MACMILLAN PUBLISHING COMPANY
NOT FOR
AUG 0 7 1991
RESALE
EMPLOYEE BOOKSTORE

MEANS REPAIR & REMODELING COST DATA
Commercial/Residential

1991

TABLE OF CONTENTS

Senior Editor
Allan B. Cleveland

Chief Editor
Phillip R. Waier

Contributing Editors
John H. Chiang
Donald D. Denzer
Jeffrey M. Goldman
Patricia Jackson
Alan E. Lew
Melville J. Mossman
John J. Moylan
Jeannene D. Murphy
Kenneth M. Randall
Kornelis Smit
Rory Woolsey
David M. Zuniga

Technical Coordinators
Marion Schofield
Wayne D. Anderson

Graphics
Carl W. Linde

First Printing

Publisher
Roger J. Grant

FOREWORD

THE COMPANY AND THE EDITORS

Since 1942, R.S. Means Company, Inc. has been actively engaged in construction cost publishing and consulting throughout North America. The primary objective of the company is to provide the construction industry professional — the contractor, the owner, the architect, the engineer, the facilities manager — with current and comprehensive construction cost data.

A thoroughly experienced and highly qualified staff of professionals at R.S. Means works daily at collecting, analyzing and disseminating reliable cost information for your needs. These staff members have years of practical construction experience and engineering training prior to joining the firm. Each contributes to the maintenance of a complete, continually-updated construction cost data system.

With the constant flow of new construction methods and materials, the construction professional often cannot find enough time to examine and evaluate all the diverse construction cost possibilities. R.S. Means performs this function by analyzing all facets of the industry. Data is collected and organized into a format that is instantly accessible. The data is useful for all phases of construction cost determination — from the preliminary budget to the detailed unit price estimate.

The Means organization is always prepared to assist you and help in the solution of construction problems through the services of its four major divisions; Construction and Cost Data Publishing, Computer Data and Software Services, Consulting Services and Educational Seminars.

DEVELOPMENT OF COST DATA

The staff at R.S. Means Company, Inc. continuously monitors developments in the construction industry in order to ensure reliable, thorough and up-to-date cost information. While *overall* construction costs may vary relative to general economic conditions, price fluctuations within the industry are dependent upon many other factors. Individual price variations may, in fact, be opposite to overall economic trends. Therefore, costs are monitored and updated and new items are added in response to industry changes.

All costs represent U.S. national averages and are given in U.S. dollars. The Means City Cost Indexes can be used to convert costs to a particular location. The City Cost Indexes for Canada can be used to convert U.S. national averages to local costs in Canadian dollars.

Material Costs are determined by contacting manufacturers, dealers, distributors, and contractors throughout the United States. If current material costs are available for a specific location, adjustments can be made to reflect differences from the national average. Material costs do not include sales tax.

Labor Costs are based on the average of wage rates from 30 major U.S. cities. Rates are determined from agreements or prevailing wages for construction trades for the current year. Rates are listed on the inside back cover of this book. If wage rates in your area vary from those used in this book, or if rate increases are expected within a given year, labor costs should be adjusted accordingly.

Labor costs reflect productivity based on actual working conditions. These figures include time spent during a normal work day on items other than actual installation such as material receiving and handling, mobilization, site movement, breaks, and cleanup. Productivity data is developed over an extended period so as not to be influenced by abnormal variations, and reflects a typical average.

Equipment Costs as presented include not only rental costs, but also operating costs. Equipment prices are obtained from industry sources throughout North America — contractors, suppliers, dealers, manufacturers and distributors.

FACTORS AFFECTING COSTS

Quality: The prices for materials and the workmanship upon which productivity is based are in line with U.S. Government specifications and represent good sound construction.

Overtime: No allowance has been made for overtime. If premium time or work during other than normal working hours is anticipated, adjustments to labor costs should be made accordingly.

Productivity: The productivity, daily output and man-hour figures for each line item are based on working an eight hour day in daylight hours. For other than normal work hours, productivity may decrease.

Size of Project: The size and type of construction project can have a significant impact on cost. Economy of scale can reduce costs for large projects. Conversely, costs may be higher for small projects due to higher percentage overhead costs, small quantity material purchases and minimum labor and/or equipment charges. Costs in this book are intended for the size and type of project as described in the "How To Use This Book" pages. Costs for projects of a significantly different size or type should be adjusted accordingly.

Location: Material prices are for metropolitan areas. Beyond a 20 mile radius of large cities, extra trucking or other transportation charges will increase the material costs slightly. This material increase may be offset by lower wage rates. Both of these factors should be considered when preparing an estimate, especially if the job site is remote. Highly specialized subcontract items may require high travel and per diem expenses for mechanics.

Other factors affecting costs are season of year, contractor management, weather, local union restrictions, building code requirements, and the availability of adequate energy, skilled labor, and building materials. General business conditions influence the "in-place" cost of all items. Substitute materials and construction methods may have to be employed, and these may increase the installed cost and/or life cycle costs. Such factors are difficult to evaluate and cannot be predicted on the basis of the job's location in a particular section of the country. Thus, there may be a significant, but unavoidable cost variation where these factors are concerned.

CSI MASTERFORMAT

Unit price data in this book is organized according to the MASTERFORMAT system of classification and numbering as developed by the Construction Specifications Institute (CSI) and Construction Specifications Canada. This system, widely accepted in the industry, is used extensively by architects and engineers for construction specifications, by contractors for estimating and record keeping, and by manufacturers and suppliers for categorization of construction materials and products. R.S. Means has organized unit price data in this system to help construction professionals categorize all aspects of the construction process.

HOW TO USE THIS BOOK

HOW THE BOOK IS ARRANGED

This book is divided into four sections: Unit Price, Assemblies, Reference and an Appendix.

Unit Price Section: All cost information has been divided into the 16 CSI divisions. These divisions are patterned after the MASTERFORMAT created and adopted by the Construction Specifications Institute, Inc. This system is widely used by most segments of the building construction industry.

A listing of all divisions and an outline of their subdivisions is shown in the Table of Contents page at the beginning of the Unit Price Section.

Each unit price line item has been assigned a 10 digit code. A graphic explanation of the numbering system is shown on the "How to Use Unit Price" page.

Descriptions Each line item number is followed by a description of the item. Sub-items and additional sizes are indented beneath appropriate line items. The first line or two of the main (bold face) item often contain descriptive information that pertains to all line items beneath this bold face listing.

Crew The first column after the description lists the typical crew needed to install the item. Typical crews are defined and listed at the beginning of the Unit Price Section. When an installation is done by one trade and requires no power equipment, the appropriate trade is listed in the crew column. For example, "2 Carp" indicates that the installation is done with 2 carpenters. If, however, the installing crew is listed as "C-2", it is made up of 1 carpenter foreman, 4 carpenters, 1 building laborer, plus an allowance for power tools. The crew costs are listed both with bare labor rates, and with the installing contractor's overhead and profit. For each, the total cost per eight-hour day and the composite hourly cost for the crew are listed.

Crew Equipment Cost The power equipment required for each crew is included in the crew cost. The daily cost for crew equipment is based on dividing the weekly bare rental rate by 5 (number of working days per week), and then adding the hourly operating cost times 8 (hours per day). This "Crew Equipment Cost" is listed in Division 016.

Daily Output To the right of every "Crew" code listing, a "Daily Output" figure is given. This is the number of units that the listed crew will install in a normal 8-hour day.

Man-hours The column following "Daily Output" is "Man-hours". This figure represents the man-hours required to install one "unit" of work. Unit man-hours are calculated by dividing the total daily crew hours (as seen in the Crew Tables) by the Daily Output.

Unit To the right of the "Man-hour" column is the "Unit" column. The abbreviated designations indicate the unit upon which the price, production and crew are based. See the Appendix for a complete list of abbreviations.

Material The first column under the "Bare Cost" heading lists the unit material cost for the line item. This figure is the "bare" material cost with no overhead and profit allowances included. This cost is developed by contacting manufacturers, dealers, distributors and contractors throughout the United States. Prices shown reflect the national average of these quoted prices for January of the current year.

Labor The second column under the "Bare Cost" heading is the unit labor cost. This cost is derived by dividing the daily labor cost by the daily output. The wage rates used are 30 city union averages and are listed on the inside back cover.

Equipment The third column under the "Bare Cost" heading lists the unit equipment cost. This figure is the daily crew equipment cost divided by the daily output.

Total The last column under the "Bare Cost" heading lists the total bare cost of the item. This is the arithmetic total of the three previous columns: "Material", "Labor", and "Equipment".

Total Incl. O&P The figure in this column is the sum of three components: the bare material cost plus 10%; the bare labor cost plus overhead and profit (per the billing rate table on the inside back cover); and the bare equipment cost plus 10%. A sample calculation is shown on the "How to Use Unit Price" page preceding the Unit Price Section.

Assemblies Section: This section uses an "Assemblies" format that groups all the functional elements of a building into 12 "Uniformat" Construction Divisions. At the top of each "Assembly" cost table is an illustration, a brief description, and the design criteria used to develop the cost. Each of the components and its contributing cost to the system is shown. For a complete breakdown and explanation of a typical "Assemblies" page, see "How to Use Assemblies Cost Tables" at the beginning of this section.

Material These cost figures include a standard 10% markup for "handling". They are national average material costs as of January of the current year and include delivery to the jobsite.

Installation The installation costs include labor and equipment, plus a markup for the installing contractor's overhead and profit.

Reference Section: Following selected item descriptions in the "Unit Price" pages are large numbers in circles. These numbers refer the reader to information and data in this Reference Section. This material includes estimating procedures, alternate pricing methods, technical data and cost derivations. These derivations illustrate the development of costs and may be of value in listing materials for purchase. This section also includes information on design and economy in construction.

Appendix: Included in this section are Historical and City Cost Indexes, a Project Schedule (CPM) example, a list of abbreviations and a comprehensive index.

Historical Cost Index This index provides data to adjust construction costs over time.

City Cost Indexes These indexes provide data to adjust the "national average" costs in this book to 162 major cities throughout the U.S. and Canada.

Abbreviations/Index A listing of the abbreviations used throughout this book, along with the terms they represent is included. Following the abbreviations list is an index for all sections.

Minimum Labor/Equipment Charge When preparing a repair and remodeling estimate, it is a good idea to carefully evaluate the "Minimum" Labor and Equipment figures listed within the Unit Price section of the book.

An estimate that has a bottom line Labor or Equipment cost LESS THAN the "Minimum" amounts listed at the bottom of specific sections of cost figures should be adjusted upward to the "Minimum" figures shown. The "Minimum" figures are included as a guide for that area of construction. Bid figures for any job must be adjusted and allowances made for the contractor to recover the expense connected with losing a portion of a normal work day. This allowance is usually needed to allow for set-up cost, clean-up and/or travel time.

Materials add a cost to an estimate over and above Labor and Equipment costs; therefore, Material costs should never be confused with this adjustment to "Minimum" Labor and Equipment figures. When Labor and Equipment costs have been developed for a specific item, ALWAYS ADJUST FIGURES UP TO THE "MINIMUM" AMOUNT SHOWN. An example is shown below:

EXAMPLE: Establish the bid price to replace a casement window unit. Assume installation of a 2′ x 5′ Metal Clad Window with Insulating Glass.

| | | **086 100** ǀ **Wood Windows** | CREW | DAILY OUTPUT | MAN-HOURS | UNIT | BARE COSTS | | | | TOTAL INCL O&P | |
							MAT.	LABOR	EQUIP.	TOTAL		
120	0010	CASEMENT WINDOW including frame, screen, and exterior trim										120
	2080	Metal clad, deluxe, insulating glass, 2′-0″ x 5′-0″ high	1 Carp	8	1	Ea.	182	22		204	235	
	9000	Minimum labor/equipment charge		3	2.670	Job		59		59	96	

SOLUTION:

STEP ONE: Develop Material Cost and Labor Cost separately.

Material Cost for Line Number [086-120-2080] = $182

Labor Cost for Line Number [086-120-2080] = $22.00

Total Incl. O&P for Line Number [086-120-2080] = [Material + 10%] + [Labor + 63.3%] = [$182 + $18] + [$22 + $14] = [$200] + [$36] = $236

STEP TWO: Evaluate Minimum Labor and Equipment Figure

Minimum Labor/Equipment Figure for this section = $59 (Compare to $22)

Minimum Labor/Equipment Figure Incl. O&P for this section = $96

Add Material Cost plus a 10% mark-up to the Minimum Labor/Equipment Cost listed in the Total Incl. O&P Column.

[$182 + 10%] + [$96] = [$200 + $96] = **$296**

ANSWER: $296 is the correct figure to use. This sum takes into consideration the Minimum Labor/Equipment figure (with O&P included), plus Material Cost (with 10% for O&P).

PROJECT SIZE

The book is aimed primarily at residential, industrial, and commercial repair/remodeling projects costing $5,000 to $500,000. With reasonable exercise of judgment the figures can be used for any repair/remodeling project but do not apply to civil engineering structures such as bridges, dams, highways or the like.

ESTIMATING PRECISION

When creating an engineering estimate, ignore the cents column. Only give the total per unit cost to the nearest dollar.

The cents will average in a column of figures. An engineering estimate of $257,323.37 is ridiculous. A figure of $257,325 is certainly more sensible and $257,000 is better and just as likely to be right.

If you follow this simple instruction, the time saved is tremendous with an added important advantage. Using round figures, your mind is left free to exercise judgment and common sense rather than being overcome and befuddled by a mass of computations.

When you have finished, roughly check the big items for location of decimal point. That is important. A large error can creep in if you write down $300 when it should be $3,000. Also check the list to be sure you have not omitted any large item. A common error is to overlook, let us say, heating or to forget finished flooring or painting or otherwise commit a gross omission. No amount of accuracy in prices can compensate for such an oversight.

It is important to keep the bare costs and costs that already include the sub's O & P separate since different mark-ups will have to be applied to each category. Organize your estimating procedures to minimize confusion and simplify checking to insure against omissions and/or duplications.

ESTIMATING GUIDELINES

Here are a few simplified rules for handling numbers and measurements when preparing a repair and remodeling estimate. These suggestions are made to enable the estimator to perform unit price and systems estimating in a logical, easy to check, and thorough manner.

1. The estimator should always visit the job site and analyze available job site data to be certain that any contingencies that pertain to the project are accounted for. Division 010-032-0010 through 2350 in the Unit Price section lists some of the factors that must be added to construction costs for particular job requirements.

2. Use preprinted forms for orderly sequence of dimensions and locations and for recording telephone dimensions.

3. Be consistent with listing dimensions, for example: Length x Width x Height. This helps in re-checking to insure that, say, the total length of partitions is appropriate for the building area.

4. Use printed (rather than measured) dimensions where given.

5. Add up multiple printed dimensions for a single entry where possible.

6. Measure all other dimensions carefully.

7. Use each set of dimensions to calculate multiple related quantities.

8. Convert foot and inch measurements to decimal feet when listing. Memorize decimal equivalents to .01 parts of a foot (1/8" equals approximately .01').

9. Do not "round off" quantities until final summary.

10. Mark drawings with different colors as items are taken off.

11. Keep similar items together; different items separate.

12. Identify location and drawing numbers to aid in future checking for completeness.

13. Measure or list everything on the drawings or mentioned in the specifications.

14. It may be necessary to list items not called for to make the job complete.

15. Be alert for: notes on plans such as N.T.S. (not to scale); changes in scale throughout the drawings; reduced size drawings; discrepancies between the specifications and the drawings.

16. Develop a consistent pattern of performing an estimate, for example:

   a. Start the quantity take-off at the lower floors and move on to the next highest floor.

   b. Proceed from the main section of the building to the wings.

   c. Move consistently through the plans south to north or vice versa.

   d. Move consistently clockwise or counterclockwise.

   e. Take off floor plan quantities first, elevations next, then the detail drawings.

17. List all gross dimensions that can be either used again for different quantities, or used as a rough-check of other quantities for verification (exterior perimeter, gross floor area, individual floor areas, etc.).

18. Utilize design symmetry or repetition:

   Repetitive floors
   Repetitive wings
   Symmetrical design around a
      center line
   Similar room layouts
   Note: Great care must be exercised, otherwise entire wings or floors can be omitted rather easily. This is one area to keep checking as the estimate progresses. Refer back to total floor area listed on the plan's posted sheet or gross dimensions.

19. Do not convert units until the final total is obtained. For instance, when estimating concrete work, keep all units to the nearest cubic foot, then summarize and convert to cubic yards.

20. When figuring alternates, it is best to total all items involved in the basic system, then total all items involved in the alternates. Thus you work with positive numbers in all cases. When adds and deducts are used, it often is confusing whether to add or subtract a given item, especially on a complicated or involved alternate.

ROUNDING OF COSTS

In general, all unit prices in excess of $5.00 have been rounded to make them easier to use and still maintain adequate precision of the results. The rounding rules are as follows:

Price from $5.01 to $20.00 rounded to the nearest 5¢

Price from $20.01 to $100.00 rounded to the nearest $1

Price from $100.01 to $1,000.00 rounded to the nearest $5

Price from $1,000.01 to $10,000.00 rounded to the nearest $25

Price from $10,000.01 to $50,000.00 rounded to the nearest $100

Price over $50,000.01 rounded to the nearest $500

HOW TO USE UNIT PRICE PAGES

Important
Prices in this section are listed in two ways: as bare costs and as costs including overhead and profit of the installing contractor. In most cases, if the work is to be subcontracted, it is best for a general contractor to add an additional 10% to the figures found in the column titled **"TOTAL INCL. O&P"**.

Unit
The unit of measure listed here reflects the material being used in the line item. For example: headers over openings are defined in linear feet (L.F.).

Productivity
The daily output represents typical total daily amount of work that the designated crew will produce. Man-hours are a unit of measure for the labor involved in performing a task. To derive the total man-hours for a task, multiply the quantity of the item involved times the man-hour figure shown.

Line Number Determination
Each line item is identified by a unique ten-digit number.

MASTERFORMAT
Division
061 128 2050
Subdivision

MASTERFORMAT
Mediumscope
061 100
061 **128** 2050
Major Classification

061 128 **2050**
Individual Line Number

Description
The meaning of this line item is 2" x 8" headers over openings will be installed by an F-2 crew at a rate of .047 man-hours per linear foot.

⑧⑥ Circle Reference Number
These reference numbers refer to charts, tables, estimating data, cost derivations and other information which may be useful to the user of this book. This information is located in the Reference Section of this book.

Crew F-2

Crew No.	Bare Costs		Incl. Subs O & P		Cost Per Man-Hour	
	Hr.	Daily	Hr.	Daily	Bare Costs	Incl. O&P
Crew F-2						
2 Carpenters	$22.00	$352.00	$35.95	$575.20	$22.00	$35.95
Power Tools		16.00		17.60	1.00	1.10
16 M.H., Daily Totals		$368.00		$592.80	$23.00	$37.05

Bare Costs are developed as follows for line no. 061-128-2050

Mat. is **Bare Material Cost ($.54)**

Labor for Crew F2 = Man-hour Cost **($22.00)** × Man-hour Units **(.047)** ≃ **$1.04**

Equip. for Crew F2 = Equip. Hour Cost **($1.00)** × Man-hour Units **(.047)** ≃ **$.05**

Total = Mat. Cost **($.54)** + Labor Cost **($1.04)** + Equip. Cost **($.05)** = **$1.63** each.

(**Note:** When a Crew is indicated Equipment and Labor costs are derived from the Crew Tables. See example above.)

Total Costs Including O&P are developed as follows:

Mat. is **Bare Material Cost** + 10% = **$.54** + **$.05** = **$.59**

Labor for Crew F2 = Man-hour Cost **($35.95)** × Man-hour Units **(.047)** ≃ **$1.69**

Equip. for Crew F2 = Equip. Hour Cost **($1.10)** × Man-hour Units **(.047)** ≃ **$.05**

Total = Mat. Cost **($.59)** + Labor Cost **($1.69)** + Equip. Cost **($.05)** = **$2.33**

(**Note:** Where a crew is indicated, Equipment and Labor costs are derived from the Crew Tables. See example at top of this page. **"Total"** line costs are rounded.)

061 | Rough Carpentry

061 100 | Wood Framing

			CREW	DAILY OUTPUT	MAN-HOURS	UNIT	BARE COSTS				TOTAL INCL O&P	
							MAT.	LABOR	EQUIP.	TOTAL		
126	1000	Canopy or soffit framing , 1" x 4"	F-2	900	.018	L.F.	.19	.39	.02	.60	.87	**126**
	1040	1" x 8"		750	.021		.38	.47	.02	.87	1.21	
	1100	2" x 4"		620	.026		.27	.57	.03	.87	1.26	
	1140	2" x 8"		500	.032		.56	.70	.03	1.29	1.81	
	1200	3" x 4"		500	.032		.62	.70	.03	1.35	1.87	
	1240	3" x 10"		300	.053		1.54	1.17	.05	2.76	3.67	
	9000	Minimum labor/equipment charge	1 Carp	4	2	Job		44		44	72	
128	0010	FRAMING, WALLS										**128**
	0020											
	2000	Headers over openings, 2" x 6" ⑧⑥	F-2	360	.044	L.F.	.40	.98	.04	1.42	2.09	
	2050	2" x 8"		340	.047		.54	1.04	.05	1.63	2.33	
	2110	2" x 10" ⑧⑦		320	.050		.84	1.10	.05	1.99	2.77	
	2150	2" x 12"		300	.053		1.02	1.17	.05	2.24	3.10	
	2200	4" x 12"		190	.084		2.51	1.85	.08	4.44	5.90	
	2250	6" x 12"		140	.114		3.58	2.51	.11	6.20	8.15	
	5000	Plates, untreated, 2" x 3"		850	.019		.21	.41	.02	.64	.93	
	5020	2" x 4"		800	.020		.27	.44	.02	.73	1.04	
	5040	2" x 6"		750	.021		.41	.47	.02	.90	1.24	
	5120	Studs, 8' high wall, 2" x 3"		1,200	.013		.21	.29	.01	.51	.72	
	5140	2" x 4"		1,100	.015		.27	.32	.01	.60	.84	
	5160	2" x 6"		1,000	.016		.40	.35	.02	.77	1.03	
	5180	3" x 4"		800	.020		.62	.44	.02	1.08	1.42	
	8200	For 12' high wall, deduct						5%				
	8220	For stub wall, 6' high, add						20%				
	8240	3' high, add						40%				
	8250	For second story & above, add						5%				
	8300	Dormer & gable, add						15%				
	9000	Minimum labor/equipment charge	1 Carp	4	2	Job		44		44	72	
130	0010	FURRING Wood strips, on walls, 1" x 2", on wood	1 Carp	550	.015	L.F.	.10	.32		.42	.63	**130**
	0300	On masonry		495	.016		.11	.36		.47	.70	
	0700	On metal lath		200	.040		.11	.88		.99	1.56	
	9000	Minimum labor/equipment charge		4	2	Job		44		44	72	

UNIT PRICE SECTION

TABLE OF CONTENTS

CREWS

Crew No.	Bare Costs Hr.	Daily	Incl. Subs O & P Hr.	Daily	Cost Per Man-Hour Bare Costs	Incl. O&P
Crew A-1	Hr.	Daily	Hr.	Daily	Bare Costs	Incl. O&P
1 Building Laborer	$17.50	$140.00	$28.60	$228.80	17.50	28.60
1 Gas Eng. Power Tool		54.40		59.85	6.80	7.48
8 M.H., Daily Totals		$194.40		$288.65	$24.30	$36.08
Crew A-1A	Hr.	Daily	Hr.	Daily	Bare Costs	Incl. O&P
1 Laborer	$17.50	$140.00	$28.60	$228.80	17.50	28.60
1 Power Equipment		29.80		32.80	3.72	4.10
8 M.H., Daily Totals		$169.80		$261.60	$21.22	$32.70
Crew A-2	Hr.	Daily	Hr.	Daily	Bare Costs	Incl. O&P
2 Building Laborers	$17.50	$280.00	$28.60	$457.60	17.70	28.73
1 Truck Driver (light)	18.10	144.80	29.00	232.00		
1 Light Truck, 1.5 Ton		146.00		160.60	6.08	6.69
24 M.H., Daily Totals		$570.80		$850.20	$23.78	$35.42
Crew A-3	Hr.	Daily	Hr.	Daily	Bare Costs	Incl. O&P
1 Truck Driver (heavy)	$18.40	$147.20	$29.50	$236.00	18.40	29.50
1 Dump Truck, 12 Ton		298.80		328.70	37.35	41.08
8 M.H., Daily Totals		$446.00		$564.70	$55.75	$70.58
Crew A-4	Hr.	Daily	Hr.	Daily	Bare Costs	Incl. O&P
2 Carpenters	$22.00	$352.00	$35.95	$575.20	21.60	35.00
1 Painter, Ordinary	20.80	166.40	33.10	264.80		
24 M.H., Daily Totals		$518.40		$840.00	$21.60	$35.00
Crew A-5	Hr.	Daily	Hr.	Daily	Bare Costs	Incl. O&P
2 Building Laborers	$17.50	$280.00	$28.60	$457.60	17.56	28.64
.25 Truck Driver (light)	18.10	36.20	29.00	58.00		
.25 Light Truck, 1.5 Ton		36.50		40.15	2.02	2.23
18 M.H., Daily Totals		$352.70		$555.75	$19.58	$30.87
Crew A-6	Hr.	Daily	Hr.	Daily	Bare Costs	Incl. O&P
1 Chief Of Party	$21.40	$171.20	$33.60	$268.80	20.30	31.87
1 Instrument Man	19.20	153.60	30.15	241.20		
16 M.H., Daily Totals		$324.80		$510.00	$20.30	$31.87
Crew A-7	Hr.	Daily	Hr.	Daily	Bare Costs	Incl. O&P
1 Chief Of Party	$21.40	$171.20	$33.60	$268.80	19.23	30.53
1 Instrument Man	19.20	153.60	30.15	241.20		
1 Rodman/Chainman	17.10	136.80	27.85	222.80		
24 M.H., Daily Totals		$461.60		$732.80	$19.23	$30.53
Crew A-8	Hr.	Daily	Hr.	Daily	Bare Costs	Incl. O&P
1 Chief Of Party	$21.40	$171.20	$33.60	$268.80	18.70	29.86
1 Instrument Man	19.20	153.60	30.15	241.20		
2 Rodmen/Chainmen	17.10	273.60	27.85	445.60		
32 M.H., Daily Totals		$598.40		$955.60	$18.70	$29.86
Crew A-9	Hr.	Daily	Hr.	Daily	Bare Costs	Incl. O&P
1 Asbestos Foreman	$25.20	$201.60	$40.35	$322.80	24.76	39.65
7 Asbestos Workers	24.70	1383.20	39.55	2214.80		
4 Airless Sprayers		101.60		111.75		
3 HDPA Vacs., 16 Gal.		89.40		98.35	2.98	3.28
64 M.H., Daily Totals		$1775.80		$2747.70	$27.74	$42.93
Crew A-10	Hr.	Daily	Hr.	Daily	Bare Costs	Incl. O&P
1 Asbestos Foreman	$25.20	$201.60	$40.35	$322.80	24.76	39.65
7 Asbestos Workers	24.70	1383.20	39.55	2214.80		
2 HEPA Vacs., 16 Gal.		59.60		65.55	.93	1.02
64 M.H., Daily Totals		$1644.40		$2603.15	$25.69	$40.67

Crew No.	Bare Costs Hr.	Daily	Incl. Subs O & P Hr.	Daily	Cost Per Man-Hour Bare Costs	Incl. O&P
Crew A-11	Hr.	Daily	Hr.	Daily	Bare Costs	Incl. O&P
1 Asbestos Foreman	$25.20	$201.60	$40.35	$322.80	24.76	39.65
7 Asbestos Workers	24.70	1383.20	39.55	2214.80		
4 Airless Sprayers		101.60		111.75		
2 HEPA Vacs., 16 Gal.		59.60		65.55		
2 Chipping Hammers		18.40		20.25	2.80	3.08
64 M.H., Daily Totals		$1764.40		$2735.15	$27.56	$42.73
Crew A-12	Hr.	Daily	Hr.	Daily	Bare Costs	Incl. O&P
1 Asbestos Foreman	$25.20	$201.60	$40.35	$322.80	24.76	39.65
7 Asbestos Workers	24.70	1383.20	39.55	2214.80		
4 Airless Sprayers		101.60		111.75		
2 HEPA Vacs., 16 Gal.		59.60		65.55		
1 Large Prod. Vac. Loader		440.80		484.90	9.40	10.34
64 M.H., Daily Totals		$2186.80		$3199.80	$34.16	$49.99
Crew B-1	Hr.	Daily	Hr.	Daily	Bare Costs	Incl. O&P
1 Labor Foreman (outside)	$19.50	$156.00	$31.85	$254.80	18.16	29.68
2 Building Laborers	17.50	280.00	28.60	457.60		
24 M.H., Daily Totals		$436.00		$712.40	$18.16	$29.68
Crew B-2	Hr.	Daily	Hr.	Daily	Bare Costs	Incl. O&P
1 Labor Foreman (outside)	$19.50	$156.00	$31.85	$254.80	17.90	29.25
4 Building Laborers	17.50	560.00	28.60	915.20		
40 M.H., Daily Totals		$716.00		$1170.00	$17.90	$29.25
Crew B-3	Hr.	Daily	Hr.	Daily	Bare Costs	Incl. O&P
1 Labor Foreman (outside)	$19.50	$156.00	$31.85	$254.80	18.96	30.56
2 Building Laborers	17.50	280.00	28.60	457.60		
1 Equip. Oper. (med.)	22.50	180.00	35.35	282.80		
2 Truck Drivers (heavy)	18.40	294.40	29.50	472.00		
F.E. Loader, T.M., 2.5 C.Y.		807.80		888.60		
2 Dump Trucks, 16 Ton		730.40		803.45	32.04	35.25
48 M.H., Daily Totals		$2448.60		$3159.25	$51.00	$65.81
Crew B-4	Hr.	Daily	Hr.	Daily	Bare Costs	Incl. O&P
1 Labor Foreman (outside)	$19.50	$156.00	$31.85	$254.80	17.98	29.29
4 Building Laborers	17.50	560.00	28.60	915.20		
1 Truck Driver (heavy)	18.40	147.20	29.50	236.00		
1 Tractor, 4 x 2, 195 H.P.		268.40		295.25		
1 Platform Trailer		130.60		143.65	8.31	9.14
48 M.H., Daily Totals		$1262.20		$1844.90	$26.29	$38.43
Crew B-5	Hr.	Daily	Hr.	Daily	Bare Costs	Incl. O&P
1 Labor Foreman (outside)	$19.50	$156.00	$31.85	$254.80	19.81	31.83
4 Building Laborers	17.50	560.00	28.60	915.20		
2 Equip. Oper. (med.)	22.50	360.00	35.35	565.60		
1 Mechanic	24.00	192.00	37.70	301.60		
1 Air Compr., 250 C.F.M.		88.80		97.70		
Air Tools & Accessories		27.60		30.35		
2-50 Ft. Air Hoses, 1.5" Dia.		10.80		11.90		
F.E. Loader, T.M., 2.5 C.Y.		807.80		888.60	14.60	16.07
64 M.H., Daily Totals		$2203.00		$3065.75	$34.41	$47.90
Crew B-6	Hr.	Daily	Hr.	Daily	Bare Costs	Incl. O&P
2 Building Laborers	$17.50	$280.00	$28.60	$457.60	18.80	30.26
1 Equip. Oper. (light)	21.40	171.20	33.60	268.80		
1 Backhoe Loader, 48 H.P.		190.40		209.45	7.93	8.72
24 M.H., Daily Totals		$641.60		$935.85	$26.73	$38.98

CREWS

Crew No.	Bare Costs		Incl. Subs O & P		Cost Per Man-Hour	
Crew B-7	Hr.	Daily	Hr.	Daily	Bare Costs	Incl. O&P
1 Labor Foreman (outside)	$19.50	$156.00	$31.85	$254.80	$18.66	$30.26
4 Building Laborers	17.50	560.00	28.60	915.20		
1 Equip. Oper. (med.)	22.50	180.00	35.35	282.80		
1 Chipping Machine		178.80		196.70		
F.E. Loader, T.M., 2.5 C.Y.		807.80		888.60		
2 Chain Saws		83.20		91.50	22.28	24.51
48 M.H., Daily Totals		$1965.80		$2629.60	$40.94	$54.77
Crew B-7A	Hr.	Daily	Hr.	Daily	Bare Costs	Incl. O&P
2 Laborers	$17.50	$280.00	$28.60	$457.60	$18.80	$30.26
1 Equip. Oper. (light)	21.40	171.20	33.60	268.80		
1 Rake w/Tractor		186.80		205.50		
2 Chain Saws		40.00		44.00	9.45	10.39
24 M.H., Daily Totals		$678.00		$975.90	$28.25	$40.65
Crew B-8	Hr.	Daily	Hr.	Daily	Bare Costs	Incl. O&P
1 Labor Foreman (outside)	$19.50	$156.00	$31.85	$254.80	$19.43	$31.11
2 Building Laborers	17.50	280.00	28.60	457.60		
2 Equip. Oper. (med.)	22.50	360.00	35.35	565.60		
1 Equip. Oper. Oiler	19.20	153.60	30.15	241.20		
2 Truck Drivers (heavy)	18.40	294.40	29.50	472.00		
1 Hyd. Crane, 25 Ton		486.40		535.05		
F.E. Loader, T.M., 2.5 C.Y.		807.80		888.60		
2 Dump Trucks, 16 Ton		730.40		803.45	31.63	34.79
64 M.H., Daily Totals		$3268.60		$4218.30	$51.06	$65.90
Crew B-9	Hr.	Daily	Hr.	Daily	Bare Costs	Incl. O&P
1 Labor Foreman (outside)	$19.50	$156.00	$31.85	$254.80	$17.90	$29.25
4 Building Laborers	17.50	560.00	28.60	915.20		
1 Air Compr., 250 C.F.M.		88.80		97.70		
Air Tools & Accessories		27.60		30.35		
2-50 Ft. Air Hoses, 1.5" Dia.		10.80		11.90	3.18	3.49
40 M.H., Daily Totals		$843.20		$1309.95	$21.08	$32.74
Crew B-10	Hr.	Daily	Hr.	Daily	Bare Costs	Incl. O&P
1 Equip. Oper. (med.)	$22.50	$180.00	$35.35	$282.80	$20.83	$33.10
.5 Building Laborer	17.50	70.00	28.60	114.40		
12 M.H., Daily Totals		$250.00		$397.20	$20.83	$33.10
Crew B-10A	Hr.	Daily	Hr.	Daily	Bare Costs	Incl. O&P
1 Equip. Oper. (med.)	$22.50	$180.00	$35.35	$282.80	$20.83	$33.10
.5 Building Laborer	17.50	70.00	28.60	114.40		
1 Roll. Compact., 2K Lbs.		74.40		81.85	6.20	6.82
12 M.H., Daily Totals		$324.40		$479.05	$27.03	$39.92
Crew B-10B	Hr.	Daily	Hr.	Daily	Bare Costs	Incl. O&P
1 Equip. Oper. (med.)	$22.50	$180.00	$35.35	$282.80	$20.83	$33.10
.5 Building Laborer	17.50	70.00	28.60	114.40		
1 Dozer, 200 H.P.		775.80		853.40	64.65	71.11
12 M.H., Daily Totals		$1025.80		$1250.60	$85.48	$104.21
Crew B-10C	Hr.	Daily	Hr.	Daily	Bare Costs	Incl. O&P
1 Equip. Oper. (med.)	$22.50	$180.00	$35.35	$282.80	$20.83	$33.10
.5 Building Laborer	17.50	70.00	28.60	114.40		
1 Dozer, 200 H.P.		775.80		853.40		
1 Vibratory Roller, Towed		90.40		99.45	72.18	79.40
12 M.H., Daily Totals		$1116.20		$1350.05	$93.01	$112.50

Crew No.	Bare Costs		Incl. Subs O & P		Cost Per Man-Hour	
Crew B-10D	Hr.	Daily	Hr.	Daily	Bare Costs	Incl. O&P
1 Equip. Oper. (med.)	$22.50	$180.00	$35.35	$282.80	$20.83	$33.10
.5 Building Laborer	17.50	70.00	28.60	114.40		
1 Dozer, 200 H.P.		775.80		853.40		
1 Sheepsft. Roller, Towed		119.80		131.80	74.63	82.10
12 M.H., Daily Totals		$1145.60		$1382.40	$95.46	$115.20
Crew B-10E	Hr.	Daily	Hr.	Daily	Bare Costs	Incl. O&P
1 Equip. Oper. (med.)	$22.50	$180.00	$35.35	$282.80	$20.83	$33.10
.5 Building Laborer	17.50	70.00	28.60	114.40		
1 Tandem Roller, 5 Ton		118.40		130.25	9.86	10.85
12 M.H., Daily Totals		$368.40		$527.45	$30.69	$43.95
Crew B-10F	Hr.	Daily	Hr.	Daily	Bare Costs	Incl. O&P
1 Equip. Oper. (med.)	$22.50	$180.00	$35.35	$282.80	$20.83	$33.10
.5 Building Laborer	17.50	70.00	28.60	114.40		
1 Tandem Roller, 10 Ton		187.60		206.35	15.63	17.19
12 M.H., Daily Totals		$437.60		$603.55	$36.46	$50.29
Crew B-10G	Hr.	Daily	Hr.	Daily	Bare Costs	Incl. O&P
1 Equip. Oper. (med.)	$22.50	$180.00	$35.35	$282.80	$20.83	$33.10
.5 Building Laborer	17.50	70.00	28.60	114.40		
1 Sheepsft. Roll., 130 H.P.		487.00		535.70	40.58	44.64
12 M.H., Daily Totals		$737.00		$932.90	$61.41	$77.74
Crew B-10H	Hr.	Daily	Hr.	Daily	Bare Costs	Incl. O&P
1 Equip. Oper. (med.)	$22.50	$180.00	$35.35	$282.80	$20.83	$33.10
.5 Building Laborer	17.50	70.00	28.60	114.40		
1 Diaphr. Water Pump, 2"		14.40		15.85		
1-20 Ft. Suction Hose, 2"		4.40		4.85		
2-50 Ft. Disch. Hoses, 2"		5.20		5.70	2.00	2.20
12 M.H., Daily Totals		$274.00		$423.60	$22.83	$35.30
Crew B-10I	Hr.	Daily	Hr.	Daily	Bare Costs	Incl. O&P
1 Equip. Oper. (med.)	$22.50	$180.00	$35.35	$282.80	$20.83	$33.10
.5 Building Laborer	17.50	70.00	28.60	114.40		
1 Diaphr. Water Pump, 4"		50.20		55.20		
1-20 Ft. Suction Hose, 4"		11.20		12.30		
2-50 Ft. Disch. Hoses, 4"		8.00		8.80	5.78	6.35
12 M.H., Daily Totals		$319.40		$473.50	$26.61	$39.45
Crew B-10J	Hr.	Daily	Hr.	Daily	Bare Costs	Incl. O&P
1 Equip. Oper. (med.)	$22.50	$180.00	$35.35	$282.80	$20.83	$33.10
.5 Building Laborer	17.50	70.00	28.60	114.40		
1 Centr. Water Pump, 3"		27.20		29.90		
1-20 Ft. Suction Hose, 3"		6.80		7.50		
2-50 Ft. Disch. Hoses, 3"		7.20		7.90	3.43	3.77
12 M.H., Daily Totals		$291.20		$442.50	$24.26	$36.87
Crew B-10K	Hr.	Daily	Hr.	Daily	Bare Costs	Incl. O&P
1 Equip. Oper. (med.)	$22.50	$180.00	$35.35	$282.80	$20.83	$33.10
.5 Building Laborer	17.50	70.00	28.60	114.40		
1 Centr. Water Pump, 6"		146.20		160.80		
1-20 Ft. Suction Hose, 6"		23.40		25.75		
2-50 Ft. Disch. Hoses, 6"		26.00		28.60	16.30	17.92
12 M.H., Daily Totals		$445.60		$612.35	$37.13	$51.02

CREWS

Crew No.	Bare Costs Hr.	Daily	Incl. Subs O&P Hr.	Daily	Cost Per Man-Hour Bare Costs	Incl. O&P
Crew B-10L	Hr.	Daily	Hr.	Daily	Bare Costs	Incl. O&P
1 Equip. Oper. (med.)	$22.50	$180.00	$35.35	$282.80	$20.83	$33.10
.5 Building Laborer	17.50	70.00	28.60	114.40		
1 Dozer, 75 H.P.		272.80		300.10	22.73	25.00
12 M.H., Daily Totals		$522.80		$697.30	$43.56	$58.10
Crew B-10M	Hr.	Daily	Hr.	Daily	Bare Costs	Incl. O&P
1 Equip. Oper. (med.)	$22.50	$180.00	$35.35	$282.80	$20.83	$33.10
.5 Building Laborer	17.50	70.00	28.60	114.40		
1 Dozer, 300 H.P.		861.80		948.00	71.81	79.00
12 M.H., Daily Totals		$1111.80		$1345.20	$92.64	$112.10
Crew B-10N	Hr.	Daily	Hr.	Daily	Bare Costs	Incl. O&P
1 Equip. Oper. (med.)	$22.50	$180.00	$35.35	$282.80	$20.83	$33.10
.5 Building Laborer	17.50	70.00	28.60	114.40		
F.E. Loader, T.M., 1.5 C.Y.		343.20		377.50	28.60	31.45
12 M.H., Daily Totals		$593.20		$774.70	$49.43	$64.55
Crew B-10O	Hr.	Daily	Hr.	Daily	Bare Costs	Incl. O&P
1 Equip. Oper. (med.)	$22.50	$180.00	$35.35	$282.80	$20.83	$33.10
.5 Building Laborer	17.50	70.00	28.60	114.40		
F.E. Loader, T.M., 2.25 C.Y.		436.40		480.05	36.36	40.00
12 M.H., Daily Totals		$686.40		$877.25	$57.19	$73.10
Crew B-10P	Hr.	Daily	Hr.	Daily	Bare Costs	Incl. O&P
1 Equip. Oper. (med.)	$22.50	$180.00	$35.35	$282.80	$20.83	$33.10
.5 Building Laborer	17.50	70.00	28.60	114.40		
F.E. Loader, T.M., 2.5 C.Y.		807.80		888.60	67.31	74.05
12 M.H., Daily Totals		$1057.80		$1285.80	$88.14	$107.15
Crew B-10Q	Hr.	Daily	Hr.	Daily	Bare Costs	Incl. O&P
1 Equip. Oper. (med.)	$22.50	$180.00	$35.35	$282.80	$20.83	$33.10
.5 Building Laborer	17.50	70.00	28.60	114.40		
F.E. Loader, T.M., 5 C.Y.		1006.00		1106.60	83.83	92.21
12 M.H., Daily Totals		$1256.00		$1503.80	$104.66	$125.31
Crew B-10R	Hr.	Daily	Hr.	Daily	Bare Costs	Incl. O&P
1 Equip. Oper. (med.)	$22.50	$180.00	$35.35	$282.80	$20.83	$33.10
.5 Building Laborer	17.50	70.00	28.60	114.40		
F.E. Loader, W.M., 1 C.Y.		225.60		248.15	18.80	20.67
12 M.H., Daily Totals		$475.60		$645.35	$39.63	$53.77
Crew B-10S	Hr.	Daily	Hr.	Daily	Bare Costs	Incl. O&P
1 Equip. Oper. (med.)	$22.50	$180.00	$35.35	$282.80	$20.83	$33.10
.5 Building Laborer	17.50	70.00	28.60	114.40		
F.E. Loader, W.M., 1.5 C.Y.		309.60		340.55	25.80	28.37
12 M.H., Daily Totals		$559.60		$737.75	$46.63	$61.47
Crew B-10T	Hr.	Daily	Hr.	Daily	Bare Costs	Incl. O&P
1 Equip. Oper. (med.)	$22.50	$180.00	$35.35	$282.80	$20.83	$33.10
.5 Building Laborer	17.50	70.00	28.60	114.40		
F.E. Loader, W.M., 2.5 C.Y.		449.20		494.10	37.43	41.17
12 M.H., Daily Totals		$699.20		$891.30	$58.26	$74.27
Crew B-10U	Hr.	Daily	Hr.	Daily	Bare Costs	Incl. O&P
1 Equip. Oper. (med.)	$22.50	$180.00	$35.35	$282.80	$20.83	$33.10
.5 Building Laborer	17.50	70.00	28.60	114.40		
F.E. Loader, W.M., 5.5 C.Y.		927.40		1020.15	77.28	85.01
12 M.H., Daily Totals		$1177.40		$1417.35	$98.11	$118.11

Crew No.	Bare Costs Hr.	Daily	Incl. Subs O&P Hr.	Daily	Cost Per Man-Hour Bare Costs	Incl. O&P
Crew B-10V	Hr.	Daily	Hr.	Daily	Bare Costs	Incl. O&P
1 Equip. Oper. (med.)	$22.50	$180.00	$35.35	$282.80	$20.83	$33.10
.5 Building Laborer	17.50	70.00	28.60	114.40		
1 Dozer, 700 H.P.		2483.20		2731.50	206.93	227.62
12 M.H., Daily Totals		$2733.20		$3128.70	$227.76	$260.72
Crew B-10W	Hr.	Daily	Hr.	Daily	Bare Costs	Incl. O&P
1 Equip. Oper. (med.)	$22.50	$180.00	$35.35	$282.80	$20.83	$33.10
.5 Building Laborer	17.50	70.00	28.60	114.40		
1 Dozer, 105 H.P.		399.20		439.10	33.26	36.59
12 M.H., Daily Totals		$649.20		$836.30	$54.09	$69.69
Crew B-10X	Hr.	Daily	Hr.	Daily	Bare Costs	Incl. O&P
1 Equip. Oper. (med.)	$22.50	$180.00	$35.35	$282.80	$20.83	$33.10
.5 Building Laborer	17.50	70.00	28.60	114.40		
1 Dozer, 410 H.P.		1156.80		1272.50	96.40	106.04
12 M.H., Daily Totals		$1406.80		$1669.70	$117.23	$139.14
Crew B-10Y	Hr.	Daily	Hr.	Daily	Bare Costs	Incl. O&P
1 Equip. Oper. (med.)	$22.50	$180.00	$35.35	$282.80	$20.83	$33.10
.5 Building Laborer	17.50	70.00	28.60	114.40		
1 Vibratory Drum Roller		296.80		326.50	24.73	27.20
12 M.H., Daily Totals		$546.80		$723.70	$45.56	$60.30
Crew B-11	Hr.	Daily	Hr.	Daily	Bare Costs	Incl. O&P
1 Equipment Oper. (med.)	$22.50	$180.00	$35.35	$282.80	$20.00	$31.97
1 Building Laborer	17.50	140.00	28.60	228.80		
16 M.H., Daily Totals		$320.00		$511.60	$20.00	$31.97
Crew B-11A	Hr.	Daily	Hr.	Daily	Bare Costs	Incl. O&P
1 Equipment Oper. (med.)	$22.50	$180.00	$35.35	$282.80	$20.00	$31.97
1 Building Laborer	17.50	140.00	28.60	228.80		
1 Dozer, 200 H.P.		775.80		853.40	48.48	53.33
16 M.H., Daily Totals		$1095.80		$1365.00	$68.48	$85.30
Crew B-11B	Hr.	Daily	Hr.	Daily	Bare Costs	Incl. O&P
1 Equipment Oper. (med.)	$22.50	$180.00	$35.35	$282.80	$20.00	$31.97
1 Building Laborer	17.50	140.00	28.60	228.80		
1 Dozer, 200 H.P.		775.80		853.40		
1 Air Powered Tamper		13.20		14.50		
1 Air Compr. 365 C.F.M.		201.80		222.00		
2-50 Ft. Air Hoses, 1.5" Dia.		10.80		11.90	62.60	68.86
16 M.H., Daily Totals		$1321.60		$1613.40	$82.60	$100.83
Crew B-11C	Hr.	Daily	Hr.	Daily	Bare Costs	Incl. O&P
1 Equipment Oper. (med.)	$22.50	$180.00	$35.35	$282.80	$20.00	$31.97
1 Building Laborer	17.50	140.00	28.60	228.80		
1 Backhoe Loader, 48 H.P.		190.40		209.45	11.90	13.09
16 M.H., Daily Totals		$510.40		$721.05	$31.90	$45.06
Crew B-11K	Hr.	Daily	Hr.	Daily	Bare Costs	Incl. O&P
1 Equipment Oper. (med.)	$22.50	$180.00	$35.35	$282.80	$20.00	$31.97
1 Building Laborer	17.50	140.00	28.60	228.80		
1 Trencher, 8' D., 16" W.		420.00		462.00	26.25	28.87
16 M.H., Daily Totals		$740.00		$973.60	$46.25	$60.84
Crew B-11L	Hr.	Daily	Hr.	Daily	Bare Costs	Incl. O&P
1 Equipment Oper. (med.)	$22.50	$180.00	$35.35	$282.80	$20.00	$31.97
1 Building Laborer	17.50	140.00	28.60	228.80		
1 Grader, 30,000 Lbs.		521.00		573.10	32.56	35.81
16 M.H., Daily Totals		$841.00		$1084.70	$52.56	$67.78

CREWS

Crew B-11M	Bare Costs Hr.	Daily	Incl. Subs O & P Hr.	Daily	Cost Per Man-Hour Bare Costs	Incl. O&P
1 Equipment Oper. (med.)	$22.50	$180.00	$35.35	$282.80	$20.00	$31.97
1 Building Laborer	17.50	140.00	28.60	228.80		
1 Backhoe Loader, 80 H.P.		286.60		315.25	17.91	19.70
16 M.H., Daily Totals		$606.60		$826.85	$37.91	$51.67

Crew B-12	Hr.	Daily	Hr.	Daily	Bare Costs	Incl. O&P
1 Equip. Oper. (crane)	$23.30	$186.40	$36.60	$292.80	$21.25	$33.37
1 Equip. Oper. Oiler	19.20	153.60	30.15	241.20		
16 M.H., Daily Totals		$340.00		$534.00	$21.25	$33.37

Crew B-12A	Hr.	Daily	Hr.	Daily	Bare Costs	Incl. O&P
1 Equip. Oper. (crane)	$23.30	$186.40	$36.60	$292.80	$21.25	$33.37
1 Equip. Oper. Oiler	19.20	153.60	30.15	241.20		
1 Hyd. Excavator, 1 C.Y.		566.60		623.25	35.41	38.95
16 M.H., Daily Totals		$906.60		$1157.25	$56.66	$72.32

Crew B-12B	Hr.	Daily	Hr.	Daily	Bare Costs	Incl. O&P
1 Equip. Oper. (crane)	$23.30	$186.40	$36.60	$292.80	$21.25	$33.37
1 Equip. Oper. Oiler	19.20	153.60	30.15	241.20		
1 Hyd. Excavator, 1.5 C.Y.		679.60		747.55	42.47	46.72
16 M.H., Daily Totals		$1019.60		$1281.55	$63.72	$80.09

Crew B-12C	Hr.	Daily	Hr.	Daily	Bare Costs	Incl. O&P
1 Equip. Oper. (crane)	$23.30	$186.40	$36.60	$292.80	$21.25	$33.37
1 Equip. Oper. Oiler	19.20	153.60	30.15	241.20		
1 Hyd. Excavator, 2 C.Y.		947.00		1041.70	59.18	65.10
16 M.H., Daily Totals		$1287.00		$1575.70	$80.43	$98.47

Crew B-12D	Hr.	Daily	Hr.	Daily	Bare Costs	Incl. O&P
1 Equip. Oper. (crane)	$23.30	$186.40	$36.60	$292.80	$21.25	$33.37
1 Equip. Oper. Oiler	19.20	153.60	30.15	241.20		
1 Hyd. Excavator, 3.5 C.Y.		2017.00		2218.70	126.06	138.66
16 M.H., Daily Totals		$2357.00		$2752.70	$147.31	$172.03

Crew B-12E	Hr.	Daily	Hr.	Daily	Bare Costs	Incl. O&P
1 Equip. Oper. (crane)	$23.30	$186.40	$36.60	$292.80	$21.25	$33.37
1 Equip. Oper. Oiler	19.20	153.60	30.15	241.20		
1 Hyd. Excavator, .5 C.Y.		336.40		370.05	21.02	23.12
16 M.H., Daily Totals		$676.40		$904.05	$42.27	$56.49

Crew B-12F	Hr.	Daily	Hr.	Daily	Bare Costs	Incl. O&P
1 Equip. Oper. (crane)	$23.30	$186.40	$36.60	$292.80	$21.25	$33.37
1 Equip. Oper. Oiler	19.20	153.60	30.15	241.20		
1 Hyd. Excavator, .75 C.Y.		457.20		502.90	28.57	31.43
16 M.H., Daily Totals		$797.20		$1036.90	$49.82	$64.80

Crew B-12G	Hr.	Daily	Hr.	Daily	Bare Costs	Incl. O&P
1 Equip. Oper. (crane)	$23.30	$186.40	$36.60	$292.80	$21.25	$33.37
1 Equip. Oper. Oiler	19.20	153.60	30.15	241.20		
1 Power Shovel, .5 C.Y.		376.40		414.05		
1 Clamshell Bucket, .5 C.Y		46.80		51.50	26.45	29.09
16 M.H., Daily Totals		$763.20		$999.55	$47.70	$62.46

Crew B-12H	Hr.	Daily	Hr.	Daily	Bare Costs	Incl. O&P
1 Equip. Oper. (crane)	$23.30	$186.40	$36.60	$292.80	$21.25	$33.37
1 Equip. Oper. Oiler	19.20	153.60	30.15	241.20		
1 Power Shovel, 1 C.Y.		435.80		479.40		
1 Clamshell Bucket, 1 C.Y.		59.60		65.55	30.96	34.05
16 M.H., Daily Totals		$835.40		$1078.95	$52.21	$67.42

Crew B-12I	Bare Costs Hr.	Daily	Incl. Subs O & P Hr.	Daily	Cost Per Man-Hour Bare Costs	Incl. O&P
1 Equip. Oper. (crane)	$23.30	$186.40	$36.60	$292.80	$21.25	$33.37
1 Equip. Oper. Oiler	19.20	153.60	30.15	241.20		
1 Power Shovel, .75 C.Y.		401.20		441.30		
1 Dragline Bucket, .75 C.Y.		29.60		32.55	26.92	29.61
16 M.H., Daily Totals		$770.80		$1007.85	$48.17	$62.98

Crew B-12J	Hr.	Daily	Hr.	Daily	Bare Costs	Incl. O&P
1 Equip. Oper. (crane)	$23.30	$186.40	$36.60	$292.80	$21.25	$33.37
1 Equip. Oper. Oiler	19.20	153.60	30.15	241.20		
1 Gradall, 3 Ton, .5 C.Y.		555.60		611.15	34.72	38.19
16 M.H., Daily Totals		$895.60		$1145.15	$55.97	$71.56

Crew B-12K	Hr.	Daily	Hr.	Daily	Bare Costs	Incl. O&P
1 Equip. Oper. (crane)	$23.30	$186.40	$36.60	$292.80	$21.25	$33.37
1 Equip. Oper. Oiler	19.20	153.60	30.15	241.20		
1 Gradall, 3 Ton, 1 C.Y.		768.00		844.80	48.00	52.80
16 M.H., Daily Totals		$1108.00		$1378.80	$69.25	$86.17

Crew B-12L	Hr.	Daily	Hr.	Daily	Bare Costs	Incl. O&P
1 Equip. Oper. (crane)	$23.30	$186.40	$36.60	$292.80	$21.25	$33.37
1 Equip. Oper. Oiler	19.20	153.60	30.15	241.20		
1 Power Shovel, .5 C.Y.		376.40		414.05		
1 F.E. Attachment, .5 C.Y.		47.00		51.70	26.46	29.10
16 M.H., Daily Totals		$763.40		$999.75	$47.71	$62.47

Crew B-12M	Hr.	Daily	Hr.	Daily	Bare Costs	Incl. O&P
1 Equip. Oper. (crane)	$23.30	$186.40	$36.60	$292.80	$21.25	$33.37
1 Equip. Oper. Oiler	19.20	153.60	30.15	241.20		
1 Power Shovel, .75 C.Y.		401.20		441.30		
1 F.E. Attachment, .75 C.Y.		86.80		95.50	30.50	33.55
16 M.H., Daily Totals		$828.00		$1070.80	$51.75	$66.92

Crew B-12N	Hr.	Daily	Hr.	Daily	Bare Costs	Incl. O&P
1 Equip. Oper. (crane)	$23.30	$186.40	$36.60	$292.80	$21.25	$33.37
1 Equip. Oper. Oiler	19.20	153.60	30.15	241.20		
1 Power Shovel, 1 C.Y.		435.80		479.40		
1 F.E. Attachment, 1 C.Y.		123.40		135.75	34.95	38.44
16 M.H., Daily Totals		$899.20		$1149.15	$56.20	$71.81

Crew B-12O	Hr.	Daily	Hr.	Daily	Bare Costs	Incl. O&P
1 Equip. Oper. (crane)	$23.30	$186.40	$36.60	$292.80	$21.25	$33.37
1 Equip. Oper. Oiler	19.20	153.60	30.15	241.20		
1 Power Shovel, 1.5 C.Y.		659.00		724.90		
1 F.E. Attachment, 1.5 C.Y.		137.80		151.60	49.80	54.78
16 M.H., Daily Totals		$1136.80		$1410.50	$71.05	$88.15

Crew B-12P	Hr.	Daily	Hr.	Daily	Bare Costs	Incl. O&P
1 Equip. Oper. (crane)	$23.30	$186.40	$36.60	$292.80	$21.25	$33.37
1 Equip. Oper. Oiler	19.20	153.60	30.15	241.20		
1 Crawler Crane, 40 Ton		659.00		724.90		
1 Dragline Bucket, 1.5 C.Y.		42.20		46.40	43.82	48.20
16 M.H., Daily Totals		$1041.20		$1305.30	$65.07	$81.57

Crew B-12Q	Hr.	Daily	Hr.	Daily	Bare Costs	Incl. O&P
1 Equip. Oper. (crane)	$23.30	$186.40	$36.60	$292.80	$21.25	$33.37
1 Equip. Oper. Oiler	19.20	153.60	30.15	241.20		
1 Hyd. Excavator, 5/8 C.Y.		351.40		386.55	21.96	24.15
16 M.H., Daily Totals		$691.40		$920.55	$43.21	$57.52

Crew No.	Bare Costs Hr.	Daily	Incl. Subs O & P Hr.	Daily	Cost Per Man-Hour Bare Costs	Incl. O&P
Crew B-12R	Hr.	Daily	Hr.	Daily	Bare Costs	Incl. O&P
1 Equip. Oper. (crane)	$23.30	$186.40	$36.60	$292.80	$21.25	$33.37
1 Equip. Oper. Oiler	19.20	153.60	30.15	241.20		
1 Hyd. Excavator, 1.5 C.Y.		679.60		747.55	42.47	46.72
16 M.H., Daily Totals		$1019.60		$1281.55	$63.72	$80.09
Crew B-12S	Hr.	Daily	Hr.	Daily	Bare Costs	Incl. O&P
1 Equip. Oper. (crane)	$23.30	$186.40	$36.60	$292.80	$21.25	$33.37
1 Equip. Oper. Oiler	19.20	153.60	30.15	241.20		
1 Hyd. Excavator, 2.5 C.Y.		1614.20		1775.60	100.88	110.97
16 M.H., Daily Totals		$1954.20		$2309.60	$122.13	$144.34
Crew B-12T	Hr.	Daily	Hr.	Daily	Bare Costs	Incl. O&P
1 Equip. Oper. (crane)	$23.30	$186.40	$36.60	$292.80	$21.25	$33.37
1 Equip. Oper. Oiler	19.20	153.60	30.15	241.20		
1 Crawler Crane, 75 Ton		871.20		958.30		
1 F.E. Attachment, 3 C.Y.		255.00		280.50	70.38	77.42
16 M.H., Daily Totals		$1466.20		$1772.80	$91.63	$110.79
Crew B-12V	Hr.	Daily	Hr.	Daily	Bare Costs	Incl. O&P
1 Equip. Oper. (crane)	$23.30	$186.40	$36.60	$292.80	$21.25	$33.37
1 Equip. Oper. Oiler	19.20	153.60	30.15	241.20		
1 Crawler Crane, 75 Ton		871.20		958.30		
1 Dragline Bucket, 3 C.Y.		75.40		82.95	59.16	65.07
16 M.H., Daily Totals		$1286.60		$1575.25	$80.41	$98.44
Crew B-13	Hr.	Daily	Hr.	Daily	Bare Costs	Incl. O&P
1 Labor Foreman (outside)	$19.50	$156.00	$31.85	$254.80	$18.85	$30.42
4 Building Laborers	17.50	560.00	28.60	915.20		
1 Equip. Oper. (crane)	23.30	186.40	36.60	292.80		
1 Equip. Oper. Oiler	19.20	153.60	30.15	241.20		
1 Hyd. Crane, 25 Ton		486.40		535.05	8.68	9.55
56 M.H., Daily Totals		$1542.40		$2239.05	$27.53	$39.97
Crew B-14	Hr.	Daily	Hr.	Daily	Bare Costs	Incl. O&P
1 Labor Foreman (outside)	$19.50	$156.00	$31.85	$254.80	$18.48	$29.97
4 Building Laborers	17.50	560.00	28.60	915.20		
1 Equip. Oper. (light)	21.40	171.20	33.60	268.80		
1 Backhoe Loader, 48 H.P.		190.40		209.45	3.96	4.36
48 M.H., Daily Totals		$1077.60		$1648.25	$22.44	$34.33
Crew B-15	Hr.	Daily	Hr.	Daily	Bare Costs	Incl. O&P
1 Equipment Oper. (med)	$22.50	$180.00	$35.35	$282.80	$19.44	$31.04
.5 Building Laborer	17.50	70.00	28.60	114.40		
2 Truck Drivers (heavy)	18.40	294.40	29.50	472.00		
2 Dump Trucks, 16 Ton		730.40		803.45		
1 Dozer, 200 H.P.		775.80		853.40	53.79	59.17
28 M.H., Daily Totals		$2050.60		$2526.05	$73.23	$90.21
Crew B-16	Hr.	Daily	Hr.	Daily	Bare Costs	Incl. O&P
1 Labor Foreman (outside)	$19.50	$156.00	$31.85	$254.80	$18.22	$29.63
2 Building Laborers	17.50	280.00	28.60	457.60		
1 Truck Driver (heavy)	18.40	147.20	29.50	236.00		
1 Dump Truck, 16 Ton		365.20		401.70	11.41	12.55
32 M.H., Daily Totals		$948.40		$1350.10	$29.63	$42.18

Crew No.	Bare Costs Hr.	Daily	Incl. Subs O & P Hr.	Daily	Cost Per Man-Hour Bare Costs	Incl. O&P
Crew B-17	Hr.	Daily	Hr.	Daily	Bare Costs	Incl. O&P
2 Building Laborers	$17.50	$280.00	$28.60	$457.60	$18.70	$30.07
1 Equip. Oper. (light)	21.40	171.20	33.60	268.80		
1 Truck Driver (heavy)	18.40	147.20	29.50	236.00		
1 Backhoe Loader, 48 H.P.		190.40		209.45		
1 Dump Truck, 12 Ton		298.80		328.70	15.28	16.81
32 M.H., Daily Totals		$1087.60		$1500.55	$33.98	$46.88
Crew B-18	Hr.	Daily	Hr.	Daily	Bare Costs	Incl. O&P
1 Labor Foreman (outside)	$19.50	$156.00	$31.85	$254.80	$18.16	$29.68
2 Building Laborers	17.50	280.00	28.60	457.60		
1 Vibrating Compactor		43.40		47.75	1.80	1.98
24 M.H., Daily Totals		$479.40		$760.15	$19.96	$31.66
Crew B-19	Hr.	Daily	Hr.	Daily	Bare Costs	Incl. O&P
1 Pile Driver Foreman	$24.20	$193.60	$41.65	$333.20	$22.35	$37.22
4 Pile Drivers	22.20	710.40	38.20	1222.40		
2 Equip. Oper. (crane)	23.30	372.80	36.60	585.60		
1 Equip. Oper. Oiler	19.20	153.60	30.15	241.20		
1 Crane, 40 Ton & Access.		659.00		724.90		
60 L.F. Leads, 15K Ft. Lbs.		60.00		66.00		
1 Hammer, 15K Ft. Lbs.		258.40		284.25		
1 Air Compr., 600 C.F.M.		269.20		296.10		
2-50 Ft. Air Hoses, 3" Dia.		20.40		22.45	19.79	21.77
64 M.H., Daily Totals		$2697.40		$3776.10	$42.14	$58.99
Crew B-20	Hr.	Daily	Hr.	Daily	Bare Costs	Incl. O&P
1 Labor Foreman (out)	$19.50	$156.00	$31.85	$254.80	$19.88	$32.36
1 Skilled Worker	22.65	181.20	36.65	293.20		
1 Building Laborer	17.50	140.00	28.60	228.80		
24 M.H., Daily Totals		$477.20		$776.80	$19.88	$32.36
Crew B-21	Hr.	Daily	Hr.	Daily	Bare Costs	Incl. O&P
1 Labor Foreman (out)	$19.50	$156.00	$31.85	$254.80	$20.37	$32.97
1 Skilled Worker	22.65	181.20	36.65	293.20		
1 Building Laborer	17.50	140.00	28.60	228.80		
.5 Equip. Oper. (crane)	23.30	93.20	36.60	146.40		
.5 S.P. Crane, 5 Ton		104.00		114.40	3.71	4.08
28 M.H., Daily Totals		$674.40		$1037.60	$24.08	$37.05
Crew B-22	Hr.	Daily	Hr.	Daily	Bare Costs	Incl. O&P
1 Labor Foreman (out)	$19.50	$156.00	$31.85	$254.80	$20.56	$33.21
1 Skilled Worker	22.65	181.20	36.65	293.20		
1 Building Laborer	17.50	140.00	28.60	228.80		
.75 Equip. Oper. (crane)	23.30	139.80	36.60	219.60		
.75 S.P. Crane, 5 Ton		156.00		171.60	5.20	5.72
30 M.H., Daily Totals		$773.00		$1168.00	$25.76	$38.93
Crew B-23	Hr.	Daily	Hr.	Daily	Bare Costs	Incl. O&P
1 Labor Foreman (outside)	$19.50	$156.00	$31.85	$254.80	$17.90	$29.25
4 Building Laborers	17.50	560.00	28.60	915.20		
1 Drill Rig		395.60		435.15		
1 Light Truck, 3 Ton		148.40		163.25	13.60	14.96
40 M.H., Daily Totals		$1260.00		$1768.40	$31.50	$44.21
Crew B-24	Hr.	Daily	Hr.	Daily	Bare Costs	Incl. O&P
1 Cement Finisher	$21.65	$173.20	$33.85	$270.80	$20.38	$32.80
1 Building Laborer	17.50	140.00	28.60	228.80		
1 Carpenter	22.00	176.00	35.95	287.60		
24 M.H., Daily Totals		$489.20		$787.20	$20.38	$32.80

Crew B-25

Crew No.	Bare Costs Hr.	Daily	Incl. Subs O & P Hr.	Daily	Cost Per Man-Hour Bare Costs	Incl. O&P
1 Labor Foreman	$19.50	$156.00	$31.85	$254.80	$19.04	$30.73
7 Laborers	17.50	980.00	28.60	1601.60		
3 Equip. Oper. (med.)	22.50	540.00	35.35	848.40		
1 Asphalt Paver, 130 H.P.		1086.40		1195.05		
1 Tandem Roller, 10 Ton		187.60		206.35		
1 Roller, Pneumatic Wheel		236.20		259.80	17.16	18.87
88 M.H., Daily Totals		$3186.20		$4366.00	$36.20	$49.60

Crew B-25B

Crew No.	Bare Costs Hr.	Daily	Incl. Subs O & P Hr.	Daily	Cost Per Man-Hour Bare Costs	Incl. O&P
1 Labor Foreman	$19.50	$156.00	$31.85	$254.80	$19.33	$31.12
7 Laborers	17.50	980.00	28.60	1601.60		
4 Equip. Oper. (medium)	22.50	720.00	35.35	1131.20		
1 Asphalt Paver, 130 H.P.		1086.40		1195.05		
2 Rollers, Steel Wheel		375.20		412.70		
1 Roller, Pneumatic Wheel		236.20		259.80	17.68	19.45
96 M.H., Daily Totals		$3553.80		$4855.15	$37.01	$50.57

Crew B-26

Crew No.	Bare Costs Hr.	Daily	Incl. Subs O & P Hr.	Daily	Cost Per Man-Hour Bare Costs	Incl. O&P
1 Labor Foreman (outside)	$19.50	$156.00	$31.85	$254.80	$19.55	$31.78
6 Building Laborers	17.50	840.00	28.60	1372.80		
2 Equip. Oper. (med.)	22.50	360.00	35.35	565.60		
1 Rodman (reinf.)	23.90	191.20	41.65	333.20		
1 Cement Finisher	21.65	173.20	33.85	270.80		
1 Grader, 30,000 Lbs.		521.00		573.10		
1 Paving Mach. & Equip.		1175.80		1293.40	19.28	21.21
88 M.H., Daily Totals		$3417.20		$4663.70	$38.83	$52.99

Crew B-27

Crew No.	Bare Costs Hr.	Daily	Incl. Subs O & P Hr.	Daily	Cost Per Man-Hour Bare Costs	Incl. O&P
1 Labor Foreman (outside)	$19.50	$156.00	$31.85	$254.80	$18.00	$29.41
3 Building Laborers	17.50	420.00	28.60	686.40		
1 Berm Machine		58.20		64.00	1.81	2.00
32 M.H., Daily Totals		$634.20		$1005.20	$19.81	$31.41

Crew B-28

Crew No.	Bare Costs Hr.	Daily	Incl. Subs O & P Hr.	Daily	Cost Per Man-Hour Bare Costs	Incl. O&P
2 Carpenters	$22.00	$352.00	$35.95	$575.20	$20.50	$33.50
1 Building Laborer	17.50	140.00	28.60	228.80		
24 M.H., Daily Totals		$492.00		$804.00	$20.50	$33.50

Crew B-29

Crew No.	Bare Costs Hr.	Daily	Incl. Subs O & P Hr.	Daily	Cost Per Man-Hour Bare Costs	Incl. O&P
1 Labor Foreman (outside)	$19.50	$156.00	$31.85	$254.80	$18.85	$30.42
4 Building Laborers	17.50	560.00	28.60	915.20		
1 Equip. Oper. (crane)	23.30	186.40	36.60	292.80		
1 Equip. Oper. Oiler	19.20	153.60	30.15	241.20		
1 Gradall, 3 Ton, 1/2 C.Y.		555.60		611.15	9.92	10.91
56 M.H., Daily Totals		$1611.60		$2315.15	$28.77	$41.33

Crew B-30

Crew No.	Bare Costs Hr.	Daily	Incl. Subs O & P Hr.	Daily	Cost Per Man-Hour Bare Costs	Incl. O&P
1 Equip. Oper. (med.)	$22.50	$180.00	$35.35	$282.80	$19.76	$31.45
2 Truck Drivers (heavy)	18.40	294.40	29.50	472.00		
1 Hyd. Excavator, 1.5 C.Y.		679.60		747.55		
2 Dump Trucks, 16 Ton		730.40		803.45	58.75	64.62
24 M.H., Daily Totals		$1884.40		$2305.80	$78.51	$96.07

Crew B-31

Crew No.	Bare Costs Hr.	Daily	Incl. Subs O & P Hr.	Daily	Cost Per Man-Hour Bare Costs	Incl. O&P
1 Labor Foreman (outside)	$19.50	$156.00	$31.85	$254.80	$18.80	$30.72
3 Building Laborers	17.50	420.00	28.60	686.40		
1 Carpenter	22.00	176.00	35.95	287.60		
1 Air Compr., 250 C.F.M.		88.80		97.70		
1 Sheeting Driver		9.20		10.10		
2-50 Ft. Air Hoses, 1.5" Dia.		10.80		11.90	2.72	2.99
40 M.H., Daily Totals		$860.80		$1348.50	$21.52	$33.71

Crew B-32

Crew No.	Bare Costs Hr.	Daily	Incl. Subs O & P Hr.	Daily	Cost Per Man-Hour Bare Costs	Incl. O&P
1 Highway Laborer	$17.50	$140.00	$28.60	$228.80	$21.25	$33.66
3 Equip. Oper. (med.)	22.50	540.00	35.35	848.40		
1 Grader, 30,000 Lbs.		521.00		573.10		
1 Tandem Roller, 10 Ton		187.60		206.35		
1 Dozer, 200 H.P.		775.80		853.40	46.38	51.02
32 M.H., Daily Totals		$2164.40		$2710.05	$67.63	$84.68

Crew B-32A

Crew No.	Bare Costs Hr.	Daily	Incl. Subs O & P Hr.	Daily	Cost Per Man-Hour Bare Costs	Incl. O&P
1 Laborer	$17.50	$140.00	$28.60	$228.80	$20.83	$33.10
2 Equip. Oper. (medium)	22.50	360.00	35.35	565.60		
1 Grader, 30,000 Lbs.		521.00		573.10		
1 Roller, Vibr., 29,000 Lbs.		341.60		375.75	35.94	39.53
24 M.H., Daily Totals		$1362.60		$1743.25	$56.77	$72.63

Crew B-32B

Crew No.	Bare Costs Hr.	Daily	Incl. Subs O & P Hr.	Daily	Cost Per Man-Hour Bare Costs	Incl. O&P
1 Laborer	$17.50	$140.00	$28.60	$228.80	$20.83	$33.10
2 Equip. Oper. (medium)	22.50	360.00	35.35	565.60		
1 Dozer, 200 H.P.		775.80		853.40		
1 Roller, Vibr., 29,000 Lbs.		341.60		375.75	46.55	51.21
24 M.H., Daily Totals		$1617.40		$2023.55	$67.38	$84.31

Crew B-32C

Crew No.	Bare Costs Hr.	Daily	Incl. Subs O & P Hr.	Daily	Cost Per Man-Hour Bare Costs	Incl. O&P
1 Labor Foreman	$19.50	$156.00	$31.85	$254.80	$20.33	$32.51
2 Laborers	17.50	280.00	28.60	457.60		
3 Equip. Oper. (medium)	22.50	540.00	35.35	848.40		
1 Grader, 30,000 Lbs.		521.00		573.10		
1 Roller, Steel Wheel		187.60		206.35		
1 Dozer, 200 H.P.		775.80		853.40	30.92	34.01
48 M.H., Daily Totals		$2460.40		$3193.65	$51.25	$66.52

Crew B-33

Crew No.	Bare Costs Hr.	Daily	Incl. Subs O & P Hr.	Daily	Cost Per Man-Hour Bare Costs	Incl. O&P
1 Equip. Oper. (med.)	$22.50	$180.00	$35.35	$282.80	$21.07	$33.42
.5 Building Laborer	17.50	70.00	28.60	114.40		
.25 Equip. Oper. (med.)	22.50	45.00	35.35	70.70		
14 M.H., Daily Totals		$295.00		$467.90	$21.07	$33.42

Crew B-33A

Crew No.	Bare Costs Hr.	Daily	Incl. Subs O & P Hr.	Daily	Cost Per Man-Hour Bare Costs	Incl. O&P
1 Equip. Oper. (med.)	$22.50	$180.00	$35.35	$282.80	$21.07	$33.42
.5 Building Laborer	17.50	70.00	28.60	114.40		
.25 Equip. Oper. (med.)	22.50	45.00	35.35	70.70		
1 Scraper, Towed, 7 C.Y.		61.60		67.75		
1 Dozer, 300 H.P.		861.80		948.00		
.25 Dozer, 300 H.P.		215.45		237.00	81.34	89.48
14 M.H., Daily Totals		$1433.85		$1720.65	$102.41	$122.90

CREWS

Crew B-33B

Crew No.	Bare Costs Hr.	Bare Costs Daily	Incl. Subs O&P Hr.	Incl. Subs O&P Daily	Cost Per Man-Hour Bare Costs	Cost Per Man-Hour Incl. O&P
1 Equip. Oper. (med.)	$22.50	$180.00	$35.35	$282.80	$21.07	$33.42
.5 Building Laborer	17.50	70.00	28.60	114.40		
.25 Equip. Oper. (med.)	22.50	45.00	35.35	70.70		
1 Scraper, Towed, 10 C.Y.		170.00		187.00		
1 Dozer, 300 H.P.		861.80		948.00		
.25 Dozer, 300 H.P.		215.45		237.00	89.08	98.00
14 M.H., Daily Totals		$1542.25		$1839.90	$110.15	$131.42

Crew B-33C

Crew No.	Bare Costs Hr.	Bare Costs Daily	Incl. Subs O&P Hr.	Incl. Subs O&P Daily	Cost Per Man-Hour Bare Costs	Cost Per Man-Hour Incl. O&P
1 Equip. Oper. (med.)	$22.50	$180.00	$35.35	$282.80	$21.07	$33.42
.5 Building Laborer	17.50	70.00	28.60	114.40		
.25 Equip. Oper. (med.)	22.50	45.00	35.35	70.70		
1 Scraper, Towed, 12 C.Y.		170.00		187.00		
1 Dozer, 300 H.P.		861.80		948.00		
.25 Dozer, 300 H.P.		215.45		237.00	89.08	98.00
14 M.H., Daily Totals		$1542.25		$1839.90	$110.15	$131.42

Crew B-33D

Crew No.	Bare Costs Hr.	Bare Costs Daily	Incl. Subs O&P Hr.	Incl. Subs O&P Daily	Cost Per Man-Hour Bare Costs	Cost Per Man-Hour Incl. O&P
1 Equip. Oper. (med.)	$22.50	$180.00	$35.35	$282.80	$21.07	$33.42
.5 Building Laborer	17.50	70.00	28.60	114.40		
.25 Equip. Oper. (med.)	22.50	45.00	35.35	70.70		
1 S.P. Scraper, 14 C.Y.		1382.60		1520.85		
.25 Dozer, 300 H.P.		215.45		237.00	114.14	125.56
14 M.H., Daily Totals		$1893.05		$2225.75	$135.21	$158.98

Crew B-33E

Crew No.	Bare Costs Hr.	Bare Costs Daily	Incl. Subs O&P Hr.	Incl. Subs O&P Daily	Cost Per Man-Hour Bare Costs	Cost Per Man-Hour Incl. O&P
1 Equip. Oper. (med.)	$22.50	$180.00	$35.35	$282.80	$21.07	$33.42
.5 Building Laborer	17.50	70.00	28.60	114.40		
.25 Equip. Oper. (med.)	22.50	45.00	35.35	70.70		
1 S.P. Scraper, 24 C.Y.		1650.60		1815.65		
.25 Dozer, 300 H.P.		215.45		237.00	133.28	146.61
14 M.H., Daily Totals		$2161.05		$2520.55	$154.35	$180.03

Crew B-33F

Crew No.	Bare Costs Hr.	Bare Costs Daily	Incl. Subs O&P Hr.	Incl. Subs O&P Daily	Cost Per Man-Hour Bare Costs	Cost Per Man-Hour Incl. O&P
1 Equip. Oper. (med.)	$22.50	$180.00	$35.35	$282.80	$21.07	$33.42
.5 Building Laborer	17.50	70.00	28.60	114.40		
.25 Equip. Oper. (med.)	22.50	45.00	35.35	70.70		
1 Elev. Scraper, 11 C.Y.		589.20		648.10		
.25 Dozer, 300 H.P.		215.45		237.00	57.47	63.22
14 M.H., Daily Totals		$1099.65		$1353.00	$78.54	$96.64

Crew B-33G

Crew No.	Bare Costs Hr.	Bare Costs Daily	Incl. Subs O&P Hr.	Incl. Subs O&P Daily	Cost Per Man-Hour Bare Costs	Cost Per Man-Hour Incl. O&P
1 Equip. Oper. (med.)	$22.50	$180.00	$35.35	$282.80	$21.07	$33.42
.5 Building Laborer	17.50	70.00	28.60	114.40		
.25 Equip. Oper. (med.)	22.50	45.00	35.35	70.70		
1 Elev. Scraper, 20 C.Y.		774.80		852.30		
.25 Dozer, 300 H.P.		215.45		237.00	70.73	77.80
14 M.H., Daily Totals		$1285.25		$1557.20	$91.80	$111.22

Crew B-34

Crew No.	Bare Costs Hr.	Bare Costs Daily	Incl. Subs O&P Hr.	Incl. Subs O&P Daily	Cost Per Man-Hour Bare Costs	Cost Per Man-Hour Incl. O&P
1 Truck Driver (heavy)	$18.40	$147.20	$29.50	$236.00	$18.40	$29.50
8 M.H., Daily Totals		$147.20		$236.00	$18.40	$29.50

Crew B-34A

Crew No.	Bare Costs Hr.	Bare Costs Daily	Incl. Subs O&P Hr.	Incl. Subs O&P Daily	Cost Per Man-Hour Bare Costs	Cost Per Man-Hour Incl. O&P
1 Truck Driver (heavy)	$18.40	$147.20	$29.50	$236.00	$18.40	$29.50
1 Dump Truck, 12 Ton		298.80		328.70	37.35	41.08
8 M.H., Daily Totals		$446.00		$564.70	$55.75	$70.58

Crew B-34B

Crew No.	Bare Costs Hr.	Bare Costs Daily	Incl. Subs O&P Hr.	Incl. Subs O&P Daily	Cost Per Man-Hour Bare Costs	Cost Per Man-Hour Incl. O&P
1 Truck Driver (heavy)	$18.40	$147.20	$29.50	$236.00	$18.40	$29.50
1 Dump Truck, 16 Ton		365.20		401.70	45.65	50.21
8 M.H., Daily Totals		$512.40		$637.70	$64.05	$79.71

Crew B-34C

Crew No.	Bare Costs Hr.	Bare Costs Daily	Incl. Subs O&P Hr.	Incl. Subs O&P Daily	Cost Per Man-Hour Bare Costs	Cost Per Man-Hour Incl. O&P
1 Truck Driver (heavy)	$18.40	$147.20	$29.50	$236.00	$18.40	$29.50
1 Truck Tractor, 40 Ton		348.80		383.70		
1 Dump Trailer, 16.5 C.Y.		108.60		119.45	57.17	62.89
8 M.H., Daily Totals		$604.60		$739.15	$75.57	$92.39

Crew B-34D

Crew No.	Bare Costs Hr.	Bare Costs Daily	Incl. Subs O&P Hr.	Incl. Subs O&P Daily	Cost Per Man-Hour Bare Costs	Cost Per Man-Hour Incl. O&P
1 Truck Driver (heavy)	$18.40	$147.20	$29.50	$236.00	$18.40	$29.50
1 Truck Tractor, 40 Ton		348.80		383.70		
1 Dump Trailer, 20 C.Y.		109.60		120.55	57.30	63.03
8 M.H., Daily Totals		$605.60		$740.25	$75.70	$92.53

Crew B-34E

Crew No.	Bare Costs Hr.	Bare Costs Daily	Incl. Subs O&P Hr.	Incl. Subs O&P Daily	Cost Per Man-Hour Bare Costs	Cost Per Man-Hour Incl. O&P
1 Truck Driver (heavy)	$18.40	$147.20	$29.50	$236.00	$18.40	$29.50
1 Truck, Off Hwy., 25 Ton		549.00		603.90	68.62	75.48
8 M.H., Daily Totals		$696.20		$839.90	$87.02	$104.98

Crew B-34F

Crew No.	Bare Costs Hr.	Bare Costs Daily	Incl. Subs O&P Hr.	Incl. Subs O&P Daily	Cost Per Man-Hour Bare Costs	Cost Per Man-Hour Incl. O&P
1 Truck Driver (heavy)	$18.40	$147.20	$29.50	$236.00	$18.40	$29.50
1 Truck, Off Hwy., 22 C.Y.		836.00		919.60	104.50	114.95
8 M.H., Daily Totals		$983.20		$1155.60	$122.90	$144.45

Crew B-34G

Crew No.	Bare Costs Hr.	Bare Costs Daily	Incl. Subs O&P Hr.	Incl. Subs O&P Daily	Cost Per Man-Hour Bare Costs	Cost Per Man-Hour Incl. O&P
1 Truck Driver (heavy)	$18.40	$147.20	$29.50	$236.00	$18.40	$29.50
1 Truck, Off Hwy., 34 C.Y.		1096.80		1206.50	137.10	150.81
8 M.H., Daily Totals		$1244.00		$1442.50	$155.50	$180.31

Crew B-34H

Crew No.	Bare Costs Hr.	Bare Costs Daily	Incl. Subs O&P Hr.	Incl. Subs O&P Daily	Cost Per Man-Hour Bare Costs	Cost Per Man-Hour Incl. O&P
1 Truck Driver (heavy)	$18.40	$147.20	$29.50	$236.00	$18.40	$29.50
1 Truck, Off Hwy., 42 C.Y.		1368.80		1505.70	171.10	188.21
8 M.H., Daily Totals		$1516.00		$1741.70	$189.50	$217.71

Crew B-34J

Crew No.	Bare Costs Hr.	Bare Costs Daily	Incl. Subs O&P Hr.	Incl. Subs O&P Daily	Cost Per Man-Hour Bare Costs	Cost Per Man-Hour Incl. O&P
1 Truck Driver (heavy)	$18.40	$147.20	$29.50	$236.00	$18.40	$29.50
1 Truck, Off Hwy., 60 C.Y.		1860.40		2046.45	232.55	255.80
8 M.H., Daily Totals		$2007.60		$2282.45	$250.95	$285.30

Crew B-34K

Crew No.	Bare Costs Hr.	Bare Costs Daily	Incl. Subs O&P Hr.	Incl. Subs O&P Daily	Cost Per Man-Hour Bare Costs	Cost Per Man-Hour Incl. O&P
1 Truck Driver (heavy)	$18.40	$147.20	$29.50	$236.00	$18.40	$29.50
1 Truck Tractor, 240 H.P.		439.00		482.90		
1 Low Bed Trailer		262.20		288.40	87.65	96.41
8 M.H., Daily Totals		$848.40		$1007.30	$106.05	$125.91

Crew B-35

Crew No.	Bare Costs Hr.	Bare Costs Daily	Incl. Subs O&P Hr.	Incl. Subs O&P Daily	Cost Per Man-Hour Bare Costs	Cost Per Man-Hour Incl. O&P
1 Laborer Foreman (out)	$19.50	$156.00	$31.85	$254.80	$21.26	$33.85
1 Skilled Worker	22.65	181.20	36.65	293.20		
1 Welder (plumber)	25.45	203.60	39.25	314.00		
1 Laborer	17.50	140.00	28.60	228.80		
1 Equip. Oper. (crane)	23.30	186.40	36.60	292.80		
1 Equip. Oper. Oiler	19.20	153.60	30.15	241.20		
1 Electric Welding Mach.		39.00		42.90		
1 Hyd. Excavator, .75 C.Y.		457.20		502.90	10.33	11.37
48 M.H., Daily Totals		$1517.00		$2170.60	$31.59	$45.22

CREWS

Crew B-36

Crew B-36	Hr.	Daily	Hr.	Daily	Bare Costs	Incl. O&P
1 Labor Foreman (outside)	$19.50	$156.00	$31.85	$254.80	$19.90	$31.95
2 Highway Laborers	17.50	280.00	28.60	457.60		
2 Equip. Oper. (med.)	22.50	360.00	35.35	565.60		
1 Dozer, 200 H.P.		775.80		853.40		
1 Aggregate Spreader		58.60		64.45		
1 Tandem Roller, 10 Ton		187.60		206.35	25.55	28.10
40 M.H., Daily Totals		$1818.00		$2402.20	$45.45	$60.05

Crew B-36A

Crew B-36A	Hr.	Daily	Hr.	Daily	Bare Costs	Incl. O&P
1 Labor Foreman	$19.50	$156.00	$31.85	$254.80	$20.64	$32.92
2 Laborers	17.50	280.00	28.60	457.60		
4 Equip. Oper. (medium)	22.50	720.00	35.35	1131.20		
1 Dozer, 200 H.P.		775.80		853.40		
1 Aggregate Spreader		58.60		64.45		
1 Roller, Steel Wheel		187.60		206.35		
1 Roller, Pneumatic Wheel		236.20		259.80	22.46	24.71
56 M.H., Daily Totals		$2414.20		$3227.60	$43.10	$57.63

Crew B-37

Crew B-37	Hr.	Daily	Hr.	Daily	Bare Costs	Incl. O&P
1 Labor Foreman (outside)	$19.50	$156.00	$31.85	$254.80	$18.48	$29.97
4 Building Laborers	17.50	560.00	28.60	915.20		
1 Equip. Oper. (light)	21.40	171.20	33.60	268.80		
1 Tandem Roller, 5 Ton		118.40		130.25	2.46	2.71
48 M.H., Daily Totals		$1005.60		$1569.05	$20.94	$32.68

Crew B-38

Crew B-38	Hr.	Daily	Hr.	Daily	Bare Costs	Incl. O&P
1 Labor Foreman (outside)	$19.50	$156.00	$31.85	$254.80	$19.68	$31.60
2 Building Laborers	17.50	280.00	28.60	457.60		
1 Equip. Oper. (light)	21.40	171.20	33.60	268.80		
1 Equip. Oper. (medium)	22.50	180.00	35.35	282.80		
1 Backhoe Loader, 48 H.P.		190.40		209.45		
1 Demol., Hyd., 1000 Lb.		337.80		371.60		
1 F.E. Loader, 170 H.P.		563.00		619.30		
1 Pavt. Rem. Bucket		38.20		42.00	28.23	31.05
40 M.H., Daily Totals		$1916.60		$2506.35	$47.91	$62.65

Crew B-39

Crew B-39	Hr.	Daily	Hr.	Daily	Bare Costs	Incl. O&P
1 Labor Foreman (outside)	$19.50	$156.00	$31.85	$254.80	$18.48	$29.97
4 Building Laborers	17.50	560.00	28.60	915.20		
1 Equipment Oper. (light)	21.40	171.20	33.60	268.80		
1 Air Compr., 250 C.F.M.		88.80		97.70		
Air Tools & Accessories		27.60		30.35		
2-50 Ft. Air Hoses, 1.5" Dia.		10.80		11.90	2.65	2.91
48 M.H., Daily Totals		$1014.40		$1578.75	$21.13	$32.88

Crew B-40

Crew B-40	Hr.	Daily	Hr.	Daily	Bare Costs	Incl. O&P
1 Pile Driver Foreman	$24.20	$193.60	$41.65	$333.20	$22.35	$37.22
4 Pile Drivers	22.20	710.40	38.20	1222.40		
2 Equip. Oper. (crane)	23.30	372.80	36.60	585.60		
1 Equip. Oper. Oiler	19.20	153.60	30.15	241.20		
1 Crane, 40 Ton		659.00		724.90		
Vibratory Hammer & Gen.		1059.80		1165.80	26.85	29.54
64 M.H., Daily Totals		$3149.20		$4273.10	$49.20	$66.76

Crew B-41

Crew B-41	Hr.	Daily	Hr.	Daily	Bare Costs	Incl. O&P
1 Labor Foreman (outside)	$19.50	$156.00	$31.85	$254.80	$18.20	$29.62
4 Building Laborers	17.50	560.00	28.60	915.20		
.25 Equip. Oper. (crane)	23.30	46.60	36.60	73.20		
.25 Equip. Oper. Oiler	19.20	38.40	30.15	60.30		
.25 Crawler Crane, 40 Ton		164.75		181.25	3.74	4.11
44 M.H., Daily Totals		$965.75		$1484.75	$21.94	$33.73

Crew B-42

Crew B-42	Hr.	Daily	Hr.	Daily	Bare Costs	Incl. O&P
1 Labor Foreman (outside)	$19.50	$156.00	$31.85	$254.80	$19.51	$32.11
4 Building Laborers	17.50	560.00	28.60	915.20		
1 Equip. Oper. (crane)	23.30	186.40	36.60	292.80		
1 Equip. Oper. Oiler	19.20	153.60	30.15	241.20		
1 Welder	24.10	192.80	43.90	351.20		
1 Hyd. Crane, 25 Ton		486.40		535.05		
1 Gas Welding Machine		68.40		75.25		
1 Horz. Boring Csg. Mch.		424.40		466.85	15.30	16.83
64 M.H., Daily Totals		$2228.00		$3132.35	$34.81	$48.94

Crew B-43

Crew B-43	Hr.	Daily	Hr.	Daily	Bare Costs	Incl. O&P
1 Labor Foreman (outside)	$19.50	$156.00	$31.85	$254.80	$19.08	$30.73
3 Building Laborers	17.50	420.00	28.60	686.40		
1 Equip. Oper. (crane)	23.30	186.40	36.60	292.80		
1 Equip. Oper. Oiler	19.20	153.60	30.15	241.20		
1 Drill Rig & Augers		744.65		819.15	15.51	17.06
48 M.H., Daily Totals		$1660.65		$2294.35	$34.59	$47.79

Crew B-44

Crew B-44	Hr.	Daily	Hr.	Daily	Bare Costs	Incl. O&P
1 Pile Driver Foreman	$24.20	$193.60	$41.65	$333.20	$22.13	$37.03
4 Pile Drivers	22.20	710.40	38.20	1222.40		
2 Equip. Oper. (crane)	23.30	372.80	36.60	585.60		
1 Building Laborer	17.50	140.00	28.60	228.80		
1 Crane, 40 Ton, & Access.		1153.25		1268.60		
45 L.F. Leads, 15K Ft. Lbs.		45.00		49.50	18.72	20.59
64 M.H., Daily Totals		$2615.05		$3688.10	$40.85	$57.62

Crew B-45

Crew B-45	Hr.	Daily	Hr.	Daily	Bare Costs	Incl. O&P
1 Equip. Oper. (med.)	$22.50	$180.00	$35.35	$282.80	$20.45	$32.42
1 Truck Driver (heavy)	18.40	147.20	29.50	236.00		
1 Dist. Tank Truck, 3K Gal.		308.20		339.00		
1 Tractor, 4 x 2, 250 H.P.		315.00		346.50	38.95	42.84
16 M.H., Daily Totals		$950.40		$1204.30	$59.40	$75.26

Crew B-46

Crew B-46	Hr.	Daily	Hr.	Daily	Bare Costs	Incl. O&P
1 Pile Driver Foreman	$24.20	$193.60	$41.65	$333.20	$20.18	$33.97
2 Pile Drivers	22.20	355.20	38.20	611.20		
3 Building Laborers	17.50	420.00	28.60	686.40		
1 Chain Saw, 36" Long		41.60		45.75	.86	.95
48 M.H., Daily Totals		$1010.40		$1676.55	$21.04	$34.92

Crew B-47

Crew B-47	Hr.	Daily	Hr.	Daily	Bare Costs	Incl. O&P
1 Blast Foreman	$19.50	$156.00	$31.85	$254.80	$19.46	$31.35
1 Driller	17.50	140.00	28.60	228.80		
1 Equip. Oper. (light)	21.40	171.20	33.60	268.80		
1 Crawler Type Drill, 4"		217.00		238.70		
1 Air Compr., 600 C.F.M.		269.20		296.10		
2-50 Ft. Air Hoses, 3" Dia.		20.40		22.45	21.10	23.21
24 M.H., Daily Totals		$973.80		$1309.65	$40.56	$54.56

CREWS

Crew No.	Bare Costs Hr.	Daily	Incl. Subs O & P Hr.	Daily	Cost Per Man-Hour Bare Costs	Incl. O&P
Crew B-47A	Hr.	Daily	Hr.	Daily	Bare Costs	Incl. O&P
1 Drilling Foreman	$19.50	$156.00	$31.85	$254.80	$20.66	$32.86
1 Equip. Oper. (heavy)	23.30	186.40	36.60	292.80		
1 Oiler	19.20	153.60	30.15	241.20		
1 Quarry Drill		451.40		496.55	18.80	20.68
24 M.H., Daily Totals		$947.40		$1285.35	$39.46	$53.54
Crew B-48	Hr.	Daily	Hr.	Daily	Bare Costs	Incl. O&P
1 Labor Foreman (outside)	$19.50	$156.00	$31.85	$254.80	$19.41	$31.14
3 Building Laborers	17.50	420.00	28.60	686.40		
1 Equip. Oper. (crane)	23.30	186.40	36.60	292.80		
1 Equip. Oper. Oiler	19.20	153.60	30.15	241.20		
1 Equip. Oper. (light)	21.40	171.20	33.60	268.80		
1 Centr. Water Pump, 6"		146.20		160.80		
1-20 Ft. Suction Hose, 6"		23.40		25.75		
1-50 Ft. Disch. Hose, 6"		13.00		14.30		
1 Drill Rig & Augers		744.65		819.15	16.55	18.21
56 M.H., Daily Totals		$2014.45		$2764.00	$35.96	$49.35
Crew B-49	Hr.	Daily	Hr.	Daily	Bare Costs	Incl. O&P
1 Labor Foreman (outside)	$19.50	$156.00	$31.85	$254.80	$20.25	$32.83
3 Building Laborers	17.50	420.00	28.60	686.40		
2 Equip. Oper. (crane)	23.30	372.80	36.60	585.60		
2 Equip. Oper. Oilers	19.20	307.20	30.15	482.40		
1 Equip. Oper. (light)	21.40	171.20	33.60	268.80		
2 Pile Drivers	22.20	355.20	38.20	611.20		
1 Hyd. Crane, 25 Ton		486.40		535.05		
1 Centr. Water Pump, 6"		146.20		160.80		
1-20 Ft. Suction Hose, 6"		23.40		25.75		
1-50 Ft. Disch. Hose, 6"		13.00		14.30		
1 Drill Rig & Augers		744.65		819.15	16.06	17.67
88 M.H., Daily Totals		$3196.05		$4444.25	$36.31	$50.50
Crew B-50	Hr.	Daily	Hr.	Daily	Bare Costs	Incl. O&P
2 Pile Driver Foremen	$24.20	$387.20	$41.65	$666.40	$21.42	$35.83
6 Pile Drivers	22.20	1065.60	38.20	1833.60		
2 Equip. Oper. (crane)	23.30	372.80	36.60	585.60		
1 Equip. Oper. Oiler	19.20	153.60	30.15	241.20		
3 Building Laborers	17.50	420.00	28.60	686.40		
1 Crane, 40 Ton		659.00		724.90		
60 L.F. Leads, 15K Ft. Lbs.		60.00		66.00		
1 Hammer, 15K Ft. Lbs.		258.40		284.25		
1 Air Compr., 600 C.F.M.		269.20		296.10		
2-50 Ft. Air Hoses, 3" Dia.		20.40		22.45		
1 Chain Saw, 36" Long		41.60		45.75	11.68	12.85
112 M.H., Daily Totals		$3707.80		$5452.65	$33.10	$48.68
Crew B-51	Hr.	Daily	Hr.	Daily	Bare Costs	Incl. O&P
1 Labor Foreman (outside)	$19.50	$156.00	$31.85	$254.80	$17.93	$29.20
4 Building Laborers	17.50	560.00	28.60	915.20		
1 Truck Driver (light)	18.10	144.80	29.00	232.00		
1 Light Truck, 1.5 Ton		146.00		160.60	3.04	3.34
48 M.H., Daily Totals		$1006.80		$1562.60	$20.97	$32.54

Crew No.	Bare Costs Hr.	Daily	Incl. Subs O & P Hr.	Daily	Cost Per Man-Hour Bare Costs	Incl. O&P
Crew B-52	Hr.	Daily	Hr.	Daily	Bare Costs	Incl. O&P
1 Carpenter Foreman	$24.00	$192.00	$39.20	$313.60	$20.47	$33.32
1 Carpenter	22.00	176.00	35.95	287.60		
3 Building Laborers	17.50	420.00	28.60	686.40		
1 Cement Finisher	21.65	173.20	33.85	270.80		
.5 Rodman (reinf.)	23.90	95.60	41.65	166.60		
.5 Equip. Oper. (med.)	22.50	90.00	35.35	141.40		
.5 F.E. Ldr., T.M., 2.5 C.Y.		403.90		444.30	7.21	7.93
56 M.H., Daily Totals		$1550.70		$2310.70	$27.68	$41.25
Crew B-53	Hr.	Daily	Hr.	Daily	Bare Costs	Incl. O&P
1 Equip. Oper. (light)	$21.40	$171.20	$33.60	$268.80	$21.40	$33.60
1 Trencher, Chain, 12 H.P.		53.20		58.50	6.65	7.31
8 M.H., Daily Totals		$224.40		$327.30	$28.05	$40.91
Crew B-54	Hr.	Daily	Hr.	Daily	Bare Costs	Incl. O&P
1 Equip. Oper. (light)	$21.40	$171.20	$33.60	$268.80	$21.40	$33.60
1 Trencher, Chain, 40 H.P.		135.20		148.70	16.90	18.58
8 M.H., Daily Totals		$306.40		$417.50	$38.30	$52.18
Crew B-55	Hr.	Daily	Hr.	Daily	Bare Costs	Incl. O&P
2 Building Laborers	$17.50	$280.00	$28.60	$457.60	$17.70	$28.73
1 Truck Driver (light)	18.10	144.80	29.00	232.00		
1 Flatbed Truck w/Auger		395.60		435.15		
1 Truck, 3 Ton		148.40		163.25	22.66	24.93
24 M.H., Daily Totals		$968.80		$1288.00	$40.36	$53.66
Crew B-56	Hr.	Daily	Hr.	Daily	Bare Costs	Incl. O&P
1 Building Laborer	$17.50	$140.00	$28.60	$228.80	$19.45	$31.10
1 Equip. Oper. (light)	21.40	171.20	33.60	268.80		
1 Crawler Type Drill, 4"		217.00		238.70		
1 Air Compr., 600 C.F.M.		269.20		296.10		
1-50 Ft. Air Hose, 3" Dia.		10.20		11.20	31.02	34.12
16 M.H., Daily Totals		$807.60		$1043.60	$50.47	$65.22
Crew B-57	Hr.	Daily	Hr.	Daily	Bare Costs	Incl. O&P
1 Labor Foreman (outside)	$19.50	$156.00	$31.85	$254.80	$19.73	$31.56
2 Building Laborers	17.50	280.00	28.60	457.60		
1 Equip. Oper. (crane)	23.30	186.40	36.60	292.80		
1 Equip. Oper. (light)	21.40	171.20	33.60	268.80		
1 Equip. Oper. Oiler	19.20	153.60	30.15	241.20		
1 Power Shovel, 1 C.Y.		435.80		479.40		
1 Clamshell Bucket, 1 C.Y.		59.60		65.55		
1 Centr. Water Pump, 6"		146.20		160.80		
1-20 Ft. Suction Hose, 6"		23.40		25.75		
20-50 Ft. Disch. Hoses, 6"		260.00		286.00	19.27	21.19
48 M.H., Daily Totals		$1872.20		$2532.70	$39.00	$52.75
Crew B-58	Hr.	Daily	Hr.	Daily	Bare Costs	Incl. O&P
2 Building Laborers	$17.50	$280.00	$28.60	$457.60	$18.80	$30.26
1 Equip. Oper. (light)	21.40	171.20	33.60	268.80		
1 Backhoe Loader, 48 H.P.		190.40		209.45		
1 Small Helicopter		2259.00		2484.90	102.05	112.26
24 M.H., Daily Totals		$2900.60		$3420.75	$120.85	$142.52
Crew B-59	Hr.	Daily	Hr.	Daily	Bare Costs	Incl. O&P
1 Truck Driver (heavy)	$18.40	$147.20	$29.50	$236.00	$18.40	$29.50
1 Truck, 30 Ton		268.40		295.25		
1 Water Tank, 5000 Gal.		184.80		203.30	56.65	62.31
8 M.H., Daily Totals		$600.40		$734.55	$75.05	$91.81

CREWS

Crew No.	Bare Costs		Incl. Subs O & P		Cost Per Man-Hour	
Crew B-60	Hr.	Daily	Hr.	Daily	Bare Costs	Incl. O&P
1 Labor Foreman (outside)	$19.50	$156.00	$31.85	$254.80	$19.97	$31.85
2 Building Laborers	17.50	280.00	28.60	457.60		
1 Equip. Oper. (crane)	23.30	186.40	36.60	292.80		
2 Equip. Oper. (light)	21.40	342.40	33.60	537.60		
1 Equip. Oper. Oiler	19.20	153.60	30.15	241.20		
1 Crawler Crane, 40 Ton		659.00		724.90		
45 L.F. Leads, 15K Ft. Lbs.		45.00		49.50		
1 Backhoe Loader, 48 H.P.		190.40		209.45	15.97	17.56
56 M.H., Daily Totals		$2012.80		$2767.85	$35.94	$49.41
Crew B-61	Hr.	Daily	Hr.	Daily	Bare Costs	Incl. O&P
1 Labor Foreman (outside)	$19.50	$156.00	$31.85	$254.80	$18.68	$30.25
3 Building Laborers	17.50	420.00	28.60	686.40		
1 Equip. Oper. (light)	21.40	171.20	33.60	268.80		
1 Cement Mixer, 2 C.Y.		230.60		253.65		
1 Air Compr., 160 C.F.M.		85.40		93.95	7.90	8.69
40 M.H., Daily Totals		$1063.20		$1557.60	$26.58	$38.94
Crew B-62	Hr.	Daily	Hr.	Daily	Bare Costs	Incl. O&P
2 Building Laborers	$17.50	$280.00	$28.60	$457.60	$18.80	$30.26
1 Equip. Oper. (light)	21.40	171.20	33.60	268.80		
1 Loader, Skid Steer		88.80		97.70	3.70	4.07
24 M.H., Daily Totals		$540.00		$824.10	$22.50	$34.33
Crew B-63	Hr.	Daily	Hr.	Daily	Bare Costs	Incl. O&P
4 Building Laborers	$17.50	$560.00	$28.60	$915.20	$18.28	$29.60
1 Equip. Oper. (light)	21.40	171.20	33.60	268.80		
1 Loader, Skid Steer		88.80		97.70	2.22	2.44
40 M.H., Daily Totals		$820.00		$1281.70	$20.50	$32.04
Crew B-64	Hr.	Daily	Hr.	Daily	Bare Costs	Incl. O&P
1 Building Laborer	$17.50	$140.00	$28.60	$228.80	$17.80	$28.80
1 Truck Driver (light)	18.10	144.80	29.00	232.00		
1 Power Mulcher (small)		88.60		97.45		
1 Light Truck, 1.5 Ton		146.00		160.60	14.66	16.12
16 M.H., Daily Totals		$519.40		$718.85	$32.46	$44.92
Crew B-65	Hr.	Daily	Hr.	Daily	Bare Costs	Incl. O&P
1 Building Laborer	$17.50	$140.00	$28.60	$228.80	$17.80	$28.80
1 Truck Driver (light)	18.10	144.80	29.00	232.00		
1 Power Mulcher (large)		202.60		222.85		
1 Light Truck, 1.5 Ton		146.00		160.60	21.78	23.96
16 M.H., Daily Totals		$633.40		$844.25	$39.58	$52.76
Crew B-66	Hr.	Daily	Hr.	Daily	Bare Costs	Incl. O&P
1 Equip. Oper. (light)	$21.40	$171.20	$33.60	$268.80	$21.40	$33.60
1 Backhoe Ldr. w/Attchmt.		162.60		178.85	20.32	22.35
8 M.H., Daily Totals		$333.80		$447.65	$41.72	$55.95
Crew B-67	Hr.	Daily	Hr.	Daily	Bare Costs	Incl. O&P
1 Millwright	$22.95	$183.60	$35.95	$287.60	$22.17	$34.77
1 Equip. Oper. (light)	21.40	171.20	33.60	268.80		
1 Forklift		183.00		201.30	11.43	12.58
16 M.H., Daily Totals		$537.80		$757.70	$33.60	$47.35
Crew B-68	Hr.	Daily	Hr.	Daily	Bare Costs	Incl. O&P
2 Millwrights	$22.95	$367.20	$35.95	$575.20	$22.43	$35.16
1 Equip. Oper. (light)	21.40	171.20	33.60	268.80		
1 Forklift		183.00		201.30	7.62	8.38
24 M.H., Daily Totals		$721.40		$1045.30	$30.05	$43.54

Crew No.	Bare Costs		Incl. Subs O & P		Cost Per Man-Hour	
Crew B-69	Hr.	Daily	Hr.	Daily	Bare Costs	Incl. O&P
1 Labor Foreman (outside)	$19.50	$156.00	$31.85	$254.80	$19.08	$30.73
3 Highway Laborers	17.50	420.00	28.60	686.40		
1 Equip Oper. (crane)	23.30	186.40	36.60	292.80		
1 Equip Oper. Oiler	19.20	153.60	30.15	241.20		
1 Truck Crane, 80 Ton		1103.40		1213.75	22.98	25.28
48 M.H., Daily Totals		$2019.40		$2688.95	$42.06	$56.01
Crew B-69A	Hr.	Daily	Hr.	Daily	Bare Costs	Incl. O&P
1 Labor Foreman	$19.50	$156.00	$31.85	$254.80	$19.35	$31.14
3 Laborers	17.50	420.00	28.60	686.40		
1 Equip. Oper. (medium)	22.50	180.00	35.35	282.80		
1 Concrete Finisher	21.65	173.20	33.85	270.80		
1 Curb Paver		394.80		434.30	8.22	9.04
48 M.H., Daily Totals		$1324.00		$1929.10	$27.57	$40.18
Crew B-69B	Hr.	Daily	Hr.	Daily	Bare Costs	Incl. O&P
1 Labor Foreman	$19.50	$156.00	$31.85	$254.80	$19.35	$31.14
3 Laborers	17.50	420.00	28.60	686.40		
1 Equip. Oper. (medium)	22.50	180.00	35.35	282.80		
1 Cement Finisher	21.65	173.20	33.85	270.80		
1 Curb/Gutter Paver		858.00		943.80	17.87	19.66
48 M.H., Daily Totals		$1787.20		$2438.60	$37.22	$50.80
Crew B-70	Hr.	Daily	Hr.	Daily	Bare Costs	Incl. O&P
1 Labor Foreman (outside)	$19.50	$156.00	$31.85	$254.80	$19.92	$31.95
3 Highway Laborers	17.50	420.00	28.60	686.40		
3 Equip. Oper. (med.)	22.50	540.00	35.35	848.40		
1 Motor Grader, 30,000 Lb.		521.00		573.10		
1 Grader Attach., Ripper		51.60		56.75		
1 Road Sweeper, S.P.		148.60		163.45		
1 F.E. Loader, 1-3/4 C.Y.		309.60		340.55	18.40	20.24
56 M.H., Daily Totals		$2146.80		$2923.45	$38.32	$52.19
Crew B-71	Hr.	Daily	Hr.	Daily	Bare Costs	Incl. O&P
1 Labor Foreman (outside)	$19.50	$156.00	$31.85	$254.80	$19.92	$31.95
3 Highway Laborers	17.50	420.00	28.60	686.40		
3 Equip. Oper. (med.)	22.50	540.00	35.35	848.40		
1 Pvmt. Profiler, 450 H.P.		3455.60		3801.15		
1 Road Sweeper, S.P.		148.60		163.45		
1 F.E. Loader, 1-3/4 C.Y.		309.60		340.55	69.88	76.87
56 M.H., Daily Totals		$5029.80		$6094.75	$89.80	$108.82
Crew B-72	Hr.	Daily	Hr.	Daily	Bare Costs	Incl. O&P
1 Labor Foreman (outside)	$19.50	$156.00	$31.85	$254.80	$20.25	$32.38
3 Highway Laborers	17.50	420.00	28.60	686.40		
4 Equip. Oper. (med.)	22.50	720.00	35.35	1131.20		
1 Pvmt. Profiler, 450 H.P.		3455.60		3801.15		
1 Hammermill, 250 H.P.		935.20		1028.70		
1 Windrow Loader		938.80		1032.70		
1 Mix Paver 165 H.P.		1230.20		1353.20		
1 Roller, Pneu. Tire, 12 T.		236.20		259.80	106.18	116.80
64 M.H., Daily Totals		$8092.00		$9547.95	$126.43	$149.18

Crew No.	Bare Costs		Incl. Subs O & P		Cost Per Man-Hour	
Crew B-73	Hr.	Daily	Hr.	Daily	Bare Costs	Incl. O&P
1 Labor Foreman (outside)	$19.50	$156.00	$31.85	$254.80	$20.87	$33.22
2 Highway Laborers	17.50	280.00	28.60	457.60		
5 Equip. Oper. (med.)	22.50	900.00	35.35	1414.00		
1 Road Mixer, 310 H.P.		871.40		958.55		
1 Roller, Tandem, 12 Ton		187.60		206.35		
1 Hammermill, 250 H.P.		935.20		1028.70		
1 Motor Grader, 30,000 Lb.		521.00		573.10		
.5 F.E. Loader, 1-3/4 C.Y.		154.80		170.30		
.5 Truck, 30 Ton		134.20		147.60		
.5 Water Tank, 5000 Gal.		92.40		101.65	45.25	49.78
64 M.H., Daily Totals		$4232.60		$5312.65	$66.12	$83.00

Crew B-74	Hr.	Daily	Hr.	Daily	Bare Costs	Incl. O&P
1 Labor Foreman (outside)	$19.50	$156.00	$31.85	$254.80	$20.47	$32.60
1 Highway Laborer	17.50	140.00	28.60	228.80		
4 Equip. Oper. (med.)	22.50	720.00	35.35	1131.20		
2 Truck Drivers (heavy)	18.40	294.40	29.50	472.00		
1 Motor Grader, 30,000 Lb.		521.00		573.10		
1 Grader Attach., Ripper		51.60		56.75		
2 Stabilizers, 310 H.P.		1198.00		1317.80		
1 Flatbed Truck, 3 Ton		148.40		163.25		
1 Chem. Spreader, Towed		89.00		97.90		
1 Vibr. Roller, 29,000 Lb.		341.60		375.75		
1 Water Tank, 5000 Gal.		184.80		203.30		
1 Truck, 30 Ton		268.40		295.25	43.79	48.17
64 M.H., Daily Totals		$4113.20		$5169.90	$64.26	$80.77

Crew B-75	Hr.	Daily	Hr.	Daily	Bare Costs	Incl. O&P
1 Labor Foreman (outside)	$19.50	$156.00	$31.85	$254.80	$20.77	$33.05
1 Highway Laborer	17.50	140.00	28.60	228.80		
4 Equip. Oper. (med.)	22.50	720.00	35.35	1131.20		
1 Truck Driver (heavy)	18.40	147.20	29.50	236.00		
1 Motor Grader, 30,000 Lb.		521.00		573.10		
1 Grader Attach., Ripper		51.60		56.75		
2 Stabilizers, 310 H.P.		1198.00		1317.80		
1 Dist. Truck, 3000 Gal.		308.20		339.00		
1 Vibr. Roller, 29,000 Lb.		341.60		375.75	43.22	47.54
56 M.H., Daily Totals		$3583.60		$4513.20	$63.99	$80.59

Crew B-76	Hr.	Daily	Hr.	Daily	Bare Costs	Incl. O&P
1 Dock Builder Foreman	$24.20	$193.60	$41.65	$333.20	$22.33	$37.33
5 Dock Builders	22.20	888.00	38.20	1528.00		
2 Equip. Oper. (crane)	23.30	372.80	36.60	585.60		
1 Equip. Oper. Oiler	19.20	153.60	30.15	241.20		
1 Crawler Crane, 50 Ton		758.00		833.80		
1 Barge, 400 Ton		384.00		422.40		
1 Hammer, 15K. Ft. Lbs.		258.40		284.25		
60 L.F. Leads, 15K. Ft. Lbs.		60.00		66.00		
1 Air Compr., 600 C.F.M.		269.20		296.10		
2-50 Ft. Air Hoses, 3" Dia.		20.40		22.45	24.30	26.73
72 M.H., Daily Totals		$3358.00		$4613.00	$46.63	$64.06

Crew B-77	Hr.	Daily	Hr.	Daily	Bare Costs	Incl. O&P
1 Labor Foreman	$19.50	$156.00	$31.85	$254.80	$18.02	$29.33
3 Laborers	17.50	420.00	28.60	686.40		
1 Truck Driver (light)	18.10	144.80	29.00	232.00		
1 Crack Cleaner, 25 H.P.		71.20		78.30		
1 Crack Filler, Trailer Mtd.		126.40		139.05		
1 Flatbed Truck, 3 Ton		148.40		163.25	8.65	9.51
40 M.H., Daily Totals		$1066.80		$1553.80	$26.67	$38.84

Crew B-78	Hr.	Daily	Hr.	Daily	Bare Costs	Incl. O&P
1 Labor Foreman	$19.50	$156.00	$31.85	$254.80	$17.93	$29.20
4 Laborers	17.50	560.00	28.60	915.20		
1 Truck Driver (light)	18.10	144.80	29.00	232.00		
1 Paint Striper, S.P.		180.40		198.45		
1 Flatbed Truck, 3 Ton		148.40		163.25		
1 Pickup Truck, 3/4 Ton		107.40		118.15	9.08	9.99
48 M.H., Daily Totals		$1297.00		$1881.85	$27.01	$39.19

Crew B-79	Hr.	Daily	Hr.	Daily	Bare Costs	Incl. O&P
1 Labor Foreman	$19.50	$156.00	$31.85	$254.80	$18.02	$29.33
3 Laborers	17.50	420.00	28.60	686.40		
1 Truck Driver (light)	18.10	144.80	29.00	232.00		
1 Thermo. Striper, T.M.		225.50		248.05		
1 Flatbed Truck, 3 Ton		148.40		163.25		
2 Pickup Trucks, 3/4 Ton		214.80		236.30	14.71	16.19
40 M.H., Daily Totals		$1309.50		$1820.80	$32.73	$45.52

Crew B-80	Hr.	Daily	Hr.	Daily	Bare Costs	Incl. O&P
1 Labor Foreman	$19.50	$156.00	$31.85	$254.80	$19.12	$30.76
1 Laborer	17.50	140.00	28.60	228.80		
1 Truck Driver (light)	18.10	144.80	29.00	232.00		
1 Equip. Oper. (light)	21.40	171.20	33.60	268.80		
1 Flatbed Truck, 3 Ton		148.40		163.25		
1 Post Driver, T.M.		256.80		282.50	12.66	13.92
32 M.H., Daily Totals		$1017.20		$1430.15	$31.78	$44.68

Crew B-81	Hr.	Daily	Hr.	Daily	Bare Costs	Incl. O&P
1 Laborer	$17.50	$140.00	$28.60	$228.80	$19.46	$31.15
1 Equip. Oper. (med.)	22.50	180.00	35.35	282.80		
1 Truck Driver (heavy)	18.40	147.20	29.50	236.00		
1 Hydromulcher, T.M.		239.80		263.80		
1 Tractor Truck, 4x2		268.40		295.25	21.17	23.29
24 M.H., Daily Totals		$975.40		$1306.65	$40.63	$54.44

Crew B-82	Hr.	Daily	Hr.	Daily	Bare Costs	Incl. O&P
1 Highway Laborer	$17.50	$140.00	$28.60	$228.80	$19.45	$31.10
1 Equip. Oper. (light)	21.40	171.20	33.60	268.80		
1 Horiz. Borer, 6 H.P.		37.00		40.70	2.31	2.54
16 M.H., Daily Totals		$348.20		$538.30	$21.76	$33.64

Crew B-83	Hr.	Daily	Hr.	Daily	Bare Costs	Incl. O&P
1 Tugboat Captain	$22.50	$180.00	$35.35	$282.80	$20.00	$31.97
1 Tugboat Hand	17.50	140.00	28.60	228.80		
1 Tugboat, 250 H.P.		456.40		502.05	28.52	31.37
16 M.H., Daily Totals		$776.40		$1013.65	$48.52	$63.34

Crew B-84	Hr.	Daily	Hr.	Daily	Bare Costs	Incl. O&P
1 Equip. Oper. (med.)	$22.50	$180.00	$35.35	$282.80	$22.50	$35.35
1 Rotary Mower/Tractor		227.40		250.15	28.42	31.26
8 M.H., Daily Totals		$407.40		$532.95	$50.92	$66.61

Crew B-85	Hr.	Daily	Hr.	Daily	Bare Costs	Incl. O&P
3 Highway Laborers	$17.50	$420.00	$28.60	$686.40	$18.68	$30.13
1 Equip. Oper. (med.)	22.50	180.00	35.35	282.80		
1 Truck Driver (heavy)	18.40	147.20	29.50	236.00		
1 Aerial Lift Truck		515.00		566.50		
1 Brush Chipper, 130 H.P.		178.80		196.70		
1 Pruning Saw, Rotary		15.20		16.70	17.72	19.49
40 M.H., Daily Totals		$1456.20		$1985.10	$36.40	$49.62

CREWS

Crew No.	Bare Costs Hr.	Bare Costs Daily	Incl. Subs O & P Hr.	Incl. Subs O & P Daily	Cost Per Man-Hour Bare Costs	Cost Per Man-Hour Incl. O&P
Crew B-86						
1 Equip. Oper. (med.)	$22.50	$180.00	$35.35	$282.80	$22.50	$35.35
1 Stump Chipper, S.P.		152.40		167.65	19.05	20.95
8 M.H., Daily Totals		$332.40		$450.45	$41.55	$56.30
Crew B-86A						
1 Equip. Oper. (medium)	$22.50	$180.00	$35.35	$282.80	$22.50	$35.35
1 Grader, 30,000 Lbs.		521.00		573.10	65.12	71.63
8 M.H., Daily Totals		$701.00		$855.90	$87.62	$106.98
Crew B-86B						
1 Equip. Oper. (medium)	$22.50	$180.00	$35.35	$282.80	$22.50	$35.35
1 Dozer, 200 H.P.		775.80		853.40	96.97	106.67
8 M.H., Daily Totals		$955.80		$1136.20	$119.47	$142.02
Crew B-87						
1 Common Laborer	$17.50	$140.00	$28.60	$228.80	$21.50	$34.00
4 Equip. Oper. (med.)	22.50	720.00	35.35	1131.20		
2 Feller Bunchers, 50 H.P.		626.80		689.50		
1 Log Chipper, 22" Tree		1820.20		2002.20		
1 Dozer, 105 H.P.		272.80		300.10		
1 Chainsaw, Gas, 36" Long		41.60		45.75	69.03	75.93
40 M.H., Daily Totals		$3621.40		$4397.55	$90.53	$109.93
Crew B-88						
1 Common Laborer	$17.50	$140.00	$28.60	$228.80	$21.78	$34.38
6 Equip. Oper. (med.)	22.50	1080.00	35.35	1696.80		
2 Feller Bunchers, 50 H.P.		626.80		689.50		
1 Log Chipper, 22" Tree		1820.20		2002.20		
2 Log Skidders, 50 H.P.		639.60		703.55		
1 Dozer, 105 H.P.		272.80		300.10		
1 Chainsaw, Gas, 36" Long		41.60		45.75	60.73	66.80
56 M.H., Daily Totals		$4621.00		$5666.70	$82.51	$101.18
Crew B-89						
1 Equip. Oper. (light)	$21.40	$171.20	$33.60	$268.80	$19.75	$31.30
1 Truck Driver (light)	18.10	144.80	29.00	232.00		
1 Truck, Stake Body, 3 Ton		148.40		163.25		
1 Concrete Saw		92.80		102.10		
1 Water Tank, 65 Gal.		6.60		7.25	15.48	17.03
16 M.H., Daily Totals		$563.80		$773.40	$35.23	$48.33
Crew B-89A						
1 Skilled Worker	$22.65	$181.20	$36.65	$293.20	$20.07	$32.62
1 Laborer	17.50	140.00	28.60	228.80		
1 Core Drill (large)		48.80		53.70	3.05	3.35
16 M.H., Daily Totals		$370.00		$575.70	$23.12	$35.97
Crew B-90						
1 Labor Foreman (outside)	$19.50	$156.00	$31.85	$254.80	$18.95	$30.48
3 Highway Laborers	17.50	420.00	28.60	686.40		
2 Equip. Oper. (light)	21.40	342.40	33.60	537.60		
2 Truck Drivers (heavy)	18.40	294.40	29.50	472.00		
1 Road Mixer, 310 H.P.		871.40		958.55		
1 Dist. Truck, 2000 Gal.		285.00		313.50	18.06	19.87
64 M.H., Daily Totals		$2369.20		$3222.85	$37.01	$50.35

Crew No.	Bare Costs Hr.	Bare Costs Daily	Incl. Subs O & P Hr.	Incl. Subs O & P Daily	Cost Per Man-Hour Bare Costs	Cost Per Man-Hour Incl. O&P
Crew B-90A						
1 Labor Foreman	$19.50	$156.00	$31.85	$254.80	$20.64	$32.92
2 Laborers	17.50	280.00	28.60	457.60		
4 Equip. Oper. (medium)	22.50	720.00	35.35	1131.20		
2 Graders, 30,000 Lbs.		1042.00		1146.20		
1 Roller, Steel Wheel		187.60		206.35		
1 Roller, Pneumatic Wheel		236.20		259.80	26.17	28.79
56 M.H., Daily Totals		$2621.80		$3455.95	$46.81	$61.71
Crew B-90B						
1 Labor Foreman	$19.50	$156.00	$31.85	$254.80	$20.33	$32.51
2 Laborers	17.50	280.00	28.60	457.60		
3 Equip. Oper. (medium)	22.50	540.00	35.35	848.40		
1 Roller, Steel Wheel		187.60		206.35		
1 Roller, Pneumatic Wheel		236.20		259.80		
1 Road Mixer, 310 H.P.		871.40		958.55	26.98	29.68
48 M.H., Daily Totals		$2271.20		$2985.50	$47.31	$62.19
Crew B-91						
1 Labor Foreman (outside)	$19.50	$156.00	$31.85	$254.80	$20.36	$32.49
2 Highway Laborers	17.50	280.00	28.60	457.60		
4 Equip. Oper. (med.)	22.50	720.00	35.35	1131.20		
1 Truck Driver (heavy)	18.40	147.20	29.50	236.00		
1 Dist. Truck, 3000 Gal.		308.20		339.00		
1 Aggreg. Spreader, S.P.		561.40		617.55		
1 Roller, Pneu. Tire, 12 Ton		236.20		259.80		
1 Roller, Steel, 10 Ton		187.60		206.35	20.20	22.22
64 M.H., Daily Totals		$2596.60		$3502.30	$40.56	$54.71
Crew B-92						
1 Labor Foreman (outside)	$19.50	$156.00	$31.85	$254.80	$18.00	$29.41
3 Highway Laborers	17.50	420.00	28.60	686.40		
1 Crack Cleaner, 25 H.P.		71.20		78.30		
1 Air Compressor		64.40		70.85		
1 Tar Kettle, T.M.		17.80		19.60		
1 Flatbed Truck, 3 Ton		148.40		163.25	9.43	10.37
32 M.H., Daily Totals		$877.80		$1273.20	$27.43	$39.78
Crew B-93						
1 Equip. Oper. (med.)	$22.50	$180.00	$35.35	$282.80	$22.50	$35.35
1 Feller Buncher, 50 H.P.		313.40		344.75	39.17	43.09
8 M.H., Daily Totals		$493.40		$627.55	$61.67	$78.44
Crew C-1						
3 Carpenters	$22.00	$528.00	$35.95	$862.80	$20.87	$34.11
1 Building Laborer	17.50	140.00	28.60	228.80		
Power Tools		24.00		26.40	.75	.82
32 M.H., Daily Totals		$692.00		$1118.00	$21.62	$34.93
Crew C-1A						
1 Carpenter	$22.00	$176.00	$35.95	$287.60	$22.00	$35.95
1 Circular Saw, 7"		8.00		8.80	1.00	1.10
8 M.H., Daily Totals		$184.00		$296.40	$23.00	$37.05
Crew C-2						
1 Carpenter Foreman (out)	$24.00	$192.00	$39.20	$313.60	$21.58	$35.26
4 Carpenters	22.00	704.00	35.95	1150.40		
1 Building Laborer	17.50	140.00	28.60	228.80		
Power Tools		32.00		35.20	.66	.73
48 M.H., Daily Totals		$1068.00		$1728.00	$22.24	$35.99

CREWS

Crew C-3

Crew No.	Bare Costs		Incl. Subs O & P		Cost Per Man-Hour	
	Hr.	Daily	Hr.	Daily	Bare Costs	Incl. O&P
1 Rodman Foreman	$25.90	$207.20	$45.15	$361.20	$22.23	$37.81
4 Rodmen (reinf.)	23.90	764.80	41.65	1332.80		
1 Equip. Oper. (light)	21.40	171.20	33.60	268.80		
2 Building Laborers	17.50	280.00	28.60	457.60		
Stressing Equipment		36.00		39.60		
Grouting Equipment		115.20		126.70	2.36	2.59
64 M.H., Daily Totals		$1574.40		$2586.70	$24.59	$40.40

Crew C-4

Crew No.	Hr.	Daily	Hr.	Daily	Bare Costs	Incl. O&P
1 Rodman Foreman	$25.90	$207.20	$45.15	$361.20	$24.40	$42.52
3 Rodmen (reinf.)	23.90	573.60	41.65	999.60		
Stressing Equipment		36.00		39.60	1.12	1.23
32 M.H., Daily Totals		$816.80		$1400.40	$25.52	$43.75

Crew C-5

Crew No.	Hr.	Daily	Hr.	Daily	Bare Costs	Incl. O&P
1 Rodman Foreman	$25.90	$207.20	$45.15	$361.20	$23.42	$39.78
4 Rodmen (reinf.)	23.90	764.80	41.65	1332.80		
1 Equip. Oper. (crane)	23.30	186.40	36.60	292.80		
1 Equip. Oper. Oiler	19.20	153.60	30.15	241.20		
1 Hyd. Crane, 25 Ton		486.40		535.05	8.68	9.55
56 M.H., Daily Totals		$1798.40		$2763.05	$32.10	$49.33

Crew C-6

Crew No.	Hr.	Daily	Hr.	Daily	Bare Costs	Incl. O&P
1 Labor Foreman (outside)	$19.50	$156.00	$31.85	$254.80	$18.52	$30.01
4 Building Laborers	17.50	560.00	28.60	915.20		
1 Cement Finisher	21.65	173.20	33.85	270.80		
2 Gas Engine Vibrators		62.00		68.20	1.29	1.42
48 M.H., Daily Totals		$951.20		$1509.00	$19.81	$31.43

Crew C-7

Crew No.	Hr.	Daily	Hr.	Daily	Bare Costs	Incl. O&P
1 Labor Foreman (outside)	$19.50	$156.00	$31.85	$254.80	$18.89	$30.50
5 Building Laborers	17.50	700.00	28.60	1144.00		
1 Cement Finisher	21.65	173.20	33.85	270.80		
1 Equip. Oper. (med.)	22.50	180.00	35.35	282.80		
2 Gas Engine Vibrators		62.00		68.20		
1 Concrete Bucket, 1 C.Y.		16.80		18.50		
1 Hyd. Crane, 55 Ton		727.20		799.90	12.59	13.85
64 M.H., Daily Totals		$2015.20		$2839.00	$31.48	$44.35

Crew C-8

Crew No.	Hr.	Daily	Hr.	Daily	Bare Costs	Incl. O&P
1 Labor Foreman (outside)	$19.50	$156.00	$31.85	$254.80	$19.68	$31.52
3 Building Laborers	17.50	420.00	28.60	686.40		
2 Cement Finishers	21.65	346.40	33.85	541.60		
1 Equip. Oper. (med.)	22.50	180.00	35.35	282.80		
1 Concrete Pump (small)		514.60		566.05	9.18	10.10
56 M.H., Daily Totals		$1617.00		$2331.65	$28.86	$41.62

Crew C-9

Crew No.	Hr.	Daily	Hr.	Daily	Bare Costs	Incl. O&P
1 Cement Finisher	$21.65	$173.20	$33.85	$270.80	$21.65	$33.85
1 Gas Finishing Mach.		34.80		38.30	4.35	4.78
8 M.H., Daily Totals		$208.00		$309.10	$26.00	$38.63

Crew C-10

Crew No.	Hr.	Daily	Hr.	Daily	Bare Costs	Incl. O&P
1 Building Laborer	$17.50	$140.00	$28.60	$228.80	$20.26	$32.10
2 Cement Finishers	21.65	346.40	33.85	541.60		
2 Gas Finishing Mach.		69.60		76.55	2.90	3.18
24 M.H., Daily Totals		$556.00		$846.95	$23.16	$35.28

Crew C-11

Crew No.	Bare Costs		Incl. Subs O & P		Cost Per Man-Hour	
	Hr.	Daily	Hr.	Daily	Bare Costs	Incl. O&P
1 Struc. Steel Foreman	$26.10	$208.80	$47.55	$380.40	$23.68	$41.96
6 Struc. Steel Workers	24.10	1156.80	43.90	2107.20		
1 Equip. Oper. (crane)	23.30	186.40	36.60	292.80		
1 Equip. Oper. Oiler	19.20	153.60	30.15	241.20		
1 Truck Crane, 150 Ton		1387.80		1526.60	19.27	21.20
72 M.H., Daily Totals		$3093.40		$4548.20	$42.95	$63.16

Crew C-12

Crew No.	Hr.	Daily	Hr.	Daily	Bare Costs	Incl. O&P
1 Carpenter Foreman (out)	$24.00	$192.00	$39.20	$313.60	$21.80	$35.37
3 Carpenters	22.00	528.00	35.95	862.80		
1 Building Laborer	17.50	140.00	28.60	228.80		
1 Equip. Oper. (crane)	23.30	186.40	36.60	292.80		
1 Hyd. Crane, 12 Ton		362.00		398.20	7.54	8.29
48 M.H., Daily Totals		$1408.40		$2096.20	$29.34	$43.66

Crew C-13

Crew No.	Hr.	Daily	Hr.	Daily	Bare Costs	Incl. O&P
1 Struc. Steel Worker	$24.10	$192.80	$43.90	$351.20	$23.40	$41.25
1 Welder	24.10	192.80	43.90	351.20		
1 Carpenter	22.00	176.00	35.95	287.60		
1 Gas Welding Machine		68.40		75.25	2.85	3.13
24 M.H., Daily Totals		$630.00		$1065.25	$26.25	$44.38

Crew C-14

Crew No.	Hr.	Daily	Hr.	Daily	Bare Costs	Incl. O&P
1 Carpenter Foreman (out)	$24.00	$192.00	$39.20	$313.60	$21.41	$35.24
5 Carpenters	22.00	880.00	35.95	1438.00		
4 Building Laborers	17.50	560.00	28.60	915.20		
4 Rodmen (reinf.)	23.90	764.80	41.65	1332.80		
2 Cement Finishers	21.65	346.40	33.85	541.60		
1 Equip. Oper. (crane)	23.30	186.40	36.60	292.80		
1 Equip. Oper. Oiler	19.20	153.60	30.15	241.20		
1 Crane, 80 Ton, & Tools		1103.40		1213.75		
Power Tools		24.00		26.40		
2 Gas Finishing Mach.		69.60		76.55	8.31	9.14
144 M.H., Daily Totals		$4280.20		$6391.90	$29.72	$44.38

Crew C-15

Crew No.	Hr.	Daily	Hr.	Daily	Bare Costs	Incl. O&P
1 Carpenter Foreman (out)	$24.00	$192.00	$39.20	$313.60	$20.85	$34.02
2 Carpenters	22.00	352.00	35.95	575.20		
3 Building Laborers	17.50	420.00	28.60	686.40		
2 Cement Finishers	21.65	346.40	33.85	541.60		
1 Rodman (reinf.)	23.90	191.20	41.65	333.20		
Power Tools		16.00		17.60		
1 Gas Finishing Mach.		34.80		38.30	.70	.77
72 M.H., Daily Totals		$1552.40		$2505.90	$21.55	$34.79

Crew C-16

Crew No.	Hr.	Daily	Hr.	Daily	Bare Costs	Incl. O&P
1 Labor Foreman (outside)	$19.50	$156.00	$31.85	$254.80	$20.62	$33.77
3 Building Laborers	17.50	420.00	28.60	686.40		
2 Cement Finishers	21.65	346.40	33.85	541.60		
1 Equip. Oper. (med.)	22.50	180.00	35.35	282.80		
2 Rodmen (reinf.)	23.90	382.40	41.65	666.40		
1 Concrete Pump (small)		514.60		566.05	7.14	7.86
72 M.H., Daily Totals		$1999.40		$2998.05	$27.76	$41.63

Crew C-17

Crew No.	Hr.	Daily	Hr.	Daily	Bare Costs	Incl. O&P
2 Skilled Worker Foremen	$24.65	$394.40	$39.90	$638.40	$23.05	$37.30
8 Skilled Workers	22.65	1449.60	36.65	2345.60		
80 M.H., Daily Totals		$1844.00		$2984.00	$23.05	$37.30

Crew No.	Bare Costs		Incl. Subs O & P		Cost Per Man-Hour	

Crew C-17A

Crew C-17A	Hr.	Daily	Hr.	Daily	Bare Costs	Incl. O&P
2 Skilled Worker Foremen	$24.65	$394.40	$39.90	$638.40	$23.05	$37.30
8 Skilled Workers	22.65	1449.60	36.65	2345.60		
.125 Equip. Oper. (crane)	23.30	23.30	36.60	36.60		
.125 Crane, 80 Ton, & Tools		143.45		157.80		
.125 Hand Held Pwr. Tools		1.05		1.15		
.125 Walk Behind Pwr. Tools		4.50		5.00	1.83	2.02
81 M.H., Daily Totals		$2016.30		$3184.55	$24.88	$39.32

Crew C-17B	Hr.	Daily	Hr.	Daily	Bare Costs	Incl. O&P
2 Skilled Worker Foremen	$24.65	$394.40	$39.90	$638.40	$23.05	$37.30
8 Skilled Workers	22.65	1449.60	36.65	2345.60		
.25 Equip. Oper. (crane)	23.30	46.60	36.60	73.20		
.25 Crane, 80 Ton, & Tools		275.85		303.45		
.25 Hand Held Power Tools		2.00		2.20		
.25 Walk Behind Power Tools		8.70		9.55	3.49	3.84
82 M.H., Daily Totals		$2177.15		$3372.40	$26.54	$41.14

Crew C-17C	Hr.	Daily	Hr.	Daily	Bare Costs	Incl. O&P
2 Skilled Worker Foremen	$24.65	$394.40	$39.90	$638.40	$23.05	$37.30
8 Skilled Workers	22.65	1449.60	36.65	2345.60		
.375 Equip. Oper. (crane)	23.30	69.90	36.60	109.80		
.375 Crane, 80 Ton & Tools		419.30		461.20		
.375 Hand Held Power Tools		3.05		3.35		
.375 Walk Bhnd Power Tools		13.20		14.55	5.24	5.77
83 M.H., Daily Totals		$2349.45		$3572.90	$28.29	$43.07

Crew C-17D	Hr.	Daily	Hr.	Daily	Bare Costs	Incl. O&P
2 Skilled Worker Foremen	$24.65	$394.40	$39.90	$638.40	$23.05	$37.30
8 Skilled Workers	22.65	1449.60	36.65	2345.60		
.5 Equip. Oper. (crane)	23.30	93.20	36.60	146.40		
.5 Crane, 80 Ton & Tools		551.70		606.85		
.5 Hand Held Power Tools		4.00		4.40		
.5 Walk Behind Power Tools		17.40		19.15	6.82	7.50
84 M.H., Daily Totals		$2510.30		$3760.80	$29.87	$44.80

Crew C-17E	Hr.	Daily	Hr.	Daily	Bare Costs	Incl. O&P
2 Skilled Worker Foremen	$24.65	$394.40	$39.90	$638.40	$23.05	$37.30
8 Skilled Workers	22.65	1449.60	36.65	2345.60		
1 Hyd. Jack with Rods		54.40		59.85	.68	.74
80 M.H., Daily Totals		$1898.40		$3043.85	$23.73	$38.04

Crew C-18	Hr.	Daily	Hr.	Daily	Bare Costs	Incl. O&P
.125 Labor Foreman (out)	$19.50	$19.50	$31.85	$31.85	$17.72	$28.96
1 Building Laborer	17.50	140.00	28.60	228.80		
1 Concrete Cart, 10 C.F.		44.00		48.40	4.88	5.37
9 M.H., Daily Totals		$203.50		$309.05	$22.60	$34.33

Crew C-19	Hr.	Daily	Hr.	Daily	Bare Costs	Incl. O&P
.125 Labor Foreman (out)	$19.50	$19.50	$31.85	$31.85	$17.72	$28.96
1 Building Laborer	17.50	140.00	28.60	228.80		
1 Concrete Cart, 18 C.F.		105.80		116.40	11.75	12.93
9 M.H., Daily Totals		$265.30		$377.05	$29.47	$41.89

Crew C-20	Hr.	Daily	Hr.	Daily	Bare Costs	Incl. O&P
1 Labor Foreman (outside)	$19.50	$156.00	$31.85	$254.80	$18.89	$30.50
5 Building Laborers	17.50	700.00	28.60	1144.00		
1 Cement Finisher	21.65	173.20	33.85	270.80		
1 Equip. Oper. (med.)	22.50	180.00	35.35	282.80		
2 Gas Engine Vibrators		62.00		68.20		
1 Concrete Pump (small)		514.60		566.05	9.00	9.91
64 M.H., Daily Totals		$1785.80		$2586.65	$27.89	$40.41

Crew C-21	Hr.	Daily	Hr.	Daily	Bare Costs	Incl. O&P
1 Labor Foreman (outside)	$19.50	$156.00	$31.85	$254.80	$18.89	$30.50
5 Building Laborers	17.50	700.00	28.60	1144.00		
1 Cement Finisher	21.65	173.20	33.85	270.80		
1 Equip. Oper. (med.)	22.50	180.00	35.35	282.80		
2 Gas Engine Vibrators		62.00		68.20		
1 Concrete Conveyer		118.80		130.70	2.82	3.10
64 M.H., Daily Totals		$1390.00		$2151.30	$21.71	$33.60

Crew C-22	Hr.	Daily	Hr.	Daily	Bare Costs	Incl. O&P
1 Rodman Foreman	$25.90	$207.20	$45.15	$361.20	$24.15	$41.92
4 Rodmen (reinf.)	23.90	764.80	41.65	1332.80		
.125 Equip. Oper. (med.)	23.30	23.30	36.60	36.60		
.125 Equip. Oper. Oiler	19.20	19.20	30.15	30.15		
.125 Hyd. Crane, 25 Ton		63.25		69.55	1.50	1.65
42 M.H., Daily Totals		$1077.75		$1830.30	$25.65	$43.57

Crew C-23	Hr.	Daily	Hr.	Daily	Bare Costs	Incl. O&P
2 Skilled Worker Foremen	$24.65	$394.40	$39.90	$638.40	$22.77	$36.64
6 Skilled Workers	22.65	1087.20	36.65	1759.20		
1 Equip. Oper. (crane)	23.30	186.40	36.60	292.80		
1 Equip. Oper. Oiler	19.20	153.60	30.15	241.20		
1 Crane, 90 Ton		1071.40		1178.55	13.39	14.73
80 M.H., Daily Totals		$2893.00		$4110.15	$36.16	$51.37

Crew C-24	Hr.	Daily	Hr.	Daily	Bare Costs	Incl. O&P
2 Skilled Worker Foremen	$24.65	$394.40	$39.90	$638.40	$22.77	$36.64
6 Skilled Workers	22.65	1087.20	36.65	1759.20		
1 Equip. Oper. (crane)	23.30	186.40	36.60	292.80		
1 Equip. Oper. Oiler	19.20	153.60	30.15	241.20		
1 Truck Crane, 150 Ton		1387.80		1526.60	17.34	19.08
80 M.H., Daily Totals		$3209.40		$4458.20	$40.11	$55.72

Crew D-1	Hr.	Daily	Hr.	Daily	Bare Costs	Incl. O&P
1 Bricklayer	$22.75	$182.00	$36.50	$292.00	$20.20	$32.40
1 Bricklayer Helper	17.65	141.20	28.30	226.40		
16 M.H., Daily Totals		$323.20		$518.40	$20.20	$32.40

Crew D-2	Hr.	Daily	Hr.	Daily	Bare Costs	Incl. O&P
3 Bricklayers	$22.75	$546.00	$36.50	$876.00	$20.82	$33.46
2 Bricklayer Helpers	17.65	282.40	28.30	452.80		
.5 Carpenter	22.00	88.00	35.95	143.80		
44 M.H., Daily Totals		$916.40		$1472.60	$20.82	$33.46

Crew D-3	Hr.	Daily	Hr.	Daily	Bare Costs	Incl. O&P
3 Bricklayers	$22.75	$546.00	$36.50	$876.00	$20.77	$33.35
2 Bricklayer Helpers	17.65	282.40	28.30	452.80		
.25 Carpenter	22.00	44.00	35.95	71.90		
42 M.H., Daily Totals		$872.40		$1400.70	$20.77	$33.35

Crew D-4

	Bare Costs Hr.	Daily	Incl. Subs O & P Hr.	Daily	Cost Per Man-Hour Bare Costs	Incl. O&P
1 Bricklayer	$22.75	$182.00	$36.50	$292.00	$19.86	$31.67
2 Bricklayer Helpers	17.65	282.40	28.30	452.80		
1 Equip. Oper. (light)	21.40	171.20	33.60	268.80		
1 Grout Pump		88.80		97.70		
1 Hoses & Hopper		23.20		25.50		
1 Accessories		9.80		10.80	3.80	4.18
32 M.H., Daily Totals		$757.40		$1147.60	$23.66	$35.85

Crew D-5

	Bare Costs Hr.	Daily	Incl. Subs O & P Hr.	Daily	Cost Per Man-Hour Bare Costs	Incl. O&P
1 Bricklayer	$22.75	$182.00	$36.50	$292.00	$22.75	$36.50
1 Power Tool		32.00		35.20	4.00	4.40
8 M.H., Daily Totals		$214.00		$327.20	$26.75	$40.90

Crew D-6

	Bare Costs Hr.	Daily	Incl. Subs O & P Hr.	Daily	Cost Per Man-Hour Bare Costs	Incl. O&P
3 Bricklayers	$22.75	$546.00	$36.50	$876.00	$20.27	$32.54
3 Bricklayer Helpers	17.65	423.60	28.30	679.20		
.25 Carpenter	22.00	44.00	35.95	71.90		
50 M.H., Daily Totals		$1013.60		$1627.10	$20.27	$32.54

Crew D-7

	Bare Costs Hr.	Daily	Incl. Subs O & P Hr.	Daily	Cost Per Man-Hour Bare Costs	Incl. O&P
1 Tile Layer	$22.15	$177.20	$34.35	$274.80	$19.80	$30.70
1 Tile Layer Helper	17.45	139.60	27.05	216.40		
16 M.H., Daily Totals		$316.80		$491.20	$19.80	$30.70

Crew D-8

	Bare Costs Hr.	Daily	Incl. Subs O & P Hr.	Daily	Cost Per Man-Hour Bare Costs	Incl. O&P
3 Bricklayers	$22.75	$546.00	$36.50	$876.00	$20.71	$33.22
2 Bricklayer Helpers	17.65	282.40	28.30	452.80		
40 M.H., Daily Totals		$828.40		$1328.80	$20.71	$33.22

Crew D-9

	Bare Costs Hr.	Daily	Incl. Subs O & P Hr.	Daily	Cost Per Man-Hour Bare Costs	Incl. O&P
3 Bricklayers	$22.75	$546.00	$36.50	$876.00	$20.20	$32.40
3 Bricklayer Helpers	17.65	423.60	28.30	679.20		
48 M.H., Daily Totals		$969.60		$1555.20	$20.20	$32.40

Crew D-10

	Bare Costs Hr.	Daily	Incl. Subs O & P Hr.	Daily	Cost Per Man-Hour Bare Costs	Incl. O&P
1 Bricklayer Foreman	$24.75	$198.00	$39.70	$317.60	$21.22	$33.88
1 Bricklayer	22.75	182.00	36.50	292.00		
2 Bricklayer Helpers	17.65	282.40	28.30	452.80		
1 Equip. Oper. (crane)	23.30	186.40	36.60	292.80		
1 Truck Crane, 12.5 Ton		419.20		461.10	10.48	11.52
40 M.H., Daily Totals		$1268.00		$1816.30	$31.70	$45.40

Crew D-11

	Bare Costs Hr.	Daily	Incl. Subs O & P Hr.	Daily	Cost Per Man-Hour Bare Costs	Incl. O&P
1 Bricklayer Foreman	$24.75	$198.00	$39.70	$317.60	$21.71	$34.83
1 Bricklayer	22.75	182.00	36.50	292.00		
1 Bricklayer Helper	17.65	141.20	28.30	226.40		
24 M.H., Daily Totals		$521.20		$836.00	$21.71	$34.83

Crew D-12

	Bare Costs Hr.	Daily	Incl. Subs O & P Hr.	Daily	Cost Per Man-Hour Bare Costs	Incl. O&P
1 Bricklayer Foreman	$24.75	$198.00	$39.70	$317.60	$20.70	$33.20
1 Bricklayer	22.75	182.00	36.50	292.00		
2 Bricklayer Helpers	17.65	282.40	28.30	452.80		
32 M.H., Daily Totals		$662.40		$1062.40	$20.70	$33.20

Crew D-13

	Bare Costs Hr.	Daily	Incl. Subs O & P Hr.	Daily	Cost Per Man-Hour Bare Costs	Incl. O&P
1 Bricklayer Foreman	$24.75	$198.00	$39.70	$317.60	$21.35	$34.22
1 Bricklayer	22.75	182.00	36.50	292.00		
2 Bricklayer Helpers	17.65	282.40	28.30	452.80		
1 Carpenter	22.00	176.00	35.95	287.60		
1 Equip. Oper. (crane)	23.30	186.40	36.60	292.80		
1 Truck Crane, 12.5 Ton		419.20		461.10	8.73	9.60
48 M.H., Daily Totals		$1444.00		$2103.90	$30.08	$43.82

Crew E-1

	Bare Costs Hr.	Daily	Incl. Subs O & P Hr.	Daily	Cost Per Man-Hour Bare Costs	Incl. O&P
1 Welder Foreman	$26.10	$208.80	$47.55	$380.40	$23.86	$41.68
1 Welder	24.10	192.80	43.90	351.20		
1 Equip. Oper. (light)	21.40	171.20	33.60	268.80		
1 Gas Welding Machine		68.40		75.25	2.85	3.13
24 M.H., Daily Totals		$641.20		$1075.65	$26.71	$44.81

Crew E-2

	Bare Costs Hr.	Daily	Incl. Subs O & P Hr.	Daily	Cost Per Man-Hour Bare Costs	Incl. O&P
1 Struc. Steel Foreman	$26.10	$208.80	$47.55	$380.40	$23.57	$41.41
4 Struc. Steel Workers	24.10	771.20	43.90	1404.80		
1 Equip. Oper. (crane)	23.30	186.40	36.60	292.80		
1 Equip. Oper. Oiler	19.20	153.60	30.15	241.20		
1 Crane, 90 Ton		1071.40		1178.55	19.13	21.04
56 M.H., Daily Totals		$2391.40		$3497.75	$42.70	$62.45

Crew E-3

	Bare Costs Hr.	Daily	Incl. Subs O & P Hr.	Daily	Cost Per Man-Hour Bare Costs	Incl. O&P
1 Struc. Steel Foreman	$26.10	$208.80	$47.55	$380.40	$24.76	$45.11
1 Struc. Steel Worker	24.10	192.80	43.90	351.20		
1 Welder	24.10	192.80	43.90	351.20		
1 Gas Welding Machine		68.40		75.25		
1 Torch, Gas & Air		62.80		69.10	5.46	6.01
24 M.H., Daily Totals		$725.60		$1227.15	$30.22	$51.12

Crew E-4

	Bare Costs Hr.	Daily	Incl. Subs O & P Hr.	Daily	Cost Per Man-Hour Bare Costs	Incl. O&P
1 Struc. Steel Foreman	$26.10	$208.80	$47.55	$380.40	$24.60	$44.81
3 Struc. Steel Workers	24.10	578.40	43.90	1053.60		
1 Gas Welding Machine		68.40		75.25	2.13	2.35
32 M.H., Daily Totals		$855.60		$1509.25	$26.73	$47.16

Crew E-5

	Bare Costs Hr.	Daily	Incl. Subs O & P Hr.	Daily	Cost Per Man-Hour Bare Costs	Incl. O&P
2 Struc. Steel Foremen	$26.10	$417.60	$47.55	$760.80	$23.93	$42.52
5 Struc. Steel Workers	24.10	964.00	43.90	1756.00		
1 Equip. Oper. (crane)	23.30	186.40	36.60	292.80		
1 Welder	24.10	192.80	43.90	351.20		
1 Equip. Oper. Oiler	19.20	153.60	30.15	241.20		
1 Crane, 90 Ton		1071.40		1178.55		
1 Gas Welding Machine		68.40		75.25		
1 Torch, Gas & Air		62.80		69.10	15.03	16.53
80 M.H., Daily Totals		$3117.00		$4724.90	$38.96	$59.05

CREWS

Crew E-6	Hr.	Daily	Hr.	Daily	Bare Costs	Incl. O&P
3 Struc. Steel Foreman	$26.10	$626.40	$47.55	$1141.20	$23.95	$42.62
9 Struc. Steel Workers	24.10	1735.20	43.90	3160.80		
1 Equip. Oper. (crane)	23.30	186.40	36.60	292.80		
1 Welder	24.10	192.80	43.90	351.20		
1 Equip. Oper. Oiler	19.20	153.60	30.15	241.20		
1 Equip. Oper. (light)	21.40	171.20	33.60	268.80		
1 Crane, 90 Ton		1071.40		1178.55		
1 Gas Welding Machine		68.40		75.25		
1 Torch, Gas & Air		62.80		69.10		
1 Air Compr., 160 C.F.M.		85.40		93.95		
2 Impact Wrenches		55.60		61.15	10.49	11.54
128 M.H., Daily Totals		$4409.20		$6934.00	$34.44	$54.16

Crew E-7	Hr.	Daily	Hr.	Daily	Bare Costs	Incl. O&P
1 Struc. Steel Foreman	$26.10	$208.80	$47.55	$380.40	$23.93	$42.52
4 Struc. Steel Workers	24.10	771.20	43.90	1404.80		
1 Equip. Oper. (crane)	23.30	186.40	36.60	292.80		
1 Equip. Oper. Oiler	19.20	153.60	30.15	241.20		
1 Welder Foreman	26.10	208.80	47.55	380.40		
2 Welders	24.10	385.60	43.90	702.40		
1 Crane, 90 Ton		1071.40		1178.55		
2 Gas Welding Machines		136.80		150.50	15.10	16.61
80 M.H., Daily Totals		$3122.60		$4731.05	$39.03	$59.13

Crew E-8	Hr.	Daily	Hr.	Daily	Bare Costs	Incl. O&P
1 Struc. Steel Foreman	$26.10	$208.80	$47.55	$380.40	$23.76	$42.05
4 Struc. Steel Workers	24.10	771.20	43.90	1404.80		
1 Welder Foreman	26.10	208.80	47.55	380.40		
4 Welders	24.10	771.20	43.90	1404.80		
1 Equip. Oper. (crane)	23.30	186.40	36.60	292.80		
1 Equip. Oper. Oiler	19.20	153.60	30.15	241.20		
1 Equip. Oper. (light)	21.40	171.20	33.60	268.80		
1 Crane, 90 Ton		1071.40		1178.55		
4 Gas Welding Machines		273.60		300.95	12.93	14.22
104 M.H., Daily Totals		$3816.20		$5852.70	$36.69	$56.27

Crew E-9	Hr.	Daily	Hr.	Daily	Bare Costs	Incl. O&P
2 Struc. Steel Foremen	$26.10	$417.60	$47.55	$760.80	$23.95	$42.62
5 Struc. Steel Workers	24.10	964.00	43.90	1756.00		
1 Welder Foreman	26.10	208.80	47.55	380.40		
5 Welders	24.10	964.00	43.90	1756.00		
1 Equip. Oper. (crane)	23.30	186.40	36.60	292.80		
1 Equip. Oper. Oiler	19.20	153.60	30.15	241.20		
1 Equip. Oper. (light)	21.40	171.20	33.60	268.80		
1 Crane, 90 Ton		1071.40		1178.55		
5 Gas Welding Machines		342.00		376.20		
1 Torch, Gas & Air		62.80		69.10	11.53	12.68
128 M.H., Daily Totals		$4541.80		$7079.85	$35.48	$55.30

Crew E-10	Hr.	Daily	Hr.	Daily	Bare Costs	Incl. O&P
1 Welder Foreman	$26.10	$208.80	$47.55	$380.40	$25.10	$45.72
1 Welder	24.10	192.80	43.90	351.20		
4 Gas Welding Machines		273.60		300.95		
1 Truck, 3 Ton		148.40		163.25	26.37	29.01
16 M.H., Daily Totals		$823.60		$1195.80	$51.47	$74.73

Crew E-11	Hr.	Daily	Hr.	Daily	Bare Costs	Incl. O&P
2 Painters, Struc. Steel	$21.45	$343.20	$40.65	$650.40	$20.45	$35.87
1 Building Laborer	17.50	140.00	28.60	228.80		
1 Equip. Oper. (light)	21.40	171.20	33.60	268.80		
1 Air Compressor 250 C.F.M.		88.80		97.70		
1 Sand Blaster		23.20		25.50		
1 Sand Blasting Accessories		9.80		10.80	3.80	4.18
32 M.H., Daily Totals		$776.20		$1282.00	$24.25	$40.05

Crew E-12	Hr.	Daily	Hr.	Daily	Bare Costs	Incl. O&P
1 Welder Foreman	$26.10	$208.80	$47.55	$380.40	$23.75	$40.57
1 Equip. Oper. (light)	21.40	171.20	33.60	268.80		
1 Gas Welding Machine		68.40		75.25	4.27	4.70
16 M.H., Daily Totals		$448.40		$724.45	$28.02	$45.27

Crew E-13	Hr.	Daily	Hr.	Daily	Bare Costs	Incl. O&P
1 Welder Foreman	$26.10	$208.80	$47.55	$380.40	$24.53	$42.90
.5 Equip. Oper. (light)	21.40	85.60	33.60	134.40		
1 Gas Welding Machine		68.40		75.25	5.70	6.27
12 M.H., Daily Totals		$362.80		$590.05	$30.23	$49.17

Crew E-14	Hr.	Daily	Hr.	Daily	Bare Costs	Incl. O&P
1 Welder Foreman	$26.10	$208.80	$47.55	$380.40	$26.10	$47.55
1 Gas Welding Machine		68.40		75.25	8.55	9.40
8 M.H., Daily Totals		$277.20		$455.65	$34.65	$56.95

Crew E-15	Hr.	Daily	Hr.	Daily	Bare Costs	Incl. O&P
2 Painters, Struc. Steel	$21.45	$343.20	$40.65	$650.40	$21.45	$40.65
1 Paint Sprayer, 17 C.F.M.		25.40		27.95	1.58	1.74
16 M.H., Daily Totals		$368.60		$678.35	$23.03	$42.39

Crew F-1	Hr.	Daily	Hr.	Daily	Bare Costs	Incl. O&P
1 Carpenter	$22.00	$176.00	$35.95	$287.60	$22.00	$35.95
Power Tools		8.00		8.80	1.00	1.10
8 M.H., Daily Totals		$184.00		$296.40	$23.00	$37.05

Crew F-2	Hr.	Daily	Hr.	Daily	Bare Costs	Incl. O&P
2 Carpenters	$22.00	$352.00	$35.95	$575.20	$22.00	$35.95
Power Tools		16.00		17.60	1.00	1.10
16 M.H., Daily Totals		$368.00		$592.80	$23.00	$37.05

Crew F-3	Hr.	Daily	Hr.	Daily	Bare Costs	Incl. O&P
4 Carpenters	$22.00	$704.00	$35.95	$1150.40	$22.26	$36.08
1 Equip. Oper. (crane)	23.30	186.40	36.60	292.80		
1 Hyd. Crane, 12 Ton		362.00		398.20		
Power Tools		16.00		17.60	9.45	10.39
40 M.H., Daily Totals		$1268.40		$1859.00	$31.71	$46.47

Crew F-4	Hr.	Daily	Hr.	Daily	Bare Costs	Incl. O&P
4 Carpenters	$22.00	$704.00	$35.95	$1150.40	$21.75	$35.09
1 Equip. Oper. (crane)	23.30	186.40	36.60	292.80		
1 Equip. Oper. Oiler	19.20	153.60	30.15	241.20		
1 Hyd. Crane, 55 Ton		727.20		799.90		
Power Tools		16.00		17.60	15.48	17.03
48 M.H., Daily Totals		$1787.20		$2501.90	$37.23	$52.12

Crew F-5	Hr.	Daily	Hr.	Daily	Bare Costs	Incl. O&P
1 Carpenter Foreman	$24.00	$192.00	$39.20	$313.60	$22.50	$36.76
3 Carpenters	22.00	528.00	35.95	862.80		
Power Tools		16.00		17.60	.50	.55
32 M.H., Daily Totals		$736.00		$1194.00	$23.00	$37.31

CREWS

Crew No.	Bare Costs Hr.	Daily	Incl. Subs O & P Hr.	Daily	Bare Costs	Incl. O&P
Crew F-6	Hr.	Daily	Hr.	Daily	Bare Costs	Incl. O&P
2 Carpenters	$22.00	$352.00	$35.95	$575.20	$20.46	$33.14
2 Building Laborers	17.50	280.00	28.60	457.60		
1 Equip. Oper. (crane)	23.30	186.40	36.60	292.80		
1 Hyd. Crane, 12 Ton		362.00		398.20		
Power Tools		16.00		17.60	9.45	10.39
40 M.H., Daily Totals		$1196.40		$1741.40	$29.91	$43.53
Crew F-7	Hr.	Daily	Hr.	Daily	Bare Costs	Incl. O&P
2 Carpenters	$22.00	$352.00	$35.95	$575.20	$19.75	$32.27
2 Building Laborers	17.50	280.00	28.60	457.60		
Power Tools		16.00		17.60	.50	.55
32 M.H., Daily Totals		$648.00		$1050.40	$20.25	$32.82
Crew G-1	Hr.	Daily	Hr.	Daily	Bare Costs	Incl. O&P
1 Roofer Foreman	$22.35	$178.80	$39.25	$314.00	$19.07	$33.50
4 Roofers, Composition	20.35	651.20	35.75	1144.00		
2 Roofer Helpers	14.90	238.40	26.15	418.40		
Application Equipment		108.20		119.00	1.93	2.12
56 M.H., Daily Totals		$1176.60		$1995.40	$21.00	$35.62
Crew G-2	Hr.	Daily	Hr.	Daily	Bare Costs	Incl. O&P
1 Plasterer	$21.80	$174.40	$34.90	$279.20	$19.06	$30.71
1 Plasterer Helper	17.90	143.20	28.65	229.20		
1 Building Laborer	17.50	140.00	28.60	228.80		
Grouting Equipment		230.40		253.45	9.60	10.56
24 M.H., Daily Totals		$688.00		$990.65	$28.66	$41.27
Crew G-3	Hr.	Daily	Hr.	Daily	Bare Costs	Incl. O&P
2 Sheet Metal Workers	$25.00	$400.00	$39.25	$628.00	$21.25	$33.92
2 Building Laborers	17.50	280.00	28.60	457.60		
Power Tools		25.10		27.65	.78	.86
32 M.H., Daily Totals		$705.10		$1113.25	$22.03	$34.78
Crew G-4	Hr.	Daily	Hr.	Daily	Bare Costs	Incl. O&P
1 Labor Foreman (outside)	$19.50	$156.00	$31.85	$254.80	$18.16	$29.68
2 Building Laborers	17.50	280.00	28.60	457.60		
1 Light Truck, 1.5 Ton		146.00		160.60		
1 Air Compr., 160 C.F.M.		85.40		93.95	9.64	10.60
24 M.H., Daily Totals		$667.40		$966.95	$27.80	$40.28
Crew G-5	Hr.	Daily	Hr.	Daily	Bare Costs	Incl. O&P
1 Roofer Foreman	$22.35	$178.80	$39.25	$314.00	$18.57	$32.61
2 Roofers, Composition	20.35	325.60	35.75	572.00		
2 Roofer Helpers	14.90	238.40	26.15	418.40		
Application Equipment		108.20		119.00	2.70	2.97
40 M.H., Daily Totals		$851.00		$1423.40	$21.27	$35.58
Crew H-1	Hr.	Daily	Hr.	Daily	Bare Costs	Incl. O&P
2 Glaziers	$22.55	$360.80	$35.75	$572.00	$23.32	$39.82
2 Struc. Steel Workers	24.10	385.60	43.90	702.40		
32 M.H., Daily Totals		$746.40		$1274.40	$23.32	$39.82
Crew H-2	Hr.	Daily	Hr.	Daily	Bare Costs	Incl. O&P
2 Glaziers	$22.55	$360.80	$35.75	$572.00	$20.86	$33.36
1 Building Laborer	17.50	140.00	28.60	228.80		
24 M.H., Daily Totals		$500.80		$800.80	$20.86	$33.36

Crew No.	Bare Costs Hr.	Daily	Incl. Subs O & P Hr.	Daily	Bare Costs	Incl. O&P
Crew J-1	Hr.	Daily	Hr.	Daily	Bare Costs	Incl. O&P
3 Plasterers	$21.80	$523.20	$34.90	$837.60	$20.24	$32.40
2 Plasterer Helpers	17.90	286.40	28.65	458.40		
1 Mixing Machine, 6 C.F.		33.20		36.50	.83	.91
40 M.H., Daily Totals		$842.80		$1332.50	$21.07	$33.31
Crew J-2	Hr.	Daily	Hr.	Daily	Bare Costs	Incl. O&P
3 Plasterers	$21.80	$523.20	$34.90	$837.60	$20.52	$32.74
2 Plasterer Helpers	17.90	286.40	28.65	458.40		
1 Lather	21.95	175.60	34.45	275.60		
1 Mixing Machine, 6 C.F.		33.20		36.50	.69	.76
48 M.H., Daily Totals		$1018.40		$1608.10	$21.21	$33.50
Crew J-3	Hr.	Daily	Hr.	Daily	Bare Costs	Incl. O&P
1 Terrazzo Worker	$22.10	$176.80	$34.25	$274.00	$19.72	$30.57
1 Terrazzo Helper	17.35	138.80	26.90	215.20		
1 Terrazzo Grinder, Electric		34.80		38.30		
1 Terrazzo Mixer		53.80		59.20	5.53	6.09
16 M.H., Daily Totals		$404.20		$586.70	$25.25	$36.66
Crew J-4	Hr.	Daily	Hr.	Daily	Bare Costs	Incl. O&P
1 Tile Layer	$22.15	$177.20	$34.35	$274.80	$19.80	$30.70
1 Tile Layer Helper	17.45	139.60	27.05	216.40		
16 M.H., Daily Totals		$316.80		$491.20	$19.80	$30.70
Crew K-1	Hr.	Daily	Hr.	Daily	Bare Costs	Incl. O&P
1 Carpenter	$22.00	$176.00	$35.95	$287.60	$20.05	$32.47
1 Truck Driver (light)	18.10	144.80	29.00	232.00		
1 Truck w/Power Equip.		390.30		429.30	24.39	26.83
16 M.H., Daily Totals		$711.10		$948.90	$44.44	$59.30
Crew K-2	Hr.	Daily	Hr.	Daily	Bare Costs	Incl. O&P
1 Struc. Steel Foreman	$26.10	$208.80	$47.55	$380.40	$22.76	$40.15
1 Struc. Steel Worker	24.10	192.80	43.90	351.20		
1 Truck Driver (light)	18.10	144.80	29.00	232.00		
1 Truck w/Power Equip.		390.30		429.30	16.26	17.88
24 M.H., Daily Totals		$936.70		$1392.90	$39.02	$58.03
Crew L-1	Hr.	Daily	Hr.	Daily	Bare Costs	Incl. O&P
1 Electrician	$25.15	$201.20	$38.40	$307.20	$25.30	$38.82
1 Plumber	25.45	203.60	39.25	314.00		
16 M.H., Daily Totals		$404.80		$621.20	$25.30	$38.82
Crew L-2	Hr.	Daily	Hr.	Daily	Bare Costs	Incl. O&P
1 Carpenter	$22.00	$176.00	$35.95	$287.60	$19.55	$31.90
1 Helper	17.10	136.80	27.85	222.80		
16 M.H., Daily Totals		$312.80		$510.40	$19.55	$31.90
Crew L-3	Hr.	Daily	Hr.	Daily	Bare Costs	Incl. O&P
1 Carpenter	$22.00	$176.00	$35.95	$287.60	$23.53	$37.38
.5 Electrician	25.15	100.60	38.40	153.60		
.5 Sheet Metal Worker	25.00	100.00	39.25	157.00		
16 M.H., Daily Totals		$376.60		$598.20	$23.53	$37.38
Crew L-4	Hr.	Daily	Hr.	Daily	Bare Costs	Incl. O&P
2 Skilled Workers	$22.65	$362.40	$36.65	$586.40	$20.80	$33.71
1 Helper	17.10	136.80	27.85	222.80		
24 M.H., Daily Totals		$499.20		$809.20	$20.80	$33.71

CREWS

Crew L-5	Bare Costs Hr.	Bare Costs Daily	Incl. Subs O & P Hr.	Incl. Subs O & P Daily	Cost Per Man-Hour Bare Costs	Cost Per Man-Hour Incl. O&P
1 Struc. Steel Foreman	$26.10	$208.80	$47.55	$380.40	$24.27	$43.37
5 Struc. Steel Workers	24.10	964.00	43.90	1756.00		
1 Equip. Oper. (crane)	23.30	186.40	36.60	292.80		
1 Hyd. Crane, 25 Ton		486.40		535.05	8.68	9.55
56 M.H., Daily Totals		$1845.60		$2964.25	$32.95	$52.92

Crew L-6	Hr.	Daily	Hr.	Daily	Bare Costs	Incl. O&P
1 Plumber	$25.45	$203.60	$39.25	$314.00	$25.35	$38.96
.5 Electrician	25.15	100.60	38.40	153.60		
12 M.H., Daily Totals		$304.20		$467.60	$25.35	$38.96

Crew L-7	Hr.	Daily	Hr.	Daily	Bare Costs	Incl. O&P
2 Carpenters	$22.00	$352.00	$35.95	$575.20	$21.16	$34.20
1 Building Laborer	17.50	140.00	28.60	228.80		
.5 Electrician	25.15	100.60	38.40	153.60		
28 M.H., Daily Totals		$592.60		$957.60	$21.16	$34.20

Crew L-8	Hr.	Daily	Hr.	Daily	Bare Costs	Incl. O&P
2 Carpenters	$22.00	$352.00	$35.95	$575.20	$22.69	$36.61
.5 Plumber	25.45	101.80	39.25	157.00		
20 M.H., Daily Totals		$453.80		$732.20	$22.69	$36.61

Crew L-9	Hr.	Daily	Hr.	Daily	Bare Costs	Incl. O&P
1 Labor Foreman (inside)	$18.00	$144.00	$29.40	$235.20	$19.92	$33.26
2 Building Laborers	17.50	280.00	28.60	457.60		
1 Struc. Steel Worker	24.10	192.80	43.90	351.20		
.5 Electrician	25.15	100.60	38.40	153.60		
36 M.H., Daily Totals		$717.40		$1197.60	$19.92	$33.26

Crew M-1	Hr.	Daily	Hr.	Daily	Bare Costs	Incl. O&P
3 Elevator Constructors	$25.35	$608.40	$39.15	$939.60	$24.08	$37.18
1 Elevator Apprentice	20.28	162.24	31.30	250.40		
Hand Tools		76.00		83.60	2.37	2.61
32 M.H., Daily Totals		$846.64		$1273.60	$26.45	$39.79

Crew M-2	Hr.	Daily	Hr.	Daily	Bare Costs	Incl. O&P
2 Millwrights	$22.95	$367.20	$35.95	$575.20	$22.95	$35.95
Power Tools		16.00		17.60	1.00	1.10
16 M.H., Daily Totals		$383.20		$592.80	$23.95	$37.05

Crew Q-1	Hr.	Daily	Hr.	Daily	Bare Costs	Incl. O&P
1 Plumber	$25.45	$203.60	$39.25	$314.00	$22.90	$35.32
1 Plumber Apprentice	20.36	162.88	31.40	251.20		
16 M.H., Daily Totals		$366.48		$565.20	$22.90	$35.32

Crew Q-2	Hr.	Daily	Hr.	Daily	Bare Costs	Incl. O&P
2 Plumbers	$25.45	$407.20	$39.25	$628.00	$23.75	$36.63
1 Plumber Apprentice	20.36	162.88	31.40	251.20		
24 M.H., Daily Totals		$570.08		$879.20	$23.75	$36.63

Crew Q-3	Hr.	Daily	Hr.	Daily	Bare Costs	Incl. O&P
1 Plumber Foreman (ins)	$25.95	$207.60	$40.00	$320.00	$24.30	$37.47
2 Plumbers	25.45	407.20	39.25	628.00		
1 Plumber Apprentice	20.36	162.88	31.40	251.20		
32 M.H., Daily Totals		$777.68		$1199.20	$24.30	$37.47

Crew Q-4	Hr.	Daily	Hr.	Daily	Bare Costs	Incl. O&P
1 Plumber Foreman (ins)	$25.95	$207.60	$40.00	$320.00	$24.30	$37.47
1 Plumber	25.45	203.60	39.25	314.00		
1 Welder (plumber)	25.45	203.60	39.25	314.00		
1 Plumber Apprentice	20.36	162.88	31.40	251.20		
1 Electric Welding Mach.		39.00		42.90	1.21	1.34
32 M.H., Daily Totals		$816.68		$1242.10	$25.51	$38.81

Crew Q-5	Hr.	Daily	Hr.	Daily	Bare Costs	Incl. O&P
1 Steamfitter	$25.50	$204.00	$39.30	$314.40	$22.95	$35.37
1 Steamfitter Apprentice	20.40	163.20	31.45	251.60		
16 M.H., Daily Totals		$367.20		$566.00	$22.95	$35.37

Crew Q-6	Hr.	Daily	Hr.	Daily	Bare Costs	Incl. O&P
2 Steamfitters	$25.50	$408.00	$39.30	$628.80	$23.80	$36.68
1 Steamfitter Apprentice	20.40	163.20	31.45	251.60		
24 M.H., Daily Totals		$571.20		$880.40	$23.80	$36.68

Crew Q-7	Hr.	Daily	Hr.	Daily	Bare Costs	Incl. O&P
1 Steamfitter Foreman (ins)	$26.00	$208.00	$40.10	$320.80	$24.35	$37.53
2 Steamfitters	25.50	408.00	39.30	628.80		
1 Steamfitter Apprentice	20.40	163.20	31.45	251.60		
32 M.H., Daily Totals		$779.20		$1201.20	$24.35	$37.53

Crew Q-8	Hr.	Daily	Hr.	Daily	Bare Costs	Incl. O&P
1 Steamfitter Foreman (ins)	$26.00	$208.00	$40.10	$320.80	$24.35	$37.53
1 Steamfitter	25.50	204.00	39.30	314.40		
1 Welder (steamfitter)	25.50	204.00	39.30	314.40		
1 Steamfitter Apprentice	20.40	163.20	31.45	251.60		
1 Electric Welding Mach.		39.00		42.90	1.21	1.34
32 M.H., Daily Totals		$818.20		$1244.10	$25.56	$38.87

Crew Q-9	Hr.	Daily	Hr.	Daily	Bare Costs	Incl. O&P
1 Sheet Metal Worker	$25.00	$200.00	$39.25	$314.00	$22.50	$35.32
1 Sheet Metal Apprentice	20.00	160.00	31.40	251.20		
16 M.H., Daily Totals		$360.00		$565.20	$22.50	$35.32

Crew Q-10	Hr.	Daily	Hr.	Daily	Bare Costs	Incl. O&P
2 Sheet Metal Workers	$25.00	$400.00	$39.25	$628.00	$23.33	$36.63
1 Sheet Metal Apprentice	20.00	160.00	31.40	251.20		
24 M.H., Daily Totals		$560.00		$879.20	$23.33	$36.63

Crew Q-11	Hr.	Daily	Hr.	Daily	Bare Costs	Incl. O&P
1 Sheet Metal Foreman (ins)	$25.50	$204.00	$40.00	$320.00	$23.87	$37.47
2 Sheet Metal Workers	25.00	400.00	39.25	628.00		
1 Sheet Metal Apprentice	20.00	160.00	31.40	251.20		
32 M.H., Daily Totals		$764.00		$1199.20	$23.87	$37.47

Crew Q-12	Hr.	Daily	Hr.	Daily	Bare Costs	Incl. O&P
1 Sprinkler Installer	$26.40	$211.20	$40.75	$326.00	$23.76	$36.67
1 Sprinkler Apprentice	21.12	168.96	32.60	260.80		
16 M.H., Daily Totals		$380.16		$586.80	$23.76	$36.67

Crew Q-13	Hr.	Daily	Hr.	Daily	Bare Costs	Incl. O&P
1 Sprinkler Foreman (ins)	$26.90	$215.20	$41.55	$332.40	$25.20	$38.91
2 Sprinkler Installers	26.40	422.40	40.75	652.00		
1 Sprinkler Apprentice	21.12	168.96	32.60	260.80		
32 M.H., Daily Totals		$806.56		$1245.20	$25.20	$38.91

CREWS

Crew Q-14

	Bare Costs Hr.	Daily	Incl. Subs O & P Hr.	Daily	Cost Per Man-Hour Bare Costs	Incl. O&P
1 Asbestos Worker	$24.70	$197.60	$39.55	$316.40	$22.23	$35.60
1 Asbestos Apprentice	19.76	158.08	31.65	253.20		
16 M.H., Daily Totals		$355.68		$569.60	$22.23	$35.60

Crew Q-15

	Bare Costs Hr.	Daily	Incl. Subs O & P Hr.	Daily	Cost Per Man-Hour Bare Costs	Incl. O&P
1 Plumber	$25.45	$203.60	$39.25	$314.00	$22.90	$35.32
1 Plumber Apprentice	20.36	162.88	31.40	251.20		
1 Electric Welding Mach.		39.00		42.90	2.43	2.68
16 M.H., Daily Totals		$405.48		$608.10	$25.33	$38.00

Crew Q-16

	Bare Costs Hr.	Daily	Incl. Subs O & P Hr.	Daily	Cost Per Man-Hour Bare Costs	Incl. O&P
2 Plumbers	$25.45	$407.20	$39.25	$628.00	$23.75	$36.63
1 Plumber Apprentice	20.36	162.88	31.40	251.20		
1 Electric Welding Mach.		39.00		42.90	1.62	1.78
24 M.H., Daily Totals		$609.08		$922.10	$25.37	$38.41

Crew Q-17

	Bare Costs Hr.	Daily	Incl. Subs O & P Hr.	Daily	Cost Per Man-Hour Bare Costs	Incl. O&P
1 Steamfitter	$25.50	$204.00	$39.30	$314.40	$22.95	$35.37
1 Steamfitter Apprentice	20.40	163.20	31.45	251.60		
1 Electric Welding Mach.		39.00		42.90	2.43	2.68
16 M.H., Daily Totals		$406.20		$608.90	$25.38	$38.05

Crew Q-18

	Bare Costs Hr.	Daily	Incl. Subs O & P Hr.	Daily	Cost Per Man-Hour Bare Costs	Incl. O&P
2 Steamfitters	$25.50	$408.00	$39.30	$628.80	$23.80	$36.68
1 Steamfitter Apprentice	20.40	163.20	31.45	251.60		
1 Electric Welding Mach.		39.00		42.90	1.62	1.78
24 M.H., Daily Totals		$610.20		$923.30	$25.42	$38.46

Crew Q-19

	Bare Costs Hr.	Daily	Incl. Subs O & P Hr.	Daily	Cost Per Man-Hour Bare Costs	Incl. O&P
1 Steamfitter	$25.50	$204.00	$39.30	$314.40	$23.68	$36.38
1 Steamfitter Apprentice	20.40	163.20	31.45	251.60		
1 Electrician	25.15	201.20	38.40	307.20		
24 M.H., Daily Totals		$568.40		$873.20	$23.68	$36.38

Crew Q-20

	Bare Costs Hr.	Daily	Incl. Subs O & P Hr.	Daily	Cost Per Man-Hour Bare Costs	Incl. O&P
1 Sheet Metal Worker	$25.00	$200.00	$39.25	$314.00	$23.03	$35.94
1 Sheet Metal Apprentice	20.00	160.00	31.40	251.20		
.5 Electrician	25.15	100.60	38.40	153.60		
20 M.H., Daily Totals		$460.60		$718.80	$23.03	$35.94

Crew Q-21

	Bare Costs Hr.	Daily	Incl. Subs O & P Hr.	Daily	Cost Per Man-Hour Bare Costs	Incl. O&P
2 Steamfitters	$25.50	$408.00	$39.30	$628.80	$24.13	$37.11
1 Steamfitter Apprentice	20.40	163.20	31.45	251.60		
1 Electrician	25.15	201.20	38.40	307.20		
32 M.H., Daily Totals		$772.40		$1187.60	$24.13	$37.11

Crew Q-22

	Bare Costs Hr.	Daily	Incl. Subs O & P Hr.	Daily	Cost Per Man-Hour Bare Costs	Incl. O&P
1 Plumber	$25.45	$203.60	$39.25	$314.00	$22.90	$35.32
1 Plumber Apprentice	20.36	162.88	31.40	251.20		
1 Truck Crane, 12 Ton		362.00		398.20	22.62	24.88
16 M.H., Daily Totals		$728.48		$963.40	$45.52	$60.20

Crew R-1

	Bare Costs Hr.	Daily	Incl. Subs O & P Hr.	Daily	Cost Per Man-Hour Bare Costs	Incl. O&P
1 Electrician Foreman	$25.65	$205.20	$39.15	$313.20	$22.55	$35.00
3 Electricians	25.15	603.60	38.40	921.60		
2 Helpers	17.10	273.60	27.85	445.60		
48 M.H., Daily Totals		$1082.40		$1680.40	$22.55	$35.00

Crew R-2

	Bare Costs Hr.	Daily	Incl. Subs O & P Hr.	Daily	Cost Per Man-Hour Bare Costs	Incl. O&P
1 Electrician Foreman	$25.65	$205.20	$39.15	$313.20	$22.65	$35.23
3 Electricians	25.15	603.60	38.40	921.60		
2 Helpers	17.10	273.60	27.85	445.60		
1 Equip. Oper. (crane)	23.30	186.40	36.60	292.80		
1 S.P. Crane, 5 Ton		208.00		228.80	3.71	4.08
56 M.H., Daily Totals		$1476.80		$2202.00	$26.36	$39.31

Crew R-3

	Bare Costs Hr.	Daily	Incl. Subs O & P Hr.	Daily	Cost Per Man-Hour Bare Costs	Incl. O&P
1 Electrician Foreman	$25.65	$205.20	$39.15	$313.20	$24.98	$38.34
1 Electrician	25.15	201.20	38.40	307.20		
.5 Equip. Oper. (crane)	23.30	93.20	36.60	146.40		
.5 S.P. Crane, 5 Ton		104.00		114.40	5.20	5.72
20 M.H., Daily Totals		$603.60		$881.20	$30.18	$44.06

Crew R-4

	Bare Costs Hr.	Daily	Incl. Subs O & P Hr.	Daily	Cost Per Man-Hour Bare Costs	Incl. O&P
1 Struc. Steel Foreman	$26.10	$208.80	$47.55	$380.40	$24.71	$43.53
3 Struc. Steel Workers	24.10	578.40	43.90	1053.60		
1 Electrician	25.15	201.20	38.40	307.20		
1 Gas Welding Machine		68.40		75.25	1.71	1.88
40 M.H., Daily Totals		$1056.80		$1816.45	$26.42	$45.41

Crew R-5

	Bare Costs Hr.	Daily	Incl. Subs O & P Hr.	Daily	Cost Per Man-Hour Bare Costs	Incl. O&P
1 Electrician Foreman	$25.65	$205.20	$39.15	$313.20	$22.26	$34.63
4 Electrician Linemen	25.15	804.80	38.40	1228.80		
2 Electrician Operators	25.15	402.40	38.40	614.40		
4 Electrician Groundmen	17.10	547.20	27.85	891.20		
1 Crew Truck		161.20		177.30		
1 Tool Van		291.80		321.00		
1 Pick-up Truck		107.40		118.15		
.2 Crane, 55 Ton		145.45		160.00		
.2 Crane, 12 Ton		72.40		79.65		
.2 Auger, Truck Mtd.		297.50		327.25		
1 Tractor w/Winch		251.40		276.55	15.08	16.58
88 M.H., Daily Totals		$3286.75		$4507.50	$37.34	$51.21

Crew R-6

	Bare Costs Hr.	Daily	Incl. Subs O & P Hr.	Daily	Cost Per Man-Hour Bare Costs	Incl. O&P
1 Electrician Foreman	$25.65	$205.20	$39.15	$313.20	$22.26	$34.63
4 Electrician Linemen	25.15	804.80	38.40	1228.80		
2 Electrician Operators	25.15	402.40	38.40	614.40		
4 Electrician Groundmen	17.10	547.20	27.85	891.20		
1 Crew Truck		161.20		177.30		
1 Tool Van		291.80		321.00		
1 Pick-up Truck		107.40		118.15		
.2 Crane, 55 Ton		145.45		160.00		
.2 Crane, 12 Ton		72.40		79.65		
.2 Auger, Truck Mtd.		297.50		327.25		
1 Tractor w/Winch		251.40		276.55		
3 Cable Trailers		370.80		407.90		
.5 Tensioning Rig		122.20		134.40		
.5 Cable Pulling Rig		723.70		796.05	28.90	31.79
88 M.H., Daily Totals		$4503.45		$5845.85	$51.16	$66.42

Crew R-7

	Bare Costs Hr.	Daily	Incl. Subs O & P Hr.	Daily	Cost Per Man-Hour Bare Costs	Incl. O&P
1 Electrician Foreman	$25.65	$205.20	$39.15	$313.20	$18.52	$29.73
5 Electrician Groundmen	17.10	684.00	27.85	1114.00		
1 Crew Truck		161.20		177.30	3.35	3.69
48 M.H., Daily Totals		$1050.40		$1604.50	$21.87	$33.42

CREWS

Crew No.	Bare Costs		Incl. Subs O & P		Cost Per Man-Hour	
Crew R-8	Hr.	Daily	Hr.	Daily	Bare Costs	Incl. O&P
1 Electrician Foreman	$25.65	$205.20	$39.15	$313.20	$22.55	$35.00
3 Electrician Linemen	25.15	603.60	38.40	921.60		
2 Electrician Groundmen	17.10	273.60	27.85	445.60		
1 Pick-up Truck		107.40		118.15		
1 Crew Truck		161.20		177.30	5.59	6.15
48 M.H., Daily Totals		$1351.00		$1975.85	$28.14	$41.15
Crew R-9	Hr.	Daily	Hr.	Daily	Bare Costs	Incl. O&P
1 Electrician Foreman	$25.65	$205.20	$39.15	$313.20	$21.18	$33.21
1 Electrician Lineman	25.15	201.20	38.40	307.20		
2 Electrician Operators	25.15	402.40	38.40	614.40		
4 Electrician Groundmen	17.10	547.20	27.85	891.20		
1 Pick-up Truck		107.40		118.15		
1 Crew Truck		161.20		177.30	4.19	4.61
64 M.H., Daily Totals		$1624.60		$2421.45	$25.37	$37.82
Crew R-10	Hr.	Daily	Hr.	Daily	Bare Costs	Incl. O&P
1 Electrician Foreman	$25.65	$205.20	$39.15	$313.20	$23.89	$36.76
4 Electrician Linemen	25.15	804.80	38.40	1228.80		
1 Electrician Groundman	17.10	136.80	27.85	222.80		
1 Crew Truck		161.20		177.30		
3 Tram Cars		436.80		480.50	12.45	13.70
48 M.H., Daily Totals		$1744.80		$2422.60	$36.34	$50.46

Crew No.	Bare Costs		Incl. Subs O & P		Cost Per Man-Hour	
Crew R-11	Hr.	Daily	Hr.	Daily	Bare Costs	Incl. O&P
1 Electrician Foreman	$25.65	$205.20	$39.15	$313.20	$22.97	$35.60
4 Electricians	25.15	804.80	38.40	1228.80		
1 Helper	17.10	136.80	27.85	222.80		
1 Common Laborer	17.50	140.00	28.60	228.80		
1 Crew Truck		161.20		177.30		
1 Crane, 12 Ton		362.00		398.20	9.34	10.27
56 M.H., Daily Totals		$1810.00		$2569.10	$32.31	$45.87
Crew R-12	Hr.	Daily	Hr.	Daily	Bare Costs	Incl. O&P
1 Carpenter Foreman	$22.50	$180.00	$36.75	$294.00	$20.64	$34.01
4 Carpenters	22.00	704.00	35.95	1150.40		
4 Common Laborers	17.50	560.00	28.60	915.20		
1 Equip. Oper. (med.)	22.50	180.00	35.35	282.80		
1 Steel Worker	24.10	192.80	43.90	351.20		
1 Dozer, 200 H.P.		775.80		853.40		
1 Pick-up Truck		107.40		118.15	10.03	11.04
88 M.H., Daily Totals		$2700.00		$3965.15	$30.67	$45.05

		010 000 |Overhead	CREW	DAILY OUTPUT	MAN-HOURS	UNIT	BARE COSTS				TOTAL INCL O&P	
							MAT.	LABOR	EQUIP.	TOTAL		
004	0011	ARCHITECTURAL FEES ⑩										004
	0020	For new construction										
	0060	Minimum				Project					4.90%	
	0090	Maximum									16%	
	0100	For alteration work, to $500,000, add to fee									50%	
	0150	Over $500,000, add to fee				↓					25%	
016	0011	CONSTRUCTION MANAGEMENT FEES										016
	0060	For work to $10,000				Project					10%	
	0070	To $25,000									9%	
	0090	To $100,000									6%	
	0100	To $500,000									5%	
	0110	To $1,000,000									4%	
020	0010	CONTINGENCIES Allowance to add at conceptual stage									20%	020
	0050	Schematic stage									15%	
	0100	Preliminary working drawing stage									10%	
	0150	Final working drawing stage				↓					2%	
028	0010	ENGINEERING FEES Educational planning consultant, minimum				Contrct					4.10%	028
	0100	Maximum									10.10%	
	0400	Elevator & conveying systems, minimum ⑪									2.50%	
	0500	Maximum									5%	
	1000	Mechanical (plumbing & HVAC), minimum				↓					4.10%	
	1100	Maximum									10.10%	
	1200	Structural, minimum				Project					1%	
	1300	Maximum				"					2.50%	
032	0010	FACTORS To be added to construction costs for particular job ⑭⑶										032
	0200											
	0500	Cut & patch to match existing construction, add, minimum				Costs	2%	3%				
	0550	Maximum					5%	9%				
	0800	Dust protection, add, minimum					1%	2%				
	0850	Maximum					4%	11%				
	1100	Equipment usage curtailment, add, minimum					1%	1%				
	1150	Maximum					3%	10%				
	1400	Material handling & storage limitation, add, minimum					1%	1%				
	1450	Maximum					6%	7%				
	1700	Protection of existing work, add, minimum					2%	2%				
	1750	Maximum					5%	7%				
	2000	Shift work requirements, add, minimum						5%				
	2050	Maximum						30%				
	2300	Temporary shoring and bracing, add, minimum					2%	5%				
	2350	Maximum				↓	5%	12%				
036	0011	FIELD PERSONNEL										036
	0020											
	0180	Project manager, minimum				Week		955		955	1,450	
	0200	Average						1,065		1,065	1,620	
	0220	Maximum						1,210		1,210	1,840	
	0240	Superintendent, minimum				↓		905		905	1,380	
	0260	Average						1,005		1,005	1,530	
	0280	Maximum						1,135		1,135	1,725	
040	0010	INSURANCE Builders risk, standard, minimum				Job					.22%	040
	0050	Maximum									.59%	
	0200	All-risk type, minimum ②									.25%	
	0250	Maximum				↓					.62%	
	0400	Contractor's equipment floater, minimum				Value					.90%	
	0450	Maximum				"					1.60%	
	0600	Public liability, average				Job					1.55%	
	0810	Workers compensation & employer's liability										
	2000	Range of 35 trades in 50 states, excl. wrecking, minimum				Payroll		2.20%				
	2100	Average				"		15.10%				

For expanded coverage of these items see *Means Building Construction Cost Data 1991*

010 000	Overhead		CREW	DAILY OUTPUT	MAN-HOURS	UNIT	BARE COSTS				TOTAL INCL O&P	
							MAT.	LABOR	EQUIP.	TOTAL		
040	2200	Maximum (7)				Payroll		162.50%				040
042	0010	**JOB CONDITIONS** Modifications to total										042
	0020	project cost summaries										
	0100	Economic conditions, favorable, deduct				Project					2%	
	0200	Unfavorable, add									5%	
	0300	Hoisting conditions, favorable, deduct									2%	
	0400	Unfavorable, add									5%	
	0500	General Contractor management, experienced, deduct									2%	
	0600	Inexperienced, add									10%	
	0700	Labor availability, surplus, deduct									1%	
	0800	Shortage, add									10%	
	0900	Material storage area, available, deduct									1%	
	1000	Not available, add									2%	
	1100	Subcontractor availability, surplus, deduct									5%	
	1200	Shortage, add									12%	
	1300	Work space, available, deduct									2%	
	1400	Not available, add				↓					5%	
048	0010	**MAIN OFFICE EXPENSE** Average for General Contractors										048
	0020	As a percentage of their annual volume										
	0030	Annual volume to $50,000, minimum				% Vol.				20%		
	0040	Maximum								30%		
	0060	To $100,000, minimum								17%		
	0070	Maximum								22%		
	0080	To $250,000, minimum								16%		
	0090	Maximum								19%		
	0110	To $500,000, minimum								14%		
	0120	Maximum								16%		
	0130	To $1,000,000, minimum								10%		
	0140	Maximum				↓				8%		
052	0010	**MARK-UP** For General Contractors for change										052
	0100	of scope of job as bid										
	0200	Extra work, by subcontractors, add				%					10%	
	0250	By General Contractor, add									15%	
	0400	Omitted work, by subcontractors, deduct									5%	
	0450	By General Contractor, deduct									7.50%	
	0600	Overtime work, by subcontractors, add									15%	
	0650	By General Contractor, add									10%	
	1000	Installing contractors, on his own labor, minimum						52.7%				
	1100	Maximum						89.6%				
058	0010	**OVERHEAD** As percent of direct costs, minimum (4)								5%		058
	0050	Average								15%		
	0100	Maximum (6)				↓				30%		
062	0010	**OVERHEAD & PROFIT** Allowance to add to items in this										062
	0020	book that do not include Subs O&P, average				%					30%	
	0100	Allowance to add to items in this book that (4)										
	0110	do include Subs O&P, minimum				%					5%	
	0150	Average									10%	
	0200	Maximum									15%	
	0290	Typical, by size of project, under $50,000								40%		
	0310	$50,000 to $100,000								35%		
	0320	$100,000 to $500,000								25%		
	0330	$500,000 to $1,000,000				↓				20%		
070	0010	**PERMITS** Rule of thumb, most cities, minimum				Job					.50%	070
	0100	Maximum				"					2%	
082	0010	**SMALL TOOLS** As % of contractor's work, minimum (4)				Total					.50%	082
	0100	Maximum				"					2%	

For expanded coverage of these items see *Means Building Construction Cost Data 1991*

010 | Overhead

		010 000	Overhead	CREW	DAILY OUTPUT	MAN-HOURS	UNIT	BARE COSTS MAT.	BARE COSTS LABOR	BARE COSTS EQUIP.	BARE COSTS TOTAL	TOTAL INCL O&P	
086	0010		**TAXES** Sales tax, State, County & City, average ⑧				%	4.44%					086
	0050		Maximum					7.50%					
	0200		Social Security, on first $51,300 of wages ⑤						7.65%				
	0300		Unemployment, MA, combined Federal and State, minimum						2.60%				
	0350		Average						6.20%				
	0400		Maximum						6.80%				

013 | Submittals

		013 300	Survey Data	CREW	DAILY OUTPUT	MAN-HOURS	UNIT	BARE COSTS MAT.	BARE COSTS LABOR	BARE COSTS EQUIP.	BARE COSTS TOTAL	TOTAL INCL O&P	
306	0010		**SURVEYING** Conventional, topographical, minimum	A-7	3.30	7.270	Acre	13.25	140		153.25	235	306
	0100		Maximum	A-8	.60	53.330		41	995		1,036	1,650	
	0300		Lot location and lines, minimum, for large quantities	A-7	2	12		22	230		252	390	
	0320		Average	"	1.25	19.200		41	370		411	630	
	0400		Maximum, for small quantities	A-8	1	32		66	600		666	1,025	
	1100		Crew for building layout, 2 man crew	A-6	1	16	Day		325		325	510	
	1200		3 man crew	A-7	1	24			460		460	735	
	1300		4 man crew	A-8	1	32			600		600	955	

		013 400	Shop Drawings										
408	0010		**RENDERINGS** Water color, matted, 20″ x 30″, eye level,										408
	0020		1 building, minimum				Ea.	400			400	440	
	0050		Average					925			925	1,025	
	0100		Maximum					1,450			1,450	1,600	
	1000		5 buildings, minimum					650			650	715	
	1100		Maximum					1,825			1,825	2,000	

014 | Quality Control

		014 100	Testing Services	CREW	DAILY OUTPUT	MAN-HOURS	UNIT	BARE COSTS MAT.	BARE COSTS LABOR	BARE COSTS EQUIP.	BARE COSTS TOTAL	TOTAL INCL O&P	
108	0011		**TESTING**										108
	1800		Compressive strength, cylinders, delivered to lab				Ea.					10	
	1900		Picked up by lab, minimum									8.40	
	1950		Average									12.60	
	2000		Maximum									21	
	2200		Compressive strength, cores (not incl. drilling)									38	
	2300		Patching core holes									22	
	4730		Soil testing										
	4750		Moisture content				Ea.					7.90	
	4800		Permeability, variable or constant head, undisturbed									195	
	4850		Recompacted									265	
	4900		Proctor compaction, 4″ standard mold									125	
	4950		6″ modified mold									145	
	5100		Shear tests, triaxial, minimum									145	
	5150		Maximum									475	
	5550		Technician for inspection, per day, earthwork									210	

For expanded coverage of these items see *Means Building Construction Cost Data 1991*

3

014 100	Testing Services	CREW	DAILY OUTPUT	MAN-HOURS	UNIT	BARE COSTS				TOTAL INCL O&P	
						MAT.	LABOR	EQUIP.	TOTAL		
108	5650	Bolting				Ea.					255	108
	5750	Roofing									210	
	5790	Welding				↓					255	
	5820	Non-destructive testing, dye penetrant				Day					300	
	5840	Magnetic particle									300	
	5860	Radiography									465	
	5880	Ultrasonic				↓					320	
	6000	Welding certification, minimum				Ea.					85	
	6100	Maximum				"					260	
	7000	Underground storage tank										
	7500	Hydrostatic tank tightness test per tank, min.				Ea.					500	
	7510	Maximum				"					1,000	
	7600	Vadose zone (soil gas) sampling, 10-40 samples, min.				Day					1,500	
	7610	Maximum				"					2,500	
	7700	Ground water monitoring incl. drilling 3 wells, min.				Total					5,000	
	7710	Maximum				"					7,000	

015 | Construction Facilities and Temporary Controls

015 100	Temporary Utilities	CREW	DAILY OUTPUT	MAN-HOURS	UNIT	BARE COSTS				TOTAL INCL O&P	
						MAT.	LABOR	EQUIP.	TOTAL		
104	0010	**TEMPORARY UTILITIES**										104
	0100	Heat, incl. fuel and operation, per week, 12 hrs. per day	1 Skwk	8.75	.914	CSF Flr	16.75	21		37.75	52	
	0200	24 hrs. per day	"	4.50	1.780		22	40		62	89	
	0350	Lighting, incl. service lamps, wiring & outlets, minimum	1 Elec	34	.235		2.21	5.90		8.11	11.45	
	0360	Maximum	"	17	.471		5.10	11.85		16.95	24	
	0400	Power for temporary lighting only, per month, minimum/month								1.03	1.02	
	0450	Maximum/month								2.61	2.61	
	0600	Power for job duration incl. elevator, etc., minimum								47	50	
	0650	Maximum				↓				105	100	

015 200	Temporary Construction	CREW	DAILY OUTPUT	MAN-HOURS	UNIT	MAT.	LABOR	EQUIP.	TOTAL	TOTAL INCL O&P	
204	0010	**PROTECTION** Stair tread, 2″ x 12″ planks, 1 use	1 Carp	75	.107	Tread	.98	2.35		3.33	4.91	204
	0100	Exterior plywood, ½″ thick, 1 use		65	.123		.47	2.71		3.18	4.94	
	0200	¾″ thick, 1 use	↓	60	.133	↓	.86	2.93		3.79	5.75	
208	0010	**TEMPORARY CONSTRUCTION** See division 010-094 & 015-300										208

015 250	Construction Aids	CREW	DAILY OUTPUT	MAN-HOURS	UNIT	MAT.	LABOR	EQUIP.	TOTAL	TOTAL INCL O&P	
254	0014	**SCAFFOLDING, STEEL TUBULAR** Rent, 1 use per mo., no plank										254
	0090	Building exterior, 1 to 5 stories	3 Carp	16.80	1.430	C.S.F.	13	31		44	66	
	0200	To 12 stories ⑭	4 Carp	15	2.130		12.20	47		59.20	90	
	0310	13 to 20 stories	5 Carp	16.75	2.390		11.40	53		64.40	98	
	0460	Building interior walls, (area) up to 16′ high	3 Carp	22.70	1.060		12.50	23		35.50	52	
	0560	16′ to 40′ high		18.70	1.280	↓	12.75	28		40.75	60	
	0800	Building interior floor area, up to 30′ high	↓	90	.267	C.C.F.	3.90	5.85		9.75	13.85	
	0900	Over 30′ high	4 Carp	100	.320	"	4.40	7.05		11.45	16.35	
255	0011	**SCAFFOLDING SPECIALTIES**										255
	0050											
	1200	Sidewalk bridge, heavy duty steel posts & beams, including										
	1210	parapet protection & waterproofing										

For expanded coverage of these items see *Means Building Construction Cost Data 1991*

		015 250	Construction Aids	CREW	DAILY OUTPUT	MAN-HOURS	UNIT	BARE COSTS				TOTAL INCL O&P	
								MAT.	LABOR	EQUIP.	TOTAL		
255	1220		8' to 10' wide, 2 posts	3 Carp	15	1.600	L.F.	10.80	35		45.80	69	255
	1230		3 posts	"	10	2.400	"	16.50	53		69.50	105	
	1500		Sidewalk bridge using tubular steel										
	1510		scaffold frames, including planking	3 Carp	55	.436	L.F.	3.71	9.60		13.31	19.75	
	1600		For 2 uses per month, deduct from all above					50%					
	1700		For 1 use every 2 months, add to all above					100%					
	1900		Catwalks, 32" wide, no guardrails, 6' span, buy				Ea.	115			115	125	
	2000		10' span, buy				"	180			180	200	
	2800		Hand winch-operated masons										
	2810		scaffolding, no plank moving required										
	2900		98' long, 10'-6" high, buy				Ea.	4,950			4,950	5,450	
	3000		Rent per month					300			300	330	
	3100		28'-6" high, buy					8,500			8,500	9,350	
	3200		Rent per month					450			450	495	
	3400		196' long, 28'-6" high, buy					16,500			16,500	18,200	
	3500		Rent per month					885			885	975	
	3600		64'-6" high, buy					28,000			28,000	30,800	
	3700		Rent per month					1,775			1,775	1,950	
	3720		Putlog, standard, 8' span, with hangers, buy					85			85	94	
	3730		Rent per month					12.35			12.35	13.60	
	3750		12' span, buy					125			125	140	
	3760		Trussed type, 14' span, buy					195			195	215	
	3770		Rent per month					18			18	19.80	
	3790		20' span, buy					230			230	255	
	3795		Rent per month					26			26	29	
	3800		Rolling ladders with handrails, 30" wide, buy, 2 step					155			155	170	
	4000		7 step					335			335	370	
	4100		Rolling towers, buy, 5' wide, 7' long, 9' high					925			925	1,025	
	4200		For 5' high added sections, add				↓	180			180	200	
	4300		Complete incl. wheels, railings, etc.										
	4400		up to 20' high, rent per month				Ea.	105			105	115	
256	0010		**SWING STAGING** For masonry, 5' wide to 7' long, hand operated										256
	0020		cable type, with 150' cables, buy				Ea.	1,800			1,800	1,975	
	0030		Rent per week				"	15.45			15.45	17	
	0600		Lightweight (not for masons) 20' long for 200' height,										
	0610		manual type, buy				Ea.	3,300			3,300	3,625	
	0620		Rent per month					365			365	400	
	0700		Powered, electric or air, to 300' high, buy					10,800			10,800	11,900	
	0710		Rent per month					915			915	1,000	
	0800		Over 300' per 100 L.F., add per month					53			53	58	
	1000		Single bosun's chair or work basket 3' x 3.5', electric, buy					6,500			6,500	7,150	
	1010		Rent per month				↓	445			445	490	
	2100		Catwalks, no handrails, 3 joists, 2" x 4"	2 Carp	55	.291	L.F.	3.10	6.40		9.50	13.85	
	2150		3 joists, 3" x 6"	"	40	.400	"	5.65	8.80		14.45	21	
		015 300	**Barriers And Enclosures**										
302	0010		**BARRICADES** 5' high, 3 rail @ 2" x 8", fixed	2 Carp	30	.533	L.F.	9.40	11.75		21.15	30	302
	0150		Movable	"	20	.800	"	10.15	17.60		27.75	40	
	0300		Stock units, 6' high, 8' wide, plain, buy				Ea.	370			370	405	
	0350		With reflective tape, buy				"	460			460	505	
	0400		Break-a-way 3" PVC pipe barricade										
	0410		with 3 ea. 1' x 4' reflectorized panels, buy				Ea.	260			260	285	
	0500		Plywood with steel legs, 32" wide					53			53	58	
	0600		Telescoping Christmas tree, 9' high, 5 flags, buy					105			105	115	
	0800		Traffic cones, PVC, 18" high				↓	5.85			5.85	6.45	
	0850		28" high					11.75			11.75	12.95	

For expanded coverage of these items see *Means Building Construction Cost Data 1991*

015 300 | Barriers And Enclosures

			CREW	DAILY OUTPUT	MAN-HOURS	UNIT	BARE COSTS MAT.	LABOR	EQUIP.	TOTAL	TOTAL INCL O&P	
302	0900	Barrels, 55 gal., with flasher	1 Clab	96	.083	Ea.	31	1.46		32.46	36	302
	1000	Guardrail, wooden, 3' high, 1" x 6", on 2" x 4" posts	2 Carp	200	.080	L.F.	1.45	1.76		3.21	4.47	
	1100	2" x 6", on 4" x 4" posts	"	165	.097		2.41	2.13		4.54	6.15	
	1200	Portable metal with base pads, buy					11.50			11.50	12.65	
	1250	Typical installation, assume 10 reuses	2 Carp	600	.027	↓	1.35	.59		1.94	2.44	
304	0010	FENCING Chain link, 5' high	2 Clab	100	.160	L.F.	4.60	2.80		7.40	9.65	304
	0100	6' high		75	.213		5.85	3.73		9.58	12.55	
	0200	Rented chain link, 6' high, to 500'		100	.160		1.89	2.80		4.69	6.65	
	0250	Over 1000' (up to 12 mo.)	↓	110	.145		1.58	2.55		4.13	5.90	
	0350	Plywood, painted, 2" x 4" frame, 4' high	A-4	135	.178		3.83	3.84		7.67	10.45	
	0400	4" x 4" frame, 8' high	"	110	.218		6.50	4.71		11.21	14.80	
	0500	Wire mesh on 4" x 4" posts, 4' high	2 Carp	100	.160		4.15	3.52		7.67	10.30	
	0550	8' high	"	80	.200	↓	6.25	4.40		10.65	14.05	
306	0010	WINTER PROTECTION Reinforced plastic on wood										306
	0100	framing to close openings	2 Clab	750	.021	S.F.	.28	.37		.65	.92	
	0200	Tarpaulins hung over scaffolding, 8 uses, not incl. scaffolding		1,500	.011		.14	.19		.33	.46	
	0300	Prefab fiberglass panels, steel frame, 8 uses	↓	1,200	.013	↓	.53	.23		.76	.96	

015 400 | Security

			CREW	DAILY OUTPUT	MAN-HOURS	UNIT	MAT.	LABOR	EQUIP.	TOTAL	TOTAL INCL O&P	
480	0010	WATCHMAN Service, monthly basis, uniformed man, minimum				Hr.					6.70	480
	0100	Maximum									12.70	
	0200	Man and command dog (man dog), minimum									9.60	
	0300	Maximum									14.35	
	0500	Sentry dog, leased, with job patrol (yard dog), 1 dog				Week					180	
	0600	2 dogs				"					250	
	0800	Purchase, trained sentry dog, minimum				Ea.					870	
	0900	Maximum				"					1,650	

015 500 | Access Roads

			CREW	DAILY OUTPUT	MAN-HOURS	UNIT	MAT.	LABOR	EQUIP.	TOTAL	TOTAL INCL O&P	
552	0010	ROADS AND SIDEWALKS Temporary										552
	2200	Sidewalks, 2" x 12" planks, 2 uses	1 Carp	350	.023	S.F.	.32	.50		.82	1.17	
	2300	Exterior plywood, 2 uses, ½" thick		750	.011		.22	.23		.45	.63	
	2400	⅝" thick		650	.012		.28	.27		.55	.75	
	2500	¾" thick	↓	600	.013	↓	.32	.29		.61	.83	

015 600 | Temporary Controls

			CREW	DAILY OUTPUT	MAN-HOURS	UNIT	MAT.	LABOR	EQUIP.	TOTAL	TOTAL INCL O&P	
602	0010	TARPAULINS Cotton duck, 10 oz. to 13.13 oz. per S.Y., minimum				S.F.	.26			.26	.29	602
	0050	Maximum					.45			.45	.50	
	0100	Polyvinyl coated nylon, 14 oz. to 18 oz., minimum					.36			.36	.40	
	0150	Maximum					.46			.46	.51	
	0200	Reinforced polyethylene 3 mils thick, white					.05			.05	.06	
	0300	4 mils thick, white, clear or black					.06			.06	.07	
	0400	5.5 mils thick, clear					.07			.07	.08	
	0500	White, fire retardant					.12			.12	.13	
	0600	7.5 mils, oil resistant, fire retardant					.13			.13	.14	
	0700	8.5 mils, black					.16			.16	.18	
	0720	Steel reinforced polyethylene, 4 mils thick					.41			.41	.45	
	0740	Mylar polyester, non-reinforced, 7 mils thick				↓	.86			.86	.95	

015 800 | Project Signs

			CREW	DAILY OUTPUT	MAN-HOURS	UNIT	MAT.	LABOR	EQUIP.	TOTAL	TOTAL INCL O&P	
804	0010	SIGNS Hi-intensity reflectorized, no posts, buy				S.F.	9.25			9.25	10.20	804

015 900 | Field Offices And Sheds

			CREW	DAILY OUTPUT	MAN-HOURS	UNIT	MAT.	LABOR	EQUIP.	TOTAL	TOTAL INCL O&P	
904	0010	OFFICE Trailer, furnished, no hookups, 20' x 8', buy	2 Skwk	1	16	Ea.	4,000	360		4,360	4,975	904
	0250	Rent per month				"	150			150	165	

For expanded coverage of these items see *Means Building Construction Cost Data 1991*

015 | Construction Facilities and Temporary Controls

		015 900	Field Offices And Sheds	CREW	DAILY OUTPUT	MAN-HOURS	UNIT	BARE COSTS				TOTAL INCL O&P	
								MAT.	LABOR	EQUIP.	TOTAL		
904	0300		32' x 8', buy	2 Skwk	.70	22.860	Ea.	6,000	520		6,520	7,425	904
	0350		Rent per month					200			200	220	
	0400		50' x 10', buy	2 Skwk	.60	26.670		11,300	605		11,905	13,400	
	0450		Rent per month					365			365	400	
	0500		50' x 12', buy	2 Skwk	.50	32		12,400	725		13,125	14,800	
	0550		Rent per month					390			390	430	
	0700		For air conditioning, rent per month, add				↓	37			37	41	
	0800		For delivery, add per mile				Mile	1.58			1.58	1.74	
	1200		Storage vans, trailer mounted, 16' x 8', buy	2 Skwk	1.80	8.890	Ea.	2,500	200		2,700	3,075	
	1250		Rent per month					89			89	98	
	1300		28' x 10', buy	2 Skwk	1.40	11.430		2,925	260		3,185	3,625	
	1350		Rent per month					89			89	98	

016 | Material and Equipment

		016 400	Equipment Rental	UNIT	HOURLY OPER. COST.	RENT PER DAY	RENT PER WEEK	RENT PER MONTH	CREW EQUIPMENT COST	
406	0010	CONCRETE EQUIPMENT RENTAL								406
	0100	without operators								
	0200	Bucket, concrete lightweight, ½ C.Y.		Ea.	.11	20	59	180	12.70	
	0600	Cart, concrete, operator walking, 10 C.F.			1	60	180	540	44	
	0700	Operator riding, 18 C.F.			2.35	145	435	1,300	105.80	
	0800	Conveyer for concrete, portable, gas, 16" wide, 26' long			2.73	115	350	1,050	91.85	
	0900	46' long			2.90	140	425	1,275	108.20	
	1000	56' long			3.10	155	470	1,400	118.80	
	1200	Finisher, concrete floor, gas, riding trowel, 48" diameter			2.30	73	220	660	62.40	
	1300	Gas, manual, 3 blade, 36" trowel			.70	33	100	300	25.60	
	1400	4 blade, 48" trowel			1.10	43	130	390	34.80	
	1500	Float, hand-operated (Bull float) 48" wide			.12	6	18	55	4.55	
	1600	Grinder, concrete and terrazzo, electric, floor			1.10	43	130	390	34.80	
	1700	Wall grinder			.39	23	70	210	17.10	
	1800	Mixer, powered, mortar and concrete, gas, 6 C.F., 18 H.P.			.90	43	130	390	33.20	
	1900	10 C.F., 25 H.P.			1.18	54	160	480	41.45	
	2120	Pump, concrete, truck mounted, 4" line, 80' boom ⑯46			8.70	740	2,225	6,675	514.60	
	2140	5" line, 110' boom			10.55	900	2,700	8,100	624.40	
	2160	Mud jack, 47 C.F. per hr.			2.62	60	180	545	56.95	
	2180	225 C.F. per hr.			5.30	315	940	2,825	230.40	
	2600	Saw, concrete, manual, gas, 18 H.P.			2.55	57	170	510	54.40	
	2650	Self-propelled, gas, 30 H.P.			4.85	90	270	810	92.80	
	2700	Vibrators, concrete, electric, 60 cycle, 2 H.P.			.27	14.55	44	130	10.95	
	2800	3 H.P.			.43	16.65	50	150	13.45	
	2900	Gas engine, 5 H.P.			.64	26	77	230	20.50	
	3000	8 H.P.		↓	1	38	115	345	31	
408	0010	EARTHWORK EQUIPMENT RENTAL Without operators ⑬		Ea.	9.55	435	1,300	3,900	336.40	408
	0100	Backhoe, diesel hydraulic, crawler mounted, ½ C.Y. cap.								
	0120	⅝ C.Y. capacity			13.30	410	1,225	3,675	351.40	
	0400	Backhoe-loader, wheel type, 40 to 45 H.P., ⅝ C.Y. capacity			6.20	190	565	1,700	162.60	
	0450	45 H.P. to 60 H.P., ¾ C.Y. capacity			7.30	220	660	2,000	190.40	
	0470	112 H.P.,1-¾ C.Y. loader, ½ C.Y. backhoe			13.15	320	970	2,900	299.20	
	1200	Compactor, roller, 2 drum, 2000 lb., operator walking			1.55	105	310	930	74.40	
	1250	Rammer compactor, gas, 1000 lb. blow			.35	35	105	315	23.80	
	1300	Vibratory plate, gas, 13" plate, 1000 lb. blow			.45	28	84	250	20.40	
	1350	24" plate, 5000 lb. blow			1.55	52	155	465	43.40	

For expanded coverage of these items see *Means Building Construction Cost Data 1991*

016 400 | Equipment Rental

		Description	UNIT	HOURLY OPER. COST.	RENT PER DAY	RENT PER WEEK	RENT PER MONTH	CREW EQUIPMENT COST	
408	1950	Hammer, pavement demo., hyd., gas, self-prop., 1000 to 1250 lb. ⑰	Ea.	15.35	360	1,075	3,225	337.80	**408**
	4110	Tractor, crawler, with bulldozer, torque converter, diesel 75 H.P.		9.10	335	1,000	3,000	272.80	
	4400	Loader, crawler, torque conv., diesel, 1-½ C.Y., 80 H.P.		10.40	435	1,300	3,900	343.20	
	4610	Tractor loader, wheel, torque conv., 4 x 4, 1 to 1-¼ C.Y., 65 H.P.		7.70	275	820	2,450	225.60	
	4900	Trencher, chain, boom type, gas, operator walking, 12 H.P.		1.65	65	200	600	53.20	
	4910	Operator riding, 40 H.P.		5.40	155	460	1,375	135.20	
	5250	Truck, dump, tandem, 12 ton payload		15.10	295	890	2,675	298.80	
	5300	Three axle dump, 16 ton payload		16.90	385	1,150	3,450	365.20	
	5550	Off highway rear dump, 25 ton capacity		16.75	690	2,075	6,225	549	
	5600	35 ton capacity		27	1,025	3,100	9,300	836	
420	0010	**GENERAL EQUIPMENT RENTAL**							**420**
	0200	Air compressor, portable, gas engine, 60 C.F.M.	Ea.	4.55	47	140	420	64.40	
	0300	160 C.F.M.		6.05	62	185	555	85.40	
	0800	For silenced models, small sizes, add		3%	5%	5%	5%		
	0900	Large sizes, add		4%	7%	7%	7%		
	0920	Air tools and accessories							
	0930	Breaker, pavement, 60 lb.	Ea.	.10	22	65	195	13.80	
	0940	80 lb.		.12	21	71	195	15.15	
	0950	Drills, hand (jackhammer) 65 lb.		.18	22	65	195	14.45	
	0960	Wagon, swing boom, 4" drifter		9	240	725	2,175	217	
	0980	Dust control per drill		.16	9	27	81	6.70	
	0990	Hammer, chipping, 12 lb.		.11	14	42	125	9.30	
	1000	Hose, air with couplings, 50' long, ¾" diameter		.39	3	8	25	4.70	
	1100	1" diameter		1	4	11	34	10.20	
	1550	Spade, 25 lb.		.05	6.25	20	60	4.40	
	1555								
	1560	Tamper, single, 35 lb.	Ea.	.10	21	62	185	13.20	
	1570	Triple, 140 lb.		1.65	35	110	330	35.20	
	1580	Wrenches, impact, air powered, up to ¾" bolt		.15	17	50	150	11.20	
	1590	Up to 1-¼" bolt		.35	42	125	375	27.80	
	1700	Carts, brick, hand powered, 1000 lb. capacity		1.15	16	47	140	18.60	
	1800	Gas engine, 1500 lb., 7-½' lift		1.32	76	230	685	56.55	
	2020	Forklift, wheeled, for brick, 18', 3000 lb., 2 wheel drive, gas		8.25	165	500	1,500	166	
	2040	28', 4000 lb., 4 wheel drive, diesel		6	225	675	2,025	183	
	2100	Generator, electric, gas engine, 1.5 KW to 3 KW		.95	29	87	260	25	
	2200	5 KW		1.18	47	140	420	37.45	
	2900	Heaters, space, oil or electric, 50 MBH		.10	20	61	185	13	
	3000	100 MBH		.06	17	51	155	10.70	
	3100	300 MBH		.11	42	125	375	25.90	
	3150	500 MBH		.17	57	170	515	35.35	
	3200	Hose, water, suction with coupling, 20' long, 2" diameter		.05	7	20	60	4.40	
	3210	3" diameter		.05	11	32	95	6.80	
	3250	Discharge hose with coupling, 50' long, 2" diameter		.05	4	11	35	2.60	
	3260	3" diameter		.05	5	16	50	3.60	
	3300	Ladders, extension type, 16' to 36' long			8.30	24	72	4.80	
	3400	40' to 60' long			19.75	60	180	12	
	3410	Level, laser type, for pipe laying, self leveling			99	295	890	59	
	3430	Manual leveling			78	235	700	47	
	3440	Rotary beacon with rod and sensor			99	295	890	59	
	3460	Builders level with tripod and rod			23	70	210	14	
	3700	Mixer, powered, plaster and mortar, 6 C.F., 7 H.P.		.70	38	115	345	28.60	
	3800	10 C.F., 9 H.P.		1.10	53	160	480	40.80	
	3900	Paint sprayers complete, 8 CFM		.05	30	90	270	18.40	
	4000	17 CFM		.05	42	125	375	25.40	
	4100	Pump, centrifugal gas pump, 1-½", 4 MGPH		.37	19.05	57	170	14.35	
	4500	Submersible electric pump, 1-¼", 55 GPM		.27	30	90	270	20.15	
	5100	Diaphragm pump, gas, single, 1-½" diameter		.43	16	48	145	13.05	
	5500	Trash pump, self-priming, gas, 2" diameter		.95	27	80	240	23.60	

For expanded coverage of these items see *Means Building Construction Cost Data 1991*

016 400	Equipment Rental	UNIT	HOURLY OPER. COST.	RENT PER DAY	RENT PER WEEK	RENT PER MONTH	CREW EQUIPMENT COST	
420								**420**
5700	Salamanders, L.P. gas fired, 100,000 B.T.U.	Ea.	.60	12	36	110	12	
5720	Sandblaster, portable, open top, 3 C.F. capacity		.15	37	110	330	23.20	
5740	Accessories for above		.05	16	47	140	9.80	
5800	Saw, chain, gas engine, 18″ long		.40	28	84	250	20	
6000	Masonry, table mounted, 14″ diameter, 5 H.P.		1.50	33	100	300	32	
6100	Circular, hand held, electric, 7″ diameter		.15	11	34	100	8	
6200	12″ diameter		.25	22	66	200	15.20	
6300	Steam cleaner, 100 gallons per hour		.32	43	130	390	28.55	
6310	200 gallons per hour		.60	45	135	405	31.80	
6350	Torch, cutting, acetylene-oxygen, 150′ hose		6.40	19	58	175	62.80	
6360	Hourly operating cost includes tips and gas		6.75	16.65	50	150	64	
6410	Toilet, portable chemical			8.30	25	75	5	
6420	Recycle flush type			10.80	31	93	6.20	
6430	Toilet, fresh water flush, garden hose,			16.20	49	145	9.80	
6440	Hoisted, non-flush, for high rise			8.65	26	78	5.20	
6450	Toilet, trailers, minimum			20	60	180	12	
6460	Maximum			81	245	730	49	
7020	Transit with tripod			27	80	240	16	
7050	Trench box, 8,000 lbs. 8′ x 16′		1.40	140	425	1,275	96.20	
7070	12,000 lbs., 10′ x 20′		1.75	220	650	1,950	144	
7100	Truck, pickup, ¾ ton, 2 wheel drive		9.05	58	175	525	107.40	
7200	4 wheel drive		10.25	60	175	530	117	
7650	Vacuum, H.E.P.A., 16 gal., wet/dry							
7655	55 gal, wet/dry	Ea.	.70	45	135	405	32.60	
7690	Large production vacuum loader		10.80	600	1,800	5,400	446.40	
7700	Welder, electric, 200 amp		.80	18	54	160	17.20	
7800	300 amp		1.75	42	125	375	39	
8100	Wheelbarrow, any size			6.65	20	60	4	
460	**LIFTING & HOISTING EQUIPMENT RENTAL**							**460**
0010								
0100	without operators							
0600	Crawler, cable, ½ C.Y., 15 tons at 12′ radius	Ea.	15.80	415	1,250	3,750	376.40	
0900	1-½ C.Y., 40 tons at 12′ radius ⑰		25.50	760	2,275	6,825	659	
1100	3 C.Y., 75 tons at 12′ radius		36.40	965	2,900	8,700	871.20	
1400	200 ton capacity, 150′ boom		105	2,350	7,075	21,200	2,255	
1600	Truck mounted, cable operated, 6 x 4, 20 tons at 10′ radius		11.55	590	1,800	5,400	452.40	
2000	8 x 4, 60 tons at 15′ radius		37.60	825	2,475	7,425	795.80	
2400	Truck mounted, hydraulic, 12 ton capacity		19	350	1,050	3,150	362	
2550	33 ton capacity		20.40	885	2,340	7,025	631.20	
2700	80 ton capacity		31.05	1,425	4,275	12,800	1,103	
2800	Self-propelled, 4 x 4, with telescoping boom, 5 ton		7.25	250	750	2,250	208	
3100	25 ton capacity		17.65	640	1,925	5,775	526.20	
3600	Hoists, chain type, overhead, manual, ¾ ton		.05	5.20	14.55	42.65	3.30	
3900	10 ton		.25	22	66	200	15.20	
4000	Hoist and tower, 5000 lb. cap., portable electric, 40′ high		3.45	150	455	1,375	118.60	
4100	For each added 10′ section, add			7.30	21	62	4.20	
4200	Hoist and single tubular tower, 5000 lb. electric, 100′ high		4.55	210	625	1,875	161.40	
4300	For each added 6′-6″ section, add		.05	18	52	155	10.80	
4400	Hoist and double tubular tower, 5000 lb., 100′ high		4.85	225	680	2,050	174.80	
4500	For each added 6′-6″ section, add		.05	11	33	99	7	
4550	Hoist and tower, mast type, 6000 lb., 100′ high		4.65	245	740	2,225	185.20	
4570	For each added 10′ section, add		.10	8.30	25	75	5.80	
4600	Hoist and tower, personnel, electric, 2000 lb., 100′ @ 125 FPM ⑬		8.65	605	1,825	5,450	434.20	
4700	3000 lb., 100′ @ 200 FPM		9.30	660	1,975	5,925	469.40	
4800	3000 lb., 150′ @ 300 FPM		9.95	710	2,125	6,400	504.60	
4900	4000 lb., 100′ @ 300 FPM		10.65	765	2,275	6,875	540.20	
5000	6000 lb., 100′ @ 275 FPM		11.30	825	2,475	7,400	585.40	
5100	For added heights up to 500′, add	L.F.		1.05	3	9	.60	
5200	Jacks, hydraulic, 20 ton	Ea.	.11	2	6	18	2.10	

For expanded coverage of these items see *Means Building Construction Cost Data 1991*

016 | Material and Equipment

016 400	Equipment Rental		UNIT	HOURLY OPER. COST.	RENT PER DAY	RENT PER WEEK	RENT PER MONTH	CREW EQUIPMENT COST	
460	5500	100 ton	Ea.	.15	18	54	160	12	460

017 | Contract Closeout

017 100	Final Cleaning		CREW	DAILY OUTPUT	MAN-HOURS	UNIT	BARE COSTS MAT.	LABOR	EQUIP.	TOTAL	TOTAL INCL O&P	
104	0011	CLEANING UP After job completion, minimum				Project					.30%	104
	0040	Maximum				"					1%	
	0050	Cleanup of floor area, continuous, per day	A-5	8	2.250	M.S.F.	1.54	40	4.56	46.10	71	
	0100	Final	"	11.50	1.570	"	1.64	28	3.17	32.81	50	

020 | Subsurface Investigation and Demolition

020 120	Std Penetration Tests		CREW	DAILY OUTPUT	MAN-HOURS	UNIT	BARE COSTS MAT.	LABOR	EQUIP.	TOTAL	TOTAL INCL O&P	
123	0010	BORINGS Initial field stake out and determination of elevations	A-6	1	16	Day		325		325	510	123
	0100	Drawings showing boring details				Total		165		165	255	
	0200	Report and recommendations from P.E.						365		365	560	
	0300	Mobilization and demobilization, minimum	B-55	4	6			105	135	240	320	
	0350	For over 100 miles, per added mile		450	.053	Mile		.94	1.21	2.15	2.86	
	0600	Auger holes in earth, no samples, 2-½" diameter		78.60	.305	L.F.		5.40	6.90	12.30	16.40	
	0650	4" diameter		67.50	.356			6.30	8.05	14.35	19.10	
	0800	Cased borings in earth, with samples, 2-½" diameter		55.50	.432			7.65	9.80	17.45	23	
	0850	4" diameter		32.60	.736			13.05	16.70	29.75	40	
	1000	Drilling in rock, "BX" core, no sampling	B-56	34.90	.458			8.90	14.20	23.10	30	
	1050	With casing & sampling		31.70	.505			9.80	15.65	25.45	33	
	1200	"NX" core, no sampling		25.92	.617			12	19.15	31.15	40	
	1250	With casing and sampling		25	.640			12.45	19.85	32.30	42	
	1400	Drill rig and crew with light duty rig	B-55	1	24	Day		425	545	970	1,300	
	1450	With heavy duty rig	B-56	1	16	"		310	495	805	1,050	
	1500	For inner city borings add, minimum									10%	
	1510	Maximum									20%	
125	0010	DRILLING, CORE Reinforced concrete slab, up to 6" thick slab										125
	0020	Including layout and set up										
	0100	1" diameter core	B-89A	43	.372	Ea.	2.14	7.45	1.13	10.72	15.75	
	0150	Each added inch thick, add		350	.046		.36	.92	.14	1.42	2.04	
	0300	3" diameter core		37	.432		4.75	8.70	1.32	14.77	21	
	0350	Each added inch thick, add		222	.072		.79	1.45	.22	2.46	3.46	
	0500	4" diameter core		34	.471		6.35	9.45	1.44	17.24	24	
	0550	Each added inch thick, add		212	.075		1.06	1.52	.23	2.81	3.89	
	0700	6" diameter core		27	.593		7.75	11.90	1.81	21.46	30	
	0750	Each added inch thick, add		180	.089		1.28	1.78	.27	3.33	4.61	
	0900	8" diameter core		19	.842		10.60	16.90	2.57	30.07	42	
	0950	Each added inch thick, add		118	.136		1.77	2.72	.41	4.90	6.85	
	1100	10" diameter core	A-1	17	.471		14.30	8.25	3.20	25.75	33	
	1150	Each added inch thick, add	"	106	.075		2.38	1.32	.51	4.21	5.35	

For expanded coverage of these items see *Means Site Work Cost Data 1991*

020 120 | Std Penetration Tests

		CREW	DAILY OUTPUT	MAN-HOURS	UNIT	BARE COSTS MAT.	LABOR	EQUIP.	TOTAL	TOTAL INCL O&P	
125	1300	12″ diameter core	A-1	15	.533	Ea.	17.25	9.35	3.63	30.23	38	**125**
	1350	Each added inch thick, add		89	.090		2.88	1.57	.61	5.06	6.40	
	1500	14″ diameter core		13	.615		21	10.75	4.18	35.93	45	
	1550	Each added inch thick, add		74	.108		3.56	1.89	.74	6.19	7.80	
	1700	18″ diameter core		6.30	1.270		27	22	8.65	57.65	76	
	1750	Each added inch thick, add		36	.222		4.77	3.89	1.51	10.17	13.25	
	1760	For horizontal holes, add to above								30%	30%	
	1770	Prestressed hollow core plank, 6″ thick										
	1780	1″ diameter core	B-89A	65	.246	Ea.	1.38	4.94	.75	7.07	10.40	
	1790	Each added inch thick, add		375	.043		.22	.86	.13	1.21	1.78	
	1800	3″ diameter core		55	.291		3.13	5.85	.89	9.87	13.90	
	1810	Each added inch thick, add		225	.071		.53	1.43	.22	2.18	3.14	
	1820	4″ diameter core		54	.296		4.17	5.95	.90	11.02	15.25	
	1830	Each added inch thick, add		210	.076		.70	1.53	.23	2.46	3.51	
	1840	6″ diameter core		46	.348		5.05	7	1.06	13.11	18.05	
	1850	Each added inch thick, add		175	.091		.84	1.84	.28	2.96	4.21	
	1860	8″ diameter core		31	.516		6.90	10.35	1.57	18.82	26	
	1870	Each added inch thick, add		135	.119		1.14	2.38	.36	3.88	5.50	
	1880	10″ diameter core	A-1	28	.286		9.30	5	1.94	16.24	21	
	1890	Each added inch thick, add		115	.070		1.14	1.22	.47	2.83	3.76	
	1900	12″ diameter core		25	.320		11.30	5.60	2.18	19.08	24	
	1910	Each added inch thick, add		95	.084		1.88	1.47	.57	3.92	5.10	
	1950	Minimum charge for above, 3″ diameter core	B-89A	6.35	2.520	Total		51	7.70	58.70	91	
	2000	4″ diameter core		6.15	2.600			52	7.95	59.95	94	
	2050	6″ diameter core		5.50	2.910			58	8.85	66.85	105	
	2100	8″ diameter core		5.10	3.140			63	9.55	72.55	115	
	2150	10″ diameter core		4.50	3.560			71	10.85	81.85	130	
	2200	12″ diameter core		3.70	4.320			87	13.20	100.20	155	
	2250	14″ diameter core		3.20	5			100	15.25	115.25	180	
	2300	18″ diameter core		3	5.330			105	16.25	121.25	190	

020 550 | Site Demolition

		CREW	DAILY OUTPUT	MAN-HOURS	UNIT	BARE COSTS MAT.	LABOR	EQUIP.	TOTAL	TOTAL INCL O&P	
554	0010	**SITE DEMOLITION** No hauling, abandon catch basin or manhole	B-6	7	3.430	Ea.		64	27	91	135	**554**
	0020	Remove existing catch basin or manhole		4	6			115	48	163	235	
	0030	Catch basin or manhole frames and covers stored		13	1.850			35	14.65	49.65	72	
	0040	Remove and reset		7	3.430			64	27	91	135	
	0100	Delineators, remove only	B-80	175	.183			3.50	2.32	5.82	8.15	
	0110	Remove and reset	"	100	.320			6.10	4.05	10.15	14.30	
	0400	Minimum labor/equipment charge	B-6	4	6	Job		115	48	163	235	
	0600	Fencing, barbed wire, 3 strand	2 Clab	430	.037	L.F.		.65		.65	1.06	
	0650	5 strand		280	.057			1		1	1.63	
	0700	Chain link, remove only, 8′ to 10′ high		310	.052			.90		.90	1.47	
	0800	Guide rail, remove only		85	.188			3.29		3.29	5.40	
	0850	Remove and reset		35	.457			8		8	13.05	
	0860	Guide posts, remove only	B-55	90	.267	Ea.		4.72	6.05	10.77	14.30	
	0870	Remove and reset	"	50	.480	"		8.50	10.90	19.40	26	
	0890	Minimum labor/equipment charge	2 Clab	4	4	Job		70		70	115	
	0900	Hydrants, fire, remove only	2 Plum	4.70	3.400	Ea.		87		87	135	
	0950	Remove and reset		1.40	11.430	"		290		290	450	
	0990	Minimum labor/equipment charge		2	8	Job		205		205	315	
	1000	Masonry walls, block or tile, solid, remove	B-5	1,800	.036	C.F.		.70	.52	1.22	1.70	
	1100	Cavity		2,200	.029			.58	.43	1.01	1.39	
	1200	Brick, solid		900	.071			1.41	1.04	2.45	3.41	
	1300	With block		1,130	.057			1.12	.83	1.95	2.71	
	1400	Stone, with mortar		900	.071			1.41	1.04	2.45	3.41	
	1500	Dry set		1,500	.043			.85	.62	1.47	2.04	
	1600	Median barrier, precast concrete, remove and store	B-3	430	.112	L.F.		2.12	3.58	5.70	7.35	
	1610	Remove and reset	"	390	.123	"		2.33	3.94	6.27	8.10	

020 550 | Site Demolition

			CREW	DAILY OUTPUT	MAN-HOURS	UNIT	BARE COSTS				TOTAL INCL O&P	
							MAT.	LABOR	EQUIP.	TOTAL		
554	1650	Minimum labor/equipment charge	A-1	4	2	Job		35	13.60	48.60	72	554
	1710	Pavement removal, bituminous, 3″ thick	B-38	690	.058	S.Y.		1.14	1.64	2.78	3.63	
	1750	4″ to 6″ thick		420	.095			1.87	2.69	4.56	5.95	
	1800	Bituminous driveways		680	.059			1.16	1.66	2.82	3.69	
	1900	Concrete to 6″ thick, mesh reinforced		255	.157			3.09	4.43	7.52	9.85	
	2000	Rod reinforced		200	.200			3.94	5.65	9.59	12.55	
	2100	Concrete 7″ to 24″ thick, plain		13.10	3.050	C.Y.		60	86	146	190	
	2200	Reinforced		9.50	4.210	″		83	120	203	265	
	2250	Minimum labor/equipment charge	↓	6	6.670	Job		130	190	320	420	
	2300	With hand held air equipment, bituminous	B-39	1,900	.025	S.F.		.47	.07	.54	.83	
	2320	Concrete to 6″ thick, no reinforcing		1,200	.040			.74	.11	.85	1.32	
	2340	Mesh reinforced		830	.058			1.07	.15	1.22	1.90	
	2360	Rod reinforced	↓	765	.063	↓		1.16	.17	1.33	2.06	
	2390	Minimum labor/equipment charge	B-38	6	6.670	Job		130	190	320	420	
	2400	Curbs, concrete, plain	B-6	325	.074	L.F.		1.39	.59	1.98	2.88	
	2500	Reinforced		220	.109			2.05	.87	2.92	4.25	
	2600	Granite curbs		355	.068			1.27	.54	1.81	2.64	
	2700	Bituminous curbs		830	.029	↓		.54	.23	.77	1.13	
	2790	Minimum labor/equipment charge		6	4	Job		75	32	107	155	
	2900	Pipe removal, concrete, no excavation, 12″ diameter		175	.137	L.F.		2.58	1.09	3.67	5.35	
	2930	15″ diameter		150	.160			3.01	1.27	4.28	6.25	
	2960	24″ diameter		120	.200			3.76	1.59	5.35	7.80	
	3000	36″ diameter		90	.267			5	2.12	7.12	10.40	
	3200	Steel, welded connections, 4″ diameter		160	.150			2.82	1.19	4.01	5.85	
	3300	10″ diameter		80	.300	↓		5.65	2.38	8.03	11.70	
	3390	Minimum labor/equipment charge	↓	3	8	Job		150	63	213	310	
	3500	Railroad track removal, ties and track	B-14	110	.436	L.F.		8.05	1.73	9.78	15	
	3600	Ballast		500	.096	C.Y.		1.77	.38	2.15	3.30	
	3700	Remove and re-install ties & track using new bolts & spikes		50	.960	L.F.		17.75	3.81	21.56	33	
	3800	Turnouts using new bolts and spikes		1	48	Ea.		885	190	1,075	1,650	
	3890	Minimum labor/equipment charge	↓	5	9.600	Job		175	38	213	330	
	4000	Sidewalk removal, bituminous, 2-½″ thick	B-6	325	.074	S.Y.		1.39	.59	1.98	2.88	
	4050	Brick, set in mortar		185	.130			2.44	1.03	3.47	5.05	
	4100	Concrete, plain		160	.150			2.82	1.19	4.01	5.85	
	4200	Mesh reinforced	↓	150	.160	↓		3.01	1.27	4.28	6.25	
	4290	Minimum labor/equipment charge	B-39	12	4	Job		74	10.60	84.60	130	

020 600 | Building Demolition

			CREW	DAILY OUTPUT	MAN-HOURS	UNIT	MAT.	LABOR	EQUIP.	TOTAL	TOTAL INCL O&P	
604	0010	**BUILDING DEMOLITION** Large urban projects, incl. disposal, steel	B-8	21,500	.003	C.F.		.06	.09	.15	.20	604
	0050	Concrete		15,300	.004			.08	.13	.21	.28	
	0080	Masonry		20,100	.003			.06	.10	.16	.21	
	0100	Mixture of types, average	↓	20,100	.003			.06	.10	.16	.21	
	0500	Small bldgs, or single bldgs, no salvage included, steel	B-3	14,800	.003			.06	.10	.16	.21	
	0600	Concrete		11,300	.004			.08	.14	.22	.28	
	0650	Masonry		14,800	.003			.06	.10	.16	.21	
	0700	Wood	↓	14,800	.003	↓		.06	.10	.16	.21	
	1000	Single family, one story house, wood, minimum				Ea.					1,950	
	1020	Maximum									3,350	
	1200	Two family, two story house, wood, minimum									2,525	
	1220	Maximum									5,025	
	1300	Three family, three story house, wood, minimum									3,250	
	1320	Maximum				↓					6,450	
	1400	Gutting building, see division 020-716										
608	0010	**DISPOSAL ONLY** Urban buildings with salvage value allowed										608
	0020	Including loading and 5 mile haul to dump										
	0200	Steel frame	B-3	430	.112	C.Y.		2.12	3.58	5.70	7.35	
	0300	Concrete frame	″	365	.132	″		2.49	4.21	6.70	8.65	

For expanded coverage of these items see *Means Site Work Cost Data 1991*

020 600 | Building Demolition

				CREW	DAILY OUTPUT	MAN-HOURS	UNIT	BARE COSTS				TOTAL INCL O&P	
								MAT.	LABOR	EQUIP.	TOTAL		
608	0400	Masonry construction		B-3	445	.108	C.Y.		2.05	3.46	5.51	7.10	608
	0500	Wood frame		"	247	.194	"		3.69	6.25	9.94	12.80	
612	0010	**DUMP CHARGES** Typical urban city, fees only											612
	0100	Building construction materials					C.Y.					25	
	0200	Demolition lumber, trees, brush										32	
	0300	Rubbish only					↓					24	
	0500	Reclamation station, usual charge					Ton					55	
620	0010	**RUBBISH HANDLING** The following are to be added to the											620
	0020	selective demolition prices											
	0400	Chute, circular, prefabricated steel, 18″ diameter		B-1	40	.600	L.F.	9.65	10.90		20.55	28	
	0440	30″ diameter		"	30	.800	"	18.70	14.55		33.25	44	
	0600	Dumpster, (debris box container), 5 C.Y., rent per week					Ea.					170	
	0700	10 C.Y. capacity										210	
	0800	30 C.Y. capacity										290	
	0840	40 C.Y. capacity					↓					345	
	1000	Dust partition, 6 mil polyethylene, 4′ x 8′ panels, 1″ x 3″ frame		2 Carp	2,000	.008	S.F.	.17	.18		.35	.47	
	1080	2″ x 4″ frame		"	2,000	.008	"	.29	.18		.47	.61	
	2000	Load, haul to chute & dumping into chute, 50′ haul		2 Clab	24	.667	C.Y.		11.65		11.65	19.05	
	2040	100′ haul			16.50	.970			16.95		16.95	28	
	2080	Over 100′ haul, add per 100 L.F.			35.50	.451			7.90		7.90	12.90	
	2120	In elevators, per 10 floors, add		↓	140	.114			2		2	3.27	
	3000	Loading & trucking, including 2 mile haul, chute loaded		B-16	32	1			18.25	11.40	29.65	42	
	3040	Hand loaded, 50′ haul		2 Clab	21.50	.744			13		13	21	
	3080	Machine loaded		B-6	80	.300			5.65	2.38	8.03	11.70	
	3120	Wheeled 50′ and ramp dump loaded		2 Clab	24	.667			11.65		11.65	19.05	
	5000	Haul, per mile, up to 8 C.Y. truck		B-34B	1,165	.007			.13	.31	.44	.55	
	5100	Over 8 C.Y. truck		"	1,550	.005	↓		.09	.24	.33	.41	

020 700 | Selective Demolition

				CREW	DAILY OUTPUT	MAN-HOURS	UNIT	MAT.	LABOR	EQUIP.	TOTAL	TOTAL INCL O&P	
702	0010	**CEILING DEMOLITION**											702
	0200	Drywall, on wood frame		2 Clab	800	.020	S.F.		.35		.35	.57	
	0220	On metal frame			760	.021			.37		.37	.60	
	0240	On suspension system, including system			720	.022			.39		.39	.63	
	1000	Plaster, lime and horse hair, on wood lath, incl. lath			700	.023			.40		.40	.65	
	1020	On metal lath			570	.028			.49		.49	.80	
	1100	Gypsum, on gypsum lath			720	.022			.39		.39	.63	
	1120	On metal lath			500	.032			.56		.56	.91	
	1500	Tile, wood fiber, 12″ x 12″, glued			900	.018			.31		.31	.51	
	1540	Stapled			1,500	.011			.19		.19	.30	
	1580	On suspension system, incl. system			760	.021			.37		.37	.60	
	2000	Wood, tongue and groove, 1″ x 4″			1,000	.016			.28		.28	.46	
	2040	1″ x 8″			1,100	.015			.25		.25	.42	
	2400	Plywood or wood fiberboard, 4′ x 8′ sheets		↓	1,200	.013	↓		.23		.23	.38	
	9000	Minimum labor/equipment charge		1 Clab	2	4	Job		70		70	115	
704	0010	**CUTOUT DEMOLITION** Conc., elev. slab, light reinf., under 6 C.F.		B-9	65	.615	C.F.		11	1.96	12.96	20	704
	0050	Light reinforcing, over 6 C.F.			75	.533	"		9.55	1.70	11.25	17.45	
	0200	Slab on grade to 6″ thick, not reinforced, under 8 S.F.			85	.471	S.F.		8.40	1.50	9.90	15.40	
	0250	Not reinforced, over 8 S.F.			175	.229	"		4.09	.73	4.82	7.50	
	0600	Walls, not reinforced, under 6 C.F.			60	.667	C.F.		11.95	2.12	14.07	22	
	0650	Not reinforced, over 6 C.F.			65	.615			11	1.96	12.96	20	
	1000	Concrete, elevated slab, bar reinforced, under 6 C.F.			45	.889			15.90	2.83	18.73	29	
	1050	Bar reinforced, over 6 C.F.			50	.800	↓		14.30	2.54	16.84	26	
	1200	Slab on grade to 6″ thick, bar reinforced, under 8 S.F.			75	.533	S.F.		9.55	1.70	11.25	17.45	
	1250	Bar reinforced, over 8 S.F.			105	.381	"		6.80	1.21	8.01	12.50	
	1400	Walls, bar reinforced, under 6 C.F.			50	.800	C.F.		14.30	2.54	16.84	26	
	1450	Bar reinforced, over 6 C.F.		↓	55	.727	"		13	2.31	15.31	24	

For expanded coverage of these items see *Means Site Work Cost Data 1991*

020 700 | Selective Demolition

		CREW	DAILY OUTPUT	MAN-HOURS	UNIT	MAT.	LABOR	EQUIP.	TOTAL	TOTAL INCL O&P
704										**704**
2000	Brick, to 4 S.F. opening, not including toothing									
2040	4" thick	B-9	30	1.330	Ea.		24	4.24	28.24	44
2060	8" thick		18	2.220			40	7.05	47.05	73
2080	12" thick		10	4			72	12.70	84.70	130
2400	Concrete block, to 4 S.F. opening, 2" thick		35	1.140			20	3.63	23.63	37
2420	4" thick		30	1.330			24	4.24	28.24	44
2440	8" thick		27	1.480			27	4.71	31.71	49
2460	12" thick		24	1.670			30	5.30	35.30	55
2600	Gypsum block, to 4 S.F. opening, 2" thick		80	.500			8.95	1.59	10.54	16.35
2620	4" thick		70	.571			10.25	1.82	12.07	18.70
2640	8" thick		55	.727			13	2.31	15.31	24
2800	Terra cotta, to 4 S.F. opening, 4" thick		70	.571			10.25	1.82	12.07	18.70
2840	8" thick		65	.615			11	1.96	12.96	20
2880	12" thick		50	.800			14.30	2.54	16.84	26
4000	For toothing masonry, see Division 045-290									
4010										
6000	Walls, interior, not including re-framing,									
6010	openings to 5 S.F.									
6100	Drywall to ⅝" thick	A-1	24	.333	Ea.		5.85	2.27	8.12	12.05
6200	Paneling to ¾" thick		20	.400			7	2.72	9.72	14.45
6300	Plaster, on gypsum lath		20	.400			7	2.72	9.72	14.45
6340	On wire lath		14	.571			10	3.89	13.89	21
7000	Wood frame, not including re-framing, openings to 5 S.F.									
7200	Floors, sheathing and flooring to 2" thick	A-1	5	1.600	Ea.		28	10.90	38.90	58
7310	Roofs, sheathing to 1" thick, not including roofing		6	1.330			23	9.05	32.05	48
7410	Walls, sheathing to 1" thick, not including siding		7	1.140			20	7.75	27.75	41
8500	Minimum labor/equipment charge		4	2	Job		35	13.60	48.60	72
706	**DOOR DEMOLITION**									**706**
0010										
0020										
0200	Doors, exterior, 1-¾" thick, single, 3' x 7' high	1 Clab	16	.500	Ea.		8.75		8.75	14.30
0220	Double, 6' x 7' high		12	.667			11.65		11.65	19.05
0500	Interior, 1-⅜" thick, single, 3' x 7' high		20	.400			7		7	11.45
0520	Double, 6' x 7' high		16	.500			8.75		8.75	14.30
0700	Bi-folding, 3' x 6'-8" high		20	.400			7		7	11.45
0720	6' x 6'-8" high		18	.444			7.80		7.80	12.70
0900	Bi-passing, 3' x 6'-8" high		16	.500			8.75		8.75	14.30
0940	6' x 6'-8" high		14	.571			10		10	16.35
1500	Remove and reset, minimum	1 Carp	8	1			22		22	36
1520	Maximum	"	6	1.330			29		29	48
2000	Frames, including trim, metal	A-1	8	1			17.50	6.80	24.30	36
2200	Wood		14	.571			10	3.89	13.89	21
2201	Alternate pricing method		200	.040	L.F.		.70	.27	.97	1.44
2950	Minimum labor/equipment charge	1 Clab	4	2	Job		35		35	57
3000	Special doors, counter doors	F-2	6	2.670	Ea.		59	2.67	61.67	99
3100	Double acting		10	1.600			35	1.60	36.60	59
3200	Floor door (trap type)		8	2			44	2	46	74
3300	Glass, sliding, including frames		12	1.330			29	1.33	30.33	49
3400	Overhead, commercial, 12' x 12' high		4	4			88	4	92	150
3440	20' x 16' high		3	5.330			115	5.35	120.35	200
3500	Residential, 9' x 7' high		8	2			44	2	46	74
3540	16' x 7' high		7	2.290			50	2.29	52.29	85
3600	Remove and reset, minimum		4	4			88	4	92	150
3620	Maximum		2.50	6.400			140	6.40	146.40	235
3700	Roll-up grille		5	3.200			70	3.20	73.20	120
3800	Revolving door		2	8			175	8	183	295
3900	Swing door		3	5.330			115	5.35	120.35	200
9000	Minimum labor/equipment charge	1 Carp	4	2	Job		44		44	72

For expanded coverage of these items see *Means Site Work Cost Data 1991*

020 700	Selective Demolition	CREW	DAILY OUTPUT	MAN-HOURS	UNIT	BARE COSTS				TOTAL INCL O&P
						MAT.	LABOR	EQUIP.	TOTAL	
708 0010	**ELECTRICAL DEMOLITION**									708
0020	Conduit to 15' high, including fittings & hangers									
0100	Rigid galvanized steel, ½" to 1" diameter	1 Elec	242	.033	L.F.		.83		.83	1.27
0120	1-¼" to 2"		200	.040			1.01		1.01	1.54
0140	2" to 4"		151	.053			1.33		1.33	2.03
0200	Electric metallic tubing (EMT) ½" to 1"		394	.020			.51		.51	.78
0220	1-¼" to 1-½"		326	.025			.62		.62	.94
0240	2" to 3"		236	.034			.85		.85	1.30
0400	Wiremold raceway, including fittings & hangers									
0420	No. 3000	1 Elec	250	.032	L.F.		.80		.80	1.23
0440	No. 4000	"	217	.037	"		.93		.93	1.42
0500	Channels, steel, including fittings & hangers									
0520	¾" x 1-½"	1 Elec	308	.026	L.F.		.65		.65	1
0540	1-½" x 1-½"	"	269	.030	"		.75		.75	1.14
0600	Copper bus duct, indoor, 3 ph, incl. removal of									
0610	hangers & supports									
0620	225 amp	1 Elec	67	.119	L.F.		3		3	4.59
0640	400 amp		53	.151			3.80		3.80	5.80
0660	600 amp		43	.186			4.68		4.68	7.15
0720	3000 amp		10	.800			20		20	31
0800	Plug-in switches, 600V 3 ph, incl. disconnecting									
0820	wire, pipe terminations, 30 amp	1 Elec	15.50	.516	Ea.		13		13	19.80
0840	60 amp		13.90	.576			14.45		14.45	22
0850	100 amp		10.40	.769			19.35		19.35	30
0860	200 amp		6.20	1.290			32		32	50
0940	1200 amp		1	8			200		200	305
0960	1600 amp		.85	9.410			235		235	360
1010	Safety switches, 250 or 600V, incl. disconnection									
1050	of wire & pipe terminations									
1100	30 amp	1 Elec	12.30	.650	Ea.		16.35		16.35	25
1120	60 amp		8.80	.909			23		23	35
1140	100 amp		7.30	1.100			28		28	42
1160	200 amp		5	1.600			40		40	61
1210	Panel boards, incl. removal of all breakers,									
1220	pipe terminations & wire connections									
1230	3 wire, 120/240V, 100A, to 20 circuits	1 Elec	2.60	3.080	Ea.		77		77	120
1240	200 amps, to 42 circuits		1.30	6.150			155		155	235
1260	4 wire, 120/208V, 125A, to 20 circuits		2.40	3.330			84		84	130
1270	200 amps, to 42 circuits		1.20	6.670			170		170	255
1300	Transformer, dry type, 1 ph, incl. removal of									
1320	supports, wire & pipe terminations									
1340	1 KVA	1 Elec	7.70	1.040	Ea.		26		26	40
1360	5 KVA		4.70	1.700			43		43	65
1420	75 KVA		1.25	6.400			160		160	245
1440	3 Phase to 600V, primary									
1460	3 KVA	1 Elec	3.85	2.080	Ea.		52		52	80
1480	15 KVA		2.10	3.810			96		96	145
1500	30 KVA		1.74	4.600			115		115	175
1530	112.5 KVA		1.16	6.900			175		175	265
1560	500 KVA		.58	13.790			345		345	530
1570	750 KVA		.45	17.780			445		445	685
1600	Pull boxes & cabinets, sheet metal, incl. removal									
1620	of supports and pipe terminations									
1640	6" x 6" x 4"	1 Elec	31.10	.257	Ea.		6.45		6.45	9.90
1660	12" x 12" x 4"		23.30	.343			8.65		8.65	13.20
1720	Junction boxes, 4" sq. & oct.		80	.100			2.52		2.52	3.84
1740	Handy box		107	.075			1.88		1.88	2.87
1760	Switch box		107	.075			1.88		1.88	2.87

For expanded coverage of these items see *Means Site Work Cost Data 1991*

020 700 | Selective Demolition

			CREW	DAILY OUTPUT	MAN-HOURS	UNIT	BARE COSTS MAT.	BARE COSTS LABOR	BARE COSTS EQUIP.	BARE COSTS TOTAL	TOTAL INCL O&P	
708	1780	Receptacle & switch plates	1 Elec	257	.031	Ea.		.78		.78	1.20	708
	1800	Wire, THW-THWN-THHN, removed from										
	1810	in place conduit, to 15' high										
	1830	#14	1 Elec	65	.123	C.L.F.		3.10		3.10	4.73	
	1840	#12		55	.145			3.66		3.66	5.60	
	1850	#10		45.50	.176			4.42		4.42	6.75	
	1880	#4		26.50	.302			7.60		7.60	11.60	
	1890	#3		25	.320			8.05		8.05	12.30	
	1910	1/0		16.60	.482			12.10		12.10	18.50	
	1920	2/0		14.60	.548			13.80		13.80	21	
	1930	3/0		12.50	.640			16.10		16.10	25	
	1980	400 MCM		8.50	.941			24		24	36	
	1990	500 MCM		8.10	.988			25		25	38	
	2000	Interior fluorescent fixtures, incl. supports										
	2010	& whips, to 15' high										
	2100	Recessed drop-in 2' x 2', 2 lamp	2 Elec	35	.457	Ea.		11.50		11.50	17.55	
	2120	2' x 4', 2 lamp		33	.485			12.20		12.20	18.60	
	2140	2' x 4', 4 lamp		30	.533			13.40		13.40	20	
	2160	4' x 4', 4 lamp		20	.800			20		20	31	
	2180	Surface mount, acrylic lens & hinged frame										
	2200	1' x 4', 2 lamp	2 Elec	44	.364	Ea.		9.15		9.15	13.95	
	2220	2' x 2', 2 lamp		44	.364			9.15		9.15	13.95	
	2260	2' x 4', 4 lamp		33	.485			12.20		12.20	18.60	
	2280	4' x 4', 4 lamp		23	.696			17.50		17.50	27	
	2300	Strip fixtures, surface mount										
	2320	4' long, 1 lamp	2 Elec	53	.302	Ea.		7.60		7.60	11.60	
	2340	4' long, 2 lamp		50	.320			8.05		8.05	12.30	
	2360	8' long, 1 lamp		42	.381			9.60		9.60	14.65	
	2380	8' long, 2 lamp		40	.400			10.05		10.05	15.35	
	2400	Pendant mount, industrial, incl. removal										
	2410	of chain or rod hangers, to 15' high										
	2420	4' long, 2 lamp	2 Elec	35	.457	Ea.		11.50		11.50	17.55	
	2440	8' long, 2 lamp	"	27	.593	"		14.90		14.90	23	
	2460	Interior incandescent, surface, ceiling										
	2470	or wall mount, to 12' high										
	2480	Metal cylinder type, 75 Watt	2 Elec	62	.258	Ea.		6.50		6.50	9.90	
	2500	150 Watt	"	62	.258	"		6.50		6.50	9.90	
	2520	Metal halide, high bay										
	2540	400 Watt	2 Elec	15	1.070	Ea.		27		27	41	
	2560	1000 Watt		12	1.330			34		34	51	
	2580	150 Watt, low bay		20	.800			20		20	31	
	2600	Exterior fixtures, incandescent, wall mount										
	2620	100 Watt	2 Elec	50	.320	Ea.		8.05		8.05	12.30	
	2640	Quartz, 500 Watt		33	.485			12.20		12.20	18.60	
	2660	1500 Watt		27	.593			14.90		14.90	23	
	2680	Wall pack, mercury vapor										
	2700	175 Watt	2 Elec	25	.640	Ea.		16.10		16.10	25	
	2720	250 Watt	"	25	.640	"		16.10		16.10	25	
	9000	Minimum labor/equipment charge	1 Elec	4	2	Job		50		50	77	
712	0010	**FLOORING DEMOLITION**										712
	0200	Brick with mortar	A-1	300	.027	S.F.		.47	.18	.65	.96	
	0400	Carpet, bonded, including surface scraping	2 Clab	2,000	.008			.14		.14	.23	
	0480	Tackless	"	9,000	.002			.03		.03	.05	
	0600	Composition, acrylic or epoxy	A-1	200	.040			.70	.27	.97	1.44	
	0800	Resilient, sheet goods (linoleum)	2 Clab	1,400	.011			.20		.20	.33	
	0820	For gym floors		900	.018			.31		.31	.51	
	0900	Tile, 12" x 12"		1,000	.016			.28		.28	.46	

For expanded coverage of these items see *Means Site Work Cost Data 1991*

			CREW	DAILY OUTPUT	MAN-HOURS	UNIT	MAT.	LABOR	EQUIP.	TOTAL	TOTAL INCL O&P	
	020 700	**Selective Demolition**						BARE COSTS				
712	2000	Tile, ceramic, thin set	A-1	400	.020	S.F.		.35	.14	.49	.72	712
	2020	Mud set		350	.023			.40	.16	.56	.82	
	2200	Marble, slate, thin set		400	.020			.35	.14	.49	.72	
	2220	Mud set		350	.023			.40	.16	.56	.82	
	2600	Terrazzo, thin set		250	.032			.56	.22	.78	1.15	
	2620	Mud set		225	.036			.62	.24	.86	1.28	
	2640	Cast in place		175	.046			.80	.31	1.11	1.65	
	3000	Wood, block, on end		400	.020			.35	.14	.49	.72	
	3200	Parquet		450	.018			.31	.12	.43	.64	
	3400	Strip flooring, interior, 2-1/4" x 25/32" thick		325	.025			.43	.17	.60	.89	
	3500	Exterior, porch flooring, 1" x 4"		220	.036			.64	.25	.89	1.31	
	3800	Subfloor, tongue and groove, 1" x 6"		325	.025			.43	.17	.60	.89	
	3820	1" x 8"		430	.019			.33	.13	.46	.67	
	3840	1" x 10"		520	.015			.27	.10	.37	.56	
	4000	Plywood, nailed		600	.013			.23	.09	.32	.48	
	4100	Glued and nailed		400	.020			.35	.14	.49	.72	
	9000	Minimum labor/equipment charge		4	2	Job		35	13.60	48.60	72	
714	0010	**FRAMING DEMOLITION**										714
	1020	Concrete, average reinforcing, beams, 8" x 10"	B-9	120	.333	L.F.		5.95	1.06	7.01	10.90	
	1040	10" x 12"		110	.364			6.50	1.16	7.66	11.90	
	1060	12" x 14"		90	.444			7.95	1.41	9.36	14.55	
	1200	Columns, 8" x 8"		120	.333			5.95	1.06	7.01	10.90	
	1240	10" x 10"		120	.333			5.95	1.06	7.01	10.90	
	1280	12" x 12"		110	.364			6.50	1.16	7.66	11.90	
	1320	14" x 14"		100	.400			7.15	1.27	8.42	13.10	
	1400	Girders, 14" x 16"		55	.727			13	2.31	15.31	24	
	1440	16" x 18"		40	1			17.90	3.18	21.08	33	
	1600	Slabs, elevated, 6" thick		600	.067	S.F.		1.19	.21	1.40	2.18	
	1640	8" thick		450	.089			1.59	.28	1.87	2.91	
	1680	10" thick		360	.111			1.99	.35	2.34	3.64	
	1900	Add for heavy reinforcement								20%	20%	
	1910	Minimum labor/equipment charge	A-1	1	8	Job		140	54	194	290	
	2000	Steel framing, beams, 4" x 6"	B-13	500	.112	L.F.		2.11	.97	3.08	4.48	
	2020	4" x 8"		400	.140			2.64	1.22	3.86	5.60	
	2080	8" x 12"		250	.224			4.22	1.95	6.17	8.95	
	2200	Columns, 6" x 6"		400	.140			2.64	1.22	3.86	5.60	
	2240	8" x 8"		350	.160			3.02	1.39	4.41	6.40	
	2280	10" x 10"		320	.175			3.30	1.52	4.82	7	
	2400	Girders, 10" x 12"		225	.249			4.69	2.16	6.85	9.95	
	2440	10" x 14"		200	.280			5.30	2.43	7.73	11.20	
	2480	10" x 16"		165	.339			6.40	2.95	9.35	13.55	
	2520	10" x 24"		125	.448			8.45	3.89	12.34	17.90	
	2950	Minimum labor/equipment charge	A-1	1	8	Job		140	54	194	290	
	3000	Wood framing, beams, 6" x 8"	B-2	275	.145	L.F.		2.60		2.60	4.25	
	3040	6" x 10"		220	.182			3.25		3.25	5.30	
	3080	6" x 12"		185	.216			3.87		3.87	6.30	
	3120	8" x 12"		140	.286			5.10		5.10	8.35	
	3160	10" x 12"		110	.364			6.50		6.50	10.65	
	3400	Fascia boards, 1" x 6"	1 Clab	500	.016			.28		.28	.46	
	3440	1" x 8"		450	.018			.31		.31	.51	
	3480	1" x 10"		400	.020			.35		.35	.57	
	3520	For trim boards, see division 020-720										
	3800	Headers over openings, 2 @ 2" x 6"	1 Clab	110	.073	L.F.		1.27		1.27	2.08	
	3840	2 @ 2" x 8"		100	.080			1.40		1.40	2.29	
	3880	2 @ 2" x 10"		90	.089			1.56		1.56	2.54	
	4200	Joists, 2" x 4"	2 Clab	1,000	.016			.28		.28	.46	
	4230	Joists, 2" x 6"	"	970	.016			.29		.29	.47	

For expanded coverage of these items see *Means Site Work Cost Data 1991*

020 700	Selective Demolition	CREW	DAILY OUTPUT	MAN-HOURS	UNIT	MAT.	LABOR	EQUIP.	TOTAL	TOTAL INCL O&P	
714	4240	2" x 8"	2 Clab	940	.017	L.F.		.30		.30	.49	714
	4250	2" x 10"		910	.018			.31		.31	.50	
	4280	2" x 12"		880	.018			.32		.32	.52	
	5400	Posts, 4" x 4"		800	.020			.35		.35	.57	
	5440	6" x 6"		400	.040			.70		.70	1.14	
	5480	8" x 8"		300	.053			.93		.93	1.52	
	5500	10" x 10"		240	.067			1.17		1.17	1.90	
	5800	Rafters, ordinary, 2" x 6"		850	.019			.33		.33	.54	
	5840	2" x 8"		720	.022			.39		.39	.63	
	5900	Hip & valley, 2" x 6"		500	.032			.56		.56	.91	
	5940	2" x 8"		420	.038			.67		.67	1.09	
	6200	Stairs and stringers, minimum		40	.400	Riser		7		7	11.45	
	6240	Maximum		26	.615	"		10.75		10.75	17.60	
	6600	Studs, 2" x 4"		2,000	.008	L.F.		.14		.14	.23	
	6640	2" x 6"		1,600	.010	"		.18		.18	.29	
	9000	Minimum labor/equipment charge	1 Clab	4	2	Job		35		35	57	
	9500	See Div. 020-620 for rubbish handling										
716	0010	**GUTTING** Building interior, including disposal										716
	0500	Residential building										
	0560	Minimum	B-16	400	.080	SF Flr.		1.46	.91	2.37	3.38	
	0580	Maximum		360	.089			1.62	1.01	2.63	3.75	
	1000	Commercial building, minimum		350	.091			1.67	1.04	2.71	3.86	
	1020	Maximum		250	.128			2.33	1.46	3.79	5.40	
	3000	Minimum labor/equipment charge		4	8	Job		145	91	236	340	
717	0010	**HAZARDOUS WASTE CLEANUP/PICKUP/DISPOSAL**										717
	0100	For contractor equipment, i.e. dozer,										
	0110	front end loader, dump truck, etc., see div. 016-408										
	1000	Solid pickup										
	1100	55 gal. drums				Ea.					150	
	1120	Bulk material, minimum				Ton					130	
	1130	Maximum				"					300	
	1200	Transportation to disposal site										
	1220	Truckload = 80 drums or 25 C.Y. or 18 tons										
	1260	Minimum				Mile					3	
	1270	Maximum				"					4	
	3000	Liquid pickup, vacuum truck, stainless steel tank										
	3100	Minimum charge, 4 hours										
	3110	1 compartment, 2200 gallon				Hr.					88	
	3120	2 compartment, 5000 gallon				"					99	
	3400	Transportation in 6900 gallon bulk truck				Mile					4	
	3410	In teflon lined truck				"					4.62	
	5000	Heavy sludge or dry vacuumable material				Hr.					99	
	6000	Dumpsite disposal charge, minimum				Ton					115	
	6020	Maximum				"					250	
718	0010	**HVAC DEMOLITION**										718
	0100	Air conditioner, split unit, 3 ton	Q-5	2	8	Ea.		185		185	285	
	0150	Package unit, 3 ton	Q-6	3	8			190		190	295	
	0300	Boiler, electric	Q-19	2	12			285		285	435	
	0340	Gas or oil, steel, under 150 MBH	Q-6	3	8			190		190	295	
	0380	Over 150 MBH	"	2	12			285		285	440	
	1000	Ductwork, 4" high, 8" wide	1 Clab	200	.040	L.F.		.70		.70	1.14	
	1020	10" wide		190	.042			.74		.74	1.20	
	1040	14" wide		180	.044			.78		.78	1.27	
	1100	6" high, 8" wide		165	.048			.85		.85	1.39	
	1120	12" wide		150	.053			.93		.93	1.52	
	1140	18" wide		135	.059			1.04		1.04	1.69	

For expanded coverage of these items see *Means Site Work Cost Data 1991*

020 700 | Selective Demolition

			CREW	DAILY OUTPUT	MAN-HOURS	UNIT	MAT.	LABOR	EQUIP.	TOTAL	TOTAL INCL O&P	
718	1200	10″ high, 12″ wide	1 Clab	125	.064	L.F.		1.12		1.12	1.83	718
	1220	18″ wide		115	.070			1.22		1.22	1.99	
	1240	24″ wide		110	.073			1.27		1.27	2.08	
	1300	12″ high, 18″ wide		85	.094			1.65		1.65	2.69	
	1320	24″ wide		75	.107			1.87		1.87	3.05	
	1340	48″ wide		71	.113			1.97		1.97	3.22	
	1400	18″ high, 24″ wide		67	.119			2.09		2.09	3.41	
	1420	36″ wide		63	.127			2.22		2.22	3.63	
	1440	48″ wide		59	.136			2.37		2.37	3.87	
	1500	30″ high, 36″ wide		56	.143			2.50		2.50	4.08	
	1520	48″ wide		53	.151			2.64		2.64	4.31	
	1540	72″ wide		50	.160			2.80		2.80	4.57	
	1550	Duct heater, electric strip	1 Elec	8	1	Ea.		25		25	38	
	1850	Minimum labor/equipment charge	1 Clab	3	2.670	Job		47		47	76	
	2200	Furnace, electric	Q-20	2	10	Ea.		230		230	360	
	2300	Gas or oil, under 120 MBH	Q-9	4	4			90		90	140	
	2340	Over 120 MBH	"	3	5.330			120		120	190	
	2800	Heat pump, package unit, 3 ton	Q-5	2.40	6.670			155		155	235	
	2840	Split unit, 3 ton		2	8			185		185	285	
	3000	Mechanical equipment, light items		.90	17.780	Ton		410		410	630	
	3600	Heavy items		1.10	14.550	"		335		335	515	
	9000	Minimum labor/equipment charge	Q-6	3	8	Job		190		190	295	
720	0010	**MILLWORK AND TRIM DEMOLITION**										720
	1000	Cabinets, wood, base cabinets	2 Clab	80	.200	L.F.		3.50		3.50	5.70	
	1020	Wall cabinets	"	80	.200	"		3.50		3.50	5.70	
	1060	Remove and reset, base cabinets	2 Carp	18	.889	Ea.		19.55		19.55	32	
	1070	Wall cabinets	"	20	.800	"		17.60		17.60	29	
	1100	Steel, painted, base cabinets	2 Clab	60	.267	L.F.		4.67		4.67	7.60	
	1120	Wall cabinets		60	.267	"		4.67		4.67	7.60	
	1200	Casework, large area		320	.050	S.F.		.88		.88	1.43	
	1220	Selective		200	.080	"		1.40		1.40	2.29	
	1500	Counter top, minimum		200	.080	L.F.		1.40		1.40	2.29	
	1510	Maximum		120	.133			2.33		2.33	3.81	
	1550	Remove and reset, minimum	2 Carp	50	.320			7.05		7.05	11.50	
	1560	Maximum	"	40	.400			8.80		8.80	14.35	
	2000	Paneling, 4′ x 8′ sheets, ¼″ thick	2 Clab	2,000	.008	S.F.		.14		.14	.23	
	2100	Boards, 1″ x 4″		700	.023			.40		.40	.65	
	2120	1″ x 6″		750	.021			.37		.37	.61	
	2140	1″ x 8″		800	.020			.35		.35	.57	
	3000	Trim, baseboard, to 6″ wide		1,200	.013	L.F.		.23		.23	.38	
	3040	12″ wide		1,000	.016			.28		.28	.46	
	3080	Remove and reset, minimum	2 Carp	400	.040			.88		.88	1.44	
	3090	Maximum	"	300	.053			1.17		1.17	1.92	
	3100	Ceiling trim	2 Clab	1,000	.016			.28		.28	.46	
	3120	Chair rail		1,200	.013			.23		.23	.38	
	3140	Railings with balusters		240	.067			1.17		1.17	1.90	
	3160	Wainscoting		700	.023	S.F.		.40		.40	.65	
	9000	Minimum labor/equipment charge	1 Clab	4	2	Job		35		35	57	
724	0010	**PLUMBING DEMOLITION**										724
	1020	Fixtures, including 10′ piping										
	1100	Bath tubs, cast iron	1 Plum	4	2	Ea.		51		51	78	
	1120	Fiberglass		6	1.330			34		34	52	
	1140	Steel		5	1.600			41		41	63	
	1200	Lavatory, wall hung		10	.800			20		20	31	
	1220	Counter top		8	1			25		25	39	
	1300	Sink, steel or cast iron, single		8	1			25		25	39	
	1320	Double		7	1.140			29		29	45	
	1400	Water closet, floor mounted		8	1			25		25	39	

For expanded coverage of these items see *Means Site Work Cost Data 1991*

020 700 | Selective Demolition

		CREW	DAILY OUTPUT	MAN-HOURS	UNIT	MAT.	LABOR	EQUIP.	TOTAL	TOTAL INCL O&P	
724	1420	Wall mounted	1 Plum	7	1.140	Ea.		29		29	45	724
	1500	Urinal, floor mounted		4	2			51		51	78	
	1520	Wall mounted		7	1.140			29		29	45	
	1600	Water fountains, free standing		8	1			25		25	39	
	1620	Recessed		6	1.330			34		34	52	
	2000	Piping, metal, to 2″ diameter		200	.040	L.F.		1.02		1.02	1.57	
	2050	To 4″ diameter		150	.053			1.36		1.36	2.09	
	2100	To 8″ diameter	2 Plum	100	.160			4.07		4.07	6.30	
	2150	To 16″ diameter	"	60	.267			6.80		6.80	10.45	
	2240	Toilet partitions, see division 020-732										
	2250	Water heater, 40 gal.	1 Plum	6	1.330	Ea.		34		34	52	
	6000	Remove and reset fixtures, minimum		6	1.330			34		34	52	
	6100	Maximum		4	2			51		51	78	
	9000	Minimum labor/equipment charge		2	4	Job		100		100	155	
726	0010	**ROOFING AND SIDING DEMOLITION**										726
	1000	Deck, roof, concrete plank	B-13	1,680	.033	S.F.		.63	.29	.92	1.33	
	1100	Gypsum plank		3,900	.014			.27	.12	.39	.57	
	1150	Metal decking		3,500	.016			.30	.14	.44	.64	
	1200	Wood, boards, tongue and groove, 2″ x 6″	2 Clab	960	.017			.29		.29	.48	
	1220	2″ x 10″		1,040	.015			.27		.27	.44	
	1280	Standard planks, 1″ x 6″		1,080	.015			.26		.26	.42	
	1320	1″ x 8″		1,160	.014			.24		.24	.39	
	1340	1″ x 12″		1,200	.013			.23		.23	.38	
	2000	Gutters, aluminum or wood, edge hung	1 Clab	240	.033	L.F.		.58		.58	.95	
	2100	Built-in		100	.080	"		1.40		1.40	2.29	
	2500	Roof accessories, plumbing vent flashing		14	.571	Ea.		10		10	16.35	
	2600	Adjustable metal chimney flashing		9	.889	"		15.55		15.55	25	
	3000	Roofing, built-up, 5 ply roof, no gravel	B-2	1,600	.025	S.F.		.45		.45	.73	
	3100	Gravel removal, minimum		5,000	.008			.14		.14	.23	
	3120	Maximum		2,000	.020			.36		.36	.59	
	3400	Roof insulation board		3,900	.010			.18		.18	.30	
	4000	Shingles, asphalt strip		3,500	.011			.20		.20	.33	
	4100	Slate		2,500	.016			.29		.29	.47	
	4300	Wood		2,200	.018			.33		.33	.53	
	4500	Skylight to 10 S.F.	1 Clab	8	1	Ea.		17.50		17.50	29	
	5000	Siding, metal, horizontal		444	.018	S.F.		.32		.32	.51	
	5020	Vertical		400	.020			.35		.35	.57	
	5200	Wood, boards, vertical		400	.020			.35		.35	.57	
	5220	Clapboards, horizontal		380	.021		.	.37		.37	.60	
	5240	Shingles		350	.023			.40		.40	.65	
	5260	Textured plywood		725	.011			.19		.19	.32	
	9000	Minimum labor/equipment charge		2	4	Job		70		70	115	
728	0010	**SAW CUTTING** Asphalt over 1000 L.F., 3″ deep	B-89	775	.021	L.F.	.22	.41	.32	.95	1.24	728
	0020	Each additional inch of depth		1,250	.013		.05	.25	.20	.50	.68	
	0400	Concrete slabs, mesh reinforcing, per inch of depth		960	.017		.26	.33	.26	.85	1.10	
	0420	Rod reinforcing, per inch of depth		550	.029		.35	.57	.45	1.37	1.80	
	0800	Concrete walls, plain, per inch of depth	A-1A	100	.080		.24	1.40	.30	1.94	2.88	
	0820	Rod reinforcing, per inch of depth		60	.133		.35	2.33	.50	3.18	4.75	
	1200	Masonry walls, brick, per inch of depth		146	.055		.24	.96	.20	1.40	2.05	
	1220	Block walls, solid, per inch of depth		122	.066		.24	1.15	.24	1.63	2.40	
	5000	Wood sheathing to 1″ thick, on walls	1 Carp	200	.040			.88		.88	1.44	
	5020	On roof	"	250	.032			.70		.70	1.15	
	9000	Minimum labor/equipment charge	A-1	2	4	Job		70	27	97	145	
	9950	See also Div. 020-125 core drilling										
730	0010	**TORCH CUTTING** Steel, 1″ thick plate	A-1A	95	.084	L.F.		1.47	.31	1.78	2.75	730
	0040	1″ diameter bar	"	210	.038	Ea.		.67	.14	.81	1.25	
	1000	Oxygen lance cutting, reinforced concrete walls										
	1040	12″ to 16″ thick walls	A-1A	10	.800	L.F.		14	2.98	16.98	26	

For expanded coverage of these items see *Means Site Work Cost Data 1991*

020 700	Selective Demolition	CREW	DAILY OUTPUT	MAN-HOURS	UNIT	BARE COSTS MAT.	BARE COSTS LABOR	BARE COSTS EQUIP.	BARE COSTS TOTAL	TOTAL INCL O&P	
730 1080	24" thick walls	A-1A	6	1.330	L.F.		23	4.97	27.97	44	**730**
1090	Minimum labor/equipment charge	A-1	1	8	Job		140	54	194	290	
1100	See also division 051-240										
732 0010	**WALLS AND PARTITIONS DEMOLITION**										**732**
0100	Brick, 4" to 12" thick	B-9	220	.182	C.F.		3.25	.58	3.83	5.95	
0200	Concrete block, 4" thick		1,000	.040	S.F.		.72	.13	.85	1.31	
0280	8" thick		810	.049			.88	.16	1.04	1.62	
1000	Drywall, nailed	1 Clab	1,000	.008			.14		.14	.23	
1020	Glued and nailed		900	.009			.16		.16	.25	
1500	Fiberboard, nailed		900	.009			.16		.16	.25	
1520	Glued and nailed		800	.010			.18		.18	.29	
2000	Movable walls, metal, 5' high		300	.027			.47		.47	.76	
2020	8' high		400	.020			.35		.35	.57	
2200	Metal or wood studs, finish 2 sides, fiberboard	B-1	520	.046			.84		.84	1.37	
2250	Lath and plaster		260	.092			1.68		1.68	2.74	
2300	Plasterboard (drywall)		520	.046			.84		.84	1.37	
2350	Plywood		450	.053			.97		.97	1.58	
3000	Plaster, lime and horsehair, on wood lath	1 Clab	400	.020			.35		.35	.57	
3020	On metal lath		335	.024			.42		.42	.68	
3400	Gypsum or perlite, on gypsum lath		410	.020			.34		.34	.56	
3420	On metal lath		300	.027			.47		.47	.76	
3800	Toilet partitions, slate or marble		5	1.600	Ea.		28		28	46	
3820	Hollow metal		8	1	"		17.50		17.50	29	
9000	Minimum labor/equipment charge		4	2	Job		35		35	57	
734 0010	**WINDOW DEMOLITION**										**734**
0020											
0200	Aluminum, including trim, to 12 S.F.	A-1A	16	.500	Ea.		8.75	1.86	10.61	16.35	
0240	To 25 S.F.		11	.727			12.75	2.71	15.46	24	
0280	To 50 S.F.		5	1.600			28	5.95	33.95	52	
0320	Storm windows, to 12 S.F.		27	.296			5.20	1.10	6.30	9.70	
0360	To 25 S.F.		21	.381			6.65	1.42	8.07	12.45	
0400	To 50 S.F.		16	.500			8.75	1.86	10.61	16.35	
0600	Glass, minimum	1 Clab	200	.040	S.F.		.70		.70	1.14	
0620	Maximum	"	150	.053	"		.93		.93	1.52	
1000	Steel, including trim, to 12 S.F.	A-1A	13	.615	Ea.		10.75	2.29	13.04	20	
1020	To 25 S.F.		9	.889			15.55	3.31	18.86	29	
1040	To 50 S.F.		4	2			35	7.45	42.45	65	
2000	Wood, including trim, to 12 S.F.	1 Clab	22	.364			6.35		6.35	10.40	
2020	To 25 S.F.		18	.444			7.80		7.80	12.70	
2060	To 50 S.F.		13	.615			10.75		10.75	17.60	
5020	Remove and reset window, minimum	1 Carp	6	1.330			29		29	48	
5040	Average		4	2			44		44	72	
5080	Maximum		2	4			88		88	145	
9000	Minimum labor/equipment charge	1 Clab	4	2	Job		35		35	57	

020 750	Concrete Removal										
754 0010	**FOOTINGS AND FOUNDATIONS DEMOLITION**										**754**
0200	Floors, concrete slab on grade,										
0240	4" thick, plain concrete	B-9	500	.080	S.F.		1.43	.25	1.68	2.62	
0280	Reinforced, wire mesh		470	.085			1.52	.27	1.79	2.79	
0300	Rods		400	.100			1.79	.32	2.11	3.27	
0400	6" thick, plain concrete		375	.107			1.91	.34	2.25	3.49	
0420	Reinforced, wire mesh		340	.118			2.11	.37	2.48	3.85	
0440	Rods		300	.133			2.39	.42	2.81	4.37	
1000	Footings, concrete, 1' thick, 2' wide	B-5	300	.213	L.F.		4.23	3.12	7.35	10.20	
1080	1'-6" thick, 2' wide	"	250	.256	"		5.05	3.74	8.79	12.25	

For expanded coverage of these items see *Means Site Work Cost Data 1991*

020 750	Concrete Removal	CREW	DAILY OUTPUT	MAN-HOURS	UNIT	BARE COSTS				TOTAL INCL O&P	
						MAT.	LABOR	EQUIP.	TOTAL		
754											**754**
1120	3' wide	B-5	200	.320	L.F.		6.35	4.68	11.03	15.35	
1140	2' thick, 3' wide	"	175	.366			7.25	5.35	12.60	17.50	
1200	Average reinforcing, add								10%	10%	
1220	Heavy reinforcing, add								20%	20%	
2000	Walls, block, 4" thick	A-1	200	.040	S.F.		.70	.27	.97	1.44	
2040	6" thick		190	.042			.74	.29	1.03	1.52	
2080	8" thick		180	.044			.78	.30	1.08	1.60	
2100	12" thick		175	.046			.80	.31	1.11	1.65	
2200	For horizontal reinforcing, add								10%	10%	
2220	For vertical reinforcing, add								20%	20%	
2400	Concrete, plain concrete, 6" thick	B-9	160	.250			4.48	.80	5.28	8.20	
2420	8" thick		140	.286			5.10	.91	6.01	9.35	
2440	10" thick		120	.333			5.95	1.06	7.01	10.90	
2500	12" thick		100	.400			7.15	1.27	8.42	13.10	
2600	For average reinforcing, add								10%	10%	
2620	For heavy reinforcing, add								20%	20%	
9000	Minimum labor/equipment charge	A-1	2	4	Job		70	27	97	145	
758	**0010** **MASONRY DEMOLITION**										**758**
1000	Chimney, 16" x 16", soft old mortar	A-1	24	.333	V.L.F.		5.85	2.27	8.12	12.05	
1020	Hard mortar		18	.444			7.80	3.02	10.82	16.05	
1080	20" x 20", soft old mortar		12	.667			11.65	4.53	16.18	24	
1100	Hard mortar		10	.800			14	5.45	19.45	29	
1140	20" x 32", soft old mortar		10	.800			14	5.45	19.45	29	
1160	Hard mortar		8	1			17.50	6.80	24.30	36	
1200	48" x 48", soft old mortar		5	1.600			28	10.90	38.90	58	
1220	Hard mortar		4	2			35	13.60	48.60	72	
2000	Columns, 8" x 8", soft old mortar		48	.167			2.92	1.13	4.05	6	
2020	Hard mortar		40	.200			3.50	1.36	4.86	7.20	
2060	16" x 16", soft old mortar		16	.500			8.75	3.40	12.15	18.05	
2100	Hard mortar		14	.571			10	3.89	13.89	21	
2140	24" x 24", soft old mortar		8	1			17.50	6.80	24.30	36	
2160	Hard mortar		6	1.330			23	9.05	32.05	48	
2200	36" x 36", soft old mortar		4	2			35	13.60	48.60	72	
2220	Hard mortar		3	2.670			47	18.15	65.15	96	
3000	Copings, precast or masonry, to 8" wide										
3020	Soft old mortar	A-1	180	.044	L.F.		.78	.30	1.08	1.60	
3040	Hard mortar	"	160	.050	"		.88	.34	1.22	1.80	
3100	To 12" wide										
3120	Soft old mortar	A-1	160	.050	L.F.		.88	.34	1.22	1.80	
3140	Hard mortar	"	140	.057	"		1	.39	1.39	2.06	
4000	Fireplace, brick, 30" x 24" opening										
4020	Soft old mortar	A-1	2	4	Ea.		70	27	97	145	
4040	Hard mortar		1.25	6.400			110	44	154	230	
4100	Stone, soft old mortar		1.50	5.330			93	36	129	190	
4120	Hard mortar		1	8			140	54	194	290	
5000	Veneers, brick, soft old mortar		140	.057	S.F.		1	.39	1.39	2.06	
5020	Hard mortar		125	.064			1.12	.44	1.56	2.31	
5100	Granite and marble, 2" thick		180	.044			.78	.30	1.08	1.60	
5120	4" thick		170	.047			.82	.32	1.14	1.70	
5140	Stone, 4" thick		180	.044			.78	.30	1.08	1.60	
5160	8" thick		175	.046			.80	.31	1.11	1.65	
5400	Alternate pricing method, stone, 4" thick		60	.133	C.F.		2.33	.91	3.24	4.81	
5420	8" thick		85	.094	"		1.65	.64	2.29	3.40	
9000	Minimum labor/equipment charge		2	4	Job		70	27	97	145	

For expanded coverage of these items see *Means Site Work Cost Data 1991*

020 800	Asbestos Removal	CREW	DAILY OUTPUT	MAN-HOURS	UNIT	BARE COSTS				TOTAL INCL O&P
						MAT.	LABOR	EQUIP.	TOTAL	
810 0010	**ASBESTOS ABATEMENT EQUIPMENT** and supplies, buy									**810**
0200	Air filtration device, 2000 C.F.M.				Ea.	2,325			2,325	2,550
0250	Large volume air sampling pump, minimum					515			515	565
0260	Maximum					880			880	970
0300	Airless sprayer unit, 2 gun					4,000			4,000	4,400
0350	Light stand, 500 watt					265			265	290
0400	Personal respirators									
0410	Negative pressure, ½ face, dual operation, min.				Ea.	14			14	15.40
0420	Maximum					18			18	19.80
0450	P.A.P.R., full face, minimum					450			450	495
0460	Maximum					650			650	715
0470	Supplied air, full face, incl. air line, minimum					350			350	385
0480	Maximum					500			500	550
0500	Personnel sampling pump, minimum					400			400	440
0510	Maximum					600			600	660
1500	Power panel, 20 unit, incl. G.F.I.					1,600			1,600	1,750
1600	Shower unit, including pump and filters					2,500			2,500	2,750
1700	Supplied air system (type C)					9,500			9,500	10,500
1750	Vacuum cleaner, HEPA, 16 gal., stainless steel, wet/dry					1,250			1,250	1,375
1760	55 gallon					2,050			2,050	2,250
1800	Vacuum loader, 9-18 ton/hr					80,000			80,000	88,000
1900	Water atomizer unit, including 55 gal. drum					200			200	220
2000	Worker protection, whole body, foot and head cover, gloves					10			10	11
2500	Respirator, single use					10			10	11
2550	Cartridge for respirator					2.85			2.85	3.14
2570	Glove bag, 7 mil, 50" x 64"					12			12	13.20
2580	10 mil, 44" x 60"					6.25			6.25	6.90
3000	HEPA vacuum for work area, minimun					1,000			1,000	1,100
3050	Maximum					4,000			4,000	4,400
6000	Disposable polyethelene bags, 6 mil, 3 C.F.					.60			.60	.66
6300	Disposable fiber drums, 3 C.F.					5.75			5.75	6.35
6400	Pressure sensitive caution lables, 3" x 5"					.12			.12	.13
6450	11" x 17"					.15			.15	.17
820 0010	**ASBESTOS ABATEMENT WORK AREA** Containment and preparation.									**820**
0100	Pre-cleaning, HEPA vacuum and wet wipe	A-10	14,400	.004	S.F.		.11		.11	.18
0200	Protect carpeted area, 2 layers 6 mil poly on ¾" plywood	"	1,000	.064		.86	1.58	.06	2.50	3.55
0300	Separation barrier, 2" x 4" @ 16", ½" plywood ea. side, 8' high	F-2	400	.040		.73	.88	.04	1.65	2.28
0310	12' high		320	.050		.85	1.10	.05	2	2.79
0320	16' high		200	.080		1.05	1.76	.08	2.89	4.12
0400	Personnel decontam. chamber, 2" x 4" @ 16", ¾" ply ea. side		280	.057		1.41	1.26	.06	2.73	3.67
0450	Waste decontam. chamber, 2" x 4" studs @ 16", ¾" ply ea side		360	.044		1.41	.98	.04	2.43	3.20
0500	Cover surfaces with polyethelene sheeting,									
0501	Including glue and tape									
0550	Floors, each layer, 6 mil	A-10	8,000	.008	S.F.	.04	.20	.01	.25	.37
0551	4 mil		9,000	.007		.03	.18	.01	.22	.32
0560	Walls, each layer, 6 mil		6,000	.011		.04	.26	.01	.31	.47
0561	4 mil		7,000	.009		.03	.23	.01	.27	.40
0570	For heights above 12', add				%		20%			
0575	For heights above 20', add				"		30%			
0580	For fire retardant poly, add				S.F.	100%				
0590	For large open areas, deduct				%	10%	20%			
0600	Seal floor penetrations with foam firestop, to 36 Sq. In.	F-2	200	.080	S.F.	2.33	1.76	.08	4.17	5.50
0610	36 Sq. In. to 72 Sq. In.		125	.128		5.70	2.82	.13	8.65	11
0615	72 Sq. In. to 144 Sq. In.		80	.200		11.40	4.40	.20	16	19.95
0620	Seal wall penetrations with foam firestop to 36 Sq. In.		180	.089	Ea.	2.33	1.96	.09	4.38	5.85
0630	36 Sq. In. to 72 Sq. In.		100	.160		5.70	3.52	.16	9.38	12.20
0640	72 Sq. In. to 144 Sq. In.		60	.267		11.40	5.85	.27	17.52	22

For expanded coverage of these items see *Means Site Work Cost Data 1991*

020 800 | Asbestos Removal

			CREW	DAILY OUTPUT	MAN-HOURS	UNIT	MAT.	LABOR	EQUIP.	TOTAL	TOTAL INCL O&P	
820	0800	Caulk seams with latex	1 Carp	230	.035	L.F.	.12	.77		.89	1.38	820
830	0010	**DEMOLITION IN ASBESTOS CONTAMINATED AREA**										830
	0200	Ceiling, including suspension system, plaster and lath	A-9	2,100	.030	S.F.		.75	.09	.84	1.31	
	0210	Finished plaster, leaving wire lath		585	.109			2.71	.33	3.04	4.70	
	0220	Suspended acoustical tile		3,500	.018			.45	.05	.50	.79	
	0230	Splined tile grid system		3,000	.021			.53	.06	.59	.92	
	0240	Metal pan grid system		1,500	.043			1.06	.13	1.19	1.83	
	0250	Gypsum board		2,500	.026			.63	.08	.71	1.10	
	0260	Lighting fixtures up to 2' x 4'		72	.889	Ea.		22	2.65	24.65	38	
	0400	Partitions, non load bearing							.25			
	0410	Plaster, lath, and studs	A-9	690	.093	S.F.		2.30	.28	2.58	3.98	
	0450	Gypsum board and studs	"	1,390	.046	"		1.14	.14	1.28	1.98	
	9000	For type C respirator equipment, add				%					10%	
840	0010	**BULK ASBESTOS REMOVAL**										840
	0020	Includes disposable tools and 4 suits and respirators/day/man										
	0100	Beams, W 10 x 19	A-9	235	.272	L.F.		6.75	.81	7.56	11.70	
	0110	W 12 x 22		210	.305			7.55	.91	8.46	13.10	
	0120	W 14 x 26		180	.356			8.80	1.06	9.86	15.25	
	0130	W 16 x 31		160	.400			9.90	1.19	11.09	17.15	
	0140	W 18 x 40		140	.457			11.30	1.36	12.66	19.65	
	0150	W 24 x 55		110	.582			14.40	1.74	16.14	25	
	0160	W 30 x 108		85	.753			18.65	2.25	20.90	32	
	0170	W 36 x 150		72	.889			22	2.65	24.65	38	
	0200	Boiler insulation		480	.133	S.F.		3.30	.40	3.70	5.70	
	0210	With metal lath add				%				50%		
	0300	Boiler breeching or flue insulation	A-9	520	.123	S.F.		3.05	.37	3.42	5.30	
	0310	For active boiler, add				%				100%		
	0400	Duct or AHU insulation	A-9	720	.089	S.F.		2.20	.27	2.47	3.82	
	0500	Duct vibration isolation joints, up to 24 Sq. In. duct		56	1.140	Ea.		28	3.41	31.41	49	
	0520	25 Sq. In. to 48 Sq. In. duct		48	1.330			33	3.98	36.98	57	
	0530	49 Sq. In. to 76 Sq. In. duct		40	1.600			40	4.78	44.78	69	
	0600	Pipe insulation up to 4" diameter pipe		900	.071	L.F.		1.76	.21	1.97	3.05	
	0610	4" to 8" diameter pipe		800	.080			1.98	.24	2.22	3.43	
	0620	10" to 12" diameter pipe		700	.091			2.26	.27	2.53	3.93	
	0630	14" to 16" diameter pipe		550	.116			2.88	.35	3.23	5	
	0700	With glove bag up to 3" diameter pipe		100	.640		4.80	15.85	1.91	22.56	33	
	1000	Pipe fitting insulation up to 4" diameter pipe		320	.200	Ea.		4.95	.60	5.55	8.60	
	1100	6" to 8" diameter pipe		304	.211			5.20	.63	5.83	9.05	
	1110	10" to 12" diameter pipe		192	.333			8.25	.99	9.24	14.30	
	1120	14" to 16" diameter pipe		128	.500			12.40	1.49	13.89	21	
	1130	Over 16" diameter pipe		176	.364	S.F.		9	1.09	10.09	15.60	
	1200	With glove bag, up to 8" diameter pipe		40	1.600	Ea.	12	40	4.78	56.78	82	
	2000	Scrape foam fireproofing from flat surface		2,400	.027	S.F.		.66	.08	.74	1.14	
	2100	Irregular surfaces		1,200	.053			1.32	.16	1.48	2.29	
	3000	Remove cementitious material from flat surface		800	.080			1.98	.24	2.22	3.43	
	3100	Irregular surface		400	.160			3.96	.48	4.44	6.85	
	4000	Scrape acoustical coating from ceiling		3,200	.020			.50	.06	.56	.86	
	5000	Remove VAT from floor by hand		2,400	.027			.66	.08	.74	1.14	
	5100	By machine	A-11	4,800	.013			.33	.04	.37	.57	
	5150	For 2 layers, add				%				50%		
	6000	Remove contaminated soil from crawl space by hand	A-9	400	.160	C.F.		3.96	.48	4.44	6.85	
	6100	With large production vacuum loader	A-12	700	.091	"		2.26	.86	3.12	4.57	
	7000	Radiator backing, not including radiator removal	A-9	1,200	.053	S.F.		1.32	.16	1.48	2.29	
	9000	For type C respirator equipment, add				%					10%	
850	0010	**WASTE PACKAGING, HANDLING, & DISPOSAL**										850
	0100	Collect and bag bulk material, 3 C.F. bags, by hand	A-9	400	.160	Ea.	.60	3.96	.48	5.04	7.55	

For expanded coverage of these items see *Means Site Work Cost Data 1991*

020 800 | Asbestos Removal

			CREW	DAILY OUTPUT	MAN-HOURS	UNIT	MAT.	LABOR	EQUIP.	TOTAL	TOTAL INCL O&P	
850	0200	Large production vacuum loader	A-12	880	.073	Ea.	.60	1.80	.68	3.08	4.30	850
	1000	Double bag and decontaminate	A-9	960	.067		.60	1.65	.20	2.45	3.52	
	2000	Containerize bagged material, per bag	"	800	.080		2	1.98	.24	4.22	5.65	
	3000	Cart bags 50' to dumpster	2 Asbe	400	.040			.99		.99	1.58	
	5000	Disposal charges, not including haul, minimum				C.Y.					40	
	5020	Maximum				"					150	
	9000	For type C respirator equipment, add				%					10%	
860	0010	**DECONTAMINATION CONTAINMENT AREA DEMOLITION** and clean-up										860
	0100	Spray exposed substrate with surfactant (bridging)										
	0200	Flat surfaces	A-9	6,000	.011	S.F.	.25	.26	.03	.54	.74	
	0250	Irregular surfaces		4,000	.016	"	.30	.40	.05	.75	1.02	
	0300	Pipes, beams, and columns		2,000	.032	L.F.	.50	.79	.10	1.39	1.92	
	1000	Spray encapsulate polyethelene sheeting		8,000	.008	S.F.	.20	.20	.02	.42	.56	
	1100	Roll down polyethelene sheeting		8,000	.008	"		.20	.02	.22	.34	
	1500	Bag polyethelene sheeting		400	.160	Ea.	.60	3.96	.48	5.04	7.55	
	2000	Fine clean exposed substrate, with nylon brush		2,400	.027	S.F.		.66	.08	.74	1.14	
	2500	Wet wipe substrate		4,800	.013			.33	.04	.37	.57	
	2600	Vacuum surfaces, fine brush		6,400	.010			.25	.03	.28	.43	
	3000	Structural demolition										
	3100	Wood stud walls	A-9	2,800	.023	S.F.		.57	.07	.64	.98	
	3500	Window manifolds, not incl. window replacement		4,200	.015			.38	.05	.43	.65	
	3600	Plywood carpet protection		2,000	.032			.79	.10	.89	1.37	
	5000	HEPA vacuum and shampoo carpeting		4,800	.013		.03	.33	.04	.40	.60	
	9000	Final cleaning of protected surfaces		8,000	.008			.20	.02	.22	.34	
870	0010	**ENCAPSULATION WITH SEALANTS**										870
	0100	Ceilings and walls, minimum	A-9	21,000	.003	S.F.	.21	.08	.01	.30	.36	
	0110	Maximum		10,600	.006		.32	.15	.02	.49	.61	
	0200	Columns and beams, minimum		13,300	.005		.21	.12	.01	.34	.44	
	0210	Maximum		5,325	.012		.32	.30	.04	.66	.87	
	0300	Pipes to 12" diameter including minor repairs, minimum		800	.080	L.F.	.31	1.98	.24	2.53	3.77	
	0310	Maximum		400	.160	"	.98	3.96	.48	5.42	7.95	
890	0010	**OSHA TESTING**										890
	0100	Certified technician, minimum				Day		275		275		
	0110	Maximum				"		450		450		
	0200	Personal sampling, PCM analysis, minimum				Ea.	2.40	23		25.40		
	0210	Maximum				"	2.40	48		50.40		
	0300	Industrial hygenist, minimum				Day		350		350		
	0310	Maximum				"		500		500		
	1000	Cleaned area samples				Ea.	2.40	23		25.40		
	1100	PCM analysis, minimum					2.40	23		25.40		
	1110	Maximum					2.40	48		50.40		
	1200	TEM analysis, minimum								400		
	1210	Maximum								1,000		

021 | Site Preparation

021 200 | Structure Moving

			CREW	DAILY OUTPUT	MAN-HOURS	UNIT	MAT.	LABOR	EQUIP.	TOTAL	TOTAL INCL O&P	
204	0010	**MOVING BUILDINGS** One day move, up to 24' wide										204
	0020	Reset on new foundation, patch & hook-up, average move				Total					7,500	

For expanded coverage of these items see *Means Site Work Cost Data 1991*

021 | Site Preparation

021 200 | Structure Moving

			CREW	DAILY OUTPUT	MAN-HOURS	UNIT	BARE COSTS MAT.	BARE COSTS LABOR	BARE COSTS EQUIP.	BARE COSTS TOTAL	TOTAL INCL O&P	
204	0040	Wood or steel frame bldg., based on ground floor area	B-4	185	.259	S.F.		4.67	2.16	6.83	9.95	204
	0060	Masonry bldg., based on ground floor area	"	137	.350			6.30	2.91	9.21	13.45	
	0200	For 24' to 42' wide, add									15%	
	0220	For each additional day on road, add	B-4	1	48	Day		865	400	1,265	1,850	
	0240	Construct new basement, move building, 1 day										
	0300	move, patch & hook-up, based on ground floor area	B-3	155	.310	S.F.	5.05	5.85	9.90	20.80	26	

021 520 | Shores

			CREW	DAILY OUTPUT	MAN-HOURS	UNIT	BARE COSTS MAT.	BARE COSTS LABOR	BARE COSTS EQUIP.	BARE COSTS TOTAL	TOTAL INCL O&P	
524	0010	SHORING Existing building, with timber, no salvage allowance	B-51	2.20	21.820	M.B.F.	465	390	66	921	1,225	524
	1000	With 35 ton screw jacks, per box and jack		3.60	13.330	Jack	28	240	41	309	465	
	1090	Minimum labor/equipment charge		2	24	Ea.		430	73	503	780	
	1100	Masonry openings in walls, see div. 020-704										

021 560 | Underpinning

			CREW	DAILY OUTPUT	MAN-HOURS	UNIT	BARE COSTS MAT.	BARE COSTS LABOR	BARE COSTS EQUIP.	BARE COSTS TOTAL	TOTAL INCL O&P	
564	0010	UNDERPINNING FOUNDATIONS Including excavation,										564
	0020	forming, reinforcing, concrete and equipment										
	0100	5' to 16' below grade, 100 to 500 C.Y.	B-52	2.30	24.350	C.Y.	155	500	175	830	1,175	
	0200	Over 500 C.Y.		2.50	22.400		135	460	160	755	1,075	
	0400	16' to 25' below grade, 100 to 500 C.Y.		2	28		175	575	200	950	1,350	
	0500	Over 500 C.Y.		2.10	26.670		150	545	190	885	1,275	
	0700	26' to 40' below grade, 100 to 500 C.Y.		1.60	35		190	715	250	1,155	1,650	
	0800	Over 500 C.Y.		1.80	31.110		175	635	225	1,035	1,475	
	0900	For under 50 C.Y., add					10%	40%				

021 610 | Sheet Piling

			CREW	DAILY OUTPUT	MAN-HOURS	UNIT	BARE COSTS MAT.	BARE COSTS LABOR	BARE COSTS EQUIP.	BARE COSTS TOTAL	TOTAL INCL O&P	
614	0010	SHEET PILING Steel, not incl. wales, 22 psf, 15' excav., left in place	B-40	10.81	5.920	Ton	715	130	160	1,005	1,175	614
	0100	Pull & salvage		7.22	8.860	"	180	200	240	620	790	
	1200	15' deep excavation, 22 psf, left in place		983	.065	S.F.	7.80	1.46	1.75	11.01	12.95	
	1300	Pull & salvage		656	.098	"	1.95	2.18	2.62	6.75	8.65	
	2100	Rent steel sheet piling and wales, first month				Ton	210			210	230	
	2200	Per added month				"	17.50			17.50	19.25	
	3900	Wood, solid sheeting, incl. wales, braces and spacers, ⑳										
	3910	pull & salvage, 8' deep excavation	B-31	330	.121	S.F.	.95	2.28	.33	3.56	5.15	
	4520	Left in place, 8' deep, 55 S.F./hr.		440	.091	"	1.92	1.71	.25	3.88	5.15	
	4990	Minimum labor/equipment charge		2	20	Job		375	54	429	675	
	5000	For treated lumber add cost of treatment to lumber										
	5010	See division 063-102										

022 | Earthwork

022 200 | Excav, Backfill, Compact

			CREW	DAILY OUTPUT	MAN-HOURS	UNIT	BARE COSTS MAT.	BARE COSTS LABOR	BARE COSTS EQUIP.	BARE COSTS TOTAL	TOTAL INCL O&P	
204	0010	BACKFILL By hand, no compaction, light soil	1 Clab	14	.571	C.Y.		10		10	16.35	204
	0100	Heavy soil		11	.727			12.75		12.75	21	
	0300	Compaction in 6" layers, hand tamp, add to above ⑱		20.60	.388			6.80		6.80	11.10	
	0400	Roller compaction operator walking, add	B-10A	90	.133			2.78	.83	3.61	5.30	
	0500	Air tamp, add	B-9	190	.211			3.77	.67	4.44	6.90	
	0600	Vibrating plate, add	A-1	60	.133			2.33	.91	3.24	4.81	
	0800	Compaction in 12" layers, hand tamp, add to above	1 Clab	34	.235			4.12		4.12	6.70	
	1000	Air tamp, add	B-9	285	.140			2.51	.45	2.96	4.60	
	1100	Vibrating plate, add	A-1	90	.089			1.56	.60	2.16	3.21	
	1200	Dozer, backfilling, bulk, 75 H.P., 50' haul, no compaction	B-10L	1,100	.011			.23	.25	.48	.63	

For expanded coverage of these items see *Means Site Work Cost Data 1991*

		022 200	Excav, Backfill, Compact	CREW	DAILY OUTPUT	MAN-HOURS	UNIT	MAT.	LABOR	EQUIP.	TOTAL	TOTAL INCL O&P	
204	1300	Dozer backfilling, bulk, up to 300' haul, no compaction		B-10B	1,200	.010	C.Y.		.21	.65	.86	1.04	204
	1400	Air tamped		B-11B	240	.067			1.33	4.17	5.50	6.70	
	2350	Spreading in 8" layers, small dozer		B-10B	1,060	.011			.24	.73	.97	1.18	
	2450	Compacting with vibrating plate, 8" lifts		A-1	73	.110			1.92	.75	2.67	3.95	
212	0010	BORROW Buy and load at pit, haul 2 miles round trip											212
	0020	and spread, with 200 H.P. dozer, no compaction											
	0100	Bank run gravel		B-15	600	.047	C.Y.	3.95	.91	2.51	7.37	8.55	
	0200	Common borrow			600	.047		3.40	.91	2.51	6.82	7.95	
	0300	Crushed stone, 1-½"			600	.047		14.95	.91	2.51	18.37	21	
	0320	¾"			600	.047		15.20	.91	2.51	18.62	21	
	0340	½"			600	.047		15.75	.91	2.51	19.17	22	
	0360	⅜"			600	.047		16.25	.91	2.51	19.67	22	
	0400	Sand, washed, concrete			600	.047		14.70	.91	2.51	18.12	20	
	0500	Dead or bank sand			600	.047		3.30	.91	2.51	6.72	7.85	
	0600	Select structural fill			600	.047		7.35	.91	2.51	10.77	12.30	
	0700	Screened loam			600	.047		15.80	.91	2.51	19.22	22	
	0800	Topsoil, weed free			600	.047		12.10	.91	2.51	15.52	17.50	
	0900	For 5 mile haul, add		B-34B	200	.040			.74	1.83	2.57	3.19	
222	0010	COMPACTION Steel wheel tandem roller, 5 tons ⑱		B-10E	8	1.500	Hr.		31	14.80	45.80	66	222
	0050	Air tamp, 8" lifts, common fill		B-9	250	.160	C.Y.		2.86	.51	3.37	5.25	
	0060	Select fill		"	300	.133			2.39	.42	2.81	4.37	
	0600	Vibratory plate, 8" lifts, common fill		A-1	200	.040			.70	.27	.97	1.44	
	0700	Select fill		"	216	.037			.65	.25	.90	1.34	
	9000	Minimum labor/equipment charge		1 Clab	4	2	Job		35		35	57	
238	0010	EXCAVATING, BULK BANK MEASURE Common earth piled ⑯											238
	0020	For loading onto trucks, add									15%	15%	
	0200	Backhoe, hydraulic, crawler mtd., 1 C.Y. cap. = 75 C.Y./hr. ⑰		B-12A	600	.027	C.Y.		.57	.94	1.51	1.93	
	1200	Front end loader, track mtd., 1-½ C.Y. cap. = 70 C.Y./hr.		B-10N	560	.021			.45	.61	1.06	1.38	
	1500	Wheel mounted, ¾ C.Y. cap. = 45 C.Y./hr.		B-10R	360	.033			.69	.63	1.32	1.79	
	9000	Minimum labor/equipment charge		B-10L	2	6	Job		125	135	260	350	
242	0010	EXCAVATING, BULK, DOZER Open site											242
	2000	75 H.P., 50' haul, sand & gravel		B-10L	460	.026	C.Y.		.54	.59	1.13	1.52	
	2200	150' haul, sand & gravel			230	.052			1.09	1.19	2.28	3.03	
	2400	300' haul, sand & gravel			120	.100			2.08	2.27	4.35	5.80	
	3000	105 H.P., 50' haul, sand & gravel		B-10W	700	.017			.36	.57	.93	1.19	
	3200	150' haul, sand & gravel			310	.039			.81	1.29	2.10	2.70	
	3300	300' haul, sand & gravel			140	.086			1.79	2.85	4.64	5.95	
	4000	200 H.P., 50' haul, sand & gravel		B-10B	1,400	.009			.18	.55	.73	.89	
	4200	150' haul, sand & gravel			595	.020			.42	1.30	1.72	2.10	
	4400	300' haul, sand & gravel			310	.039			.81	2.50	3.31	4.03	
	5040	Clay		B-10M	1,025	.012			.24	.84	1.08	1.31	
	5400	300' haul, sand & gravel		"	470	.026			.53	1.83	2.36	2.86	
250	0010	EXCAVATING, STRUCTURAL Hand, pits to 6' deep, sandy soil		1 Clab	8	1			17.50		17.50	29	250
	0100	Heavy soil or clay			4	2			35		35	57	
	0300	Pits 6' to 12' deep, sandy soil			5	1.600			28		28	46	
	0500	Heavy soil or clay			3	2.670			47		47	76	
	0700	Pits 12' to 18' deep, sandy soil			4	2			35		35	57	
	0900	Heavy soil or clay			2	4			70		70	115	
	1100	Hand loading trucks from stock pile, sandy soil			12	.667			11.65		11.65	19.05	
	1300	Heavy soil or clay			8	1			17.50		17.50	29	
	1500	For wet or muck hand excavation, add to above					%				50%	50%	
	9000	Minimum labor/equipment charge		1 Clab	4	2	Job		35		35	57	
254	0010	EXCAVATING, TRENCH or continuous footing, common earth											254
	0020	No sheeting or dewatering included											
	1400	By hand with pick and shovel to 6' deep, light soil		1 Clab	8	1	C.Y.		17.50		17.50	29	
	1500	Heavy soil		"	4	2			35		35	57	
	1700	For tamping backfilled trenches, air tamp, add		A-1	100	.080			1.40	.54	1.94	2.89	
	1900	Vibrating plate, add		"	90	.089			1.56	.60	2.16	3.21	

For expanded coverage of these items see *Means Site Work Cost Data 1991*

022 200 | Excav, Backfill, Compact

		CREW	DAILY OUTPUT	MAN-HOURS	UNIT	BARE COSTS MAT.	BARE COSTS LABOR	BARE COSTS EQUIP.	BARE COSTS TOTAL	TOTAL INCL O&P	
254 2100	Trim sides and bottom for concrete pours, common earth	A-1	600	.013	S.F.		.23	.09	.32	.48	**254**
2300	Hardpan	"	180	.044	"		.78	.30	1.08	1.60	
9000	Minimum labor/equipment charge	1 Clab	4	2	Job		35		35	57	
258 0010	**EXCAVATING, UTILITY TRENCH** Common earth										**258**
0050	Trenching with chain trencher, 12 H.P., operator walking										
0100	4" wide trench, 12" deep	B-53	800	.010	L.F.		.21	.07	.28	.41	
0150	18" deep		750	.011			.23	.07	.30	.44	
0200	24" deep		700	.011			.24	.08	.32	.47	
0300	6" wide trench, 12" deep		650	.012			.26	.08	.34	.50	
0350	18" deep		600	.013			.29	.09	.38	.55	
0400	24" deep		550	.015			.31	.10	.41	.60	
0450	36" deep		450	.018			.38	.12	.50	.73	
0600	8" wide trench, 12" deep		475	.017			.36	.11	.47	.69	
0650	18" deep		400	.020			.43	.13	.56	.82	
0700	24" deep		350	.023			.49	.15	.64	.94	
0750	36" deep		300	.027			.57	.18	.75	1.09	
0900	Minimum labor/equipment charge		2	4	Job		86	27	113	165	
1000	Backfill by hand including compaction, add										
1050	4" wide trench, 12" deep	A-1	800	.010	L.F.		.18	.07	.25	.36	
1100	18" deep		530	.015			.26	.10	.36	.54	
1150	24" deep		400	.020			.35	.14	.49	.72	
1300	6" wide trench, 12" deep		540	.015			.26	.10	.36	.53	
1350	18" deep		405	.020			.35	.13	.48	.71	
1400	24" deep		270	.030			.52	.20	.72	1.07	
1450	36" deep		180	.044			.78	.30	1.08	1.60	
1600	8" wide trench, 12" deep		400	.020			.35	.14	.49	.72	
1650	18" deep		265	.030			.53	.21	.74	1.09	
1700	24" deep		200	.040			.70	.27	.97	1.44	
1750	36" deep		135	.059			1.04	.40	1.44	2.14	
9000	Minimum labor/equipment charge		4	2	Job		35	13.60	48.60	72	
262 0010	**FILL** Spread dumped material, by dozer, no compaction	B-10B	1,000	.012	C.Y.		.25	.78	1.03	1.25	**262**
0100	By hand	1 Clab	12	.667	"		11.65		11.65	19.05	
9000	Minimum labor/equipment charge	"	4	2	Job		35		35	57	
266 0010	**HAULING** Earth 6 C.Y. dump truck, ¼ mile round trip, 5.0 loads/hr.	B-34A	240	.033	C.Y.		.61	1.25	1.86	2.35	**266**
0200	4 mile round trip, 1.8 loads/hr.	"	85	.094			1.73	3.52	5.25	6.65	
0300	12 C.Y. dump truck, 1 mile round trip, 2.7 loads/hr. ⑯	B-34B	260	.031			.57	1.40	1.97	2.45	
0500	4 mile round trip, 1.6 loads/hr.	"	150	.053			.98	2.43	3.41	4.25	
0600	16.5 C.Y. dump trailer, 1 mile round trip, 2.6 loads/hr.	B-34C	340	.024			.43	1.35	1.78	2.17	
1100	4 mile round trip, 1.6 loads/hr.	"	210	.038			.70	2.18	2.88	3.52	
1150	20 C.Y. dump trailer, 1 mile round trip, 2.5 loads/hr.	B-34D	400	.020			.37	1.15	1.52	1.85	
1240	4 mile round trip, 1.5 loads/hr.	"	240	.033			.61	1.91	2.52	3.08	
1300	Hauling in medium traffic, add								20%	20%	
1400	Heavy traffic, add								30%	30%	
1600	Grading at dump, or embankment if required, by dozer	B-10B	1,000	.012			.25	.78	1.03	1.25	
1800	Spotter at fill or cut, if required	1 Clab	8	1	Hr.		17.50		17.50	29	
274 0010	**MOBILIZATION AND DEMOBILIZATION** Up to 25 miles							160			**274**
0020	Dozer or loader, 105 H.P.	B-34K	4	2	Ea.		37	175	212	250	
0900	Shovel, backhoe or dragline, ¾ C.Y.		3.60	2.220			41	195	236	280	
1200	Tractor shovel or front end loader, 1 C.Y.		4.50	1.780			33	155	188	225	
286 0010	**LOAM OR TOPSOIL** Remove and stockpile on site										**286**
0700	Furnish and place, truck dumped @ $17.00 per C.Y., 4" deep	B-10S	1,300	.009	S.Y.	1.87	.19	.24	2.30	2.63	
0800	6" deep		820	.015	"	2.82	.30	.38	3.50	4	
0810	Minimum labor/equipment charge		2	6	Ea.		125	155	280	370	
0900	Fine grading and seeding, incl. lime, fertilizer & seed,										
1000	With equipment	B-14	1,000	.048	S.Y.	.18	.89	.19	1.26	1.85	

For expanded coverage of these items see *Means Site Work Cost Data 1991*

		022 200 Excav, Backfill, Compact	CREW	DAILY OUTPUT	MAN-HOURS	UNIT	MAT.	LABOR	EQUIP.	TOTAL	TOTAL INCL O&P	
								BARE COSTS				
286	2000	Minimum labor/equipment charge	1 Clab	4	2	Job		35		35	57	**286**

022 300   Pavement Base

			CREW	DAILY OUTPUT	MAN-HOURS	UNIT	MAT.	LABOR	EQUIP.	TOTAL	INCL O&P	
304	0010	**BASE** Prepare and roll sub-base, small areas to 2500 S.Y.	B-32A	1,500	.016	S.Y.		.33	.58	.91	1.16	**304**
	0100	Large areas over 2500 S.Y.	B-32	3,700	.009	"		.18	.40	.58	.73	
	9000	Minimum labor/equipment charge	1 Clab	4	2	Job		35		35	57	
308	0010	**BASE COURSE** For roadways and large paved areas										**308**
	0050	¾" stone compacted to 3" deep	B-36	4,000	.010	S.Y.	2.15	.20	.26	2.61	2.97	
	0100	6" deep		3,900	.010		4.30	.20	.26	4.76	5.35	
	0200	9" deep		2,875	.014		6.48	.28	.36	7.12	7.95	
	0300	12" deep		2,350	.017		8.55	.34	.43	9.32	10.45	
	0301	Crushed 1-½" stone base, compacted to 4" deep		5,225	.008		2.58	.15	.20	2.93	3.30	
	0302	6" deep		3,900	.010		3.90	.20	.26	4.36	4.91	
	0303	8" deep		3,000	.013		5.19	.27	.34	5.80	6.50	
	0304	12" deep		1,800	.022		7.80	.44	.57	8.81	9.90	
	0310	Minimum labor/equipment charge		1	40	Job		795	1,025	1,820	2,400	
	0350	Bank run gravel, spread and compacted										
	0370	6" deep	B-32	6,000	.005	S.Y.	1.54	.11	.25	1.90	2.14	
	0390	9" deep		44,000	.001		2.31	.02	.03	2.36	2.60	
	0400	12" deep		3,600	.009		3.08	.19	.41	3.68	4.14	
	8900	For small and irregular areas, add							50%			
	9000	Minimum labor/equipment charge	1 Clab	3	2.670	Job		47		47	76	

022 400   Soil Stabilization

			CREW	DAILY OUTPUT	MAN-HOURS	UNIT	MAT.	LABOR	EQUIP.	TOTAL	INCL O&P	
408	0010	**GROUTING, PRESSURE** Cement and sand, 1:1 mix, minimum	B-47	124	.194	Bag	6.80	3.77	4.09	14.66	18.05	**408**
	0100	Maximum		51	.471	"	7.80	9.15	9.95	26.90	34	
	0200	Cement and sand, 1:1 mix, minimum		250	.096	C.F.	4.55	1.87	2.03	8.45	10.25	
	0300	Maximum		100	.240		6	4.67	5.05	15.72	19.70	
	0400	Cement grout, minimum (1 bag = 1 C.F.)		137	.175		6.30	3.41	3.70	13.41	16.50	
	0500	Maximum		57	.421		7.45	8.20	8.90	24.55	31	
	0700	Alternate pricing method: (Add for materials)										
	0710	5 man crew and equipment	B-47	1	24	Day		465	505	970	1,300	

022 700   Slope/Erosion Control

			CREW	DAILY OUTPUT	MAN-HOURS	UNIT	MAT.	LABOR	EQUIP.	TOTAL	INCL O&P	
704	0010	**EROSION CONTROL** Jute mesh, 100 S.Y. per roll, 4' wide, stapled	B-1	2,500	.010	S.Y.	.62	.17		.79	.96	**704**
	0100	Plastic netting, stapled, 2" x 1" mesh, 20 mil		2,500	.010		.33	.17		.50	.64	
	0200	Polypropylene mesh, stapled, 6.5 oz./S.Y.		2,500	.010		1.63	.17		1.80	2.07	
	0300	Tobacco netting, or jute mesh #2, stapled		2,500	.010		.03	.17		.20	.31	
716	0010	**STONE WALL** Including excavation, concrete footing and										**716**
	0020	stone 3' below grade. Price is exposed face area.										
	0200	Decorative random stone, to 6' high, 1'-6" thick, dry set	D-1	35	.457	S.F.	7.15	9.25		16.40	23	
	0300	Mortar set		40	.400		8.65	8.10		16.75	22	
	0500	Cut stone, to 6' high, 1'-6" thick, dry set		35	.457		11.45	9.25		20.70	27	
	0600	Mortar set		40	.400		12.30	8.10		20.40	26	
	0800	Retaining wall, random stone, 6' to 10' high, 2' thick, dry set		45	.356		8.90	7.20		16.10	21	
	0900	Mortar set		50	.320		10.90	6.45		17.35	22	
	1100	Cut stone, 6' to 10' high, 2' thick, dry set		45	.356		14.60	7.20		21.80	28	
	1200	Mortar set		50	.320		15.75	6.45		22.20	28	
	9000	Minimum labor/equipment charge		2	8	Job		160		160	260	

022 800   Soil Treatment

			CREW	DAILY OUTPUT	MAN-HOURS	UNIT	MAT.	LABOR	EQUIP.	TOTAL	INCL O&P	
804	0010	**TERMITE PRETREATMENT**	1 Skwk	1,508	.005	SF Flr.	.11	.12		.23	.32	**804**
	0100	Commercial, minimum		2,496	.003		.05	.07		.12	.17	
	0200	Maximum		1,645	.005		.08	.11		.19	.27	
	0390	Minimum labor/equipment charge		4	2	Job		45		45	73	

For expanded coverage of these items see *Means Site Work Cost Data 1991*

022 800	**Soil Treatment**	CREW	DAILY OUTPUT	MAN-HOURS	UNIT	BARE COSTS				TOTAL INCL O&P	
						MAT.	LABOR	EQUIP.	TOTAL		
804	0400	Insecticides for termite control, minimum				Gal.	10.50			10.50	11.55	804
	0500	Maximum				"	17.25			17.25	19	

023 550	**Pile Driving**	CREW	DAILY OUTPUT	MAN-HOURS	UNIT	BARE COSTS				TOTAL INCL O&P	
						MAT.	LABOR	EQUIP.	TOTAL		
554	0010	**MOBILIZATION** Set up & remove, air compressor, 600 C.F.M.	A-5	3.30	5.450	Ea.		96	11.05	107.05	170	554
	0100	1200 C.F.M.	"	2.20	8.180			145	16.60	161.60	255	
	0200	Crane, with pile leads and pile hammer, 75 ton	B-19	.60	107			2,375	2,100	4,475	6,300	
	0300	150 ton	"	.36	178			3,975	3,525	7,500	10,500	
558	0011	**PILING SPECIAL COSTS** pile caps, see also division 033-130										558
	0500	Cutoffs, concrete piles, plain	1 Pile	5.50	1.450	Ea.		32		32	56	
	0600	With steel thin shell		38	.211			4.67		4.67	8.05	
	0700	Steel pile or "H" piles		19	.421			9.35		9.35	16.10	
	0800	Wood piles		38	.211			4.67		4.67	8.05	
	1000	Testing, any type piles, test load is twice the design load										
	1050	50 ton design load, 100 ton test				Ea.					10,500	
	1100	100 ton design load, 200 ton test									12,000	
	1200	200 ton design load, 400 ton test									21,000	

023 600	**Driven Piles**	CREW	DAILY OUTPUT	MAN-HOURS	UNIT	BARE COSTS				TOTAL INCL O&P	
						MAT.	LABOR	EQUIP.	TOTAL		
604	0010	**PILES, CONCRETE** 200 piles, 60' long										604
	0020	unless specified otherwise, not incl. pile caps or mobilization										
	0800	Cast in place friction pile, 50' long, fluted,										
	0810	tapered steel, 4000 psi concrete, no reinforcing										
	0900	12" diameter, 7 ga.	B-19	600	.107	V.L.F.	12.20	2.38	2.11	16.69	19.70	
	1200	18" diameter, 7 ga.	"	480	.133	"	18.35	2.98	2.64	23.97	28	
	1300	End bearing, fluted, constant diameter,										
	1320	4000 psi concrete, no reinforcing										
	1340	12" diameter, 7 ga.	B-19	600	.107	V.L.F.	12.50	2.38	2.11	16.99	20	
	1400	18" diameter, 7 ga.		480	.133		19.60	2.98	2.64	25.22	29	
	3100	Precast, prestressed, 40' long, 10" thick, square		700	.091		6.35	2.04	1.81	10.20	12.40	
	3200	12" thick, square		680	.094		8.30	2.10	1.86	12.26	14.70	
	3400	14" thick, square		600	.107		12.70	2.38	2.11	17.19	20	
	4000	18" thick, square		520	.123		17.85	2.75	2.44	23.04	27	
608	0010	**PILES, STEEL** Not including mobilization or demobilization										608
	0100	Step tapered, round, concrete filled										
	0110	8" tip, 60 ton capacity, 30' depth	B-19	760	.084	V.L.F.	5.05	1.88	1.67	8.60	10.55	
	0120	60' depth		740	.086		5.65	1.93	1.71	9.29	11.30	
	0250	"H" Sections, 8" x 8", 36 lb. per L.F.		640	.100		9	2.24	1.98	13.22	15.80	
	1300	14" x 14", 102 lb. per L.F.		510	.125		24	2.80	2.48	29.28	34	
	2600	Pipe piles, 8" diameter, 29 lb. per L.F., no concrete		500	.128		8.75	2.86	2.53	14.14	17.20	
	2700	Concrete filled		460	.139		9.65	3.11	2.75	15.51	18.85	
	3500	14" diameter, 46 lb. per L.F., no concrete		430	.149		13.65	3.33	2.95	19.93	24	
	3600	Concrete filled		355	.180		16.35	4.03	3.57	23.95	29	
	4100	18" diameter, 59 lb. per L.F., no concrete		355	.180		17.35	4.03	3.57	24.95	30	
	4200	Concrete filled		310	.206		23	4.61	4.09	31.70	37	
612	0010	**PILES, WOOD** Untreated, friction or end bearing, not including										612
	0050	mobilization or demobilization										
	0100	Up to 30' long, 12" butts, 8" points	B-19	625	.102	V.L.F.	4	2.29	2.03	8.32	10.45	
	0200	30' to 39' long, 12" butts, 8" points	"	700	.091	"	4	2.04	1.81	7.85	9.80	

For expanded coverage of these items see *Means Site Work Cost Data 1991*

023 600 | Driven Piles

			DAILY	MAN-			BARE COSTS			TOTAL	
		CREW	OUTPUT	HOURS	UNIT	MAT.	LABOR	EQUIP.	TOTAL	INCL O&P	
612	0800	Treated piles, 12 lb. creosote per C.F.,										612
	0810	friction or end bearing, ASTM class B										
	1000	Up to 30' long, 12" butts, 8" points ㉒	B-19	625	.102	V.L.F.	6.65	2.29	2.03	10.97	13.35	
	1100	30' to 39' long, 12" butts, 8" points	"	700	.091	"	6.65	2.04	1.81	10.50	12.70	

023 800 | Caissons

			DAILY	MAN-			BARE COSTS			TOTAL	
		CREW	OUTPUT	HOURS	UNIT	MAT.	LABOR	EQUIP.	TOTAL	INCL O&P	
804	0010	CAISSONS Incl. excav., concrete, 50 lbs. reinf. per C.Y., but not incl.										804
	0020	mobilization, boulder removal, disposal										
	0100	Open style, machine drilled, to 50' deep, in stable ground, no										
	0110	casings or ground water, 18" diam., .065 C.Y./L.F.	B-43	200	.240	V.L.F.	8.35	4.58	3.72	16.65	21	
	0200	24" diameter, .116 C.Y./L.F. ⑲		190	.253	"	14.85	4.82	3.92	23.59	28	
	0210	4' bell diameter		20	2.400	Ea.	47	46	37	130	165	
	0500	48" diameter, .465 C.Y./L.F.		100	.480	V.L.F.	58	9.15	7.45	74.60	87	
	0510	9' bell diameter		2	24	Ea.	460	460	370	1,290	1,650	
	1210	Open style, machine drilled, to 25' deep, in wet ground,										
	1220	pulled casing and pumping										
	1400	24" diameter, .116 C.Y./L.F.	B-48	125	.448	V.L.F.	14.85	8.70	7.40	30.95	38	
	1410	4' bell diameter	"	19.80	2.830	Ea.	47	55	47	149	190	
	1700	48" diameter, .465 C.Y./L.F.	B-49	55	1.600	V.L.F.	58	32	26	116	145	
	1710	9' bell diameter	"	3.30	26.670	Ea.	460	540	430	1,430	1,850	
	2310	Open style, machine drilled, to 25' deep, in soft rocks and										
	2320	medium hard shales										
	2500	24" diameter, .116 C.Y./L.F.	B-49	30	2.930	V.L.F.	14.90	59	47	120.90	165	
	2510	4' bell diameter		10.90	8.070	Ea.	47	165	130	342	460	
	2800	48" diameter, .465 C.Y./L.F.		10	8.800	V.L.F.	60	180	140	380	510	
	2810	9' bell diameter		1.10	80	Ea.	410	1,625	1,275	3,310	4,500	
	3600	For rock excavation, sockets, add, minimum		120	.733	C.F.		14.85	11.80	26.65	37	
	3650	Average		95	.926			18.75	14.90	33.65	47	
	3700	Maximum		48	1.830			37	29	66	93	
	3900	For 50' to 100' deep, add				V.L.F.				7%	7%	
	4000	For 100' to 150' deep, add								25%	25%	
	4100	For 150' to 200' deep, add								30%	30%	
	4200	For casings left in place, add				Lb.	.52			.52	.57	
	4300	For other than 50 lb. reinf. per C.Y., add or deduct				"	.51			.51	.56	
	4400	For steel "I" beam cores, add	B-49	8.30	10.600	Ton	545	215	170	930	1,125	
	4500	Load and haul excess excavation, 2 miles	B-34B	178	.045	C.Y.		.83	2.05	2.88	3.58	
	4600	For mobilization, 50 mile radius, rig to 36"	B-43	2	24	Ea.		460	370	830	1,150	
	4650	Rig to 84"	B-48	1.75	32			620	530	1,150	1,575	
	4700	For low headroom, add								50%		
	4750	For difficult access, add								25%		

025 100 | Walk/Rd/Parkng Paving

			DAILY	MAN-			BARE COSTS			TOTAL	
		CREW	OUTPUT	HOURS	UNIT	MAT.	LABOR	EQUIP.	TOTAL	INCL O&P	
104	0010	ASPHALTIC CONCRETE PAVEMENT for highways										104
	0020	and large paved areas										
	0080	Binder course, 1-½" thick	B-25	7,725	.011	S.Y.	2.17	.22	.20	2.59	2.96	
	0120	2" thick		6,345	.014		2.91	.26	.24	3.41	3.89	
	0160	3" thick ㉖		4,905	.018		4.36	.34	.31	5.01	5.70	
	0300	Wearing course, 1" thick	B-25B	10,575	.009		1.55	.18	.16	1.89	2.17	
	0340	1-½" thick		7,725	.012		2.35	.24	.22	2.81	3.22	
	0380	2" thick		6,345	.015		3.12	.29	.27	3.68	4.20	

For expanded coverage of these items see *Means Site Work Cost Data 1991*

025 100 | Walk/Rd/Parkng Paving

			CREW	DAILY OUTPUT	MAN-HOURS	UNIT	BARE COSTS MAT.	LABOR	EQUIP.	TOTAL	TOTAL INCL O&P	
104	0420	2-½" thick	B-25B	5,480	.018	S.Y.	3.89	.34	.31	4.54	5.15	104
	0460	3" thick	"	4,900	.020	"	4.70	.38	.35	5.43	6.15	
	0800	Alternate method of figuring paving costs										
	0810	Binder course, 1-½" thick	B-25	630	.140	Ton	28.35	2.66	2.40	33.41	38	
	0811	2" thick		690	.128		28.35	2.43	2.19	32.97	38	
	0812	3" thick		800	.110		28.35	2.10	1.89	32.34	37	
	0813	4" thick	↓	900	.098		28.35	1.86	1.68	31.89	36	
	0850	Wearing course, 1" thick	B-25B	575	.167		28.80	3.23	2.95	34.98	40	
	0851	1-½" thick		630	.152		28.80	2.95	2.69	34.44	39	
	0852	2" thick		690	.139		28.80	2.69	2.46	33.95	39	
	0853	2-½" thick		745	.129		28.80	2.49	2.28	33.57	38	
	0854	3" thick	↓	800	.120	↓	28.80	2.32	2.12	33.24	38	
120	0010	**CONCRETE PAVEMENT** Including joints, finishing, and curing										120
	0020	Fixed form, 12' pass, unreinforced, 6" thick	B-26	3,000	.029	S.Y.	12.50	.57	.57	13.64	15.30	
	0100	8" thick		2,700	.033		15.90	.64	.63	17.17	19.20	
	0400	12" thick	↓	1,800	.049	↓	22.85	.96	.94	24.75	28	
	0450	Minimum labor/equipment charge	1 Cefi	1	8	Job		175		175	270	
	0700	Finishing, broom finish small areas	2 Cefi	135	.119	S.Y.		2.57		2.57	4.01	
	1000	Curing, with sprayed membrane by hand ㉖	2 Clab	1,500	.011	"	.16	.19		.35	.48	
128	0010	**SIDEWALKS, DRIVEWAYS, & PATIOS** No base										128
	0020	Asphaltic concrete, 2" thick	B-37	720	.067	S.Y.	3.17	1.23	.16	4.56	5.65	
	0100	2-½" thick	"	660	.073	"	3.97	1.34	.18	5.49	6.75	
	0300	Concrete, 3000 psi, cast in place with 6 x 6 - #10/10 mesh,										
	0310	broomed finish, no base, 4" thick	B-24	600	.040	S.F.	.91	.82		1.73	2.31	
	1700	Redwood, prefabricated, 4' x 4' sections	F-2	316	.051		2.97	1.11	.05	4.13	5.15	
	1750	Redwood planks, 1" thick, on sleepers	"	240	.067	↓	1.95	1.47	.07	3.49	4.62	
	9000	Minimum labor/equipment charge	D-1	2	8	Job		160		160	260	

025 150 | Unit Pavers

			CREW	DAILY OUTPUT	MAN-HOURS	UNIT	BARE COSTS MAT.	LABOR	EQUIP.	TOTAL	TOTAL INCL O&P	
154	0010	**ASPHALT BLOCKS** Premold, 6"x12"x1-¼", w/bed & neopr. adhesive	D-1	135	.119	S.F.	2.54	2.39		4.93	6.65	154
	0100	3" thick		130	.123		3.68	2.49		6.17	8.05	
	0300	Hexagonal tile, 8" wide, 1-¼" thick		135	.119		2.81	2.39		5.20	6.95	
	0400	2" thick		130	.123		3.46	2.49		5.95	7.80	
	0500	Square, 8" x 8", 1-¼" thick		135	.119		2.80	2.39		5.19	6.90	
	0600	2" thick	↓	130	.123	↓	3.46	2.49		5.95	7.80	
	9000	Minimum labor/equipment charge	1 Bric	2	4	Job		91		91	145	
158	0010	**BRICK PAVING** 4" x 8" x 1-½", without joints (4.5 brick/S.F.)	D-1	110	.145	S.F.	2.32	2.94		5.26	7.25	158
	0100	Grouted, ⅜" joint (3.9 brick/S.F.)		90	.178		2.05	3.59		5.64	8	
	0200	4" x 8" x 2-¼", without joints (4.5 bricks/S.F.)		110	.145		2.51	2.94		5.45	7.45	
	0300	Grouted, ⅜" joint (3.9 brick/S.F.)		90	.178		2.19	3.59		5.78	8.15	
	1500	Brick on 4" thick sand bed laid flat, 4.5 per S.F.		110	.145		2.45	2.94		5.39	7.40	
	2000	Laid on edge, 7.2 per S.F.		70	.229		3.54	4.62		8.16	11.30	
	2500	For 4" thick concrete bed and joints, add	↓	595	.027		.73	.54		1.27	1.67	
	2800	For steam cleaning, add	A-1	950	.008	↓	.05	.15	.06	.26	.36	
	9000	Minimum labor/equipment charge	1 Bric	2	4	Job		91		91	145	
166	0010	**STONE PAVERS**										166
	1300	Slate, natural cleft, irregular, ¾" thick	D-1	92	.174	S.F.	1.48	3.51		4.99	7.25	
	1350	Random rectangular, gauged, ½" thick		105	.152		3.31	3.08		6.39	8.60	
	1400	Random rectangular, butt joint, gauged, ¼" thick	↓	150	.107		3.53	2.15		5.68	7.35	
	1450	For sand rubbed finish, add					2.27			2.27	2.50	
	1550	Granite blocks, 3-½" x 3-½" x 3-½"	D-1	92	.174		4.38	3.51		7.89	10.45	
	1600	4" to 12" long, 3" to 5" wide, 3" to 5" thick	"	98	.163	↓	3.85	3.30		7.15	9.55	

For expanded coverage of these items see *Means Site Work Cost Data 1991*

025 | Paving and Surfacing

025 250 | Curbs

			CREW	DAILY OUTPUT	MAN-HOURS	UNIT	BARE COSTS MAT.	LABOR	EQUIP.	TOTAL	TOTAL INCL O&P	
254	0010	**CURBS** Asphaltic, machine formed, 8" wide, 6" high, 40 L.F./ton	B-27	1,000	.032	L.F.	.70	.58	.06	1.34	1.78	254
	0100	8" wide, 8" high, 30 L.F. per ton		900	.036		.91	.64	.06	1.61	2.12	
	0150	Asphaltic berm, 12"W, 3"-6"H, 35 L.F./ton, before pavement		700	.046		.94	.82	.08	1.84	2.47	
	0200	12"W, 1-½"to 4" H, 60 L.F. per ton, laid with pavement	B-2	1,050	.038		.51	.68		1.19	1.67	
	0300	Concrete, 6" x 18", wood forms, straight	C-2	500	.096		3.02	2.07	.06	5.15	6.80	
	0400	6" x 18", radius	"	200	.240		3.23	5.20	.16	8.59	12.20	
	0550	Precast, 6" x 18", straight	B-29	700	.080		5.15	1.51	.79	7.45	9	
	0600	6" x 18", radius		325	.172		7.75	3.25	1.71	12.71	15.65	
	1000	Granite, split face, straight, 5" x 16"		500	.112		12.35	2.11	1.11	15.57	18.20	
	1100	6" x 18", see also division 044-651		450	.124		16.30	2.35	1.23	19.88	23	
	1300	Radius curbing, 6" x 18", over 10' radius		260	.215		19.95	4.06	2.14	26.15	31	
	1400	Corners, 2' radius		80	.700	Ea.	68	13.20	6.95	88.15	105	
	1600	Edging, 4-½" x 12", straight		300	.187	L.F.	6.30	3.52	1.85	11.67	14.65	
	1800	Curb inlets, (guttermouth) straight		41	1.370	Ea.	150	26	13.55	189.55	220	
256	0010	**EDGING** Redwood, untreated, 1" x 4"	F-2	500	.032	L.F.	.87	.70	.03	1.60	2.15	256
	0100	2" x 4"		330	.048		.81	1.07	.05	1.93	2.69	
	0200	Steel edge strips, ¼" x 5" including stakes		330	.048		2.50	1.07	.05	3.62	4.55	
	0300	³⁄₁₆" x 4"		330	.048		1.86	1.07	.05	2.98	3.85	
	0500	Brick edging, set vertically, 3 brick per L.F.	D-1	135	.119		1.52	2.39		3.91	5.50	
	0600	Set horizontally, 1-½ brick per L.F.	"	370	.043		.77	.87		1.64	2.25	
	9000	Minimum labor/equipment charge	1 Carp	4	2	Job		44		44	72	

025 450 | Surfacing

			CREW	DAILY OUTPUT	MAN-HOURS	UNIT	BARE COSTS MAT.	LABOR	EQUIP.	TOTAL	TOTAL INCL O&P	
458	0010	**SEALCOATING** 2 coal tar pitch emulsion over 10,000 S.Y.	B-45	5,000	.003	S.Y.	.33	.07	.12	.52	.60	458
	0100	Under 1000 S.Y.	B-1	1,050	.023	"	.33	.42		.75	1.04	
	9000	Minimum labor/equipment charge	1 Clab	2	4	Job		70		70	115	

025 800 | Pavement Marking

			CREW	DAILY OUTPUT	MAN-HOURS	UNIT	BARE COSTS MAT.	LABOR	EQUIP.	TOTAL	TOTAL INCL O&P	
804	0010	**PAINTING LINES** On pavement, reflectorized white or yellow, 4" wide	B-78	20,000	.002	L.F.	.03	.04	.02	.09	.12	804
	0200	6" wide	"	11,000	.004	"	.04	.08	.04	.16	.21	
	0760	Arrows	B-79	660	.061	S.F.	1.21	1.09	.89	3.19	4.09	
	0800	Parking stall, paint, white	B-78	860	.056	Stall	.68	1	.51	2.19	2.94	
	1000	Street letters and numbers	"	1,600	.030	S.F.	.21	.54	.27	1.02	1.41	

026 | Piped Utilities

026 010 | Piped Utilities

			CREW	DAILY OUTPUT	MAN-HOURS	UNIT	BARE COSTS MAT.	LABOR	EQUIP.	TOTAL	TOTAL INCL O&P	
012	0010	**BEDDING** For pipe and conduit, not incl. compaction										012
	0050	Crushed or screened bank run gravel	B-6	150	.160	C.Y.	11.05	3.01	1.27	15.33	18.40	
	0100	Crushed stone ¾" to ½"		150	.160		12.90	3.01	1.27	17.18	20	
	0200	Sand, dead or bank,		150	.160		3.45	3.01	1.27	7.73	10.05	
	0500	Compacting bedding in trench	A-1	90	.089			1.56	.60	2.16	3.21	
014	0010	**EXCAVATION AND BACKFILL** See division 022-204 & 254										014
	0100	Hand excavate and trim for pipe bells after trench excavation										
	0200	8" pipe	1 Clab	155	.052	L.F.		.90		.90	1.47	
	0300	18" pipe		130	.062	"		1.08		1.08	1.76	
	9000	Minimum labor/equipment charge		4	2	Job		35		35	57	

For expanded coverage of these items see *Means Site Work Cost Data 1991*

026 050 | Manholes And Cleanouts

			CREW	DAILY OUTPUT	MAN-HOURS	UNIT	BARE COSTS				TOTAL INCL O&P	
							MAT.	LABOR	EQUIP.	TOTAL		
054	0010	UTILITY VAULTS Precast concrete, 6" thick										054
	0050	5' x 10' x 6' high, I.D.	B-13	2	28	Ea.	1,600	530	245	2,375	2,875	
	0350	Hand hole, precast concrete, 1-½" thick										
	0400	1'-0" x 2'-0" x 1'-9", I.D., light duty	B-1	4	6	Ea.	255	110		365	460	
	0450	4'-6" x 3'-2" x 2'-0", O.D., heavy duty	B-6	3	8	"	540	150	63	753	905	

026 650 | Water Systems

			CREW	DAILY OUTPUT	MAN-HOURS	UNIT	BARE COSTS				TOTAL INCL O&P	
							MAT.	LABOR	EQUIP.	TOTAL		
686	0010	PIPING, WATER DISTRIBUTION SYSTEMS Pipe laid in trench,										686
	0020	excavation and backfill not included										
	1400	Ductile Iron pipe, class 250 water piping, 18' lengths ⑫⑨										
	1410	Mechanical joint, 4" diameter	B-20	144	.167	L.F.	6.50	3.31		9.81	12.55	
	9000	Minimum labor/equipment charge	1 Clab	4	2	Job		35		35	57	

026 700 | Water Wells

			CREW	DAILY OUTPUT	MAN-HOURS	UNIT	BARE COSTS				TOTAL INCL O&P	
							MAT.	LABOR	EQUIP.	TOTAL		
704	0010	WELLS Domestic water, drilled and cased, including casing										704
	0100	4" to 6" diameter	B-23	160	.250	V.L.F.	7.75	4.48	3.40	15.63	19.60	
	0200	8" diameter ⑫⑨	"	127	.315	"	9.75	5.65	4.28	19.68	25	
	1400	Remove & reset pump, minimum	B-21	4	7	Ea.		145	26	171	260	
	1420	Maximum	"	2	14	"		285	52	337	520	
	1500	Pumps, installed in wells to 100' deep, 4" submersible										
	1510	½ H.P.	Q-1	3.22	4.970	Ea.	305	115		420	510	
	1520	¾ H.P.		2.66	6.020		370	140		510	620	
	1600	1 H.P.	↓	2.29	6.990		380	160		540	665	
	1700	1-½ H.P.	Q-22	1.60	10		510	230	225	965	1,175	
	1800	2 H.P.		1.33	12.030		550	275	270	1,095	1,325	
	1900	3 H.P.		1.14	14.040		725	320	320	1,365	1,650	
	2000	5 H.P.	↓	1.14	14.040		760	320	320	1,400	1,675	
	5000	Wells to 180 ft. deep, 4" submersible, 1 HP	B-21	1.10	25.450		400	520	95	1,015	1,375	
	5500	2 HP		1.10	25.450		660	520	95	1,275	1,675	
	6000	3 HP		1	28		680	570	105	1,355	1,775	
	7000	5 HP		.90	31.110	↓	710	635	115	1,460	1,925	
	9000	Minimum labor/equipment charge	↓	1.80	15.560	Job		315	58	373	575	

026 850 | Gas Distribution System

			CREW	DAILY OUTPUT	MAN-HOURS	UNIT	BARE COSTS				TOTAL INCL O&P	
							MAT.	LABOR	EQUIP.	TOTAL		
852	0010	GAS SERVICE & DISTRIBUTION Not including excavation										852
	0050	or backfill										
	0100	Polyethylene, 60 psi, coils, ½" diameter, SDR 7	B-20	450	.053	L.F.	.15	1.06		1.21	1.90	
	0150	1-¼" diameter, SDR 10	↓	400	.060	"	.55	1.19		1.74	2.55	
	9000	Minimum labor/equipment charge	↓	2	12	Job		240		240	390	

027 | Sewerage & Drainage

027 100 | Subdrainage Systems

			CREW	DAILY OUTPUT	MAN-HOURS	UNIT	BARE COSTS				TOTAL INCL O&P	
							MAT.	LABOR	EQUIP.	TOTAL		
106	0010	PIPING, SUBDRAINAGE, BITUMINOUS										106
	0021	Not including excavation and backfill										
	2000	Perforated underdrain, 3" diameter	B-20	800	.030	L.F.	1.20	.60		1.80	2.29	
	2020	4" diameter		760	.032		1.50	.63		2.13	2.67	
	2040	5" diameter	↓	720	.033	↓	2.60	.66		3.26	3.94	

For expanded coverage of these items see *Means Site Work Cost Data 1991*

027 150 | Sewage Systems

		CREW	DAILY OUTPUT	MAN-HOURS	UNIT	BARE COSTS MAT.	LABOR	EQUIP.	TOTAL	TOTAL INCL O&P	
152	0010	**CATCH BASINS OR MANHOLES** Including footing & excavation,										**152**
	0020	not including frame and cover										
	0050	Brick, 4' inside diameter, 4' deep	D-1	1	16	Ea.	385	325		710	940	
	0100	6' deep		.70	22.860		535	460		995	1,325	
	0150	8' deep		.50	32		680	645		1,325	1,775	
	0200	For depths over 8', add		4	4	V.L.F.	88	81		169	225	
	1110	Precast, 4' I.D., 4' deep	B-6	4.10	5.850	Ea.	258	110	46	414	510	
	1120	6' deep		3	8		358	150	63	571	705	
	1130	8' deep		2	12		446	225	95	766	960	
	1140	For depths over 8', add		16	1.500	V.L.F.	63	28	11.90	102.90	130	
	1600	Frames and covers, C.I., 24" square, 500 lb.		7.80	3.080	Ea.	172.20	58	24	254.20	310	
	1700	26" D shape, 600 lb.		7	3.430	"	194.50	64	27	285.50	350	
160	0010	**PIPING, DRAINAGE & SEWAGE, BITUMINOUS FIBER**										**160**
	2000	Plain, 2" diameter	2 Clab	400	.040	L.F.	.93	.70		1.63	2.17	
	2080	4" diameter	"	380	.042	"	1.48	.74		2.22	2.83	
162	0010	**PIPING, DRAINAGE & SEWAGE, CONCRETE**										**162**
	0020	Not including excavation or backfill										
	1000	Non-reinforced pipe, extra strength, B&S or T&G joints										
	1010	6" diameter	B-20	150	.160	L.F.	2.95	3.18		6.13	8.45	
	1020	8" diameter	B-21	200	.140		3.25	2.85	.52	6.62	8.75	
	1040	12" diameter	"	173	.162		4.40	3.30	.60	8.30	10.85	
	2000	Reinforced culvert, class 3, no gaskets										
	2010	12" diameter	B-21	190	.147	L.F.	5.20	3	.55	8.75	11.20	
	2020	15" diameter	"	155	.181	"	6.50	3.68	.67	10.85	13.85	
164	0010	**PIPING, DRAINAGE & SEWAGE, CORRUGATED METAL**										**164**
	0020	Not including excavation or backfill										
	2000	Corrugated Metal Pipe, galv. or aluminum,										
	2020	Bituminous coated with paved invert, 20' to 30' lengths										
	2040	8" diameter 16 ga.	B-20	330	.073	L.F.	5.10	1.45		6.55	7.95	
	2080	12" diameter 16 ga.	"	210	.114	"	7.80	2.27		10.07	12.30	
168	0010	**PIPING, DRAINAGE & SEWAGE, POLYVINYL CHLORIDE**										**168**
	0020	Not including excavation or backfill										
	2000	10' lengths, S.D.R. 35, 4" diameter	B-20	375	.064	L.F.	.68	1.27		1.95	2.82	
	2040	6" diameter		350	.069		1.65	1.36		3.01	4.04	
	2080	8" diameter		335	.072		2.35	1.42		3.77	4.91	
	2120	10" diameter	B-21	330	.085		3.79	1.73	.32	5.84	7.30	
172	0010	**PIPING, DRAINAGE & SEWAGE, VITRIFIED CLAY** C700										**172**
	0020	Not including excavation or backfill, 4' & 5' lengths										
	4030	Extra strength, compression joints, C425										
	5000	4" diameter	B-20	265	.091	L.F.	1.05	1.80		2.85	4.09	
	5020	6" diameter	"	200	.120		1.75	2.39		4.14	5.80	
	5040	8" diameter	B-21	200	.140		2.75	2.85	.52	6.12	8.20	
	5060	10" diameter	"	190	.147		4.50	3	.55	8.05	10.40	
	9000	Minimum labor/equipment charge	1 Clab	4	2	Job		35		35	57	

027 400 | Septic Systems

		CREW	DAILY OUTPUT	MAN-HOURS	UNIT	MAT.	LABOR	EQUIP.	TOTAL	TOTAL INCL O&P	
404	0010	**SEPTIC TANKS** Not incl. excav. or piping, precast, 1,000 gallon	B-21	8	3.500	Ea.	430	71	13	514	605	**404**
	0100	2,000 gallon		5	5.600		810	115	21	946	1,100	
	0600	Fiberglass, 1,000 gallon		6	4.670		485	95	17.35	597.35	705	
	0700	1,500 gallon		4	7		610	145	26	781	930	
	1000	Distribution boxes, concrete, 7 outlets	2 Clab	16	1		51	17.50		68.50	85	
	1420	Leaching pit, 6', dia, 3' deep complete									575	
	2200	Excavation for septic tank, ¾ C.Y. backhoe	B-12F	145	.110	C.Y.		2.34	3.15	5.49	7.15	
	2400	4' trench for disposal field, ¾ C.Y. backhoe	"	335	.048	L.F.		1.01	1.36	2.37	3.10	
	2600	Gravel fill, run of bank	B-6	150	.160	C.Y.	10.85	3.01	1.27	15.13	18.20	
	2800	Crushed stone, ¾"	"	150	.160	"	14.35	3.01	1.27	18.63	22	

For expanded coverage of these items see *Means Site Work Cost Data 1991*

028 100 | Irrigation Systems

		CREW	DAILY OUTPUT	MAN-HOURS	UNIT	MAT.	LABOR	EQUIP.	TOTAL	TOTAL INCL O&P	
104	0010 **SPRINKLER IRRIGATION SYSTEM** For lawns										**104**
	0100 Golf course with fully automatic system	C-17	.05	1,600	9 Holes	73,750	36,900		110,650	141,000	
	0200 24' diam. head at 15' O.C incl. piping, minimum	B-20	70	.343	Head	16	6.80		22.80	29	
	0300 Maximum		40	.600		37	11.95		48.95	60	
	0500 60' diameter head, automatic operation, minimum		28	.857		50	17.05		67.05	83	
	0600 Maximum		23	1.040		135	21		156	180	
	0800 Residential system, custom, 1" supply		2,619	.009	S.F.	.22	.18		.40	.54	
	0900 1-½" supply		2,311	.010	"	.20	.21		.41	.56	

028 300 | Fences And Gates

		CREW	DAILY OUTPUT	MAN-HOURS	UNIT	MAT.	LABOR	EQUIP.	TOTAL	TOTAL INCL O&P	
308	0010 **FENCE, CHAIN LINK INDUSTRIAL** 6' high plus 3 strands										**308**
	0020 barbed wire, 2" line post @ 10' O.C., 1-⅝" top rail										
	0200 9 ga. wire, galv. steel	B-80	200	.160	L.F.	5.60	3.06	2.03	10.69	13.30	
	0300 Aluminized steel		200	.160		7.05	3.06	2.03	12.14	14.90	
	0800 6 ga. wire, 6' high but omit barbed wire, galv. steel		180	.178		8.40	3.40	2.25	14.05	17.20	
	0900 Aluminized steel		180	.178		10	3.40	2.25	15.65	18.95	
	1100 Add for corner posts, 3" diam., galv. steel		40	.800	Ea.	49	15.30	10.15	74.45	90	
	1200 Aluminized steel		40	.800		62	15.30	10.15	87.45	105	
	1400 Gate for 6' high fence, 1-⅝" frame, 3' wide, galv. steel		10	3.200		72	61	41	174	220	
	1500 Aluminized steel		10	3.200		89	61	41	191	240	
	3100 Overhead slide gate, chain link, 6' high, to 18' wide		38	.842	L.F.	50	16.10	10.65	76.75	93	
	3110 Cantilever type		48	.667	"	34	12.75	8.45	55.20	67	
	3150 Tennis courts, 11 ga. wire, 1-¾" mesh, 2-½" line posts										
	3170 1-⅝" top rail, 3" corner and gate posts										
	3190 10' high, galvanized steel	B-80	190	.168	L.F.	8.50	3.22	2.13	13.85	16.90	
	3210 Vinyl covered 9 ga. wire		190	.168	"	9.45	3.22	2.13	14.80	17.95	
	3240 Corner posts for above, 3" diameter, 10 ft. high		30	1.070	Ea.	95	20	13.50	128.50	150	
	3300 Residential, 11 ga. wire, 1-⅝" line post @ 10' O.C.										
	3310 1-⅜" top rail										
	3350 4' high, galvanized steel	B-80	475	.067	L.F.	3.71	1.29	.85	5.85	7.10	
	3400 Aluminized		475	.067	"	4.35	1.29	.85	6.49	7.80	
	3600 Gate, 3' wide, 1-⅜" frame, galv. steel		10	3.200	Ea.	40	61	41	142	185	
	3700 Aluminized		10	3.200	"	50	61	41	152	200	
	3900 3' high, galvanized steel		620	.052	L.F.	3.06	.99	.65	4.70	5.70	
	4000 Aluminized		620	.052	"	3.71	.99	.65	5.35	6.40	
	4200 Gate, 3' wide, 1-⅜" frame, galv. steel		12	2.670	Ea.	32	51	34	117	155	
	4300 Aluminized		12	2.670	"	37	51	34	122	160	
	8000 Components only, fabric, galvanized, 4' high		585	.055	L.F.	1.10	1.05	.69	2.84	3.65	
	8040 6' high		430	.074	"	1.55	1.42	.94	3.91	5.05	
	8200 Posts, line post, 4' high		34	.941	Ea.	5	18	11.90	34.90	48	
	8240 6' high		26	1.230	"	6.65	24	15.60	46.25	62	
	8400 Top rails, 1-⅜" O.D.		1,000	.032	L.F.	.69	.61	.41	1.71	2.19	
	8440 1-⅝" O.D.		1,000	.032	"	.85	.61	.41	1.87	2.37	
	9000 Minimum labor/equipment charge		2	16	Job		305	205	510	715	
324	0010 **FENCE, WOOD** Picket, No. 2 cedar picket, Gothic, 2 rail, 3' high	B-1	160	.150	L.F.	3.79	2.73		6.52	8.60	**324**
	0050 Gate, 3'-6" wide		9	2.670	Ea.	33	48		81	115	
	0600 Open rail, rustic, No. 1 cedar, 2 rail, 3' high		160	.150	L.F.	2.99	2.73		5.72	7.75	
	0650 Gate, 3' wide		9	2.670	Ea.	38	48		86	120	
	1200 Stockade, No. 2 cedar, treated wood rails, 6' high		160	.150	L.F.	4.69	2.73		7.42	9.60	
	1250 Gate, 3' wide		9	2.670	Ea.	39	48		87	120	
	3300 Board, shadow box, 1" x 6", treated pine, 6' high		160	.150	L.F.	7.15	2.73		9.88	12.30	
	3400 No. 1 cedar, 6' high		150	.160		13.80	2.91		16.71	19.95	
	3900 Basket weave, No. 1 cedar, 6' high		160	.150		15.40	2.73		18.13	21	
	3950 Gate, 3'-6" wide		8	3	Ea.	96	55		151	195	
	4000 Treated pine, 6' high		150	.160	L.F.	7.75	2.91		10.66	13.30	
	4200 Gate, 3'-6" wide		9	2.670	Ea.	44	48		92	130	
	8000 Posts only, 4' high		40	.600		10.25	10.90		21.15	29	
	8040 6' high		30	.800		14.65	14.55		29.20	40	

For expanded coverage of these items see *Means Site Work Cost Data 1991*

028 300 | Fences And Gates

			CREW	DAILY OUTPUT	MAN-HOURS	UNIT	BARE COSTS MAT.	LABOR	EQUIP.	TOTAL	TOTAL INCL O&P	
324	9000	Minimum labor/equipment charge	1 Clab	2	4	Job		70		70	115	324

028 400 | Walk/Road/Parkg Appurt

			CREW	DAILY OUTPUT	MAN-HOURS	UNIT	MAT.	LABOR	EQUIP.	TOTAL	TOTAL INCL O&P	
404	0010	GUIDE RAIL Corrugated steel, galv. steel posts, 6'-3" O.C.	B-80	850	.038	L.F.	9.35	.72	.48	10.55	11.95	404
	0200	End sections, galvanized, flared		50	.640	Ea.	38	12.25	8.10	58.35	70	
	0300	Wrap around end		50	.640	"	48	12.25	8.10	68.35	81	
	0400	Timber guide rail, 4" x 8" with 6" x 8" wood posts, treated	↓	960	.033	L.F.	7	.64	.42	8.06	9.20	
408	0010	PARKING BARRIERS Timber with saddles, treated type										408
	1000	Wheel stops, precast concrete incl. dowels, 6" x 10" x 6'-0"	B-2	120	.333	Ea.	22	5.95		27.95	34	
416	0010	STEPS Incl. excav., borrow & concrete base, where applicable										416
	0100	Bricks	B-24	35	.686	LF Rsr	22	14		36	47	
	0200	Railroad ties	2 Clab	25	.640		7.90	11.20		19.10	27	
	0300	Bluestone treads, 12" x 2" or 12" x 1-½"	B-24	30	.800	↓	13	16.30		29.30	41	
	0490	Minimum labor/equipment charge	D-1	2	8	Job		160		160	260	
	0500	Concrete, cast in place, see division 033-130										
	0600	Precast concrete, see division 034-804										

028 700 | Site/Street Furnishings

			CREW	DAILY OUTPUT	MAN-HOURS	UNIT	MAT.	LABOR	EQUIP.	TOTAL	TOTAL INCL O&P	
704	0010	BENCHES Park, precast concrete, w/backs, wood rails, 4' long	2 Clab	5	3.200	Ea.	230	56		286	345	704
	0100	8' long		4	4		395	70		465	550	
	0300	Fiberglass, backless, one piece, 4' long		10	1.600		305	28		333	380	
	0400	8' long		7	2.290		645	40		685	775	
	0500	Steel barstock pedestals w/backs, 2" x 3" wood rails, 4' long		10	1.600		525	28		553	625	
	0510	8' long		7	2.290		625	40		665	755	
	0520	3" x 8" wood plank, 4' long		10	1.600		565	28		593	665	
	0530	8' long		7	2.290		660	40		700	790	
	0540	Backless, 4" x 4" wood plank, 4' square		10	1.600		550	28		578	650	
	0550	8' long		7	2.290		525	40		565	645	
	0600	Aluminum pedestals, with backs, aluminum slats, 8' long		8	2		140	35		175	210	
	0610	15' long		5	3.200		235	56		291	350	
	0620	Portable, aluminum slats, 8' long		8	2		160	35		195	235	
	0630	15' long		5	3.200		260	56		316	375	
	0800	Cast iron pedestals, back & arms, wood slats, 4' long		8	2		450	35		485	550	
	0820	8' long		5	3.200		755	56		811	920	
	0840	Backless, wood slats, 4' long		8	2		395	35		430	490	
	0860	8' long		5	3.200		660	56		716	815	
	1700	Steel frame, fir seat, 10' long		10	1.600	↓	130	28		158	190	
	9000	Minimum labor/equipment charge		2	8	Job		140		140	230	
716	0010	PLANTERS Concrete, sandblasted, precast, 48" diameter, 24" high		15	1.070	Ea.	410	18.65		428.65	480	716
	0100	Fluted, precast, 7' diameter, 36" high		10	1.600		1,725	28		1,753	1,950	
	0300	Fiberglass, circular, 36" diameter, 24" high		15	1.070		460	18.65		478.65	535	
	0400	60" diameter, 24" high	↓	10	1.600	↓	815	28		843	940	
	9000	Minimum labor/equipment charge	1 Clab	2	4	Job		70		70	115	

029 | Landscaping

029 100 | Shrub/Tree Transplanting

			CREW	DAILY OUTPUT	MAN-HOURS	UNIT	MAT.	LABOR	EQUIP.	TOTAL	TOTAL INCL O&P	
104	0010	TREE GUYING Including stakes, guy wire and wrap										104
	0100	Less than 3" caliper, 2 stakes	2 Clab	35	.457	Ea.	6.65	8		14.65	20	

For expanded coverage of these items see *Means Site Work Cost Data 1991*

029 | Landscaping

		029 100	Shrub/Tree Transplanting	CREW	DAILY OUTPUT	MAN-HOURS	UNIT	MAT.	LABOR	EQUIP.	TOTAL	TOTAL INCL O&P	
104	0200		3″ to 4″ caliper, 3 stakes	2 Clab	21	.762	Ea.	12.20	13.35		25.55	35	104
	1000		Including arrowhead anchor, cable, turnbuckles and wrap										
	1100		Less than 3″ caliper, 3″ anchors	2 Clab	20	.800	Ea.	38	14		52	65	
	1200		3″ to 6″ caliper, 4″ anchors		15	1.070		48	18.65		66.65	83	
	1300		6″ caliper, 6″ anchors		12	1.330		71	23		94	115	
	1400		8″ caliper, 8″ anchors		9	1.780		85	31		116	145	

029 200 | Soil Preparation

		Description	CREW	DAILY OUTPUT	MAN-HOURS	UNIT	MAT.	LABOR	EQUIP.	TOTAL	TOTAL INCL O&P	
208	0010	PLANT BED PREPARATION										208
	0100	Backfill planting pit, by hand, on site topsoil	2 Clab	18	.889	C.Y.		15.55		15.55	25	
	0200	Prepared planting mix	"	24	.667			11.65		11.65	19.05	
	0300	Skid steer loader, on site topsoil	B-62	340	.071			1.33	.26	1.59	2.42	
	0400	Prepared planting mix	"	410	.059			1.10	.22	1.32	2.01	
	1000	Excavate planting pit, by hand, sandy soil	2 Clab	16	1			17.50		17.50	29	
	1100	Heavy soil or clay	"	8	2			35		35	57	
	1200	½ C.Y. backhoe, sandy soil	B-11C	150	.107			2.13	1.27	3.40	4.81	
	1300	Heavy soil or clay	"	115	.139			2.78	1.66	4.44	6.25	
	2000	Mix planting soil, incl. loam, manure, peat, by hand	2 Clab	60	.267		23	4.67		27.67	33	
	2100	Skid steer loader	B-62	150	.160		23	3.01	.59	26.60	31	
	3000	Pile sod, skid steer loader	"	2,800	.009	S.Y.		.16	.03	.19	.29	
	3100	By hand	2 Clab	400	.040			.70		.70	1.14	
	4000	Remove sod, F.E. loader	B-10S	2,000	.006			.13	.15	.28	.37	
	4100	Sod cutter	B-12K	3,200	.005			.11	.24	.35	.43	
	4200	By hand	2 Clab	240	.067			1.17		1.17	1.90	

029 300 | Lawns & Grasses

		Description	CREW	DAILY OUTPUT	MAN-HOURS	UNIT	MAT.	LABOR	EQUIP.	TOTAL	TOTAL INCL O&P	
304	0010	SEEDING Mechanical seeding, $2.00/lb., 215 lb./acre ㉙	A-1	.31	25.810	Acre	435	450	175	1,060	1,400	304
	0100	$2.00/lb., 44 lb./M.S.Y.		1,550	.005	S.Y.	.08	.09	.04	.21	.28	
	0600	Limestone hand push spreader, 50 lbs. per M.S.F.		200	.040	M.S.F.	1.55	.70	.27	2.52	3.15	
	9000	Minimum labor/equipment charge	1 Clab	4	2	Job		35		35	57	
312	0010	SODDING In East, 1 inch deep, incl. fine grade, on level ground	B-14	1,000	.048	S.Y.	1.51	.89	.19	2.59	3.31	312
	0200	On slopes		800	.060		1.61	1.11	.24	2.96	3.83	
	1200	In Midwest on level ground, prepared area, over 400 S.Y.		840	.057		.98	1.06	.23	2.27	3.04	
	1230	100 S.Y. area		800	.060		1.36	1.11	.24	2.71	3.56	
	1260	50 S.Y. area		750	.064		1.95	1.18	.25	3.38	4.35	
	1300	On slopes, 400 S.Y. area		720	.067		1.03	1.23	.26	2.52	3.42	
	1700	Polyurethane with ceramic chips for median strip, minimum		1,153	.042	S.F.	.58	.77	.17	1.52	2.07	
	1800	Maximum		875	.055	"	.82	1.01	.22	2.05	2.78	
	9000	Minimum labor/equipment charge	1 Clab	2	4	Job		70		70	115	

029 500 | Trees/Plants/Grnd Cover

		Description	CREW	DAILY OUTPUT	MAN-HOURS	UNIT	MAT.	LABOR	EQUIP.	TOTAL	TOTAL INCL O&P	
504	0010	GROUND COVER Plants, pachysandra, in prepared beds	B-1	10	2.400	C	12.70	44		56.70	85	504
	0200	Vinca minor, 1 yr, bare root		10	2.400	"	47	44		91	125	
	0600	Stone chips, in 50 lb. bags, Georgia marble		520	.046	Bag	2.75	.84		3.59	4.40	
	0700	Onyx gemstone		260	.092		10.50	1.68		12.18	14.30	
	0800	Quartz		260	.092		4.68	1.68		6.36	7.90	
	0900	Pea gravel, truckload lots		28	.857	C.Y.	16.70	15.55		32.25	44	
520	0010	PLANTING Moving shrubs on site, 12″ ball ㉚		28	.857	Ea.		15.55		15.55	25	520
	0100	24″ ball		22	1.090			19.80		19.80	32	
	0300	Moving trees on site, 36″ ball	B-6	3.75	6.400			120	51	171	250	
	0400	60″ ball	"	1	24			450	190	640	935	
524	0010	SHRUBS Broadleaf evergreen, planted in prepared beds	B-1	96	.250	Ea.	16.45	4.54		20.99	26	524
	0100	Andromeda, 15″-18″, container		96	.250		16.45	4.54		20.99	26	
	0200	Azalea, 15″-18″, container		96	.250		16.30	4.54		20.84	25	
	0300	Barberry, 9″-12″, container		130	.185		5.80	3.35		9.15	11.85	
	0400	Boxwood, 15″-18″, B & B		96	.250		22	4.54		26.54	32	
	0500	Euonymus, emerald gaiety, 12″ to 15″, container		115	.209		8.40	3.79		12.19	15.45	

For expanded coverage of these items see *Means Site Work Cost Data 1991*

029 500	Trees/Plants/Grnd Cover	CREW	DAILY OUTPUT	MAN-HOURS	UNIT	BARE COSTS MAT.	LABOR	EQUIP.	TOTAL	TOTAL INCL O&P	
524	0600	Holly, 15"-18", B & B	B-1	96	.250	Ea.	16.60	4.54		21.14	26	**524**
	0900	Mount laurel, 18"-24", B & B		80	.300		18.75	5.45		24.20	30	
	1000	Privet, 18" to 24" high		130	.185		2.25	3.35		5.60	7.95	
	1100	Rhododendron, 18"-24", container		48	.500		18.50	9.10		27.60	35	
	1200	Rosemary, 1 gal container		600	.040		3.60	.73		4.33	5.15	
	2000	Deciduous, amelanchier, 2'-3', B & B		57	.421		14.65	7.65		22.30	29	
	2100	Azalea, 15"-18", B & B		96	.250		13.65	4.54		18.19	22	
	2300	Bayberry, 2'-3', B & B		57	.421		17.55	7.65		25.20	32	
	2600	Cotoneaster, 15"-18", B & B	↓	80	.300		8.25	5.45		13.70	18	
	2800	Dogwood, 3'-4', B & B	B-17	40	.800		14.30	14.95	12.25	41.50	53	
	2900	Euonymus, alatus compacta, 15" to 18", container	B-1	80	.300		9.20	5.45		14.65	19.05	
	3200	Forsythia, 2'-3', container	"	60	.400		9.70	7.25		16.95	23	
	3300	Hibiscus, 3'-4', B & B	B-17	75	.427		25	8	6.50	39.50	48	
	3400	Honeysuckle, 3'-4', B & B	B-1	60	.400		25	7.25		32.25	39	
	3500	Hydrangea, 2'-3', B & B	"	57	.421		10.20	7.65		17.85	24	
	3600	Lilac, 3'-4', B & B	B-17	40	.800		20	14.95	12.25	47.20	60	
	3900	Privet, bare root, 18"-24"	B-1	80	.300		2.50	5.45		7.95	11.65	
	4100	Quince, 2'-3', B & B	"	57	.421		16.39	7.65		24.04	31	
	4200	Russian olive, 3'-4', B & B	B-17	75	.427		15.25	8	6.50	29.75	37	
	4400	Spirea, 3'-4', B & B	B-1	70	.343		16.20	6.25		22.45	28	
	4500	Viburnum, 3'-4', B & B	B-17	40	.800	↓	16.20	14.95	12.25	43.40	55	
528	0010	**SHRUBS AND TREES** Evergreen, in prepared beds, B & B										**528**
	0100	Arborvitae pyramidal, 4'-5'	B-17	30	1.070	Ea.	26	19.95	16.30	62.25	79	
	0150	Globe, 12"-15"	B-1	96	.250		8.65	4.54		13.19	16.95	
	0300	Cedar, blue, 8'-10'	B-17	18	1.780		105	33	27	165	200	
	0500	Hemlock, canadian, 2-½'-3'	B-1	36	.667		18.35	12.10		30.45	40	
	0600	Juniper, andora, 18"-24"		80	.300		9.20	5.45		14.65	19.05	
	0620	Wiltoni, 15"-18"	↓	80	.300		13.60	5.45		19.05	24	
	0640	Skyrocket, 4-½'-5'	B-17	55	.582		33	10.90	8.90	52.80	64	
	0660	Blue pfitzer, 2'-2-½'	B-1	44	.545		14.30	9.90		24.20	32	
	0680	Ketleerie, 2-½'-3'		50	.480		22	8.70		30.70	38	
	0700	Pine, black, 2-½'-3'		50	.480		25	8.70		33.70	42	
	0720	Mugo, 18"-24"	↓	60	.400		28	7.25		35.25	43	
	0740	White, 4'-5'	B-17	75	.427		31	8	6.50	45.50	54	
	0800	Spruce, blue, 18"-24"	B-1	60	.400		20	7.25		27.25	34	
	0840	Norway, 4'-5'	B-17	75	.427		39	8	6.50	53.50	63	
	0900	Yew, denisforma, 12"-15"	B-1	60	.400		10.35	7.25		17.60	23	
	1000	Capitata, 18"-24"		30	.800		15.30	14.55		29.85	41	
	1100	Hicksi, 2'-2-½'	↓	30	.800	↓	16.35	14.55		30.90	42	
536	0010	**TREES** Deciduous, in prep. beds, balled & burlapped (B&B)										**536**
	0100	Ash, 2" caliper	B-17	8	4	Ea.	76	75	61	212	270	
	0200	Beech, 5'-6'		50	.640		39	11.95	9.80	60.75	73	
	0300	Birch, 6'-8', 3 stems ㉚		20	1.600		45	30	24	99	125	
	0500	Crabapple, 6'-8'		20	1.600		40	30	24	94	120	
	0600	Dogwood, 4'-5'		40	.800		36	14.95	12.25	63.20	77	
	0700	Eastern redbud 4'-5'		40	.800		51	14.95	12.25	78.20	94	
	0800	Elm, 8'-10'		20	1.600		60	30	24	114	140	
	0900	Ginkgo, 6'-7'		24	1.330		51	25	20	96	120	
	1000	Hawthorn, 8'-10', 1" caliper		20	1.600		55	30	24	109	135	
	1100	Honeylocust, 10'-12', 1-½" caliper		10	3.200		43	60	49	152	195	
	1300	Larch, 8'		32	1		26	18.70	15.30	60	75	
	1400	Linden, 8'-10', 1" caliper		20	1.600		49	30	24	103	130	
	1500	Magnolia, 4'-5'		20	1.600		46	30	24	100	125	
	1600	Maple, red, 8'-10', 1-½" caliper		10	3.200		45	60	49	154	200	
	1700	Mountain ash, 8'-10', 1" caliper		16	2		50	37	31	118	150	
	1800	Oak, 2-½"-3" caliper		3	10.670		170	200	165	535	685	
	2100	Planetree, 9'-11', 1-¼" caliper	↓	10	3.200	↓	69	60	49	178	225	

For expanded coverage of these items see *Means Site Work Cost Data 1991*

029 500 Trees/Plants/Grnd Cover		CREW	DAILY OUTPUT	MAN-HOURS	UNIT	BARE COSTS				TOTAL INCL O&P	
						MAT.	LABOR	EQUIP.	TOTAL		
536 2200	Plum, 6'-8', 1" caliper	B-17	20	1.600	Ea.	67	30	24	121	150	**536**
2300	Poplar, 9'-11', 1-¼" caliper		10	3.200		41	60	49	150	195	
2500	Sumac, 2'-3'		75	.427		11.50	8	6.50	26	33	
2700	Tulip, 5'-6'		40	.800		35	14.95	12.25	62.20	76	
2800	Willow, 6'-8', 1" caliper	↓	20	1.600	↓	25	30	24	79	105	
9000	Minimum labor equipment charge	1 Clab	4	2	Job		35		35	57	

031 100 Struct C.I.P. Formwork		CREW	DAILY OUTPUT	MAN-HOURS	UNIT	BARE COSTS				TOTAL INCL O&P	
						MAT.	LABOR	EQUIP.	TOTAL		
132 0010	**EXPANSION JOINT** Keyed cold joint, 24 ga., incl. stakes, 3-½" high	1 Carp	200	.040	L.F.	.82	.88		1.70	2.34	**132**
0050	4-½" high		200	.040		.89	.88		1.77	2.42	
0100	5-½" high		195	.041		1.08	.90		1.98	2.66	
0150	7-½" high	↓	190	.042		1.38	.93		2.31	3.03	
0300	Poured asphalt, plain, ½" x 1"	1 Clab	450	.018		.24	.31		.55	.77	
0350	1" x 2"		400	.020		.90	.35		1.25	1.56	
0500	Neoprene, liquid, cold applied, ½" x 1"		450	.018		1.39	.31		1.70	2.04	
0550	1" x 2"		400	.020		5.15	.35		5.50	6.25	
0700	Polyurethane, 1 part, ½" x 1"		400	.020		1.64	.35		1.99	2.38	
0750	1" x 2"		350	.023		6.50	.40		6.90	7.80	
0900	Rubberized asphalt, hot or cold applied, ½" x 1"		450	.018		.29	.31		.60	.83	
0950	1" x 2"		400	.020		.92	.35		1.27	1.58	
1100	Hot applied, fuel resistant, ½" x 1"		450	.018		.35	.31		.66	.89	
1150	1" x 2"	↓	400	.020		.97	.35		1.32	1.64	
2000	Premolded, bituminous fiber, ½" x 6"	1 Carp	375	.021		.48	.47		.95	1.29	
2050	1" x 12"		300	.027		2.15	.59		2.74	3.32	
2250	Cork with resin binder, ½" x 6"		375	.021		.95	.47		1.42	1.81	
2300	1" x 12"		300	.027		3.60	.59		4.19	4.92	
2500	Neoprene sponge, closed cell, ½" x 6"		375	.021		1.30	.47		1.77	2.20	
2550	1" x 12"		300	.027		5	.59		5.59	6.45	
2750	Polyethylene foam, ½" x 6"		375	.021		.90	.47		1.37	1.76	
2800	1" x 12"		300	.027		3.40	.59		3.99	4.70	
3000	Polyethylene backer rod, ⅜" diameter		460	.017		.07	.38		.45	.70	
3050	¾" diameter		460	.017		.11	.38		.49	.75	
3100	1" diameter	↓	460	.017	↓	.15	.38		.53	.79	
3110											
3500	Polyurethane foam, with polybutylene, ½" x ½"	1 Carp	475	.017	L.F.	.67	.37		1.04	1.34	
3550	1" x 1"		450	.018		1.33	.39		1.72	2.10	
3750	Polyurethane foam, regular, closed cell, ½" x 6"		375	.021		.61	.47		1.08	1.44	
3800	1" x 12"		300	.027		2.25	.59		2.84	3.43	
4000	Polyvinyl chloride foam, closed cell, ½" x 6"		375	.021		1.42	.47		1.89	2.33	
4050	1" x 12"		300	.027		4.90	.59		5.49	6.35	
4250	Rubber, gray sponge, ½" x 6"		375	.021		2.40	.47		2.87	3.41	
4300	1" x 12"	↓	300	.027	↓	9.55	.59		10.14	11.45	
4500	Lead wool for joints, 1 ton lots				Lb.	1.75			1.75	1.93	
4550	Retail				"	1.95			1.95	2.15	
5000	For installation in walls, add				L.F.		75%				
5250	For installation in boxouts, add				"		25%				
138 0010	**FORMS IN PLACE, BEAMS AND GIRDERS** ㉜										**138**
0030											
0500	Beam and Girder, exterior spandrel, 12" wide, 1 use ㉝	C-2	225	.213	SFCA	2.19	4.60	.14	6.93	10.10	
0650	4 use	"	310	.155	"	.74	3.34	.10	4.18	6.40	

For expanded coverage of these items see *Means Concrete Cost Data 1991*

		031 100	Struct C.I.P. Formwork	CREW	DAILY OUTPUT	MAN-HOURS	UNIT	MAT.	LABOR	EQUIP.	TOTAL	TOTAL INCL O&P	
138	1000	Exterior spandrel, 18" wide, 1 use		C-2	250	.192	SFCA	1.82	4.14	.13	6.09	8.90	**138**
	1150	4 use			315	.152		.67	3.29	.10	4.06	6.25	
	1500	Exterior spandrel, 24" wide, 1 use			265	.181		1.82	3.91	.12	5.85	8.50	
	1650	4 use			325	.148		.67	3.19	.10	3.96	6.05	
	2000	Interior beams, 12" wide, 1 use			300	.160		1.75	3.45	.11	5.31	7.70	
	2150	4 use			377	.127		.71	2.75	.08	3.54	5.35	
	2500	Interior beams, 24" wide, 1 use			320	.150		1.71	3.24	.10	5.05	7.30	
	2650	4 use			395	.122		.64	2.62	.08	3.34	5.05	
	3000	Beam and Girder, encasing steel frame, hung, 1 use			325	.148		1.72	3.19	.10	5.01	7.20	
	3150	4 use			430	.112		.65	2.41	.07	3.13	4.74	
	9000	Minimum labor/equipment charge		F-2	2	8	Job		175	8	183	295	
142	0010	**FORMS IN PLACE, COLUMNS**											**142**
	0020												
	0500	Round fiberglass, 4 use per mo., rent, 12" diameter		C-1	160	.200	L.F.	1.60	4.18	.15	5.93	8.75	
	0650	24" diameter			135	.237		2.60	4.95	.18	7.73	11.15	
	1500	Round fiber tube, 1 use, 8" diameter			155	.206		1.76	4.31	.15	6.22	9.15	
	1700	16" diameter			140	.229		5.50	4.77	.17	10.44	14.05	
	2200	For seamless type, add						15%					
	3000	Round, steel, 4 use per mo., rent, regular duty, 12" diam.		C-1	145	.221		2.25	4.61	.17	7.03	10.20	
	3150	24" diameter			85	.376		2.77	7.85	.28	10.90	16.20	
	4000	Column capitals, 4 use per mo., 24" col, 4' cap diameter			12	2.670	Ea.	15.40	56	2	73.40	110	
	5000	Plywood, 8" x 8" columns, 1 use (32)			165	.194	SFCA	1.85	4.05	.15	6.05	8.80	
	5150	4 use			215	.149		.65	3.11	.11	3.87	5.90	
	6000	16" x 16" plywood columns, 1 use (31)			185	.173		1.75	3.61	.13	5.49	7.95	
	6150	4 use			235	.136		.59	2.84	.10	3.53	5.40	
	9000	Minimum labor/equipment charge		F-2	2	8	Job		175	8	183	295	
150	0010	**FORMS IN PLACE, ELEVATED SLABS**											**150**
	0020												
	1000	Flat plate to 15' high, 1 use		C-2	470	.102	S.F.	1.60	2.20	.07	3.87	5.45	
	1150	4 use			560	.086		.58	1.85	.06	2.49	3.73	
	1500	15' to 20' high ceilings, 4 use (33)			495	.097		.68	2.09	.06	2.83	4.24	
	2000	Flat slab with drop panels, to 15' high, 1 use			445	.108		1.80	2.33	.07	4.20	5.85	
	2150	4 use			544	.088		.74	1.90	.06	2.70	3.99	
	2250	15' to 20' high ceilings, 4 use			480	.100		.97	2.16	.07	3.20	4.67	
	3500	Floor slab, with 20" metal pans, 1 use (36)			350	.137		2.63	2.96	.09	5.68	7.85	
	3650	4 use			415	.116		.92	2.50	.08	3.50	5.15	
	4500	With 30" fiberglass domes, 1 use			405	.119		2.68	2.56	.08	5.32	7.20	
	4550	4 use			470	.102		1.06	2.20	.07	3.33	4.85	
	5000	Box out for slab openings, over 16" deep, 1 use			190	.253	SFCA	2.10	5.45	.17	7.72	11.40	
	5050	2 use			240	.200	"	1.24	4.32	.13	5.69	8.55	
	5500	Shallow slab box outs, to 10 S.F.			42	1.140	Ea.	7.82	25	.76	33.58	50	
	5550	Over 10 S.F. (use perimeter)			400	.120	L.F.	.91	2.59	.08	3.58	5.30	
	6000	Bulkhead forms for slab, with keyway, 1 use, 2 piece			500	.096		1.08	2.07	.06	3.21	4.65	
	6100	3 piece (see also edge forms)			460	.104		1.45	2.25	.07	3.77	5.35	
	6500	Curb forms, wood, 6" to 12" high, on elevated slabs, 1 use		C-1	180	.178	SFCA	1.61	3.71	.13	5.45	8	
	6550	2 use			205	.156		.84	3.26	.12	4.22	6.35	
	6600	3 use			220	.145		.60	3.04	.11	3.75	5.75	
	6650	4 use			225	.142		.49	2.97	.11	3.57	5.50	
	7000	Edge forms to 6" high, on elevated slab, 4 use			500	.064	L.F.	.27	1.34	.05	1.66	2.54	
	7100	7" to 12" high, 4 use			350	.091	SFCA	.47	1.91	.07	2.45	3.71	
	7500	Depressed area forms to 12" high, 4 use			300	.107	L.F.	.48	2.23	.08	2.79	4.26	
	7550	12" to 24" high, 4 use			175	.183		.65	3.82	.14	4.61	7.10	
	8000	Perimeter deck and rail for elevated slabs, straight			90	.356		6.30	7.40	.27	13.97	19.35	
	8050	Curved			65	.492		8.66	10.30	.37	19.33	27	
	8500	Void forms, round fiber, 3" diameter			450	.071		.40	1.48	.05	1.93	2.92	
	8650	8" diameter, void			375	.085		1.22	1.78	.06	3.06	4.32	

For expanded coverage of these items see *Means Concrete Cost Data 1991*

031 100 | Struct C.I.P. Formwork

		Description	CREW	DAILY OUTPUT	MAN-HOURS	UNIT	MAT.	LABOR	EQUIP.	TOTAL	TOTAL INCL O&P	
150	9000	Minimum labor/equipment charge	F-2	2	8	Job		175	8	183	295	150
154	0010	**FORMS IN PLACE, EQUIPMENT FOUNDATIONS** 1 use	C-2	160	.300	SFCA	1.64	6.50	.20	8.34	12.60	154
	0150	4 use	"	205	.234	"	.58	5.05	.16	5.79	9.05	
	9000	Minimum labor/equipment charge	F-1	3	2.670	Job		59	2.67	61.67	99	
158	0010	**FORMS IN PLACE, FOOTINGS** Continuous wall, 1 use	C-1	375	.085	SFCA	1.05	1.78	.06	2.89	4.14	158
	0150	4 use	"	485	.066	"	.35	1.38	.05	1.78	2.70	
	1500	Keyway, 4 use, tapered wood, 2" x 4"	1 Carp	530	.015	L.F.	.07	.33		.40	.62	
	1550	2" x 6"	"	500	.016	"	.10	.35		.45	.68	
	3000	Pile cap, square or rectangular, 1 use	C-1	290	.110	SFCA	1.42	2.30	.08	3.80	5.40	
	3150	4 use		380	.084		.51	1.76	.06	2.33	3.50	
	5000	Spread footings, 1 use ③③		305	.105		1.05	2.19	.08	3.32	4.83	
	5150	4 use		415	.077		.38	1.61	.06	2.05	3.11	
	9000	Minimum labor/equipment charge	F-1	3	2.670	Job		59	2.67	61.67	99	
162	0010	**FORMS IN PLACE, GRADE BEAM** 1 use	C-2	530	.091	SFCA	1.30	1.95	.06	3.31	4.69	162
	0150	4 use	"	605	.079	"	.50	1.71	.05	2.26	3.41	
	9000	Minimum labor/equipment charge	F-2	2	8	Job		175	8	183	295	
170	0010	**FORMS IN PLACE, SLAB ON GRADE**										170
	1000	Bulkhead forms with keyway, 1 use, 2 piece	C-1	510	.063	L.F.	.44	1.31	.05	1.80	2.67	
	2000	Curb forms, wood, 6" to 12" high, on grade, 1 use		215	.149	SFCA	1.34	3.11	.11	4.56	6.65	
	2150	4 use		275	.116	"	.44	2.43	.09	2.96	4.55	
	3000	Edge forms, to 6" high, 4 use, on grade		600	.053	L.F.	.23	1.11	.04	1.38	2.11	
	3050	7" to 12" high, 4 use, on grade		435	.074	SFCA	.70	1.54	.06	2.30	3.34	
	3500	For depressed slabs, 4 use, to 12" high		300	.107	L.F.	.51	2.23	.08	2.82	4.29	
	3550	To 24" high		175	.183		.65	3.82	.14	4.61	7.10	
	4000	For slab blockouts, 1 use to 12" high		200	.160		.51	3.34	.12	3.97	6.15	
	4050	To 24" high		120	.267		.65	5.55	.20	6.40	10.05	
	5000	Screed, 24 ga. metal key joint										
	5020	Wood, incl. wood stakes, 1" x 3"	C-1	900	.036	L.F.	.31	.74	.03	1.08	1.58	
	5050	2" x 4"		900	.036	"	.87	.74	.03	1.64	2.20	
	6000	Trench forms in floor, 1 use		160	.200	SFCA	1.67	4.18	.15	6	8.85	
	6150	4 use		185	.173	"	.54	3.61	.13	4.28	6.65	
	9000	Minimum labor/equipment charge	F-1	2	4	Job		88	4	92	150	
182	0010	**FORMS IN PLACE, WALLS**										182
	0100	Box out for wall openings, to 16" thick, to 10 S.F.	C-2	24	2	Ea.	14.70	43	1.33	59.03	88	
	0150	Over 10 S.F. (use perimeter)	"	280	.171	L.F.	1.35	3.70	.11	5.16	7.65	
	0250	Brick shelf, 4" wide, add to wall forms, use wall area										
	0260	above shelf, 1 use	C-2	240	.200	SFCA	1.44	4.32	.13	5.89	8.80	
	0350	4 use		300	.160	"	.58	3.45	.11	4.14	6.40	
	0500	Bulkhead forms for walls, with keyway, 1 use, 2 piece		265	.181	L.F.	1.75	3.91	.12	5.78	8.45	
	0550	3 piece		175	.274	"	2.20	5.90	.18	8.28	12.30	
	0700	Buttress forms, to 8' high, 1 use		350	.137	SFCA	1.64	2.96	.09	4.69	6.75	
	0850	4 use		480	.100	"	.60	2.16	.07	2.83	4.26	
	1000	Corbel (haunch) forms, up to 12" wide, add to wall forms, 1 use		150	.320	L.F.	1.71	6.90	.21	8.82	13.40	
	1150	4 use		180	.267	"	.60	5.75	.18	6.53	10.25	
	2000	Job built plyform wall forms, to 8' high, 1 use ③③		370	.130	SFCA	1.50	2.80	.09	4.39	6.30	
	2150	4 use		505	.095		.62	2.05	.06	2.73	4.10	
	2400	Over 8' to 16' high, 1 use ③④		280	.171		1.69	3.70	.11	5.50	8.05	
	2550	4 use		395	.122		.64	2.62	.08	3.34	5.05	
	2700	Over 16' high, 1 use		235	.204		1.87	4.41	.14	6.42	9.40	
	2850	4 use		330	.145		.75	3.14	.10	3.99	6.05	
	3000	For architectural finish, add		1,820	.026		.41	.57	.02	1	1.40	
	3101											
	3300	For battered walls, 1 side battered, add				SFCA	10%	10%				
	3350	2 sides battered, add				"	15%	15%				

For expanded coverage of these items see *Means Concrete Cost Data 1991*

031 100 | Struct C.I.P. Formwork

		CREW	DAILY OUTPUT	MAN-HOURS	UNIT	MAT.	LABOR	EQUIP.	TOTAL	TOTAL INCL O&P	
182											**182**
4000	Radial wall forms, smooth curved, 1 use	C-2	245	.196	SFCA	1.80	4.23	.13	6.16	9.05	
4150	4 use		335	.143		.71	3.09	.10	3.90	5.95	
4600	Retaining wall forms, battered, to 8' high, 1 use		300	.160		1.75	3.45	.11	5.31	7.70	
4750	4 use		390	.123		.70	2.66	.08	3.44	5.20	
4900	Over 8' to 16' high, 1 use		240	.200		1.93	4.32	.13	6.38	9.30	
5050	4 use		320	.150		.71	3.24	.10	4.05	6.20	
5200	For elevated walls, add to above						10%				
5750	Liners for forms (add to wall forms), A.B.S. plastic										
5800	Aged wood, 4" wide, 1 use	1 Carp	250	.032	SFCA	4.15	.70		4.85	5.70	
5820	2 use		400	.020		2.30	.44		2.74	3.25	
5840	4 use		750	.011		1.35	.23		1.58	1.87	
5850											
5900	Fractured rope rib, 1 use	1 Carp	250	.032	SFCA	6.50	.70		7.20	8.30	
6000	4 use		750	.011		2.10	.23		2.33	2.69	
6100	Ribbed look, ½" & ¾" deep, 1 use		300	.027		4.50	.59		5.09	5.90	
6200	4 use		800	.010		1.40	.22		1.62	1.90	
6300	Rustic brick pattern, 1 use		250	.032		4.20	.70		4.90	5.75	
6400	4 use		750	.011		1.37	.23		1.60	1.89	
6500	Striated, random, ⅜" x ⅜" deep, 1 use		300	.027		4.45	.59		5.04	5.85	
6600	4 use		800	.010		1.45	.22		1.67	1.95	
7500	Lintel or sill forms, 1 use		30	.267		1.70	5.85		7.55	11.45	
7560	4 use		37	.216		.55	4.76		5.31	8.35	
7800	Modular prefabricated plywood, to 8' high, 1 use per month	C-2	910	.053		.83	1.14	.04	2.01	2.81	
7860	4 use per month		970	.049		.28	1.07	.03	1.38	2.09	
8000	To 16' high, 1 use per month		550	.087		1.12	1.88	.06	3.06	4.37	
8060	4 use per month		610	.079		.40	1.70	.05	2.15	3.27	
8600	Pilasters, 1 use		270	.178		1.75	3.84	.12	5.71	8.35	
8660	4 use		385	.125		.76	2.69	.08	3.53	5.35	
9100	Minimum labor/equipment charge	F-2	2	8	Job		175	8	183	295	
186	**GAS STATION FORMS** Curb fascia, with template,										**186**
0010											
0050	12 ga. steel, left in place, 9" high	1 Carp	50	.160	L.F.	6.90	3.52		10.42	13.35	
1000	Sign or light bases, 18" diameter, 9" high		9	.889	Ea.	32	19.55		51.55	67	
1050	30" diameter, 13" high		8	1	"	56	22		78	98	
1990	Minimum labor/equipment charge		2	4	Job		88		88	145	
1995											
2000	Island forms, 10' long, 9" high, 3'- 6" wide	C-1	10	3.200	Ea.	140	67	2.40	209.40	265	
2050	4' wide		9	3.560		155	74	2.67	231.67	295	
2500	20' long, 9" high, 4' wide		6	5.330		255	110	4	369	465	
2550	5' wide		5	6.400		290	135	4.80	429.80	545	
9000	Minimum labor/equipment charge	F-1	3	2.670	Job		59	2.67	61.67	99	
192	**SHORES** Erect and strip, by hand, horizontal members										**192**
0010											
1000	Vertical members to 10' high	2 Carp	55	.291	Ea.		6.40		6.40	10.45	
1050	To 13' high		50	.320			7.05		7.05	11.50	
1100	To 16' high		45	.356			7.80		7.80	12.75	
1500	Reshoring		1,400	.011	S.F.	.14	.25		.39	.56	
198	**WATERSTOP** Polyvinyl chloride, ribbed 3/16" thick, 4" wide	1 Carp	155	.052	L.F.	1.29	1.14		2.43	3.27	**198**
0010											
0050	3/16" thick, 6" wide		145	.055		1.65	1.21		2.86	3.80	
0500	Ribbed, PVC, with center bulb, 3/16" thick, 9" wide		135	.059		2	1.30		3.30	4.33	
0550	⅜" thick		130	.062		3.20	1.35		4.55	5.75	

032 | Concrete Reinforcement

032 100 | Reinforcing Steel

			CREW	DAILY OUTPUT	MAN-HOURS	UNIT	BARE COSTS				TOTAL INCL O&P
							MAT.	LABOR	EQUIP.	TOTAL	
107	0010	**REINFORCING IN PLACE** A615 Grade 60										107
	0100	Beams & Girders, #3 to #7	4 Rodm	3,200	.010	Lb.	.28	.24		.52	.72
	0150	#8 to #18		5,400	.006		.27	.14		.41	.54
	0200	Columns, #3 to #7		3,000	.011		.28	.25		.53	.75
	0250	#8 to #18		4,600	.007		.27	.17		.44	.59
	0400	Elevated slabs, #4 to #7		5,800	.006		.28	.13		.41	.54
	0500	Footings, #4 to #7		4,200	.008		.27	.18		.45	.61
	0550	#8 to #18		7,200	.004		.27	.11		.38	.48
	0600	Slab on grade, #3 to #7		4,200	.008		.28	.18		.46	.63
	0700	Walls, #3 to #7		6,000	.005		.28	.13		.41	.53
	0750	#8 to #18	↓	8,000	.004	↓	.27	.10		.37	.46
	2000	Unloading & sorting, add to above	C-5	100	.560	Ton		13.10	4.86	17.96	28
	2200	Crane cost for handling, add to above, minimum		135	.415			9.70	3.60	13.30	20
	2210	Average		92	.609			14.25	5.30	19.55	30
	2220	Maximum	↓	35	1.600	↓		37	13.90	50.90	79
	2400	Dowels, 2 feet long, deformed, #3 bar	2 Rodm	140	.114	Ea.	.82	2.73		3.55	5.65
	2410	#4 bar		125	.128		.97	3.06		4.03	6.40
	2420	#5 bar		110	.145		1.18	3.48		4.66	7.35
	2430	#6 bar		105	.152	↓	1.59	3.64		5.23	8.10
	2450	Longer and heavier dowels		450	.036	Lb.	.39	.85		1.24	1.91
	2500	Smooth dowels, 12" long, ¼" or ⅜" diameter		140	.114	Ea.	.60	2.73		3.33	5.40
	2520	⅝" diameter		125	.128		1.05	3.06		4.11	6.50
	2530	¾" diameter	↓	110	.145	↓	1.29	3.48		4.77	7.50
	2550										
	2700	Dowel caps, 5" long, ½" to ¾" diameter	2 Rodm	800	.020	Ea.	.06	.48		.54	.90
	2720	1-¼" diameter	"	750	.021	"	.08	.51		.59	.98
	9000	Minimum labor/equipment charge	1 Rodm	4	2	Job		48		48	83

032 200 | Welded Wire Fabric

			CREW	DAILY OUTPUT	MAN-HOURS	UNIT	BARE COSTS				TOTAL INCL O&P
							MAT.	LABOR	EQUIP.	TOTAL	
207	0010	**WELDED WIRE FABRIC** Rolls, 6 x 6 #10/10 (W1.4/W1.4) 21 lb.	2 Rodm	35	.457	C.S.F.	7.10	10.95		18.05	27	207
	0300	6 x 6 - #6/6 (W2.9/W2.9) 42 lb. per C.S.F.		29	.552		13.80	13.20		27	38
	0500	4 x 4 - #10/10 (W1.4/W1.4) 31 lb. per C.S.F.		31	.516		10.90	12.35		23.25	33
	0900	2 x 2 - #12 galv. for gunite reinforcing	↓	6.50	2.460	↓	17.95	59		76.95	120
	0950	Material prices for above include 10% lap									
	9000	Minimum labor/equipment charge	1 Rodm	4	2	Job		48		48	83

033 | Cast-In-Place Concrete

033 100 | Structural Concrete

			CREW	DAILY OUTPUT	MAN-HOURS	UNIT	BARE COSTS				TOTAL INCL O&P
							MAT.	LABOR	EQUIP.	TOTAL	
118	0010	**CONCRETE ADMIXTURES & SURFACE TREATMENTS**										118
	0020										
	0040	Abrasives, aluminum oxide, over 20 tons				Lb.	.75			.75	.83
	0070	Under 1 ton					.80			.80	.88
	0100	Silicon carbide, black, over 20 tons					1			1	1.10
	0120	Under 1 ton				↓	1.05			1.05	1.16
	0200	Air entraining agent, .7 to 1.5 oz. per bag, 55 gallon lots				Gal.	4.10			4.10	4.51
	0220	5 gallon lots					5.10			5.10	5.60
	0300	Bonding agent, acrylic latex (200-250 S.F. per gallon)					20			20	22
	0320	Epoxy resin (70-80 S.F. per gallon)				↓	49			49	54
	0400	Calcium chloride, 100 lb. bags, FOB plant, truckload lots				Ton	291			291	320
	0420	Less than truckload lots				Bag	16.75			16.75	18.45

For expanded coverage of these items see *Means Concrete Cost Data 1991*

		033 100 | Structural Concrete		DAILY OUTPUT	MAN-HOURS	UNIT	BARE COSTS				TOTAL INCL O&P	
			CREW				MAT.	LABOR	EQUIP.	TOTAL		
118	0500	Carbon black, liquid, 2 to 8 lbs. per bag of cement				Lb.	2.50			2.50	2.75	**118**
	0600	Colors, integral, 2 to 10 lb. per bag of cement, minimum					.71			.71	.78	
	0610	Average					1.35			1.35	1.49	
	0620	Maximum					3.20			3.20	3.52	
	0700	Curing compound, 200 to 400 S.F. per gallon, 55 gal. lots				Gal.	6.70			6.70	7.35	
	0720	5 gallon lots					7.45			7.45	8.20	
	0800	Premium grade, 450 S.F. per gallon, 55 gallon lots					13			13	14.30	
	0820	5 gallon lots					13.75			13.75	15.15	
	0900	Dustproofing compound, 200-600 S.F./gal., 55 gallon lots					6.80			6.80	7.50	
	0920	5 gallon lots					7.80			7.80	8.60	
	1000	Epoxy dustproof coating, colors, 300-400 S.F. per coat,										
	1010	or transparent, 400-600 S.F. per coat				Gal.	49			49	54	
	1100	Hardeners, metallic, 55 lb. bags, natural (grey)				Lb.	.40			.40	.44	
	1200	Colors, average					.65			.65	.72	
	1300	Non-metallic, 55 lb. bags, natural (grey), minimum					.33			.33	.36	
	1310	Maximum					.56			.56	.62	
	1320	Non-metallic, colors, mininum					.36			.36	.40	
	1340	Maximum					.58			.58	.64	
	1400	Non-metallic, non-slip, 100 lb. bags, minimum					.38			.38	.42	
	1420	Maximum					.70			.70	.77	
	1500	Solution type, 300 to 400 S.F. per gallon				Gal.	5.05			5.05	5.55	
	1510											
	1600	Sealer, hardener and dustproofer, clear, 450 S.F., minimum				Gal.	6.50			6.50	7.15	
	1620	Maximum					18			18	19.80	
	1700	Colors (300-400 S.F. per gallon)					16.50			16.50	18.15	
	1750											
	1800	Set accelerator for below freezing, 1 to 1-½ gal. per C.Y.				Gal.	4.20			4.20	4.62	
	1900	Set retarder, 2 to 4 fl. oz. per bag of cement				"	13.25			13.25	14.60	
	2000	Waterproofing, integral 1 lb. per bag of cement				Lb.	.80			.80	.88	
	2100	Powdered metallic, 40 lbs. per 100 S.F., minimum					.80			.80	.88	
	2120	Maximum					1.60			1.60	1.76	
	2200	Water reducing admixture, average				Gal.	7.50			7.50	8.25	
126	0010	**CONCRETE, READY MIX** Regular weight, 2000 psi (43)				C.Y.	47.80			47.80	53	**126**
	0100	2500 psi					49.45			49.45	54	
	0150	3000 psi (42)					51.05			51.05	56	
	0200	3500 psi					52.70			52.70	58	
	0250	3750 psi (43)					53.50			53.50	59	
	0300	4000 psi					54.30			54.30	60	
	0350	4500 psi (42)					55.90			55.90	61	
	0400	5000 psi					57.50			57.50	63	
	1000	For high early strength cement, add					10%					
	2000	For all lightweight aggregate, add					45%					
	3000	For integral colors, 2500 psi, 5 bag mix										
	3100	Red, yellow or brown, 1.8 lb. per bag, add				C.Y.	12			12	13.20	
	3200	9.4 lb. per bag, add					63			63	69	
	3400	Black, 1.8 lb. per bag, add					12.75			12.75	14.05	
	3500	7.5 lb. per bag, add					53			53	58	
	3700	Green, 1.8 lb. per bag, add (41)					25			25	28	
	3800	7.5 lb. per bag, add					105			105	115	
130	0010	**CONCRETE IN PLACE** Including forms (4 uses), reinforcing										**130**
	0050	steel, including finishing unless otherwise indicated										
	0100	Average for concrete framed building, (35)										
	0110	including finishing	C-17B	15.75	5.210	C.Y.	117	120	18.20	255.20	345	
	0130	Average for substructure only, simple design, incl. finishing		29.07	2.820		75.70	65	9.85	150.55	200	
	0150	Average for superstructure only, including finishing		13.42	6.110		116.10	140	21	277.10	380	
	0200	Base, granolithic, 1" x 5" high, straight (50)	C-10	175	.137	L.F.	.15	2.78	.40	3.33	5	
	0220	Cove	"	140	.171	"	.15	3.47	.50	4.12	6.20	

For expanded coverage of these items see *Means Concrete Cost Data 1991*

033 100 | Structural Concrete

			CREW	DAILY OUTPUT	MAN-HOURS	UNIT	MAT.	LABOR	EQUIP.	TOTAL	TOTAL INCL O&P	
130	0300	Beams, 5 kip per L.F., 10' span	(49)	C-17A	6.28	12.900	C.Y.	187	295	24	506	715	130
	0350	25' span			7.40	10.950		149	250	20	419	595	
	0500	Chimney foundations, industrial, minimum			26.70	3.030		96	70	5.60	171.60	225	
	0510	Maximum			19.70	4.110		110	95	7.55	212.55	285	
	0700	Columns, square, 12" x 12", minimum reinforcing	(49)		4.60	17.610		205	405	32	642	920	
	0720	Average reinforcing			4.10	19.760		294	455	36	785	1,100	
	0740	Maximum reinforcing	(49)	C-17B	3.84	21.350		425	490	75	990	1,350	
	0800	16" x 16", minimum reinforcing		C-17A	6	13.500		178	310	25	513	725	
	0820	Average reinforcing		"	4.97	16.300		284	375	30	689	955	
	0840	Maximum reinforcing		C-17B	8.34	9.830		468	225	34	727	920	
	1200	Columns, round, tied, 16" diameter, minimum reinforcing	(47)		13.02	6.300		263	145	22	430	550	
	1220	Average reinforcing			8.30	9.880		375	230	35	640	820	
	1240	Maximum reinforcing			6.05	13.550		600	315	47	962	1,225	
	1300	20" diameter, minimum reinforcing			17.35	4.730		243	110	16.50	369.50	460	
	1320	Average reinforcing			10.43	7.860		390	180	27	597	750	
	1340	Maximum reinforcing			7.47	10.980		530	255	38	823	1,025	
	3800	Footings, spread under 1 C.Y.	(49)		35.95	2.280		84	53	7.95	144.95	185	
	3850	Over 5 C.Y.		C-17C	73.91	1.120		70	26	5.90	101.90	125	
	3900	Footings, strip, 18" x 9", plain		C-17B	29.24	2.800		66	65	9.80	140.80	190	
	3950	36" x 12", reinforced			51.42	1.590		70	37	5.55	112.55	145	
	4000	Foundation mat, under 10 C.Y.			32.32	2.540		107	59	8.85	174.85	220	
	4050	Over 20 C.Y.			47.37	1.730		103	40	6.05	149.05	185	
	4200	Grade walls, 8" thick, 8' high	(49)	C-15	9.50	7.580		115	160	5.35	280.35	390	
	4250	14' high		C-20	7.30	8.770		172	165	79	416	545	
	4260	12" thick, 8' high		C-17A	13.50	6		135	140	11.05	286.05	385	
	4270	14' high		C-20	11.60	5.520		126	105	50	281	360	
	4300	15" thick, 8' high		C-17B	18.40	4.460		89	105	15.55	209.55	280	
	4350	12' high		C-20	14.80	4.320		107	82	39	228	290	
	4750	Slab on grade, incl. troweled finish, not incl. forms											
	4760	or reinforcing, over 10,000 S.F., 4" thick slab		C-8	3,520	.016	S.F.	.65	.31	.15	1.11	1.38	
	4820	6" thick slab			3,610	.016		.98	.31	.14	1.43	1.73	
	4840	8" thick slab			3,275	.017		1.30	.34	.16	1.80	2.14	
	4900	12" thick slab			2,875	.019		1.95	.38	.18	2.51	2.96	
	4950	15" thick slab			2,560	.022		2.44	.43	.20	3.07	3.59	
	5200	Lift slab in place above the foundation, incl. forms,											
	5210	reinforcing, concrete and columns, minimum		C-17E	745	.107	S.F.	2.34	2.48	.07	4.89	6.65	
	5250	Average	(48)		675	.119		3.77	2.73	.08	6.58	8.65	
	5300	Maximum			430	.186		4.06	4.29	.13	8.48	11.55	
	5900	Pile caps, incl. forms and reinf., sq. or rect., under 5 C.Y.	(49)	C-17C	48.80	1.700	C.Y.	76	39	8.95	123.95	155	
	5950	Over 10 C.Y.			84.32	.984		70	23	5.15	98.15	120	
	6000	Triangular or hexagonal, under 5 C.Y.	(47)		47.95	1.730		73	40	9.10	122.10	155	
	6050	Over 10 C.Y.			77.85	1.070		70	25	5.60	100.60	125	
	6200	Retaining walls, gravity, 4' high see division 022-708		C-17B	19.10	4.290		64	99	15	178	245	
	6250	10' high			27.10	3.030		66	70	10.55	146.55	195	
	6300	Cantilever, level backfill loading, 8' high			16.30	5.030		121	115	17.60	253.60	340	
	6350	16' high			17	4.820		110	110	16.85	236.85	320	
	6800	Stairs, not including safety treads, free standing		C-15	120	.600	LF Nose	4.85	12.50	.42	17.77	26	
	6850	Cast on ground			180	.400	"	3.66	8.35	.28	12.29	17.95	
	7000	Stair landings, free standing			285	.253	S.F.	2	5.25	.18	7.43	11	
	7050	Cast on ground			685	.105	"	1.18	2.19	.07	3.44	4.96	
	9000	Minimum labor/equipment charge		F-2	1	16	Job		350	16	366	595	
134	0010	CURING With burlap, 4 uses assumed, 7.5 oz.		2 Clab	55	.291	C.S.F.	2.57	5.10		7.67	11.15	134
	0100	12 oz.			55	.291		3.70	5.10		8.80	12.40	
	0200	With waterproof curing paper, 2 ply, reinforced			95	.168		5.15	2.95		8.10	10.50	
	0300	With sprayed membrane curing compound			95	.168		1.96	2.95		4.91	6.95	
	9000	Minimum labor/equipment charge		1 Clab	5	1.600	Job		28		28	46	

For expanded coverage of these items see *Means Concrete Cost Data 1991*

033 100 | Structural Concrete

		CREW	DAILY OUTPUT	MAN-HOURS	UNIT	MAT.	LABOR	EQUIP.	TOTAL	TOTAL INCL O&P	
160	0010	**GUNITE**									160
	0020	Applied in 1″ layers, no mesh included	C-8	1,550	.036	S.F.	.75	.71	.33	1.79	2.33	
	0300	Typical in place, including mesh, 2″ thick, minimum	C-16	775	.093		1.40	1.92	.66	3.98	5.40	
	0350	Maximum		395	.182		2.10	3.76	1.30	7.16	9.90	
	0500	4″ thick, minimum		595	.121		2	2.50	.86	5.36	7.25	
	0550	Maximum		275	.262		3	5.40	1.87	10.27	14.20	
	0900	Prepare old walls, no scaffolding, minimum	C-10	1,000	.024		.49	.49	.07	1.05	1.39	
	0950	Maximum	″	275	.087		1.49	1.77	.25	3.51	4.72	
	1100	For high finish requirement or close tolerance, add, minimum						50%				
	1150	Maximum						110%				
	9000	Minimum labor/equipment charge	C-10	1	24	Job		485	70	555	845	
168	0010	**PATCHING CONCRETE**										168
	0050											
	0100	Floors, ¼″ thick, small areas, regular grout	1 Cefi	170	.047	S.F.	.82	1.02		1.84	2.49	
	0150	Epoxy grout	″	100	.080	″	3.50	1.73		5.23	6.55	
	2000	Walls, including chipping, cleaning and epoxy grout										
	2100	Minimum	1 Cefi	65	.123	S.F.	.14	2.66		2.80	4.32	
	2150	Average		50	.160		.23	3.46		3.69	5.65	
	2200	Maximum		40	.200		.37	4.33		4.70	7.15	
	9000	Minimum labor/equipment charge		4.50	1.780	Job		38		38	60	
172	0010	**PLACING CONCRETE** and vibrating, including labor & equipment										172
	0020											
	0050	Beams, elevated, small beams, pumped	C-20	40	1.600	C.Y.		30	14.40	44.40	65	
	0100	With crane and bucket	C-7	45	1.420			27	17.90	44.90	63	
	0400	Columns, square or round, 12″ thick, pumped	C-20	45	1.420			27	12.80	39.80	57	
	0450	With crane and bucket	C-7	40	1.600			30	20	50	71	
	0800	24″ thick, pumped	C-20	92	.696			13.15	6.25	19.40	28	
	0850	With crane and bucket	C-7	70	.914			17.25	11.50	28.75	41	
	1400	Elevated slabs, less than 6″ thick, pumped	C-20	110	.582			11	5.25	16.25	24	
	1450	With crane and bucket	C-7	95	.674			12.75	8.50	21.25	30	
	1600	Slabs over 10″ thick, pumped	C-20	150	.427			8.05	3.84	11.89	17.25	
	1650	With crane and bucket	C-7	130	.492			9.30	6.20	15.50	22	
	1900	Footings, continuous, shallow, direct chute	C-6	120	.400			7.40	.52	7.92	12.60	
	1950	Pumped	C-20	100	.640			12.10	5.75	17.85	26	
	2000	With crane and bucket	C-7	90	.711			13.45	8.95	22.40	32	
	2100	Deep continuous footings, direct chute	C-6	155	.310			5.75	.40	6.15	9.75	
	2150	Pumped	C-20	120	.533			10.10	4.81	14.91	22	
	2200	With crane and bucket	C-7	110	.582			11	7.35	18.35	26	
	2400	Footings, spread, under 1 C.Y., direct chute	C-6	55	.873			16.15	1.13	17.28	27	
	2450	Pumped	C-20	50	1.280			24	11.55	35.55	52	
	2500	With crane and bucket	C-7	45	1.420			27	17.90	44.90	63	
	2600	Spread footings, over 5 C.Y., direct chute	C-6	110	.436			8.10	.56	8.66	13.70	
	2650	Pumped	C-20	105	.610			11.50	5.50	17	25	
	2700	With crane and bucket	C-7	100	.640			12.10	8.05	20.15	28	
	2900	Foundation mats, over 20 C.Y., direct chute	C-6	350	.137			2.54	.18	2.72	4.31	
	2950	Pumped	C-20	325	.197			3.72	1.77	5.49	7.95	
	3000	With crane and bucket	C-7	300	.213			4.03	2.69	6.72	9.45	
	3200	Grade beams, direct chute	C-6	150	.320			5.95	.41	6.36	10.05	
	3250	Pumped	C-20	130	.492			9.30	4.44	13.74	19.90	
	3300	With crane and bucket	C-7	120	.533			10.10	6.70	16.80	24	
	3500	High rise, for more than 5 stories, pumped, add per story	C-20	2,100	.030			.58	.27	.85	1.23	
	3510	With crane and bucket, add per story	C-7	2,100	.030			.58	.38	.96	1.35	
	3700	Pile caps, under 5 C.Y., direct chute	C-6	90	.533			9.90	.69	10.59	16.75	
	3750	Pumped	C-20	85	.753			14.25	6.80	21.05	30	
	3800	With crane and bucket	C-7	80	.800			15.10	10.10	25.20	35	
	4300	Slab on grade, 4″ thick, direct chute	C-6	110	.436			8.10	.56	8.66	13.70	

For expanded coverage of these items see *Means Concrete Cost Data 1991*

033 100 | Structural Concrete

		CREW	DAILY OUTPUT	MAN-HOURS	UNIT	MAT.	LABOR	EQUIP.	TOTAL	TOTAL INCL O&P	
172	4350 Pumped	C-20	120	.533	C.Y.		10.10	4.81	14.91	22	**172**
	4400 With crane and bucket	C-7	110	.582			11	7.35	18.35	26	
	4600 Slab over 6″ thick, direct chute	C-6	165	.291			5.40	.38	5.78	9.15	
	4650 Pumped	C-20	165	.388			7.35	3.49	10.84	15.70	
	4700 With crane and bucket	C-7	145	.441			8.35	5.55	13.90	19.60	
	4900 Walls, 8″ thick, direct chute	C-6	90	.533			9.90	.69	10.59	16.75	
	4950 Pumped	C-20	85	.753			14.25	6.80	21.05	30	
	5000 With crane and bucket	C-7	80	.800			15.10	10.10	25.20	35	
	5050 12″ thick, direct chute	C-6	100	.480			8.90	.62	9.52	15.10	
	5100 Pumped	C-20	95	.674			12.75	6.05	18.80	27	
	5200 With crane and bucket	C-7	90	.711			13.45	8.95	22.40	32	
	5210 Minimum labor/equipment charge	C-6	1	48	Job		890	62	952	1,500	
	5320										
	5600 Wheeled concrete dumping, add to placing costs above										
	5610 Walking cart, 50′ haul, add	A-1	31	.258	C.Y.		4.52	1.75	6.27	9.30	
	5620 150′ haul, add		27	.296			5.20	2.01	7.21	10.70	
	5700 250′ haul, add		23	.348			6.10	2.37	8.47	12.55	
	5800 Riding cart, 50′ haul, add	B-9	320	.125			2.24	.40	2.64	4.09	
	5810 150′ haul, add		230	.174			3.11	.55	3.66	5.70	
	5900 250′ haul, add		200	.200			3.58	.64	4.22	6.55	
	9000 Minimum labor/equipment charge	A-1	2	4	Job		70	27	97	145	
196	0010 **WINTER PROTECTION** For heated ready mix, add, minimum				C.Y.	3			3	3.30	**196**
	0050 Maximum				″	4			4	4.40	
	0100 Protecting concrete and temporary heat, add, minimum	2 Clab	6,000	.003	S.F.	.06	.05		.11	.14	
	0150 Maximum, see also division 010-094	″	2,000	.008	″	.46	.14		.60	.73	

033 450 | Concrete Finishing

		CREW	DAILY OUTPUT	MAN-HOURS	UNIT	MAT.	LABOR	EQUIP.	TOTAL	TOTAL INCL O&P	
454	0010 **FINISHING FLOORS** Monolithic, screed finish	1 Cefi	900	.009	S.F.		.19		.19	.30	**454**
	0050 Darby finish	″	750	.011			.23		.23	.36	
	0100 Float finish	C-9	725	.011			.24	.05	.29	.43	
	0150 Broom finish		675	.012			.26	.05	.31	.46	
	0200 Steel trowel finish, for resilient tile		625	.013			.28	.06	.34	.49	
	0250 For finish floor		550	.015			.31	.06	.37	.56	
	0370 Minimum labor/equipment charge	1 Cefi	4	2	Job		43		43	68	
	0380										
	0400 Integral topping and finish, using 1:1:2 mix, ³⁄₁₆″ thick (54)	C-10	1,000	.024	S.F.	.10	.49	.07	.66	.96	
	0450 ½″ thick		950	.025		.13	.51	.07	.71	1.03	
	0500 ¾″ thick		850	.028		.19	.57	.08	.84	1.21	
	0600 1″ thick		750	.032		.25	.65	.09	.99	1.41	
	0800 Granolithic topping, laid after, 1:1:1-½ mix, ½″ thick (50)		590	.041		.14	.82	.12	1.08	1.59	
	0820 ¾″ thick		580	.041		.21	.84	.12	1.17	1.69	
	0850 1″ thick		575	.042		.27	.85	.12	1.24	1.77	
	0950 2″ thick		500	.048		.55	.97	.14	1.66	2.30	
	9100 Minimum labor/equipment charge		2	12	Job		245	35	280	425	
458	0010 **FINISHING WALLS** Break ties and patch voids	1 Cefi	540	.015	S.F.	.01	.32		.33	.51	**458**
	0050 Burlap rub with grout		450	.018		.05	.38		.43	.66	
	0300 Bush hammer, green concrete		170	.047		.01	1.02		1.03	1.60	
	0350 Cured concrete		110	.073		.02	1.57		1.59	2.48	
	0700 Sandblast, light penetration	C-10	1,100	.022		.13	.44	.06	.63	.91	
	0750 Heavy penetration		375	.064		.40	1.30	.19	1.89	2.70	
	9000 Minimum labor/equipment charge		2	12	Job		245	35	280	425	

For expanded coverage of these items see *Means Concrete Cost Data 1991*

034 | Precast Concrete

034 100 | Structural Precast

			CREW	DAILY OUTPUT	MAN-HOURS	UNIT	BARE COSTS MAT.	LABOR	EQUIP.	TOTAL	TOTAL INCL O&P	
104	0010	BEAMS Rectangular, to 20' spans, small beams	C-11	320	.225	L.F.	47	5.35	4.34	56.69	66	104
	0050	Large beams		240	.300		150	7.10	5.80	162.90	185	
112	0010	COLUMNS Rectangular to 12' high, small columns		125	.576		63	13.65	11.10	87.75	105	112
	0050	Large columns		100	.720		99	17.05	13.90	129.95	155	
136	0010	SLABS Prestressed roof and floor members, grouted, 4" deep		4,875	.015	S.F.	3.15	.35	.28	3.78	4.40	136
	0050	6" deep		5,850	.012		3.30	.29	.24	3.83	4.41	
	0100	8" deep ㊝		7,200	.010		3.42	.24	.19	3.85	4.39	
	0150	10" deep		8,800	.008		3.87	.19	.16	4.22	4.78	
	0200	12" deep		9,500	.008		4.14	.18	.15	4.47	5.05	

034 500 | Architectural Precast

			CREW	DAILY OUTPUT	MAN-HOURS	UNIT	BARE COSTS MAT.	LABOR	EQUIP.	TOTAL	TOTAL INCL O&P	
504	0011	WALL PANELS										504
	2200	Fiberglass reinforced cement with urethane core										
	2210	R20, 8' x 8', minimum	E-2	750	.075	S.F.	7.60	1.76	1.43	10.79	13	
	2220	Maximum	"	600	.093	"	13.50	2.20	1.79	17.49	21	

034 800 | Precast Specialties

			CREW	DAILY OUTPUT	MAN-HOURS	UNIT	BARE COSTS MAT.	LABOR	EQUIP.	TOTAL	TOTAL INCL O&P	
802	0010	LINTELS										802
	0800	Precast concrete, 4" x 8", stock units to 5' long	D-1	175	.091	L.F.	4.60	1.85		6.45	8	
	0850	To 12' long	D-4	190	.168		6	3.35	.64	9.99	12.65	
	1000	6" wide, 8" high, solid, stock units to 5' long		185	.173		7.05	3.44	.66	11.15	13.95	
	1050	To 12' long		190	.168		8.55	3.35	.64	12.54	15.45	
	1200	8" wide, 8" high, stock units to 5' long		185	.173		10.75	3.44	.66	14.85	18.05	
	1250	To 12' long		190	.168		11.60	3.35	.64	15.59	18.80	
	1400	10" wide, 8" high, stock units to 14' long		180	.178		13.85	3.53	.68	18.06	22	
	1450	12" wide, 8" high, stock units to 19' long		185	.173		14.90	3.44	.66	19	23	
804	0010	STAIRS Concrete treads on steel stringers, 3' wide	C-12	75	.640	Riser	58	13.95	4.83	76.78	92	804
	0300	Front entrance, 5' wide with 48" platform, 2 risers		16	3	Flight	290	65	23	378	450	
	0350	5 risers		12	4		335	87	30	452	545	
	0500	6' wide, 2 risers		15	3.200		305	70	24	399	475	
	0550	5 risers		11	4.360		335	95	33	463	560	
	1200	Basement entrance stairs, steel bulkhead doors, minimum	B-51	22	2.180		485	39	6.65	530.65	605	
	1250	Maximum	"	11	4.360		725	78	13.25	816.25	940	

035 | Cementitious Decks

035 200 | Lightweight Concrete

			CREW	DAILY OUTPUT	MAN-HOURS	UNIT	BARE COSTS MAT.	LABOR	EQUIP.	TOTAL	TOTAL INCL O&P	
212	0010	INSULATING Lightweight cellular concrete roof fill										212
	0020	Portland cement and foaming agent	C-8	50	1.120	C.Y.	81	22	10.30	113.30	135	
	0100	Poured vermiculite or perlite, field mix, ㊱	C-8	50	1.120	C.Y.	73	22	10.30	105.30	125	
	0110	1:6 roof fill, add formboards above										
	0200	Ready mix, 1:6 mix, roof fill, 2" thick		10,000	.006	S.F.	.45	.11	.05	.61	.73	
	0250	3" thick		7,700	.007	"	.68	.14	.07	.89	1.05	

For expanded coverage of these items see *Means Concrete Cost Data 1991*

041 000 | Mortar

			CREW	DAILY OUTPUT	MAN-HOURS	UNIT	BARE COSTS MAT.	LABOR	EQUIP.	TOTAL	TOTAL INCL O&P	
008	0010	**CEMENT** Gypsum 80 lb. bag, T.L. lots				Bag	10.85			10.85	11.95	008
	0050	L.T.L. lots					11.05			11.05	12.15	
	0100	Masonry, 70 lb. bag, T.L. lots					5.60			5.60	6.15	
	0150	L.T.L. lots					5.85			5.85	6.45	
	0200	White masonry cement, 70 lb. bag, T.L. lots					13.85			13.85	15.25	
	0250	L.T.L. lots					14.10			14.10	15.50	
016	0010	**GROUTING** Bond bms. & lintels, 8″ dp., pumped, not incl. block										016
	0020	8″ thick, 0.2 C.F. per L.F.	D-4	1,750	.018	L.F.	.65	.36	.07	1.08	1.38	
	0050	10″ thick, 0.25 C.F. per L.F.		1,500	.021		.82	.42	.08	1.32	1.67	
	0060	12″ thick, 0.3 C.F. per L.F.		1,300	.025		.98	.49	.09	1.56	1.96	
	0200	Concrete block cores, solid, 4″ thk., by hand, .067 C.F./S.F.	D-8	1,200	.033	S.F.	.22	.69		.91	1.35	
	0250	8″ thick, pumped .258 C.F. per S.F.	D-4	850	.038		.84	.75	.14	1.73	2.27	
	0300	10″ thick, .340 C.F. per S.F.		825	.039		1.11	.77	.15	2.03	2.61	
	0350	12″ thick, .422 C.F. per S.F.		800	.040		1.38	.79	.15	2.32	2.95	
	0500	Cavity walls, 2″ space, pumped, .167 C.F. per S.F.		2,000	.016		.54	.32	.06	.92	1.16	
	0550	3″ space, shoring not incl., .250 C.F. per S.F.		1,500	.021		.82	.42	.08	1.32	1.67	
	0600	4″ space, .333 C.F. per S.F.		1,250	.026		1.09	.51	.10	1.70	2.12	
	0700	6″ space, .500 C.F. per S.F.		950	.034		1.63	.67	.13	2.43	3	
	0800	Door frames, 3′ x 7′ opening, 2.5 C.F. per opening		60	.533	Opng.	8.15	10.60	2.03	20.78	28	
	0850	6′ x 7′ opening, 3.5 C.F. per opening		45	.711	″	11.40	14.10	2.71	28.21	38	
	9000	Minimum labor/equipment charge	1 Bric	2	4	Job		91		91	145	
020	0010	**LIME** Masons, hydrated, 50 lb. bag, T.L. lots				Bag	4.10			4.10	4.51	020
	0050	L.T.L. lots					4.35			4.35	4.79	
	0200	Finish, double hydrated, 50 lb. bag, T.L. lots					5.35			5.35	5.90	
	0250	L.T.L. lots					6.05			6.05	6.65	

041 500 | Masonry Accessories

			CREW	DAILY OUTPUT	MAN-HOURS	UNIT	MAT.	LABOR	EQUIP.	TOTAL	TOTAL INCL O&P	
504	0010	**ANCHOR BOLTS** Hooked type with nut, ½″ diam., 8″ long	1 Bric	200	.040	Ea.	.34	.91		1.25	1.83	504
	0030	12″ long		190	.042		.44	.96		1.40	2.02	
	0060	¾″ diameter, 8″ long		160	.050		1.38	1.14		2.52	3.34	
	0070	12″ long		150	.053		1.69	1.21		2.90	3.81	
512	0010	**JOINT REINFORCING** Steel bars, placed horizontal, #3 & #4 bars		450	.018	Lb.	.30	.40		.70	.98	512
	0020	#5 & #6 bars		800	.010		.30	.23		.53	.69	
	0050	Placed vertical, #3 & #4 bars		350	.023		.30	.52		.82	1.16	
	0060	#5 & #6 bars		650	.012		.30	.28		.58	.78	
	0500	Wire strips, ladder type, galvanized										
	0600	9 ga. sides, 9 ga. ties, 4″ wall	1 Bric	30	.267	C.L.F.	9	6.05		15.05	19.65	
	0750	10″ wall	″	20	.400	″	10.50	9.10		19.60	26	
	1000	Wire strips, truss type, galvanized										
	1010	To 8″ wide	1 Bric	3,000	.003	L.F.	.18	.06		.24	.30	
	1030	12″ wide		2,000	.004	″	.19	.09		.28	.35	
	1100	9 ga. sides, 9 ga. ties, 4″ wall	1 Bric	30	.267	C.L.F.	12.50	6.05		18.55	23	
	1250	10″ wall		20	.400	″	14.50	9.10		23.60	31	
	3500	For hot dip galvanizing, add					65%					
520	0010	**WALL TIES** To brick veneer, galv., corrugated, ⅞″ x 7″, 22 gauge	1 Bric	10.50	.762	C	4.20	17.35		21.55	32	520
	0150	16 gauge		10.50	.762		11.40	17.35		28.75	40	
	0200	Buck anchors, galv., corrugated, 16 gauge, 2″ bend, 8″ x 2″		10.50	.762		43.70	17.35		61.05	76	
	0250	8″ x 3″		10.50	.762		57.65	17.35		75	91	
	0600	Cavity wall, 6″ long, Z type, galvanized, ¼″ diameter		10.50	.762		17.10	17.35		34.45	47	
	0650	³⁄₁₆″ diameter		10.50	.762		7.15	17.35		24.50	36	
	0800	8″ long, ¼″ diameter, galvanized		10.50	.762		20.40	17.35		37.75	50	
	0850	³⁄₁₆″ diameter		10.50	.762		8.10	17.35		25.45	37	
	1000	Rectangular type, ¼″ diameter, galv., 2″ x 6″		10.50	.762		23.70	17.35		41.05	54	
	1050	2″ x 8″ or 4″ x 6″		10.50	.762		26.90	17.35		44.25	57	
	1100	³⁄₁₆″ diameter, galv., 2″ x 6″		10.50	.762		15.30	17.35		32.65	45	
	1150	2″ x 8″ or 4″ x 6″		10.50	.762		17.40	17.35		34.75	47	

For expanded coverage of these items see *Means Concrete Cost Data 1991*

041 500	Masonry Accessories	CREW	DAILY OUTPUT	MAN-HOURS	UNIT	BARE COSTS MAT.	LABOR	EQUIP.	TOTAL	TOTAL INCL O&P	
520	1500	Rigid partition anchors, plain, 8" long, 1" x ⅛"	1 Bric	10.50	.762	C	48	17.35		65.35	81	520
	1550	X ¼"		10.50	.762		94	17.35		111.35	130	
	1580	1-½" x ⅛"		10.50	.762		66	17.35		83.35	100	
	1600	X ¼"		10.50	.762		160	17.35		177.35	205	
	1650	2" x ⅛"		10.50	.762		85	17.35		102.35	120	
	1700	X ¼"		10.50	.762		225	17.35		242.35	275	

042 050	Chimneys	CREW	DAILY OUTPUT	MAN-HOURS	UNIT	BARE COSTS MAT.	LABOR	EQUIP.	TOTAL	TOTAL INCL O&P	
054	0010	CHIMNEY For foundation, add to prices below, see div 033-130-0500										054
	0100	Brick @ $250/M, 16" x 16", 8" flue, scaff. not incl.	D-1	18.20	.879	V.L.F.	11.70	17.75		29.45	41	
	0150	16" x 20" with one 8" x 12" flue ⓖ⑤		16	1		14.60	20		34.60	48	
	0200	16" x 24" with two 8" x 8" flues		14	1.140		17.25	23		40.25	56	
	0250	20" x 20" with one 12" x 12" flue		13.70	1.170		17.10	24		41.10	57	
	0300	20" x 24" with two 8" x 12" flues		12	1.330		23.20	27		50.20	69	
	0350	20" x 32" with two 12" x 12" flues		10	1.600		28.85	32		60.85	84	

042 100	Brick Masonry	CREW	DAILY OUTPUT	MAN-HOURS	UNIT	BARE COSTS MAT.	LABOR	EQUIP.	TOTAL	TOTAL INCL O&P	
108	0010	COLUMNS Brick @ $250 per M, 8" x 8", 9 brick, scaff. not incl.	D-1	56	.286	V.L.F.	2.57	5.75		8.32	12.10	108
	0100	12" x 8", 13.5 brick		37	.432		3.85	8.75		12.60	18.25	
	0200	12" x 12", 20.3 brick		25	.640		5.80	12.95		18.75	27	
	0300	16" x 12", 27 brick		19	.842		7.70	17		24.70	36	
	0400	16" x 16", 36 brick		14	1.140		10.25	23		33.25	48	
	0500	20" x 16", 45 brick		11	1.450		12.85	29		41.85	61	
	0600	20" x 20", 56.3 brick		9	1.780		16.05	36		52.05	75	
	0700	24" x 20", 67.5 brick		7	2.290		19.25	46		65.25	95	
	0800	24" x 24", 81 brick		6	2.670		23.10	54		77.10	110	
	1000	36" x 36", 182.3 brick		3	5.330		52	110		162	230	
	9000	Minimum labor/equipment charge		2	8	Job		160		160	260	
110	0010	COMMON BRICK Standard size, material only, minimum ⓖ①				M	220			220	240	110
	0050	Average (select common)				"	250			250	275	
116	0010	COPING For 12" wall, stock units, aluminum	D-1	80	.200	L.F.	9	4.04		13.04	16.40	116
	0050	Precast concrete, special order, 8" wide		100	.160		12.50	3.23		15.73	18.95	
	0100	10" wide		90	.178		16.40	3.59		19.99	24	
	0150	14" wide		80	.200		22.75	4.04		26.79	32	
	0300	Limestone for 12" wall, 4" thick		90	.178		11.50	3.59		15.09	18.40	
	0350	6" thick		80	.200		13.75	4.04		17.79	22	
	0500	Marble to 4" thick, no wash, 9" wide		90	.178		16.80	3.59		20.39	24	
	0550	12" wide		80	.200		24.20	4.04		28.24	33	
	0700	Terra cotta, 9" wide		90	.178		3.95	3.59		7.54	10.10	
	0750	12" wide		80	.200		6.75	4.04		10.79	13.90	
	9000	Minimum labor/equipment charge		2	8	Job		160		160	260	
120	0010	CORNICES Brick cornice on existing building										120
	0020	Not including scaffolding										
	0110	Face bricks @ $280 per M, 12 brick/S.F., minimum	D-1	30	.533	S.F.Face	3.97	10.75		14.72	22	
	0150	15 brick/S.F., maximum		23	.696	"	4.81	14.05		18.86	28	
	9000	Minimum labor/equipment charge		1.50	10.670	Job		215		215	345	

For expanded coverage of these items see *Means Concrete Cost Data 1991*

042 100 | Brick Masonry

			CREW	DAILY OUTPUT	MAN-HOURS	UNIT	BARE COSTS MAT.	LABOR	EQUIP.	TOTAL	TOTAL INCL O&P	
124	0010	**FACE BRICK** T.L. lots, material only, ⑥1										124
	0020	Including truck delivery, red brick only										
	0300	Boston, standard size, 8″ x 2-⅔″ x 4″, minimum				M	280			280	310	
	0350	Maximum					385			385	425	
	0450	Econo size, 8″ x 4″ x 4″, minimum					490			490	540	
	0500	Maximum					530			530	585	
	0550	Jumbo, 12″ x 4″ x 6″, minimum					985			985	1,075	
	0600	Maximum					1,240			1,240	1,375	
	0650	Norwegian, 12″ x 3-⅛″ x 4″, minimum					595			595	655	
	0700	Maximum					640			640	705	
	0850	Standard glazed, plain colors, 8″ x 2-⅔″ x 4″, minimum					830			830	915	
	0900	Maximum					990			990	1,100	
	1000	Deep trim shades, 8″ x 2-⅔″ x 4″, minimum					980			980	1,075	
	1050	Maximum					1,200			1,200	1,325	
	1080	Utility 12″ x 4″ x 4″					795			795	875	
	1120	8″ x 8″ x 4″					1,100			1,100	1,200	
	1140	8″ x 16″ x 4″				↓	2,385			2,385	2,625	
	1150											
	1200	Chicago, standard size, 8″ x 2-⅔″ x 4″, minimum				M	340			340	375	
	1250	Maximum					365			365	400	
	1260	Engineer size, 8″ x 3-⅛″ x 4″, minimum					395			395	435	
	1270	Maximum					425			425	470	
	1350	King size, 10″ x 2-¾″ x 4″, minimum					350			350	385	
	1360	Maximum					385			385	425	
	1460	Jumbo glazed, 12″ x 4″ x 6″, minimum					1,295			1,295	1,425	
	1480	Maximum					1,540			1,540	1,700	
	1710	New York City, standard size, industrial, 8″ x 2-⅔″ x 4″					340			340	375	
	1730	Office building, standard size, 8″ x 2-⅔″ x 4″					365			365	400	
	1770	Standard size, double glazed, 8″ x 2-⅔″ x 4″					890			890	980	
	1800	Jumbo, unglazed, 12″ x 4″ x 6″					965			965	1,050	
	1850	Jumbo, colored glazed ceramic, 12″ x 4″ x 6″					1,340			1,340	1,475	
	2050	Utility brick, glazed, 12″ x 4″ x 4″					940			940	1,025	
	2100	8″ x 8″ x 4″					1,380			1,380	1,525	
	2150	8″ x 16″ x 4″				↓	2,750			2,750	3,025	
	2160											
	2170	For less than truck load lots, add				M	5%			35	39	
	2180	For buff or gray brick, add					10%					
	3050	Used brick, minimum					240			240	265	
	3100	Maximum					325			325	360	
	3150	Add for brick to match existing work, minimum					5%					
	3200	Maximum				↓	50%					
134	0010	**LINTELS** See division 051-232										134
162	0010	**SIMULATED BRICK** Aluminum, baked on colors	1 Carp	200	.040	S.F.	1.75	.88		2.63	3.36	162
	0050	Fiberglass panels		200	.040		1.85	.88		2.73	3.47	
	0100	Urethane pieces cemented in mastic		150	.053		3.95	1.17		5.12	6.25	
	0150	Vinyl siding panels	↓	200	.040	↓	1.50	.88		2.38	3.09	
184	0016	**WALLS**										184
	0800	Common 8″ x 2-⅔″ x 4″ at $220/M, 4″ wall, as face brick	D-8	215	.186	S.F.	1.72	3.85		5.57	8.05	
	0850	4″ thick, as back up, 6.75 bricks per S.F. ⑤9		240	.167		1.72	3.45		5.17	7.45	
	0900	8″ thick wall, 13.50 brick per S.F.		135	.296		3.51	6.15		9.66	13.70	
	1000	12″ thick wall, 20.25 bricks per S.F. ⑥1		95	.421		5.35	8.70		14.05	19.90	
	1050	16″ thick wall, 27.00 bricks per S.F.		75	.533		7.15	11.05		18.20	26	
	1200	Reinf., straight hard, 8″ x 2-⅔″ x 4″ at $250/M, 4″ wall		205	.195		2.03	4.04		6.07	8.70	
	1250	8″ thick wall, 13.50 brick per S.F.		130	.308		4.19	6.35		10.54	14.85	
	1300	12″ thick wall, 20.25 bricks per S.F.		90	.444		6.30	9.20		15.50	22	
	1350	16″ thick wall, 27.00 bricks per S.F.	↓	70	.571	↓	8.50	11.85		20.35	28	

For expanded coverage of these items see *Means Concrete Cost Data 1991*

042 100 | Brick Masonry

		CREW	DAILY OUTPUT	MAN-HOURS	UNIT	BARE COSTS				TOTAL INCL O&P	
						MAT.	LABOR	EQUIP.	TOTAL		
184	9000	Minimum labor/ equipment charge	D-1	2	8	Job		160		160	260	**184**
194	0010	WINDOW SILL Bluestone, natural cleft, 12" wide, 1-½" thick	D-1	85	.188	S.F.	10.35	3.80		14.15	17.50	**194**
	0050	2" thick		85	.188	"	10.75	3.80		14.55	17.95	
	0100	Cut stone, 5" x 8" plain		85	.188	L.F.	8.15	3.80		11.95	15.05	
	0200	Face brick on edge, brick @ $280 per M, 8" wide		80	.200		1.65	4.04		5.69	8.30	
	0400	Marble, 12" wide, ¾" thick		85	.188		12.40	3.80		16.20	19.75	
	0600	Precast concrete, special order, 6" wide		85	.188		13.50	3.80		17.30	21	
	0650	10" wide		85	.188		22	3.80		25.80	30	
	0700	14" wide		85	.188		30	3.80		33.80	39	
	0900	Slate, colored, unfading, honed, 12" wide, 1" thick		85	.188		22.50	3.80		26.30	31	
	0950	2" thick	↓	85	.188		32.50	3.80		36.30	42	
	9000	Minimum labor/equipment charge	1 Bric	2	4	Job		91		91	145	

042 200 | Concrete Unit Masonry

		CREW	DAILY OUTPUT	MAN-HOURS	UNIT	MAT.	LABOR	EQUIP.	TOTAL	TOTAL INCL O&P	
216 0010	CONCRETE BLOCK, BACK-UP, Scaffolding not included										**216**
0020	Sand aggregate, tooled joint 1 side										
0050	Not-reinforced, 8" x 16", 2" thick, 2000 psi	D-8	475	.084	S.F.	.67	1.74		2.41	3.54	
0200	Regular block, 4" thick		440	.091		.90	1.88		2.78	4.01	
0300	6" thick		420	.095		1.09	1.97		3.06	4.36	
0350	8" thick		400	.100		1.37	2.07		3.44	4.83	
0400	10" thick	↓	390	.103		1.87	2.12		3.99	5.45	
0450	12" thick	D-9	370	.130	↓	1.91	2.62		4.53	6.30	
9000	Minimum labor/equipment charge	D-1	2	8	Job		160		160	260	
220 0010	CONCRETE BLOCK, DECORATIVE Scaffolding not included										**220**
1000	Fluted high strength										
1100	Flutes 1 side, 8" x 16" x 4" thick	D-8	345	.116	S.F.	2.18	2.40		4.58	6.25	
1150	Flutes 2 sides, 8" x 16" x 4" thick		335	.119		2.58	2.47		5.05	6.80	
1200	8" thick	↓	300	.133	↓	3.34	2.76		6.10	8.10	
1250	For special colors, add					.25			.25	.28	
1400	Deep grooved, smooth face										
1450	8" x 16" x 4" thick	D-8	345	.116	S.F.	1.40	2.40		3.80	5.40	
1500	8" thick	"	300	.133	"	2.58	2.76		5.34	7.25	
1600											
4000	Slump block										
4100	4" face height x 16" x 4" thick	D-1	165	.097	S.F.	1.62	1.96		3.58	4.92	
4150	6" thick		160	.100		2.07	2.02		4.09	5.50	
4200	8" thick		155	.103		2.71	2.09		4.80	6.30	
4250	10" thick		140	.114		4.33	2.31		6.64	8.45	
4300	12" thick	↓	130	.123	↓	4.53	2.49		7.02	8.95	
5000	Split rib profile units, 1" deep ribs, 8 ribs	D-8	345	.116	S.F.	1.87	2.40		4.27	5.90	
5100	8" x 16" x 4" thick										
5150	6" thick		325	.123		2.33	2.55		4.88	6.65	
5200	8" thick		305	.131		2.98	2.72		5.70	7.65	
5250	12" thick	D-9	275	.175	↓	3.61	3.53		7.14	9.65	
9000	Minimum labor/equipment charge	D-1	2	8	Job		160		160	260	
224 0010	CONCRETE BLOCK, INTERLOCKING Scaffolding not incl., mortar incl.										**224**
0100	Not including grout or reinforcing										
0200	8" x 16" units, 8" thick, 2000 psi	D-1	245	.065	S.F.	1.44	1.32		2.76	3.70	
0300	12" thick	"	220	.073		2.12	1.47		3.59	4.69	
0400	Including grout & reinforcing, 8" thick (70)	D-4	245	.131		2.44	2.59	.50	5.53	7.35	
0450	12" thick	"	220	.145	↓	3.83	2.89	.55	7.27	9.45	
9000	Minimum labor/equipment charge	D-1	2	8	Job		160		160	260	
232 0010	CONCRETE BLOCK, PARTITIONS Scaffolding not included										**232**
1000	Lightweight block, tooled joints, 2 sides, hollow										

For expanded coverage of these items see *Means Concrete Cost Data 1991*

042 200 | Concrete Unit Masonry

			CREW	DAILY OUTPUT	MAN-HOURS	UNIT	BARE COSTS				TOTAL INCL O&P	
							MAT.	LABOR	EQUIP.	TOTAL		
232	1100	Not reinforced, 8″ x 16″ x 4″ thick	D-8	440	.091	S.F.	.92	1.88		2.80	4.03	232
	1150	6″ thick		410	.098		1.23	2.02		3.25	4.59	
	1200	8″ thick		385	.104		1.61	2.15		3.76	5.20	
	1250	10″ thick		370	.108		2.39	2.24		4.63	6.20	
	1300	12″ thick	D-9	350	.137		2.44	2.77		5.21	7.10	
	4000	Regular block, tooled joints, 2 sides, hollow										
	4100	Not reinforced, 8″ x 16″ x 4″ thick	D-8	430	.093	S.F.	.83	1.93		2.76	4	
	4150	6″ thick		400	.100		1.02	2.07		3.09	4.44	
	4200	8″ thick		375	.107		1.30	2.21		3.51	4.97	
	4250	10″ thick		360	.111		1.80	2.30		4.10	5.65	
	4300	12″ thick	D-9	340	.141		1.85	2.85		4.70	6.60	
	9000	Minimum labor/equipment charge	D-1	2	8	Job		160		160	260	
248	0010	**GLAZED CONCRETE BLOCK** No scaffolding or reinforcing incl.										248
	0100	Single face, 8″ x 16″ units, 2″ thick	D-8	360	.111	S.F.	5.20	2.30		7.50	9.40	
	0200	4″ thick		345	.116		5.40	2.40		7.80	9.80	
	0250	6″ thick		330	.121		5.75	2.51		8.26	10.35	
	0300	8″ thick		310	.129		6.35	2.67		9.02	11.30	
	0350	10″ thick		295	.136		6.95	2.81		9.76	12.15	
	0400	12″ thick	D-9	280	.171		7.50	3.46		10.96	13.80	
	0700	Double face, 8″ x 16″ units, 4″ thick	D-8	310	.129		8.10	2.67		10.77	13.20	
	0750	6″ thick		290	.138		8.55	2.86		11.41	14	
	0800	8″ thick		270	.148		9.20	3.07		12.27	15.05	
	1500	Cove base, 8″ x 16″, 2″ thick		315	.127	L.F.	4.60	2.63		7.23	9.30	
	1550	4″ thick		285	.140		5	2.91		7.91	10.15	
	1600	6″ thick		265	.151		5.35	3.13		8.48	10.90	
	1650	8″ thick		245	.163		5.90	3.38		9.28	11.90	
	9000	Minimum labor/equipment charge	D-1	2	8	Job		160		160	260	
252	0010	**INSULATION** See also division 072-108										252
	0100	Inserts, styrofoam, plant installed, add to block prices										
	0200	8″ x 16″ units, 6″ thick				S.F.	.65			.65	.72	
	0250	8″ thick					.65			.65	.72	
	0300	10″ thick					.79			.79	.87	
	0350	12″ thick					.79			.79	.87	
	0500	8″ x 8″ units, 8″ thick					.50			.50	.55	
	0550	12″ thick					.55			.55	.61	
	9000	Minimum labor/equipment charge	D-1	2	8	Job		160		160	260	

042 300 | Reinforced Unit Masonry

			CREW	DAILY OUTPUT	MAN-HOURS	UNIT	MAT.	LABOR	EQUIP.	TOTAL	TOTAL INCL O&P	
304	0010	**CONCRETE BLOCK BOND BEAM** Scaffolding not included										304
	0020	Not including grout or reinforcing										
	0100	Regular block, 8″ high, 8″ thick	D-8	565	.071	L.F.	1.07	1.47		2.54	3.53	
	0150	12″ thick	D-9	510	.094	″	1.57	1.90		3.47	4.78	
	9000	Minimum labor/equipment charge	D-1	2	8	Job		160		160	260	
310	0010	**CONCRETE BLOCK, EXTERIOR** Not including scaffolding										310
	0020	Reinforced, tooled joints 2 sides, styrofoam inserts										
	0100	Regular, 8″ x 16″ x 6″ thick	D-8	390	.103	S.F.	1.77	2.12		3.89	5.35	
	0200	8″ thick		365	.110		2.05	2.27		4.32	5.90	
	0250	10″ thick		355	.113		2.70	2.33		5.03	6.70	
	0300	12″ thick	D-9	330	.145		2.76	2.94		5.70	7.75	
	9000	Minimum labor/equipment charge	D-1	2	8	Job		160		160	260	
320	0010	**CONCRETE BLOCK FOUNDATION WALL** Scaffolding not included										320
	0050	Sand aggregate, trowel cut joints, not reinf., parged ½″ thick										
	0200	Regular, 8″ x 16″ x 6″ thick	D-8	450	.089	S.F.	1.25	1.84		3.09	4.33	
	0250	8″ thick	″	430	.093	″	1.60	1.93		3.53	4.85	

For expanded coverage of these items see *Means Concrete Cost Data 1991*

042 300 | Reinforced Unit Masonry

		CREW	DAILY OUTPUT	MAN-HOURS	UNIT	MAT.	LABOR	EQUIP.	TOTAL	TOTAL INCL O&P	
320	0300	10" thick	D-8	420	.095	S.F.	2.11	1.97		4.08	5.50	320
	0350	12" thick	D-9	395	.122	"	2.17	2.45		4.62	6.35	
	1000	Reinforced										
	1100	Regular, 8" x 16" block, 6" thick	D-8	445	.090	S.F.	1.35	1.86		3.21	4.48	
	1150	8" thick		425	.094		1.70	1.95		3.65	5	
	1200	10" thick		415	.096		2.22	2		4.22	5.65	
	1250	12" thick	D-9	390	.123		2.29	2.49		4.78	6.50	
	9000	Minimum labor/equipment charge	D-1	2	8	Job		160		160	260	
330	0010	**CONCRETE BLOCK, LINTELS** Scaffolding not included										330
	0100	Including grout and reinforcing										
	0200	8" x 8" x 8", 1 #4 bar	D-4	300	.107	L.F.	2.33	2.12	.41	4.86	6.40	
	0250	2 #4 bars		295	.108		2.53	2.15	.41	5.09	6.65	
	1000	12" x 8" x 8", 1 #4 bar		275	.116		3.25	2.31	.44	6	7.75	
	1150	2 #5 bars		270	.119		3.67	2.35	.45	6.47	8.30	
	9000	Minimum labor/equipment charge	D-1	2	8	Job		160		160	260	

042 450 | Structural Facing Tile

		CREW	DAILY OUTPUT	MAN-HOURS	UNIT	MAT.	LABOR	EQUIP.	TOTAL	TOTAL INCL O&P	
454	0010	**STRUCTURAL FACING TILE** Scaffolding not incl., 6T, 5-1/3" x 12"										454
	0020	Functional colors, 2.3 pieces per S.F., 2" thick, glazed 1 side	D-8	225	.178	S.F.	3.57	3.68		7.25	9.85	
	0100	4" thick, glazed 1 side		220	.182		4.45	3.77		8.22	10.95	
	0150	Glazed 2 sides		195	.205		6.90	4.25		11.15	14.40	
	0250	6" thick, glazed 1 side		210	.190		6.35	3.94		10.29	13.30	
	0300	Glazed 2 sides		185	.216		8.75	4.48		13.23	16.80	
	0400	8" thick, glazed 1 side		180	.222		7.60	4.60		12.20	15.75	
	9000	Minimum labor/equipment charge	D-1	2	8	Job		160		160	260	

042 500 | Ceramic Veneer

		CREW	DAILY OUTPUT	MAN-HOURS	UNIT	MAT.	LABOR	EQUIP.	TOTAL	TOTAL INCL O&P	
510	0010	**TERRA COTTA** Coping, split type, not glazed, 9" wide	D-1	90	.178	L.F.	4.10	3.59		7.69	10.25	510
	0100	13" wide		80	.200		6.50	4.04		10.54	13.65	
	0200	Split type, glazed, 9" wide		90	.178		4.75	3.59		8.34	11	
	0250	13" wide		80	.200		7.90	4.04		11.94	15.15	
	0350											
	0500	Partition or back-up blocks, scored, in C.L. lots										
	0700	Non-load bearing, 12" x 12", 3" thick	D-8	550	.073	S.F.	2.35	1.51		3.86	5	
	0850	8" thick		400	.100		4.05	2.07		6.12	7.80	
	1000	Load bearing, 12" x 12", 4" thick, in walls		500	.080		2.70	1.66		4.36	5.65	
	1400	8" thick, in walls		400	.100		4.60	2.07		6.67	8.40	
	9000	Minimum labor/equipment charge	D-1	2	8	Job		160		160	260	

042 550 | Masonry Veneer

		CREW	DAILY OUTPUT	MAN-HOURS	UNIT	MAT.	LABOR	EQUIP.	TOTAL	TOTAL INCL O&P	
554	0010	**BRICK VENEER** Scaffolding not included, truck load lots										554
	0015	Material costs include a 3% brick waste allowance										
	2000	Standard, sel. common, 8" x 2-2/3" x 4", $250 per M (6.75/S.F.)	D-8	230	.174	S.F.	1.92	3.60		5.52	7.90	
	2020	Stnd, 8" x 2-2/3" x 4", running bond, red, (6.75/S.F.), $280/M		220	.182		2.13	3.77		5.90	8.40	
	2050	Full header every 6th course (7.88/S.F.)		185	.216		2.48	4.48		6.96	9.90	
	2100	English, full header every 2nd course (10.13/S.F.) (59)		140	.286		3.19	5.90		9.09	13	
	2150	Flemish, alternate header every course (9.00/S.F.) (61)		150	.267		2.84	5.50		8.34	12	
	2200	Flemish, alt. header every 6th course (7.13/S.F.)		205	.195		2.25	4.04		6.29	8.95	
	2250	Full headers throughout (13.50/S.F.)		105	.381		4.25	7.90		12.15	17.35	
	2300	Rowlock course (13.50/S.F.)		100	.400		4.25	8.30		12.55	17.95	
	2350	Rowlock stretcher (4.50/S.F.)		310	.129		1.42	2.67		4.09	5.85	
	2400	Soldier course (6.75/S.F.)		200	.200		2.13	4.14		6.27	9	
	2450	Sailor course (4.50/S.F.)		290	.138		1.42	2.86		4.28	6.15	
	2600	Running bond, buff or gray face at $355 per M (6.75/S.F.)		220	.182		2.67	3.77		6.44	9	
	2700	Glazed face, brick at $825 per M, running bond		210	.190		6.35	3.94		10.29	13.30	
	2750	Full header every 6th course (7.88/S.F.)		170	.235		6.95	4.87		11.82	15.45	

For expanded coverage of these items see *Means Concrete Cost Data 1991*

042 | Unit Masonry

042 550 | Masonry Veneer

		CREW	DAILY OUTPUT	MAN-HOURS	UNIT	MAT.	LABOR	EQUIP.	TOTAL	TOTAL INCL O&P	
554	3000	Jumbo, 12" x 4" x 6" running bond @ $985 per M (3.00/S.F.)	D-8	435	.092	S.F.	3.24	1.90		5.14	6.60	554
	3050	Norman, 12" x 2-⅔" x 4" running bond, $550/M (4.50/S.F.)		320	.125		2.72	2.59		5.31	7.15	
	3100	Norwegian, 12" x 3-⅛" x 4" at $595 per M (3.75/S.F.)		375	.107		2.46	2.21		4.67	6.25	
	3150	Economy, 8" x 4" x 4" at $490 per M (4.50/S.F.)		310	.129		2.41	2.67		5.08	6.95	
	3200	Engineer, 8" x 3-⅛" x 4" at $395 per M (5.63/S.F.)		260	.154		2.45	3.19		5.64	7.80	
	3250	Roman, 12" x 2" x 4" at $575 per M (6.00/S.F.)		250	.160		3.78	3.31		7.09	9.50	
	3300	SCR, 12" x 2-⅔" x 6" at $690 per M (4.50/S.F.)		310	.129		3.47	2.67		6.14	8.10	
	3350	Utility, 12" x 4" x 4" at $795 per M (3.00/S.F.)		450	.089		2.58	1.84		4.42	5.80	
	3400	For cavity wall construction, add						15%				
	3450	For stacked bond, add						10%				
	3500	For interior veneer construction, add						15%				
	3550	For curved walls, add						30%				
	9000	Minimum labor/equipment charge	D-1	2	8	Job		160		160	260	

042 700 | Glass Unit Masonry

		CREW	DAILY OUTPUT	MAN-HOURS	UNIT	MAT.	LABOR	EQUIP.	TOTAL	TOTAL INCL O&P	
704	0010	GLASS BLOCK Scaffolding not included										704
	0100	4" thick, plain, under 1000 S.F., 6" x 6" block	D-8	115	.348	S.F.	19.60	7.20		26.80	33	
	0150	8" x 8" block		160	.250		14.70	5.20		19.90	24	
	0200	12" x 12" block		175	.229		18.50	4.73		23.23	28	
	0300	Plain, 1000 to 5000 S.F., 6" x 6" block		135	.296		17.55	6.15		23.70	29	
	0350	8" x 8" block		190	.211		14.05	4.36		18.41	22	
	0400	12" x 12" block		215	.186		15.90	3.85		19.75	24	
	0700	For solar reflective blocks, add					100%					
	0800	Plain, 4" x 8" blocks, under 1000 S.F.	D-8	145	.276		23.55	5.70		29.25	35	
	0850	Over 5000 S.F.		170	.235		20.55	4.87		25.42	30	
	1000	3-⅛" thick thinline, plain, under 1000 S.F., 6" x 6" block		115	.348		17.35	7.20		24.55	31	
	1050	8" x 8" block		160	.250		11.05	5.20		16.25	20	
	1400	For cleaning block after installation (both sides), add		1,000	.040		.08	.83		.91	1.42	
	9000	Minimum labor/equipment charge	D-1	2	8	Job		160		160	260	

044 | Stone

044 100 | Rough Stone

		CREW	DAILY OUTPUT	MAN-HOURS	UNIT	MAT.	LABOR	EQUIP.	TOTAL	TOTAL INCL O&P	
104	0011	ROUGH STONE WALL Dry										104
	0020											
	0100	Random fieldstone, under 18" thick	D-12	60	.533	C.F.	9.75	11.05		20.80	28	
	0150	Over 18" thick	"	63	.508	"	12.90	10.50		23.40	31	
	9000	Minimum labor/equipment charge	D-1	2	8	Job		160		160	260	

044 200 | Cut Stone

		CREW	DAILY OUTPUT	MAN-HOURS	UNIT	MAT.	LABOR	EQUIP.	TOTAL	TOTAL INCL O&P	
204	0010	BLUESTONE Cut to size										204
	0020											
	0500	Sills, natural cleft, 10" wide to 6' long, 1-½" thick	D-11	70	.343	L.F.	8.85	7.45		16.30	22	
	0550	2" thick		63	.381		9.40	8.25		17.65	24	
	0600	Smooth finish, 1-½" thick		70	.343		11.95	7.45		19.40	25	
	0650	2" thick		63	.381		12.70	8.25		20.95	27	
	0800	Thermal finish, 1-½" thick		70	.343		12.50	7.45		19.95	26	
	0850	2" thick		63	.381		13.60	8.25		21.85	28	
	1000	Stair treads, natural cleft, 12" wide, 6' long, 1-½" thick	D-10	115	.348		12.80	7.40	3.65	23.85	30	
	1050	2" thick	"	105	.381		13.65	8.10	3.99	25.74	32	

For expanded coverage of these items see *Means Concrete Cost Data 1991*

044 200 | Cut Stone

			CREW	DAILY OUTPUT	MAN-HOURS	UNIT	BARE COSTS MAT.	LABOR	EQUIP.	TOTAL	TOTAL INCL O&P	
204	1100	Smooth finish, 1-½" thick	D-10	115	.348	L.F.	17.30	7.40	3.65	28.35	35	204
	1150	2" thick		105	.381		18.45	8.10	3.99	30.54	38	
	1300	Thermal finish, 1-½" thick		115	.348		18.60	7.40	3.65	29.65	36	
	1350	2" thick		105	.381		19.80	8.10	3.99	31.89	39	
	9000	Minimum labor/equipment charge	D-1	2.50	6.400	Job		130		130	205	

044 550 | Marble

			CREW	DAILY OUTPUT	MAN-HOURS	UNIT	MAT.	LABOR	EQUIP.	TOTAL	TOTAL INCL O&P	
554	0011	MARBLE Ashlar, split face, 4" + or - thick, random										554
	0040	lengths 1' to 4' & heights 2" to 7-½", average	D-8	175	.229	S.F.	8.45	4.73		13.18	16.90	
	0100	Base, polished, ¾" or ⅞" thick, polished, 6" high	D-10	65	.615	L.F.	10.40	13.05	6.45	29.90	39	
	0260											
	0300	Carvings or bas relief, from templates, average	D-10	80	.500	S.F.	100	10.60	5.25	115.85	135	
	0350	Maximum	"	80	.500	"	235	10.60	5.25	250.85	280	
	1000	Facing, polished finish, cut to size, ¾" to ⅞" thick										
	1050	Average	D-10	130	.308	S.F.	16.50	6.55	3.22	26.27	32	
	1100	Maximum	"	130	.308	"	38.40	6.55	3.22	48.17	56	
	2500	Flooring, polished tiles, 12" x 12" x ⅜" thick										
	2510	Thin set, average	D-11	90	.267	S.F.	7.20	5.80		13	17.20	
	2600	Maximum		90	.267		21.50	5.80		27.30	33	
	2700	Mortar bed, average		65	.369		7.20	8		15.20	21	
	2740	Maximum		65	.369		21.50	8		29.50	37	
	3500	Thresholds, 3' long, ⅞" thick, 4" to 5" wide, plain	D-12	24	1.330	Ea.	12.50	28		40.50	58	
	3550	Beveled		24	1.330	"	13.50	28		41.50	59	
	3700	Window stools, polished, ⅞" thick, 5" wide		85	.376	L.F.	10.50	7.80		18.30	24	
	9000	Minimum labor/equipment charge	D-1	2	8	Job		160		160	260	

044 600 | Limestone

			CREW	DAILY OUTPUT	MAN-HOURS	UNIT	MAT.	LABOR	EQUIP.	TOTAL	TOTAL INCL O&P	
604	0012	LIMESTONE, Cut to size										604
	0020	Veneer facing panels										
	0100	Sawn finish, 2" thick, to 3' x 5' panels	D-10	130	.308	S.F.	9.65	6.55	3.22	19.42	25	
	0150	Smooth finish, 2" thick, to 3' x 5' panels		130	.308		11.75	6.55	3.22	21.52	27	
	0300	3" thick, to 4' x 9' panels		225	.178		13.60	3.77	1.86	19.23	23	
	0350	4" thick, to 5' x 11' panels		275	.145		15.85	3.09	1.52	20.46	24	
	0700	Light stick, textured finish, 4-½" thick to 5' x 12'		275	.145		15.60	3.09	1.52	20.21	24	
	0750	5" thick, to 5' x 14' panels		275	.145		18.60	3.09	1.52	23.21	27	
	1000	Medium ribbed, textured finish, 4-½" thick, to 5' x 12'		275	.145		16.50	3.09	1.52	21.11	25	
	1050	5" thick, to 5' x 14' panels		275	.145		19	3.09	1.52	23.61	28	
	1200	Deep ribbed, textured finish, 4-½" thick, to 5' x 10'		275	.145		17.50	3.09	1.52	22.11	26	
	1250	5" thick, to 5' x 14' panels		275	.145		19.85	3.09	1.52	24.46	28	
	1400	Sugar cube, textured finish, 4-½" thick, to 5' x 12'		275	.145		16.50	3.09	1.52	21.11	25	
	1450	5" thick, to 5' x 14' panels		275	.145		19	3.09	1.52	23.61	28	
	2000	Coping, smooth finish, top & 2 sides		30	1.330	C.F.	47	28	13.95	88.95	110	
	2100	Sills, lintels, jambs, smooth finish, average		20	2		51	42	21	114	145	
	2150	Detailed		20	2		65	42	21	128	160	
	2300	Steps, extra hard, 14" wide, 6" rise		50	.800	L.F.	32	17	8.40	57.40	72	
	9000	Minimum labor/equipment charge	D-1	2	8	Job		160		160	260	

044 650 | Granite

			CREW	DAILY OUTPUT	MAN-HOURS	UNIT	MAT.	LABOR	EQUIP.	TOTAL	TOTAL INCL O&P	
651	0010	GRANITE Cut to size										651
	0050	Veneer, polished face, ¾" to 1-½" thick										
	0150	Low price, gray, light gray, etc.	D-10	130	.308	S.F.	11.40	6.55	3.22	21.17	27	
	0180	Medium price, pink, brown, etc.		130	.308		19.15	6.55	3.22	28.92	35	
	0220	High price, red, black, etc.		130	.308		26.90	6.55	3.22	36.67	44	
	2500	Steps, copings, etc., finished on more than one surface										
	2550	Minimum	D-10	50	.800	C.F.	60	17	8.40	85.40	100	
	2600	Maximum	"	50	.800	"	90	17	8.40	115.40	135	

For expanded coverage of these items see *Means Concrete Cost Data 1991*

044 650 | Granite

			CREW	DAILY OUTPUT	MAN-HOURS	UNIT	BARE COSTS				TOTAL INCL O&P
							MAT.	LABOR	EQUIP.	TOTAL	
651	3500	Curbing, city street type, 6" x 18", split face,										651
	3510	sawn top, radius nosing, 4' to 7' lengths	D-10	75	.533	L.F.	17.75	11.30	5.60	34.65	44	
	3600	Highway type, 5" x 16", split face,										
	3610	sawn top, 4' to 7' lengths	D-10	300	.133	L.F.	14.50	2.83	1.40	18.73	22	
	3800	Radius curbs, over 5' radius, add					50%					
	3850	Under 5' radius, add					100%					
	9000	Minimum labor/equipment charge	D-1	2	8	Job		160		160	260	

044 700 | Sandstone

			CREW	DAILY OUTPUT	MAN-HOURS	UNIT	BARE COSTS				TOTAL INCL O&P
							MAT.	LABOR	EQUIP.	TOTAL	
704	0011	SANDSTONE OR BROWNSTONE										704
	0100	Sawed face veneer, 2-½" thick, to 2' x 4' panels	D-10	130	.308	S.F.	10.80	6.55	3.22	20.57	26	
	0150	4" thick, to 3'-6" x 8' panels		100	.400		10.50	8.50	4.19	23.19	30	
	0300	Split face, random sizes		100	.400		7.65	8.50	4.19	20.34	27	
	9000	Minimum labor/equipment charge	D-1	2.50	6.400	Job		130		130	205	

044 750 | Slate

			CREW	DAILY OUTPUT	MAN-HOURS	UNIT	BARE COSTS				TOTAL INCL O&P
							MAT.	LABOR	EQUIP.	TOTAL	
754	0010	SLATE Pennsylvania, blue gray to gray black; Vermont,										754
	0050	Unfading green, mottled green & purple, gray & purple										
	0100	Virginia, blue black										
	0150											
	3100	Stair landings, 1" thick, black, clear	D-1	65	.246	S.F.	9.75	4.97		14.72	18.70	
	3200	Ribbon	"	65	.246	"	9.50	4.97		14.47	18.45	
	3500	Stair treads, sand finish, 1" thick x 12" wide										
	3550	Under 3 L.F.	D-10	85	.471	L.F.	13.80	10	4.93	28.73	37	
	3600	3 L.F. to 6 L.F.	"	120	.333	"	15.25	7.05	3.49	25.79	32	
	3650											
	3700	Ribbon, sand finish, 1" thick x 12" wide										
	3750	To 6 L.F.	D-10	120	.333	L.F.	10.90	7.05	3.49	21.44	27	
	4000	Stools or sills, sand finish, 1" thick, 6" wide	D-12	160	.200		6.75	4.14		10.89	14.05	
	4200	10" wide		90	.356		10.30	7.35		17.65	23	
	4400	2" thick, 6" wide		140	.229		10.50	4.73		15.23	19.15	
	4600	10" wide		90	.356		14.10	7.35		21.45	27	
	4800	For lengths over 3', add					25%					
	9000	Minimum labor/equipment charge	D-1	2.50	6.400	Job		130		130	205	

045 100 | Masonry Cleaning

			CREW	DAILY OUTPUT	MAN-HOURS	UNIT	BARE COSTS				TOTAL INCL O&P
							MAT.	LABOR	EQUIP.	TOTAL	
102	0010	CLEANING MASONRY No staging included										102
	0200	Chemical cleaning, brush and wash, minimum	D-1	800	.020	S.F.	.05	.40		.45	.71	
	0220	Average		400	.040		.09	.81		.90	1.40	
	0240	Maximum		330	.048		.12	.98		1.10	1.70	
	0400	High pressure water only, minimum	B-9	2,000	.020			.36	.06	.42	.65	
	0420	Average		1,500	.027			.48	.08	.56	.87	
	0440	Maximum		1,000	.040			.72	.13	.85	1.31	
	0800	High pressure water and chemical, minimum		1,800	.022		.05	.40	.07	.52	.79	
	0820	Average		1,200	.033		.10	.60	.11	.81	1.20	
	0840	Maximum		800	.050		.16	.90	.16	1.22	1.82	
	1200	Sandblast, wet system, minimum		1,750	.023		.19	.41	.07	.67	.96	
	1220	Average		1,100	.036		.26	.65	.12	1.03	1.48	

For expanded coverage of these items see *Means Concrete Cost Data 1991*

045 100 | Masonry Cleaning

			DAILY OUTPUT	MAN-HOURS	UNIT	BARE COSTS				TOTAL INCL O&P	
						MAT.	LABOR	EQUIP.	TOTAL		
102	1240	Maximum	B-9	700	.057	S.F.	.28	1.02	.18	1.48	2.18	102
	1400	Dry system, minimum		2,500	.016		.16	.29	.05	.50	.70	
	1420	Average		1,750	.023		.22	.41	.07	.70	.99	
	1440	Maximum		1,000	.040		.28	.72	.13	1.13	1.62	
	1800	For walnut shells, add					.38			.38	.42	
	1820	For corn chips, add					.38			.38	.42	
	2000	Steam cleaning, minimum	B-9	3,000	.013			.24	.04	.28	.44	
	2020	Average		2,500	.016			.29	.05	.34	.52	
	2040	Maximum		1,500	.027			.48	.08	.56	.87	
	2060											
	4000	Add for masking doors and windows				S.F.					.80	
	4200	Add for pedestrian protection				Job					10%	
	4400	Add for wire cut face brick				S.F.					.28	
	9000	Minimum labor/equipment charge	D-4	2	16	Job		320	61	381	575	

045 200 | Masonry Restoration

			CREW	DAILY OUTPUT	MAN-HOURS	UNIT	MAT.	LABOR	EQUIP.	TOTAL	INCL O&P	
210	0010	**CAULKING MASONRY** No staging included, ½" x ½" joint										210
	0050	Re-caulk only, oil base	1 Bric	225	.036	L.F.	.26	.81		1.07	1.58	
	0100	Butyl		205	.039		.40	.89		1.29	1.86	
	0200	Polysulfide		200	.040		.58	.91		1.49	2.10	
	0300	Silicone		195	.041		.73	.93		1.66	2.30	
	0350											
	1000	Cut out and re-caulk, oil base	1 Bric	145	.055	L.F.	.29	1.26		1.55	2.33	
	1050	Butyl		130	.062		.44	1.40		1.84	2.73	
	1100	Polysulfide		125	.064		.64	1.46		2.10	3.04	
	1150	Silicone		125	.064		.76	1.46		2.22	3.17	
	9000	Minimum labor/equipment charge		4	2	Job		46		46	73	
240	0010	**NEEDLE MASONRY** Includes shoring										240
	0020											
	0400	Block, concrete, 8" thick	B-9	7.10	5.630	Ea.	38	100	17.90	155.90	225	
	0420	12" thick		6.70	5.970		44	105	19	168	245	
	0800	Brick, 4" thick with 8" backup block		5.70	7.020		44	125	22	191	280	
	0810											
	1000	Brick, solid, 8" thick	B-9	6.20	6.450	Ea.	38	115	21	174	255	
	1040	12" thick		4.90	8.160		44	145	26	215	315	
	1080	16" thick		4.50	8.890		50	160	28	238	345	
	2000	Add for additional floors of shoring	B-1	6	4		38	73		111	160	
	9000	Minimum labor/equipment charge	"	2	12	Job		220		220	355	
270	0010	**POINTING MASONRY**										270
	0020											
	0300	Cut and repoint brick, hard mortar, running bond	1 Bric	80	.100	S.F.	.20	2.28		2.48	3.87	
	0320	Common bond		77	.104		.20	2.36		2.56	4.01	
	0360	Flemish bond		70	.114		.21	2.60		2.81	4.40	
	0400	English bond		65	.123		.21	2.80		3.01	4.72	
	0600	Soft old mortar, running bond		100	.080		.20	1.82		2.02	3.14	
	0620	Common bond		96	.083		.20	1.90		2.10	3.26	
	0640	Flemish bond		90	.089		.21	2.02		2.23	3.47	
	0680	English bond		82	.098		.21	2.22		2.43	3.79	
	0700	Stonework, hard mortar		140	.057	L.F.	.28	1.30		1.58	2.39	
	0720	Soft old mortar		160	.050	"	.28	1.14		1.42	2.13	
	1000	Repoint, mask and grout method, running bond		95	.084	S.F.	.28	1.92		2.20	3.38	
	1020	Common bond		90	.089		.28	2.02		2.30	3.55	
	1040	Flemish bond		86	.093		.28	2.12		2.40	3.70	
	1060	English bond		77	.104		.28	2.36		2.64	4.10	
	2000	Scrub coat, sand grout on walls, minimum		120	.067		.28	1.52		1.80	2.74	
	2020	Maximum		98	.082		.30	1.86		2.16	3.31	

For expanded coverage of these items see *Means Concrete Cost Data 1991*

045 200 | Masonry Restoration

		CREW	DAILY OUTPUT	MAN-HOURS	UNIT	MAT.	LABOR	EQUIP.	TOTAL	TOTAL INCL O&P	
270	9000	Minimum labor/equipment charge	1 Bric	3	2.670	Job		61		61	97	270
290	0010	**TOOTHING MASONRY**										
	0020											290
	0500	Brickwork, soft old mortar	1 Clab	40	.200	V.L.F.		3.50		3.50	5.70	
	0520	Hard mortar		30	.267			4.67		4.67	7.60	
	0700	Blockwork, soft old mortar		70	.114			2		2	3.27	
	0720	Hard mortar		50	.160			2.80		2.80	4.57	
	9000	Minimum labor/equipment charge	↓	4	2	Job		35		35	57	

045 650 | Fire Brick

		CREW	DAILY OUTPUT	MAN-HOURS	UNIT	MAT.	LABOR	EQUIP.	TOTAL	TOTAL INCL O&P	
654	0010	**FIRE CLAY** Gray, high duty, 100 lb. bag				Bag	30			30	33	654
	0050	100 lb. drum, premixed (400 brick per drum)				Drum	45			45	50	
656	0010	**FIREPLACE** For prefabricated fireplace, see div. 103-054										656
	0020											
	0100	Brick fireplace, not incl. foundations or chimneys										
	0110	30″ x 29″ opening, incl. chamber, plain brickwork	D-1	.40	40	Ea.	330	810		1,140	1,650	
	0200	Fireplace box only (110 brick)	"	2	8		120	160		280	390	
	0300	For elaborate brickwork and details, add			80		35%	35%				
	0400	For hearth, brick & stone, add	D-1	2	8		100	160		260	370	
	0410	For steel angle, damper, cleanouts, add		4	4		75	81		156	210	
	0600	Plain brickwork, incl. metal circulator		.50	32		585	645		1,230	1,675	
	0800	Face brick only, standard size, 8″ x 2-⅔″ x 4″	↓	.30	53.330	M	330	1,075		1,405	2,100	
	0900	Stone fireplace, fieldstone, add				S.F.Face					9.25	
	1000	Cut stone, add				"					9	
	9000	Minimum labor/equipment charge	D-1	2	8	Job		160		160	260	

050 500 | Metal Fastening

		CREW	DAILY OUTPUT	MAN-HOURS	UNIT	MAT.	LABOR	EQUIP.	TOTAL	TOTAL INCL O&P	
510	0010	**CURB EDGING** Steel angle w/anchors, on forms, 1″ x 1″, 1.5#/L.F.	E-4	350	.091	L.F.	2.35	2.25	.20	4.80	6.90	510
	0300	4″ x 4″ angles, 8#/L.F.	"	275	.116	"	5.65	2.86	.25	8.76	11.70	
	9000	Minimum labor/equipment charge	A-1	4	2	Job		35	13.60	48.60	72	
515	0010	**DRILLING** And layout for anchors, per										515
	0050	inch of depth, concrete or brick walls										
	0100	¼″ diameter	1 Carp	75	.107	Ea.	.07	2.35		2.42	3.91	
	0200	⅜″ diameter		63	.127		.09	2.79		2.88	4.66	
	0300	½″ diameter		50	.160		.11	3.52		3.63	5.85	
	0400	⅝″ diameter		48	.167		.15	3.67		3.82	6.15	
	0500	¾″ diameter		45	.178		.18	3.91		4.09	6.60	
	0600	⅞″ diameter		43	.186		.26	4.09		4.35	6.95	
	0700	1″ diameter		40	.200		.33	4.40		4.73	7.55	
	0800	1-¼″ diameter		38	.211		.56	4.63		5.19	8.20	
	0900	1-½″ diameter	↓	35	.229	↓	.90	5.05		5.95	9.20	
	1000	For ceiling installations add						40%				
	1100	Drilling & layout for drywall or plaster walls										
	1200	Holes, ¼″ diameter	1 Carp	150	.053	Ea.	.04	1.17		1.21	1.96	
	1300	⅜″ diameter		140	.057		.05	1.26		1.31	2.11	
	1400	½″ diameter	↓	130	.062		.06	1.35		1.41	2.28	

	050 500	Metal Fastening	CREW	DAILY OUTPUT	MAN-HOURS	UNIT	MAT.	LABOR	EQUIP.	TOTAL	TOTAL INCL O&P	
									BARE COSTS			
515	1500	¾″ diameter	1 Carp	120	.067	Ea.	.09	1.47		1.56	2.49	**515**
	1600	1″ diameter		110	.073		.17	1.60		1.77	2.80	
	1700	1-¼″ diameter		100	.080		.28	1.76		2.04	3.18	
	1800	1-½″ diameter		90	.089		.45	1.96		2.41	3.69	
	1900	For ceiling installations add						40%				
	2000	Minimum labor/equipment charge	1 Carp	4	2	Job		44		44	72	
520	0010	**EXPANSION ANCHORS & shields**										**520**
	0100	Bolt anchors for concrete, brick or stone, no layout and drilling										
	0200	Expansion shields, zinc, ¼″ diameter, 1″ long, single	1 Carp	90	.089	Ea.	.51	1.96		2.47	3.75	
	0300	1-⅜″ long, double		85	.094		.58	2.07		2.65	4.02	
	0400	⅜″ diameter, 2″ long, single		85	.094		.90	2.07		2.97	4.37	
	0500	2″ long, double		80	.100		1.08	2.20		3.28	4.78	
	0600	½″ diameter, 2-½″ long, single		80	.100		1.39	2.20		3.59	5.10	
	0700	2-½″ long, double		75	.107		1.41	2.35		3.76	5.40	
	0800	⅝″ diameter, 2-⅝″ long, single		75	.107		1.98	2.35		4.33	6	
	0900	3″ long, double		70	.114		2.14	2.51		4.65	6.45	
	1000	¾″ diameter, 2-¾″ long, single		70	.114		2.91	2.51		5.42	7.30	
	1100	4″ long, double		65	.123		4	2.71		6.71	8.80	
	1300	1″ diameter, 6″ long, double		60	.133		17.55	2.93		20.48	24	
	1500	Self drilling, steel, ¼″ diameter bolt		26	.308		.62	6.75		7.37	11.75	
	1600	⅜″ diameter bolt		23	.348		.94	7.65		8.59	13.55	
	1700	½″ diameter bolt		20	.400		1.41	8.80		10.21	15.90	
	1800	⅝″ diameter bolt		18	.444		2.49	9.80		12.29	18.70	
	1900	¾″ diameter bolt		16	.500		4.23	11		15.23	23	
	2000	⅞″ diameter bolt		14	.571		6.20	12.55		18.75	27	
	2100	Hollow wall anchors for gypsum board,										
	2200	plaster, tile or wall board										
	2300	⅛″ diameter, short				Ea.	.18			.18	.20	
	2400	Long					.19			.19	.21	
	2500	3/16″ diameter, short					.38			.38	.42	
	2600	Long					.41			.41	.45	
	2700	¼″ diameter, short					.46			.46	.51	
	2800	Long					.54			.54	.59	
	3000	Toggle bolts, bright steel, ⅛″ diameter, 2″ long	1 Carp	85	.094		.24	2.07		2.31	3.64	
	3100	4″ long		80	.100		.29	2.20		2.49	3.91	
	3200	3/16″ diameter, 3″ long		80	.100		.27	2.20		2.47	3.89	
	3300	6″ long		75	.107		.40	2.35		2.75	4.27	
	3400	¼″ diameter, 3″ long		75	.107		.30	2.35		2.65	4.16	
	3500	6″ long		70	.114		.44	2.51		2.95	4.59	
	3600	⅜″ diameter, 3″ long		70	.114		.67	2.51		3.18	4.84	
	3700	6″ long		60	.133		.95	2.93		3.88	5.85	
	3800	½″ diameter, 4″ long		60	.133		1.97	2.93		4.90	6.95	
	3900	6″ long		50	.160		2.62	3.52		6.14	8.65	
	4000	Nailing anchors										
	4100	Nylon anchor, standard nail, ¼″ diameter, 1″ long				C	10.80			10.80	11.90	
	4200	1-½″ long					14.05			14.05	15.45	
	4300	2″ long					21.25			21.25	23	
	4400	Zamac anchor, stainless nail, ¼″ diameter, 1″ long					20.55			20.55	23	
	4500	1-½″ long					24.95			24.95	27	
	4600	2″ long					37.90			37.90	42	
	8000	Wedge anchors, not including layout or drilling										
	8050	Carbon steel, ¼″ diameter, 1-¾″ long	1 Carp	150	.053	Ea.	.37	1.17		1.54	2.32	
	8100	3 ¼″ long		145	.055		.52	1.21		1.73	2.55	
	8150	⅜″ diameter, 2-¼″ long		150	.053		.49	1.17		1.66	2.45	
	8200	5″ long		145	.055		.88	1.21		2.09	2.95	
	8250	½″ diameter, 2-¾″ long		140	.057		.91	1.26		2.17	3.05	
	8300	7″ long		130	.062		1.60	1.35		2.95	3.97	
	8350	⅝″ diameter, 3-½″ long		130	.062		1.69	1.35		3.04	4.07	

		050 500 |Metal Fastening	CREW	DAILY OUTPUT	MAN-HOURS	UNIT	BARE COSTS				TOTAL INCL O&P	
							MAT.	LABOR	EQUIP.	TOTAL		
520	8400	8-½″ long	1 Carp	115	.070	Ea.	2.86	1.53		4.39	5.65	**520**
	8450	¾″ diameter, 4-¼″ long		115	.070		2.48	1.53		4.01	5.25	
	8500	10″ long		100	.080		5.05	1.76		6.81	8.45	
	8550	1″ diameter, 6″ long		100	.080		7.60	1.76		9.36	11.25	
	8600	12″ long		80	.100		10.95	2.20		13.15	15.65	
	8650	1-¼″ diameter, 9″ long		70	.114		14.85	2.51		17.36	20	
	8700	12″ long		60	.133		16.55	2.93		19.48	23	
	8750	For type 303 stainless steel, add					350%					
	8800	For type 316 stainless steel, add					450%					
	9000	Minimum labor/equipment charge	1 Carp	4	2	Job		44		44	72	
530	0010	LAG SCREWS Steel, ¼″ diameter, 2″ long		140	.057	Ea.	.04	1.26		1.30	2.10	**530**
	0200	½″ diameter, 3″ long		95	.084		.21	1.85		2.06	3.26	
	0300	⅝″ diameter, 3″ long		85	.094		.36	2.07		2.43	3.78	
540	0010	MACHINERY ANCHORS Standard, flush mounted,										**540**
	0020	incl. stud w/fiber plug, nut & washer, anchor & anchor bolt										
	0200	Material only, ½″ diameter stud & bolt				Ea.	20.95			20.95	23	
	0300	⅝″ diameter					22.85			22.85	25	
	0500	¾″ diameter					24.85			24.85	27	
	0600	⅞″ diameter					28.75			28.75	32	
	0800	1″ diameter					31.75			31.75	35	
	0900	1-¼″ diameter					40.40			40.40	44	
550	0010	STUDS .22 caliber stud driver, buy, minimum					230			230	255	**550**
	0100	Maximum					375			375	415	
	0300	Powder charges for above, low velocity				C	12			12	13.20	
	0400	Standard velocity					20			20	22	
	0600	Drive pins & studs, ¼″ & ⅜″ diam., to 3″ long, minimum					30			30	33	
	0700	Maximum					80			80	88	
	0800	Pneumatic stud driver for ⅛″ diameter studs				Ea.	850			850	935	
	0900	Drive pins for above, ½″ to ¾″ long				M	40			40	44	
565	0010	WELDED STUDS ¼″ diameter, 2-¹¹⁄₁₆″ long	E-10	1,030	.016	Ea.	.21	.39	.41	1.01	1.39	**565**
	0100	4-⅛″ long		1,030	.016		.22	.39	.41	1.02	1.40	
	0200	⅜″ diameter, 4-⅛″ long		1,030	.016		.31	.39	.41	1.11	1.50	
	0300	6-⅛″ long		1,030	.016		.40	.39	.41	1.20	1.60	
	9000	Minimum labor/equipment charge	1 Sswk	2	4	Job		96		96	175	
575	0010	WELDING Field. Cost per welder, no operating engineer	E-14	8	1	Hr.	2.55	26	8.55	37.10	60	**575**
	0200	With ½ operating engineer	E-13	8	1.500		2.55	37	8.55	48.10	77	
	0300	With 1 operating engineer	E-12	8	2		2.55	48	8.55	59.10	93	
	0500	With no operating engineer, minimum	E-14	13.30	.602	Ton	1.90	15.70	5.15	22.75	36	
	0600	Maximum	″	2.50	3.200		7.60	84	27	118.60	190	
	0800	With one operating engineer per welder, minimum	E-12	13.30	1.200		1.90	29	5.15	36.05	57	
	0900	Maximum	″	2.50	6.400		7.60	150	27	184.60	300	
	1200	Continuous fillet, stick welding, incl. equipment										
	1300	Single pass, ⅛″ thick, 0.1#/L.F.	E-14	240	.033	L.F.	.09	.87	.29	1.25	2	
	1400	³⁄₁₆″ thick, 0.2#/L.F.		120	.067		.17	1.74	.57	2.48	3.99	
	1500	¼″ thick, 0.3#/L.F.		80	.100		.26	2.61	.86	3.73	6	
	1610	⁵⁄₁₆″ thick, 0.4#/L.F.		60	.133		.34	3.48	1.14	4.96	7.95	
	1800	3 passes, ⅜″ thick, 0.5#/L.F.		48	.167		.43	4.35	1.43	6.21	9.95	
	2010	4 passes, ½″ thick, 0.7#/L.F.		34	.235		.60	6.15	2.01	8.76	14.05	
	2200	5 to 6 passes, ¾″ thick, 1.3#/L.F.		19	.421		1.11	11	3.60	15.71	25	
	2400	8 to 11 passes, 1″ thick, 2.4#/L.F.		10	.800		2.04	21	6.85	29.89	48	
	2600	For all position welding, add, minimum						20%				
	2700	Maximum						300%				
	2900	For semi-automatic welding, deduct, minimum						5%				
	3000	Maximum						15%				
	4000	Cleaning and welding plates, bars, or rods										
	4010	to existing beams, columns, or trusses	E-14	12	.667	L.F.	.40	17.40	5.70	23.50	38	

050 | Metal Materials, Finishes and Fastenings

		050 500	Metal Fastening	CREW	DAILY OUTPUT	MAN-HOURS	UNIT	BARE COSTS				TOTAL INCL O&P	
								MAT.	LABOR	EQUIP.	TOTAL		
575	9000		Minimum labor/equipment charge	E-14	4	2	Job		52	17.10	69.10	115	575

051 | Structural Metal Framing

		051 100	Bracing	CREW	DAILY OUTPUT	MAN-HOURS	UNIT	BARE COSTS				TOTAL INCL O&P	
								MAT.	LABOR	EQUIP.	TOTAL		
108	0010	**BRACING**											108
	0300	Let-in, "T" shaped, 20 ga. galv. steel, studs at 16" O.C.		F-1	580	.014	L.F.	.60	.30	.01	.91	1.17	
	0400	Studs at 24" O.C.			600	.013		.60	.29	.01	.90	1.15	
	0500	16 ga. galv. steel straps, studs at 16" O.C.			600	.013		.54	.29	.01	.84	1.08	
	0600	Studs at 24" O.C.			620	.013		.54	.28	.01	.83	1.07	
110	0010	**PIPE SUPPORT** Framing, under 10#/L.F.		E-4	3,900	.008	Lb.	.75	.20	.02	.97	1.22	110
	0200	10.1 to 15#/L.F.			4,300	.007		.65	.18	.02	.85	1.07	
	0400	15.1 to 20#/L.F.			4,800	.007		.55	.16	.01	.72	.92	
	0600	Over 20#/L.F.			5,400	.006		.50	.15	.01	.66	.83	

		051 200	Structural Steel	CREW	DAILY OUTPUT	MAN-HOURS	UNIT	BARE COSTS				TOTAL INCL O&P	
								MAT.	LABOR	EQUIP.	TOTAL		
212	0010	**CANOPY FRAMING** 6" and 8" members		E-4	3,000	.011	Lb.	.80	.26	.02	1.08	1.38	212
	9000	Minimum labor/equipment charge		1 Sswk	1	8	Job		195		195	350	
215	0010	**CEILING SUPPORTS**											215
	1000	Entrance door/folding partition supports		E-4	60	.533	L.F.	10.50	13.10	1.14	24.74	37	
	1100	Linear accelerator door supports			14	2.290		45	56	4.89	105.89	155	
	1200	Lintels or shelf angles, hung, exterior hot dipped galv.			267	.120		8	2.95	.26	11.21	14.45	
	1250	Two coats primer paint instead of galv.			267	.120		6.50	2.95	.26	9.71	12.80	
	1400	Monitor support, ceiling hung, expansion bolted			4	8	Ea.	200	195	17.10	412.10	595	
	1450	Hung from pre-set inserts			6	5.330		200	130	11.40	341.40	470	
	1600	Motor supports for overhead doors			4	8		115	195	17.10	327.10	505	
	1700	Partition support for heavy folding partitions, without pocket			24	1.330	L.F.	25	33	2.85	60.85	90	
	1750	Supports at pocket only			12	2.670		50	66	5.70	121.70	180	
	2000	Rolling grilles & fire door supports			34	.941		20	23	2.01	45.01	66	
	2100	Spider-leg light supports, expansion bolted to ceiling slab			8	4	Ea.	68	98	8.55	174.55	265	
	2150	Hung from pre-set inserts			12	2.670	"	68	66	5.70	139.70	200	
	2400	Toilet partition support			36	.889	L.F.	30	22	1.90	53.90	75	
	2500	X-ray travel gantry support			12	2.670	"	85	66	5.70	156.70	220	
220	0010	**COLUMNS** Aluminum, extruded, stock units, 6" diameter		E-4	240	.133	L.F.	8	3.28	.29	11.57	15.10	220
	0100	8" diameter		"	170	.188		11.10	4.63	.40	16.13	21	
	0500	For square columns, add to column prices above						50%					
	0800	Steel, concrete filled, extra strong pipe, 3-½" diameter		E-2	660	.085		10	2	1.62	13.62	16.30	
	0830	4" diameter			780	.072		12	1.69	1.37	15.06	17.70	
	0890	5" diameter			1,020	.055		16	1.29	1.05	18.34	21	
	1500	Steel pipe, extra strong, no concrete, 3" to 5" O.D.			12,960	.004	Lb.	.80	.10	.08	.98	1.15	
	3300	Square structural tubing, 4" to 6" square, light section			11,270	.005	"	.75	.12	.10	.97	1.14	
	9000	Minimum labor/equipment charge		1 Sswk	1	8	Job		195		195	350	
230	0010	**LIGHTWEIGHT FRAMING**											230
	0200	For steel studs see division 092-612											
	0400	Angle framing, 4" and larger		E-4	3,000	.011	Lb.	.72	.26	.02	1	1.29	
	0450	Less than 4" angles		"	1,800	.018		.75	.44	.04	1.23	1.67	
	1400	Tie rod, not upset, 1-½" to 4" diameter, with turnbuckle		2 Sswk	800	.020		.75	.48		1.23	1.70	
	1420	No turnbuckle		"	700	.023		.68	.55		1.23	1.75	

051 | Structural Metal Framing

051 200 | Structural Steel

			CREW	DAILY OUTPUT	MAN-HOURS	UNIT	BARE COSTS MAT.	LABOR	EQUIP.	TOTAL	TOTAL INCL O&P	
230	9000	Minimum labor/equipment charge	2 Sswk	2	8	Job		195		195	350	230
232	0010	**LINTELS** Plain steel angles, under 500 lb.	1 Bric	500	.016	Lb.	.50	.36		.86	1.13	232
	0100	500 to 1000 lb.		600	.013	"	.46	.30		.76	.99	
	2000	Steel angles, 3-½" x 3", ¼" thick, 2'-6" long		50	.160	Ea.	6.75	3.64		10.39	13.25	
	2100	4'-6" long		45	.178		12.15	4.04		16.19	19.85	
	2600	4" x 3-½", ¼" thick, 5'-0" long		40	.200		15.50	4.55		20.05	24	
	2700	9'-0" long		35	.229		27.90	5.20		33.10	39	
	3500	For precast concrete lintels, see div. 034-802										
	9000	Minimum labor/equipment charge	1 Bric	4	2	Job		46		46	73	
240	0010	**STEEL CUTTING** Hand burning, including preparation,										240
	0020	torch cutting, and grinding, no staging										
	0100	Steel to ½" thick	A-1	70	.114	L.F.	.40	2	.78	3.18	4.56	
	0150	¾" thick		50	.160		.56	2.80	1.09	4.45	6.40	
	0200	1" thick		45	.178		.62	3.11	1.21	4.94	7.10	
	9000	Minimum labor/equipment charge		2	4	Job		70	27	97	145	
255	0010	**STRUCTURAL STEEL PROJECTS** Bolted, unless mentioned otherwise										255
	1300	Industrial bldgs., 1 story, beams & girders, steel bearing	E-5	12.90	6.200	Ton	1,050	150	93	1,293	1,525	
	1400	Masonry bearing		10	8		1,050	190	120	1,360	1,625	
	1600	1 story with roof trusses, steel bearing		10.60	7.550		1,250	180	115	1,545	1,825	
	1700	Masonry bearing		8.30	9.640		1,250	230	145	1,625	1,950	
260	0010	**STRUCTURAL STEEL** Bolted, incl. fabrication, not incl. trucking										260
	0050	Beams, W 6 x 9	E-2	720	.078	L.F.	6	1.83	1.49	9.32	11.45	
	0100	W 8 x 10		720	.078		7	1.83	1.49	10.32	12.55	
	0150	W 10 x 15		720	.078		8	1.83	1.49	11.32	13.65	
	0200	Columns, W 6 x 15		540	.104		8	2.44	1.98	12.42	15.30	
	0250	W 8 x 31		540	.104		17	2.44	1.98	21.42	25	
	0500	Girders, W 12 x 22		900	.062		12	1.47	1.19	14.66	17.10	
	0550	W 14 x 26		900	.062		14	1.47	1.19	16.66	19.30	
	0600	W 16 x 31		900	.062		16	1.47	1.19	18.66	21	
	0700	Joists (bar joists, H or K series), span to 30'	E-7	30,000	.003	Lb.	.35	.06	.04	.45	.55	
	0750	Span to 50'	"	20,000	.004	"	.33	.10	.06	.49	.60	
	9000	Minimum labor/equipment charge	E-2	2	28	Job		660	535	1,195	1,750	

(76)

052 | Metal Joists

052 100 | Steel Joists

			CREW	DAILY OUTPUT	MAN-HOURS	UNIT	BARE COSTS MAT.	LABOR	EQUIP.	TOTAL	TOTAL INCL O&P	
108	0011	**LIGHTGAGE JOISTS**										108
	0050											
	0120	Punched, double nailable, 6" deep, 16 gauge, galv.	E-4	1,800	.018	L.F.	1.30	.44	.04	1.78	2.27	
	0140	14 gauge		1,800	.018		1.65	.44	.04	2.13	2.66	
	0160	8" deep, 16 gauge		1,550	.021		1.55	.51	.04	2.10	2.68	
	0180	14 gauge		1,550	.021		1.95	.51	.04	2.50	3.12	
	0200	10" deep, 14 gauge		1,350	.024		2.25	.58	.05	2.88	3.60	
	0220	12 gauge		1,350	.024		3.25	.58	.05	3.88	4.70	
	0240	12" deep, 14 gauge		1,200	.027		2.55	.66	.06	3.27	4.07	
	0260	12 gauge		1,200	.027		3.75	.66	.06	4.47	5.40	
	9000	Minimum labor/equipment charge	1 Sswk	2	4	Job		96		96	175	

053 100 | Steel Deck

104		CREW	DAILY OUTPUT	MAN-HOURS	UNIT	BARE COSTS MAT.	LABOR	EQUIP.	TOTAL	TOTAL INCL O&P	104
0010	METAL DECKING Steel floor panels, over 15,000 S.F.										
0200	Cellular units, galvanized, 2″ deep, 20-20 gauge	E-4	1,460	.022	S.F.	2.45	.54	.05	3.04	3.73	
0400	3″ deep, galvanized, 20-20 gauge		1,375	.023		2.55	.57	.05	3.17	3.91	
1000	4-½″ deep, galvanized, 20-18 gauge	↓	1,100	.029		4.70	.72	.06	5.48	6.55	
1500	For acoustical deck, add					1			1	1.10	
1900	For congested site, add						50%				
2100	Open type, galv., 1-½″ deep, 22 ga., under 50 square	E-4	4,500	.007		.71	.17	.02	.90	1.12	
2400	Over 50 square		4,900	.007		.68	.16	.01	.85	1.06	
5200	Non-cellular composite deck, galv., 2″ deep, 22 gauge		3,860	.008		.77	.20	.02	.99	1.24	
5300	20 gauge		3,600	.009		.93	.22	.02	1.17	1.44	
5400	18 gauge		3,380	.009		1.22	.23	.02	1.47	1.79	
5500	16 gauge		3,200	.010		1.53	.25	.02	1.80	2.15	
5700	3″ deep, galv., 22 gauge		3,200	.010		.84	.25	.02	1.11	1.39	
5800	20 gauge		3,000	.011		1.01	.26	.02	1.29	1.61	
5900	18 gauge		2,850	.011		1.32	.28	.02	1.62	1.98	
6000	16 gauge	↓	2,700	.012		1.66	.29	.03	1.98	2.39	
6100	Slab form, steel 28 gauge, 9/16″ deep, uncoated	E-1	4,000	.006		.32	.14	.02	.48	.62	
6200	Galvanized	"	4,000	.006	↓	.39	.14	.02	.55	.70	
9000	Minimum labor/equipment charge	1 Sswk	1	8	Job		195		195	350	

055 100 | Metal Stairs

104		CREW	DAILY OUTPUT	MAN-HOURS	UNIT	BARE COSTS MAT.	LABOR	EQUIP.	TOTAL	TOTAL INCL O&P	104
0010	STAIR Steel, safety nosing, steel stringers										
0020	Grating tread and pipe railing, 3′-6″ wide	E-4	35	.914	Riser	70	22	1.95	93.95	120	
0100	4′-0″ wide		30	1.070		75	26	2.28	103.28	135	
0200	Cement fill metal pan and picket rail, 3′-6″ wide		35	.914		60	22	1.95	83.95	110	
0300	4′-0″ wide		30	1.070		64	26	2.28	92.28	120	
0350	Wall rail, both sides, 3′-6″ wide		53	.604		45	14.85	1.29	61.14	78	
0500	Checkered plate tread, industrial, 3′-6″ wide		28	1.140		80	28	2.44	110.44	140	
0550	Circular for tanks, 3′-0″ wide		33	.970		69	24	2.07	95.07	120	
0600	For isolated stairs, add						100%				
0800	Custom steel stairs, minimum	E-4	35	.914		85	22	1.95	108.95	135	
0810	Average		30	1.070		130	26	2.28	158.28	195	
0900	Maximum	↓	20	1.600		180	39	3.42	222.42	275	
1100	For 4′ wide stairs, add					10%	5%				
1300	For 5′ wide stairs, add				↓	20%	10%				
1500	Landing, steel pan, conventional	E-4	160	.200	S.F.	30	4.92	.43	35.35	42	
1810	Spiral aluminum, 5′-0″ diameter, stock units		45	.711	Riser	205	17.50	1.52	224.02	260	
1820	Custom units		45	.711		310	17.50	1.52	329.02	375	
1830	Stock units, 4′-0″ diameter, safety treads		50	.640		230	15.75	1.37	247.12	285	
1840	Oak treads		50	.640		245	15.75	1.37	262.12	300	
1850	5′-0″ diameter, safety treads		45	.711		250	17.50	1.52	269.02	310	
1860	Oak treads		45	.711		270	17.50	1.52	289.02	330	
1870	6′-0″ diameter, safety treads		40	.800		285	19.70	1.71	306.41	350	
1880	Oak treads		40	.800		320	19.70	1.71	341.41	390	
1900	Spiral, cast iron, 4′-0″ diameter, ornamental, minimum		45	.711		160	17.50	1.52	179.02	210	
1920	Maximum		25	1.280		235	31	2.74	268.74	320	
2000	Spiral, steel, industrial checkered plate, 4′ diameter		45	.711		80	17.50	1.52	99.02	120	
2200	Stock units, 6′-0″ diameter		40	.800		125	19.70	1.71	146.41	175	
3900	Industrial ships ladder, 3′ W, grating treads, 2 line pipe rail	↓	30	1.070	↓	54	26	2.28	82.28	110	

055 100 | Metal Stairs

		CREW	DAILY OUTPUT	MAN-HOURS	UNIT	BARE COSTS MAT.	BARE COSTS LABOR	BARE COSTS EQUIP.	BARE COSTS TOTAL	TOTAL INCL O&P	
104	4000	Aluminum	E-4	30	1.070	Riser	72	26	2.28	100.28	130	104
	9000	Minimum labor/equipment charge	"	2	16	Job		395	34	429	755	

055 150 | Ladders

		CREW	DAILY OUTPUT	MAN-HOURS	UNIT	MAT.	LABOR	EQUIP.	TOTAL	INCL O&P	
154	0010	**FIRE ESCAPE**										154
	0200	2' wide balcony, 1" x ¼" bars 1-½" O.C.				L.F.	28			28	31	
	0400	1st story cantilever, standard				Ea.	1,225			1,225	1,350	
	0500	Cable counterweight				"	1,130			1,130	1,250	
	0700	Platform & stair, 36" x 40"				Flight	550			550	605	
	0900	For 3'-6" wide escapes, add to above					100%					
156	0010	**FIRE ESCAPE LADDERS** One story, collapsible				Ea.	560			560	615	156
	0100	Portable				"	145			145	160	
158	0010	**LADDER** Steel, bolted to concrete, with cage	E-4	50	.640	V.L.F.	45	15.75	1.37	62.12	80	158
	0100	Without cage		85	.376		22	9.25	.80	32.05	42	
	0300	Aluminum, bolted to concrete, with cage		50	.640		62	15.75	1.37	79.12	98	
	0400	Without cage		85	.376		36	9.25	.80	46.05	57	
	9000	Minimum labor/equipment charge		2	16	Job		395	34	429	755	

055 200 | Handrails & Railings

		CREW	DAILY OUTPUT	MAN-HOURS	UNIT	MAT.	LABOR	EQUIP.	TOTAL	INCL O&P	
202	0010	**BUMPER RAILS** For garages, 12 ga. rail, 6" wide, with steel										202
	0020	posts 12'-6" O.C., Minimum	E-4	190	.168	L.F.	9.40	4.14	.36	13.90	18.30	
	0030	Average		165	.194		11.40	4.77	.41	16.58	22	
	0100	Maximum		140	.229		14.10	5.60	.49	20.19	26	
	9000	Minimum labor/equipment charge	A-1	1	8	Job		140	54	194	290	
203	0010	**RAILING, PIPE** Aluminum, 2 rail, 1-¼" diam., satin finish	E-4	160	.200	L.F.	9.50	4.92	.43	14.85	19.90	203
	0030	Clear anodized		160	.200		12.40	4.92	.43	17.75	23	
	0040	Dark anodized		160	.200		13.05	4.92	.43	18.40	24	
	0080	1-½" diameter, satin finish		160	.200		10.05	4.92	.43	15.40	20	
	0090	Clear anodized		160	.200		13.10	4.92	.43	18.45	24	
	0100	Dark anodized		160	.200		13.80	4.92	.43	19.15	25	
	0140	Aluminum, 3 rail, 1-¼" diam., satin finish		137	.234		14.10	5.75	.50	20.35	27	
	0150	Clear anodized		137	.234		19.10	5.75	.50	25.35	32	
	0160	Dark anodized		137	.234		19.50	5.75	.50	25.75	32	
	0200	1-½" diameter, satin finish		137	.234		14.95	5.75	.50	21.20	27	
	0210	Clear anodized		137	.234		19.40	5.75	.50	25.65	32	
	0220	Dark anodized		137	.234		21.05	5.75	.50	27.30	34	
	0500	Steel, 2 rail, primed, 1-¼" diameter		160	.200		6.85	4.92	.43	12.20	16.95	
	0520	1-½" diameter		160	.200		7.50	4.92	.43	12.85	17.70	
	0540	Galvanized, 1-¼" diameter		160	.200		9.40	4.92	.43	14.75	19.75	
	0560	1-½" diameter		160	.200		10.60	4.92	.43	15.95	21	
	0580	Steel, 3 rail, primed, 1-¼" diameter		137	.234		10.15	5.75	.50	16.40	22	
	0600	1-½" diameter		137	.234		10.85	5.75	.50	17.10	23	
	0620	Galvanized, 1-¼" diameter		137	.234		14.10	5.75	.50	20.35	27	
	0640	1-½" diameter		137	.234		15.35	5.75	.50	21.60	28	
	0700	Stainless steel, 2 rail, 1-¼" diam. #4 finish		137	.234		22.05	5.75	.50	28.30	35	
	0720	High polish		137	.234		35.40	5.75	.50	41.65	50	
	0740	Mirror polish		137	.234		43.40	5.75	.50	49.65	59	
	0760	Stainless steel, 3 rail, 1-½" diam., #4 finish		120	.267		34.10	6.55	.57	41.22	50	
	0770	High polish		120	.267		54.20	6.55	.57	61.32	72	
	0780	Mirror finish		120	.267		65	6.55	.57	72.12	84	
	0900	Wall rail, alum. pipe, 1-¼" diam., satin finish		213	.150		6.50	3.70	.32	10.52	14.25	
	0905	Clear anodized		213	.150		8.15	3.70	.32	12.17	16.05	
	0910	Dark anodized		213	.150		8.85	3.70	.32	12.87	16.85	
	0915	1-½" diameter, satin finish		213	.150		7.40	3.70	.32	11.42	15.25	
	0920	Clear anodized		213	.150		9.50	3.70	.32	13.52	17.55	
	0925	Dark anodized		213	.150		10.05	3.70	.32	14.07	18.15	

055 200 | Handrails & Railings

			CREW	DAILY OUTPUT	MAN-HOURS	UNIT	MAT.	LABOR	EQUIP.	TOTAL	TOTAL INCL O&P	
203	0930	Steel pipe, 1-¼" diameter, primed	E-4	213	.150	L.F.	4.05	3.70	.32	8.07	11.55	203
	0935	Galvanized		213	.150		5.65	3.70	.32	9.67	13.30	
	0940	1-½" diameter, primed		213	.150		4.30	3.70	.32	8.32	11.80	
	0945	Galvanized		213	.150		5.85	3.70	.32	9.87	13.55	
	0955	Stainless steel pipe, 1-½" diam., #4 finish		107	.299		21.10	7.35	.64	29.09	37	
	0960	High polish		107	.299		34.05	7.35	.64	42.04	52	
	0965	Mirror polish		107	.299		42.10	7.35	.64	50.09	60	
	9000	Minimum labor/equipment charge	1 Sswk	2	4	Job		96		96	175	
206	0010	RAILINGS, INDUSTRIAL Welded, 2 rail, 3'-6" high, 1-½" pipe	E-4	255	.125	L.F.	10.50	3.09	.27	13.86	17.45	206
	0200	For 4" high kick plate, 10 gauge, add					2.15			2.15	2.37	
	0500	For curved rails, add					30%					
	9000	Minimum labor/equipment charge	1 Sswk	2	4	Job		96		96	175	
208	0010	RAILINGS, ORNAMENTAL Aluminum, bronze or stainless, minimum		24	.333	L.F.	59	8.05		67.05	80	208
	0100	Maximum		9	.889		350	21		371	425	
	0200	Aluminum pipe rail, minimum		15	.533		40	12.85		52.85	67	
	0300	Maximum		8	1		125	24		149	180	
	0400	Hand-forged wrought iron, minimum		12	.667		58	16.05		74.05	93	
	0500	Maximum		8	1		220	24		244	285	
	0600	Composite metal and wood or glass, minimum		6	1.330		125	32		157	195	
	0700	Maximum		5	1.600		300	39		339	400	
	9000	Minimum labor/equipment charge		2	4	Job		96		96	175	

055 400 | Castings

			CREW	DAILY OUTPUT	MAN-HOURS	UNIT	MAT.	LABOR	EQUIP.	TOTAL	TOTAL INCL O&P	
404	0010	CONSTRUCTION CASTINGS										404
	0020	Manhole covers and frames see div. 027-152										
	0100	Column bases, cast iron, 16" x 16", approx. 65 lbs.	E-4	46	.696	Ea.	60	17.10	1.49	78.59	99	
	0200	32" x 32", approx. 256 lbs.		23	1.390	"	240	34	2.97	276.97	330	
	0600	Miscellaneous C.I. castings, light sections		3,200	.010	Lb.	.70	.25	.02	.97	1.24	
	1300	Special low volume items		3,200	.010	"	1.60	.25	.02	1.87	2.23	

055 500 | Metal Specialties

			CREW	DAILY OUTPUT	MAN-HOURS	UNIT	MAT.	LABOR	EQUIP.	TOTAL	TOTAL INCL O&P	
504	0010	LAMP POSTS Only, 7' high, stock units, aluminum	1 Carp	16	.500	Ea.	37	11		48	59	504
	0100	Mild steel, plain		16	.500	"	26	11		37	47	
	9000	Minimum labor/equipment charge		4	2	Job		44		44	72	
508	0010	WINDOW GUARDS Expanded metal, steel angle frame, permanent	E-4	350	.091	S.F.	11.50	2.25	.20	13.95	16.95	508
	0020	Steel bars, ½" x ½", spaced 5" O.C.	"	290	.110	"	7.90	2.71	.24	10.85	13.90	
	0030	Hinge mounted, add				Opng.	19.50			19.50	21	
	0040	Removable type, add				"	13.50			13.50	14.85	
	0050	For galvanized guards, add				S.F.	35%					
	0070	For pivoted or projected type, add					110%	40%				
	0100	Mild steel, stock units, economy	E-4	405	.079		3.05	1.94	.17	5.16	7.10	
	0200	Deluxe		405	.079		6.40	1.94	.17	8.51	10.75	
	0400	Woven wire, stock units, ⅜" channel frame, 3' x 5' opening		40	.800	Opng.	85	19.70	1.71	106.41	130	
	0500	4' x 6' opening		38	.842		135	21	1.80	157.80	190	
	0800	Basket guards for above, add					125			125	140	
	1000	Swinging guards for above, add					35			35	39	
	9000	Minimum labor/equipment charge	1 Sswk	2	4	Job		96		96	175	

057 | Ornamental Metal

		057 250	Ornamental Metal	CREW	DAILY OUTPUT	MAN-HOURS	UNIT	BARE COSTS				TOTAL INCL O&P	
								MAT.	LABOR	EQUIP.	TOTAL		
252	0010	WEATHERVANES Residential types, minimum		1 Carp	8	1	Ea.	25	22		47	63	252
	0100	Maximum			2	4	"	600	88		688	805	
	9000	Minimum labor/equipment charge		↓	4	2	Job		44		44	72	

058 | Expansion Control

		058 100	Exp. Cover Assemblies	CREW	DAILY OUTPUT	MAN-HOURS	UNIT	BARE COSTS				TOTAL INCL O&P	
								MAT.	LABOR	EQUIP.	TOTAL		
104	0010	EXPANSION JOINT ASSEMBLIES Custom units											104
	0020												
	0200	Floor cover assemblies, 1" space, aluminum		1 Sswk	38	.211	L.F.	12.40	5.05		17.45	23	
	0300	Bronze or stainless			38	.211		20	5.05		25.05	31	
	0500	2" space, aluminum			38	.211		16.80	5.05		21.85	28	
	0600	Bronze or stainless			38	.211		27	5.05		32.05	39	
	0800	Wall and ceiling assemblies, 1" space, aluminum			38	.211		7.30	5.05		12.35	17.25	
	0900	Bronze or stainless			38	.211		24	5.05		29.05	36	
	1100	2" space, aluminum			38	.211		9.70	5.05		14.75	19.90	
	1200	Bronze or stainless			38	.211		34	5.05		39.05	47	
	1400	Floor to wall assemblies, 1" space, aluminum			38	.211		8.90	5.05		13.95	19.05	
	1500	Bronze or stainless			38	.211		29	5.05		34.05	41	
	1700	Gym floor angle covers, aluminum, 3" x 3" angle			46	.174		10	4.19		14.19	18.65	
	1800	3" x 4" angle			46	.174		11	4.19		15.19	19.75	
	2000	Roof closures, aluminum, 1" space, flat roof, low profile			57	.140		23	3.38		26.38	31	
	2100	High profile			57	.140		24	3.38		27.38	33	
	2300	Roof to wall, 1" space, low profile			57	.140		8.10	3.38		11.48	15.05	
	2400	High profile			57	.140		10	3.38		13.38	17.15	
	9000	Minimum labor/equipment charge		↓	2	4	Job		96		96	175	

060 | Fasteners and Adhesives

		060 500	Fasteners & Adhesives	CREW	DAILY OUTPUT	MAN-HOURS	UNIT	BARE COSTS				TOTAL INCL O&P	
								MAT.	LABOR	EQUIP.	TOTAL		
504	0010	NAILS Prices of material only, copper, plain					Lb.	3.75			3.75	4.13	504
	0400	Stainless steel, plain						5			5	5.50	
	0600	Common, 3d to 60d, plain						.70			.70	.77	
	0700	Galvanized						.90			.90	.99	
	0800	Aluminum						4.40			4.40	4.84	
	1000	Annular or spiral thread, 4d to 60d, plain						1			1	1.10	
	1200	Galvanized						1.10			1.10	1.21	
	1400	Drywall nails, plain						.98			.98	1.08	
	1600	Galvanized						1.20			1.20	1.32	
	1800	Finish nails, 4d to 10d, plain						.75			.75	.83	
	2000	Galvanized						.95			.95	1.05	
	2100	Aluminum						4.40			4.40	4.84	
	2300	Flooring nails, hardened steel, 2d to 10d, plain						1			1	1.10	
	2400	Galvanized						1.10			1.10	1.21	

68   For expanded coverage of these items see *Means Interior Cost Data 1991*

		060 500	Fasteners & Adhesives	CREW	DAILY OUTPUT	MAN-HOURS	UNIT	MAT.	LABOR	EQUIP.	TOTAL	TOTAL INCL O&P	
504	2500		Gypsum lath nails, 1-⅛", 13 ga. flathead, blued				Lb.	1.20			1.20	1.32	**504**
	2600		Masonry nails, hardened steel, ¾" to 3" long, plain					1			1	1.10	
	2700		Galvanized					1.15			1.15	1.27	
	2900		Roofing nails, threaded, galvanized					.90			.90	.99	
	3100		Aluminum					4.70			4.70	5.15	
	3300		Compressed lead head, threaded, galvanized					.98			.98	1.08	
	3600		Siding nails, plain shank, galvanized					.95			.95	1.05	
	3800		Aluminum					3.95			3.95	4.35	
	5000		Add to prices above for cement coating					.05			.05	.06	
	5200		Zinc or tin plating					.10			.10	.11	
512	0010		**TIMBER CONNECTORS** Add up cost of each part for total										**512**
	0020		cost of connection										
	0100		Connector plates, steel, with bolts, straight	2 Carp	75	.213	Ea.	13.80	4.69		18.49	23	
	0110		Tee	"	50	.320		20.50	7.05		27.55	34	
	0200		Bolts, machine, sq. hd. with nut & washer, ½" diameter, 4" long	1 Carp	140	.057		.85	1.26		2.11	2.99	
	0300		7-½" long		130	.062		1.05	1.35		2.40	3.37	
	0500		¾" diameter, 7-½" long		130	.062		1.70	1.35		3.05	4.08	
	0600		15" long		95	.084		2.05	1.85		3.90	5.30	
	0800		Drilling bolt holes in timber, ½" diameter		450	.018	Inch		.39		.39	.64	
	0900		1" diameter		350	.023	"		.50		.50	.82	
	1100		Framing anchors, 2 or 3 dimensional, 10 gauge, no nails incl.		175	.046	Ea.	.38	1.01		1.39	2.06	
	1200												
	1250		Holdowns, 3 gauge base, 10 gauge body	1 Carp	8	1	Ea.	12.10	22		34.10	49	
	1300		Joist and beam hangers, 18 ga. galv., for 2" x 4" joist		175	.046		.38	1.01		1.39	2.06	
	1400		2" x 6" to 2" x 10" joist		165	.048		.55	1.07		1.62	2.35	
	1600		16 ga. galv., 3" x 6" to 3" x 10" joist		160	.050		1.15	1.10		2.25	3.06	
	1700		3" x 10" to 3" x 14" joist		160	.050		1.85	1.10		2.95	3.83	
	1800		4" x 6" to 4" x 10" joist		155	.052		1.95	1.14		3.09	4	
	1900		4" x 10" to 4" x 14" joist		155	.052		1.95	1.14		3.09	4	
	2000		Two-2" x 6" to two-2" x 10" joists		150	.053		1.10	1.17		2.27	3.13	
	2100		Two-2" x 10" to two-2" x 14" joists		150	.053		1.65	1.17		2.82	3.73	
	2300		³⁄₁₆" thick for 6" x 8" joist		145	.055		4.10	1.21		5.31	6.50	
	2400		6" x 10" joist		140	.057		4.85	1.26		6.11	7.40	
	2500		6" x 12" joist		135	.059		6.80	1.30		8.10	9.60	
	2700		¼" thick, 6" x 14" joist		130	.062		7.25	1.35		8.60	10.20	
	2800		Joist anchors, ¼" x 1-¼" x 18"		140	.057		2.80	1.26		4.06	5.15	
	2850												
	2900		Plywood clips, extruded aluminum H clip, for ¾" panels				Ea.	.06			.06	.07	
	3000		Galvanized 18 ga. back-up clip					.05			.05	.06	
	3200		Post framing, 16 ga. galv. for 4" x 4" base, 2 piece	1 Carp	130	.062		2.95	1.35		4.30	5.45	
	3300		Cap		130	.062		1.85	1.35		3.20	4.25	
	3500		Rafter anchors, 18 ga. galv., 1-½" wide, 5-¼" long		145	.055		.40	1.21		1.61	2.42	
	3600		10-¾" long		145	.055		.70	1.21		1.91	2.75	
	3800		Shear plates, 2-⅝" diameter		120	.067		1.10	1.47		2.57	3.60	
	3900		4" diameter		115	.070		2.35	1.53		3.88	5.10	
	4000		Sill anchors, (embedded in concrete or block), 18-⅝" long		115	.070		.60	1.53		2.13	3.16	
	4100		Spike grids, 4" x 4", flat or curved		120	.067		3.60	1.47		5.07	6.35	
	4400		Split rings, 2-½" diameter		120	.067		.70	1.47		2.17	3.16	
	4500		4" diameter		110	.073		1.25	1.60		2.85	3.99	
	4700		Strap ties, 16 ga., 1-⅜" wide, 12" long		180	.044		.85	.98		1.83	2.53	
	4800		24" long		160	.050		1.20	1.10		2.30	3.12	
	5000		Toothed rings, 2-⅝" or 4" diameter		90	.089		.75	1.96		2.71	4.02	
	5200		Truss plates, nailed, 20 gauge, up to 32' span		17	.471	Truss	6.15	10.35		16.50	24	
	5400		Washers, 2" x 2" x ⅛"				Ea.	.20			.20	.22	
	5500		3" x 3" x ³⁄₁₆"				"	.50			.50	.55	
	9000		Minimum labor/equipment charge	1 Carp	4	2	Job		44		44	72	
516	0010		**WOOD SCREWS #8, 1" long, steel**				C	2.95			2.95	3.25	**516**
	0100		Brass				"	14			14	15.40	

For expanded coverage of these items see *Means Interior Cost Data 1991*

060 | Fasteners and Adhesives

		060 500 Fasteners & Adhesives	CREW	DAILY OUTPUT	MAN-HOURS	UNIT	BARE COSTS MAT.	LABOR	EQUIP.	TOTAL	TOTAL INCL O&P	
516	0600	#10, 2" long, steel				C	6.30			6.30	6.95	516
	0700	Brass					33.40			33.40	37	
	1500	#12, 3" long, steel				↓	13.30			13.30	14.65	

061 | Rough Carpentry

		061 100 Wood Framing	CREW	DAILY OUTPUT	MAN-HOURS	UNIT	BARE COSTS MAT.	LABOR	EQUIP.	TOTAL	TOTAL INCL O&P	
102	0011	BLOCKING										102
	2600	Miscellaneous, to wood construction										
	2620	2" x 4"	F-1	.17	47.060	M.B.F.	400	1,025	47	1,472	2,175	
	2660	2" x 8"	"	.27	29.630	"	405	650	30	1,085	1,550	
	2720	To steel construction										
	2740	2" x 4"	F-1	.14	57.140	M.B.F.	400	1,250	57	1,707	2,550	
	2780	2" x 8"		.21	38.100	"	405	840	38	1,283	1,850	
	9000	Minimum labor/equipment charge		4	2	Job		44	2	46	74	
104	0010	BRACING Let-in, with 1" x 6" boards, studs @ 16" O.C.		150	.053	L.F.	.26	1.17	.05	1.48	2.27	104
	0200	Studs @ 24" O.C.		230	.035	"	.26	.77	.03	1.06	1.58	
106	0010	BRIDGING Wood, for joists 16" O.C., 1" x 3"		130	.062	Pr.	.31	1.35	.06	1.72	2.62	106
	0100	2" x 3" bridging	↓	130	.062		.57	1.35	.06	1.98	2.91	
	0300	Steel, galvanized, 18 ga., for 2" x 10" joists at 12" O.C.	1 Carp	130	.062		.78	1.35		2.13	3.07	
	0400	24" O.C.		140	.057		1.19	1.26		2.45	3.36	
	0900	Compression type, 16" O.C., 2" x 8" joists	↓	200	.040		1.15	.88		2.03	2.70	
	1000	2" x 12" joists	↓	200	.040	↓	1.23	.88		2.11	2.79	
110	0010	FRAMING, BEAMS & GIRDERS										110
	0020											
	1000	Single, 2" x 6" ⑧⑥	F-2	700	.023	L.F.	.39	.50	.02	.91	1.28	
	1020	2" x 8"		650	.025		.55	.54	.02	1.11	1.52	
	1040	2" x 10" ⑧⑦		600	.027		.82	.59	.03	1.44	1.89	
	1060	2" x 12"		550	.029		1.03	.64	.03	1.70	2.21	
	1080	2" x 14"		500	.032		1.20	.70	.03	1.93	2.51	
	1100	3" x 8"		550	.029		1.17	.64	.03	1.84	2.37	
	1120	3" x 10"		500	.032		1.59	.70	.03	2.32	2.94	
	1140	3" x 12"		450	.036		1.80	.78	.04	2.62	3.30	
	1160	3" x 14"	↓	400	.040		2.23	.88	.04	3.15	3.93	
	1180	4" x 8"	F-3	1,000	.040		1.65	.89	.38	2.92	3.68	
	1200	4" x 10"		950	.042		2.12	.94	.40	3.46	4.29	
	1220	4" x 12"		900	.044		2.55	.99	.42	3.96	4.88	
	1240	4" x 14"	↓	850	.047	↓	2.98	1.05	.44	4.47	5.45	
	1260											
	2000	Double, 2" x 6"	F-2	625	.026	L.F.	.80	.56	.03	1.39	1.83	
	2020	2" x 8"		575	.028		1.10	.61	.03	1.74	2.24	
	2040	2" x 10"		550	.029		1.69	.64	.03	2.36	2.94	
	2060	2" x 12"		525	.030		2.06	.67	.03	2.76	3.40	
	2080	2" x 14"	↓	475	.034	↓	2.42	.74	.03	3.19	3.91	
	2100											
	3000	Triple, 2" x 6"	F-2	550	.029	L.F.	1.18	.64	.03	1.85	2.38	
	3020	2" x 8"		525	.030		1.65	.67	.03	2.35	2.95	
	3040	2" x 10"		500	.032		2.54	.70	.03	3.27	3.98	
	3060	2" x 12"		475	.034		3.10	.74	.03	3.87	4.66	
	3080	2" x 14"	↓	450	.036	↓	3.63	.78	.04	4.45	5.30	
	9000	Minimum labor/equipment charge	F-1	2	4	Job		88	4	92	150	

For expanded coverage of these items see *Means Interior Cost Data 1991*

061 100 | Wood Framing

			CREW	DAILY OUTPUT	MAN-HOURS	UNIT	MAT.	LABOR	EQUIP.	TOTAL	TOTAL INCL O&P	
112	0010	FRAMING, CEILINGS										112
	0020											
	6000	Suspended, 2" x 3"	F-2	1,000	.016	L.F.	.22	.35	.02	.59	.83	
	6050	2" x 4"		900	.018		.27	.39	.02	.68	.96	
	6100	2" x 6"		800	.020		.40	.44	.02	.86	1.18	
	6150	2" x 8"		650	.025		.54	.54	.02	1.10	1.50	
	9000	Minimum labor/equipment charge	1 Carp	4	2	Job		44		44	72	
114	0010	FRAMING, JOISTS										114
	0020											
	2000	Joists, 2" x 4"	F-2	1,250	.013	L.F.	.27	.28	.01	.56	.77	
	2100	2" x 6"		1,250	.013		.40	.28	.01	.69	.91	
	2150	2" x 8" ⑧⑥		1,100	.015		.54	.32	.01	.87	1.13	
	2200	2" x 10"		900	.018		.84	.39	.02	1.25	1.58	
	2250	2" x 12"		875	.018		1.02	.40	.02	1.44	1.80	
	2300	2" x 14"		770	.021		1.21	.46	.02	1.69	2.10	
	2350	3" x 6" ⑧⑦		925	.017		.92	.38	.02	1.32	1.65	
	2400	3" x 10"		780	.021		1.54	.45	.02	2.01	2.45	
	2450	3" x 12"		600	.027		1.85	.59	.03	2.47	3.03	
	2500	4" x 6"		800	.020		1.25	.44	.02	1.71	2.12	
	2550	4" x 10"		600	.027		2.09	.59	.03	2.71	3.29	
	2600	4" x 12"		450	.036		2.51	.78	.04	3.33	4.08	
	2605	Sister joist, 2" x 6"		800	.020		.40	.44	.02	.86	1.18	
	2610	2" x 8"		640	.025		.54	.55	.03	1.12	1.52	
	2615	2" x 10"		535	.030		.84	.66	.03	1.53	2.03	
	2620	2" x 12"		455	.035		1.03	.77	.04	1.84	2.43	
	9000	Minimum labor/equipment charge	1 Carp	4	2	Job		44		44	72	
116	0010	FRAMING, MISCELLANEOUS										116
	0020											
	2000	Firestops, 2" x 4"	F-2	780	.021	L.F.	.27	.45	.02	.74	1.06	
	2100	2" x 6"		600	.027		.41	.59	.03	1.03	1.44	
	5000	Nailers, treated, wood construction, 2" x 4"		800	.020		.37	.44	.02	.83	1.15	
	5100	2" x 6"		750	.021		.53	.47	.02	1.02	1.37	
	5120	2" x 8"		700	.023		.73	.50	.02	1.25	1.65	
	5200	Steel construction, 2" x 4"		750	.021		.37	.47	.02	.86	1.20	
	5220	2" x 6"		700	.023		.53	.50	.02	1.05	1.43	
	5240	2" x 8"		650	.025		.73	.54	.02	1.29	1.71	
	7000	Rough bucks, treated, for doors or windows, 2" x 6"		400	.040		.53	.88	.04	1.45	2.06	
	7100	2" x 8"		380	.042		.73	.93	.04	1.70	2.36	
	8000	Stair stringers, 2" x 10"		130	.123		.87	2.71	.12	3.70	5.50	
	8100	2" x 12"		130	.123		1.04	2.71	.12	3.87	5.70	
	8150	3" x 10"		125	.128		1.59	2.82	.13	4.54	6.50	
	8200	3" x 12"		125	.128		1.91	2.82	.13	4.86	6.85	
	9000	Minimum labor/equipment charge	1 Carp	4	2	Job		44		44	72	
118	0010	FRAMING, COLUMNS										118
	0020											
	0100	4" x 4"	F-2	390	.041	L.F.	.85	.90	.04	1.79	2.46	
	0150	4" x 6"		275	.058		1.26	1.28	.06	2.60	3.55	
	0200	4" x 8"		220	.073		1.64	1.60	.07	3.31	4.49	
	0250	6" x 6"		215	.074		2.15	1.64	.07	3.86	5.15	
	0300	6" x 8"		175	.091		2.86	2.01	.09	4.96	6.55	
	0350	6" x 10"		150	.107		3.60	2.35	.11	6.06	7.90	
	9000	Minimum labor/equipment charge	F-1	2	4	Job		88	4	92	150	
120	0010	FRAMING, ROOFS										120
	6065											

For expanded coverage of these items see *Means Interior Cost Data 1991*

061 100 | Wood Framing

			CREW	DAILY OUTPUT	MAN-HOURS	UNIT	BARE COSTS MAT.	LABOR	EQUIP.	TOTAL	TOTAL INCL O&P	
120	6070	Fascia boards, 2" x 8"	F-2	225	.071	L.F.	.54	1.56	.07	2.17	3.22	120
	6080	2" x 10"		180	.089		.84	1.96	.09	2.89	4.21	
	7000	Rafters, to 4 in 12 pitch, 2" x 6"		1,000	.016		.40	.35	.02	.77	1.03	
	7060	2" x 8"		950	.017		.54	.37	.02	.93	1.21	
	7300	Hip and valley rafters, 2" x 6"		760	.021		.40	.46	.02	.88	1.22	
	7360	2" x 8"		720	.022		.54	.49	.02	1.05	1.41	
	7540	Hip and valley jacks, 2" x 6"		600	.027		.40	.59	.03	1.02	1.43	
	7600	2" x 8"		490	.033		.54	.72	.03	1.29	1.80	
	7780	For slopes steeper than 4 in 12, add						30%				
	7790	For dormers or complex roofs, add						50%				
	7800	Rafter tie, 1" x 4", #3	F-2	800	.020	L.F.	.17	.44	.02	.63	.93	
	7810											
	7820	Ridge board, #2 or better, 1" x 6"	F-2	600	.027	L.F.	.28	.59	.03	.90	1.30	
	7840	1" x 8"		550	.029		.38	.64	.03	1.05	1.50	
	7860	1" x 10"		500	.032		.48	.70	.03	1.21	1.72	
	7880	2" x 6"		500	.032		.39	.70	.03	1.12	1.62	
	7900	2" x 8"		450	.036		.54	.78	.04	1.36	1.91	
	7920	2" x 10"		400	.040		.82	.88	.04	1.74	2.38	
	7940	Roof cants, split, 4" x 4"		650	.025		1	.54	.02	1.56	2.01	
	7960	6" x 6"		600	.027		2.20	.59	.03	2.82	3.41	
	7980	Roof curbs, untreated, 2" x 6"		520	.031		.39	.68	.03	1.10	1.57	
	8000	2" x 12"		400	.040		1.02	.88	.04	1.94	2.60	
	8020	Sister rafters, 2" x 6"		800	.020		.40	.44	.02	.86	1.18	
	8040	2" x 8"		640	.025		.56	.55	.03	1.14	1.55	
	8060	2" x 10"		535	.030		.87	.66	.03	1.56	2.07	
	8080	2" x 12"		455	.035		1.02	.77	.04	1.83	2.42	
	9000	Minimum labor/equipment charge	1 Carp	4	2	Job		44		44	72	
122	0010	**FRAMING, SILLS**										122
	1810											
	2000	Ledgers, nailed, 2" x 4"	F-2	755	.021	L.F.	.27	.47	.02	.76	1.09	
	2050	2" x 6"		600	.027		.41	.59	.03	1.03	1.44	
	2100	Bolted, not including bolts, 3" x 6"		325	.049		.92	1.08	.05	2.05	2.83	
	2150	3" x 12"		233	.069		1.85	1.51	.07	3.43	4.58	
	2600	Mud sills, redwood, construction grade, 2" x 4"		895	.018		.70	.39	.02	1.11	1.43	
	2620	2" x 6"		780	.021		1.03	.45	.02	1.50	1.89	
	4000	Sills, 2" x 4"		600	.027		.27	.59	.03	.89	1.29	
	4050	2" x 6"		550	.029		.41	.64	.03	1.08	1.53	
	4080	2" x 8"		500	.032		.55	.70	.03	1.28	1.80	
	4200	Treated, 2" x 4"		550	.029		.37	.64	.03	1.04	1.49	
	4220	2" x 6"		500	.032		.53	.70	.03	1.26	1.77	
	4240	2" x 8"		450	.036		.73	.78	.04	1.55	2.12	
	4400	4" x 4"		450	.036		1.11	.78	.04	1.93	2.54	
	4420	4" x 6"		350	.046		1.67	1.01	.05	2.73	3.53	
	4460	4" x 8"		300	.053		2.18	1.17	.05	3.40	4.38	
	4480	4" x 10"		260	.062		2.78	1.35	.06	4.19	5.35	
	9000	Minimum labor/equipment charge	1 Carp	4	2	Job		44		44	72	
124	0010	**FRAMING, SLEEPERS**										124
	0020											
	0100	On concrete, treated, 1" x 2"	F-2	2,350	.007	L.F.	.12	.15	.01	.28	.38	
	0150	1" x 3"		2,000	.008		.19	.18	.01	.38	.51	
	0200	2" x 4"		1,500	.011		.37	.23	.01	.61	.81	
	0250	2" x 6"		1,300	.012		.55	.27	.01	.83	1.07	
	9000	Minimum labor/equipment charge	1 Carp	4	2	Job		44		44	72	
126	0010	**FRAMING, SOFFITS & CANOPIES**										126
	0020											

For expanded coverage of these items see *Means Interior Cost Data 1991*

		061 100 |Wood Framing	CREW	DAILY OUTPUT	MAN-HOURS	UNIT	BARE COSTS				TOTAL INCL O&P	
							MAT.	LABOR	EQUIP.	TOTAL		
126	1000	Canopy or soffit framing , 1″ x 4″	F-2	900	.018	L.F.	.19	.39	.02	.60	.87	**126**
	1040	1″ x 8″		750	.021		.38	.47	.02	.87	1.21	
	1100	2″ x 4″		620	.026		.27	.57	.03	.87	1.26	
	1140	2″ x 8″		500	.032		.56	.70	.03	1.29	1.81	
	1200	3″ x 4″		500	.032		.62	.70	.03	1.35	1.87	
	1240	3″ x 10″		300	.053		1.54	1.17	.05	2.76	3.67	
	9000	Minimum labor/equipment charge	1 Carp	4	2	Job		44		44	72	
128	0010	**FRAMING, WALLS**										**128**
	0020											
	2000	Headers over openings, 2″ x 6″ ⑧⑥	F-2	360	.044	L.F.	.40	.98	.04	1.42	2.09	
	2050	2″ x 8″		340	.047		.54	1.04	.05	1.63	2.33	
	2110	2″ x 10″ ⑧⑦		320	.050		.84	1.10	.05	1.99	2.77	
	2150	2″ x 12″		300	.053		1.02	1.17	.05	2.24	3.10	
	2200	4″ x 12″		190	.084		2.51	1.85	.08	4.44	5.90	
	2250	6″ x 12″		140	.114		3.58	2.51	.11	6.20	8.15	
	5000	Plates, untreated, 2″ x 3″		850	.019		.21	.41	.02	.64	.93	
	5020	2″ x 4″		800	.020		.27	.44	.02	.73	1.04	
	5040	2″ x 6″		750	.021		.41	.47	.02	.90	1.24	
	5120	Studs, 8′ high wall, 2″ x 3″		1,200	.013		.21	.29	.01	.51	.72	
	5140	2″ x 4″		1,100	.015		.27	.32	.01	.60	.84	
	5160	2″ x 6″		1,000	.016		.40	.35	.02	.77	1.03	
	5180	3″ x 4″		800	.020		.62	.44	.02	1.08	1.42	
	8200	For 12′ high wall, deduct						5%				
	8220	For stub wall, 6′ high, add						20%				
	8240	3′ high, add						40%				
	8250	For second story & above, add						5%				
	8300	Dormer & gable, add						15%				
	9000	Minimum labor/equipment charge	1 Carp	4	2	Job		44		44	72	
130	0010	**FURRING** Wood strips, on walls, 1″ x 2″, on wood	1 Carp	550	.015	L.F.	.10	.32		.42	.63	**130**
	0300	On masonry		495	.016		.11	.36		.47	.70	
	0400	On concrete		260	.031		.11	.68		.79	1.23	
	0600	1″ x 3″, wood strips, on walls, on wood		550	.015		.14	.32		.46	.68	
	0700	On masonry		495	.016		.15	.36		.51	.75	
	0800	On concrete		260	.031		.15	.68		.83	1.27	
	0850	1″ x 3″, wood strips, on ceilings, on wood		350	.023		.14	.50		.64	.98	
	0900	On masonry		320	.025		.15	.55		.70	1.06	
	0950	On concrete		210	.038		.15	.84		.99	1.53	
	9000	Minimum labor/equipment charge		4	2	Job		44		44	72	
132	0010	**GROUNDS** For casework, 1″ x 2″ wood strips, on wood		330	.024	L.F.	.10	.53		.63	.98	**132**
	0100	On masonry		285	.028		.11	.62		.73	1.13	
	0200	On concrete		250	.032		.11	.70		.81	1.27	
	0400	For plaster, ¾″ deep, on wood		450	.018		.10	.39		.49	.75	
	0500	On masonry		225	.036		.11	.78		.89	1.40	
	0600	On concrete		175	.046		.11	1.01		1.12	1.76	
	0700	On metal lath		200	.040		.11	.88		.99	1.56	
	9000	Minimum labor/equipment charge		4	2	Job		44		44	72	
134	0010	**INSULATION** See division 072										**134**
136	0010	**LAMINATED** See division 061-804										**136**
138	0010	**PARTITIONS** Wood stud with single bottom plate and										**138**
	0020	double top plate, no waste, std. & better lumber										
	0180	2″ x 4″ studs, 8′ high, studs 12″ O.C.	F-2	80	.200	L.F.	3.02	4.40	.20	7.62	10.75	
	0200	16″ O.C.		100	.160		2.48	3.52	.16	6.16	8.65	
	0300	24″ O.C.		125	.128		1.93	2.82	.13	4.88	6.85	
	0380	10′ high, studs 12″ O.C.		80	.200		3.58	4.40	.20	8.18	11.35	

For expanded coverage of these items see *Means Interior Cost Data 1991*

061 100 | Wood Framing

		CREW	DAILY OUTPUT	MAN-HOURS	UNIT	MAT.	LABOR	EQUIP.	TOTAL	TOTAL INCL O&P	
138	0400	16" O.C.	F-2	100	.160	L.F.	2.89	3.52	.16	6.57	9.10	138
	0500	24" O.C.		125	.128		2.20	2.82	.13	5.15	7.15	
	0580	12' high, studs 12" O.C.		65	.246		4.13	5.40	.25	9.78	13.65	
	0600	16" O.C.		80	.200		3.30	4.40	.20	7.90	11.05	
	0700	24" O.C.		100	.160		2.48	3.52	.16	6.16	8.65	
	0702											
	0780	2" x 6" studs, 8' high, studs 12" O.C.	F-2	70	.229	L.F.	4.48	5.05	.23	9.76	13.40	
	0800	16" O.C.		90	.178		3.66	3.91	.18	7.75	10.60	
	0900	24" O.C.		115	.139		2.85	3.06	.14	6.05	8.30	
	0980	10' high, studs 12" O.C.		70	.229		5.29	5.05	.23	10.57	14.30	
	1000	16" O.C.		90	.178		4.27	3.91	.18	8.36	11.30	
	1100	24" O.C.		115	.139		3.25	3.06	.14	6.45	8.75	
	1180	12' high, studs 12" O.C.		55	.291		6.10	6.40	.29	12.79	17.50	
	1200	16" O.C.		70	.229		4.88	5.05	.23	10.16	13.85	
	1300	24" O.C.		90	.178		3.66	3.91	.18	7.75	10.60	
	1400	For horizontal blocking, 2" x 4", add		600	.027		.27	.59	.03	.89	1.29	
	1500	2" x 6", add		600	.027		.40	.59	.03	1.02	1.43	
	1600	For openings, add		250	.064			1.41	.06	1.47	2.37	
	1700	Headers for above openings, material only, add				B.F.	.55			.55	.61	
	9000	Minimum labor/equipment charge	1 Carp	4	2	Job		44		44	72	
140	0010	**ROUGH HARDWARE** Average % of carpentry material, minimum					.50%					140
	0200	Maximum					1.50%					

061 150 | Sheathing

		CREW	DAILY OUTPUT	MAN-HOURS	UNIT	MAT.	LABOR	EQUIP.	TOTAL	TOTAL INCL O&P	
154	0010	**SHEATHING** Plywood on roof, CDX										154
	0030	5/16" thick	F-2	1,600	.010	S.F.	.24	.22	.01	.47	.63	
	0050	3/8" thick		1,525	.010		.26	.23	.01	.50	.68	
	0100	1/2" thick (83)		1,400	.011		.32	.25	.01	.58	.77	
	0200	5/8" thick (87)		1,300	.012		.40	.27	.01	.68	.90	
	0300	3/4" thick		1,200	.013		.46	.29	.01	.76	1	
	0500	Plywood on walls with exterior CDX, 3/8" thick		1,200	.013		.26	.29	.01	.56	.78	
	0600	1/2" thick		1,125	.014		.32	.31	.01	.64	.88	
	0700	5/8" thick		1,050	.015		.40	.34	.02	.76	1	
	0800	3/4" thick		975	.016		.46	.36	.02	.84	1.12	
	1000	For shear wall construction, add						20%				
	1200	For structural 1 exterior plywood, add					10%					
	1400	With boards, on roof 1" x 6" boards, laid horizontal	F-2	725	.022		.67	.49	.02	1.18	1.56	
	1500	Laid diagonal		650	.025		.69	.54	.02	1.25	1.67	
	1700	1" x 8" boards, laid horizontal		875	.018		.67	.40	.02	1.09	1.42	
	1800	Laid diagonal		725	.022		.69	.49	.02	1.20	1.58	
	2000	For steep roofs, add						40%				
	2200	For dormers, hips and valleys, add					5%	50%				
	2400	Boards on walls, 1" x 6" boards, laid regular	F-2	650	.025		.67	.54	.02	1.23	1.65	
	2500	Laid diagonal		585	.027		.69	.60	.03	1.32	1.77	
	2700	1" x 8" boards, laid regular		765	.021		.67	.46	.02	1.15	1.51	
	2800	Laid diagonal		650	.025		.69	.54	.02	1.25	1.67	
	2850	Gypsum, weatherproof, 1/2" thick		1,050	.015		.33	.34	.02	.69	.92	
	2900	Sealed, 4/10" thick		1,100	.015		.31	.32	.01	.64	.88	
	3000	Wood fiber, regular, no vapor barrier, 1/2" thick		1,200	.013		.35	.29	.01	.65	.88	
	3100	5/8" thick		1,200	.013		.46	.29	.01	.76	1	
	3300	No vapor barrier, in colors, 1/2" thick		1,200	.013		.49	.29	.01	.79	1.03	
	3400	5/8" thick		1,200	.013		.60	.29	.01	.90	1.15	
	3600	With vapor barrier one side, white, 1/2" thick		1,200	.013		.48	.29	.01	.78	1.02	
	3700	Vapor barrier 2 sides		1,200	.013		.75	.29	.01	1.05	1.32	
	3800	Asphalt impregnated, 25/32" thick		1,200	.013		.30	.29	.01	.60	.82	
	3850	Intermediate, 1/2" thick		1,200	.013		.26	.29	.01	.56	.78	
	9000	Minimum labor/equipment charge	1 Carp	2	4	Job		88		88	145	

For expanded coverage of these items see *Means Interior Cost Data 1991*

061 160 | Subfloor

			CREW	DAILY OUTPUT	MAN-HOURS	UNIT	MAT.	LABOR	EQUIP.	TOTAL	TOTAL INCL O&P	
							BARE COSTS					
161	0010	**FLOORING, WOOD** See division 095-604										161
164	0010	**SUBFLOOR** Plywood, CDX, ½" thick	F-2	1,500	.011	SF Flr.	.32	.23	.01	.56	.75	164
	0100	⅝" thick		1,350	.012		.39	.26	.01	.66	.87	
	0200	¾" thick		1,250	.013		.46	.28	.01	.75	.98	
	0300	1-⅛" thick, 2-4-1 including underlayment ⑧③		1,050	.015		1	.34	.02	1.36	1.66	
	0500	With boards, 1" x 10" S4S, laid regular		1,100	.015		.68	.32	.01	1.01	1.29	
	0600	Laid diagonal		900	.018		.70	.39	.02	1.11	1.43	
	0800	1" x 8" S4S, laid regular		1,000	.016		.67	.35	.02	1.04	1.33	
	0900	Laid diagonal		850	.019		.69	.41	.02	1.12	1.46	
	1100	Wood fiber, T&G, 2' x 8' planks, 1" thick		1,000	.016		.97	.35	.02	1.34	1.66	
	1200	1-⅜" thick		900	.018		1.28	.39	.02	1.69	2.07	
	9000	Minimum labor/equipment charge	F-1	4	2	Job		44	2	46	74	
168	0010	**UNDERLAYMENT** Plywood, underlayment grade, ⅜" thick	F-2	1,500	.011	SF Flr.	.32	.23	.01	.56	.75	168
	0100	½" thick		1,450	.011		.36	.24	.01	.61	.81	
	0200	⅝" thick ⑧③		1,400	.011		.44	.25	.01	.70	.90	
	0300	¾" thick		1,300	.012		.51	.27	.01	.79	1.02	
	0500	Particle board, ⅜" thick		1,500	.011		.26	.23	.01	.50	.69	
	0600	½" thick		1,450	.011		.28	.24	.01	.53	.72	
	0800	⅝" thick		1,400	.011		.31	.25	.01	.57	.76	
	0900	¾" thick		1,300	.012		.35	.27	.01	.63	.85	
	1100	Hardboard, underlayment grade, 4' x 4', .215" thick		1,500	.011		.28	.23	.01	.52	.71	
	9000	Minimum labor/equipment charge	F-1	4	2	Job		44	2	46	74	

061 200 | Structural Panels

			CREW	DAILY OUTPUT	MAN-HOURS	UNIT	MAT.	LABOR	EQUIP.	TOTAL	TOTAL INCL O&P	
208	0010	**STRESSED SKIN PLYWOOD ROOF PANELS** ⅜" group 1 top										208
	0020	skin, ⅜" exterior AD bottom skin										
	0030	1150f stringers, 4' x 8' panels										
	0100	4-¼" deep	F-3	2,075	.019	SF Roof	2.05	.43	.18	2.66	3.16	
	0200	6-⅛" deep		1,725	.023		2.25	.52	.22	2.99	3.56	
	0300	8-⅛" deep		1,475	.027		2.65	.60	.26	3.51	4.18	
	0500	⅜" top skin, no bottom skin, 5-¾" deep		1,725	.023		1.80	.52	.22	2.54	3.06	
	0600	7-¾" deep		1,475	.027		2	.60	.26	2.86	3.46	
	0800	For 3-½" factory fiberglass insulation, add					.25			.25	.28	
	1000	For ½" thick top skin, add					.13			.13	.14	
	1500	Floor panels, substitute ⅝" underlayment as										
	1510	top skin, add to roof panels above				SF Flr.	.16			.16	.18	
	2000	Curved roof panels, ⅜" structural 1 top skin,										
	2010	⅜" exterior AC bottom skin, laminated ribs										
	2200	8' radius, 2-¼" deep, tie rods not req'd.	F-3	1,150	.035	SF Flr.	4.25	.77	.33	5.35	6.30	
	2400	10' radius, 1-½" deep, tie rods are included		950	.042		3.35	.94	.40	4.69	5.65	
	2600	10' radius, 3-⅜" deep, tie rods not req'd.		1,150	.035		5.85	.77	.33	6.95	8.05	
	2800	12' radius, 2" deep, tie rods are included		950	.042		3.50	.94	.40	4.84	5.80	
	3000	12' radius, 4-½" deep, tie rods not req'd.		1,150	.035		6.10	.77	.33	7.20	8.35	
	3200	Cost of tie rods included where required										
	4000	Folded plate roofs, structural 1 top skin with intermediate										
	4010	rafters and end chord. Cost of tie rods included										
	4200	Slope 7 in 12, 4' fold, 2" thick, 32' span	F-3	850	.047	SF Flr.	2.80	1.05	.44	4.29	5.25	
	4400	Slope 8-½ in 12, 5' fold, 4" thick, 56' span		950	.042		6.05	.94	.40	7.39	8.60	
	4600	Slope 10 in 12, 8' fold, 4" thick, 52' span		950	.042		3.65	.94	.40	4.99	6	
	4800	Slope 10 in 12, 8' fold, 4" thick, 72' span		1,025	.039		6.40	.87	.37	7.64	8.85	
	6000	Box beams, structural 1 web										
	6200	24" deep, 2-2" x 4" flanges, 2 webs @ ⅜"	F-3	295	.136	L.F.	7.50	3.02	1.28	11.80	14.55	
	6400	24" deep, 3-2" x 4" flanges, 2 webs @ ½"		260	.154		9.40	3.42	1.45	14.27	17.50	
	6600	48" deep, 3-2" x 6" flanges, 2 webs @ ¾"		140	.286		16.55	6.35	2.70	25.60	31	
	6800	48" deep, 6-2" x 6" flanges, 4 webs @ ⅜",										
	6810	including 2 interior webs	F-3	115	.348	L.F.	35	7.75	3.29	46.04	55	

For expanded coverage of these items see *Means Interior Cost Data 1991*

061 200 | Structural Panels

		CREW	DAILY OUTPUT	MAN-HOURS	UNIT	BARE COSTS				TOTAL INCL O&P	
						MAT.	LABOR	EQUIP.	TOTAL		
208	7000	For exterior AC outer webs, add				L.F.	.80			.80	.88	208
	7200	For medium density overlaid outer webs, add				"	1.35			1.35	1.49	

061 250 | Wood Decking

		CREW	DAILY OUTPUT	MAN-HOURS	UNIT	MAT.	LABOR	EQUIP.	TOTAL	TOTAL INCL O&P	
258	0010	ROOF DECKS For laminated decks, see division 061-808										258
	0200	For cementitious decks, see division 035										
	0400	Cedar planks, 3.65 B.F. per S.F., 3″ thick	F-2	320	.050	S.F.	3.86	1.10	.05	5.01	6.10	
	0500	4.65 B.F. per S.F., 4″ thick		250	.064		4.88	1.41	.06	6.35	7.75	
	0700	Douglas fir, 3″ thick		320	.050		1.85	1.10	.05	3	3.89	
	0800	4″ thick		250	.064		2.51	1.41	.06	3.98	5.15	
	1000	Hemlock, 3″ thick		320	.050		1.85	1.10	.05	3	3.89	
	1100	4″ thick		250	.064		2.51	1.41	.06	3.98	5.15	
	1300	Western white spruce, 3″ thick		320	.050		1.82	1.10	.05	2.97	3.85	
	1400	4″ thick		250	.064		2.42	1.41	.06	3.89	5.05	
	9000	Minimum labor/equipment charge	F-1	2	4	Job		88	4	92	150	

061 280 | Mineral Fbr. Cem. Panel

		CREW	DAILY OUTPUT	MAN-HOURS	UNIT	MAT.	LABOR	EQUIP.	TOTAL	TOTAL INCL O&P	
281	0010	MINERAL FIBER CEMENT PANELS Including panels, fasteners,										281
	0100	accessories, trim & sealant										
	0130	Architectural, textured finish, ⅛″ thick, minimum	G-3	500	.064	S.F.	5.50	1.36	.05	6.91	8.30	
	0140	Maximum		500	.064		6.90	1.36	.05	8.31	9.80	
	0150	¼″ thick, minimum		500	.064		6.45	1.36	.05	7.86	9.35	
	0200	Maximum		500	.064		8.45	1.36	.05	9.86	11.55	
	0250	⅜″ thick, minimum		500	.064		8.50	1.36	.05	9.91	11.60	
	0300	Maximum		500	.064		11.20	1.36	.05	12.61	14.55	
	0350	⅝″ thick, minimum		300	.107		11.50	2.27	.08	13.85	16.35	
	0400	Maximum		300	.107		16.30	2.27	.08	18.65	22	
	2000	Flat sheets, ⅛″ thick		1,200	.027		3.85	.57	.02	4.44	5.15	
	2100	¼″ thick		1,020	.031		5.55	.67	.02	6.24	7.20	
	2200	⅜″ thick		885	.036		6.95	.77	.03	7.75	8.90	
	2300	⅝″ thick		795	.040		9.50	.86	.03	10.39	11.85	
	3000	Glasweld, mineral enamel coating, ⅛″ thick		600	.053		3.35	1.13	.04	4.52	5.55	
	3100	¼″ thick		322	.099		4.25	2.11	.08	6.44	8.15	
	4000	Sandwich panel, Glasweld face and back										
	4100	1″ thick, perlite core	G-3	322	.099	S.F.	6.90	2.11	.08	9.09	11.05	
	4200	Polyurethane core		322	.099		7.30	2.11	.08	9.49	11.50	
	4500	2″ thick, perlite core		322	.099		7.60	2.11	.08	9.79	11.80	
	4600	Polyurethane core		322	.099		8	2.11	.08	10.19	12.25	

061 300 | Heavy Timber Const

		CREW	DAILY OUTPUT	MAN-HOURS	UNIT	MAT.	LABOR	EQUIP.	TOTAL	TOTAL INCL O&P	
304	0010	FRAMING, HEAVY Mill timber, beams, single 6″ x 10″	F-2	220	.073	L.F.	3.25	1.60	.07	4.92	6.25	304
	0100	Single 8″ x 16″		115	.139	"	7.04	3.06	.14	10.24	12.90	
	0200	Built from 2″ lumber, multiple 2″ x 14″		900	.018	B.F.	.52	.39	.02	.93	1.23	
	0210	Built from 3″ lumber, multiple 3″ x 6″		700	.023		.62	.50	.02	1.14	1.53	
	0220	Multiple 3″ x 8″		800	.020		.62	.44	.02	1.08	1.42	
	0230	Multiple 3″ x 10″		900	.018		.62	.39	.02	1.03	1.34	
	0240	Multiple 3″ x 12″		1,000	.016		.62	.35	.02	.99	1.27	
	0245											
	0250	Built from 4″ lumber, multiple 4″ x 6″	F-2	800	.020	B.F.	.63	.44	.02	1.09	1.43	
	0260	Multiple 4″ x 8″		900	.018		.63	.39	.02	1.04	1.35	
	0270	Multiple 4″ x 10″		1,000	.016		.63	.35	.02	1	1.28	
	0280	Multiple 4″ x 12″		1,100	.015		.63	.32	.01	.96	1.23	
	0281											
	0290	Columns, structural grade, 1500f, 4″ x 4″	F-2	450	.036	L.F.	.85	.78	.04	1.67	2.26	
	0300	6″ x 6″		225	.071		2.20	1.56	.07	3.83	5.05	
	0400	8″ x 8″		240	.067		3.95	1.47	.07	5.49	6.80	

(86)

For expanded coverage of these items see *Means Interior Cost Data 1991*

			DAILY OUTPUT	MAN-HOURS	UNIT	BARE COSTS				TOTAL INCL O&P	
	061 300	**Heavy Timber Const**				MAT.	LABOR	EQUIP.	TOTAL		
304	0500	10" x 10"	F-2	90	.178	L.F.	6.65	3.91	.18	10.74	13.90	**304**
	0600	12" x 12"		70	.229	"	9.75	5.05	.23	15.03	19.20	
	0800	Floor planks, 2" thick, T & G, 2" x 6"		1,050	.015	B.F.	.59	.34	.02	.95	1.21	
	0900	2" x 10"		1,100	.015		.61	.32	.01	.94	1.21	
	1100	3" thick, 3" x 6"		1,050	.015		.81	.34	.02	1.17	1.45	
	1200	3" x 10"		1,100	.015		.82	.32	.01	1.15	1.44	
	1400	Girders, structural grade, 12" x 12"		800	.020		.80	.44	.02	1.26	1.62	
	1500	10" x 16"		1,000	.016		.80	.35	.02	1.17	1.47	
	2050	Roof planks, see division 061-258										
	2300	Roof purlins, 4" thick, structural grade	F-2	1,050	.015	B.F.	.63	.34	.02	.99	1.25	
	2500	Roof trusses, add timber connectors, division 060-512	"	450	.036	"	.63	.78	.04	1.45	2.01	
	9000	Minimum labor/equipment charge	F-1	2	4	Job		88	4	92	150	
	061 500	**Wood-Metal Systems**										
508	0010	**STRUCTURAL JOISTS** Fabricated ''I'' joists with wood flanges,										**508**
	0100	Plywood webs, incl. bridging & blocking, panels 24" O.C.										
	1200	15' to 24' span, 50 psf live load	F-5	2,400	.013	SF Flr.	1.17	.30	.01	1.48	1.79	
	1300	55 psf live load		2,250	.014		1.24	.32	.01	1.57	1.89	
	1400	24' to 30' span, 45 psf live load		2,600	.012		1.41	.28	.01	1.70	2.01	
	1500	55 psf live load		2,400	.013		1.48	.30	.01	1.79	2.13	
	1600	Tubular steel open webs, 45 psf, 24" O.C., 40' span	F-3	6,250	.006		1.38	.14	.06	1.58	1.82	
	1700	55' span		5,150	.008		1.33	.17	.07	1.57	1.82	
	1800	70' span		9,250	.004		1.84	.10	.04	1.98	2.22	
	1900	85 psf live load, 26' span		2,300	.017		1.74	.39	.16	2.29	2.72	
	061 800	**Glued-Laminated Const**										
804	0010	**LAMINATED FRAMING** Not including decking										**804**
	0020	30 lb., short term live load, 15 lb. dead load										
	0200	Straight roof beams, 20' clear span, beams 8' O.C.	F-3	2,560	.016	SF Flr.	1.20	.35	.15	1.70	2.05	
	0300	Beams 16' O.C.		3,200	.013		.82	.28	.12	1.22	1.48	
	0500	40' clear span, beams 8' O.C.		3,200	.013		2.26	.28	.12	2.66	3.07	
	0600	Beams 16' O.C.		3,840	.010		2.15	.23	.10	2.48	2.85	
	0800	60' clear span, beams 8' O.C.	F-4	2,880	.017		4.43	.36	.26	5.05	5.75	
	0900	Beams 16' O.C.	"	3,840	.013		2.65	.27	.19	3.11	3.57	
	1100	Tudor arches, 30' to 40' clear span, frames 8' O.C.	F-3	1,680	.024		5.20	.53	.23	5.96	6.85	
	1200	Frames 16' O.C.	"	2,240	.018		3.90	.40	.17	4.47	5.10	
	1400	50' to 60' clear span, frames 8' O.C.	F-4	2,200	.022		5.30	.47	.34	6.11	6.95	
	1500	Frames 16' O.C.		2,640	.018		4.55	.40	.28	5.23	5.95	
	1700	Radial arches, 60' clear span, frames 8' O.C.		1,920	.025		5	.54	.39	5.93	6.80	
	1800	Frames 16' O.C.		2,880	.017		3.85	.36	.26	4.47	5.10	
	2000	100' clear span, frames 8' O.C.		1,600	.030		5.15	.65	.46	6.26	7.25	
	2100	Frames 16' O.C.		2,400	.020		4.60	.44	.31	5.35	6.10	
	2300	120' clear span, frames 8' O.C.		1,440	.033		6.90	.73	.52	8.15	9.35	
	2400	Frames 16' O.C.		1,920	.025		6.25	.54	.39	7.18	8.20	
	2600	Bowstring trusses, 20' O.C., 40' clear span	F-3	2,400	.017		3.10	.37	.16	3.63	4.18	
	2700	60' clear span	F-4	3,600	.013		2.80	.29	.21	3.30	3.77	
	2800	100' clear span		4,000	.012		3.95	.26	.19	4.40	4.98	
	2900	120' clear span		3,600	.013		4.25	.29	.21	4.75	5.35	
	3100	For premium appearance, add to S.F. prices					5%					
	3300	For industrial type, deduct					15%					
	3500	For stain and varnish, add					5%					
	3900	For ¾" laminations, add to straight					25%					
	4100	Add to curved					15%					
	4300	Alternate pricing method: (use nominal footage of										
	4310	components). Straight beams, camber less than 6"	F-3	3.50	11.430	M.B.F.	1,425	255	110	1,790	2,100	
	4400	Columns, including hardware		2	20		1,600	445	190	2,235	2,700	
	4600	Curved members, radius over 32'		2.50	16		1,625	355	150	2,130	2,525	
	4700	Radius 10' to 32'		3	13.330		1,750	295	125	2,170	2,550	

For expanded coverage of these items see *Means Interior Cost Data 1991*

061 800 | Glued-Laminated Const

			CREW	DAILY OUTPUT	MAN-HOURS	UNIT	MAT.	LABOR	EQUIP.	TOTAL	TOTAL INCL.O&P	
804	4900	For complicated shapes, add maximum				M.B.F.	100%					804
	5000											
	5100	For pressure treating, add to straight				M.B.F.	35%					
	5200	Add to curved				"	45%					
	9000	Minimum labor/equipment charge	F-3	2.50	16	Job		355	150	505	745	
808	0010	LAMINATED ROOF DECK Pine or hemlock, 3" thick	F-2	425	.038	S.F.	2.31	.83	.04	3.18	3.93	808
	0100	4" thick		325	.049		2.69	1.08	.05	3.82	4.78	
	0300	Cedar, 3" thick		425	.038		2.86	.83	.04	3.73	4.54	
	0400	4" thick		325	.049		3.41	1.08	.05	4.54	5.55	
	0600	Fir, 3" thick		425	.038		2.53	.83	.04	3.40	4.17	
	0700	4" thick		325	.049		2.91	1.08	.05	4.04	5	
	9000	Minimum labor/equipment charge	F-1	3	2.670	Job		59	2.67	61.67	99	

061 900 | Wood Trusses

			CREW	DAILY OUTPUT	MAN-HOURS	UNIT	MAT.	LABOR	EQUIP.	TOTAL	TOTAL INCL.O&P	
908	0010	ROOF TRUSSES For timber connectors, see div. 060-512										908
	0100	Fink (W) or King post type, 2'-0" O.C.										
	0200	Metal plate connected, 4 in 12 slope ⑧⑤										
	0210	24' to 29' span	F-3	3,000	.013	SF Flr.	.85	.30	.13	1.28	1.56	
	0300	30' to 43' span		3,000	.013		.83	.30	.13	1.26	1.53	
	0400	44' to 60' span		3,000	.013		.82	.30	.13	1.25	1.52	
	0600	For change in roof pitch, subtract					.03			.03	.03	
	0700	Glued and nailed, add					50%					

062 200 | Millwork Moldings

			CREW	DAILY OUTPUT	MAN-HOURS	UNIT	MAT.	LABOR	EQUIP.	TOTAL	TOTAL INCL O&P	
208	0010	MOLDINGS, BASE										208
	0020											
	0500	Base, stock pine, 9/16" x 3-1/2"	1 Carp	240	.033	L.F.	.80	.73		1.53	2.08	
	0550	9/16" x 4-1/2"		200	.040		.92	.88		1.80	2.45	
	0561	Base shoe, oak, 3/4" x 1"		240	.033		.92	.73		1.65	2.21	
	9000	Minimum labor/equipment charge		4	2	Job		44		44	72	
212	0010	MOLDINGS, CASINGS										212
	0020											
	0090	Apron, stock pine, 5/8" x 2"	1 Carp	250	.032	L.F.	.61	.70		1.31	1.82	
	0110	5/8" x 3-1/2"		220	.036		1.28	.80		2.08	2.71	
	0300	Band, stock pine, 11/16" x 1-1/8"		270	.030		.37	.65		1.02	1.47	
	0350	11/16" x 1-3/4"		250	.032		.48	.70		1.18	1.68	
	0700	Casing, stock pine, 11/16" x 2-1/2"		240	.033		.59	.73		1.32	1.85	
	0750	11/16" x 3-1/2"		215	.037		.81	.82		1.63	2.23	
	9000	Minimum labor/equipment charge		4	2	Job		44		44	72	
216	0010	MOLDINGS, CEILINGS										216
	0020											
	0600	Bed, stock pine, 9/16" x 1-3/4"	1 Carp	270	.030	L.F.	.44	.65		1.09	1.55	
	0650	9/16" x 2"		240	.033		.59	.73		1.32	1.85	
	1200	Cornice molding, stock pine, 9/16" x 1-3/4"		330	.024		.46	.53		.99	1.38	
	1300	9/16" x 2-1/4"		300	.027		.66	.59		1.25	1.68	
	2400	Cove scotia, stock pine, 9/16" x 1-3/4"		270	.030		.47	.65		1.12	1.58	
	2500	11/16" x 2-3/4"		255	.031		.77	.69		1.46	1.97	

For expanded coverage of these items see *Means Interior Cost Data 1991*

062 200 | Millwork Moldings

		CREW	DAILY OUTPUT	MAN-HOURS	UNIT	BARE COSTS				TOTAL INCL O&P	
						MAT.	LABOR	EQUIP.	TOTAL		
216	2600	Crown, stock pine, 9/16″ x 3-5/8″	1 Carp	250	.032	L.F.	1	.70		1.70	2.25	**216**
	2700	11/16″ x 4-5/8″		220	.036	″	2.03	.80		2.83	3.54
	9000	Minimum labor/equipment charge	↓	4	2	Job		44		44	72
220	0010	**MOLDINGS, EXTERIOR**										**220**
	1500	Cornice, boards, pine, 1″ x 2″	1 Carp	330	.024	L.F.	.14	.53		.67	1.02
	1700	1″ x 6″		200	.040		.55	.88		1.43	2.04
	2000	1″ x 12″		180	.044		1.60	.98		2.58	3.36
	2200	Three piece, built-up, pine, minimum		80	.100		1.20	2.20		3.40	4.91
	2300	Maximum		65	.123		3.35	2.71		6.06	8.10
	3000	Trim, exterior, sterling pine, corner board, 1″ x 4″		200	.040		.17	.88		1.05	1.62
	3100	1″ x 6″		200	.040		.27	.88		1.15	1.73
	3350	Fascia, 1″ x 6″		250	.032		.27	.70		.97	1.45
	3370	1″ x 8″		225	.036		.36	.78		1.14	1.67
	3400	Moldings, back band		250	.032		.49	.70		1.19	1.69
	3500	Casing		250	.032		.26	.70		.96	1.44
	3600	Crown		250	.032		1	.70		1.70	2.25
	3700	Porch rail with balusters		22	.364		5.35	8		13.35	18.95
	3800	Screen	↓	395	.020	↓	.19	.45		.64	.94
	3850										
	4100	Verge board, sterling pine, 1″ x 4″	1 Carp	200	.040	L.F.	.16	.88		1.04	1.61
	4200	1″ x 6″		200	.040		.25	.88		1.13	1.71
	4300	2″ x 6″		165	.048		.38	1.07		1.45	2.16
	4400	2″ x 8″		165	.048		.51	1.07		1.58	2.30
	4700	For redwood trim, add	↓			↓	100%				
	9000	Minimum labor/equipment charge	1 Carp	4	2	Job		44		44	72
224	0010	**MOLDINGS, TRIM**										**224**
	0020										
	0200	Astragal, stock pine, 11/16″ x 1-3/4″	1 Carp	255	.031	L.F.	.70	.69		1.39	1.90
	0250	1-5/16″ x 2-3/16″		240	.033		1.17	.73		1.90	2.48
	0800	Chair rail, stock pine, 5/8″ x 2-1/2″		270	.030		.66	.65		1.31	1.79
	0900	5/8″ x 3-1/2″		240	.033		1.18	.73		1.91	2.50
	1000	Closet pole, stock pine, 1-1/8″ diameter		200	.040		.83	.88		1.71	2.35
	1100	Fir, 1-5/8″ diameter		200	.040		.97	.88		1.85	2.50
	3300	Half round, stock pine, 1/4″ x 1/2″		270	.030		.10	.65		.75	1.17
	3350	1/2″ x 1″	↓	255	.031		.33	.69		1.02	1.49
	3400	Handrail, fir, single piece, stock, hardware not included									
	3450	1-1/2″ x 1-3/4″	1 Carp	80	.100	L.F.	.90	2.20		3.10	4.58
	3470	Pine, 1-1/2″ x 1-3/4″		80	.100		.86	2.20		3.06	4.54
	3500	1-1/2″ x 2-1/2″		76	.105		1.22	2.32		3.54	5.10
	3600	Lattice, stock pine, 1/4″ x 1-1/8″		270	.030		.18	.65		.83	1.26
	3700	1/4″ x 1-3/4″		250	.032		.22	.70		.92	1.39
	3800	Miscellaneous, custom, pine or cedar, 1″ x 1″		270	.030		.14	.65		.79	1.22
	3900	Nominal 1″ x 3″		240	.033		.34	.73		1.07	1.57
	4100	Birch or oak, custom, nominal 1″ x 1″		240	.033		.20	.73		.93	1.42
	4200	Nominal 1″ x 3″		215	.037		.56	.82		1.38	1.95
	4400	Walnut, custom, nominal 1″ x 1″		215	.037		.31	.82		1.13	1.68
	4500	Nominal 1″ x 3″		200	.040		.86	.88		1.74	2.38
	4700	Teak, custom, nominal 1″ x 1″		215	.037		.77	.82		1.59	2.18
	4800	Nominal 1″ x 3″		200	.040		2.04	.88		2.92	3.68
	4900	Quarter round, stock pine, 1/4″ x 1/4″		275	.029		.10	.64		.74	1.15
	4950	3/4″ x 3/4″	↓	255	.031	↓	.21	.69		.90	1.36
	5600	Wainscot moldings, 1-1/8″ x 9/16″, 2′ high, minimum		76	.105	S.F.	5.40	2.32		7.72	9.70
	5700	Maximum		65	.123	″	10.70	2.71		13.41	16.20
	9000	Minimum labor/equipment charge	↓	4	2	Job		44		44	72
228	0010	**MOLDINGS, WINDOW AND DOOR**										**228**
	0020										

For expanded coverage of these items see *Means Interior Cost Data 1991*

			CREW	DAILY OUTPUT	MAN-HOURS	UNIT	MAT.	LABOR	EQUIP.	TOTAL	TOTAL INCL O&P	
062 200		**Millwork Moldings**										
228	2800	Door moldings, stock, decorative, 1-1/8" wide, plain	1 Carp	17	.471	Set	20	10.35		30.35	39	228
	2900	Detailed		17	.471	"	50	10.35		60.35	72	
	3150	Door trim set, 1 head and 2 sides, pine, 2-1/2 wide		5.90	1.360	Opng.	8.50	30		38.50	58	
	3170	4-1/2" wide		5.30	1.510	"	37	33		70	95	
	3200	Glass beads, stock pine, 1/4" x 11/16"		285	.028	L.F.	.17	.62		.79	1.20	
	3250	3/8" x 1/2"		275	.029		.24	.64		.88	1.31	
	3270	3/8" x 7/8"		270	.030		.28	.65		.93	1.37	
	4850	Parting bead, stock pine, 3/8" x 3/4"		275	.029		.22	.64		.86	1.29	
	4870	1/2" x 3/4"		255	.031		.27	.69		.96	1.42	
	5000	Stool caps, stock pine, 11/16" x 3-1/2"		200	.040		1.25	.88		2.13	2.81	
	5100	1-1/16" x 3-1/4"		150	.053		2.45	1.17		3.62	4.61	
	5300	Threshold, oak, 3' long, inside, 5/8" x 3-5/8"		32	.250	Ea.	4.80	5.50		10.30	14.25	
	5400	Outside, 1-1/2" x 7-5/8"		16	.500	"	18	11		29	38	
	5900	Window trim sets , including casings, header, stops,										
	5910	stool and apron, 2-1/2" wide, minimum	1 Carp	13	.615	Opng.	10.50	13.55		24.05	34	
	5950	Average		10	.800		15	17.60		32.60	45	
	6000	Maximum		6	1.330		20	29		49	70	
	9000	Minimum labor/equipment charge		4	2	Job		44		44	72	
062 300		**Shelving**										
304	0010	SHELVING Pine, clear grade, no edge band, 1" x 8"	F-1	115	.070	L.F.	.96	1.53	.07	2.56	3.64	304
	0100	1" x 10"		110	.073		1.22	1.60	.07	2.89	4.03	
	0200	1" x 12"		105	.076		1.50	1.68	.08	3.26	4.47	
	0300											
	0400	For lumber edge band, by hand, add				L.F.	1.30			1.30	1.43	
	0420	By machine, add					.85			.85	.94	
	0600	Plywood, 3/4" thick with lumber edge, 12" wide	F-1	75	.107		1.04	2.35	.11	3.50	5.10	
	0700	24" wide		70	.114		1.95	2.51	.11	4.57	6.40	
	0900	Bookcase, pine, clear grade, 8" shelves, 12" O.C.		70	.114	S.F.	3.80	2.51	.11	6.42	8.40	
	1000	12" wide shelves		65	.123	"	4.50	2.71	.12	7.33	9.50	
	1200	Adjustable closet rod and shelf, 12" wide, 3' long		20	.400	Ea.	7.70	8.80	.40	16.90	23	
	1300	8' long		15	.533	"	13.90	11.75	.53	26.18	35	
	1500	Prefinished shelves with supports, stock, 8" wide		75	.107	L.F.	5.30	2.35	.11	7.76	9.80	
	1600	10" wide		70	.114	"	5.95	2.51	.11	8.57	10.80	
	1800	Custom, high quality dadoed pine shelving units, minimum				S.F.				23	28	
	1900	Maximum				"				31.50	40	
	9000	Minimum labor/equipment charge	F-1	4	2	Job		44	2	46	74	
062 400		**Plastic Laminate**										
404	0010	CONVECTOR COVERS Laminated plastic on 3/4"										404
	0020	thick particle board, 12" wide, minimum	1 Carp	16	.500	L.F.	19.80	11		30.80	40	
	0050	Average		13	.615		24.60	13.55		38.15	49	
	0100	Maximum		13	.615		34	13.55		47.55	60	
	0300	Add to above for grille, minimum		150	.053	S.F.	1.60	1.17		2.77	3.68	
	0400	Maximum		75	.107	"	4.25	2.35		6.60	8.50	
	9000	Minimum labor/equipment charge		4	2	Job		44		44	72	
408	0010	COUNTER TOP Stock, plastic lam., 24" wide w/backsplash, min.	1 Carp	30	.267	L.F.	5.20	5.85		11.05	15.30	408
	0100	Maximum		25	.320		16	7.05		23.05	29	
	0300	Custom plastic, 7/8" thick, aluminum molding, no splash		30	.267		15.25	5.85		21.10	26	
	0400	Cove splash		30	.267		19.90	5.85		25.75	31	
	0600	1-1/4" thick, no splash		28	.286		17.95	6.30		24.25	30	
	0700	Square splash		28	.286		22.50	6.30		28.80	35	
	0900	Square edge, plastic face, 7/8" thick, no splash		30	.267		19.20	5.85		25.05	31	
	1000	With splash		30	.267		25	5.85		30.85	37	
	1200	For stainless channel edge, 7/8" thick, add					2.15			2.15	2.37	
	1300	1-1/4" thick, add					2.50			2.50	2.75	

For expanded coverage of these items see *Means Interior Cost Data 1991*

062 400 | Plastic Laminate

		CREW	DAILY OUTPUT	MAN-HOURS	UNIT	BARE COSTS				TOTAL INCL O&P
						MAT.	LABOR	EQUIP.	TOTAL	
408 1500	For solid color suede finish, add				L.F.	1.85			1.85	2.04	408
1700	For end splash, add				Ea.	12			12	13.20
1900	For cut outs, standard, add, minimum	1 Carp	32	.250		2.55	5.50		8.05	11.80
2000	Maximum		8	1		3.05	22		25.05	39
2100	Postformed, including backsplash and front edge		30	.267	L.F.	8.05	5.85		13.90	18.45
2110	Mitred, add		12	.667	Ea.		14.65		14.65	24
2200	Built-in place, 25″ wide, plastic laminate		25	.320	L.F.	10.25	7.05		17.30	23
2300	Ceramic tile mosaic		25	.320		24	7.05		31.05	38
2500	Marble, stock, with splash, ½″ thick, minimum	1 Bric	17	.471		29	10.70		39.70	49
2700	¾″ thick, maximum	"	13	.615		74	14		88	105
2900	Maple, solid, laminated, 1-½″ thick, no splash	1 Carp	28	.286		30	6.30		36.30	43
3000	With square splash		28	.286		34	6.30		40.30	48
3200	Stainless steel		24	.333	S.F.	71	7.35		78.35	90
3400	Recessed cutting block with trim, 16″ x 20″ x 1″		8	1	Ea.	38	22		60	78
3600	Table tops, plastic laminate, square edge, ⅞″ thick		45	.178	S.F.	6.80	3.91		10.71	13.85
3700	1-⅛″ thick		40	.200	"	7	4.40		11.40	14.90
9000	Minimum labor/equipment charge		3.75	2.130	Job		47		47	77

062 500 | Prefin. Wood Paneling

		CREW	DAILY OUTPUT	MAN-HOURS	UNIT	MAT.	LABOR	EQUIP.	TOTAL	TOTAL INCL O&P
504 0010	PANELING, PLYWOOD										504
0020										
2400	Plywood, prefinished, ¼″ thick, 4′ x 8′ sheets									
2410	with vertical grooves. Birch faced, minimum ㊳	F-2	500	.032	S.F.	.60	.70	.03	1.33	1.85
2420	Average		420	.038		.90	.84	.04	1.78	2.40
2430	Maximum		350	.046		1.25	1.01	.05	2.31	3.07
2600	Mahogany, African		400	.040		1.55	.88	.04	2.47	3.19
2700	Philippine (Lauan)		500	.032		.50	.70	.03	1.23	1.74
2900	Oak or Cherry, minimum		500	.032		1.35	.70	.03	2.08	2.68
3000	Maximum		400	.040		2.40	.88	.04	3.32	4.12
3200	Rosewood		320	.050		9.50	1.10	.05	10.65	12.30
3400	Teak		400	.040		2.40	.88	.04	3.32	4.12
3600	Chestnut		375	.043		3.60	.94	.04	4.58	5.55
3800	Pecan		400	.040		1.55	.88	.04	2.47	3.19
3900	Walnut, minimum		500	.032		2.05	.70	.03	2.78	3.45
3950	Maximum		400	.040		3.85	.88	.04	4.77	5.70
4000	Plywood, prefinished, ¾″ thick, stock grades, minimum		320	.050		.90	1.10	.05	2.05	2.84
4100	Maximum		224	.071		4.15	1.57	.07	5.79	7.20
4300	Architectural grade, minimum		224	.071		3	1.57	.07	4.64	5.95
4400	Maximum		160	.100		4.60	2.20	.10	6.90	8.75
4600	Plywood, "A" face, birch, V.C., ½″ thick, natural		450	.036		1.40	.78	.04	2.22	2.86
4700	Select		450	.036		1.50	.78	.04	2.32	2.97
4900	Veneer core, ¾″ thick, natural		320	.050		1.50	1.10	.05	2.65	3.50
5000	Select		320	.050		1.65	1.10	.05	2.80	3.67
5200	Lumber core, ¾″ thick, natural		320	.050		2.25	1.10	.05	3.40	4.33
5250										
5500	Plywood, knotty pine, ¼″ thick, A2 grade	F-2	450	.036	S.F.	1	.78	.04	1.82	2.42
5600	A3 grade		450	.036		1.45	.78	.04	2.27	2.92
5800	¾″ thick, veneer core, A2 grade		320	.050		1.50	1.10	.05	2.65	3.50
5900	A3 grade		320	.050		1.70	1.10	.05	2.85	3.72
6100	Aromatic cedar, ¼″ thick, plywood		400	.040		1.45	.88	.04	2.37	3.08
6200	¼″ thick, particle board		400	.040		.70	.88	.04	1.62	2.25
9000	Minimum labor/equipment charge	1 Carp	2	4	Job		88		88	145

062 550 | Prefin. Hardboard Panel

554 0010	PANELING, HARDBOARD										554
0020										

For expanded coverage of these items see *Means Interior Cost Data 1991*

062 550 | Prefin. Hardboard Panel

			DAILY OUTPUT	MAN-HOURS	UNIT	BARE COSTS				TOTAL INCL O&P	
			CREW			MAT.	LABOR	EQUIP.	TOTAL		
554	0050	Not incl. furring or trim, hardboard, tempered, ⅛" thick	F-2	500	.032	S.F.	.27	.70	.03	1	1.49	554
	0100	¼" thick		500	.032		.33	.70	.03	1.06	1.55	
	0300	Tempered pegboard, ⅛" thick		500	.032		.30	.70	.03	1.03	1.52	
	0400	¼" thick		500	.032		.38	.70	.03	1.11	1.61	
	0600	Untempered hardboard, natural finish, ⅛" thick		500	.032		.24	.70	.03	.97	1.45	
	0700	¼" thick		500	.032		.31	.70	.03	1.04	1.53	
	0900	Untempered pegboard, ⅛" thick		500	.032		.27	.70	.03	1	1.49	
	1000	¼" thick		500	.032		.33	.70	.03	1.06	1.55	
	1200	Plastic faced hardboard, ⅛" thick		500	.032		.44	.70	.03	1.17	1.67	
	1300	¼" thick		500	.032		.60	.70	.03	1.33	1.85	
	1500	Plastic faced pegboard, ⅛" thick		500	.032		.43	.70	.03	1.16	1.66	
	1600	¼" thick		500	.032		.52	.70	.03	1.25	1.76	
	1800	Wood grained, plain or grooved, ¼" thick, minimum		500	.032		.40	.70	.03	1.13	1.63	
	1900	Maximum		425	.038		.75	.83	.04	1.62	2.22	
	2100	Moldings for hardboard, wood or aluminum, minimum		500	.032	L.F.	.28	.70	.03	1.01	1.50	
	2200	Maximum		425	.038	"	.75	.83	.04	1.62	2.22	
	9000	Minimum labor/equipment charge	1 Carp	2	4	Job		88		88	145	

062 600 | Board Paneling

			CREW	DAILY OUTPUT	MAN-HOURS	UNIT	MAT.	LABOR	EQUIP.	TOTAL	TOTAL INCL O&P	
604	0010	PANELING, BOARDS										604
	0020											
	6400	Wood board paneling, ¾" thick, knotty pine	F-2	300	.053	S.F.	.88	1.17	.05	2.10	2.95	
	6500	Rough sawn cedar		300	.053		1.22	1.17	.05	2.44	3.32	
	6700	Redwood, clear, 1" x 4" boards		300	.053		2.80	1.17	.05	4.02	5.05	
	6900	Aromatic cedar, closet lining, boards		275	.058		1.90	1.28	.06	3.24	4.25	
	9000	Minimum labor/equipment charge	1 Carp	2	4	Job		88		88	145	

062 700 | Misc Finish Carpentry

			CREW	DAILY OUTPUT	MAN-HOURS	UNIT	MAT.	LABOR	EQUIP.	TOTAL	TOTAL INCL O&P	
704	0010	BEAMS, DECORATIVE Rough sawn cedar, non-load bearing, 4" x 4"	2 Carp	180	.089	L.F.	1.70	1.96		3.66	5.05	704
	0100	4" x 6"		170	.094		2.95	2.07		5.02	6.65	
	0200	4" x 8"		160	.100		3.65	2.20		5.85	7.60	
	0300	4" x 10"		150	.107		4.25	2.35		6.60	8.50	
	0400	4" x 12"		140	.114		5.50	2.51		8.01	10.15	
	0500	8" x 8"		130	.123		7.60	2.71		10.31	12.80	
	1100	Beam connector plates see div 060-512										
	9000	Minimum labor/equipment charge	1 Carp	3	2.670	Job		59		59	96	
720	0010	FIREPLACE MANTEL BEAMS Rough texture wood, 4" x 8"		36	.222	L.F.	3.65	4.89		8.54	12	720
	0100	4" x 10"		35	.229	"	4.20	5.05		9.25	12.85	
	0300	Laminated hardwood, 2-¼" x 10-½" wide, 6' long		5	1.600	Ea.	88	35		123	155	
	0400	8' long		5	1.600	"	122	35		157	190	
	0600	Brackets for above, rough sawn		12	.667	Pr.	8.15	14.65		22.80	33	
	0700	Laminated		12	.667	"	12.20	14.65		26.85	37	
	9000	Minimum labor/equipment charge		4	2	Job		44		44	72	
725	0010	FIREPLACE MANTELS 6" molding, 6' x 3'-6" opening, minimum	1 Carp	5	1.600	Opng.	98	35		133	165	725
	0100	Maximum		5	1.600		121	35		156	190	
	0300	Prefabricated pine, colonial type, stock, deluxe		2	4		535	88		623	730	
	0400	Economy		3	2.670		200	59		259	315	
	9000	Minimum labor/equipment charge		3	2.670	Job		59		59	96	
730	0010	GRILLES and Panels, hardwood, sanded										730
	0020	2' x 4' to 4' x 8', custom designs, unfinished, minimum	1 Carp	38	.211	S.F.	11.10	4.63		15.73	19.75	
	0050	Average		30	.267		17.30	5.85		23.15	29	
	0100	Maximum		19	.421		26.80	9.25		36.05	45	
	0300	As above, but prefinished, minimum		38	.211		13.85	4.63		18.48	23	
	0400	Maximum		19	.421		34	9.25		43.25	53	

For expanded coverage of these items see *Means Interior Cost Data 1991*

		062 700 | Misc Finish Carpentry	CREW	DAILY OUTPUT	MAN-HOURS	UNIT	BARE COSTS				TOTAL INCL O&P	
							MAT.	LABOR	EQUIP.	TOTAL		
730	9000	Minimum labor/equipment charge	1 Carp	2	4	Job		88		88	145	730
735	0010	HARDWARE Finish, see divisions 064-108 & 087										735
	0100	Rough, see division 050										
740	0010	LOUVERS Redwood, 2'-0" opening, full circle	1 Carp	16	.500	Ea.	76	11		87	100	740
	0100	Half circle, 2'-0" diameter		16	.500		72	11		83	97	
	0200	Octagonal, 2'-0" diameter		16	.500		68	11		79	93	
	0300	Triangular, 5/12 pitch, 5'-0" at base		16	.500		70	11		81	95	
	9000	Minimum labor/equipment charge		3.50	2.290	Job		50		50	82	
760	0010	SHUTTERS, EXTERIOR Aluminum, louvered, 1'-4" wide, 3'-0" long	1 Carp	10	.800	Pr.	25	17.60		42.60	56	760
	0400	6'-8" long		9	.889		42	19.55		61.55	78	
	1000	Pine, louvered, primed, each 1'-2" wide, 3'-3" long		10	.800		31	17.60		48.60	63	
	1100	4'-7" long		10	.800		39	17.60		56.60	72	
	1500	Each 1'-6" wide, 3'-3" long		10	.800		35	17.60		52.60	67	
	1600	4'-7" long		10	.800		48	17.60		65.60	82	
	1620	Hemlock, louvered, 1'-2" wide, 5'-7" long		10	.800		44	17.60		61.60	77	
	1630	Each 1'-4" wide, 2'-2" long		10	.800		28	17.60		45.60	60	
	1670	4'-3" long		10	.800		36	17.60		53.60	68	
	1690	5'-11" long		10	.800		45	17.60		62.60	78	
	1700	Door blinds, 6'-9" long, each 1'-3" wide		9	.889		53	19.55		72.55	90	
	1710	1'-6" wide		9	.889		66	19.55		85.55	105	
	1720	Hemlock, solid raised panel, each 1'-4" wide, 3'-3" long		10	.800		45	17.60		62.60	78	
	1740	4'-3" long		10	.800		57	17.60		74.60	91	
	1770	5'-11" long		10	.800		79	17.60		96.60	115	
	1790											
	1800	Door blinds, 6'-9" long, each 1'-3" wide	1 Carp	9	.889	Pr.	88	19.55		107.55	130	
	1900	1'-6" wide		9	.889		92	19.55		111.55	135	
	2500	Polystyrene, solid raised panel, each 1'-4" wide, 3'-3" long		10	.800		40	17.60		57.60	73	
	2700	4'-7" long		10	.800		50	17.60		67.60	84	
	4500	Polystyrene, louvered, each 1'-2" wide, 3'-3" long		10	.800		30	17.60		47.60	62	
	4750	5'-3" long		10	.800		42	17.60		59.60	75	
	6000	Vinyl, louvered, each 1'-2" x 4'-7" long		10	.800		40	17.60		57.60	73	
	6200	Each 1'-4" x 6'-8" long		9	.889		64	19.55		83.55	100	
	9000	Minimum labor/equipment charge		4	2	Job		44		44	72	
775	0010	SOFFITS Wood fiber, no vapor barrier, 15/32" thick	F-2	525	.030	S.F.	.50	.67	.03	1.20	1.68	775
	0100	5/8" thick		525	.030		.56	.67	.03	1.26	1.75	
	0300	As above, 5/8" thick, with factory finish		525	.030		.72	.67	.03	1.42	1.92	
	0500	Hardboard, 3/8" thick, slotted		525	.030		.70	.67	.03	1.40	1.90	
	1000	Exterior AC plywood, 1/4" thick (87)		420	.038		.38	.84	.04	1.26	1.83	
	1100	1/2" thick		420	.038		.50	.84	.04	1.38	1.96	
	1150	For aluminum soffit, see division 076-217										
	9000	Minimum labor/equipment charge	F-2	5	3.200	Job		70	3.20	73.20	120	
778	0010	DOORS AND FRAMES See division 081 & 082										778

		063 100 | Preservative Treatment	CREW	DAILY OUTPUT	MAN-HOURS	UNIT	BARE COSTS				TOTAL INCL O&P	
							MAT.	LABOR	EQUIP.	TOTAL		
102	0010	LUMBER TREATMENT Creosoted 8 lbs. per C.F., add				M.B.F.	210			210	230	102
	0200	For every added 2#/C.F., add per increment				"	25			25	28	

For expanded coverage of these items see *Means Interior Cost Data 1991*

		063 100	Preservative Treatment	CREW	DAILY OUTPUT	MAN-HOURS	UNIT	BARE COSTS				TOTAL INCL O&P	
								MAT.	LABOR	EQUIP.	TOTAL		
102	0400		Fire retardant, wet				M.B.F.	240			240	265	102
	0500		KDAT					275			275	305	
	0700		Salt treated, water borne, .40 lb. retention					155			155	170	
	0800		Oil borne, 8 lb. retention					165			165	180	
	1000		Kiln dried lumber, 1" & 2" thick, soft woods					100			100	110	
	1100		Hard woods					105			105	115	
	1500		For small size 1" stock, add					10			10	11	
	1700		For full size rough lumber, add					20%					
104	0010		PLYWOOD TREATMENT Fire retardant, 1/4" thick				M.S.F.	220			220	240	104
	0030		3/8" thick					240			240	265	
	0050		1/2" thick					255			255	280	
	0070		5/8" thick					270			270	295	
	0100		3/4" thick					300			300	330	
	0200		For KDAT, add					60			60	66	
	0500		Salt treated water borne, .25 lb., wet, 1/4" thick					115			115	125	
	0530		3/8" thick					120			120	130	
	0550		1/2" thick					125			125	140	
	0570		5/8" thick					135			135	150	
	0600		3/4" thick					145			145	160	
	0800		For KDAT add					60			60	66	
	0900		For .40 lb., per C.F. retention, add					40			40	44	
	1000		For certification stamp, add					30			30	33	

		064 100	Custom Casework	CREW	DAILY OUTPUT	MAN-HOURS	UNIT	BARE COSTS				TOTAL INCL O&P	
								MAT.	LABOR	EQUIP.	TOTAL		
102	0010		CABINETS Corner china cabinets, stock pine,										102
	0020		80" high, unfinished, minimum	2 Carp	6.60	2.420	Ea.	240	53		293	350	
	0100		Maximum	"	4.40	3.640	"	560	80		640	745	
	0300		Built-in drawer units, pine, 18" deep, 32" high, unfinished										
	0400		Minimum	2 Carp	53	.302	L.F.	35	6.65		41.65	49	
	0500		Maximum	"	40	.400	"	65	8.80		73.80	86	
	0700		Kitchen base cabinets, hardwood, not incl. counter tops,										
	0710		24" deep, 35" high, prefinished										
	0800		One top drawer, one door below, 12" wide	2 Carp	24.80	.645	Ea.	95	14.20		109.20	130	
	0820		15" wide		24	.667		100	14.65		114.65	135	
	0840		18" wide		23.30	.687		105	15.10		120.10	140	
	0860		21" wide		22.70	.705		115	15.50		130.50	150	
	0880		24" wide		22.30	.717		125	15.80		140.80	165	
	0890												
	1000		Four drawers, 12" wide	2 Carp	24.80	.645	Ea.	105	14.20		119.20	140	
	1020		15" wide		24	.667		115	14.65		129.65	150	
	1040		18" wide		23.30	.687		120	15.10		135.10	155	
	1060		24" wide		22.30	.717		140	15.80		155.80	180	
	1200		Two top drawers, two doors below, 27" wide		22	.727		160	16		176	200	
	1220		30" wide		21.40	.748		165	16.45		181.45	210	
	1240		33" wide		20.90	.766		175	16.85		191.85	220	
	1260		36" wide		20.30	.788		180	17.35		197.35	225	
	1280		42" wide		19.80	.808		195	17.80		212.80	245	
	1300		48" wide		18.90	.847		205	18.60		223.60	255	
	1500		Range or sink base, two doors below, 30" wide		21.40	.748		105	16.45		121.45	140	
	1520		33" wide		20.90	.766		125	16.85		141.85	165	

For expanded coverage of these items see *Means Interior Cost Data 1991*

064 100 | Custom Casework

		CREW	DAILY OUTPUT	MAN-HOURS	UNIT	BARE COSTS				TOTAL INCL O&P	
						MAT.	LABOR	EQUIP.	TOTAL		
102											**102**
1540	36" wide	2 Carp	20.30	.788	Ea.	135	17.35		152.35	175	
1560	42" wide		19.80	.808		145	17.80		162.80	190	
1580	48" wide		18.90	.847		160	18.60		178.60	205	
1800	For sink front units, deduct					45			45	50	
2000	Corner base cabinets, 36" wide, standard	2 Carp	18	.889		135	19.55		154.55	180	
2100	Lazy Susan with revolving door	"	16.50	.970		185	21		206	240	
4000	Kitchen wall cabinets, hardwood, 12" deep with two doors										
4050	12" high, 30" wide	2 Carp	24.80	.645	Ea.	80	14.20		94.20	110	
4100	36" wide		24	.667		90	14.65		104.65	125	
4400	15" high, 30" wide		24	.667		85	14.65		99.65	115	
4420	33" wide		23.30	.687		90	15.10		105.10	125	
4440	36" wide		22.70	.705		100	15.50		115.50	135	
4450	42" wide		22.70	.705		100	15.50		115.50	135	
4700	24" high, 30" wide		23.30	.687		100	15.10		115.10	135	
4720	36" wide		22.70	.705		120	15.50		135.50	155	
4740	42" wide		22.30	.717		130	15.80		145.80	170	
5000	30" high, one door, 12" wide		22	.727		80	16		96	115	
5020	15" wide		21.40	.748		85	16.45		101.45	120	
5040	18" wide		20.90	.766		95	16.85		111.85	130	
5060	24" wide		20.30	.788		100	17.35		117.35	140	
5300	Two doors, 27" wide		19.80	.808		115	17.80		132.80	155	
5320	30" wide		19.30	.829		120	18.25		138.25	160	
5340	36" wide		18.80	.851		125	18.70		143.70	170	
5360	42" wide		18.50	.865		135	19.05		154.05	180	
5380	48" wide		18.40	.870		150	19.15		169.15	195	
6000	Corner wall, 30" high, 24" wide		18	.889		120	19.55		139.55	165	
6050	30" wide		17.20	.930		140	20		160	185	
6100	36" wide		16.50	.970		155	21		176	205	
6500	Revolving Lazy Susan		15.20	1.050		170	23		193	225	
7000	Broom cabinet, 84" high, 24" deep, 18" wide		10	1.600		140	35		175	210	
7500	Oven cabinets, 84" high, 24" deep, 27" wide		8	2		160	44		204	250	
7750	Valance board trim		396	.040	L.F.	5	.89		5.89	6.95	
7760											
9000	For deluxe models of all cabinets, add to above				Ea.	40%					
9500	For custom built in place, add to above				"	25%	10%				
9550	Rule of thumb, kitchen cabinets not including										
9560	appliances & counter top, minimum	2 Carp	30	.533	L.F.	60	11.75		71.75	85	
9600	Maximum	"	25	.640	"	130	14.10		144.10	165	
9700	Minimum labor/equipment charge	1 Carp	3	2.670	Job		59		59	96	
104	**CASEWORK, FRAMES**										**104**
0010											
0050	Base cabinets, counter storage, 36" high, one bay										
0100	18" wide	1 Carp	2.70	2.960	Ea.	70	65		135	185	
0400	Two bay, 36" wide		2.20	3.640		110	80		190	250	
1100	Three bay, 54" wide		1.50	5.330		140	115		255	345	
2800	Book cases, one bay, 7' high, 18" wide		2.40	3.330		85	73		158	215	
3500	Two bay, 36" wide		1.60	5		130	110		240	325	
4100	Three bay, 54" wide		1.20	6.670		215	145		360	475	
5100	Coat racks, one bay, 7' high, 24" wide		4.50	1.780		85	39		124	155	
5300	Two bay, 48" wide		2.75	2.910		125	64		189	240	
5800	Three bay, 72" wide		2.10	3.810		180	84		264	335	
6100	Wall mounted cabinet, one bay, 24" high, 18" wide		3.60	2.220		40	49		89	125	
6800	Two bay, 36" wide		2.20	3.640		65	80		145	200	
7400	Three bay, 54" wide		1.70	4.710		85	105		190	265	
8400	30" high, one bay, 18" wide		3.60	2.220		45	49		94	130	
9000	Two bay, 36" wide		2.15	3.720		65	82		147	205	
9400	Three bay, 54" wide		1.60	5		85	110		195	275	
9800	Wardrobe, 7' high, single, 24" wide		2.70	2.960		95	65		160	210	

For expanded coverage of these items see *Means Interior Cost Data 1991*

	064 100	Custom Casework	CREW	DAILY OUTPUT	MAN-HOURS	UNIT	BARE COSTS				TOTAL INCL O&P	
							MAT.	LABOR	EQUIP.	TOTAL		
104	9880	Partition & adjustable shelves, 48″ wide	1 Carp	1.70	4.710	Ea.	125	105		230	305	104
	9950	Partition, adjustable shelves & drawers, 48″ wide		1.40	5.710	″	190	125		315	415	
	9970	Minimum labor/equipment charge	↓	4	2	Job		44		44	72	
106	0010	**CABINET DOORS**										106
	2000	Glass panel, hardwood frame										
	2200	12″ wide, 18″ high	1 Carp	34	.235	Ea.	5.60	5.20		10.80	14.60	
	2600	30″ high		32	.250		7.50	5.50		13	17.25	
	4450	18″ wide, 18″ high		32	.250		6.50	5.50		12	16.15	
	4550	30″ high	↓	29	.276	↓	8.60	6.05		14.65	19.35	
	4800											
	5000	Hardwood, raised panel										
	5100	12″ wide, 18″ high	1 Carp	16	.500	Ea.	10.45	11		21.45	29	
	5200	30″ high		15	.533		16.60	11.75		28.35	37	
	5500	18″ wide, 18″ high		15	.533		15.80	11.75		27.55	37	
	5600	30″ high	↓	14	.571		25.20	12.55		37.75	48	
	5900											
	6000	Plastic laminate on particle board										
	6100	12″ wide, 18″ high	1 Carp	25	.320	Ea.	3.35	7.05		10.40	15.20	
	6140	30″ high		23	.348		6.15	7.65		13.80	19.25	
	6500	18″ wide, 18″ high		24	.333		5.20	7.35		12.55	17.70	
	6600	30″ high	↓	22	.364	↓	8.60	8		16.60	23	
	6801											
	7000	Plywood, with edge band										
	7010	12″ wide, 18″ high	1 Carp	27	.296	Ea.	7.15	6.50		13.65	18.50	
	7120	30″ high		25	.320		9.65	7.05		16.70	22	
	7650	18″ wide, 18″ high		26	.308		8.75	6.75		15.50	21	
	7750	30″ high	↓	24	.333	↓	14.60	7.35		21.95	28	
	9000	Minimum labor/equipment charge	↓	4	2	Job		44		44	72	
108	0010	**CABINET HARDWARE**										108
	1000	Catches, minimum	1 Carp	235.29	.034	Ea.	.60	.75		1.35	1.88	
	1040	Maximum	″	80	.100	″	2.95	2.20		5.15	6.85	
	2000	Door/drawer pulls, handles										
	2200	Handles and pulls, projecting, metal, minimum	1 Carp	160	.050	Ea.	.90	1.10		2	2.79	
	2240	Maximum		68.83	.116		5.70	2.56		8.26	10.45	
	2300	Wood, minimum		160	.050		.80	1.10		1.90	2.68	
	2340	Maximum		68.38	.117		2.60	2.57		5.17	7.05	
	2600	Flush, metal, minimum		160	.050		1.40	1.10		2.50	3.34	
	2640	Maximum		68.38	.117		10	2.57		12.57	15.20	
	3000	Drawer tracks/glides, minimum		47.90	.167	Pr.	5.45	3.67		9.12	12	
	3040	Maximum		23.95	.334		15.50	7.35		22.85	29	
	4000	Cabinet hinges, minimum		160	.050		1.25	1.10		2.35	3.17	
	4040	Maximum	↓	68.38	.117	↓	5.70	2.57		8.27	10.45	
110	0010	**DRAWERS**										110
	0100	Solid hardwood front										
	1000	4″ high, 12″ wide	1 Carp	17	.471	Ea.	13.85	10.35		24.20	32	
	1200	18″ wide	″	16	.500	″	17.40	11		28.40	37	
	2800	Plastic laminate on particle board front										
	3000	4″ high, 12″ wide	1 Carp	17	.471	Ea.	13.40	10.35		23.75	32	
	3200	18″ wide	″	16	.500	″	16	11		27	36	
	5400	Plywood, flush panel front										
	6000	4″ high, 12″ wide	1 Carp	17	.471	Ea.	14.50	10.35		24.85	33	
	6200	18″ wide		16	.500	″	17.90	11		28.90	38	
	9000	Minimum labor/equipment charge	↓	4	2	Job		44		44	72	
140	0010	**VANITIES**										140
	0020											

For expanded coverage of these items see *Means Interior Cost Data 1991*

064 100 | Custom Casework

		Description	CREW	DAILY OUTPUT	MAN-HOURS	UNIT	MAT.	LABOR	EQUIP.	TOTAL	TOTAL INCL O&P	
140	8000	Vanity bases, 2 doors, 30″ high, 21″ deep, 24″ wide	2 Carp	20	.800	Ea.	145	17.60		162.60	190	**140**
	8050	30″ wide		16	1		180	22		202	235	
	8100	36″ wide		13.33	1.200		230	26		256	295	
	8150	48″ wide		11.43	1.400		315	31		346	395	
	9000	For deluxe models of all vanities, add to above					40%					
	9500	For custom built in place, add to above					25%	10%				

064 300 | Stairwork & Handrails

		Description	CREW	DAILY OUTPUT	MAN-HOURS	UNIT	MAT.	LABOR	EQUIP.	TOTAL	TOTAL INCL O&P	
306	0011	**STAIRS, PREFABRICATED**										**306**
	0100	Box stairs, prefabricated, 3′-0″ wide										
	0110	Oak treads, no handrails, 2′ high	2 Carp	5	3.200	Flight	105	70		175	230	
	0200	4′ high		4	4		230	88		318	395	
	0300	6′ high		3.50	4.570		370	100		470	570	
	0400	8′ high		3	5.330		470	115		585	710	
	0600	With pine treads for carpet, 2′ high		5	3.200		80	70		150	205	
	0700	4′ high		4	4		145	88		233	305	
	0800	6′ high		3.50	4.570		225	100		325	410	
	0900	8′ high		3	5.330		275	115		390	495	
	1100	For 4′ wide stairs, add					25%					
	1500	Prefabricated stair rail with balusters, 5 risers	2 Carp	15	1.070	Ea.	135	23		158	185	
	1700	Basement stairs, prefabricated, soft wood,										
	1710	open risers, 3′ wide, 8′ high	2 Carp	4	4	Flight	90	88		178	245	
	1900	Open stairs, prefabricated prefinished poplar, metal stringers,										
	1910	treads 3′-6″ wide, no railings										
	2000	3′ high	2 Carp	5	3.200	Flight	320	70		390	465	
	2100	4′ high		4	4		395	88		483	580	
	2200	6′ high		3.50	4.570		695	100		795	930	
	2300	8′ high		3	5.330		1,080	115		1,195	1,375	
	2500	For prefab. 3 piece wood railings & balusters, add for										
	2600	3′ high stairs	2 Carp	15	1.070	Ea.	100	23		123	150	
	2700	4′ high stairs		14	1.140		130	25		155	185	
	2800	6′ high stairs		13	1.230		200	27		227	265	
	2900	8′ high stairs		12	1.330		250	29		279	325	
	3100	For 3′-6″ x 3′-6″ platform, add		4	4		115	88		203	270	
	3300	Curved stairways, 3′-3″ wide, prefabricated, oak, unfinished,										
	3310	incl. curved balustrade system, open one side										
	3400	9′ high	2 Carp	.70	22.860	Flight	4,500	505		5,005	5,775	
	3500	10′ high		.70	22.860		4,900	505		5,405	6,200	
	3700	Open two sides, 9′ high		.50	32		7,400	705		8,105	9,300	
	3800	10′ high		.50	32		8,000	705		8,705	9,950	
	4000	Residential, wood, oak treads, prefabricated		1.50	10.670		850	235		1,085	1,325	
	4200	Built in place		.44	36.360		984	800		1,784	2,400	
	4400	Spiral, oak, 4′-6″ diameter, unfinished, prefabricated,										
	4500	incl. railing, 9′ high	2 Carp	1.50	10.670	Flight	3,000	235		3,235	3,675	
	9000	Minimum labor/equipment charge	″	3	5.330	Job		115		115	190	
308	0010	**STAIR PARTS** Balusters, turned, 30″ high, pine, minimum	1 Carp	28	.286	Ea.	3.90	6.30		10.20	14.55	**308**
	0100	Maximum		26	.308		5.50	6.75		12.25	17.10	
	0300	30″ high birch balusters, minimum		28	.286		5.60	6.30		11.90	16.40	
	0400	Maximum		26	.308		6.50	6.75		13.25	18.20	
	0600	42″ high, pine balusters, minimum		27	.296		4.75	6.50		11.25	15.85	
	0700	Maximum		25	.320		6.50	7.05		13.55	18.65	
	0900	42″ high birch balusters, minimum		27	.296		6	6.50		12.50	17.25	
	1000	Maximum		25	.320		8.50	7.05		15.55	21	
	1050	Baluster, stock pine, 1-1/16″ x 1-1/16″		240	.033	L.F.	.65	.73		1.38	1.91	
	1100	1-5/8″ x 1-5/8″		220	.036	″	1.20	.80		2	2.63	
	1200	Newels, 3-1/4″ wide, starting, minimum		7	1.140	Ea.	34	25		59	78	
	1300	Maximum		6	1.330	″	250	29		279	325	

⟨84⟩

For expanded coverage of these items see *Means Interior Cost Data 1991*

064 300 | Stairwork & Handrails

		Description	CREW	DAILY OUTPUT	MAN-HOURS	UNIT	MAT.	LABOR	EQUIP.	TOTAL	TOTAL INCL O&P	
308	1500	Landing, minimum	1 Carp	5	1.600	Ea.	65	35		100	130	308
	1600	Maximum		4	2	"	280	44		324	380	
	1800	Railings, oak, built-up, minimum		60	.133	L.F.	4.25	2.93		7.18	9.45	
	1900	Maximum		55	.145		10	3.20		13.20	16.20	
	2100	Add for sub rail		110	.073		3.30	1.60		4.90	6.25	
	2110											
	2300	Risers, Beech, ¾" x 7-½" high	1 Carp	64	.125	L.F.	4.25	2.75		7	9.15	
	2400	Fir, ¾" x 7-½" high		64	.125		1.20	2.75		3.95	5.80	
	2600	Oak, ¾" x 7-½" high		64	.125		3.90	2.75		6.65	8.80	
	2800	Pine, ¾" x 7-½" high		66	.121		1.20	2.67		3.87	5.65	
	2850	Skirt board, pine, 1" x 10"		55	.145		1.50	3.20		4.70	6.85	
	2900	1" x 12"		52	.154		1.80	3.38		5.18	7.50	
	3000	Treads, oak, 1-1/16" x 9-½" wide, 3' long		18	.444	Ea.	17.25	9.80		27.05	35	
	3100	4' long		17	.471		23	10.35		33.35	42	
	3300	1-1/16" x 11-½" wide, 3' long		18	.444		21.75	9.80		31.55	40	
	3400	6' long		14	.571		43.50	12.55		56.05	68	
	3600	Beech treads, add					40%					
	3800	For mitered return nosings, add				L.F.	2.35			2.35	2.59	
	9000	Minimum labor/equipment charge	1 Carp	3	2.670	Job		59		59	96	
310	0010	RAILING Custom design, architectural grade, hardwood, minimum	1 Carp	38	.211	L.F.	10.40	4.63		15.03	19	310
	0100	Maximum		30	.267		35	5.85		40.85	48	
	0300	Stock interior railing with spindles 6" O.C., 4' long		40	.200		25	4.40		29.40	35	
	0400	8' long		48	.167		23	3.67		26.67	31	
	9000	Minimum labor/equipment charge		3	2.670	Job		59		59	96	

064 400 | Misc Ornamental Items

		Description	CREW	DAILY OUTPUT	MAN-HOURS	UNIT	MAT.	LABOR	EQUIP.	TOTAL	TOTAL INCL O&P	
402	0011	COLUMNS										402
	0050	Aluminum, round colonial, 6" diameter	2 Carp	80	.200	V.L.F.	10.60	4.40		15	18.85	
	0100	8" diameter		62.25	.257		13.25	5.65		18.90	24	
	0200	10" diameter		55	.291		17.30	6.40		23.70	29	
	0250	Fir, stock units, hollow round, 6" diameter		80	.200		10.40	4.40		14.80	18.60	
	0300	8" diameter		80	.200		11.30	4.40		15.70	19.60	
	0350	10" diameter		70	.229		14.60	5.05		19.65	24	
	0400	Solid turned, to 8' high, 3-½" diameter		80	.200		4.10	4.40		8.50	11.70	
	0500	4-½" diameter		75	.213		6.10	4.69		10.79	14.35	
	0600	5-½" diameter		70	.229		7.90	5.05		12.95	16.90	
	0800	Square columns, built-up , 5" x 5"		65	.246		5.60	5.40		11	15	
	0900	Solid, 3-½" x 3-½"		130	.123		4.50	2.71		7.21	9.35	
	1600	Hemlock, tapered, T & G, 12" diam, 10' high		100	.160		25.50	3.52		29.02	34	
	1700	16' high		65	.246		38.75	5.40		44.15	51	
	1900	10' high, 14" diameter		100	.160		53	3.52		56.52	64	
	2000	18' high		65	.246		51	5.40		56.40	65	
	2200	18" diameter, 12' high		65	.246		72.40	5.40		77.80	88	
	2300	20' high		50	.320		65.25	7.05		72.30	83	
	2500	20" diameter, 14' high		40	.400		84.70	8.80		93.50	110	
	2600	20' high		35	.457		81.60	10.05		91.65	105	
	2800	For flat pilasters, deduct					33%					
	3000	For splitting into halves, add				Ea.	52			52	57	
	4000	Rough sawn cedar posts, 4" x 4"	2 Carp	250	.064	V.L.F.	1.30	1.41		2.71	3.73	
	4100	4" x 6"		235	.068		2	1.50		3.50	4.65	
	4200	6" x 6"		220	.073		3.50	1.60		5.10	6.45	
	4300	8" x 8"		200	.080		9.10	1.76		10.86	12.90	
	9000	Minimum labor/equipment charge	1 Carp	3	2.670	Job		59		59	96	

For expanded coverage of these items see *Means Interior Cost Data 1991*

071 100 | Sheet Waterproofing

		Description	CREW	DAILY OUTPUT	MAN-HOURS	UNIT	BARE COSTS MAT.	LABOR	EQUIP.	TOTAL	TOTAL INCL O&P	
102	0010	ELASTOMERIC WATERPROOFING EPDM, plain, 1/32" thick	2 Rofc	570	.028	S.F.	.44	.57		1.01	1.49	102
	0100	1/16" thick		570	.028		.58	.57		1.15	1.64	
	0300	Nylon reinforced sheets, 1/32" thick		570	.028		.52	.57		1.09	1.57	
	0400	1/16" thick		570	.028		.61	.57		1.18	1.67	
	0600	Vulcanizing splicing tape for above, 2" wide				L.F.	.20			.20	.22	
	0700	4" wide				"	.36			.36	.40	
	0900	Adhesive bonding, 60 S.F. per gal.				Gal.	14			14	15.40	
	1000	Splicing, 75 S.F. per gal.				"	16			16	17.60	
	1200	Neoprene sheets, plain, 1/32" thick	2 Rofc	570	.028	S.F.	.76	.57		1.33	1.84	
	1300	1/16" thick		570	.028		1.25	.57		1.82	2.38	
	1500	Nylon reinforced, 1/32" thick		570	.028		.80	.57		1.37	1.88	
	1600	1/16" thick		570	.028		1.29	.57		1.86	2.42	
	1800	1/8" thick		500	.032		2.20	.65		2.85	3.56	
	1900	Adhesive, splicing, 150 S.F. per gal. per coat				Gal.	16			16	17.60	
	2100	Fiberglass reinforced, fluid applied, 1/8" thick	2 Rofc	500	.032	S.F.	1.20	.65		1.85	2.46	
	2200	Polyethylene and rubberized asphalt sheets, 1/8" thick		570	.028		.72	.57		1.29	1.79	
	2400	Polyvinyl chloride sheets, plain, 10 mils thick		570	.028		.11	.57		.68	1.12	
	2500	20 mils thick		570	.028		.22	.57		.79	1.24	
	2700	30 mils thick		570	.028		.34	.57		.91	1.38	
	2800	Mineral fiber back PVC, 45 mils thick		570	.028		.73	.57		1.30	1.81	
	3000	Adhesives, trowel grade, 40-100 S.F. per gal.				Gal.	13.95			13.95	15.35	
	3100	Brush grade, 100-250 S.F. per gal.				"	13			13	14.30	
	3300	Bitumen modified polyurethane, fluid applied, 55 mils thick	2 Rofc	665	.024	S.F.	1	.49		1.49	1.96	
	3600	Vinyl plastic, sprayed on, 25 to 40 mils thick		475	.034	"	.86	.69		1.55	2.15	
	9000	Minimum labor/equipment charge		2	8	Job		165		165	285	
104	0010	MEMBRANE WATERPROOFING On slabs, 1 ply, felt	G-1	3,000	.019	S.F.	.15	.36	.04	.55	.84	104
	0100	Fabric		2,100	.027		.18	.51	.05	.74	1.15	
	0300	2 ply, felt		2,500	.022		.29	.43	.04	.76	1.12	
	0400	Fabric		1,650	.034		.35	.65	.07	1.07	1.60	
	0600	3 ply, felt		2,100	.027		.47	.51	.05	1.03	1.47	
	0700	Fabric		1,550	.036		.57	.69	.07	1.33	1.92	
	0900	For installation on walls, add						15%				
	1000	For 1/4" backer board, add	2 Rofc	3,500	.005		.30	.09		.39	.49	
	1050	For protector board, 3/8" thick, add		3,500	.005		.62	.09		.71	.85	
	1060	1/2" thick, add		3,500	.005		.79	.09		.88	1.03	
	1070	Fiberglass fabric, black, 20/10 mesh		116	.138	Sq.	8.87	2.81		11.68	14.70	
	1080	White, 20/10 mesh		116	.138	"	9	2.81		11.81	14.85	
	1100	Fluid neoprene, 4 coats, 50 mil		200	.080	S.F.	2.55	1.63		4.18	5.65	
	1200	90 mil		120	.133		5.10	2.71		7.81	10.35	
	1300	Fluid elastomeric copolymer compound, 32 mils thick		4,400	.004		2.25	.07		2.32	2.60	
	9000	Minimum labor/equipment charge		2	8	Job		165		165	285	

071 600 | Bitum. Dampproofing

		Description	CREW	DAILY OUTPUT	MAN-HOURS	UNIT	BARE COSTS MAT.	LABOR	EQUIP.	TOTAL	TOTAL INCL O&P	
602	0010	BITUMINOUS ASPHALT COATING For foundation										602
	0020											
	0030	Brushed on, below grade, 1 coat	1 Rofc	665	.012	S.F.	.08	.24		.32	.52	
	0100	2 coat		500	.016		.16	.33		.49	.75	
	0300	Sprayed on, below grade, 1 coat, 25.6 S.F./gal.		830	.010		.16	.20		.36	.52	
	0400	2 coat, 20.5 S.F./gal.		500	.016		.21	.33		.54	.80	
	0600	Troweled on, asphalt with fibers, 1/16" thick		500	.016		.14	.33		.47	.73	
	0700	1/8" thick		400	.020		.26	.41		.67	1	
	1000	1/2" thick		350	.023		1.10	.47		1.57	2.03	
	9000	Minimum labor/equipment charge		3	2.670	Job		54		54	95	

071 750 | Water Repellent Coat

		Description	CREW	DAILY OUTPUT	MAN-HOURS	UNIT	BARE COSTS MAT.	LABOR	EQUIP.	TOTAL	TOTAL INCL O&P	
754	0010	RUBBER COATING Water base liquid, roller applied	2 Rofc	7,000	.002	S.F.	.55	.05		.60	.69	754
	0200	Silicone or stearate, sprayed on masonry, 1 coat	1 Rofc	4,000	.002	"	.10	.04		.14	.18	

071 | Waterproofing and Dampproofing

071 750 | Water Repellent Coat

			CREW	DAILY OUTPUT	MAN-HOURS	UNIT	BARE COSTS				TOTAL INCL O&P	
							MAT.	LABOR	EQUIP.	TOTAL		
754	0300	2 coats	1 Rofc	2,000	.004	S.F.	.18	.08		.26	.34	754
	9000	Minimum labor/equipment charge	"	3	2.670	Job		54		54	95	

071 920 | Vapor Retarders

			CREW	DAILY OUTPUT	MAN-HOURS	UNIT	BARE COSTS				TOTAL INCL O&P	
							MAT.	LABOR	EQUIP.	TOTAL		
922	0010	BUILDING PAPER Aluminum and kraft laminated, foil 1 side	1 Carp	19	.421	Sq.	4	9.25		13.25	19.50	922
	0100	Foil 2 sides		19	.421		7.15	9.25		16.40	23	
	0300	Asphalt, two ply, 30#, for subfloors		37	.216		5.70	4.76		10.46	14.05	
	0400	Asphalt felt sheathing paper, 15#		37	.216		2.75	4.76		7.51	10.80	
	0450	Housewrap, exterior, spun bonded polypropylene										
	0470	Small roll, 3′ wide x 165′	1 Carp	3,800	.002	S.F.	.10	.05		.15	.19	
	0480	Large roll, 9′ wide x 195′	"	4,000	.002	"	.09	.04		.13	.17	
	0500	Material only, 3′ wide roll				Ea.	44			44	48	
	0520	9′ wide roll				"	148			148	165	
	0600	Polyethylene vapor barrier, standard, .002″ thick	1 Carp	37	.216	Sq.	.68	4.76		5.44	8.50	
	0700	.004″ thick		37	.216		1.19	4.76		5.95	9.10	
	0900	.006″ thick		37	.216		1.73	4.76		6.49	9.65	
	1000	.008″ thick		37	.216		2.49	4.76		7.25	10.50	
	1200	.010″ thick		37	.216		3.40	4.76		8.16	11.50	
	1300	Clear reinforced, fire retardant, .008″ thick		37	.216		7.50	4.76		12.26	16	
	1350	Cross laminated type, .003″ thick		37	.216		4.10	4.76		8.86	12.30	
	1400	.004″ thick		37	.216		5	4.76		9.76	13.25	
	1500	Red rosin paper, 5 sq. rolls, 4 lbs. per square		37	.216		2.05	4.76		6.81	10	
	1600	5 lbs. per square		37	.216		2.35	4.76		7.11	10.35	
	1800	Reinf. waterproof, .002″ polyethylene backing, 1 side		37	.216		4.40	4.76		9.16	12.60	
	1900	2 sides		37	.216		5.70	4.76		10.46	14.05	
	2100	Roof deck vapor barrier, class 1 metal decks	1 Rofc	37	.216		5.65	4.40		10.05	13.95	
	2200	For all other decks	"	37	.216		3.55	4.40		7.95	11.65	
	2400	Waterproofed kraft with sisal or fiberglass fibers, minimum	1 Carp	37	.216		4.85	4.76		9.61	13.10	
	2500	Maximum		37	.216		12.60	4.76		17.36	22	
	9950	Minimum labor/equipment charge		4	2	Job		44		44	72	

072 | Insulation

072 100 | Building Insulation

			CREW	DAILY OUTPUT	MAN-HOURS	UNIT	BARE COSTS				TOTAL INCL O&P	
							MAT.	LABOR	EQUIP.	TOTAL		
101	0010	BLOWN-IN INSULATION Ceilings, with open access										101
	0020	Cellulose, 3-½″ thick, R13	G-4	5,000	.005	S.F.	.16	.09	.05	.30	.37	
	0030	5-³⁄₁₆″ thick, R19		3,800	.006		.21	.11	.06	.38	.48	
	0050	6-½″ thick, R22		3,000	.008		.28	.15	.08	.51	.63	
	0100	8-¹¹⁄₁₆″ thick, R30		2,600	.009		.35	.17	.09	.61	.76	
	1000	Fiberglass, 5″ thick, R11		3,800	.006		.16	.11	.06	.33	.43	
	1050	6″ thick, R13		3,000	.008		.19	.15	.08	.42	.53	
	1100	8-½″ thick, R19		2,200	.011		.23	.20	.11	.54	.69	
	1200	10″ thick, R23		1,800	.013		.29	.24	.13	.66	.86	
	1300	12″ thick, R26		1,500	.016		.38	.29	.15	.82	1.06	
	2000	Mineral wool, 4″ thick, R12		3,500	.007		.20	.12	.07	.39	.50	
	2050	6″ thick, R17		2,500	.010		.31	.17	.09	.57	.73	
	2100	9″ thick, R23		1,750	.014		.39	.25	.13	.77	.98	
	2500	Wall installation, incl. drilling & patching from outside, two 1″										
	2510	diam. holes @ 16″ O.C., top & mid-point of wall										
	2700	For masonry	G-4	415	.058	S.F.	.05	1.05	.56	1.66	2.39	
	2800	For wood siding		840	.029		.05	.52	.28	.85	1.21	
	2900	For stucco/plaster		665	.036		.05	.66	.35	1.06	1.51	

072 100 | Building Insulation

			DAILY	MAN-			BARE COSTS			TOTAL	
		CREW	OUTPUT	HOURS	UNIT	MAT.	LABOR	EQUIP.	TOTAL	INCL O&P	
101	9000	Minimum labor/equipment charge	A-1	4	2	Job		35	13.60	48.60	72	101
106	0010	**FLOOR INSULATION, NONRIGID** Including										106
	0020	spring type wire fasteners										
	2000	Fiberglass, blankets or batts, paper or foil backing										
	2100	1 side, 3-½" thick, R11	1 Carp	700	.011	S.F.	.22	.25		.47	.65	
	2150	6" thick, R19		600	.013		.37	.29		.66	.89	
	2200	8-½" thick, R30		550	.015		.54	.32		.86	1.12	
	9000	Minimum labor/equipment charge		4	2	Job		44		44	72	
108	0011	**MASONRY INSULATION**										108
	0700	Foamed in place, urethane in 2-⅝" cavity	G-2	1,035	.023	S.F.	.18	.44	.22	.84	1.16	
	0800	For each 1" added thickness, add	"	2,372	.010	"	.06	.19	.10	.35	.49	
109	0010	**PERIMETER INSULATION** Asphalt impregnated cork, ½" thick, R1.12	1 Carp	685	.012	S.F.	.44	.26		.70	.90	109
	0100	1" thick, R2.24		685	.012		.65	.26		.91	1.13	
	0600	Polystyrene, molded bead board, 1" thick, R4		685	.012		.27	.26		.53	.72	
	0700	2" thick, R8		685	.012		.39	.26		.65	.85	
	9000	Minimum labor/equipment charge		4	2	Job		44		44	72	
110	0010	**POURED INSULATION** Cellulose fiber, R3.8 per inch	1 Carp	200	.040	C.F.	.46	.88		1.34	1.94	110
	0040	Ceramic type (perlite), R3.2 per inch		200	.040		1.51	.88		2.39	3.10	
	0080	Fiberglass wool, R4 per inch		200	.040		.32	.88		1.20	1.79	
	0100	Mineral wool, R3 per inch		200	.040		.80	.88		1.68	2.32	
	0300	Polystyrene, R4 per inch		200	.040		1.64	.88		2.52	3.24	
	0400	Vermiculite or perlite, R2.7 per inch		200	.040		1.77	.88		2.65	3.38	
	9000	Minimum labor/equipment charge		4	2	Job		44		44	72	
111	0010	**REFLECTIVE** Aluminum foil on 40 lb. kraft, foil 1 side, R9	1 Carp	1,900	.004	S.F.	.03	.09		.12	.18	111
	0100	Multilayered with air spaces, 2 ply, R14		1,900	.004		.15	.09		.24	.32	
	0500	3 ply, R17		1,500	.005		.19	.12		.31	.40	
	0600	5 ply, R22		1,500	.005		.28	.12		.40	.50	
	9000	Minimum labor/equipment charge		4	2	Job		44		44	72	
115	0010	**SPRAYED** Fibrous/cementitious, finished wall, 1" thick, R3.7	G-2	2,050	.012	S.F.	.12	.22	.11	.45	.61	115
	0100	Attic, 5.2" thick, R19	"	1,550	.015	"	.28	.30	.15	.73	.95	
	0300	Foam type on roofs, incl. preparation										
	0600	Urethane, 3 lb./C.F., 1" thick, R7.7	G-2	770	.031	S.F.	.06	.59	.30	.95	1.36	
	0700	2" thick, R15.4	"	475	.051	"	.18	.96	.49	1.63	2.29	
	1000	Coating, 50-70 mils, elastomeric aromatic urethane										
	1100	Finish coat, 5 mils, elastomeric aliphatic, 200 S.F./gal.				Gal.	38			38	42	
	9000	Minimum labor/equipment charge	A-1	2	4	Job		70	27	97	145	
116	0010	**WALL INSULATION, RIGID**										116
	0040	Fiberglass, 1.5#/C.F., unfaced, 1" thick, R4.1	1 Carp	1,000	.008	S.F.	.18	.18		.36	.49	
	0060	1-½" thick, R6.2		1,000	.008		.29	.18		.47	.61	
	0080	2" thick, R8.3		1,000	.008		.38	.18		.56	.71	
	0120	3" thick, R12.4		800	.010		.59	.22		.81	1.01	
	0370	3#/C.F., unfaced, 1" thick, R4.3		1,000	.008		.40	.18		.58	.73	
	0390	1-½" thick, R6.5		1,000	.008		.59	.18		.77	.94	
	0400	2" thick, R8.7		890	.009		.88	.20		1.08	1.29	
	0420	2-½" thick, R10.9		800	.010		1.05	.22		1.27	1.51	
	0440	3" thick, R13		800	.010		1.34	.22		1.56	1.83	
	0520	Foil faced, 1" thick, R4.3		1,000	.008		.79	.18		.97	1.16	
	0540	1-½" thick, R6.5		1,000	.008		1.14	.18		1.32	1.54	
	0560	2" thick, R8.7		890	.009		1.45	.20		1.65	1.92	
	0580	2-½" thick, R10.9		800	.010		1.70	.22		1.92	2.23	
	0600	3" thick, R13		800	.010		2.15	.22		2.37	2.72	
	0670	6#/C.F., unfaced, 1" thick, R4.3		1,000	.008		.75	.18		.93	1.11	

(90)

072 100 | Building Insulation

		CREW	DAILY OUTPUT	MAN-HOURS	UNIT	MAT.	LABOR	EQUIP.	TOTAL	TOTAL INCL O&P	
116	0690	1-½" thick, R6.5	1 Carp	890	.009	S.F.	1.14	.20		1.34	1.58	**116**
	0700	2" thick, R8.7		800	.010		1.52	.22		1.74	2.03	
	0721	2-½" thick, R10.9		800	.010		1.91	.22		2.13	2.46	
	0741	3" thick, R13		730	.011		2.30	.24		2.54	2.92	
	0821	Foil faced, 1" thick, R4.3		1,000	.008		.89	.18		1.07	1.27	
	0840	1-½" thick, R6.5		890	.009		1.12	.20		1.32	1.55	
	0850	2" thick, R8.7		800	.010		1.71	.22		1.93	2.24	
	0880	2-½" thick, R10.9		800	.010		2.20	.22		2.42	2.78	
	0900	3" thick, R13	↓	730	.011	↓	2.60	.24		2.84	3.25	
	1000											
	1500	Foamglass, 1-½" thick, R2.64	1 Carp	800	.010	S.F.	1.58	.22		1.80	2.10	
	1550	2" thick, R5.26	"	730	.011	"	2.21	.24		2.45	2.82	
	1600	Isocyanurate, 4' x 8' sheet, foil faced, both sides										
	1610	½" thick, R3.9	1 Carp	800	.010	S.F.	.27	.22		.49	.66	
	1620	⅝" thick, R4.5		800	.010		.32	.22		.54	.71	
	1630	¾" thick, R5.4		800	.010		.39	.22		.61	.79	
	1640	1" thick, R7.2		800	.010		.50	.22		.72	.91	
	1650	1-½" thick, R10.8		730	.011		.75	.24		.99	1.22	
	1660	2" thick, R14.4		730	.011		1	.24		1.24	1.49	
	1670	3" thick, R21.6		730	.011		1.49	.24		1.73	2.03	
	1680	4" thick, R28.8		730	.011		2	.24		2.24	2.59	
	1700	Perlite, 1" thick, R2.77		800	.010		.40	.22		.62	.80	
	1750	2" thick, R5.55		730	.011		.75	.24		.99	1.22	
	1900	Polystyrene, extruded blue, 2.2#/C.F., ¾" thick, R4		800	.010		.41	.22		.63	.81	
	1940	1-½" thick, R8.1		730	.011		.67	.24		.91	1.13	
	1960	2" thick, R10.8		730	.011		.88	.24		1.12	1.36	
	2100	Molded bead board, white, 1" thick, R3.85		800	.010		.17	.22		.39	.55	
	2120	1-½" thick, R5.6		730	.011		.27	.24		.51	.69	
	2140	2" thick, R7.7		730	.011		.34	.24		.58	.77	
	2350	Sheathing, insulating foil faced fiberboard, ⅜" thick		670	.012		.22	.26		.48	.67	
	2510	Urethane, no paper backing, ½" thick, R2.9		800	.010		.28	.22		.50	.67	
	2520	1" thick, R5.8		800	.010		.50	.22		.72	.91	
	2540	1-½" thick, R8.7		730	.011		.75	.24		.99	1.22	
	2560	2" thick, R11.7		730	.011		1	.24		1.24	1.49	
	2710	Fire resistant, ½" thick, R2.9		800	.010		.35	.22		.57	.74	
	2720	1" thick, R5.8		800	.010		.69	.22		.91	1.12	
	2740	1-½" thick, R8.7		730	.011		.97	.24		1.21	1.46	
	2760	2" thick, R11.7		730	.011	↓	1.24	.24		1.48	1.76	
	9000	Minimum labor/equipment charge	↓	4	2	Job		44		44	72	
118	0010	**WALL OR CEILING INSUL., NON-RIGID**										**118**
	0040	Fiberglass, kraft faced, batts or blankets										
	0060	3-½" thick, R11, 11" wide	1 Carp	1,150	.007	S.F.	.25	.15		.40	.52	
	0080	15" wide		1,600	.005		.25	.11		.36	.45	
	0100	23" wide		1,600	.005		.27	.11		.38	.48	
	0140	6" thick, R19, 11" wide		1,000	.008		.39	.18		.57	.72	
	0160	15" wide		1,350	.006		.39	.13		.52	.64	
	0180	23" wide		1,600	.005		.39	.11		.50	.61	
	0200	9" thick, R30, 15" wide		1,150	.007		.60	.15		.75	.91	
	0220	23" wide		1,350	.006		.60	.13		.73	.87	
	0240	12" thick, R38, 15" wide		1,000	.008		.78	.18		.96	1.15	
	0260	23" wide	↓	1,350	.006	↓	.78	.13		.91	1.07	
	0400	Fiberglass, foil faced, batts or blankets										
	0420	3-½" thick, R11, 15" wide	1 Carp	1,600	.005	S.F.	.27	.11		.38	.48	
	0440	23" wide		1,600	.005		.27	.11		.38	.48	
	0460	6" thick, R19, 15" wide		1,350	.006		.39	.13		.52	.64	
	0480	23" wide		1,600	.005		.39	.11		.50	.61	
	0500	9" thick, R30, 15" wide	↓	1,150	.007	↓	.56	.15		.71	.87	

072 100 | Building Insulation

		CREW	DAILY OUTPUT	MAN-HOURS	UNIT	MAT.	LABOR	EQUIP.	TOTAL	TOTAL INCL O&P	
118	0550	23" wide	1 Carp	1,350	.006	S.F.	.56	.13		.69	.83	118
	0800	Fiberglass, unfaced, batts or blankets										
	0820	3-½" thick, R11, 15" wide	1 Carp	1,350	.006	S.F.	.21	.13		.34	.44	
	0830	23" wide		1,600	.005		.21	.11		.32	.41	
	0860	6" thick, R19, 15" wide		1,150	.007		.36	.15		.51	.65	
	0880	23" wide		1,350	.006		.36	.13		.49	.61	
	0900	9" thick, R30, 15" wide		1,000	.008		.54	.18		.72	.88	
	0920	23" wide		1,150	.007		.54	.15		.69	.84	
	0940	12" thick, R38, 15" wide		1,000	.008		.79	.18		.97	1.16	
	0960	23" wide		1,150	.007		.79	.15		.94	1.12	
	1300	Mineral fiber batts, kraft faced										
	1320	3-½" thick, R13	1 Carp	1,600	.005	S.F.	.28	.11		.39	.49	
	1340	6" thick, R19		1,600	.005		.44	.11		.55	.66	
	1380	10" thick, R30		1,350	.006		.70	.13		.83	.98	
	1900	For foil backing, add					.04			.04	.04	
	9000	Minimum labor/equipment charge	1 Carp	4	2	Job		44		44	72	

072 200 | Roof & Deck Insulation

		CREW	DAILY OUTPUT	MAN-HOURS	UNIT	MAT.	LABOR	EQUIP.	TOTAL	TOTAL INCL O&P	
203	0010	ROOF DECK INSULATION										203
	0030	Fiberboard, mineral, 1" thick, R2.78	1 Rofc	800	.010	S.F.	.27	.20		.47	.65	
	0080	1-½" thick, R4		800	.010		.44	.20		.64	.84	
	0100	2" thick, R5.26		800	.010		.61	.20		.81	1.03	
	0300	Fiberglass, in 3' x 4' or 4' x 8' sheets										
	0400	15/16" thick, R3.3	1 Rofc	1,000	.008	S.F.	.39	.16		.55	.71	
	0460	1-1/16" thick, R3.8		1,000	.008		.50	.16		.66	.84	
	0600	1-5/16" thick, R5.3		1,000	.008		.62	.16		.78	.97	
	0650	1-5/8" thick, R5.7		1,000	.008		.71	.16		.87	1.07	
	0700	1-7/8" thick, R7.7		1,000	.008		.76	.16		.92	1.12	
	0800	2-¼" thick, R8		800	.010		.76	.20		.96	1.19	
	0900	Fiberglass and urethane composite, 3' x 4' sheets										
	1000	1-11/16" thick, R11.1	1 Rofc	1,000	.008	S.F.	.64	.16		.80	.99	
	1200	2" thick, R14.3		800	.010		.75	.20		.95	1.18	
	1300	2-5/8" thick, R18.2		800	.010		.96	.20		1.16	1.41	
	1500	Foamglass, 2' x 4' sheets, rectangular										
	1510	1-½" thick R3.95	1 Rofc	800	.010	S.F.	1.63	.20		1.83	2.15	
	1520	2" thick R5.26		800	.010		2.06	.20		2.26	2.62	
	1530	3" thick R7.89		700	.011		2.49	.23		2.72	3.15	
	1540	4" thick R10.53		700	.011		4.30	.23		4.53	5.15	
	1600	Tapered 1/16", 1/8" or ¼" per foot				B.F.	1.13			1.13	1.24	
	1650	Perlite, 2' x 4' sheets										
	1655	¾" thick, R2.08	1 Rofc	800	.010	S.F.	.32	.20		.52	.71	
	1660	1" thick, R2.78		800	.010		.37	.20		.57	.76	
	1670	1-½" thick, R4.17		800	.010		.59	.20		.79	1.01	
	1680	2" thick, R5.26		700	.011		.75	.23		.98	1.23	
	1700	Perlite/urethane composite										
	1711	1-¼" thick, R5.88	1 Rofc	1,000	.008	S.F.	.71	.16		.87	1.07	
	1721	1-½" thick, R7.2		1,000	.008		.76	.16		.92	1.12	
	1730	1-¾" thick, R10		1,000	.008		.80	.16		.96	1.17	
	1740	2" thick, R12.5		800	.010		.83	.20		1.03	1.27	
	1750	2-½" thick, R14.3		750	.011		.95	.22		1.17	1.43	
	1760	3" thick, R20		700	.011		1.12	.23		1.35	1.64	
	1800	Phenolic foam, 4' x 8' sheets										
	1810	1-3/16" thick, R10	1 Rofc	1,000	.008	S.F.	.43	.16		.59	.76	
	1820	1-½" thick, R12.5		1,000	.008		.51	.16		.67	.85	
	1830	1-¾" thick, R14.6		1,000	.008		.60	.16		.76	.95	
	1840	2" thick, R16.7		800	.010		.80	.20		1	1.24	
	1850	2-½" thick, R20		800	.010		.87	.20		1.07	1.31	
	1860	3" thick, R25		800	.010		1.05	.20		1.25	1.51	

072 200 | Roof & Deck Insulation

			CREW	DAILY OUTPUT	MAN-HOURS	UNIT	BARE COSTS				TOTAL INCL O&P	
							MAT.	LABOR	EQUIP.	TOTAL		
203	1900	Polystyrene										203
	1910	Extruded, 2.3#/C.F., 1″ thick, R5.26	1 Rofc	1,500	.005	S.F.	.50	.11		.61	.74	
	1920	2″ thick, R10		1,250	.006		.94	.13		1.07	1.26	
	1930	3″ thick, R15		1,000	.008		1.40	.16		1.56	1.83	
	2010	Expanded bead board, 1″ thick, R3.57		1,500	.005		.22	.11		.33	.43	
	2100	2″ thick, R7.14		1,250	.006		.33	.13		.46	.59	
	2200	Urethane, felt both sides										
	2210	1″ thick, R6.7	1 Rofc	1,000	.008	S.F.	.56	.16		.72	.90	
	2220	1-½″ thick, R11.11		1,000	.008		.72	.16		.88	1.08	
	2230	2″ thick, R14.3		800	.010		.80	.20		1	1.24	
	2240	2-½″ thick, R20		800	.010		.94	.20		1.14	1.39	
	2250	3″ thick, R25		800	.010		1.27	.20		1.47	1.75	
	2300	Urethane and gypsum board composite										
	2310	1-⅝″ thick, R7.7	1 Rofc	1,000	.008	S.F.	.87	.16		1.03	1.24	
	2320	2″ thick, R10		800	.010		1.11	.20		1.31	1.58	
	2330	2-½″ thick, R14.3		800	.010		1.16	.20		1.36	1.63	
	2340	3″ thick, R18.2		800	.010		1.31	.20		1.51	1.80	
	9000	Minimum labor/equipment charge		3.25	2.460	Job		50		50	88	

072 400 | Exterior Insulation

			CREW	DAILY OUTPUT	MAN-HOURS	UNIT	BARE COSTS				TOTAL INCL O&P	
							MAT.	LABOR	EQUIP.	TOTAL		
402	0010	INTEGRATED SIDING Fabric reinforced synthetic exterior										402
	0020	finish, on 1″ polystyrene insulation board										
	0100	Minimum	J-1	380	.105	S.F.	1.80	2.13	.09	4.02	5.50	
	0200	Maximum	″	270	.148		2.40	3	.12	5.52	7.60	
	0300	For insulation, 2″ polystyrene, add	1 Plas	725	.011		.40	.24		.64	.83	
	0400	For heavy duty protective fabric, add					1.35			1.35	1.49	

072 500 | Fireproofing

			CREW	DAILY OUTPUT	MAN-HOURS	UNIT	BARE COSTS				TOTAL INCL O&P	
							MAT.	LABOR	EQUIP.	TOTAL		
554	0010	SPRAYED Mineral fiber or cementitious for fireproofing,										554
	0050	not incl. tamping or canvas protection										
	0100	1″ thick, on flat plate steel	G-2	3,000	.008	S.F.	.35	.15	.08	.58	.72	
	0200	Flat decking		2,400	.010		.35	.19	.10	.64	.80	
	0400	Beams		1,500	.016		.35	.31	.15	.81	1.05	
	0500	Corrugated or fluted decks		1,250	.019		.35	.37	.18	.90	1.18	
	0700	Columns, 1-⅛″ thick		1,100	.022		.38	.42	.21	1.01	1.32	
	0800	2-3/16″ thick		700	.034		.70	.65	.33	1.68	2.19	
	0850	For tamping, add						10%				
	0900	For canvas protection, add	G-2	5,000	.005		.03	.09	.05	.17	.23	
	9000	Minimum labor/equipment charge	″	3	8	Job		155	77	232	330	

073 100 | Shingles

			CREW	DAILY OUTPUT	MAN-HOURS	UNIT	BARE COSTS				TOTAL INCL O&P	
							MAT.	LABOR	EQUIP.	TOTAL		
101	0010	ALUMINUM Shingles, mill finish, .020″ thick	1 Carp	2.50	3.200	Sq.	96	70		166	220	101
	0100	.030″ thick	″	2.50	3.200		105	70		175	230	
	0300	For colors, anodized finish, add					30			30	33	
	0400	For bonderized finish, add					50			50	55	
	0600	Ridge cap, .020″ thick	1 Carp	170	.047	L.F.	1.50	1.04		2.54	3.34	
	0700	.030″ thick		170	.047		1.75	1.04		2.79	3.62	
	0900	Valley section for above, .020″ thick		170	.047		1.50	1.04		2.54	3.34	
	1000	.030″ thick		170	.047		1.75	1.04		2.79	3.62	

073 100 | Shingles

			DAILY OUTPUT	MAN-HOURS	UNIT	BARE COSTS MAT.	LABOR	EQUIP.	TOTAL	TOTAL INCL O&P	
101	1200	For 1" factory applied polystyrene insulation, add			Sq.	25			25	28	101
	9000	Minimum labor/equipment charge	1 Carp	3	2.670	Job		59		59	96	
103	0010	**MINERAL FIBER** Strip shingles, 14" x 30", 325 lb. per square		4	2	Sq.	129	44		173	215	103
	0100	12" x 24", 167 lb. per square		3.50	2.290		65	50		115	155	
	0200	Shakes, 9.35" x 16", 500 lb. per square (siding) ⑨③		2.20	3.640		205	80		285	355	
	0300	Hip & ridge shingles, 5-⅜" x 14"		1	8	C.L.F.	315	175		490	635	
	0400	Hexagonal shape, 16" x 16"		3	2.670	Sq.	130	59		189	240	
	0500	Square, 16" x 16"		3	2.670		120	59		179	230	
	2000	For steep roofs, add						50%				
	9000	Minimum labor/equipment charge	1 Carp	3	2.670	Job		59		59	96	
104	0010	**ASPHALT SHINGLES**										104
	0100	Standard strip shingles										
	0150	Inorganic, class A, 210-235 lb./square, 3 bundles/square	1 Rofc	5.50	1.450	Sq.	32	30		62	87	
	0200	Organic, class C, 235-240 lb./square, 3 bundles/square	"	5	1.600	"	32.35	33		65.35	93	
	0250	Standard, laminated multi-layered shingles										
	0300	Class A, 240-260 lb./square, 3 bundles/square	1 Rofc	4.50	1.780	Sq.	52.50	36		88.50	120	
	0350	Class C, 260-300 lb./square, 4 bundles/square	"	4	2	"	47.85	41		88.85	125	
	0400	Premium, laminated multi-layered shingles										
	0450	Class A, 260-300 lb./square, 4 bundles/square	1 Rofc	3.50	2.290	Sq.	77.80	47		124.80	165	
	0500	Class C, 300-385 lb./square, 5 bundles/square		3	2.670	"	74.20	54		128.20	175	
	0700	Hip and ridge roll		400	.020	L.F.	.57	.41		.98	1.34	
	0900	Ridge shingles		330	.024	"	.65	.49		1.14	1.58	
	1000	For steep roofs, add				Sq.		50%				
	9000	Minimum labor/equipment charge	1 Rofc	3	2.670	Job		54		54	95	
105	0010	**PORCELAIN ENAMEL** 22 ga., 10" x 10", 225 lb. per sq., minimum	1 Rots	1.30	6.150	Sq.	435	125		560	700	105
	0100	Maximum		1	8	"	540	165		705	880	
	9000	Minimum labor/equipment charge		3	2.670	Job		55		55	96	
106	0010	**SLATE** Including felt underlay & nails, Buckingham, Virginia, black										106
	0100	³⁄₁₆" thick	1 Rots	1.75	4.570	Sq.	350	93		443	550	
	0200	¼" thick		1.75	4.570		350	93		443	550	
	0900	Pennsylvania black, Bangor, #1 clear		1.75	4.570		280	93		373	470	
	1200	Vermont, unfading colors, green, mottled green		1.75	4.570		390	93		483	595	
	1300	Semi-weathering green & gray		1.75	4.570		300	93		393	495	
	1400	Purple		1.75	4.570		450	93		543	660	
	1500	Black or gray		1.75	4.570		360	93		453	560	
	1510	For steep roofs, add to above						50%				
	1520											
	2500	Slate roof repair, extensive replacement	1 Rots	1	8	Sq.	335	165		500	655	
	2600	Repair individual pieces, scattered		19	.421	Ea.	3.75	8.60		12.35	19.25	
	9000	Minimum labor/equipment charge		3	2.670	Job		55		55	96	
107	0010	**STEEL** Shingles, galvanized, 26 gauge	1 Rots	2.20	3.640	Sq.	139	74		213	285	107
	0200	24 gauge	"	2.20	3.640		168	74		242	315	
	0300	For colored galvanized shingles, add					50			50	55	
	0500	For 1" factory applied polystyrene insulation, add					30			30	33	
	9000	Minimum labor/equipment charge	1 Rots	3	2.670	Job		55		55	96	
108	0010	**WOOD** 16" No. 1 red cedar shingles, 5X, 5" exposure, on roof	1 Carp	2.50	3.200	Sq.	105	70		175	230	108
	0200	7-½" exposure, on walls		2.05	3.900		76	86		162	225	
	0300	18" No. 1 red cedar perfections, 5-½" exposure, on roof ⑧⑦		2.75	2.910		111	64		175	225	
	0400											
	0600	Resquared, and rebutted, 5-½" exposure, on roof	1 Carp	3	2.670	Sq.	87	59		146	190	
	0900	7-½" exposure, on walls	"	2.45	3.270		62	72		134	185	
	1000	Add to above for fire retardant shingles, 16" long					30			30	33	
	1050	18" long					30			30	33	
	1100	Hand-split red cedar shakes, on roof, 24" long, 10" exposure	1 Carp	2.50	3.200		92	70		162	215	
	1200	18" long, 8-½" exposure	"	2	4		80	88		168	230	

073 100 | Shingles

			CREW	DAILY OUTPUT	MAN-HOURS	UNIT	MAT.	LABOR	EQUIP.	TOTAL	TOTAL INCL O&P	
108	1700	Add to above for fire retardant shakes, 24″ long				Sq.	35			35	39	108
	1800	18″ long					30			30	33	
	2000	White cedar shingles, 16″ long, extras, 5″ exposure, on roof	1 Carp	2.40	3.330		105	73		178	235	
	2100	7-½″ exposure, on walls		2	4		76	88		164	225	
	2300	For #15 organic felt underlayment on roof, 1 layer, add		64	.125		2.80	2.75		5.55	7.55	
	2400	2 layers, add		32	.250		5.60	5.50		11.10	15.15	
	2600	For steep roofs, add to above						50%				
	2700	Panelized systems, No.1 cedar shingles on ⁵⁄₁₆″ CDX plywood										
	2800	On walls, 8′ strips, 7″ or 14″ exposure	F-2	700	.023	S.F.	2.07	.50	.02	2.59	3.13	
	2900	Matching flush corners	"	400	.040	L.F.	2.07	.88	.04	2.99	3.76	
	3500	On roofs, 8′ strips, 7″ or 14″ exposure	1 Carp	3	2.670	Sq.	170	59		229	285	
	3600	Matching lap corners		200	.040	L.F.	1.10	.88		1.98	2.65	
	3700	Matching rake corners		200	.040		1.40	.88		2.28	2.98	
	3800	Matching valley sheets		200	.040		5.70	.88		6.58	7.70	
	9000	Minimum labor/equipment charge		3	2.670	Job		59		59	96	

073 200 | Roofing Tile

			CREW	DAILY OUTPUT	MAN-HOURS	UNIT	MAT.	LABOR	EQUIP.	TOTAL	TOTAL INCL O&P	
201	0010	ALUMINUM Tiles, .019″ thick, mission tile	1 Carp	2.50	3.200	Sq.	220	70		290	355	201
	0200	Spanish tiles		3	2.670	"	185	59		244	300	
	9000	Minimum labor/equipment charge		3	2.670	Job		59		59	96	
202	0010	CLAY TILE										202
	0200	Lanai tile or Classic tile, 158 pcs. per sq.	1 Rots	1.65	4.850	Sq.	275	99		374	475	
	0300	Americana, 158 pcs. per sq., most colors		1.65	4.850		275	99		374	475	
	0350	Green, gray or brown		1.65	4.850		280	99		379	480	
	0400	Blue		1.65	4.850		650	99		749	890	
	0600	Spanish tile, 171 pcs. per sq., red		1.80	4.440		270	91		361	455	
	0800	Buff, green, gray, brown		1.80	4.440		380	91		471	580	
	0900	Blue		1.80	4.440		650	91		741	875	
	1100	Mission tile, 166 pcs. per sq., machine scored finish, red		1.15	6.960		405	140		545	695	
	1700	French tile, 133 pcs. per sq., smooth finish, red		1.35	5.930		390	120		510	640	
	1750	Blue or green		1.35	5.930		815	120		935	1,100	
	1800	Norman tile, 317 pcs. per sq.		1	8		890	165		1,055	1,275	
	2200	Williamsburg tile, 158 pcs. per sq., aged cedar		1.35	5.930		290	120		410	530	
	2250	Gray or green		1.35	5.930		290	120		410	530	
	3000	For steep roofs, add to above						50%				
	9000	Minimum labor/equipment charge	1 Rots	3	2.670	Job		55		55	96	
204	0010	CONCRETE TILE Including installation of accessories										204
	0050	Earthtone colors, nailed to wood deck	1 Rots	1.35	5.930	Sq.	54	120		174	270	
	0150	Custom blues		1.35	5.930		200	120		320	435	
	0200	Custom greens		1.35	5.930		72	120		192	290	
	0500	Shakes, 13″ x 16-½″, 90 per sq., 950 lb. per sq.										
	0600	All colors, nailed to wood deck	1 Rots	1.50	5.330	Sq.	73	110		183	270	
	1500	Accessory pieces, ridge & hip, 10″ x 16-½″, 8 lbs. each				Ea.	1.25			1.25	1.38	
	1700	Rake, 6-½″ x 16-¾″, 9 lbs. each					1.25			1.25	1.38	
	1800	Mansard hip, 10″ x 16-½″, 9.2 lbs. each					1.35			1.35	1.49	
	1900	Hip starter, 10″ x 16-½″, 10.5 lbs. each					2.75			2.75	3.03	
	2000	3 or 4 way apex, 10″ each side, 11.5 lbs. each					5.75			5.75	6.35	
	9000	Minimum labor/equipment charge	1 Rots	3	2.670	Job		55		55	96	

074 100	Preformed Panels	CREW	DAILY OUTPUT	MAN-HOURS	UNIT	BARE COSTS				TOTAL INCL O&P	
						MAT.	LABOR	EQUIP.	TOTAL		
101	0010	**ALUMINUM ROOFING** Corrugated or ribbed, .0175″ thick, natural	G-3	1,000	.032	S.F.	.58	.68	.03	1.29	1.75	**101**
	0300	Painted		1,000	.032		.73	.68	.03	1.44	1.91	
	0400	Corrugated, .0215″ thick, on steel frame, natural finish		1,000	.032		.63	.68	.03	1.34	1.80	
	0600	Painted		1,000	.032		.78	.68	.03	1.49	1.97	
	0700	Corrugated, on steel frame, natural, .024″ thick		1,000	.032		.90	.68	.03	1.61	2.10	
	0900	.032″ thick		1,000	.032		1.25	.68	.03	1.96	2.49	
	1000	Painted, .024″ thick		1,000	.032		1.09	.68	.03	1.80	2.31	
	1200	.032″ thick		1,000	.032		1.55	.68	.03	2.26	2.82	
	9000	Minimum labor/equipment charge	1 Rofc	3	2.670	Job		54		54	95	
104	0010	**FIBERGLASS** Corrugated panels, roofing, 6 oz. per S.F.	G-3	1,000	.032	S.F.	1.20	.68	.03	1.91	2.43	**104**
	0100	8 oz. per S.F.		1,000	.032		1.70	.68	.03	2.41	2.98	
	0300	Corrugated siding, 4 oz. per S.F.		880	.036		.95	.77	.03	1.75	2.32	
	0400	5 oz. per S.F.		880	.036		1.08	.77	.03	1.88	2.46	
	0600	8 oz. siding, textured		880	.036		1.55	.77	.03	2.35	2.98	
	0700	Fire retardant		880	.036		1.65	.77	.03	2.45	3.09	
	0900	Flat panels, 6 oz. per S.F., clear or colors		880	.036		1.70	.77	.03	2.50	3.14	
	1100	Fire retardant, class A		880	.036		1.80	.77	.03	2.60	3.25	
	1300	8 oz. per S.F., clear or colors		880	.036		2.05	.77	.03	2.85	3.53	
	9000	Minimum labor/equipment charge	1 Rofc	2	4	Job		81		81	145	
107	0010	**STEEL ROOFING** Galv., corrugated or ribbed, on steel frame, 30 ga.	G-3	1,100	.029	S.F.	.58	.62	.02	1.22	1.65	**107**
	0100	28 gauge		1,050	.030		.61	.65	.02	1.28	1.73	
	0300	26 gauge ⑨¹		1,000	.032		.65	.68	.03	1.36	1.83	
	0400	24 gauge		950	.034		.79	.72	.03	1.54	2.04	
	0600	Colored, corrugated or ribbed, on steel frame, 28 gauge		1,050	.030		.76	.65	.02	1.43	1.90	
	0700	26 gauge		1,000	.032		.80	.68	.03	1.51	1.99	
	9000	Minimum labor/equipment charge	1 Rofc	2	4	Job		81		81	145	

074 600	Cladding/Siding	CREW	DAILY OUTPUT	MAN-HOURS	UNIT	MAT.	LABOR	EQUIP.	TOTAL	TOTAL INCL O&P	
602	0010	**ALUMINUM SIDING** .019″ thick, on steel construction, natural	G-3	775	.041	S.F.	.52	.88	.03	1.43	2.01	**602**
	0100	Painted		775	.041		.74	.88	.03	1.65	2.25	
	0400	Farm type, .021″ thick on steel frame, natural		775	.041		.62	.88	.03	1.53	2.12	
	0600	Painted		775	.041		.79	.88	.03	1.70	2.31	
	0700	Industrial type, corrugated, .024″ thick, on steel, natural ⑨⁰		775	.041		.86	.88	.03	1.77	2.39	
	0900	Painted		775	.041		1.05	.88	.03	1.96	2.60	
	1000	.032″ thick, natural		775	.041		1	.88	.03	1.91	2.54	
	1200	Painted		775	.041		1.26	.88	.03	2.17	2.83	
	1300	V-Beam, on steel frame, .032″ thick, natural		775	.041		1.10	.88	.03	2.01	2.65	
	1500	Painted		775	.041		1.30	.88	.03	2.21	2.87	
	1600	.040″ thick, natural		775	.041		1.25	.88	.03	2.16	2.82	
	1800	Painted		775	.041		1.55	.88	.03	2.46	3.15	
	1910	Minimum labor/equipment charge	1 Carp	3	2.670	Job		59		59	96	
	2000											
	3800	Horizontal, colored clapboard, 8″ or 10″ wide, plain	1 Carp	255	.031	S.F.	1	.69		1.69	2.23	
	3810	Insulated		255	.031		1.14	.69		1.83	2.38	
	3830	8″ embossed, painted		255	.031		1.05	.69		1.74	2.28	
	3840	Insulated		255	.031		1.20	.69		1.89	2.45	
	3860	12″ painted, smooth		300	.027		1.08	.59		1.67	2.15	
	3870	Insulated		300	.027		1.25	.59		1.84	2.33	
	3890	12″ embossed, painted		300	.027		1.15	.59		1.74	2.22	
	3900	Insulated		300	.027		1.30	.59		1.89	2.39	
	3920	12″ embossed, to 6″ exposure, painted		300	.027		1.17	.59		1.76	2.24	
	3930	Insulated		300	.027		1.30	.59		1.89	2.39	
	4000	Vertical board & batten, colored, non-insulated		255	.031		1.30	.69		1.99	2.56	
	4200	For simulated wood design, add					.05			.05	.06	
	4300	Corners for above, outside	1 Carp	255	.031	V.L.F.	1.63	.69		2.32	2.92	
	4500	Inside corners	″	255	.031	″	.77	.69		1.46	1.97	

074 600		Cladding/Siding	CREW	DAILY OUTPUT	MAN-HOURS	UNIT	MAT.	BARE COSTS LABOR	EQUIP.	TOTAL	TOTAL INCL O&P	
602	4520	For simulated wood design, add				S.F.	.05			.05	.06	602
	9000	Minimum labor/equipment charge	1 Carp	3	2.670	Job		59		59	96	
606	0010	STEEL SIDING Beveled, vinyl coated, 8″ wide		265	.030	S.F.	.69	.66		1.35	1.84	606
	0050	10″ wide		275	.029	″	.66	.64		1.30	1.77	
	0060	Minimum labor/equipment charge		3	2.670	Job		59		59	96	
	0070											
	0100	28 gauge	G-3	795	.040	S.F.	.52	.86	.03	1.41	1.97	
	0300	26 gauge		790	.041		.63	.86	.03	1.52	2.10	
	0400	24 gauge		785	.041		.75	.87	.03	1.65	2.25	
	0600	22 gauge		770	.042		.86	.88	.03	1.77	2.40	
	0700	Colored, corrugated/ribbed, on steel frame, 10 yr. fin., 28 ga.		800	.040		.70	.85	.03	1.58	2.16	
	0900	26 gauge		795	.040		.80	.86	.03	1.69	2.28	
	1000	24 gauge		790	.041		.92	.86	.03	1.81	2.42	
	9000	Minimum labor/equipment charge	1 Carp	3	2.670	Job		59		59	96	
607	0010	VINYL SIDING Solid PVC panels, 8″ to 10″ wide, plain		255	.031	S.F.	.60	.69		1.29	1.79	607
	0100	Insulated		255	.031		.85	.69		1.54	2.06	
	0200	Soffit and fascia		205	.039		.58	.86		1.44	2.04	
	0300	Window and door trim moldings		185	.043	L.F.	.36	.95		1.31	1.95	
	0500	Corner posts, outside corner		205	.039		.79	.86		1.65	2.27	
	0600	Inside corner		205	.039		.40	.86		1.26	1.84	
	0800	Corrugated vinyl sheets, .090″ thick		235	.034	S.F.	2.15	.75		2.90	3.59	
	0900	.120″ thick		225	.036		2.95	.78		3.73	4.52	
	1100	Flat sheets with fibers, colored, 1/16″ thick		235	.034		1.45	.75		2.20	2.82	
	1200	1/8″ thick		225	.036		2.15	.78		2.93	3.64	
	1400	Insulated sandwich panels with 1/16″ skin, 1″ thick		180	.044		3.45	.98		4.43	5.40	
	1500	1-1/2″ thick		190	.042		3.79	.93		4.72	5.70	
	9000	Minimum labor/equipment charge		3	2.670	Job		59		59	96	
609	0010	WOOD SIDING, BOARDS										609
	0020											
	3200	Wood, cedar bevel, short lengths, A grade, 1/2″ x 6″	1 Carp	250	.032	S.F.	1.44	.70		2.14	2.73	
	3300	1/2″ x 8″		275	.029		1.79	.64		2.43	3.01	
	3500	3/4″ x 10″, clear grade, 3′ to 16′		300	.027		1.59	.59		2.18	2.71	
	3600	"B" grade		300	.027		1.57	.59		2.16	2.68	
	3800	Cedar, rough sawn, 1″ x 4″, B & Btr., natural		240	.033		1.83	.73		2.56	3.21	
	3900	Stained		240	.033		1.88	.73		2.61	3.27	
	4100	1″ x 12″, board & batten, #3 & Btr., natural		260	.031		1.10	.68		1.78	2.32	
	4200	Stained		260	.031		1.14	.68		1.82	2.36	
	4400	1″ x 8″ channel siding, #3 & Btr., natural		250	.032		1.09	.70		1.79	2.35	
	4500	Stained		250	.032		1.14	.70		1.84	2.40	
	4700	Redwood, clear, beveled, vertical grain, 1/2″ x 4″		200	.040		1.26	.88		2.14	2.82	
	4800	1/2″ x 8″		250	.032		1.29	.70		1.99	2.57	
	5000	3/4″ x 10″		300	.027		1.73	.59		2.32	2.86	
	5200	Channel siding, 1″ x 10″, clear		285	.028		1.14	.62		1.76	2.26	
	5250	Redwood, T&G boards, clear, 1″ x 4″	F-2	300	.053		2.29	1.17	.05	3.51	4.50	
	5270	1″ x 8″	″	375	.043		2.06	.94	.04	3.04	3.85	
	5400	White pine, rough sawn, 1″ x 8″, natural	1 Carp	275	.029		.52	.64		1.16	1.62	
	5500	Stained		275	.029		.60	.64		1.24	1.70	
	9000	Minimum labor/equipment charge		2	4	Job		88		88	145	
611	0010	WOOD SIDING, SHEETS										611
	0020											
	0030	Siding, hardboard, 7/16″ thick, prime painted, lap,										
	0050	plain or grooved finish	F-2	750	.021	S.F.	.41	.47	.02	.90	1.24	
	0100	Board finish, 7/16″ thick, lap or grooved, primed		750	.021		.66	.47	.02	1.15	1.52	
	0200	Stained		750	.021		.69	.47	.02	1.18	1.55	
	0700	Particle board, overlaid, 3/8″ thick		750	.021		.55	.47	.02	1.04	1.40	
	0900	Plywood, medium density overlaid, 3/8″ thick		750	.021		.64	.47	.02	1.13	1.49	

(83)

074 600	Cladding/Siding		CREW	DAILY OUTPUT	MAN-HOURS	UNIT	BARE COSTS				TOTAL INCL O&P	
							MAT.	LABOR	EQUIP.	TOTAL		
611	1000	½" thick	F-2	700	.023	S.F.	.70	.50	.02	1.22	1.62	**611**
	1100	¾" thick		650	.025		.87	.54	.02	1.43	1.87	
	1600	Texture 1-11, cedar, ⅝" thick, natural		675	.024		.97	.52	.02	1.51	1.95	
	1700	Factory stained		675	.024		1.09	.52	.02	1.63	2.08	
	1900	Texture 1-11, fir, ⅝" thick, natural		675	.024		.52	.52	.02	1.06	1.45	
	2000	Factory stained		675	.024		.59	.52	.02	1.13	1.53	
	2050	Texture 1-11, S.Y.P., ⅝" thick, natural		675	.024		.57	.52	.02	1.11	1.51	
	2100	Factory stained		675	.024		.64	.52	.02	1.18	1.58	
	2200	Rough sawn cedar, ⅜" thick, natural		675	.024		1.15	.52	.02	1.69	2.15	
	2300	Factory stained		675	.024		1.25	.52	.02	1.79	2.26	
	2500	Rough sawn fir, ⅜" thick, natural		675	.024		.42	.52	.02	.96	1.34	
	2600	Factory stained		675	.024		.49	.52	.02	1.03	1.42	
	2800	Redwood, textured siding, ⅝" thick		675	.024		1.15	.52	.02	1.69	2.15	
	3000	Polyvinyl chloride coated, ⅜" thick		750	.021		.79	.47	.02	1.28	1.66	
	9000	Minimum labor/equipment charge	1 Carp	2	4	Job		88		88	145	

075 100	Built-Up Roofing		CREW	DAILY OUTPUT	MAN-HOURS	UNIT	BARE COSTS				TOTAL INCL O&P	
							MAT.	LABOR	EQUIP.	TOTAL		
101	0010	**ASPHALT** Coated felt, #30, 2 sq. per roll, not mopped	1 Rofc	58	.138	Sq.	5	2.81		7.81	10.45	**101**
	0200	#15, 4 sq. per roll, plain or perforated, not mopped		58	.138		2.50	2.81		5.31	7.70	
	0300	Roll roofing, smooth, #65		15	.533		10.25	10.85		21.10	30	
	0500	#90		15	.533		12	10.85		22.85	32	
	0520	Mineralized		15	.533		12.60	10.85		23.45	33	
	0540	D.C. (Double coverage), 19" selvage edge		10	.800		24	16.30		40.30	55	
	0580	Adhesive (lap cement)				Gal.	6.50			6.50	7.15	
	0595											
	0600	Steep, flat or dead level asphalt, 10 ton lots, bulk				Ton	259			259	285	
	0800	Packaged				"	295			295	325	
	9000	Minimum labor/equipment charge	1 Rofc	4	2	Job		41		41	71	
102	0010	**BUILT-UP ROOFING**										**102**
	0020											
	0120	Asphalt flood coat with gravel/slag surfacing, not including										
	0140	Insulation, flashing or wood nailers										
	0200	Asphalt base sheet, 3 plies #15 asphalt felt, mopped	G-1	22	2.550	Sq.	33.25	49	4.92	87.17	125	
	0350	On nailable decks		21	2.670		34.75	51	5.15	90.90	135	
	0500	4 plies #15 asphalt felt, mopped		20	2.800		36	53	5.40	94.40	140	
	0550	On nailable decks		19	2.950		34.75	56	5.70	96.45	145	
	0600											
	0700	Coated glass base sheet, 2 plies glass (type IV), mopped	G-1	22	2.550	Sq.	32.65	49	4.92	86.57	125	
	0850	3 plies glass, mopped		20	2.800		38.95	53	5.40	97.35	145	
	0950	On nailable decks		19	2.950		34.15	56	5.70	95.85	145	
	1000	3 plies glass fiber felt (type IV), mopped		22	2.550		39.05	49	4.92	92.97	135	
	1050	On nailable decks		21	2.670		34.90	51	5.15	91.05	135	
	1100	4 plies glass fiber felt (type IV), mopped		20	2.800		47.75	53	5.40	106.15	150	
	1150	On nailable decks		19	2.950		44.25	56	5.70	105.95	155	
	1200	Organic base sheet, 3 plies #15 organic felt, mopped		20	2.800		34	53	5.40	92.40	135	
	1250	On nailable decks		19	2.950		30.65	56	5.70	92.35	140	
	1300	4 plies #15 organic felt, mopped		22	2.550		37	49	4.92	90.92	130	
	2000	Asphalt flood coat, smooth surface										

075 100 | Built-Up Roofing

			CREW	DAILY OUTPUT	MAN-HOURS	UNIT	BARE COSTS				TOTAL INCL O&P	
							MAT.	LABOR	EQUIP.	TOTAL		
102	2200	Asphalt base sheet & 3 plies #15 asphalt felt, mopped	G-1	24	2.330	Sq.	29	45	4.51	78.51	115	102
	2400	On nailable decks		23	2.430		28	46	4.70	78.70	120	
	2600	4 plies #15 asphalt felt, mopped		24	2.330		29.50	45	4.51	79.01	115	
	2700	On nailable decks		23	2.430		28	46	4.70	78.70	120	
	2900	Coated glass fiber base sheet, mopped, and 2 plies of										
	2910	glass fiber felt (type IV)	G-1	25	2.240	Sq.	29.50	43	4.33	76.83	110	
	3100	On nailable decks	"	24	2.330	"	27.75	45	4.51	77.26	115	
	3150											
	3200	3 plies, mopped	G-1	23	2.430	Sq.	36.50	46	4.70	87.20	125	
	3300	On nailable decks		22	2.550		34.75	49	4.92	88.67	130	
	3500	3 plies glass fiber felt (type IV), mopped		25	2.240		38.50	43	4.33	85.83	120	
	3600	On nailable decks		24	2.330		35	45	4.51	84.51	120	
	3800	4 plies glass fiber felt (type IV), mopped		23	2.430		44.60	46	4.70	95.30	135	
	3900	On nailable decks		22	2.550		40	49	4.92	93.92	135	
	4000	Organic base sheet & 3 plies #15 organic felt, mopped		24	2.330		35.40	45	4.51	84.91	120	
	4200	On nailable decks		23	2.430		32	46	4.70	82.70	120	
	4300	4 plies #15 organic felt, mopped		22	2.550		38.50	49	4.92	92.42	135	
	4500	Coal tar pitch with gravel/slag surfacing										
	4600	4 plies #15 tarred felt, mopped	G-1	21	2.670	Sq.	60	51	5.15	116.15	160	
	4800	3 plies glass fiber felt (type IV), mopped	"	19	2.950	"	49.20	56	5.70	110.90	160	
	5000	Coated glass fiber base sheet, and 2 plies of										
	5010	glass fiber felt, (type IV), mopped	G-1	19	2.950	Sq.	50.70	56	5.70	112.40	160	
	5300	On nailable decks		18	3.110		46.50	59	6	111.50	160	
	5400	On wood decks		18	3.110		46.50	59	6	111.50	160	
	5600	4 plies glass fiber felt (type IV), mopped		21	2.670		70	51	5.15	126.15	170	
	5800	On nailable decks		20	2.800		65	53	5.40	123.40	170	
	5900	On wood decks		20	2.800		65	53	5.40	123.40	170	
	6000	4 plies #15 tarred felt, mopped		21	2.670		61.75	51	5.15	117.90	165	
103	0010	CANTS 4" x 4" treated timber, cut diagonally	1 Rofc	325	.025	L.F.	.54	.50		1.04	1.47	103
	0100	Foamglass		325	.025		.52	.50		1.02	1.45	
	0300	Mineral or fiber, trapezoidal, 1" x 4" x 48"		325	.025		.15	.50		.65	1.04	
	0400	1-½" x 5-⅝" x 48"		325	.025		.25	.50		.75	1.15	
	9000	Minimum labor/equipment charge		4	2	Job		41		41	71	
104	0010	FELT Glass fibered, #15, no mopping	1 Rofc	58	.138	Sq.	4.50	2.81		7.31	9.90	104
	0200	#43 base sheet		58	.138		6.50	2.81		9.31	12.10	
	0300	Base sheet, #45, channel vented		58	.138		16.50	2.81		19.31	23	
	0400	#50, coated		58	.138		7.10	2.81		9.91	12.75	
	0500	Cap, mineral surfaced		58	.138		13.25	2.81		16.06	19.50	
	0510											
	0600	Flashing membrane, #65	1 Rofc	16	.500	Sq.	32.50	10.20		42.70	54	
	0800	Coal tar fibered, #15, no mopping		58	.138		8.65	2.81		11.46	14.45	
	0900	Asphalt felt, #15, 4 sq. per roll, no mopping		58	.138		3.10	2.81		5.91	8.35	
	1100	#30, 2 sq. per roll		58	.138		5.90	2.81		8.71	11.40	
	1200	Double coated, #33		58	.138		6.25	2.81		9.06	11.80	
	1400	#40, base sheet		58	.138		6.65	2.81		9.46	12.25	
	1500	Tarred felt, organic, #15, 4 sq. rolls		58	.138		4.95	2.81		7.76	10.35	
	1550	#30, 2 sq. roll		58	.138		9.80	2.81		12.61	15.70	
	1700	Add for mopping above felts, per ply, asphalt, 20 lbs. per sq.		28	.286		2.55	5.80		8.35	13	
	1800	Coal tar mopping, 30 lbs. per sq.		28	.286		5.25	5.80		11.05	16	
	1900	Flood coat, with asphalt, 60 lbs. per sq.	2 Rofc	16.30	.982		7.30	20		27.30	43	
	2000	With coal tar, 75 lbs. per sq.	"	16.30	.982		10.75	20		30.75	47	
	9000	Minimum labor/equipment charge	1 Rofc	4	2	Job		41		41	71	
105	0010	WALKWAY For built-up roofs, asphalt impregnated, 3' x 6' x ½" thk.	1 Rofc	400	.020	S.F.	1.50	.41		1.91	2.36	105
	0100	3' x 3' x ¾" thick, hot application		400	.020	"	2.10	.41		2.51	3.02	
	9000	Minimum labor/equipment charge		2.75	2.910	Job		59		59	105	

075 150 | Cold Applied Roofing

		CREW	DAILY OUTPUT	MAN-HOURS	UNIT	MAT.	LABOR	EQUIP.	TOTAL	TOTAL INCL O&P	
152	0010	COLD APPLIED 3-ply system (components listed below)	G-5	50	.800	Sq.		14.85	2.16	17.01	28	152
	0100	Spunbond poly. fabric, 1.35 oz/SY, 36"W, 10.8 Sq/roll				Ea.	67.50			67.50	74	
	0200	49" wide, 14.6 Sq./roll					87.50			87.50	96	
	0300	2.10 oz./S.Y., 36" wide, 10.8 Sq./roll					94			94	105	
	0400	49" wide, 14.6 Sq./roll					120			120	130	
	0500	Base & finish coat, 3 gal./Sq., 5 gal./can				Gal.	3.50			3.50	3.85	
	0600	Coating, ceramic granules, ½ Sq./bag				Ea.	9.40			9.40	10.35	
	0700	Aluminum, 2 gal./Sq.				Gal.	7.25			7.25	8	
	0800	Emulsion, fibered or non-fibered, 4 gal./Sq.				"	4.25			4.25	4.68	

075 200 | Prepared Roll Roofing

		CREW	DAILY OUTPUT	MAN-HOURS	UNIT	MAT.	LABOR	EQUIP.	TOTAL	TOTAL INCL O&P	
204	0010	ROLL ROOFING										204
	0100	Asphalt, mineral surface										
	0200	1 ply #15 organic felt, 1 ply mineral surfaced										
	0300	selvage roofing, lap 19", nailed & mopped	G-1	27	2.070	Sq.	33	40	4.01	77.01	110	
	0400	3 plies glass fiber felt (type IV), 1 ply mineral surfaced										
	0500	selvage roofing, lapped 19", mopped	G-1	25	2.240	Sq.	51	43	4.33	98.33	135	
	0600	Coated glass fiber base sheet, 2 plies of glass fiber										
	0700	felt (type IV), 1 ply mineral surfaced selvage										
	0800	roofing, lapped 19", mopped	G-1	25	2.240	Sq.	52.50	43	4.33	99.83	140	
	0900	On nailable decks	"	24	2.330	"	50	45	4.51	99.51	140	
	1000	3 plies glass fiber felt (type III), 1 ply mineral surfaced										
	1100	selvage roofing, lapped 19", mopped	G-1	25	2.240	Sq.	47.80	43	4.33	95.13	130	

075 300 | Elastomeric Roofing

		CREW	DAILY OUTPUT	MAN-HOURS	UNIT	MAT.	LABOR	EQUIP.	TOTAL	TOTAL INCL O&P	
301	0010	ELASTOMERIC ROOFING										301
	0100	For Elastomeric waterproofing, see division 071-102										
	0300	Hypalon neoprene, fluid applied, 20 mil thick, not-reinforced	G-1	1,135	.049	S.F.	1.45	.94	.10	2.49	3.36	
	0600	Non-woven polyester, reinforced		960	.058		1.60	1.11	.11	2.82	3.84	
	0700	5 coat neoprene deck, 60 mil thick, under 10,000 S.F.		325	.172		4.35	3.29	.33	7.97	10.95	
	0900	Over 10,000 S.F.		625	.090		4.05	1.71	.17	5.93	7.65	
	1300	Vinyl plastic traffic deck, sprayed, 2 to 4 mils thick		625	.090		1.15	1.71	.17	3.03	4.46	
	1500	Vinyl and neoprene membrane traffic deck		1,550	.036		1.25	.69	.07	2.01	2.67	
	9000	Minimum labor/equipment charge	1 Rofc	2	4	Job		81		81	145	
302	0010	SINGLE-PLY MEMBRANE										302
	0020											
	0200	Chlorinated polyethylene(CPE), 40 mils, 0.31 P.S.F.										
	0300	Partially adhered with mechanical fasteners	G-5	3,500	.011	S.F.	1.50	.21	.03	1.74	2.06	
	0800	Chlorosulfonated polyethylene-hypalon (CSPE), 35 mils,										
	0900	0.25 P.S.F., fully adhered with neoprene latex	G-5	2,600	.015	S.F.	1.30	.29	.04	1.63	1.98	
	3500	Ethylene propylene diene monomer (EPDM), 45 mils, 0.28 P.S.F.										
	4000	55 mils, 0.35 P.S.F.										
	4100	Loose-laid & ballasted with stone, (10 P.S.F.)	G-5	5,100	.008	S.F.	.69	.15	.02	.86	1.04	
	4200	Partially adhered		3,500	.011		.74	.21	.03	.98	1.22	
	4300	Fully adhered with adhesive		2,600	.015		.85	.29	.04	1.18	1.49	
	4850	Vulcanizing tape for membrane, 2" x 50' roll				Ea.	27.50			27.50	30	
	4900	Batten strips, 10' sections					3.50			3.50	3.85	
	4910	Vulcanizing tape for batten strips, 6" x 100' roll					44.75			44.75	49	
	4930	Plate anchors				M	100			100	110	
	4970	Adhesive for fully adhered systems, 60 S.F./gal.				Gal.	16.65			16.65	18.30	
	6800	Neoprene, 60 mils, 0.45 P.S.F.										
	7000	Partially adhered with mechanical fasteners	G-5	3,500	.011	S.F.	1.65	.21	.03	1.89	2.23	
	7100	Fully adhered with contact adhesive	"	2,600	.015	"	1.55	.29	.04	1.88	2.26	
	7500	Polyisobutylene (PIB), 100 mils, 0.57 P.S.F.										
	7600	Loose-laid & ballasted with stone/gravel (10 P.S.F.)	G-5	5,100	.008	S.F.	1.05	.15	.02	1.22	1.44	
	7700	Partially adhered with adhesive	"	3,500	.011	"	.98	.21	.03	1.22	1.49	

075 | Membrane Roofing

075 300 | Elastomeric Roofing

			CREW	DAILY OUTPUT	MAN-HOURS	UNIT	BARE COSTS MAT.	LABOR	EQUIP.	TOTAL	TOTAL INCL O&P	
302	7900	Fully adhered with contact cement	G-5	2,600	.015	S.F.	1.09	.29	.04	1.42	1.75	302
	8200	Polyvinyl chloride (PVC)										
	8400	48 mils, 0.33 to 0.38 P.S.F.										
	8450	Loose-laid & ballasted with stone/gravel(10 P.S.F.)	G-5	5,100	.008	S.F.	.52	.15	.02	.69	.85	
	8500	Partially adhered with mechanical & solvent weld		3,500	.011		.80	.21	.03	1.04	1.29	
	8550	Fully adhered with adhesive	↓	2,600	.015		.85	.29	.04	1.18	1.49	
	8680	Uncured neoprene, 60 mils, for flashing	1 Rofc	600	.013		1.21	.27		1.48	1.81	
	8690	Separator sheet	G-5	4,000	.010	↓	.08	.19	.03	.30	.45	
	8700	Reinforced PVC, 48 mils, 0.33 P.S.F.										
	8750	Loose-laid & ballasted with stone/gravel (12 P.S.F.)	G-5	5,100	.008	S.F.	.60	.15	.02	.77	.94	
	8800	Partially adhered with mechanical fasteners		3,500	.011		.82	.21	.03	1.06	1.31	
	8850	Fully adhered with adhesive	↓	2,600	.015		.86	.29	.04	1.19	1.50	

075 350 | Modified Bit. Roofing

			CREW	DAILY OUTPUT	MAN-HOURS	UNIT	MAT.	LABOR	EQUIP.	TOTAL	TOTAL INCL O&P	
352	0010	MODIFIED BITUMEN										352
	0200	150 mils, 0.82 P.S.F.										
	0300	Loose-laid & ballasted with gravel (4 P.S.F.)	G-5	3,200	.013	S.F.	.57	.23	.03	.83	1.07	
	0400	Partially adhered with torch welding		2,500	.016		.74	.30	.04	1.08	1.38	
	0500	Fully adhered with torch welding	↓	2,000	.020	↓	.74	.37	.05	1.16	1.52	

075 600 | Roof Maint. & Repairs

			CREW	DAILY OUTPUT	MAN-HOURS	UNIT	MAT.	LABOR	EQUIP.	TOTAL	TOTAL INCL O&P	
604	0010	ROOF COATINGS Asphalt				Gal.	2.75			2.75	3.03	604
	0200	Asphalt base, fibered aluminum coating					7.25			7.25	8	
	0300	Asphalt primer, 5 gallon				↓	3.25			3.25	3.58	
	0600	Coal tar pitch, 200 lb. barrels				Ton	380			380	420	
	0700	Tar roof cement, 5 gal. lots				Gal.	4.10			4.10	4.51	
	0800	Glass fibered roof & patching cement, 5 gallon				"	6.50			6.50	7.15	
	0900	Reinforcing glass membrane, 450 S.F./roll				Ea.	46			46	51	
	1000	Neoprene roof coating, 5 gal., 2 gal./sq.				Gal.	16.90			16.90	18.60	
	1100	Roof patch & flashing cement, 5 gallon					16.95			16.95	18.65	
	1200	Roof resurtant, glass fibered, 3 gal./sq.					6.25			6.25	6.90	
	1300	Mineral rubber, 3 gal./sq.				↓	4			4	4.40	

076 | Flashing and Sheet Metal

076 100 | Sheet Metal Roofing

			CREW	DAILY OUTPUT	MAN-HOURS	UNIT	BARE COSTS MAT.	LABOR	EQUIP.	TOTAL	TOTAL INCL O&P	
101	0010	COPPER ROOFING Batten seam, over 10 sq., 16 oz., 130 lb./sq.	1 Shee	1.10	7.270	Sq.	395	180		575	720	101
	0200	18 oz., 145 lb. per sq.		1	8		440	200		640	800	
	0300	20 oz., 160 lb. per sq.		1	8		500	200		700	865	
	0400	Standing seam, over 10 squares, 16 oz., 125 lb. per sq.		1.30	6.150		395	155		550	675	
	0600	18 oz., 140 lb. per sq.		1.20	6.670		440	165		605	745	
	0700	20 oz., 150 lb. per sq.		1.10	7.270		485	180		665	820	
	0900	Flat seam, over 10 squares, 16 oz., 115 lb. per sq.		1.20	6.670		365	165		530	665	
	1000	20 oz., 145 lb. per sq.	↓	1.10	7.270		450	180		630	780	
	1200	For abnormal conditions or small areas, add				↓	25%	100%				
	1300	For lead-coated copper, add					20%					
	9000	Minimum labor/equipment charge	1 Shee	2	4	Job		100		100	155	
102	0010	LEAD ROOFING 3 lb. per S.F., batten seam	1 Shee	1.20	6.670	Sq.	470	165		635	780	102
	0100	Flat seam	"	1.30	6.150	"	425	155		580	710	

076 100 | Sheet Metal Roofing

			DAILY OUTPUT	MAN-HOURS	UNIT	BARE COSTS				TOTAL INCL O&P	
		CREW				MAT.	LABOR	EQUIP.	TOTAL		
102	9000	Minimum labor/equipment charge	1 Shee	2	4	Job		100		100	155	102
103	0010	**MONEL ROOFING** Batten seam, over 10 squares, .018″ thick	1 Shee	1.20	6.670	Sq.	400	165		565	700	103
	0100	.021″ thick		1.15	6.960		420	175		595	735	
	0300	Standing seam, .018″ thick		1.35	5.930		455	150		605	735	
	0400	.021″ thick		1.30	6.150		495	155		650	785	
	0600	Flat seam, .018″ thick		1.30	6.150		450	155		605	735	
	0700	.021″ thick		1.20	6.670		500	165		665	810	
	9000	Minimum labor/equipment charge		2.50	3.200	Job		80		80	125	
104	0010	**STAINLESS STEEL ROOFING** Type 304, batten seam, 28 gauge	1 Shee	1.20	6.670	Sq.	285	165		450	575	104
	0100	26 gauge	"	1.15	6.960		355	175		530	665	
	0200	For standing seam construction, deduct					2%					
	0500	For flat seam construction, deduct					3%					
	0800	For lead or terne coated stainless, 28 gauge, add					65			65	72	
	0900	For 26 gauge, add					85			85	94	
	9000	Minimum labor/equipment charge	1 Shee	2.75	2.910	Job		73		73	115	
105	0010	**ZINC** Copper alloy roofing, batten seam, .020″ thick	1 Shee	1.20	6.670	Sq.	525	165		690	840	105
	0100	.027″ thick		1.15	6.960		635	175		810	970	
	0300	.032″ thick		1.10	7.270		715	180		895	1,075	
	0400	.040″ thick		1.05	7.620		845	190		1,035	1,225	
	0600	For standing seam construction, deduct					2%					
	0700	For flat seam construction, deduct					3%					
	9000	Minimum labor/equipment charge	1 Shee	2.75	2.910	Job		73		73	115	

076 200 | Sheet Mtl Flash & Trim

			CREW	DAILY OUTPUT	MAN-HOURS	UNIT	MAT.	LABOR	EQUIP.	TOTAL	TOTAL INCL O&P	
201	0010	**DOWNSPOUTS** Aluminum 2″ x 3″, .020″ thick, embossed	1 Shee	190	.042	L.F.	.52	1.05		1.57	2.22	201
	0100	Enameled		190	.042		.55	1.05		1.60	2.26	
	0300	Enameled, .024″ thick, 2″ x 3″		190	.042		.53	1.05		1.58	2.23	
	0400	3″ x 4″		140	.057		.85	1.43		2.28	3.18	
	0600	Round, corrugated aluminum, 3″ diameter, .020″ thick		190	.042		.65	1.05		1.70	2.37	
	0700	4″ diameter, .025″ thick		140	.057		.95	1.43		2.38	3.29	
	0900	Wire strainer, round, 2″ diameter		155	.052	Ea.	.76	1.29		2.05	2.86	
	1000	4″ diameter		155	.052		1.05	1.29		2.34	3.18	
	1200	Rectangular, perforated, 2″ x 3″		145	.055		1.45	1.38		2.83	3.76	
	1300	3″ x 4″		145	.055		2.35	1.38		3.73	4.75	
	1500	Copper, round, 16 oz., stock, 2″ diameter		190	.042	L.F.	3.15	1.05		4.20	5.10	
	1600	3″ diameter		190	.042		4.25	1.05		5.30	6.35	
	1800	4″ diameter		145	.055		5.40	1.38		6.78	8.10	
	1900	5″ diameter		130	.062		5.55	1.54		7.09	8.50	
	2100	Rectangular, corrugated copper, stock, 2″ x 3″		190	.042		4.60	1.05		5.65	6.70	
	2200	3″ x 4″		145	.055		6.20	1.38		7.58	9	
	2400	Rectangular, plain copper, stock, 2″ x 3″		190	.042		4.75	1.05		5.80	6.90	
	2500	3″ x 4″		145	.055		5.80	1.38		7.18	8.55	
	2700	Wire strainers, rectangular, 2″ x 3″		145	.055	Ea.	2.40	1.38		3.78	4.80	
	2800	3″ x 4″		145	.055		3.80	1.38		5.18	6.35	
	3000	Round, 2″ diameter		145	.055		2.30	1.38		3.68	4.69	
	3100	3″ diameter		145	.055		3.30	1.38		4.68	5.80	
	3300	4″ diameter		145	.055		5.10	1.38		6.48	7.75	
	3400	5″ diameter		115	.070		7.05	1.74		8.79	10.50	
	3600	Lead-coated copper, round, stock, 2″ diameter		190	.042	L.F.	4.10	1.05		5.15	6.15	
	3700	3″ diameter		190	.042		4.95	1.05		6	7.10	
	3900	4″ diameter		145	.055		6.10	1.38		7.48	8.85	
	4000	5″ diameter, corrugated		130	.062		7.80	1.54		9.34	11	
	4200	6″ diameter, corrugated		105	.076		9.70	1.90		11.60	13.65	
	4300	Rectangular, corrugated, stock, 2″ x 3″		190	.042		6.70	1.05		7.75	9	

076 200	Sheet Mtl Flash & Trim	CREW	DAILY OUTPUT	MAN-HOURS	UNIT	BARE COSTS				TOTAL INCL O&P	
						MAT.	LABOR	EQUIP.	TOTAL		
201 4500	Plain, stock, 2" x 3"	1 Shee	190	.042	L.F.	5.40	1.05		6.45	7.60	**201**
4600	3" x 4"		145	.055		7	1.38		8.38	9.85	
4800	Steel, galvanized, round, corrugated, 2" or 3" diam., 28 ga.		190	.042		.45	1.05		1.50	2.15	
4900	4" diameter, 28 gauge		145	.055		.57	1.38		1.95	2.79	
5100	5" diameter, 28 gauge		130	.062		.70	1.54		2.24	3.18	
5200	26 gauge		130	.062		.92	1.54		2.46	3.43	
5400	6" diameter, 28 gauge		105	.076		1.05	1.90		2.95	4.14	
5500	26 gauge		105	.076		1.20	1.90		3.10	4.31	
5700	Rectangular, corrugated, 28 gauge, 2" x 3"		190	.042		.45	1.05		1.50	2.15	
5800	3" x 4"		145	.055		1.25	1.38		2.63	3.54	
6000	Rectangular, plain, 28 gauge, galvanized, 2" x 3"		190	.042		.45	1.05		1.50	2.15	
6100	3" x 4"		145	.055		1.25	1.38		2.63	3.54	
6300	Epoxy painted, 24 gauge, corrugated, 2" x 3"		190	.042		.80	1.05		1.85	2.53	
6400	3" x 4"		145	.055		1.33	1.38		2.71	3.63	
6600	Wire strainers, rectangular, 2" x 3"		145	.055	Ea.	1.35	1.38		2.73	3.65	
6700	3" x 4"		145	.055		2.18	1.38		3.56	4.56	
6900	Round strainers, 2" or 3" diameter		145	.055		.95	1.38		2.33	3.21	
7000	4" diameter		145	.055		1.18	1.38		2.56	3.46	
7200	5" diameter		145	.055		1.75	1.38		3.13	4.09	
7300	6" diameter		115	.070		2.15	1.74		3.89	5.10	
9000	Minimum labor/equipment charge		4	2	Job		50		50	78	
202 0010	**DRIP EDGE** Aluminum, .016" thick, 5" girth, mill finish	1 Carp	400	.020	L.F.	.13	.44		.57	.86	**202**
0100	White finish		400	.020		.15	.44		.59	.88	
0200	8" girth		400	.020		.27	.44		.71	1.02	
0300	28" girth		100	.080		1.18	1.76		2.94	4.17	
0400	Galvanized, 5" girth		400	.020		.21	.44		.65	.95	
0500	8" girth		400	.020		.26	.44		.70	1	
9000	Minimum labor/equipment charge		4	2	Job		44		44	72	
203 0010	**ELBOWS** Aluminum, 2" x 3", embossed	1 Shee	100	.080	Ea.	.51	2		2.51	3.70	**203**
0100	Enameled		100	.080		.51	2		2.51	3.70	
0200	3" x 4", .025" thick, embossed		100	.080		.80	2		2.80	4.02	
0300	Enameled		100	.080		.80	2		2.80	4.02	
0400	Round corrugated, 3", embossed, .020" thick		100	.080		.75	2		2.75	3.96	
0500	4", .025" thick		100	.080		.87	2		2.87	4.09	
0600	Copper, 16 oz. round, 2" diameter		100	.080		4.25	2		6.25	7.80	
0700	3" diameter		100	.080		5.30	2		7.30	8.95	
0800	4" diameter		100	.080		9.40	2		11.40	13.50	
0900	5" diameter		100	.080		11.65	2		13.65	15.95	
1000	2" x 3" corrugated		100	.080		9	2		11	13.05	
1100	3" x 4" corrugated		100	.080		12.70	2		14.70	17.10	
9000	Minimum labor/equipment charge		4	2	Job		50		50	78	
204 0010	**FLASHING** Aluminum, mill finish, .013" thick	1 Shee	145	.055	S.F.	.26	1.38		1.64	2.45	**204**
0030	.016" thick		145	.055		.30	1.38		1.68	2.49	
0060	.019" thick		145	.055		.65	1.38		2.03	2.88	
0100	.032" thick		145	.055		.78	1.38		2.16	3.02	
0200	.040" thick		145	.055		1.32	1.38		2.70	3.62	
0300	.050" thick		145	.055		1.60	1.38		2.98	3.92	
0400	Painted finish, add					.15			.15	.17	
0500	Fabric-backed 2 sides, .004" thick	1 Shee	330	.024		.35	.61		.96	1.34	
0700	.016" thick		330	.024		.95	.61		1.56	2	
0750	Mastic-backed, self adhesive		460	.017		1.85	.43		2.28	2.72	
0800	Mastic-coated 2 sides, .004" thick		330	.024		.40	.61		1.01	1.39	
1000	.005" thick		330	.024		.52	.61		1.13	1.52	
1100	.016" thick		330	.024		.95	.61		1.56	2	
1300	Asphalt flashing cement, 5 gallon				Gal.	6.50			6.50	7.15	

		CREW	DAILY OUTPUT	MAN-HOURS	UNIT	BARE COSTS				TOTAL INCL O&P
076 200 Sheet Mtl Flash & Trim						MAT.	LABOR	EQUIP.	TOTAL	
204 1600	Copper, 16 oz., sheets, under 6000 lbs.	1 Shee	115	.070	S.F.	2.60	1.74		4.34	5.60	204
1700	Over 6000 lbs.		155	.052		2.60	1.29		3.89	4.88
1900	20 oz. sheets, under 6000 lbs.		110	.073		3.25	1.82		5.07	6.45
2000	Over 6000 lbs.		145	.055		3.25	1.38		4.63	5.75
2200	24 oz. sheets, under 6000 lbs.		105	.076		3.90	1.90		5.80	7.30
2300	Over 6000 lbs.		135	.059		3.90	1.48		5.38	6.60
2500	32 oz. sheets, under 6000 lbs.		100	.080		5.05	2		7.05	8.70
2600	Over 6000 lbs.		130	.062		5.05	1.54		6.59	7.95
2800	Copper, paperbacked 1 side, 2 oz.		330	.024		1	.61		1.61	2.05
2900	3 oz.		330	.024		1.35	.61		1.96	2.44
3100	Paperbacked 2 sides, copper, 2 oz.		330	.024		1	.61		1.61	2.05
3150	3 oz.		330	.024		1.45	.61		2.06	2.55
3200	5 oz.		330	.024		2.25	.61		2.86	3.43
3250	7 oz.		330	.024		2.55	.61		3.16	3.76
3400	Mastic-backed 2 sides, copper, 2 oz.		330	.024		1.15	.61		1.76	2.22
3500	3 oz.		330	.024		1.65	.61		2.26	2.77
3700	5 oz.		330	.024		2.40	.61		3.01	3.59
3800	Fabric-backed 2 sides, copper, 2 oz.		330	.024		1.30	.61		1.91	2.38
4000	3 oz.		330	.024		1.70	.61		2.31	2.82
4100	5 oz.		330	.024		2.45	.61		3.06	3.65
4300	Copper-clad stainless steel, .015" thick, under 500 lbs.		115	.070		2.80	1.74		4.54	5.80
4400	Over 2000 lbs.		155	.052		2.65	1.29		3.94	4.94
4600	.018" thick, under 500 lbs.		100	.080		3.75	2		5.75	7.25
4700	Over 2000 lbs.		145	.055		2.65	1.38		4.03	5.10
4900	Fabric, asphalt-saturated cotton, specification grade	1 Rofc	35	.229	S.Y.	1.22	4.65		5.87	9.50
5000	Utility grade		35	.229		.75	4.65		5.40	9
5200	Open-mesh fabric, saturated, 40 oz. per S.Y.		35	.229		1.45	4.65		6.10	9.75
5300	Close-mesh fabric, saturated, 17 oz. per S.Y.		35	.229		1.45	4.65		6.10	9.75
5500	Fiberglass, resin-coated		35	.229		1.47	4.65		6.12	9.80
5600	Asphalt-coated, 40 oz. per S.Y.		35	.229		2.15	4.65		6.80	10.55
5800	Lead, 2.5 lb. per S.F., up to 12" wide		135	.059	S.F.	3.15	1.21		4.36	5.60
5900	Over 12" wide		135	.059		3.15	1.21		4.36	5.60
6100	Lead-coated copper, fabric-backed, 2 oz.	1 Shee	330	.024		1.25	.61		1.86	2.33
6200	5 oz.		330	.024		1.45	.61		2.06	2.55
6400	Mastic-backed 2 sides, 2 oz.		330	.024		.87	.61		1.48	1.91
6500	5 oz.		330	.024		1.40	.61		2.01	2.49
6700	Paperbacked 1 side, 2 oz.		330	.024		.80	.61		1.41	1.83
6800	3 oz.		330	.024		.98	.61		1.59	2.03
7000	Paperbacked 2 sides, 2 oz.		330	.024		.88	.61		1.49	1.92
7100	5 oz.		330	.024		1.45	.61		2.06	2.55
7300	Polyvinyl chloride, black, .010" thick	1 Rofc	285	.028		.10	.57		.67	1.11
7400	.020" thick		285	.028		.15	.57		.72	1.17
7600	.030" thick		285	.028		.25	.57		.82	1.28
7700	.056" thick		285	.028		.45	.57		1.02	1.50
7900	Black or white for exposed roofs, .060" thick		285	.028		1.50	.57		2.07	2.65
8000	Asbestos-backed for parking decks, .045" thick		285	.028		.85	.57		1.42	1.94
8050	PVC (19 mils) coated galv. steel (24 mils), 4' x 8' sheets		240	.033		1.40	.68		2.08	2.73
8060	PVC tape, 5" x 45 mils, for joint covers, 100 L.F./roll				Ea.	80			80	88
8100	Rubber, butyl, 1/32" thick	1 Rofc	285	.028	S.F.	.65	.57		1.22	1.72
8200	1/16" thick		285	.028		.95	.57		1.52	2.05
8300	Neoprene, cured, 1/16" thick		285	.028		1.45	.57		2.02	2.60
8400	1/8" thick		285	.028		2.95	.57		3.52	4.25
8500	Shower pan, bituminous membrane, 7 oz.	1 Shee	155	.052		1.05	1.29		2.34	3.18
8550	3 ply copper and fabric, 3 oz.		155	.052		1.85	1.29		3.14	4.06
8600	7 oz.		155	.052		3.40	1.29		4.69	5.75
8650	Copper, 16 oz.		100	.080		2.70	2		4.70	6.10
8700	Lead on copper and fabric, 5 oz.		155	.052		2.15	1.29		3.44	4.39
8800	7 oz.		155	.052		2.45	1.29		3.74	4.72

		076 200 Sheet Mtl Flash & Trim	CREW	DAILY OUTPUT	MAN-HOURS	UNIT	BARE COSTS MAT.	LABOR	EQUIP.	TOTAL	TOTAL INCL O&P	
204	8900	Stainless steel sheets, 32 ga., .010" thick	1 Shee	155	.052	S.F.	1.85	1.29		3.14	4.06	204
	9000	28 ga., .015" thick		155	.052		2.30	1.29		3.59	4.55	
	9100	26 ga., .018" thick		155	.052		2.85	1.29		4.14	5.15	
	9200	24 ga., .025" thick		155	.052		3.45	1.29		4.74	5.80	
	9290	For mechanically keyed flashing, add					40%					
	9300	Stainless steel, paperbacked 2 sides, .005" thick	1 Shee	330	.024		1.15	.61		1.76	2.22	
	9320	Steel sheets, galvanized, 20 gauge		130	.062		.55	1.54		2.09	3.02	
	9340	30 gauge		160	.050		.22	1.25		1.47	2.20	
	9400	Terne coated stainless steel, .015" thick, 28 ga.		155	.052		1.60	1.29		2.89	3.78	
	9500	.018" thick, 26 ga.		155	.052		1.80	1.29		3.09	4	
	9600	Zinc and copper alloy, .020" thick		155	.052		1.35	1.29		2.64	3.51	
	9700	.027" thick		155	.052		1.80	1.29		3.09	4	
	9800	.032" thick		155	.052		2.10	1.29		3.39	4.33	
	9900	.040" thick		155	.052		2.55	1.29		3.84	4.83	
	9950	Minimum labor/equipment charge		3	2.670	Job		67		67	105	
205	0010	**GUTTERS** Aluminum, stock units, 5" box, .027" thick, plain	1 Shee	120	.067	L.F.	.80	1.67		2.47	3.49	205
	0100	Enameled		120	.067		.90	1.67		2.57	3.60	
	0300	5" box type, .032" thick, plain		120	.067		1	1.67		2.67	3.71	
	0400	Enameled		120	.067		1	1.67		2.67	3.71	
	0600	5" x 6" combination fascia & gutter, .032" thick, enameled		60	.133		2.60	3.33		5.93	8.10	
	0700	Copper, half round, 16 oz., stock units, 4" wide		120	.067		4.05	1.67		5.72	7.05	
	0900	5" wide		120	.067		4.10	1.67		5.77	7.10	
	1000	6" wide		115	.070		4.55	1.74		6.29	7.75	
	1200	K type copper gutter, stock, 4" wide		120	.067		4.40	1.67		6.07	7.45	
	1300	5" wide		120	.067		4.80	1.67		6.47	7.90	
	1500	Lead coated copper, half round, stock, 4" wide		120	.067		4.70	1.67		6.37	7.80	
	1600	6" wide		115	.070		5.70	1.74		7.44	9	
	1800	K type lead coated copper, stock, 4" wide		120	.067		5.50	1.67		7.17	8.65	
	1900	5" wide		120	.067		6	1.67		7.67	9.20	
	2100	Stainless steel, half round or box, stock, 4" wide		120	.067		4.50	1.67		6.17	7.55	
	2200	5" wide		120	.067		4.90	1.67		6.57	8	
	2400	Steel, galv., half round or box, 28 ga., 5" wide, plain		120	.067		.62	1.67		2.29	3.30	
	2500	Enameled		120	.067		.66	1.67		2.33	3.34	
	2700	26 ga. galvanized steel, stock, 5" wide		120	.067		.75	1.67		2.42	3.44	
	2800	6" wide		120	.067		.95	1.67		2.62	3.66	
	3000	Vinyl, O.G., 4" wide	1 Carp	110	.073		.88	1.60		2.48	3.58	
	3100	5" wide		110	.073		1.04	1.60		2.64	3.76	
	3200	4" half round, stock units		110	.073		.63	1.60		2.23	3.31	
	3250	Joint connectors				Ea.	1.50			1.50	1.65	
	3300	Wood, clear treated cedar, fir or hemlock, 3" x 4"	1 Carp	100	.080	L.F.	3.85	1.76		5.61	7.10	
	3400	4" x 5"	"	100	.080	"	4.85	1.76		6.61	8.20	
	9000	Minimum labor/equipment charge	1 Shee	3.75	2.130	Job		53		53	84	
206	0010	**GUTTER GUARD** 6" wide strip, aluminum mesh	1 Carp	500	.016	L.F.	.39	.35		.74	1	206
	0100	Vinyl mesh		500	.016	"	.15	.35		.50	.74	
	9000	Minimum labor/equipment charge		4	2	Job		44		44	72	
207	0010	**MANSARD** Colored aluminum, with battens, .032" thick										207
	0600	Stock units, straight surfaces	1 Shee	115	.070	S.F.	1.95	1.74		3.69	4.87	
	0700	Concave or convex surfaces		75	.107	"	2.15	2.67		4.82	6.55	
	0800	For framing, to 5' high, add		115	.070	L.F.	2.15	1.74		3.89	5.10	
	0900	Soffits, to 1' wide		125	.064	S.F.	1.15	1.60		2.75	3.78	
	9000	Minimum labor/equipment charge		2.50	3.200	Job		80		80	125	
210	0010	**REGLET** Aluminum, .025" thick, in concrete parapet	1 Carp	225	.036	L.F.	.90	.78		1.68	2.27	210
	0100	Copper, 10 oz.		225	.036		1.90	.78		2.68	3.37	
	0300	16 oz.		225	.036		2.60	.78		3.38	4.14	
	0400	Galvanized steel, 24 gauge		225	.036		.61	.78		1.39	1.95	

076 | Flashing and Sheet Metal

076 200 | Sheet Mtl Flash & Trim

			CREW	DAILY OUTPUT	MAN-HOURS	UNIT	BARE COSTS MAT.	LABOR	EQUIP.	TOTAL	TOTAL INCL O&P	
210	0600	Stainless steel, .020" thick	1 Carp	225	.036	L.F.	1.15	.78		1.93	2.54	210
	0700	Zinc and copper alloy, 20 oz.	"	225	.036		1.60	.78		2.38	3.04	
	0900	Counter flashing for above, 12" wide, .032" aluminum	1 Shee	150	.053		1.30	1.33		2.63	3.52	
	1000	Copper, 10 oz.		150	.053		2.15	1.33		3.48	4.46	
	1200	16 oz.		150	.053		2.60	1.33		3.93	4.95	
	1300	Galvanized steel, .020" thick		150	.053		.53	1.33		1.86	2.67	
	1500	Stainless steel, .020" thick		150	.053		1.55	1.33		2.88	3.80	
	1600	Zinc and copper alloy, 20 oz.		150	.053		2.95	1.33		4.28	5.35	
	9000	Minimum labor/equipment charge	1 Carp	3	2.670	Job		59		59	96	
217	0010	SOFFIT Aluminum, residential, stock units, .020" thick	1 Carp	210	.038	S.F.	.65	.84		1.49	2.08	217
	0100	Baked enamel on steel, 16 or 18 gauge		105	.076		3.95	1.68		5.63	7.10	
	0300	Polyvinyl chloride, white, solid		230	.035		.50	.77		1.27	1.80	
	0400	Perforated		230	.035		.51	.77		1.28	1.81	
	0500	For colors, add					.02			.02	.02	
	9000	Minimum labor/equipment charge	1 Carp	3	2.670	Job		59		59	96	
219	0010	TERMITE SHIELDS Zinc, 10" wide, .012" thick		350	.023	L.F.	1	.50		1.50	1.92	219
	0100	.020" thick		350	.023	"	.70	.50		1.20	1.59	
	9000	Minimum labor/equipment charge		4	2	Job		44		44	72	

077 | Roof Specialties and Accessories

077 100 | Prefab Roof Specialties

			CREW	DAILY OUTPUT	MAN-HOURS	UNIT	BARE COSTS MAT.	LABOR	EQUIP.	TOTAL	TOTAL INCL O&P	
103	0010	EXPANSION JOINT Butyl, 1/16" thick, 29" wide	1 Rofc	165	.048	L.F.	3.65	.99		4.64	5.75	103
	0300	Butyl or neoprene center with foam insulation, metal flanges										
	0400	Aluminum, .032" thick for openings to 2-1/2"	1 Rofc	165	.048	L.F.	5.35	.99		6.34	7.60	
	0600	For joint openings to 3-1/2"		165	.048		5.35	.99		6.34	7.60	
	0700	Copper, 16 oz. for openings to 2-1/2"		165	.048		11.25	.99		12.24	14.10	
	0900	For joint openings to 3-1/2"		165	.048		14	.99		14.99	17.15	
	1000	Galvanized steel, 26 ga. for openings to 2-1/2"		165	.048		4.05	.99		5.04	6.20	
	1200	For joint openings to 3-1/2"		165	.048		4.25	.99		5.24	6.40	
	1300	Lead-coated copper, 16 oz. for openings to 2-1/2"		165	.048		12.25	.99		13.24	15.20	
	1500	For joint openings to 3-1/2"		165	.048		15.25	.99		16.24	18.50	
	1600	Stainless steel, .018", for openings to 2-1/2"		165	.048		6.30	.99		7.29	8.65	
	1800	For joint openings to 3-1/2"		165	.048		11.10	.99		12.09	13.95	
	1900	Neoprene, double-seal type with thick center, 4-1/2" wide		125	.064		5.60	1.30		6.90	8.45	
	1910											
	1950	Polyethylene bellows, with galv. steel flat flanges	1 Rofc	100	.080	L.F.	3.10	1.63		4.73	6.25	
	1960	With galvanized angle flanges	"	100	.080		3.40	1.63		5.03	6.60	
	2000	Roof joint with extruded aluminum cover, 2"	1 Shee	115	.070		20.25	1.74		21.99	25	
	2050											
	2100	Roof joint, plastic curbs, foam center, standard	1 Rofc	100	.080	L.F.	7.85	1.63		9.48	11.50	
	2200	Large		100	.080	"	10.50	1.63		12.13	14.40	
	2300	Transitions, regular, minimum		10	.800	Ea.	35	16.30		51.30	67	
	2350	Maximum		4	2		120	41		161	205	
	2400	Large, minimum		9	.889		40	18.10		58.10	76	
	2450	Maximum		3	2.670		130	54		184	240	
	2500	Roof to wall joint with extruded aluminum cover	1 Shee	115	.070	L.F.	17.95	1.74		19.69	22	
	2650											
	2700	Wall joint, closed cell foam on PVC cover, 9" wide	1 Rofc	125	.064	L.F.	2.75	1.30		4.05	5.30	
	2800	12" wide	"	115	.070	"	3.15	1.42		4.57	5.95	

077 100 | Prefab Roof Specialties

			CREW	DAILY OUTPUT	MAN-HOURS	UNIT	BARE COSTS MAT.	LABOR	EQUIP.	TOTAL	TOTAL INCL O&P	
103	9000	Minimum labor/equipment charge	1 Shee	3	2.670	Job		67		67	105	103
104	0010	**FASCIA** Aluminum, reverse board and batten,										104
	0100	.032" thick, colored, no furring included	1 Shee	145	.055	S.F.	1.70	1.38		3.08	4.03	
	0200	Residential type, aluminum	1 Carp	200	.040	L.F.	1.15	.88		2.03	2.70	
	0220	Vinyl	"	200	.040	"	.80	.88		1.68	2.32	
	0300	Steel, galv. and enameled, stock, no furring, long panels	1 Shee	145	.055	S.F.	1.90	1.38		3.28	4.25	
	0600	Short panels		115	.070	"	3.05	1.74		4.79	6.10	
	9000	Minimum labor/equipment charge	↓	4	2	Job		50		50	78	
105	0010	**GRAVEL STOP** Aluminum, .050" thick, 4" height, mill finish	1 Shee	145	.055	L.F.	2.35	1.38		3.73	4.75	105
	0080	Duranodic finish		145	.055		3.15	1.38		4.53	5.65	
	0100	Painted		145	.055		3.60	1.38		4.98	6.10	
	0300	6" face height, .050" thick, mill finish		135	.059		2.55	1.48		4.03	5.15	
	0350	Duranodic finish		135	.059		3.40	1.48		4.88	6.05	
	0400	Painted		135	.059		3.85	1.48		5.33	6.55	
	0600	8" face height, .050" thick, mill finish		125	.064		3.10	1.60		4.70	5.90	
	0650	Duranodic finish		125	.064		4	1.60		5.60	6.90	
	0700	Painted		125	.064		4.50	1.60		6.10	7.45	
	0900	12" face height, 2 piece, mill finish		100	.080		6.40	2		8.40	10.20	
	0950	Duranodic finish		100	.080		7.15	2		9.15	11	
	1000	Painted		100	.080		7.75	2		9.75	11.65	
	1500	Polyvinyl chloride, 6" face height		135	.059		2.65	1.48		4.13	5.25	
	1600	9" face height		125	.064		3.15	1.60		4.75	6	
	1800	Stainless steel, 24 ga., 6" face height		135	.059		6	1.48		7.48	8.90	
	1900	12" face height		100	.080		11.50	2		13.50	15.80	
	2100	20 ga., 6" face height		135	.059		6.80	1.48		8.28	9.80	
	2200	12" face height		100	.080		13.60	2		15.60	18.10	
	9000	Minimum labor/equipment charge	↓	3.50	2.290	Job		57		57	90	

077 200 | Roof Accessories

			CREW	DAILY OUTPUT	MAN-HOURS	UNIT	MAT.	LABOR	EQUIP.	TOTAL	TOTAL INCL O&P	
204	0010	**CEILING HATCHES** 2'-6" x 2'-6", single leaf, steel frame & cover	G-3	11	2.910	Ea.	285	62	2.28	349.28	415	204
	0100	Aluminum cover		11	2.910		315	62	2.28	379.28	450	
	0300	2'-6" x 3'-0", single leaf, steel frame & steel cover		11	2.910		300	62	2.28	364.28	430	
	0400	Aluminum cover	↓	11	2.910	↓	340	62	2.28	404.28	475	
	9000	Minimum labor/equipment charge	F-2	2	8	Job		175	8	183	295	
205	0010	**ROOF DRAINS** See division 151-125										205
206	0010	**ROOF HATCHES** With curb, 1" fiberglass insulation, 2'-6" x 3'-0"										206
	0500	Aluminum curb and cover	G-3	10	3.200	Ea.	275	68	2.51	345.51	415	
	0520	Galvanized steel		10	3.200		260	68	2.51	330.51	395	
	0540	Plain steel, primed		10	3.200		230	68	2.51	300.51	365	
	0600	2'-6" x 4'-6", aluminum curb & cover		9	3.560		415	76	2.79	493.79	580	
	0800	Galvanized steel		9	3.560		375	76	2.79	453.79	535	
	0900	Plain steel, primed		9	3.560		345	76	2.79	423.79	505	
	1200	2'-6" x 8'-0", aluminum curb and cover		6.60	4.850		680	105	3.81	788.81	915	
	1400	Galvanized steel		6.60	4.850		650	105	3.81	758.81	885	
	1500	Plain steel, primed	↓	6.60	4.850		600	105	3.81	708.81	830	
	1800	For plexiglass panels, add to above					185			185	205	
	2000	For galv. curb and alum. cover, deduct from aluminum				↓	15			15	16.50	
	9000	Minimum labor/equipment charge	F-2	2	8	Job		175	8	183	295	
207	0010	**SMOKE HATCHES** Unlabeled, not including hand winch operator										207
	0100											
	0200	For 3'-0" long, add to roof hatches from division 077-206				Ea.	25%	5%				
	0300	For 8'-0" long, add to roof hatches from division 077-206				"	10%	5%				

077 | Roof Specialties and Accessories

077 200 | Roof Accessories

			CREW	DAILY OUTPUT	MAN-HOURS	UNIT	BARE COSTS MAT.	LABOR	EQUIP.	TOTAL	TOTAL INCL O&P	
208	0010	**SMOKE VENTS** Metal cover, heavy duty, low profile, 4' x 4'										208
	0100	Aluminum	G-3	13	2.460	Ea.	950	52	1.93	1,003	1,125	
	0200	Galvanized steel	"	13	2.460	"	815	52	1.93	868.93	980	
	0250											
	0300	4' x 8' aluminum	G-3	8	4	Ea.	1,125	85	3.14	1,213	1,375	
	0400	Galvanized steel	"	8	4		1,050	85	3.14	1,138	1,300	
	0500	Sloped cover style, deduct				↓	10%					
	9000	Minimum labor/equipment charge	F-2	2	8	Job		175	8	183	295	
210	0010	**ROOF VENTS** Mushroom for built-up roofs, aluminum	1 Rofc	30	.267	Ea.	20	5.45		25.45	32	210
	0100	PVC, 6" high		30	.267	"	25	5.45		30.45	37	
	9000	Minimum labor/equipment charge	↓	2.75	2.910	Job		59		59	105	
212	0010	**VENTS, ONE-WAY** For insul. decks, 1 per M.S.F., plastic, min.	1 Rofc	40	.200	Ea.	3.95	4.07		8.02	11.50	212
	0100	Maximum		20	.400		24.50	8.15		32.65	41	
	0300	Aluminum	↓	30	.267		10.75	5.45		16.20	21	
	0800	Fiber board baffles, 12" wide for 16" O.C. rafter spacing	1 Carp	100	.080		.42	1.76		2.18	3.34	
	0900	For 24" O.C. rafter spacing		100	.080	↓	.69	1.76		2.45	3.63	
	9000	Minimum labor/equipment charge	↓	3	2.670	Job		59		59	96	
214	0010	**VENTILATORS** See division 157-490										214

078 | Skylights

078 100 | Plastic Skylights

			CREW	DAILY OUTPUT	MAN-HOURS	UNIT	BARE COSTS MAT.	LABOR	EQUIP.	TOTAL	TOTAL INCL O&P	
101	0010	**SKYLIGHT** Plastic roof domes, flush or curb mounted, ten or										101
	0100	more units, curb not included, "L" frames										
	0300	Nominal size under 10 S.F., double	G-3	130	.246	S.F.	20	5.25	.19	25.44	31	
	0400	Single		160	.200		15.50	4.25	.16	19.91	24	
	0600	10 S.F. to 20 S.F., double		315	.102		15.20	2.16	.08	17.44	20	
	0700	Single		395	.081		12.35	1.72	.06	14.13	16.40	
	0900	20 S.F. to 30 S.F., double		395	.081		14.30	1.72	.06	16.08	18.55	
	1000	Single		465	.069		11	1.46	.05	12.51	14.50	
	1200	30 S.F. to 65 S.F., double		465	.069		13.40	1.46	.05	14.91	17.15	
	1300	Single	↓	610	.052		11	1.11	.04	12.15	13.90	
	1500	For insulated 4" curbs, double, add					25%					
	1600	Single, add					30%					
	1800	For integral insulated 9" curbs, double, add					30%					
	1900	Single, add					40%					
	2100	Ceiling plastic domes compared with single roof domes				↓	85%	100%				
	2110											
	2120	Ventilating insulated plexiglass dome with										
	2130	curb mounting, 36" x 36"	G-3	12	2.670	Ea.	300	57	2.09	359.09	425	
	2150	52" x 52"		12	2.670		450	57	2.09	509.09	590	
	2160	28" x 52"		10	3.200		350	68	2.51	420.51	495	
	2170	36" x 52"	↓	10	3.200		380	68	2.51	450.51	530	
	2180	For electric opening system, add				↓	225			225	250	
	2200	Field fabricated, factory type, aluminum and wire glass	G-3	120	.267	S.F.	11.60	5.65	.21	17.46	22	
	2300	Insulated safety glass with aluminum frame		160	.200		69	4.25	.16	73.41	83	
	2400	Sandwich panels, fiberglass, for walls, 1-9/16" thick, to 250 S.F.		200	.160		12.25	3.40	.13	15.78	19.05	
	2500	250 S.F. and up		265	.121		11	2.57	.09	13.66	16.30	
	2700	As above, but for roofs, 2-3/4" thick, to 250 S.F.		295	.108		17.70	2.31	.09	20.10	23	
	2800	250 S.F. and up	↓	330	.097	↓	14.50	2.06	.08	16.64	19.30	

078 | Skylights

078 200 | Metal Framed Skylights

		CREW	DAILY OUTPUT	MAN-HOURS	UNIT	BARE COSTS MAT.	BARE COSTS LABOR	BARE COSTS EQUIP.	BARE COSTS TOTAL	TOTAL INCL O&P	
202	0010	SKYROOFS Translucent panels, 2-¾" thick, under 5000 S.F.	G-3	395	.081	SF Hor.	15	1.72	.06	16.78	19.30	202
	0100	Over 5000 S.F.		465	.069		13.50	1.46	.05	15.01	17.25	
	0300	Continuous vaulted, semi-circular, to 8' wide, double glazed		145	.221		41	4.69	.17	45.86	53	
	0400	Single glazed		160	.200		25	4.25	.16	29.41	34	
	0600	To 20' wide, single glazed		175	.183		30	3.89	.14	34.03	39	
	0700	Over 20' wide, single glazed		200	.160		31.50	3.40	.13	35.03	40	
	0900	Motorized opening type, single glazed, ⅓ opening		145	.221		35	4.69	.17	39.86	46	
	1000	Full opening	↓	130	.246		40	5.25	.19	45.44	53	
	1200	Pyramid type units, self-supporting, to 30' clear opening,										
	1300	square or circular, single glazed, minimum	G-3	200	.160	SF Hor.	19	3.40	.13	22.53	26	
	1310	Average		165	.194		27	4.12	.15	31.27	36	
	1400	Maximum		130	.246		39	5.25	.19	44.44	51	
	1500	Grid type, 4' to 10' modules, single glass glazed, minimum		200	.160		17.50	3.40	.13	21.03	25	
	1550	Maximum		128	.250		40	5.30	.20	45.50	53	
	1600	Preformed acrylic, minimum		299	.107		11.50	2.27	.08	13.85	16.35	
	1650	Maximum	↓	174	.184	↓	35	3.91	.14	39.05	45	
	9000	Minimum labor/equipment charge	F-2	8	2	Job		44	2	46	74	

079 | Joint Sealers

079 204 | Sealants & Caulkings

		CREW	DAILY OUTPUT	MAN-HOURS	UNIT	BARE COSTS MAT.	BARE COSTS LABOR	BARE COSTS EQUIP.	BARE COSTS TOTAL	TOTAL INCL O&P	
204	0010	CAULKING AND SEALANTS										204
	0020	Acoustical sealant, elastomeric				Gal.	14.50			14.50	15.95	
	0030	Backer rod, polyethylene, ¼" diameter	1 Bric	460	.017	L.F.	.05	.40		.45	.69	
	0050	½" diameter		460	.017		.07	.40		.47	.71	
	0070	¾" diameter		460	.017		.07	.40		.47	.71	
	0090	1" diameter	↓	460	.017	↓	.09	.40		.49	.73	
	0100	Caulking compound, oil base, bulk										
	0200	Brilliant white color				Gal.	10.85			10.85	11.95	
	0300	Aluminum pigment and other colors				"	14			14	15.40	
	0500	Bulk, in place, ¼" x ½", 154 L.F./gal.	1 Bric	260	.031	L.F.	.07	.70		.77	1.20	
	0600	½" x ½", 77 L.F./gal.		250	.032		.14	.73		.87	1.32	
	0800	¾" x ¾", 34 L.F./gal.		230	.035		.32	.79		1.11	1.62	
	0900	¾" x 1", 26 L.F./gal.		200	.040		.42	.91		1.33	1.92	
	1000	1" x 1", 19 L.F./gal.	↓	180	.044	↓	.57	1.01		1.58	2.25	
	1100	Acrylic latex based, bulk				Gal.	12.50			12.50	13.75	
	1200	Cartridges					17.25			17.25	19	
	1400	Butyl based, bulk					13.50			13.50	14.85	
	1500	Cartridges				↓	21			21	23	
	1700	Bulk, in place ¼" x ½", 154 L.F./gal.	1 Bric	230	.035	L.F.	.09	.79		.88	1.37	
	1800	½" x ½", 77 L.F./gal.	"	180	.044	"	.18	1.01		1.19	1.82	
	2000	Latex acrylic based, bulk				Gal.	12.50			12.50	13.75	
	2100	Cartridges				"	17.25			17.25	19	
	2200	Bulk in place, ¼" x ½", 154 L.F./gal.	1 Bric	230	.035	L.F.	.11	.79		.90	1.39	
	2250											
	2300	Polysulfide compounds, 1 component, bulk				Gal.	27.50			27.50	30	
	2400	Cartridges				"	34.50			34.50	38	
	2600	1 or 2 component, in place, ¼" x ¼", 308 L.F./gal.	1 Bric	145	.055	L.F.	.10	1.26		1.36	2.12	
	2700	½" x ¼", 154 L.F./gal.		135	.059		.21	1.35		1.56	2.39	
	2900	¾" x ⅜", 68 L.F./gal.		130	.062		.47	1.40		1.87	2.76	
	3000	1" x ½", 38 L.F./gal.	↓	130	.062	↓	.82	1.40		2.22	3.15	

079 | Joint Sealers

204

	079 204	Sealants & Caulkings	CREW	DAILY OUTPUT	MAN-HOURS	UNIT	BARE COSTS MAT.	LABOR	EQUIP.	TOTAL	TOTAL INCL O&P	
204	3200	Polyurethane, 1 or 2 component, bulk				Gal.	35.75			35.75	39	204
	3300	Cartridges				"	42.50			42.50	47	
	3500	1 or 2 component, in place, ¼" x ¼", 308 L.F./gal.	1 Bric	150	.053	L.F.	.11	1.21		1.32	2.07	
	3600	½" x ¼", 154 L.F./gal.		145	.055		.24	1.26		1.50	2.28	
	3800	¾" x ⅜", 68 L.F./gal.		130	.062		.54	1.40		1.94	2.84	
	3900	1" x ½", 38 L.F./gal.		110	.073		.98	1.65		2.63	3.73	
	4100	Silicone rubber, bulk				Gal.	29.75			29.75	33	
	4200	Cartridges				"	40.15			40.15	44	
	4300	Bulk in place, ¼" x ½", 154 L.F./gal.	1 Bric	235	.034	L.F.	.24	.77		1.01	1.51	
	4350											
	4400	Neoprene gaskets, closed cell, adhesive, ⅛" x ⅜"	1 Bric	240	.033	L.F.	.18	.76		.94	1.41	
	4500	¼" x ¾"		215	.037		.45	.85		1.30	1.85	
	4700	½" x 1"		200	.040		1.30	.91		2.21	2.89	
	4800	¾" x 1-½"		165	.048		2.70	1.10		3.80	4.74	
	5500	Resin epoxy coating, 2 component, heavy duty				Gal.	25			25	28	
	5800	Tapes, sealant, P.V.C. foam adhesive, 1⁄16" x ¼"				L.F.	.06			.06	.07	
	5900	1⁄16" x ½"					.07			.07	.08	
	5950	1⁄16" x 1"					.13			.13	.14	
	6000	⅛" x ½"					.11			.11	.12	
	6200	Urethane foam, 2 component, handy pack, 0.75 C.F.				Ea.	21			21	23	
	6300	50.0 C.F. pack				C.F.	14.30			14.30	15.75	
	9000	Minimum labor/equipment charge	1 Bric	4	2	Job		46		46	73	

081 | Metal Doors and Frames

103

	081 100	Steel Doors And Frames	CREW	DAILY OUTPUT	MAN-HOURS	UNIT	BARE COSTS MAT.	LABOR	EQUIP.	TOTAL	TOTAL INCL O&P	
103	0010	COMMERCIAL STEEL DOORS Flush, full panel										103
	0020	Hollow core, 1-⅜" thick, 20 ga., 2'-0" x 6'-8"	F-2	20	.800	Ea.	122	17.60	.80	140.40	165	
	0040	2'-8" x 6'-8"		18	.889		130	19.55	.89	150.44	175	
	0060	3'-0" x 6'-8" (95)		17	.941		137	21	.94	158.94	185	
	0100	3'-0" x 7'-0"		17	.941		146	21	.94	167.94	195	
	0120	For vision lite, add					41			41	45	
	0140	For narrow lite, add					50			50	55	
	0160	For bottom louver, add					122			122	135	
	0230	For baked enamel finish, add					30%	60%				
	0260	For galvanizing, add					15%					
	0290	For porcelain enamel finish, add					125%	100%				
	0300											
	0320	Half glass, 20 ga., 2'-0" x 6'-8"	F-2	20	.800	Ea.	163	17.60	.80	181.40	210	
	0340	2'-8" x 6'-8"		18	.889		173	19.55	.89	193.44	225	
	0360	3'-0" x 6'-8"		17	.941		180	21	.94	201.94	235	
	0400	3'-0" x 7'-0"		17	.941		184	21	.94	205.94	235	
	1020	Hollow core, 1-¾" thick, full panel, 20 ga., 2'-8" x 6'-8"		18	.889		160	19.55	.89	180.44	210	
	1040	3'-0" x 6'-8"		17	.941		170	21	.94	191.94	220	
	1060	3'-0" x 7'-0"		17	.941		180	21	.94	201.94	235	
	1080	4'-0" x 7'-0"		15	1.070		210	23	1.07	234.07	270	
	1100	4'-0" x 8'-0"		13	1.230		240	27	1.23	268.23	310	
	1120	18 ga., 2'-8" x 6'-8"		17	.941		175	21	.94	196.94	225	
	1140	3'-0" x 6'-8"		16	1		181	22	1	204	235	
	1160	3'-0" x 7'-0"		16	1		188	22	1	211	245	
	1180	4'-0" x 7'-0"		14	1.140		235	25	1.14	261.14	300	
	1200	4'-0" x 8'-0"		14	1.140		267	25	1.14	293.14	335	

For expanded coverage of these items see *Means Interior Cost Data 1991*

081 100 | Steel Doors And Frames

		Description	CREW	DAILY OUTPUT	MAN-HOURS	UNIT	MAT.	LABOR	EQUIP.	TOTAL	TOTAL INCL O&P	
103	1220	Half glass, 20 ga., 2'-8" x 6'-8"	F-2	20	.800	Ea.	190	17.60	.80	208.40	240	**103**
	1240	3'-0" x 6'-8"		18	.889		195	19.55	.89	215.44	245	
	1260	3'-0" x 7'-0"		18	.889		215	19.55	.89	235.44	270	
	1280	4'-0" x 7'-0"		16	1		260	22	1	283	325	
	1300	4'-0" x 8'-0"		13	1.230		294	27	1.23	322.23	370	
	1320	18 ga., 2'-8" x 6'-8"		18	.889		223	19.55	.89	243.44	280	
	1340	3'-0" x 6'-8"		17	.941		230	21	.94	251.94	290	
	1360	3'-0" x 7'-0"		17	.941		238	21	.94	259.94	295	
	1380	4'-0" x 7'-0"		15	1.070		294	23	1.07	318.07	365	
	1400	4'-0" x 8'-0"		14	1.140		330	25	1.14	356.14	405	
	1720	Insulated, 1-¾" thick, full panel, 18 ga., 3'-0" x 6'-8"		15	1.070		198	23	1.07	222.07	255	
	1740	2'-8" x 7'-0"		16	1		205	22	1	228	265	
	1760	3'-0" x 7'-0"		15	1.070		205	23	1.07	229.07	265	
	1800	4'-0" x 8'-0"		13	1.230		290	27	1.23	318.23	365	
	1820	Half glass, 18 ga., 3'-0" x 6'-8"		16	1		254	22	1	277	315	
	1840	2'-8" x 7'-0"		17	.941		265	21	.94	286.94	325	
	1860	3'-0" x 7'-0"		16	1		285	22	1	308	350	
	1900	4'-0" x 8'-0"		14	1.140		345	25	1.14	371.14	420	
	9000	Minimum labor/equipment charge	F-1	4	2	Job		44	2	46	74	
106	0010	**DOOR FRAMES** Steel channels with anchors and bar stops										**106**
	0100	6" channel @ 8.2#/L.F., 3' x 7' door, weighs 200#	E-4	13	2.460	Ea.	176	61	5.25	242.25	310	
	0200	8" channel @ 11.5#/L.F., 6' x 8' door, weighs 300#		9	3.560		264	87	7.60	358.60	460	
	0300	8' x 12' door, weighs 450#		6.50	4.920		396	120	10.50	526.50	670	
	0800	For frames without bar stops, light sections, deduct					15%					
	0900	Heavy sections, deduct					10%					
	9000	Minimum labor/equipment charge	A-1	4	2	Job		35	13.60	48.60	72	
110	0010	**FIRE DOOR** Steel, flush, "B" label, 90 minute										**110**
	0020	Full panel, 20 ga., 2'-0" x 6'-8"	F-2	20	.800	Ea.	165	17.60	.80	183.40	210	
	0040	2'-8" x 6'-8"		18	.889		170	19.55	.89	190.44	220	
	0060	3'-0" x 6'-8" ㉟		17	.941		175	21	.94	196.94	225	
	0080	3'-0" x 7'-0"		17	.941		184	21	.94	205.94	235	
	0120											
	0140	18 ga., 3'-0" x 6'-8"	F-2	16	1	Ea.	198	22	1	221	255	
	0160	2'-8" x 7'-0"		17	.941		198	21	.94	219.94	255	
	0180	3'-0" x 7'-0"		16	1		205	22	1	228	265	
	0200	4'-0" x 7'-0"		15	1.070		240	23	1.07	264.07	305	
	0220	For "A" label, 3 hour, 18 ga., use same price as "B" label										
	0240	For vision lite, add				Ea.	40			40	44	
	0520	Flush, "B" label 90 min., composite, 20 ga., 2'-0" x 6'-8"	F-2	18	.889		175	19.55	.89	195.44	225	
	0540	2'-8" x 6'-8"		17	.941		175	21	.94	196.94	225	
	0560	3'-0" x 6'-8"		16	1		180	22	1	203	235	
	0580	3'-0" x 7'-0"		16	1		185	22	1	208	240	
	0640	Flush, "A" label 3 hour, composite, 18 ga., 3'-0" x 6'-8"	F-2	15	1.070		200	23	1.07	224.07	260	
	0660	2'-8" x 7'-0"		16	1		200	22	1	223	255	
	0680	3'-0" x 7'-0"		15	1.070		210	23	1.07	234.07	270	
	0700	4'-0" x 7'-0"		14	1.140		250	25	1.14	276.14	315	
	9000	Minimum labor/equipment charge	F-1	4	2	Job		44	2	46	74	
114	0010	**RESIDENTIAL STEEL DOOR**										**114**
	0020	Prehung, insulated, exterior										
	0030	Embossed, full panel, 2'-8" x 6'-8"	F-2	16	1	Ea.	135	22	1	158	185	
	0040	3'-0" x 6'-8"		15	1.070		150	23	1.07	174.07	205	
	0060	3'-0" x 7'-0"		15	1.070		165	23	1.07	189.07	220	
	0070	5'-4" x 6'-8", double		8	2		330	44	2	376	435	
	0220	Half glass, 2'-8" x 6'-8"		17	.941		210	21	.94	231.94	265	
	0240	3'-0" x 6'-8"		16	1		210	22	1	233	270	

For expanded coverage of these items see *Means Interior Cost Data 1991*

081 100	Steel Doors And Frames	CREW	DAILY OUTPUT	MAN-HOURS	UNIT	BARE COSTS				TOTAL INCL O&P
						MAT.	LABOR	EQUIP.	TOTAL	
114 0260	3'-0" x 7'-0"	F-2	16	1	Ea.	225	22	1	248	285	**114**
0270	5'-4" x 6'-8", double		8	2		445	44	2	491	565	
0720	Raised plastic face, full panel, 2'-8" x 6'-8"		16	1		165	22	1	188	220	
0740	3'-0" x 6'-8"		15	1.070		170	23	1.07	194.07	225	
0760	3'-0" x 7'-0"		15	1.070		185	23	1.07	209.07	245	
0780	5'-4" x 6'-8", double		8	2		355	44	2	401	465	
0820	Half glass, 2'-8" x 6'-8"		17	.941		205	21	.94	226.94	260	
0840	3'-0" x 6'-8"		16	1		215	22	1	238	275	
0860	3'-0" x 7'-0"		16	1		225	22	1	248	285	
0880	5'-4" x 6'-8", double		8	2		425	44	2	471	540	
1320	Flush face, full panel, 2'-6" x 6'-8"		16	1		135	22	1	158	185	
1340	3'-0" x 6'-8"		15	1.070		145	23	1.07	169.07	200	
1360	3'-0" x 7'-0"		15	1.070		155	23	1.07	179.07	210	
1380	5'-4" x 6'-8", double		8	2		320	44	2	366	425	
1420	Half glass, 2'-8" x 6'-8"		17	.941		185	21	.94	206.94	240	
1440	3'-0" x 6'-8"		16	1		190	22	1	213	245	
1460	3'-0" x 7'-0"		16	1		210	22	1	233	270	
1480	5'-4" x 6'-8", double		8	2		405	44	2	451	520	
2300	Interior, residential, closet, bi-fold, 6'-8" x 2'-0" wide		16	1		57	22	1	80	100	
2330	3'-0" wide		16	1		65	22	1	88	110	
2360	4'-0" wide		15	1.070		110	23	1.07	134.07	160	
2400	5'-0" wide		14	1.140		115	25	1.14	141.14	170	
2420	6'-0" wide		13	1.230		126	27	1.23	154.23	185	
9000	Minimum labor/equipment charge	F-1	4	2	Job		44	2	46	74	
118 0010	STEEL FRAMES, KNOCK DOWN 18 ga., up to 5-¾" deep										**118**
0020											
0025	6'-8" high, 3'-0" wide, single	F-2	16	1	Ea.	66	22	1	89	110	
0040	6'-0" wide, double		14	1.140		88	25	1.14	114.14	140	
0100	7'-0" high, 3'-0" wide, single		16	1		66	22	1	89	110	
0140	6'-0" wide, double		14	1.140		88	25	1.14	114.14	140	
2800	18 ga. drywall, up to 4-⅞" deep, 7'-0" high, 3'-0" wide, single		16	1		66	22	1	89	110	
2840	6'-0" wide, double		14	1.140		88	25	1.14	114.14	140	
3600	16 ga., up to 5-¾" deep, 7'-0" high, 4'-0" wide, single		15	1.070		71	23	1.07	95.07	120	
3640	8'-0" wide, double		12	1.330		106	29	1.33	136.33	165	
3700	8'-0" high, 4'-0" wide, single		15	1.070		85	23	1.07	109.07	135	
3740	8'-0" wide, double		12	1.330		120	29	1.33	150.33	180	
4000	6-¾" deep, 7'-0" high, 4'-0" wide, single		15	1.070		74	23	1.07	98.07	120	
4040	8'-0" wide, double		12	1.330		105	29	1.33	135.33	165	
4100	8'-0" high, 4'-0" wide, single		15	1.070		89	23	1.07	113.07	135	
4140	8'-0" wide, double		12	1.330		108	29	1.33	138.33	170	
4400	8-¾" deep, 7'-0" high, 4'-0" wide, single		15	1.070		81	23	1.07	105.07	130	
4440	8'-0" wide, double		12	1.330		120	29	1.33	150.33	180	
4500	8'-0" high, 4'-0" wide, single		15	1.070		100	23	1.07	124.07	150	
4540	8'-0" wide, double		12	1.330		135	29	1.33	165.33	200	
4800	16 ga. drywall, up to 3-⅞" deep, 7'-0" high, 3'-0" wide, single		16	1		66	22	1	89	110	
4840	6'-0" wide, double		14	1.140		88	25	1.14	114.14	140	
4900	For welded frames, add					30			30	33	
4902											
5400	16 ga. "B" label, up to 5-¾" deep, 7'-0" high, 4'-0" wide, single	F-2	15	1.070	Ea.	83	23	1.07	107.07	130	
5440	8'-0" wide, double		12	1.330		120	29	1.33	150.33	180	
5800	6-¾" deep, 7'-0" high, 4'-0" wide, single		15	1.070		92	23	1.07	116.07	140	
5840	8'-0" wide, double		12	1.330		132	29	1.33	162.33	195	
6200	8-¾" deep, 7'-0" high, 4'-0" wide, single		15	1.070		100	23	1.07	124.07	150	
6240	8'-0" wide, double		12	1.330		136	29	1.33	166.33	200	
6300	For "A" label use same price as "B" label										
6400	For baked enamel finish, add					30%	60%				
6500	For galvanizing, add					15%					
6600	For porcelain enamel finish, add					125%	100%				

For expanded coverage of these items see *Means Interior Cost Data 1991*

081 | Metal Doors and Frames

081 100 | Steel Doors And Frames

			CREW	DAILY OUTPUT	MAN-HOURS	UNIT	BARE COSTS MAT.	BARE COSTS LABOR	BARE COSTS EQUIP.	BARE COSTS TOTAL	TOTAL INCL O&P	
118	7900	Transom lite frames, fixed, add	F-2	155	.103	S.F.	13	2.27	.10	15.37	18.10	118
	8000	Movable, add	"	130	.123	"	17	2.71	.12	19.83	23	
	9000	Minimum labor/equipment charge	F-1	4	2	Job		44	2	46	74	

081 200 | Alum Doors And Frames

			CREW	DAILY OUTPUT	MAN-HOURS	UNIT	BARE COSTS MAT.	BARE COSTS LABOR	BARE COSTS EQUIP.	BARE COSTS TOTAL	TOTAL INCL O&P	
204	0010	ALUMINUM FRAMES Entrance, 3' x 7' opening, clear finish	2 Sswk	7	2.290	Opng.	165	55		220	280	204
	0100	Bronze finish		7	2.290		210	55		265	330	
	0500	6' x 7' opening, clear finish		6	2.670		210	64		274	350	
	0520	Bronze finish		6	2.670		240	64		304	380	
	1000	With 3' high transoms, 3' x 10' opening, clear finish		6.50	2.460		250	59		309	385	
	1050	Bronze finish		6.50	2.460		290	59		349	425	
	1100	Black finish		6.50	2.460		335	59		394	475	
	1500	With 3' high transoms, 6' x 10' opening, clear finish		5.50	2.910		320	70		390	480	
	1550	Bronze finish		5.50	2.910		370	70		440	535	
	1600	Black finish		5.50	2.910		425	70		495	595	
	9000	Minimum labor/equipment charge	↓	4	4	Job		96		96	175	
212	0010	ALUMINUM DOORS & FRAMES Entrance, narrow stile, including										212
	0020	hardware & closer, clear finish, not incl. glass, 3' x 7' opening	2 Sswk	2	8	Ea.	531	195		726	935	
	0100	3' x 10' opening, 3' high transom		1.80	8.890		615	215		830	1,075	
	0200	3'-6" x 10' opening, 3' high transom		1.80	8.890	↓	650	215		865	1,100	
	0300	6' x 7' opening		1.30	12.310	Pr.	850	295		1,145	1,475	
	0400	6' x 10' opening, 3' high transom	↓	1.10	14.550	"	950	350		1,300	1,675	
	1000	Add to above for wide stile doors				Leaf	55%					
	1100	Full vision doors, with ½" glass, add					55%					
	1200	Non-standard size, add					67%					
	1300	Light bronze finish, add					36%					
	1400	Dark bronze finish, add					18%					
	1500	Black finish, add				↓	36%					
	1600	Concealed panic device, add					240			240	265	
	1700	Electric striker release, add				Opng.	170			170	185	
	1800	Floor check, add				Leaf	564			564	620	
	1900	Concealed closer, add				"	710			710	780	
	9000	Minimum labor/equipment charge	F-2	4	4	Job		88	4	92	150	

082 | Wood and Plastic Doors

082 050 | Wood And Plastic Doors

			CREW	DAILY OUTPUT	MAN-HOURS	UNIT	BARE COSTS MAT.	BARE COSTS LABOR	BARE COSTS EQUIP.	BARE COSTS TOTAL	TOTAL INCL O&P	
054	0010	WOOD FRAMES										054
	0400	Exterior frame, incl. ext. trim, pine, ⁵⁄₄ x 4-⁹⁄₁₆" deep	F-2	375	.043	L.F.	2.80	.94	.04	3.78	4.66	
	0420	5-³⁄₁₆" deep		375	.043		3.06	.94	.04	4.04	4.95	
	0440	6-⁹⁄₁₆" deep		375	.043		3.55	.94	.04	4.53	5.50	
	0600	Oak, ⁵⁄₄ x 4-⁹⁄₁₆" deep		350	.046		3.75	1.01	.05	4.81	5.80	
	0620	5-³⁄₁₆" deep		350	.046		4.05	1.01	.05	5.11	6.15	
	0640	6-⁹⁄₁₆" deep		350	.046		4.80	1.01	.05	5.86	6.95	
	0800	Walnut, ⁵⁄₄ x 4-⁹⁄₁₆" deep		350	.046		5.65	1.01	.05	6.71	7.90	
	0820	5-³⁄₁₆" deep		350	.046		6.45	1.01	.05	7.51	8.80	
	0840	6-⁹⁄₁₆" deep		350	.046		7.30	1.01	.05	8.36	9.70	
	1000	Sills, ⁵⁄₄ x 8" deep, oak, no horns		100	.160		7.50	3.52	.16	11.18	14.20	
	1020	2" horns	↓	100	.160	↓	8.80	3.52	.16	12.48	15.60	

For expanded coverage of these items see *Means Interior Cost Data 1991*

			CREW	DAILY OUTPUT	MAN-HOURS	UNIT	BARE COSTS MAT.	LABOR	EQUIP.	TOTAL	TOTAL INCL O&P	
082 050		**Wood And Plastic Doors**										
054	1040	3" horns	F-2	100	.160	L.F.	9.18	3.52	.16	12.86	16.05	054
	1100	⅝ x 10" deep, oak, no horns		90	.178		9.50	3.91	.18	13.59	17.05	
	1120	2" horns		90	.178		10.55	3.91	.18	14.64	18.20	
	1140	3" horns		90	.178		11.20	3.91	.18	15.29	18.90	
	1200	For casing, see division 062-212										
	1220											
	2000	Exterior, colonial, frame & trim, 3' opng., in-swing, minimum	F-2	22	.727	Ea.	230	16	.73	246.73	280	
	2020	Maximum		20	.800		520	17.60	.80	538.40	600	
	2100	5'-4" opening, in-swing, minimum		17	.941		420	21	.94	441.94	495	
	2120	Maximum		15	1.070		775	23	1.07	799.07	890	
	2140	Out-swing, minimum		17	.941		470	21	.94	491.94	550	
	2160	Maximum		15	1.070		825	23	1.07	849.07	945	
	2400	6'-0" opening, in-swing, minimum		16	1		430	22	1	453	510	
	2420	Maximum		10	1.600		815	35	1.60	851.60	955	
	2460	Out-swing, minimum		16	1		475	22	1	498	560	
	2480	Maximum		10	1.600		860	35	1.60	896.60	1,000	
	2600	For two sidelights, add, minimum		30	.533	Opng.	190	11.75	.53	202.28	230	
	2620	Maximum		20	.800	"	625	17.60	.80	643.40	715	
	2700	Custom birch frame, 3'-0" opening		16	1	Ea.	112	22	1	135	160	
	2750	6'-0" opening		16	1	"	160	22	1	183	215	
	3000	Interior frame, pine, 11/16" x 3-5/8" deep		375	.043	L.F.	1.15	.94	.04	2.13	2.85	
	3020	4-9/16" deep		375	.043		1.40	.94	.04	2.38	3.12	
	3040	5-3/16" deep		375	.043		1.98	.94	.04	2.96	3.76	
	3200	Oak, 11/16" x 3-5/8" deep		350	.046		2.10	1.01	.05	3.16	4	
	3220	4-9/16" deep		350	.046		2.50	1.01	.05	3.56	4.44	
	3240	5-3/16" deep		350	.046		2.85	1.01	.05	3.91	4.83	
	3400	Walnut, 11/16" x 3-5/8" deep		350	.046		2.45	1.01	.05	3.51	4.39	
	3420	4-9/16" deep		350	.046		3	1.01	.05	4.06	4.99	
	3440	5-3/16" deep		350	.046		3.25	1.01	.05	4.31	5.25	
	3600	Pocket door frame		16	1	Ea.	60	22	1	83	105	
	3800	Threshold, oak, 5/8" x 3-5/8" deep		200	.080	L.F.	2	1.76	.08	3.84	5.15	
	3820	4-5/8" deep		190	.084		2.57	1.85	.08	4.50	5.95	
	3840	5-5/8" deep		180	.089		3	1.96	.09	5.05	6.60	
	4020	For casing, see division 062-212										
	9000	Minimum labor/equipment charge	F-1	4	2	Job		44	2	46	74	
062	0010	**WOOD DOOR, ARCHITECTURAL** Flush, interior, 7 ply, hollow core,										062
	0020	Lauan face, 2'-0" x 6'-8"	F-2	17	.941	Ea.	30	21	.94	51.94	68	
	0040	2'-6" x 6'-8" ⑨⑦		17	.941		32	21	.94	53.94	70	
	0080	3'-0" x 6'-8"		17	.941		34	21	.94	55.94	72	
	0100	4'-0" x 6'-8"		16	1		44	22	1	67	85	
	0120	Birch face, 2'-0" x 6'-8"		17	.941		34	21	.94	55.94	72	
	0140	2'-6" x 6'-8"		17	.941		36	21	.94	57.94	74	
	0180	3'-0" x 6'-8"		17	.941		39	21	.94	60.94	78	
	0200	4'-0" x 6'-8"		16	1		48	22	1	71	90	
	0220	Oak face, 2'-0" x 6'-8"		17	.941		41	21	.94	62.94	80	
	0240	2'-6" x 6'-8"		17	.941		45	21	.94	66.94	84	
	0280	3'-0" x 6'-8"		17	.941		50	21	.94	71.94	90	
	0300	4'-0" x 6'-8"		16	1		60	22	1	83	105	
	0320	Walnut face, 2'-0" x 6'-8"		17	.941		100	21	.94	121.94	145	
	0340	2'-6" x 6'-8"		17	.941		100	21	.94	121.94	145	
	0380	3'-0" x 6'-8"		17	.941		111	21	.94	132.94	155	
	0400	4'-0" x 6'-8"		16	1		135	22	1	158	185	
	0420	For 7'-0" high, add					7			7	7.70	
	0440	For 8'-0" high, add					18			18	19.80	
	0460	For 8'-0" high walnut, add					36			36	40	
	0480	For prefinishing, clear, add					22			22	24	
	0500	For prefinishing, stain, add					26			26	29	

For expanded coverage of these items see *Means Interior Cost Data 1991*

082 050	Wood And Plastic Doors	CREW	DAILY OUTPUT	MAN-HOURS	UNIT	BARE COSTS				TOTAL INCL O&P
						MAT.	LABOR	EQUIP.	TOTAL	

		CREW	DAILY OUTPUT	MAN-HOURS	UNIT	MAT.	LABOR	EQUIP.	TOTAL	TOTAL INCL O&P
0600										
1320	M.D. overlay on hardboard, 2'-0" x 6'-8"	F-2	17	.941	Ea.	62	21	.94	83.94	105
1340	2'-6" x 6'-8"		17	.941		66	21	.94	87.94	105
1380	3'-0" x 6'-8"		17	.941		70	21	.94	91.94	110
1400	4'-0" x 6'-8"	↓	16	1		85	22	1	108	130
1420	For 7'-0" high, add					5			5	5.50
1440	For 8'-0" high, add					12			12	13.20
1720	H.P. plastic laminate, 2'-0" x 6'-8"	F-2	16	1		80	22	1	103	125
1740	2'-6" x 6'-8"		16	1		82	22	1	105	125
1780	3'-0" x 6'-8"		15	1.070		90	23	1.07	114.07	140
1800	4'-0" x 6'-8"	↓	14	1.140		120	25	1.14	146.14	175
1820	For 7'-0" high, add					5			5	5.50
1840	For 8'-0" high, add					14			14	15.40
2020	5 ply particle core, lauan face, 2'-6" x 6'-8"	F-2	15	1.070		46	23	1.07	70.07	90
2040	3'-0" x 6'-8"		14	1.140		49	25	1.14	75.14	96
2080	3'-0" x 7'-0"		13	1.230		54	27	1.23	82.23	105
2100	4'-0" x 7'-0"		12	1.330		70	29	1.33	100.33	125
2120	Birch face, 2'-6" x 6'-8"		15	1.070		49	23	1.07	73.07	93
2140	3'-0" x 6'-8"		14	1.140		53	25	1.14	79.14	100
2180	3'-0" x 7'-0"		13	1.230		58	27	1.23	86.23	110
2200	4'-0" x 7'-0"		12	1.330		73	29	1.33	103.33	130
2220	Oak face, 2'-6" x 6'-8"		15	1.070		58	23	1.07	82.07	105
2240	3'-0" x 6'-8"		14	1.140		64	25	1.14	90.14	115
2280	3'-0" x 7'-0"		13	1.230		69	27	1.23	97.23	120
2300	4'-0" x 7'-0"		12	1.330		88	29	1.33	118.33	145
2320	Walnut face, 2'-0" x 6'-8"		15	1.070		120	23	1.07	144.07	170
2340	2'-6" x 6'-8"		14	1.140		125	25	1.14	151.14	180
2380	3'-0" x 6'-8"		13	1.230		135	27	1.23	163.23	195
2400	4'-0" x 6'-8"	↓	12	1.330		165	29	1.33	195.33	230
2440	For 8'-0" high, add					18			18	19.80
2460	For 8'-0" high walnut, add					30			30	33
2480	For solid wood core, add					25			25	28
2720	For prefinishing, clear, add					33			33	36
2740	For prefinishing, stain, add				↓	48			48	53
2750										
3320	M.D. overlay on hardboard, 2'-6" x 6'-8"	F-2	14	1.140	Ea.	76	25	1.14	102.14	125
3340	3'-0" x 6'-8"		13	1.230		83	27	1.23	111.23	135
3380	3'-0" x 7'-0"		12	1.330		86	29	1.33	116.33	145
3400	4'-0" x 7'-0"	↓	10	1.600		102	35	1.60	138.60	170
3440	For 8'-0" height, add					12			12	13.20
3460	For solid wood core, add					25			25	28
3720	H.P. plastic laminate, 2'-6" x 6'-8"	F-2	13	1.230		85	27	1.23	113.23	140
3740	3'-0" x 6'-8"		12	1.330		92	29	1.33	122.33	150
3780	3'-0" x 7'-0"		11	1.450		107	32	1.45	140.45	170
3800	4'-0" x 7'-0"	↓	8	2		145	44	2	191	235
3840	For 8'-0" height, add					15			15	16.50
3860	For solid wood core, add					25			25	28
4000	Exterior, flush, solid wood core, birch, 1-¾" x 7'-0" x 2'-6"	F-2	15	1.070		90	23	1.07	114.07	140
4020	2'-8" wide		15	1.070		93	23	1.07	117.07	140
4040	3'-0" wide		14	1.140		95	25	1.14	121.14	145
4100	Oak faced 1-¾" x 7'-0" x 2'-6" wide		15	1.070		96	23	1.07	120.07	145
4120	2'-8" wide		15	1.070		111	23	1.07	135.07	160
4140	3'-0" wide		14	1.140		113	25	1.14	139.14	165
4200	Walnut faced 1-¾" x 7'-0" x 2'-6" wide		15	1.070		173	23	1.07	197.07	230
4220	2'-8" wide		15	1.070		178	23	1.07	202.07	235
4240	3'-0" wide	↓	14	1.140		186	25	1.14	212.14	245
4300	For 6'-8" high door, deduct from 7'-0" door				↓	10			10	11
9000	Minimum labor/equipment charge	F-1	4	2	Job		44	2	46	74

For expanded coverage of these items see *Means Interior Cost Data 1991*

082 050	Wood And Plastic Doors	CREW	DAILY OUTPUT	MAN-HOURS	UNIT	BARE COSTS MAT.	LABOR	EQUIP.	TOTAL	TOTAL INCL O&P
066 0010	**WOOD DOORS, DECORATOR**									066
3000	Solid wood, stile and rail									
3020	Mahogany, 3'-0" x 7'-0", minimum	F-2	14	1.140	Ea.	330	25	1.14	356.14	405
3030	Maximum		10	1.600		635	35	1.60	671.60	760
3040	3'-6" x 8'-0", minimum		10	1.600		485	35	1.60	521.60	595
3050	Maximum		8	2		920	44	2	966	1,075
3100	Pine, 3'-0" x 7'-0", minimum		14	1.140		180	25	1.14	206.14	240
3110	Maximum		10	1.600		350	35	1.60	386.60	445
3120	3'-6" x 8'-0", minimum		10	1.600		255	35	1.60	291.60	340
3130	Maximum		8	2		475	44	2	521	595
3200	Red oak, 3'-0" x 7'-0", minimum		14	1.140		470	25	1.14	496.14	560
3210	Maximum		10	1.600		1,150	35	1.60	1,186	1,325
3220	3'-6" x 8'-0", minimum		10	1.600		670	35	1.60	706.60	795
3230	Maximum		8	2		1,395	44	2	1,441	1,600
4000	Hand carved door, mahogany	F-2	14	1.140	Ea.	510	25	1.14	536.14	605
4020	3'-0" x 7'-0", minimum		11	1.450		1,400	32	1.45	1,433	1,600
4030	Maximum		10	1.600		820	35	1.60	856.60	960
4040	3'-6" x 8'-0", minimum		8	2		1,750	44	2	1,796	2,000
4050	Maximum		14	1.140		1,000	25	1.14	1,026	1,150
4200	Red oak, 3'-0" x 7'-0", minimum		11	1.450		2,700	32	1.45	2,733	3,025
4210	Maximum		10	1.600		1,300	35	1.60	1,336	1,500
4220	3'-6" x 8'-0", minimum									
4280	For 6'-8" high door, deduct from 7'-0" door					20			20	22
4400	For custom finish, add					95			95	105
4600	Side light, mahogany, 7'-0" x 1'-6" wide, minimum	F-2	18	.889		200	19.55	.89	220.44	255
4610	Maximum		14	1.140		630	25	1.14	656.14	735
4620	8'-0" x 1'-6" wide, minimum		14	1.140		250	25	1.14	276.14	315
4630	Maximum		10	1.600		730	35	1.60	766.60	860
4640	Side light, oak, 7'-0" x 1'-6" wide, minimum		18	.889		270	19.55	.89	290.44	330
4650	Maximum		14	1.140		725	25	1.14	751.14	840
4660	8'-0" x 1-6" wide, minimum		14	1.140		330	25	1.14	356.14	405
4670	Maximum		10	1.600		850	35	1.60	886.60	995
6520	Interior cafe doors, 2'-6" opening, stock, panel pine		16	1		98	22	1	121	145
6540	3'-0" opening		16	1		109	22	1	132	155
6550	Custom hardwood or louvered pine									
6560	2'-6" opening	F-2	16	1	Ea.	72	22	1	95	115
8000	3'-0" opening	"	16	1	"	82	22	1	105	125
8800	Pre-hung doors, see division 082-082									
9000	Minimum labor/equipment charge	F-1	4	2	Job		44	2	46	74
070 0010	**WOOD FIRE DOORS** Mineral core, 3 ply stile, "B" label, (97)									070
0040	1 hour, birch face, 2'-6" x 6'-8"	F-2	14	1.140	Ea.	116	25	1.14	142.14	170
0090	3'-0" x 7'-0"	"	12	1.330	"	125	29	1.33	155.33	185
0110										
0140	Oak face, 2'-6" x 6'-8"	F-2	14	1.140	Ea.	120	25	1.14	146.14	175
0190	3'-0" x 7'-0"		12	1.330		130	29	1.33	160.33	190
0240	Walnut face, 2'-6" x 6'-8"		14	1.140		198	25	1.14	224.14	260
0290	3'-0" x 7'-0"		12	1.330		208	29	1.33	238.33	280
0440	M.D. overlay on hardboard, 2'-6" x 6'-8"		15	1.070		104	23	1.07	128.07	155
0490	3'-0" x 7'-0"		13	1.230		118	27	1.23	146.23	175
0540	H.P. plastic laminate, 2'-6" x 6'-8"		13	1.230		187	27	1.23	215.23	250
0590	3'-0" x 7'-0"		11	1.450		205	32	1.45	238.45	280
2460	For 6'-8" high door, deduct from 7'-0" door					10			10	11
2480	For oak veneer, add					50%				
2500	For walnut veneer, add					75%				
9000	Minimum labor/equipment charge	F-1	4	2	Job		44	2	46	74
074 0010	**WOOD DOORS, PANELED** Interior, six panel, hollow core, 1-3/8" thick									074
0040	Molded hardboard, 2'-0" x 6'-8"	F-2	17	.941	Ea.	42	21	.94	63.94	81

For expanded coverage of these items see *Means Interior Cost Data 1991*

		082 050	Wood And Plastic Doors	CREW	DAILY OUTPUT	MAN-HOURS	UNIT	MAT.	LABOR	EQUIP.	TOTAL	TOTAL INCL O&P	
074	0060		2'-6" x 6'-8"	F-2	17	.941	Ea.	47	21	.94	68.94	87	074
	0080		3'-0" x 6'-8"		17	.941		52	21	.94	73.94	92	
	0140		Embossed print, molded hardboard, 2'-0" x 6'-8"		17	.941		50	21	.94	71.94	90	
	0160		2'-6" x 6'-8"		17	.941		55	21	.94	76.94	95	
	0180		3'-0" x 6'-8"		17	.941		60	21	.94	81.94	100	
	0540		Six panel, solid, 1-⅜" thick, pine, 2'-0" x 6'-8"		15	1.070		98	23	1.07	122.07	145	
	0560		2'-6" x 6'-8"		14	1.140		103	25	1.14	129.14	155	
	0580		3'-0" x 6'-8"		13	1.230		122	27	1.23	150.23	180	
	1020		Two panel, bored rail, solid, 1-⅜" thick, pine, 1'-6" x 6'-8"		16	1		135	22	1	158	185	
	1040		2'-0" x 6'-8"		15	1.070		145	23	1.07	169.07	200	
	1060		2'-6" x 6'-8"		14	1.140		150	25	1.14	176.14	205	
	1340		Two panel, solid, 1-⅜" thick, fir, 2'-0" x 6'-8"		15	1.070		130	23	1.07	154.07	185	
	1360		2'-6" x 6'-8"		14	1.140		135	25	1.14	161.14	190	
	1380		3'-0" x 6'-8"		13	1.230		140	27	1.23	168.23	200	
	1740		Five panel, solid, 1-⅜" thick, fir, 2'-0" x 6'-8"		15	1.070		150	23	1.07	174.07	205	
	1760		2'-6" x 6'-8"		14	1.140		153	25	1.14	179.14	210	
	1780		3'-0" x 6'-8"		13	1.230		160	27	1.23	188.23	220	
	9000		Minimum labor/equipment charge	F-1	4	2	Job		44	2	46	74	
078	0010		**WOOD DOORS, RESIDENTIAL**										078
	0200		Exterior, combination storm & screen, pine										
	0220		Cross buck, 6'-9" x 2'-6" wide ⑨⑦	F-2	11	1.450	Ea.	170	32	1.45	203.45	240	
	0260		2'-8" wide		10	1.600		173	35	1.60	209.60	250	
	0280		3'-0" wide		9	1.780		178	39	1.78	218.78	260	
	0300		7'-1" x 3'-0" wide		9	1.780		190	39	1.78	230.78	275	
	0400		Full lite, 6'-9" x 2'-6" wide		11	1.450		150	32	1.45	183.45	220	
	0420		2'-8" wide		10	1.600		160	35	1.60	196.60	235	
	0440		3'-0" wide		9	1.780		165	39	1.78	205.78	245	
	0500		7'-1" x 3'-0" wide		9	1.780		175	39	1.78	215.78	260	
	0700		Dutch door, pine, 1-¾" x 6'-8" x 2'-8" wide, minimum		12	1.330		314	29	1.33	344.33	395	
	0720		Maximum		10	1.600		346	35	1.60	382.60	440	
	0800		3'-0" wide, minimum		12	1.330		329	29	1.33	359.33	410	
	0820		Maximum		10	1.600		363	35	1.60	399.60	460	
	1000		Entrance door, colonial, 1-¾" x 6'-8" x 2'-8" wide ⑨④		16	1		260	22	1	283	325	
	1020		6 panel pine, 3'-0" wide		15	1.070		270	23	1.07	294.07	335	
	1100		8 panel pine, 2'-8" wide		16	1		285	22	1	308	350	
	1120		3'-0" wide		15	1.070		290	23	1.07	314.07	360	
	1200		For tempered safety glass lites, add					21			21	23	
	1220												
	1300		Flush, birch, solid core, 1-¾" x 6'-8" x 2'-8" wide ⑨④	F-2	16	1	Ea.	80	22	1	103	125	
	1320		3'-0" wide		15	1.070		85	23	1.07	109.07	135	
	1340		7'-0" x 2'-8" wide		16	1		88	22	1	111	135	
	1360		3'-0" wide		15	1.070		90	23	1.07	114.07	140	
	1380		For tempered safety glass lites, add					46			46	51	
	1550		For handcarved door, see division 082-066-4000										
	2700		Interior closet, bi-fold, w/hardware, no frame or trim incl.										
	2720		Flush, birch, 6'-6" or 6'-8" x 2'-6" wide	F-2	13	1.230	Ea.	51	27	1.23	79.23	100	
	2740		3'-0" wide		13	1.230		55	27	1.23	83.23	105	
	2760		4'-0" wide		12	1.330		89	29	1.33	119.33	145	
	2780		5'-0" wide		11	1.450		98	32	1.45	131.45	160	
	2800		6'-0" wide		10	1.600		105	35	1.60	141.60	175	
	2820		Flush, hardboard, primed, 6'-8" x 2'-6" wide		13	1.230		40	27	1.23	68.23	90	
	2840		3'-0" wide		13	1.230		43	27	1.23	71.23	93	
	2860		4'-0" wide		12	1.330		69	29	1.33	99.33	125	
	2880		5'-0" wide		11	1.450		76	32	1.45	109.45	135	
	2900		6'-0" wide		10	1.600		84	35	1.60	120.60	150	
	3000		Raised panel pine, 6'-6" or 6'-8" x 2'-6" wide		13	1.230		95	27	1.23	123.23	150	
	3020		3'-0" wide		13	1.230		115	27	1.23	143.23	170	
	3040		4'-0" wide		12	1.330		175	29	1.33	205.33	240	

For expanded coverage of these items see *Means Interior Cost Data 1991*

082 050 | Wood And Plastic Doors

		CREW	DAILY OUTPUT	MAN-HOURS	UNIT	BARE COSTS				TOTAL INCL O&P	
						MAT.	LABOR	EQUIP.	TOTAL		
078 3060	5'-0" wide	F-2	11	1.450	Ea.	200	32	1.45	233.45	275	078
3080	6'-0" wide		10	1.600		220	35	1.60	256.60	300	
3200	Louvered, pine, 6'-6" or 6'-8" x 2'-6" wide		13	1.230		70	27	1.23	98.23	125	
3220	3'-0" wide		13	1.230		76	27	1.23	104.23	130	
3240	4'-0" wide		12	1.330		125	29	1.33	155.33	185	
3260	5'-0" wide		11	1.450		145	32	1.45	178.45	215	
3280	6'-0" wide		10	1.600		150	35	1.60	186.60	225	
4400	Bi-passing closet, incl. hardware and frame, no trim incl.										
4420	Flush, lauan, 6'-8" x 4'-0" wide	F-2	12	1.330	Opng.	75	29	1.33	105.33	130	
4440	5'-0" wide		11	1.450		83	32	1.45	116.45	145	
4460	6'-0" wide		10	1.600		90	35	1.60	126.60	160	
4600	Flush, birch, 6'-8" x 4'-0" wide		12	1.330		85	29	1.33	115.33	145	
4620	5'-0" wide		11	1.450		99	32	1.45	132.45	165	
4640	6'-0" wide		10	1.600		108	35	1.60	144.60	180	
4800	Louvered, pine, 6'-8" x 4'-0" wide		12	1.330		102	29	1.33	132.33	160	
4820	5'-0" wide		11	1.450		116	32	1.45	149.45	180	
4840	6'-0" wide		10	1.600		127	35	1.60	163.60	200	
5000	Paneled, pine, 6'-8" x 4'-0" wide		12	1.330		183	29	1.33	213.33	250	
5020	5'-0" wide		11	1.450		205	32	1.45	238.45	280	
5040	6'-0" wide		10	1.600		230	35	1.60	266.60	310	
5050											
6100	Folding accordion, closet, not including frame										
6120	Vinyl, 2 layer, stock (see also division 106-552)	F-2	400	.040	S.F.	4	.88	.04	4.92	5.90	
6140	Woven mahogany and vinyl, stock		400	.040		1.60	.88	.04	2.52	3.24	
6160	Wood slats with vinyl overlay, stock		400	.040		7.95	.88	.04	8.87	10.25	
6180	Economy vinyl, stock		400	.040		1.10	.88	.04	2.02	2.69	
6200	Rigid PVC		400	.040		2.15	.88	.04	3.07	3.85	
6220	For custom folding, add to above					25%					
6230		⑨⑧									
7400	Passage doors, flush, no frame included										
7420	Lauan, hollow core, 1-3/8" x 6'-8" x 1'-6" wide	F-2	19	.842	Ea.	28	18.55	.84	47.39	62	
7440	2'-0" wide		18	.889		28	19.55	.89	48.44	64	
7460	2'-6" wide		18	.889		29	19.55	.89	49.44	65	
7480	2'-8" wide		18	.889		29	19.55	.89	49.44	65	
7500	3'-0" wide		17	.941		31	21	.94	52.94	69	
7700	Birch, hollow core, 1-3/8" x 6'-8" x 1'-6" wide		19	.842		31	18.55	.84	50.39	65	
7720	2'-0" wide		18	.889		31	19.55	.89	51.44	67	
7740	2'-6" wide		18	.889		35	19.55	.89	55.44	71	
7760	2'-8" wide		18	.889		35	19.55	.89	55.44	71	
7780	3'-0" wide		17	.941		38	21	.94	59.94	77	
8000	Pine louvered, 1-3/8" x 6'-8" x 1'-6" wide		19	.842		75	18.55	.84	94.39	115	
8020	2'-0" wide		18	.889		80	19.55	.89	100.44	120	
8040	2'-6" wide		18	.889		90	19.55	.89	110.44	130	
8060	2'-8" wide		18	.889		110	19.55	.89	130.44	155	
8080	3'-0" wide		17	.941		110	21	.94	131.94	155	
8090											
8300	Pine paneled, 1-3/8" x 6'-8" x 1'-6" wide	F-2	19	.842	Ea.	85	18.55	.84	104.39	125	
8320	2'-0" wide		18	.889		89	19.55	.89	109.44	130	
8340	2'-6" wide		18	.889		101	19.55	.89	121.44	145	
8360	2'-8" wide		18	.889		107	19.55	.89	127.44	150	
8380	3'-0" wide		17	.941		115	21	.94	136.94	160	
9000	Minimum labor/equipment charge	F-1	4	2	Job		44	2	46	74	
082 0010	**PRE-HUNG DOORS**										082
0300	Exterior, wood, combination storm & screen, 6'-9" x 2'-6" wide	F-2	15	1.070	Ea.	150	23	1.07	174.07	205	
0320	2'-8" wide		15	1.070		157	23	1.07	181.07	210	
0340	3'-0" wide		15	1.070		165	23	1.07	189.07	220	
0360	For 7'-0" high door, add					7			7	7.70	
0370	For aluminum storm doors, see division 083-900										

For expanded coverage of these items see *Means Interior Cost Data 1991*

082 050 | Wood And Plastic Doors

			CREW	DAILY OUTPUT	MAN-HOURS	UNIT	BARE COSTS MAT.	LABOR	EQUIP.	TOTAL	TOTAL INCL O&P
082	1600	Entrance door, flush, birch, solid core										**082**
	1620	4-⅝" solid jamb, 1-¾" x 6'-8" x 2'-8" wide	F-2	16	1	Ea.	180	22	1	203	235
	1640	3'-0" wide	"	16	1		200	22	1	223	255
	1680	For 7'-0" high door, add					10			10	11
	2000	Entrance door, colonial, 6 panel pine									
	2020	4-⅝" solid jamb, 1-¾" x 6'-8" x 2'-8" wide	F-2	16	1	Ea.	335	22	1	358	405
	2040	3'-0" wide	"	16	1		365	22	1	388	440
	2060	For 7'-0" high door, add					10			10	11
	2200	For 5-⅝" solid jamb, add					14			14	15.40
	2300	French door, 6'-8" x 6'-0" wide, ½" insul. glass and grille	F-2	7	2.290		885	50	2.29	937.29	1,050
	2990										
	4000	Interior, passage door, 4-⅝" solid jamb ⊛98									
	4400	Lauan, flush, solid core, 1-⅜" x 6'-8" x 2'-6" wide	F-2	20	.800	Ea.	110	17.60	.80	128.40	150
	4420	2'-8" wide		20	.800		112	17.60	.80	130.40	155
	4440	3'-0" wide		19	.842		115	18.55	.84	134.39	160
	4600	Hollow core, 1-⅜" x 6'-8" x 2'-6" wide		20	.800		78	17.60	.80	96.40	115
	4620	2'-8" wide		20	.800		83	17.60	.80	101.40	120
	4640	3'-0" wide		19	.842		88	18.55	.84	107.39	130
	4700	For 7'-0" high door, add					10			10	11
	5000	Birch, flush, solid core, 1-⅜" x 6'-8" x 2'-6" wide	F-2	20	.800		145	17.60	.80	163.40	190
	5020	2'-8" wide		20	.800		148	17.60	.80	166.40	190
	5040	3'-0" wide		19	.842		153	18.55	.84	172.39	200
	5200	Hollow core, 1-⅜" x 6'-8" x 2'-6" wide		20	.800		118	17.60	.80	136.40	160
	5220	2'-8" wide		20	.800		122	17.60	.80	140.40	165
	5240	3'-0" wide		19	.842		127	18.55	.84	146.39	170
	5280	For 7'-0" high door, add					10			10	11
	5500	Hardboard paneled, 1-⅜" x 6'-8" x 2'-6" wide	F-2	20	.800		125	17.60	.80	143.40	165
	5520	2'-8" wide		20	.800		128	17.60	.80	146.40	170
	5540	3'-0" wide		19	.842		133	18.55	.84	152.39	180
	5600										
	6000	Pine paneled, 1-⅜" x 6'-8" x 2'-6" wide	F-2	20	.800	Ea.	175	17.60	.80	193.40	220
	6020	2'-8" wide		20	.800		185	17.60	.80	203.40	235
	6040	3'-0" wide		20	.800		195	17.60	.80	213.40	245
	6200										
	6500	For 5-⅝" solid jamb, add				Ea.	7			7	7.70
	6520	For split jamb, deduct				"	22			22	24
	9000	Minimum labor/equipment charge	F-1	4	2	Job		44	2	46	74

083 050 | Access Doors

			CREW	DAILY OUTPUT	MAN-HOURS	UNIT	BARE COSTS MAT.	LABOR	EQUIP.	TOTAL	TOTAL INCL O&P
054	0010	**ACCESS DOORS**										**054**
	1000	Fire rated door with lock									
	1100	Metal, 12" x 12"	1 Carp	10	.800	Ea.	109	17.60		126.60	150
	1150	18" x 18"		9	.889		139	19.55		158.55	185
	1200	24" x 24"		9	.889		177	19.55		196.55	225
	1250	24" x 36"		8	1		231	22		253	290
	1300	24" x 48"		8	1		262	22		284	325
	1350	36" x 36"		7.50	1.070		352	23		375	425
	1400	48" x 48"		7.50	1.070		419	23		442	500
	1600	Stainless steel, 12" x 12"		10	.800		177	17.60		194.60	225

For expanded coverage of these items see *Means Interior Cost Data 1991*

083 050 | Access Doors

			CREW	DAILY OUTPUT	MAN-HOURS	UNIT	MAT.	LABOR	EQUIP.	TOTAL	TOTAL INCL O&P	
054	1650	18" x 18"	1 Carp	9	.889	Ea.	261	19.55		280.55	320	054
	1700	24" x 24"		9	.889		312	19.55		331.55	375	
	1750	24" x 36"	↓	8	1	↓	387	22		409	460	
	2000	Flush door for finishing										
	2100	Metal 8" x 8"	1 Carp	10	.800	Ea.	21	17.60		38.60	52	
	2150	12" x 12"	"	10	.800	"	24	17.60		41.60	55	
	3000	Recessed door for acoustic tile										
	3100	Metal, 12" x 12"	1 Carp	4.50	1.780	Ea.	35	39		74	100	
	3150	12" x 24"		4.50	1.780		53	39		92	120	
	3200	24" x 24"		4	2		71	44		115	150	
	3250	24" x 36"	↓	4	2	↓	95	44		139	175	
	4000	Recessed door for drywall										
	4100	Metal 12" x 12"	1 Carp	6	1.330	Ea.	42	29		71	94	
	4150	12" x 24"		5.50	1.450		64	32		96	125	
	4200	24" x 36"	↓	5	1.600	↓	115	35		150	185	
	6000	Standard door										
	6100	Metal, 8" x 8"	1 Carp	10	.800	Ea.	19	17.60		36.60	50	
	6150	12" x 12"		10	.800		22	17.60		39.60	53	
	6200	18" x 18"		9	.889		31	19.55		50.55	66	
	6250	24" x 24"		9	.889		45	19.55		64.55	81	
	6300	24" x 36"		8	1		63	22		85	105	
	6350	36" x 36"		8	1		91	22		113	135	
	6500	Stainless steel, 8" x 8"		10	.800		62	17.60		79.60	97	
	6550	12" x 12"		10	.800		85	17.60		102.60	120	
	6600	18" x 18"		9	.889		184	19.55		203.55	235	
	6650	24" x 24"		9	.889	↓	225	19.55		244.55	280	
	9000	Minimum labor/equipment charge	↓	4	2	Job		44		44	72	

083 100 | Sliding Doors

			CREW	DAILY OUTPUT	MAN-HOURS	UNIT	MAT.	LABOR	EQUIP.	TOTAL	TOTAL INCL O&P	
102	0010	**GLASS, SLIDING** Vinyl clad, 1" insulated glass, 6'-0" x 6'-10" high	2 Carp	4	4	Opng.	904	88		992	1,150	102
	0100	8'-0" x 6'-10" high		4	4		1,104	88		1,192	1,350	
	0500	3 leaf, 9'-0" x 6'-10" high		3	5.330		1,293	115		1,408	1,625	
	0600	12'-0" x 6'-10" high	↓	3	5.330	↓	1,588	115		1,703	1,950	
	9000	Minimum labor/equipment charge	1 Carp	4	2	Job		44		44	72	
104	0010	**GLASS, SLIDING** Wood, ⅝" tempered insul. glass, 6' wide, premium	2 Carp	4	4	Ea.	775	88		863	995	104
	0100	Economy		4	4		470	88		558	660	
	0150	8' wide, wood, premium		3	5.330		1,000	115		1,115	1,300	
	0200	Economy		3	5.330		600	115		715	850	
	0250	12' wide, wood, premium		2.50	6.400		1,490	140		1,630	1,875	
	0300	Economy	↓	2.50	6.400	↓	960	140		1,100	1,275	
	0350	Aluminum sliding, ⅝" tempered insulated glass, 6' wide										
	0400	Premium	2 Carp	4	4	Ea.	440	88		528	630	
	0450	Economy	"	4	4	"	290	88		378	465	
	0460											
	0500	8' wide, premium	2 Carp	3	5.330	Ea.	460	115		575	700	
	0550	Economy		3	5.330		370	115		485	600	
	0600	12' wide, premium		2.50	6.400		560	140		700	845	
	0650	Economy	↓	2.50	6.400	↓	464	140		604	740	
	1000	Replacement doors, wood										
	1050	6' wide, premium	2 Carp	4	4	Ea.	425	88		513	610	
	9000	Minimum labor/equipment charge	1 Carp	2	4	Job		88		88	145	
106	0010	**MALL FRONT** 2 fixed end panels with remaining panels sliding,										106
	0700	incl. automatic oper., see division 084-604										

For expanded coverage of these items see *Means Interior Cost Data 1991*

083 200 | Metal-Clad Doors

			CREW	DAILY OUTPUT	MAN-HOURS	UNIT	BARE COSTS MAT.	LABOR	EQUIP.	TOTAL	TOTAL INCL O&P	
202	0010	KALAMEIN Interior, flush type, 3' x 7'	2 Carp	4.30	3.720	Opng.	290	82		372	455	202
	9000	Minimum labor/equipment charge	1 Carp	2	4	Job		88		88	145	
204	0010	TIN CLAD 3 ply, 6' x 7', double sliding, manual	2 Carp	1	16	Opng.	1,640	350		1,990	2,375	204
	1000	For electric operator, add	1 Elec	2	4	"	1,690	100		1,790	2,025	
	9000	Minimum labor/equipment charge	2 Carp	1	16	Job		350		350	575	

083 250 | Cold Storage Doors

			CREW	DAILY OUTPUT	MAN-HOURS	UNIT	MAT.	LABOR	EQUIP.	TOTAL	TOTAL INCL O&P	
251	0010	COLD STORAGE Single, galvanized steel										251
	0300	Horizontal sliding, 5' x 7', manual operation, 2" thick	F-2	2	8	Ea.	1,703	175	8	1,886	2,175	
	0400	4" thick		2	8		1,981	175	8	2,164	2,475	
	0500	6" thick		2	8		2,200	175	8	2,383	2,725	
	0800	Power operation, 2" thick		1.90	8.420		3,790	185	8.40	3,983	4,475	
	0900	4" thick		1.90	8.420		4,067	185	8.40	4,260	4,775	
	1000	6" thick		1.90	8.420		4,185	185	8.40	4,378	4,925	
	1300	9' x 10', manual operation, 2" insulation		1.70	9.410		2,541	205	9.40	2,755	3,150	
	1400	4" insulation		1.70	9.410		2,981	205	9.40	3,195	3,625	
	1500	6" insulation		1.70	9.410		3,150	205	9.40	3,364	3,825	
	1800	Power operation, 2" insulation		1.60	10		4,800	220	10	5,030	5,650	
	1900	4" insulation		1.60	10		5,214	220	10	5,444	6,100	
	2000	6" insulation		1.70	9.410		5,450	205	9.40	5,664	6,350	
	2300	For stainless steel face, add					20%					
	3000	Hinged, lightweight, 3' x 7'-0", galvanized 1 face, 2" thick	2 Carp	2	8	Ea.	875	175		1,050	1,250	
	3050	4" thick		1.90	8.420		946	185		1,131	1,350	
	3300	Aluminum doors, 3' x 7'-0", 4" thick		1.90	8.420		993	185		1,178	1,400	
	3350	6" thick		1.40	11.430		1,075	250		1,325	1,600	
	3600	Stainless steel, 3' x 7'-0", 4" thick		1.90	8.420		1,135	185		1,320	1,550	
	3650	6" thick		1.40	11.430		1,385	250		1,635	1,925	
	3900	Painted, 3' x 7'-0", 4" thick		1.90	8.420		960	185		1,145	1,350	
	3950	6" thick		1.40	11.430		1,060	250		1,310	1,575	
	5000	Bi-parting, electric operated										
	5010	6' x 8' opening, galv. faces, 4" thick for cooler	2 Carp	.80	20	Opng.	4,925	440		5,365	6,125	
	5050	For freezer, 4" thick		.80	20		5,703	440		6,143	7,000	
	5300	For door buck framing and door protection, add		2.50	6.400		530	140		670	815	
	6000	Galvanized batten door, galvanized hinges, 4' x 7'		2	8		1,100	175		1,275	1,500	
	6050	6' x 8'		1.80	8.890		1,600	195		1,795	2,075	
	6500	Fire door, 3 hr., 6' x 8', single slide		.80	20		6,890	440		7,330	8,300	
	6550	Double, bi-parting		.70	22.860		8,050	505		8,555	9,675	
	9000	Minimum labor/equipment charge	1 Carp	2	4	Job		88		88	145	

083 300 | Coiling Doors

			CREW	DAILY OUTPUT	MAN-HOURS	UNIT	MAT.	LABOR	EQUIP.	TOTAL	TOTAL INCL O&P	
302	0010	COUNTER DOORS 4' high roll-up, 6' long, galv. steel or aluminum	2 Carp	2	8	Opng.	717	175		892	1,075	302
	0300	Galvanized steel, UL label		1.80	8.890		1,071	195		1,266	1,500	
	0600	Stainless steel, 4' high roll-up, 6' long		2	8		1,300	175		1,475	1,725	
	0700	10' long		1.80	8.890		1,680	195		1,875	2,175	
	9000	Minimum labor/equipment charge	1 Carp	2	4	Job		88		88	145	

083 400 | Coiling Grilles

			CREW	DAILY OUTPUT	MAN-HOURS	UNIT	MAT.	LABOR	EQUIP.	TOTAL	TOTAL INCL O&P	
402	0010	ROLL UP GRILLE Aluminum, manual operated, mill finish	2 Sswk	82	.195	S.F.	13.60	4.70		18.30	24	402
	0100	Bronze anodized		82	.195	"	23.20	4.70		27.90	34	
	0400	Steel, manual operated, 10' x 10' high		1	16	Opng.	1,100	385		1,485	1,900	
	0500	15' x 8' high		.80	20	"	1,440	480		1,920	2,450	
	1000	For safety edge bottom bar, electric, add				L.F.	18			18	19.80	
	1100	For motor operation, add	2 Sswk	5	3.200	Opng.	800	77		877	1,025	
	9000	Minimum labor/equipment charge	"	1	16	Job		385		385	700	

For expanded coverage of these items see *Means Interior Cost Data 1991*

083 600 | Sectional Overhead Drs

		CREW	DAILY OUTPUT	MAN-HOURS	UNIT	BARE COSTS MAT.	LABOR	EQUIP.	TOTAL	TOTAL INCL O&P
604	0010	**OVERHEAD, COMMERCIAL** Frames not included									604
	0020										
	1000	Stock, sectional, heavy duty, wood, 1-¾" thick, 8' x 8' high	2 Carp	2	8	Ea.	403	175		578	730
	1200	12' x 12' high		1.50	10.670		825	235		1,060	1,300
	1300	Chain hoist, 14' x 14' high		1.30	12.310		1,396	270		1,666	1,975
	1600	20' x 16' high		.60	26.670		2,800	585		3,385	4,050
	2100	For medium duty custom doors, deduct					5%	5%			
	2150	For medium duty stock doors, deduct					10%	5%			
	2300	Fiberglass and aluminum, heavy duty, sectional, 12' x 12' high	2 Carp	1.50	10.670		1,050	235		1,285	1,550
	2450	Chain hoist, 20' x 20' high	"	.50	32		3,615	705		4,320	5,125
	2900	For electric trolley operator, ⅓ H.P., to 12' x 12', add	1 Carp	2	4		360	88		448	540
	2950	Over 12' x 12', ½ H.P., add	"	1	8		400	175		575	725
	9000	Minimum labor/equipment charge	2 Carp	1.50	10.670	Job		235		235	385

		CREW	DAILY OUTPUT	MAN-HOURS	UNIT	BARE COSTS MAT.	LABOR	EQUIP.	TOTAL	TOTAL INCL O&P
606	0010	**RESIDENTIAL GARAGE DOORS** Including hardware, no frame									606
	0020										
	0050	Hinged, wood, custom, double door, 9' x 7'	2 Carp	3	5.330	Ea.	290	115		405	510
	0070	16' x 7'		2	8		550	175		725	890
	0200	Overhead, sectional, incl. hardware, fiberglass, 9' x 7', standard		8	2		355	44		399	460
	0220	Deluxe		8	2		412	44		456	525
	0300	16' x 7', standard		6	2.670		595	59		654	750
	0320	Deluxe		6	2.670		745	59		804	915
	0500	Hardboard, 9' x 7', standard		8	2		255	44		299	350
	0520	Deluxe		8	2		325	44		369	430
	0600	16' x 7', standard		6	2.670		450	59		509	590
	0620	Deluxe		6	2.670		510	59		569	655
	0700	Metal, 9' x 7', standard		8	2		270	44		314	370
	0720	Deluxe		8	2		380	44		424	490
	0800	16' x 7', standard		6	2.670		425	59		484	565
	0820	Deluxe		6	2.670		640	59		699	800
	0900	Wood, 9' x 7', standard		8	2		265	44		309	365
	0920	Deluxe		8	2		760	44		804	910
	1000	16' x 7', standard		6	2.670		530	59		589	680
	1020	Deluxe		6	2.670		1,150	59		1,209	1,350
	1800	Door hardware, sectional	1 Carp	4	2		105	44		149	185
	1810	Door tracks only		4	2		55	44		99	130
	1820	One side only		7	1.140		39	25		64	84
	3000	Swing-up, including hardware, metal, 9' x 7', standard	2 Carp	8	2		455	44		499	570
	3100	16' x 7', standard	"	6	2.670		565	59		624	715
	4000	For electric operator, economy, add	1 Carp	8	1		185	22		207	240
	4100	Deluxe, including remote control	"	8	1		300	22		322	365
	4500	For transmitter/receiver control , add to operator				Total	65			65	72
	4600	Transmitters, additional				"	20			20	22
	6000	Replace section, on sectional door, fiberglass, 9' x 7'	1 Carp	4	2	Ea.	85	44		129	165
	6020	16' x 7'		3.50	2.290		160	50		210	260
	6200	Hardboard, 9' x 7'		4	2		87	44		131	170
	6220	16' x 7'		3.50	2.290		145	50		195	240
	6300	Metal, 9' x 7'		4	2		100	44		144	180
	6320	16' x 7'		3.50	2.290		165	50		215	265
	6500	Wood, 9' x 7'		4	2		95	44		139	175
	6520	16' x 7'		3.50	2.290		180	50		230	280
	9000	Minimum labor/equipment charge		2.50	3.200	Job		70		70	115

083 720 | Special Purpose Doors

		CREW	DAILY OUTPUT	MAN-HOURS	UNIT	BARE COSTS MAT.	LABOR	EQUIP.	TOTAL	TOTAL INCL O&P
721	0010	**BULKHEAD CELLAR DOORS** Steel, not incl. sides, minimum	1 Carp	5.50	1.450	Ea.	210	32		242	285	721
	0100	Maximum		5.10	1.570		229	35		264	310
	0500	With sides and foundation plates, minimum		4.70	1.700		212	37		249	295
	0600	Maximum		4.30	1.860		268	41		309	360

For expanded coverage of these items see *Means Interior Cost Data 1991*

083 720 | Special Purpose Doors

		CREW	DAILY OUTPUT	MAN-HOURS	UNIT	BARE COSTS MAT.	LABOR	EQUIP.	TOTAL	TOTAL INCL O&P	
721	9000	Minimum labor/equipment charge	1 Carp	2	4	Job		88		88	145	721
723	0010	FLOOR, COMMERCIAL Alum. tile, steel frame, one leaf, 2'x2' opng.	2 Sswk	3.50	4.570	Opng.	350	110		460	585	723
	0050	3'-6" x 3'-6" opening		3.50	4.570		615	110		725	875	
	0500	Double leaf, 4' x 4' opening		3	5.330		902	130		1,032	1,225	
	0550	5' x 5' opening		3	5.330		1,320	130		1,450	1,675	
	9000	Minimum labor/equipment charge		2	8	Job		195		195	350	
725	0010	FLOOR, INDUSTRIAL Steel 300 psf L.L., single leaf, 2' x 2', 175#	2 Sswk	6	2.670	Opng.	476	64		540	640	725
	0050	3' x 3' opening, 300#		5.50	2.910		669	70		739	865	
	0300	Double leaf, 4' x 4' opening, 455#		5	3.200		990	77		1,067	1,225	
	0350	5' x 5' opening, 645#		4.50	3.560		1,304	86		1,390	1,600	
	1000	Aluminum, 300 psf L.L., single leaf, 2' x 2', 60#		6	2.670		440	64		504	600	
	1050	3' x 3' opening, 100#		5.50	2.910		685	70		755	880	
	1500	Double leaf, 4' x 4' opening, 160#		5	3.200		1,076	77		1,153	1,325	
	1550	5' x 5' opening, 235#		4.50	3.560		1,428	86		1,514	1,725	
	9000	Minimum labor/equipment charge		2	8	Job		195		195	350	
732	0010	ROLLING SERVICE DOORS Steel, manual, 20 ga., 8' x 8', std.	2 Sswk	1.60	10	Ea.	785	240		1,025	1,300	732
	0120	8' x 8' high, class A fire door		1.40	11.430		1,325	275		1,600	1,950	
	0130	12' x 12' high, standard		1.20	13.330		1,275	320		1,595	2,000	
	0140	12' x 12' high, class A fire door		1	16		2,260	385		2,645	3,200	
	0160	10' x 20' high, standard		.50	32		2,096	770		2,866	3,700	
	0180	10' x 20' high, class A fire door		.40	40		3,470	965		4,435	5,575	
	3000	For 18 ga. doors, add				S.F.	1			1	1.10	
	3300	For enamel finish, add				"	.90			.90	.99	
	3600	For safety edge bottom bar, pneumatic, add				L.F.	10			10	11	
	4000	For weatherstripping, extruded rubber, jambs, add					6			6	6.60	
	4100	Hood, add					9			9	9.90	
	4200	Sill, add					3.10			3.10	3.41	
	4500	Motor operators, to 14' x 14' opening	2 Sswk	5	3.200	Ea.	820	77		897	1,050	
	4700	For fire door, additional fusible link, add				"	30			30	33	
	9000	Minimum labor/equipment charge	2 Sswk	1	16	Job		385		385	700	
736	0010	SHOCK ABSORBING Rigid, no frame, 1-1/2" thick, 5' x 7'	2 Sswk	1.90	8.420	Opng.	1,158	205		1,363	1,650	736
	0100	8' x 8'		1.80	8.890		1,644	215		1,859	2,200	
	0500	Flexible, no frame, insulated, 1-13/16" thick, economy, 5' x 7'		2	8		1,425	195		1,620	1,925	
	0600	Deluxe		1.90	8.420		2,175	205		2,380	2,750	
	1000	8' x 8' opening, economy		2	8		2,160	195		2,355	2,725	
	1100	Deluxe		1.90	8.420		3,200	205		3,405	3,900	
	9000	Minimum labor/equipment charge		2	8	Job		195		195	350	

083 750 | Swing Doors

		CREW	DAILY OUTPUT	MAN-HOURS	UNIT	BARE COSTS MAT.	LABOR	EQUIP.	TOTAL	TOTAL INCL O&P	
752	0010	DOUBLE ACTING With vision panel, incl. frame, closer & hardware										752
	0020											
	1000	.063" aluminum, 7'-0" high, 4'-0" wide	2 Carp	4.20	3.810	Pr.	440	84		524	620	
	1050	6'-8" wide	"	4	4	"	578	88		666	780	
	2000	Solid core wood, 3/4" thick, metal frame, stainless steel										
	2010	base plate, 7' high opening, 4' wide	2 Carp	4	4	Pr.	695	88		783	910	
	2050	7' wide		3.80	4.210	"	1,030	93		1,123	1,275	
	9000	Minimum labor/equipment charge		2	8	Job		175		175	285	
754	0010	GLASS, SWING Tempered, 1/2" thick, incl. hardware, 3' x 7' opening	2 Glaz	2	8	Opng.	1,250	180		1,430	1,650	754
	0100	6' x 7' opening		1.40	11.430	"	2,900	260		3,160	3,600	
	9000	Minimum labor/equipment charge		2	8	Job		180		180	285	

For expanded coverage of these items see *Means Interior Cost Data 1991*

083 | Special Doors

083 800 | Sound Retardant Doors

		CREW	DAILY OUTPUT	MAN-HOURS	UNIT	BARE COSTS MAT.	LABOR	EQUIP.	TOTAL	TOTAL INCL O&P	
804	0010	ACOUSTICAL Incl. framed seals, 3' x 7', wood, 27 STC rating	F-2	1.50	10.670	Ea.	480	235	10.65	725.65	925	804
	0100	Steel, 40 STC rating		1.50	10.670		1,550	235	10.65	1,795	2,100	
	0200	45 STC rating		1.50	10.670		2,000	235	10.65	2,245	2,600	
	0300	48 STC rating		1.50	10.670		2,400	235	10.65	2,645	3,025	
	0400	52 STC rating	↓	1.50	10.670	↓	2,700	235	10.65	2,945	3,375	
	9000	Minimum labor/equipment charge	F-1	4	2	Job		44	2	46	74	

083 900 | Screen And Storm Doors

		CREW	DAILY OUTPUT	MAN-HOURS	UNIT	BARE COSTS MAT.	LABOR	EQUIP.	TOTAL	TOTAL INCL O&P	
904	0010	STORM DOORS & FRAMES Aluminum, residential,										904
	0020	combination storm and screen										
	0400	Clear anodic coating, 6'-8" x 2'-6" wide	F-2	15	1.070	Ea.	185	23	1.07	209.07	245	
	0420	2'-8" wide		14	1.140		185	25	1.14	211.14	245	
	0440	3'-0" wide	↓	14	1.140		190	25	1.14	216.14	250	
	0500	For 7'-0" door, add					5%					
	1000	Mill finish, 6'-8" x 2'-6" wide	F-2	15	1.070		172	23	1.07	196.07	230	
	1020	2'-8" wide		14	1.140		182	25	1.14	208.14	245	
	1040	3'-0" wide	↓	14	1.140		182	25	1.14	208.14	245	
	1100	For 7'-0" door, add					5%					
	1500	White painted, 6'-8" x 2'-6" wide	F-2	15	1.070		190	23	1.07	214.07	250	
	1520	2'-8" wide		14	1.140		190	25	1.14	216.14	250	
	1540	3'-0" wide	↓	14	1.140	↓	200	25	1.14	226.14	260	
	1600	For 7'-0" door, add					5%					
	2000	Wood door & screen, see division 082-078										
	2020											
	9000	Minimum labor/equipment charge	F-1	4	2	Job		44	2	46	74	

084 | Entrances and Storefronts

084 100 | Aluminum

		CREW	DAILY OUTPUT	MAN-HOURS	UNIT	BARE COSTS MAT.	LABOR	EQUIP.	TOTAL	TOTAL INCL O&P	
103	0010	BALANCED DOORS Hdwre & frame, alum. & glass, 3' x 7', econ.	2 Sswk	.90	17.780	Ea.	2,375	430		2,805	3,400	103
	0150	Premium		.70	22.860	"	4,150	550		4,700	5,575	
	9000	Minimum labor/equipment charge	↓	1	16	Job		385		385	700	
105	0010	STOREFRONT SYSTEMS Aluminum frame, clear ⅜" plate glass,										105
	0020	incl. 3' x 7' door with hardware (400 sq. ft. max. wall)										
	0500	Wall height to 12' high, commercial grade	2 Glaz	150	.107	S.F.	11	2.41		13.41	15.90	
	0600	Institutional grade		130	.123		14.30	2.78		17.08	20	
	0700	Monumental grade		115	.139		20.50	3.14		23.64	28	
	1000	6' x 7' door with hardware, commercial grade		135	.119		11.25	2.67		13.92	16.60	
	1100	Institutional grade		115	.139		15.20	3.14		18.34	22	
	1200	Monumental grade	↓	100	.160	↓	21.25	3.61		24.86	29	
	1500	For bronze anodized finish, add					18%					
	1600	For black anodized finish, add					36%					
	1700	For stainless steel framing, add to monumental					75%					
	2000	For individual doors see division 081										
	9000	Minimum labor/equipment charge	2 Glaz	1	16	Job		360		360	570	
107	0010	SWING DOORS Alum. entrance, 6' x 7', incl. hdwre & oper.	2 Sswk	.70	22.860	Opng.	4,600	550		5,150	6,075	107
	0020	For anodized finish, add				"	400			400	440	
	9000	Minimum labor/equipment charge	2 Sswk	1	16	Job		385		385	700	

For expanded coverage of these items see *Means Interior Cost Data 1991*

125

084 300 | Stainless Steel

			CREW	DAILY OUTPUT	MAN-HOURS	UNIT	BARE COSTS MAT.	BARE COSTS LABOR	BARE COSTS EQUIP.	BARE COSTS TOTAL	TOTAL INCL O&P	
301	0010	**STAINLESS STEEL AND GLASS** Entrance unit, narrow stiles										301
	0020	3' x 7' opening, including hardware, minimum	2 Sswk	1.60	10	Opng.	1,700	240		1,940	2,300	
	0050	Average		1.40	11.430		3,000	275		3,275	3,800	
	0100	Maximum		1.20	13.330		4,600	320		4,920	5,650	
	1000	For solid bronze entrance units, statuary finish, add					60%					
	1100	Without statuary finish, add					45%					
	2000	Balanced doors, 3' x 7', economy	2 Sswk	.90	17.780	Ea.	5,020	430		5,450	6,300	
	2100	Premium		.70	22.860	"	8,650	550		9,200	10,500	
	9000	Minimum labor/equipment charge		2	8	Job		195		195	350	

084 600 | Automatic Doors

			CREW	DAILY OUTPUT	MAN-HOURS	UNIT	BARE COSTS MAT.	BARE COSTS LABOR	BARE COSTS EQUIP.	BARE COSTS TOTAL	TOTAL INCL O&P	
602	0010	**SLIDING ENTRANCE** 12' x 7'-6" opng., 5' x 7' door, 2 way traf.,										602
	0020	mat activated, panic pushout, incl. operator & hardware,										
	0030	not including glass or glazing	2 Glaz	.70	22.860	Opng.	5,100	515		5,615	6,425	
	9000	Minimum labor/equipment charge		.70	22.860	Job		515		515	815	
604	0010	**SLIDING PANEL** Mall fronts, aluminum & glass, 15' x 9' high		1.30	12.310	Opng.	1,820	280		2,100	2,450	604
	0100	24' x 9' high		.70	22.860		2,900	515		3,415	4,000	
	0200	48' x 9' high, with fixed panels		.90	17.780		5,450	400		5,850	6,625	
	0500	For bronze finish, add					17%					
	9000	Minimum labor/equipment charge	2 Glaz	1	16	Job		360		360	570	

085 100 | Steel Windows

			CREW	DAILY OUTPUT	MAN-HOURS	UNIT	BARE COSTS MAT.	BARE COSTS LABOR	BARE COSTS EQUIP.	BARE COSTS TOTAL	TOTAL INCL O&P	
102	0010	**STEEL SASH** Custom units, glazing and trim not included,										102
	0100	Casement, 100% vented	2 Sswk	200	.080	S.F.	26.25	1.93		28.18	32	
	0200	50% vented		200	.080		22.25	1.93		24.18	28	
	0300	Fixed		200	.080		15.50	1.93		17.43	21	
	1000	Projected, commercial, 40% vented		200	.080		23.25	1.93		25.18	29	
	1100	Intermediate, 50% vented		200	.080		26	1.93		27.93	32	
	1500	Industrial, horizontally pivoted		200	.080		19.50	1.93		21.43	25	
	1600	Fixed		200	.080		17	1.93		18.93	22	
	2000	Industrial security sash, 50% vented		200	.080		29.75	1.93		31.68	36	
	2100	Fixed		200	.080		24.75	1.93		26.68	31	
	2500	Picture window		200	.080		10.50	1.93		12.43	15.05	
	3000	Double hung		200	.080		28.75	1.93		30.68	35	
	5000	Mullions for above, open interior face		240	.067	L.F.	5.50	1.61		7.11	9	
	5100	With interior cover		240	.067	"	10.75	1.61		12.36	14.75	
	5200	Single glazing for above, add	2 Glaz	200	.080	S.F.	3.75	1.80		5.55	7	
	6000	Double glazing for above, add		200	.080		5.90	1.80		7.70	9.35	
	6100	Triple glazing for above, add		85	.188		7.20	4.24		11.44	14.65	
	9000	Minimum labor/equipment charge	1 Sswk	2	4	Job		96		96	175	
104	0010	**STEEL WINDOWS** Stock, including frame, trim and insulating glass										104
	0050											
	1000	Custom units, double hung, 2'-8" x 4'-6" opening	2 Sswk	12	1.330	Ea.	395	32		427	495	
	1100	2'-4" x 3'-9" opening		12	1.330		325	32		357	415	
	1500	Commercial projected, 3'-9" x 5'-5" opening		10	1.600		680	39		719	820	
	1600	6'-9" x 4'-1" opening		7	2.290		910	55		965	1,100	
	2000	Intermediate projected, 2'-9" x 4'-1" opening		12	1.330		390	32		422	490	
	2100	4'-1" x 5'-5" opening		10	1.600		770	39		809	915	

For expanded coverage of these items see *Means Interior Cost Data 1991*

085 100 | Steel Windows

			CREW	DAILY OUTPUT	MAN-HOURS	UNIT	MAT.	LABOR	EQUIP.	TOTAL	TOTAL INCL O&P	
104	9000	Minimum labor/equipment charge	1 Sswk	3	2.670	Job		64		64	115	104

085 200 | Aluminum Windows

			CREW	DAILY OUTPUT	MAN-HOURS	UNIT	MAT.	LABOR	EQUIP.	TOTAL	TOTAL INCL O&P	
202	0010	ALUMINUM SASH Stock, grade c, glaze & trim not incl., casement	2 Sswk	200	.080	S.F.	16.90	1.93		18.83	22	202
	0050	Double hung		200	.080		9.80	1.93		11.73	14.30	
	0100	Fixed casement		200	.080		6.95	1.93		8.88	11.15	
	0150	Picture window		200	.080		7.80	1.93		9.73	12.10	
	0200	Projected window		200	.080		15.90	1.93		17.83	21	
	0250	Single hung		200	.080		8.85	1.93		10.78	13.25	
	0300	Sliding		200	.080		10.60	1.93		12.53	15.15	
	1000	Mullions for above, tubular		240	.067	L.F.	2.95	1.61		4.56	6.15	
	2950	Single glazing for above, add	2 Glaz	200	.080	S.F.	4.50	1.80		6.30	7.80	
	3000	Double glazing for above, add		200	.080		7.25	1.80		9.05	10.85	
	3100	Triple glazing for above, add		85	.188		8.50	4.24		12.74	16.10	
	9000	Minimum labor/equipment charge	1 Sswk	2	4	Job		96		96	175	
204	0010	ALUMINUM WINDOWS Incl. frame and glazing, grade C										204
	0050											
	1000	Stock units, casement, 3'-1" x 3'-2" opening	2 Sswk	10	1.600	Ea.	194	39		233	285	
	1050	Add for storms					39			39	43	
	1600	Projected, with screen, 3'-1" x 3'-2" opening	2 Sswk	10	1.600		134	39		173	220	
	1700	Add for storms					39			39	43	
	2000	4'-5" x 5'-3" opening	2 Sswk	8	2		190	48		238	295	
	2100	Add for storms					53			53	58	
	2500	Enamel finish windows, 3'-1" x 3'-2"	2 Sswk	10	1.600		124	39		163	205	
	2600	4'-5" x 5'-3"		8	2		212	48		260	320	
	3000	Single hung, 2' x 3' opening, enameled, standard glazed		10	1.600		92	39		131	170	
	3100	Insulating glass		10	1.600		110	39		149	190	
	3300	2'-8" x 6'-8" opening, standard glazed		8	2		194	48		242	300	
	3400	Insulating glass		8	2		248	48		296	360	
	3700	3'-4" x 5'-0" opening, standard glazed		9	1.780		124	43		167	215	
	3800	Insulating glass		9	1.780		175	43		218	270	
	4000	Sliding aluminum, 3' x 2' opening, standard glazed		10	1.600		90	39		129	170	
	4100	Insulating glass		10	1.600		100	39		139	180	
	4300	5' x 3' opening, standard glazed		9	1.780		112	43		155	200	
	4400	Insulating glass		9	1.780		160	43		203	255	
	4600	8' x 4' opening, standard glazed		6	2.670		165	64		229	300	
	4700	Insulating glass		6	2.670		265	64		329	410	
	5000	9' x 5' opening, standard glazed		4	4		250	96		346	450	
	5100	Insulating glass		4	4		400	96		496	615	
	5500	Sliding, with thermal barrier and screen, 6' x 4', 2 track		8	2		340	48		388	460	
	5700	4 track		8	2		415	48		463	545	
	6000	For above units with bronze finish, add					12%					
	6200	For installation in concrete openings, add					5%					
	6400											
	9000	Minimum labor/equipment charge	1 Sswk	3	2.670	Job		64		64	115	

085 500 | Metal Jalousie Windows

			CREW	DAILY OUTPUT	MAN-HOURS	UNIT	MAT.	LABOR	EQUIP.	TOTAL	TOTAL INCL O&P	
502	0010	JALOUSIES Aluminum incl. glazing & screens, stock, 1'-7" x 3'-2"	2 Sswk	10	1.600	Ea.	75	39		114	155	502
	0100	2'-3" x 4'-0"		10	1.600		120	39		159	200	
	0200	3'-1" x 2'-0"		10	1.600		115	39		154	195	
	0300	3'-1" x 5'-3"		10	1.600		175	39		214	265	
	1000	Mullions for above, 2'-0" long		80	.200		7.20	4.82		12.02	16.70	
	1100	5'-3" long		80	.200		12.85	4.82		17.67	23	
	9000	Minimum labor/equipment charge	1 Sswk	3.50	2.290	Job		55		55	100	
504	0013	LOUVERS See division 062-740 & 102-104										504

For expanded coverage of these items see *Means Interior Cost Data 1991*

		085 600	Metal Storm Windows	CREW	DAILY OUTPUT	MAN-HOURS	UNIT	BARE COSTS MAT.	LABOR	EQUIP.	TOTAL	TOTAL INCL O&P
601	0010	**STORM WINDOWS** Aluminum, residential										601
	0020											
	0300	Basement, mill finish, incl. fiberglass screen										
	0320	1'-10" x 1'-0" high	F-2	30	.533	Ea.	16	11.75	.53	28.28	37
	0340	2'-9" x 1'-6" high		30	.533		19	11.75	.53	31.28	41
	0360	3'-4" x 2'-0" high		30	.533		24	11.75	.53	36.28	46
	1600	Double-hung, combination, storm & screen										
	1700	Custom, clear anodic coating, 2'-0" x 3'-5" high	F-2	30	.533	Ea.	70	11.75	.53	82.28	97
	1720	2'-6" x 5'-0" high		28	.571		106	12.55	.57	119.12	140
	1740	4'-0" x 6'-0" high		25	.640		190	14.10	.64	204.74	235
	1800	White painted, 2'-0" x 3'-5" high		30	.533		66	11.75	.53	78.28	92
	1820	2'-6" x 5'-0" high		28	.571		106	12.55	.57	119.12	140
	1840	4'-0" x 6'-0" high		25	.640		190	14.10	.64	204.74	235
	2000	Average quality, clear anodic coating, 2'-0" x 3'-5" high		30	.533		57	11.75	.53	69.28	82
	2020	2'-6" x 5'-0" high		28	.571		72	12.55	.57	85.12	100
	2040	4'-0" x 6'-0" high		25	.640		80	14.10	.64	94.74	110
	2400	White painted, 2'-0" x 3'-5" high		30	.533		54	11.75	.53	66.28	79
	2420	2'-6" x 5'-0" high		28	.571		60	12.55	.57	73.12	87
	2440	4'-0" x 6'-0" high		25	.640		65	14.10	.64	79.74	95
	2600	Mill finish, 2'-0" x 3'-5" high		30	.533		50	11.75	.53	62.28	75
	2620	2'-6" x 5'-0" high		28	.571		56	12.55	.57	69.12	83
	2640	4'-0" x 6-8" high		25	.640		62	14.10	.64	76.74	92
	4000	Picture window, storm, 1 lite, white or bronze finish										
	4020	4'-6" x 4'-6" high	F-2	25	.640	Ea.	82	14.10	.64	96.74	115
	4040	5'-8" x 4'-6" high		20	.800		84	17.60	.80	102.40	120
	4400	Mill finish, 4'-6" x 4'-6" high		25	.640		72	14.10	.64	86.74	105
	4420	5'-8" x 4'-6" high		20	.800		76	17.60	.80	94.40	115
	4600	3 lite, white or bronze finish										
	4620	4'-6" x 4'-6" high	F-2	25	.640	Ea.	96	14.10	.64	110.74	130
	4640	5'-8" x 4'-6" high		20	.800		100	17.60	.80	118.40	140
	4800	Mill finish, 4'-6" x 4'-6" high		25	.640		90	14.10	.64	104.74	125
	4820	5'-8" x 4'-6" high		20	.800		93	17.60	.80	111.40	130
	5001	Sliding glass door, storm window, 6'-0" x 6'-8", fixed	1 Glaz	1.60	5		175	115		290	370
	5101	Operable	"	2.10	3.810		260	86		346	420
	6000	Sliding window, storm, 2 lite, white or bronze finish										
	6020	3'-4" x 2'-7" high	F-2	28	.571	Ea.	52	12.55	.57	65.12	78
	6040	4'-4" x 3'-3" high		25	.640		60	14.10	.64	74.74	90
	6060	5'-4" x 6'-0" high		20	.800		90	17.60	.80	108.40	130
	6400	3 lite, white or bronze finish										
	6420	4'-4" x 3'-3" high	F-2	25	.640	Ea.	80	14.10	.64	94.74	110
	6440	5'-4" x 6'-0" high		20	.800		125	17.60	.80	143.40	165
	6460	6'-0" x 6'-0" high		18	.889		130	19.55	.89	150.44	175
	6800	Mill finish, 4'-4" x 3'-3" high		25	.640		72	14.10	.64	86.74	105
	6820	5'-4" x 6'-0" high		20	.800		103	17.60	.80	121.40	145
	6840	6'-0" x 6-0" high		18	.889		110	19.55	.89	130.44	155
	8000	PVC framed										
	8100	Double-hung, combination, storm & screen										
	8120	2'-6" x 3'-5"	F-2	15	1.070	Ea.	46	23	1.07	70.07	90
	8140	4' x 6'	"	12.50	1.280	"	78	28	1.28	107.28	135
	8600	Single lite picture storm										
	8620	4'-6" x 4'-6"	F-2	12.50	1.280	Ea.	75	28	1.28	104.28	130
	8640	5'-8" x 4'-6"	"	10	1.600	"	82	35	1.60	118.60	150
	9000	Magnetic interior storm window										
	9100	³⁄₁₆" plate glass	1 Glaz	107	.075	S.F.	3.75	1.69		5.44	6.80
	9410	Minimum labor/equipment charge	F-1	4	2	Job		44	2	46	74

For expanded coverage of these items see *Means Interior Cost Data 1991*

085 | Metal Windows

085 700	Screens	CREW	DAILY OUTPUT	MAN-HOURS	UNIT	BARE COSTS MAT.	LABOR	EQUIP.	TOTAL	TOTAL INCL O&P	
701	0010	SCREENS For metal sash, aluminum or bronze mesh, flat screen	2 Sswk	1,200	.013	S.F.	2.65	.32		2.97	3.50	701
	0500	Wicket screen, inside window		1,000	.016		4.10	.39		4.49	5.20	
	0800	Security screen, aluminum frame with stainless steel cloth		1,200	.013		14.45	.32		14.77	16.50	
	0900	Steel grate, painted, on steel frame		1,600	.010		7.75	.24		7.99	8.95	
	1000	For solar louvers, add		160	.100		14.95	2.41		17.36	21	
	4000	See also division 055-508										

086 | Wood and Plastic Windows

086 100	Wood Windows	CREW	DAILY OUTPUT	MAN-HOURS	UNIT	BARE COSTS MAT.	LABOR	EQUIP.	TOTAL	TOTAL INCL O&P	
104	0010	AWNING WINDOW Including frame, screen, and exterior trim										104
	0020											
	0100	Average quality, builders model, 34" x 22", standard glazed	1 Carp	10	.800	Ea.	135	17.60		152.60	175	
	0200	Insulating glass		10	.800		150	17.60		167.60	195	
	0300	40" x 28", standard glazed		9	.889		155	19.55		174.55	200	
	0400	Insulating glass		9	.889		180	19.55		199.55	230	
	0500	48" x 36", standard glazed		8	1		180	22		202	235	
	0600	Insulating glass		8	1		216	22		238	275	
	1000	Vinyl clad, premium, insulating glass, 34" x 22"		10	.800		176	17.60		193.60	220	
	1100	40" x 22"		10	.800		192	17.60		209.60	240	
	1200	36" x 28"		9	.889		209	19.55		228.55	260	
	1300	36" x 36"		9	.889		250	19.55		269.55	305	
	1400	48" x 28"		8	1		272	22		294	335	
	1500	60" x 36"		8	1		430	22		452	510	
	2000	Metal clad, deluxe, insulating glass, 34" x 22"		10	.800		145	17.60		162.60	190	
	2100	40" x 22"		10	.800		157	17.60		174.60	200	
	2200	36" x 25"		9	.889		168	19.55		187.55	215	
	2300	40" x 30"		9	.889		170	19.55		189.55	220	
	2400	48" x 28"		8	1		190	22		212	245	
	2500	60" x 36"		8	1		290	22		312	355	
	9000	Minimum labor/equipment charge		4	2	Job		44		44	72	
108	0010	BOW-BAY WINDOW Including frame, screen and exterior trim,										108
	0020	end panels operable										
	1000	Awning type, builders model, 8' x 5' high, std. glazed, 4 panels	2 Carp	10	1.600	Ea.	670	35		705	795	
	1050	Insulating glass		10	1.600		766	35		801	900	
	1100	10'-0" x 5'-0" high, standard glazed		6	2.670		833	59		892	1,000	
	1200	Insulating glass, 6 panels		6	2.670		935	59		994	1,125	
	1300	Vinyl clad, premium, insulating glass, 6'-0" x 4'-0"		10	1.600		1,120	35		1,155	1,300	
	1340	9'-0" x 4'-0"		8	2		1,510	44		1,554	1,725	
	1380	10'-0" x 6'-0"		7	2.290		1,975	50		2,025	2,250	
	1420	12'-0" x 6'-0"		6	2.670		2,460	59		2,519	2,800	
	1600	Metal clad, deluxe, insul. glass, 6'-0" x 4'-0" high, 3 panels		10	1.600		672	35		707	795	
	1640	9'-0" x 4'-0" high, 4 panels		8	2		1,040	44		1,084	1,225	
	1680	10'-0" x 5'-0" high, 5 panels		7	2.290		1,430	50		1,480	1,650	
	1720	12'-0" x 6'-0" high, 6 panels		6	2.670		2,035	59		2,094	2,325	
	2000	Casement, builders model, bow, 8' x 5' high, std. glazed, 4 panels		10	1.600		870	35		905	1,025	
	2050	Insulating glass		10	1.600		1,055	35		1,090	1,225	
	2100	12'-0" x 6'-0" high, 6 panels, standard glazed		6	2.670		1,310	59		1,369	1,525	
	2200	Insulating glass		6	2.670		2,085	59		2,144	2,400	
	2300	Vinyl clad, premium, insulating glass, 8'-0" x 5'-0"		10	1.600		1,190	35		1,225	1,375	
	2340	10'-0" x 5'-0"		8	2		1,580	44		1,624	1,800	

For expanded coverage of these items see *Means Interior Cost Data 1991*

086 100	Wood Windows	CREW	DAILY OUTPUT	MAN-HOURS	UNIT	BARE COSTS				TOTAL INCL O&P
						MAT.	LABOR	EQUIP.	TOTAL	
108 2380	10'-0" x 6'-0"	2 Carp	7	2.290	Ea.	1,815	50		1,865	2,075	**108**
2420	12'-0" x 6'-0"		6	2.670		2,300	59		2,359	2,625	
2600	Metal clad, deluxe, insul. glass, 8'-0" x 5'-0" high, 4 panels		10	1.600		870	35		905	1,025	
2640	10'-0" x 5'-0" high, 5 panels		8	2		1,105	44		1,149	1,275	
2680	10'-0" x 6'-0" high, 5 panels		7	2.290		1,335	50		1,385	1,550	
2720	12'-0" x 6'-0" high, 6 panels		6	2.670		2,415	59		2,474	2,750	
3000	Double hung, bldrs. model, bay, 8' x 4' high, std. glazed		10	1.600		810	35		845	950	
3050	Insulating glass		10	1.600		915	35		950	1,075	
3100	9'-0" x 5'-0" high, standard glazed		6	2.670		875	59		934	1,050	
3200	Insulating glass		6	2.670		995	59		1,054	1,200	
3300	Vinyl clad, premium, insulating glass, 7'-0" x 4'-6"		10	1.600		990	35		1,025	1,150	
3340	8'-0" x 4'-6"		8	2		1,080	44		1,124	1,250	
3380	8'-0" x 5'-0"		7	2.290		1,105	50		1,155	1,300	
3420	9'-0" x 5'-0"		6	2.670		1,200	59		1,259	1,425	
3600	Metal clad, deluxe, insul. glass, 7'-0" x 4'-0" high		10	1.600		685	35		720	810	
3640	8'-0" x 4'-0" high		8	2		725	44		769	870	
3680	8'-0" x 5'-0" high		7	2.290		790	50		840	950	
3720	9'-0" x 5'-0" high		6	2.670		830	59		889	1,000	
9000	Minimum labor/equipment charge		2.50	6.400	Job		140		140	230	
120 0010	**CASEMENT WINDOW** Including frame, screen, and exterior trim										**120**
0020											
0100	Average quality, bldrs. model, 2'-0" x 3'-0" high, standard glazed	1 Carp	10	.800	Ea.	130	17.60		147.60	170	
0150	Insulating glass		10	.800		168	17.60		185.60	215	
0200	2'-0" x 4'-6" high, standard glazed		9	.889		170	19.55		189.55	220	
0250	Insulating glass		9	.889		217	19.55		236.55	270	
0300	2'-3" x 6'-0" high, standard glazed		8	1		263	22		285	325	
0350	Insulating glass		8	1		310	22		332	375	
1000	Vinyl clad, premium, insulating glass, 2'-0" x 3'-0"		10	.800		197	17.60		214.60	245	
1040	2'-0" x 4'-0"		9	.889		213	19.55		232.55	265	
1080	2'-0" x 5'-0"		8	1		247	22		269	310	
1120	2'-0" x 6'-0"		8	1		292	22		314	355	
2000	Metal clad, deluxe, insulating glass, 2'-0" x 3'-0" high		10	.800		197	17.60		214.60	245	
2040	2'-0" x 4'-0" high		9	.889		160	19.55		179.55	210	
2080	2'-0" x 5'-0" high		8	1		182	22		204	235	
2120	2'-0" x 6'-0" high		8	1		212	22		234	270	
2200	For multiple leaf units, deduct for stationary sash										
2210	2' high				Ea.	37			37	41	
2300	4'-6" high					52			52	57	
2400	6' high					51			51	56	
3000	For installation, add per leaf						15%				
9000	Minimum labor/equipment charge	1 Carp	3	2.670	Job		59		59	96	
124 0010	**DOUBLE HUNG** Including frame, screen, and exterior trim										**124**
0020											
0100	Average quality, bldrs. model, 2'-0" x 3'-0" high, standard glazed	1 Carp	10	.800	Ea.	85	17.60		102.60	120	
0150	Insulating glass		10	.800		122	17.60		139.60	165	
0200	3'-0" x 4'-0" high, standard glazed		9	.889		106	19.55		125.55	150	
0250	Insulating glass		9	.889		141	19.55		160.55	185	
0300	4'-0" x 4'-6" high, standard glazed		8	1		120	22		142	170	
0350	Insulating glass		8	1		170	22		192	225	
1000	Vinyl clad, premium, insulating glass, 2'-6" x 3'-0"		10	.800		175	17.60		192.60	220	
1100	3'-0" x 3'-6"		10	.800		211	17.60		228.60	260	
1200	3'-0" x 4'-0"		9	.889		230	19.55		249.55	285	
1300	3'-0" x 4'-6"		9	.889		242	19.55		261.55	300	
1400	3'-0" x 5'-0"		8	1		252	22		274	315	
1500	3'-6" x 6'-0"		8	1		303	22		325	370	
2000	Metal clad, deluxe, insulating glass, 2'-6" x 3'-0" high		10	.800		108	17.60		125.60	150	
2100	3'-0" x 3'-6" high		10	.800		130	17.60		147.60	170	

For expanded coverage of these items see *Means Interior Cost Data 1991*

086 100	Wood Windows	CREW	DAILY OUTPUT	MAN-HOURS	UNIT	BARE COSTS				TOTAL INCL O&P	
						MAT.	LABOR	EQUIP.	TOTAL		
124 2200	3'-0" x 4'-0" high	1 Carp	9	.889	Ea.	142	19.55		161.55	190	**124**
2300	3'-0" x 4'-6" high		9	.889		149	19.55		168.55	195	
2400	3'-0" x 5'-0" high		8	1		169	22		191	220	
2500	3'-6" x 6'-0" high		8	1		189	22		211	245	
9000	Minimum labor/equipment charge		3	2.670	Job		59		59	96	
132 0010	**PICTURE WINDOW** Including frame and exterior trim										**132**
0020											
0100	Average quality, bldrs. model, 3'-6" x 4'-0" high, standard glazed	2 Carp	12	1.330	Ea.	155	29		184	220	
0150	Insulating glass		12	1.330		192	29		221	260	
0200	4'-0" x 4'-6" high, standard glazed		11	1.450		177	32		209	245	
0250	Insulating glass		11	1.450		220	32		252	295	
0300	5'-0" x 4'-0" high, standard glazed		11	1.450		205	32		237	280	
0350	Insulating glass		11	1.450		237	32		269	315	
0400	6'-0" x 4'-6" high, standard glazed		10	1.600		259	35		294	340	
0450	Insulating glass		10	1.600		267	35		302	350	
1000	Vinyl clad, premium, insulating glass, 4'-0" x 4'-0"		12	1.330		335	29		364	415	
1100	4'-0" x 6'-0"		11	1.450		445	32		477	540	
1200	5'-0" x 6'-0"		10	1.600		567	35		602	680	
1300	6'-0" x 6'-0"		10	1.600		660	35		695	785	
2000	Metal clad, deluxe, insulating glass, 4'-0" x 4'-0" high		12	1.330		184	29		213	250	
2100	4'-0" x 6'-0" high		11	1.450		268	32		300	345	
2200	5'-0" x 6'-0" high		10	1.600		340	35		375	430	
2300	6'-0" x 6'-0" high		10	1.600		415	35		450	515	
9000	Minimum labor/equipment charge		2.75	5.820	Job		130		130	210	
140 0010	**SLIDING WINDOW** Including frame, screen, and exterior trim										**140**
0020											
0100	Average quality, bldrs. model, 3'-0" x 3'-0" high, standard glazed	1 Carp	10	.800	Ea.	95	17.60		112.60	135	
0120	Insulating glass		10	.800		131	17.60		148.60	175	
0200	4'-0" x 3'-6" high, standard glazed		9	.889		121	19.55		140.55	165	
0220	Insulating glass		9	.889		187	19.55		206.55	240	
0300	6'-0" x 5'-0" high, standard glazed		8	1		205	22		227	260	
0320	Insulating glass		8	1		305	22		327	370	
1000	Vinyl clad, premium, insulating glass, 3'-0" x 3'-0"		10	.800		245	17.60		262.60	300	
1050	4'-0" x 3'-6"		9	.889		303	19.55		322.55	365	
1100	5'-0" x 4'-0"		9	.889		366	19.55		385.55	435	
1150	6'-0" x 5'-0"		8	1		522	22		544	610	
2000	Metal clad, deluxe, insulating glass, 3'-0" x 3'-0" high		10	.800		247	17.60		264.60	300	
2050	4'-0" x 3'-6" high		9	.889		307	19.55		326.55	370	
2100	5'-0" x 4'-0" high		9	.889		369	19.55		388.55	440	
2150	6'-0" x 5'-0" high		8	1		530	22		552	620	
9000	Minimum labor/equipment charge		3	2.670	Job		59		59	96	
144 0010	**WINDOW GRILLE OR MUNTIN** Snap-in type										**144**
0020	Colonial or diamond pattern										
2000	Wood, awning window, glass size 28" x 16" high	1 Carp	30	.267	Ea.	9.25	5.85		15.10	19.75	
2060	44" x 24" high		32	.250		12.55	5.50		18.05	23	
2100	Casement, glass size, 20" x 36" high		30	.267		9.65	5.85		15.50	20	
2180	20" x 56" high		32	.250		13.55	5.50		19.05	24	
2200	Double hung, glass size, 16" x 24" high		24	.333	Set	13	7.35		20.35	26	
2280	32" x 32" high		34	.235	"	16.25	5.20		21.45	26	
2500	Picture, glass size, 48" x 48" high		30	.267	Ea.	61	5.85		66.85	77	
2580	60" x 68" high		28	.286	"	71.40	6.30		77.70	89	
2600	Sliding, glass size, 14" x 36" high		24	.333	Set	35	7.35		42.35	50	
2680	36" x 36" high		22	.364	"	55	8		63	74	
9000	Minimum labor/equipment charge		5	1.600	Job		35		35	57	

For expanded coverage of these items see *Means Interior Cost Data 1991*

086 100 | Wood Windows

		CREW	DAILY OUTPUT	MAN-HOURS	UNIT	BARE COSTS MAT.	LABOR	EQUIP.	TOTAL	TOTAL INCL O&P	
148	0010	**WOOD SASH** Including glazing but not including trim										148
	0020											
	0050	Custom, 5'-0" x 4'-0", 1" dbl. glazed, 3/16" thick lites	2 Carp	3.20	5	Ea.	257	110		367	460	
	0100	1/4" thick lites		5	3.200		273	70		343	415	
	0200	1" thick, triple glazed		5	3.200		394	70		464	550	
	0300	7'-0" x 4'-6" high, 1" double glazed, 3/16" thick lites		4.30	3.720		394	82		476	565	
	0400	1/4" thick lites		4.30	3.720		445	82		527	625	
	0500	1" thick, triple glazed		4.30	3.720		499	82		581	685	
	0600	8'-6" x 5'-0" high, 1" double glazed, 3/16" thick lites		3.50	4.570		535	100		635	755	
	0700	1/4" thick lites		3.50	4.570		588	100		688	810	
	0800	1" thick, triple glazed	↓	3.50	4.570		588	100		688	810	
	0900	Window frames only, based on perimeter length				L.F.	1.95			1.95	2.15	
	1200	Window sill, stock, per lineal foot					5.50			5.50	6.05	
	1250	Casing, stock				↓	1.95			1.95	2.15	
	3000	Replacement sash, double hung, double glazing, to 12 S.F.	1 Carp	64	.125	S.F.	12.20	2.75		14.95	17.90	
	3100	12 S.F. to 20 S.F.		94	.085		11.20	1.87		13.07	15.40	
	3200	20 S.F. and over	↓	106	.075		10.50	1.66		12.16	14.25	
	3800	Triple glazing for above, add				↓	2.36			2.36	2.60	
	7000	Sash, single lite, 2'-0" x 2'-0" high	1 Carp	20	.400	Ea.	20.15	8.80		28.95	37	
	7050	2'-6" x 2'-0" high		19	.421		24.25	9.25		33.50	42	
	7100	2'-6" x 2'-6" high		18	.444		27.30	9.80		37.10	46	
	7150	3'-0" x 2'-0" high	↓	17	.471	↓	27.30	10.35		37.65	47	
	9000	Minimum labor/equipment charge	↓	4	2	Job		44		44	72	
152	0010	**WOOD SCREENS** Over 3 S.F., 3/4" frames	2 Carp	375	.043	S.F.	2.42	.94		3.36	4.19	152
	0100	1-1/8" frames	"	375	.043	"	2.63	.94		3.57	4.43	
	9000	Minimum labor/equipment charge	1 Carp	4	2	Job		44		44	72	

087 100 | Finish Hardware

		CREW	DAILY OUTPUT	MAN-HOURS	UNIT	BARE COSTS MAT.	LABOR	EQUIP.	TOTAL	TOTAL INCL O&P	
101	0010	**AVERAGE** Percentage for hardware, total job cost, minimum									.75%	101
	0050	Maximum									3.50%	
	0500	Total hardware for building, average distribution					85%	15%				
	1000	Door hardware, apartment, interior				Door	89			89	98	
	1500	Hospital bedroom, minimum					116			116	130	
	2000	Maximum					410			410	450	
	2250	School, single exterior, incl. lever, not incl. panic device					298			298	330	
	2500	Single interior, regular use, no lever included					134			134	145	
	2600	Heavy use, incl. lever and closer					245			245	270	
	2850	Stairway, single interior				↓	160			160	175	
	3100	Double exterior, with panic device				Pr.	700			700	770	
	3300											
	3600	Toilet, public, single interior				Door	108			108	120	
108	0010	**DEADLOCKS** Mortise, heavy duty, outside key	1 Carp	9	.889	Ea.	85	19.55		104.55	125	108
	0020	Double cylinder		9	.889		95	19.55		114.55	135	
	0100	Medium duty, outside key		10	.800		67	17.60		84.60	100	
	0110	Double cylinder		10	.800		84	17.60		101.60	120	
	1000	Tubular, standard duty, outside key		10	.800		36.50	17.60		54.10	69	
	1010	Double cylinder	↓	10	.800	↓	47	17.60		64.60	80	

For expanded coverage of these items see *Means Interior Cost Data 1991*

087 100 | Finish Hardware

			CREW	DAILY OUTPUT	MAN-HOURS	UNIT	BARE COSTS				TOTAL INCL O&P	
							MAT.	LABOR	EQUIP.	TOTAL		
108	1200	Night latch, outside key	1 Carp	10	.800	Ea.	46	17.60		63.60	79	108
110	0010	**DOORSTOPS** Holder and bumper, floor or wall	1 Carp	24	.333	Ea.	10	7.35		17.35	23	110
	1300	Wall bumper		24	.333		4.25	7.35		11.60	16.65	
	1600	Floor bumper, 1″ high		24	.333		3.50	7.35		10.85	15.80	
	1900	Plunger type, door mounted		24	.333		24	7.35		31.35	38	
	9000	Minimum labor/equipment charge		6	1.330	Job		29		29	48	
112	0010	**ENTRANCE LOCKS** Cylinder, grip handle, deadlocking latch	1 Carp	9	.889	Ea.	88	19.55		107.55	130	112
	0020	Deadbolt		8	1		105	22		127	150	
	0100	Push and pull plate, dead bolt		8	1		97	22		119	145	
	0900	For handicapped lever, add					65			65	72	
116	0010	**HINGES** Full mortise, avg. freq., steel base, 4-½″x 4-½″, USP				Pr.	18.50			18.50	20	116
	0100	5″ x 5″, USP					35			35	39	
	0200	6″ x 6″, USP					52			52	57	
	0400	Brass base, 4-½″ x 4-½″, US10					45			45	50	
	0500	5″ x 5″, US10					50			50	55	
	0600	6″ x 6″, US10					90			90	99	
	0800	Stainless steel base, 4-½″ x 4-½″, US32					60			60	66	
	0900	For non removable pin, add				Ea.	3.40			3.40	3.74	
	0910	For floating pin, driven tips, add					3.75			3.75	4.13	
	0930	For hospital type tip on pin, add					11			11	12.10	
	0940	For steeple type tip on pin, add					3.75			3.75	4.13	
	1000	Full mortise, high frequency, steel base, 4-½″ x 4-½″, USP				Pr.	45			45	50	
	1100	5″ x 5″, USP					50			50	55	
	1200	6″ x 6″, USP					75			75	83	
	1400	Brass base, 4-½″ x 4-½″, US10					70			70	77	
	1500	5″ x 5″, US10					85			85	94	
	1600	6″ x 6″, US10					95			95	105	
	1800	Stainless steel base, 4-½″ x 4-½″, US32					90			90	99	
	1930	For hospital type tip on pin, add				Ea.	12.50			12.50	13.75	
	2000	Full mortise low frequency, steel base, 4-½″ x 4-½″, USP				Pr.	10.85			10.85	11.95	
	2100	5″ x 5″, USP					24.50			24.50	27	
	2200	6″ x 6″, USP					38			38	42	
	2400	Brass base, 4-½″ x 4-½″, US10					35			35	39	
	2500	5″ x 5″, US10					46			46	51	
	2800	Stainless steel base, 4-½″ x 4-½″, US32					53			53	58	
118	0010	**KICK PLATE** 6″ high, for 3′ door, stainless steel	1 Carp	15	.533	Ea.	17	11.75		28.75	38	118
	0500	Bronze		15	.533	″	35	11.75		46.75	58	
	9000	Minimum labor/equipment charge		6	1.330	Job		29		29	48	
120	0010	**LOCKSET** Standard duty, cylindrical, with sectional trim	1 Carp	12	.667	Ea.	26.50	14.65		41.15	53	120
	0020	Non-keyed, passage		12	.667		29.50	14.65		44.15	56	
	0100	Privacy		12	.667		36	14.65		50.65	64	
	0400	Keyed, single cylinder function		10	.800		58	17.60		75.60	93	
	0420	Hotel		8	1		72	22		94	115	
	1000	Heavy duty with sectional trim, non-keyed, passages		12	.667		87	14.65		101.65	120	
	1100	Privacy		12	.667		112	14.65		126.65	145	
	1400	Keyed, single cylinder function		10	.800		128	17.60		145.60	170	
	1420	Hotel		8	1		185	22		207	240	
	1600	Communicating		10	.800		150	17.60		167.60	195	
	1690	For re-core cylinder, add					22			22	24	
	1700	Residential, interior door, minimum	1 Carp	16	.500		10.25	11		21.25	29	
	1720	Maximum		8	1		28	22		50	67	
	1800	Exterior, minimum		14	.571		22	12.55		34.55	45	
	1820	Maximum		8	1		96	22		118	140	
	9000	Minimum labor/equipment charge		6	1.330	Job		29		29	48	

For expanded coverage of these items see *Means Interior Cost Data 1991*

087 100 | Finish Hardware

		CREW	DAILY OUTPUT	MAN-HOURS	UNIT	BARE COSTS MAT.	LABOR	EQUIP.	TOTAL	TOTAL INCL O&P	
125	0010	**MORTISE LOCKSET** Comm., wrought knobs & full escutcheon trim									**125**
	0020	Non-keyed, passage, minimum	1 Carp	9	.889	Ea.	80	19.55		99.55	120
	0030	Maximum		8	1		160	22		182	210
	0040	Privacy, minimum		9	.889		96	19.55		115.55	140
	0050	Maximum		8	1		190	22		212	245
	0100	Keyed, office/entrance/apartment, minimum		8	1		122	22		144	170
	0110	Maximum		7	1.140		226	25		251	290
	0120	Single cylinder, typical, minimum		8	1		105	22		127	150
	0130	Maximum		7	1.140		205	25		230	265
	0200	Hotel, minimum		7	1.140		130	25		155	185
	0210	Maximum		6	1.330		232	29		261	305
	0300	Communication, double cylinder, minimum		8	1		130	22		152	180
	0310	Maximum		7	1.140		185	25		210	245
	1000	Wrought knobs and sectional trim, non-keyed, passage, minimum		10	.800		68	17.60		85.60	105
	1010	Maximum		9	.889		145	19.55		164.55	190
	1040	Privacy, minimum		10	.800		78	17.60		95.60	115
	1050	Maximum		9	.889		160	19.55		179.55	210
	1100	Keyed, entrance, office/apartment, minimum		9	.889		115	19.55		134.55	160
	1110	Maximum		8	1		190	22		212	245
	1120	Single cylinder, typical, minimum		9	.889		110	19.55		129.55	155
	1130	Maximum		8	1		185	22		207	240
	2000	Cast knobs and full escutcheon trim									
	2010	Non-keyed, passage, minimum	1 Carp	9	.889	Ea.	145	19.55		164.55	190
	2020	Maximum		8	1		265	22		287	325
	2040	Privacy, minimum		9	.889		175	19.55		194.55	225
	2050	Maximum		8	1		285	22		307	350
	2120	Keyed, single cylinder, typical, minimum		8	1		175	22		197	230
	2130	Maximum		7	1.140		300	25		325	370
	2200	Hotel, minimum		7	1.140		200	25		225	260
	2210	Maximum		6	1.330		385	29		414	470
	2220										
	3000	Cast knob and sectional trim, non-keyed, passage, minimum	1 Carp	10	.800	Ea.	110	17.60		127.60	150
	3010	Maximum		10	.800		240	17.60		257.60	295
	3040	Privacy, minimum		10	.800		125	17.60		142.60	165
	3050	Maximum		10	.800		245	17.60		262.60	300
	3100	Keyed, office/entrance/apartment, minimum		9	.889		140	19.55		159.55	185
	3110	Maximum		9	.889		255	19.55		274.55	310
	3120	Single cylinder, typical, minimum		9	.889		145	19.55		164.55	190
	3130	Maximum		9	.889		300	19.55		319.55	360
	3190	For re-core cylinder, add					20			20	22
	3200										
	4000	Keyless, pushbutton type, with deadbolt, standard	1 Carp	9	.889	Ea.	75	19.55		94.55	115
	4100	Heavy duty	"	9	.889		105	19.55		124.55	145
	4150	Card type, 1 time zone, minimum					258			258	285
	4200	Maximum					840			840	925
	4250	3 time zones, minimum					670			670	735
	4300	Maximum					1,500			1,500	1,650
	4350	System with printer, and control console, 3 zones				Total	7,680			7,680	8,450
	4400	6 zones				"	10,100			10,100	11,100
	4450	For each door, minimum, add				Ea.	1,130			1,130	1,250
	4500	Maximum, add				"	1,625			1,625	1,800
127	0010	**PANIC DEVICE** For rim locks, single door, exit only	1 Carp	3	2.670	Ea.	235	59		294	355	**127**
	0020	Outside key and pull		5	1.600		285	35		320	370
	0200	Bar and vertical rod, exit only		5	1.600		360	35		395	455
	0210	Outside key and pull		4	2		430	44		474	545
	0400	Bar and concealed rod		4	2		395	44		439	505
	0501										

For expanded coverage of these items see *Means Interior Cost Data 1991*

087 100	Finish Hardware	CREW	DAILY OUTPUT	MAN-HOURS	UNIT	BARE COSTS				TOTAL INCL O&P	
						MAT.	LABOR	EQUIP.	TOTAL		
127 0600	Touch bar, exit only	1 Carp	6	1.330	Ea.	295	29		324	370	**127**
0610	Outside key and pull		5	1.600		375	35		410	470	
0700	Touch bar and vertical rod, exit only		5	1.600		425	35		460	525	
0710	Outside key and pull		4	2	↓	495	44		539	615	
1000	Mortise, bar, exit only		2	4	Pr.	270	88		358	440	
1600	Touch bar, exit only		4	2	Ea.	385	44		429	495	
2000	Narrow stile, rim mounted, bar, exit only		6	1.330		340	29		369	420	
2010	Outside key and pull		5	1.600		395	35		430	490	
2200	Bar and vertical rod, exit only		5	1.600		360	35		395	455	
2210	Outside key and pull		4	2		425	44		469	540	
2400	Bar and concealed rod, exit only		3	2.670		405	59		464	540	
3000	Mortise, bar, exit only		3	2.670		285	59		344	410	
3600	Touch bar, exit only		4	2	↓	415	44		459	530	
4000	Double doors, exit only		2	4	Pr.	520	88		608	715	
4500	Exit & entrance		2	4	"	550	88		638	750	
9000	Minimum labor/equipment charge		2.50	3.200	Job		70		70	115	
129 0010	**PUSH-PULL** Push plate, pull plate, aluminum		12	.667	Ea.	48	14.65		62.65	77	**129**
0500	Bronze		12	.667		60	14.65		74.65	90	
1500	Pull handle and push bar, aluminum		11	.727		89	16		105	125	
2000	Bronze		10	.800		115	17.60		132.60	155	
3000	Push plate both sides, aluminum		14	.571		11	12.55		23.55	33	
3500	Bronze		13	.615		27	13.55		40.55	52	
4000	Door pull, designer style, cast aluminum, minimum		12	.667		50	14.65		64.65	79	
5000	Maximum		8	1		215	22		237	270	
6000	Cast bronze, minimum		12	.667		55	14.65		69.65	84	
7000	Maximum		8	1		240	22		262	300	
8000	Walnut, minimum		12	.667		45	14.65		59.65	73	
9000	Maximum		8	1	↓	235	22		257	295	
9800	Minimum labor/equipment charge	↓	5	1.600	Job		35		35	57	
132 0010	**SPECIAL HINGES** Paumelle, high frequency										**132**
0020	Steel base, 6" x 4-½", US10				Pr.	80			80	88	
0100	Bronze base, 5" x 4-½", US10					105			105	115	
0200	Paumelle, average frequency, steel base, 4-½" x 3-½", US10					50			50	55	
0400	Olive knuckle, low frequency, bronze base, 6" x 4-½", US10				↓	95			95	105	
1000	Electric hinge with concealed conductor, average frequency										
1010	Steel base, 4-½" x 4-½", US26D				Pr.	85			85	94	
1100	Bronze base, 4-½" x 4-½", US26D				"	105			105	115	
1200	Electric hinge with concealed conductor, high frequency										
1210	Steel base, 4-½" x 4-½", US26D				Pr.	100			100	110	
1600	Double weight, 800 lb., steel base, removable pin, 5" x 6", USP					110			110	120	
1700	Steel base-welded pin, 5" x 6", USP					100			100	110	
1800	Triple weight, 2000 lb., steel base, welded pin, 5" x 6", USP				↓	135			135	150	
1810											
2000	Pivot reinf., high frequency, steel base, 7-¾" door plate, USP				Pr.	140			140	155	
2200	Bronze base, 7-¾" door plate, US10				"	190			190	210	
3000	Swing clear, full mortise, full or half surface, high frequency,										
3010	Steel base, 5" high, USP				Pr.	110			110	120	
3200	Swing clear, full mortise, average frequency										
3210	Steel base, 4-½" high, USP				Pr.	85			85	94	
4000	Wide throw, average frequency, steel base, 4-½" x 6", USP					62			62	68	
4200	High frequency, steel base, 4-½" x 6", USP				↓	110			110	120	
4600	Spring hinge, single acting, 6" flange, steel				Ea.	28			28	31	
4700	Brass					48			48	53	
4900	Double acting, 6" flange, steel					50			50	55	
4950	Brass					80			80	88	
8000	Continuous hinge, steel	2 Carp	64	.250	L.F.	2.95	5.50		8.45	12.25	

For expanded coverage of these items see *Means Interior Cost Data 1991*

087 100	Finish Hardware	CREW	DAILY OUTPUT	MAN-HOURS	UNIT	BARE COSTS				TOTAL INCL O&P
						MAT.	LABOR	EQUIP.	TOTAL	
134										**134**
0010	**WINDOW HARDWARE**									
1000	Handles, surface mounted, aluminum	1 Carp	24	.333	Ea.	1.55	7.35		8.90	13.70
1020	Brass		24	.333		1.80	7.35		9.15	13.95
1040	Chrome		24	.333		1.65	7.35		9	13.80
1500	Recessed, aluminum		12	.667		.90	14.65		15.55	25
1520	Brass		12	.667		1	14.65		15.65	25
1540	Chrome		12	.667		.95	14.65		15.60	25
2000	Latches, aluminum		20	.400		1.30	8.80		10.10	15.80
2020	Brass		20	.400		1.55	8.80		10.35	16.05
2040	Chrome		20	.400		1.45	8.80		10.25	15.95
9000	Minimum labor/equipment charge		6	1.330	Job		29		29	48

087 200	Operators	CREW	DAILY OUTPUT	MAN-HOURS	UNIT	MAT.	LABOR	EQUIP.	TOTAL	TOTAL INCL O&P
202										**202**
0010	**AUTOMATIC OPENERS** Swing doors, single	2 Carp	.80	20	Ea.	1,900	440		2,340	2,800
0100	Single operating pair		.50	32	Pr.	3,500	705		4,205	5,000
0400	For double simultaneous doors, one way, add		1.20	13.330		300	295		595	810
0500	Two way, add		.90	17.780		370	390		760	1,050
1000	Sliding doors, 3' wide, including track & hanger, single		.60	26.670	Opng.	3,230	585		3,815	4,500
1300	Bi-parting		.50	32		3,840	705		4,545	5,375
1450	Activating carpet, single door, one way, add		2.20	7.270		605	160		765	925
1550	Two way, add		1.30	12.310		865	270		1,135	1,400
1750	Handicap opener, button operating		8	2	Ea.	985	44		1,029	1,150
206										**206**
0010	**DOOR CLOSER** Rack and pinion	1 Carp	6.50	1.230	Ea.	68	27		95	120
0020	Adjustable backcheck, 3 way mount, all sizes, regular arm		6	1.330		73	29		102	130
0040	Hold open arm		6	1.330		76	29		105	130
0100	Fusible link		6.50	1.230		64	27		91	115
0200	Non sized, regular arm		6	1.330		76	29		105	130
0240	Hold open arm		6	1.330		85	29		114	140
0400	4 way mount, non sized, regular arm		6	1.330		95	29		124	150
0440	Hold open arm		6	1.330		105	29		134	165
1950										
2000	Backcheck and adjustable power, hinge face mount									
2010	All sizes, regular arm	1 Carp	6.50	1.230	Ea.	67	27		94	120
2040	Hold open arm		6.50	1.230		76	27		103	130
2400	Top jamb mount, all sizes, regular arm		6	1.330		65	29		94	120
2440	Hold open arm		6	1.330		75	29		104	130
2800	Top face mount, all sizes, regular arm		6.50	1.230		67	27		94	120
2840	Hold open arm		6.50	1.230		77	27		104	130
4000	Backcheck, overhead concealed, all sizes, regular arm		5.50	1.450		105	32		137	170
4040	Concealed arm		5	1.600		112	35		147	180
4400	Compact overhead, concealed, all sizes, regular arm		5.50	1.450		130	32		162	195
4440	Concealed arm		5	1.600		146	35		181	220
4800	Concealed in door, all sizes, regular arm		5.50	1.450		76	32		108	135
4840	Concealed arm		5	1.600		67	35		102	130
4900	Floor concealed, all sizes, single acting		2.20	3.640		82	80		162	220
4940	Double acting		2.20	3.640		105	80		185	245
5000	For cast aluminum cylinder, deduct					12			12	13.20
5040	For delayed action, add					12			12	13.20
5080	For fusible link arm, add					9			9	9.90
5120	For shock absorbing arm, add					17			17	18.70
5160	For spring power adjustment, add					18.50			18.50	20
6000	Closer-holder, hinge face mount, all sizes, exposed arm	1 Carp	6.50	1.230		75	27		102	125
7000	Electronic closer-holder, hinge facemount, concealed arm		5	1.600		115	35		150	185
7400	With built-in detector		5	1.600		340	35		375	430
9000	Minimum labor/equipment charge		4	2	Job		44		44	72

For expanded coverage of these items see *Means Interior Cost Data 1991*

087 300 | Weatherstripping/Seals

		CREW	DAILY OUTPUT	MAN-HOURS	UNIT	BARE COSTS MAT.	BARE COSTS LABOR	BARE COSTS EQUIP.	BARE COSTS TOTAL	TOTAL INCL O&P	
302	0010	**ASTRAGALS** One piece overlapping										**302**
	0400	Cadmium plated steel, flat, ³⁄₁₆″ x 2″	1 Carp	90	.089	L.F.	4.25	1.96		6.21	7.85	
	0600	Prime coated steel, flat, ⅛″ x 3″		90	.089		3.65	1.96		5.61	7.20	
	0800	Stainless steel, flat, ³⁄₃₂″ x 1-⅝″		90	.089		6.95	1.96		8.91	10.85	
	1000	Aluminum, flat, ⅛″ x 2″		90	.089		2.65	1.96		4.61	6.10	
	1200	Nail on, "T" extrusion		120	.067		.70	1.47		2.17	3.16	
	1300	Vinyl bulb insert		105	.076		.80	1.68		2.48	3.62	
	1600	Screw on, "T" extrusion		90	.089		2.65	1.96		4.61	6.10	
	1700	Vinyl insert		75	.107		2.25	2.35		4.60	6.30	
	2000	"L" extrusion, neoprene bulbs		75	.107		1.80	2.35		4.15	5.80	
	2100	Neoprene sponge insert		75	.107		4.30	2.35		6.65	8.55	
	2200	Magnetic		75	.107		6.60	2.35		8.95	11.10	
	2400	Spring hinged security seal, with cam		75	.107		4.60	2.35		6.95	8.90	
	2600	Spring loaded locking bolt, vinyl insert		45	.178		5.55	3.91		9.46	12.50	
	2800	Neoprene sponge strip, "Z" shaped, aluminum		60	.133		2.95	2.93		5.88	8.05	
	2900	Solid neoprene strip, nail on aluminum strip	↓	90	.089	↓	2.40	1.96		4.36	5.85	
	2910											
	3000	One piece stile protection										
	3020	Neoprene fabric loop, nail on aluminum strips	1 Carp	60	.133	L.F.	4.20	2.93		7.13	9.40	
	3110	Flush mounted aluminum extrusion, ½″ x 1-¼″		60	.133		2.30	2.93		5.23	7.30	
	3140	¾″ x 1-⅜″		60	.133		2.90	2.93		5.83	8	
	3160	1-⅛″ x 1-¾″		60	.133		4.65	2.93		7.58	9.90	
	3300	Mortise, ⁹⁄₁₆″ x ¾″		60	.133		2.45	2.93		5.38	7.50	
	3320	¹³⁄₁₆″ x 1-⅜″		60	.133		2.65	2.93		5.58	7.70	
	3600	Spring bronze strip, nail on type		105	.076		2.10	1.68		3.78	5.05	
	3620	Screw on, with retainer		75	.107		1.70	2.35		4.05	5.70	
	3800	Flexible stainless steel housing, pile insert, ½″ door		105	.076		4.75	1.68		6.43	7.95	
	3820	¾″ door		105	.076		5.35	1.68		7.03	8.60	
	4000	Extruded aluminum retainer, flush mount, pile insert		105	.076		1.65	1.68		3.33	4.55	
	4080	Mortise, felt insert		90	.089		2.95	1.96		4.91	6.45	
	4160	Mortise with spring, pile insert		90	.089		2.30	1.96		4.26	5.70	
	4400	Rigid vinyl retainer, mortise, pile insert		105	.076		1.65	1.68		3.33	4.55	
	4600	Wool pile filler strip, aluminum backing	↓	105	.076	↓	1.65	1.68		3.33	4.55	
	4610											
	5000	Two piece overlapping astragal, extruded aluminum retainer										
	5010	Pile insert	1 Carp	60	.133	L.F.	2.15	2.93		5.08	7.15	
	5020	Vinyl bulb insert		60	.133		1.40	2.93		4.33	6.35	
	5040	Vinyl flap insert		60	.133		4.35	2.93		7.28	9.55	
	5060	Solid neoprene flap insert		60	.133		4.20	2.93		7.13	9.40	
	5080	Hypalon rubber flap insert		60	.133		4.30	2.93		7.23	9.50	
	5090	Snap on cover, pile insert		60	.133		4.95	2.93		7.88	10.25	
	5400	Magnetic aluminum, surface mounted		60	.133		15.80	2.93		18.73	22	
	5500	Interlocking aluminum, ⅝″ x 1″ neoprene bulb insert		45	.178		2.35	3.91		6.26	8.95	
	5600	Adjustable aluminum, ⁹⁄₁₆″ x ²¹⁄₃₂″, pile insert	↓	45	.178		12.50	3.91		16.41	20	
	5790	For vinyl bulb, deduct					.32			.32	.35	
	5800	Magnetic, adjustable, ⁹⁄₁₆″ x ²¹⁄₃₂″	1 Carp	45	.178	↓	15.30	3.91		19.21	23	
	6000	Two piece stile protection										
	6010	Cloth backed rubber loop, 1″ gap, nail on aluminum strips	1 Carp	45	.178	L.F.	2.70	3.91		6.61	9.35	
	6040	Screw on aluminum strips		45	.178		4.20	3.91		8.11	11	
	6100	1-½″ gap, screw on aluminum extrusion		45	.178		3.80	3.91		7.71	10.55	
	6240	Vinyl fabric loop, slotted aluminum extrusion, 1″ gap		45	.178		1.35	3.91		5.26	7.85	
	6300	1-¼″ gap		45	.178	↓	3.80	3.91		7.71	10.55	
304	0010	**THRESHOLD** 3′ long door saddles, aluminum, minimum		20	.400	Ea.	22	8.80		30.80	39	**304**
	0100	Maximum		12	.667		68	14.65		82.65	99	
	0500	Bronze, minimum		20	.400		44	8.80		52.80	63	
	0600	Maximum		12	.667		120	14.65		134.65	155	
	0700	Rubber, ½″ thick, 5-½″ wide		20	.400		27	8.80		35.80	44	
	0800	2-¾″ wide	↓	20	.400	↓	12.50	8.80		21.30	28	

For expanded coverage of these items see *Means Interior Cost Data 1991*

		087 300 | Weatherstripping/Seals	CREW	DAILY OUTPUT	MAN-HOURS	UNIT	BARE COSTS MAT.	LABOR	EQUIP.	TOTAL	TOTAL INCL O&P	
304	9000	Minimum labor/equipment charge	1 Carp	4	2	Job		44		44	72	304
306	0010	WEATHERSTRIPPING Window, double hung, 3' x 5', zinc	1 Carp	7.20	1.110	Opng.	8.50	24		32.50	49	306
	0100	Bronze		7.20	1.110		15.50	24		39.50	57	
	0200	Vinyl V strip		7	1.140		3	25		28	44	
	0500	As above but heavy duty, zinc		4.60	1.740		10.50	38		48.50	74	
	0600	Bronze		4.60	1.740		18.50	38		56.50	83	
	1000	Doors, wood frame, interlocking, for 3' x 7' door, zinc		3	2.670		10	59		69	105	
	1100	Bronze		3	2.670		15	59		74	110	
	1300	6' x 7' opening, zinc		2	4		11.50	88		99.50	155	
	1400	Bronze	↓	2	4	↓	17	88		105	160	
	1700	Wood frame, spring type, bronze										
	1800	3' x 7' door	1 Carp	7.60	1.050	Opng.	6.90	23		29.90	45	
	1900	6' x 7' door	"	7	1.140	"	7.95	25		32.95	50	
	2200	Metal frame, spring type, bronze										
	2300	3' x 7' door	1 Carp	3	2.670	Opng.	30	59		89	130	
	2400	6' x 7' door	"	2.50	3.200		35	70		105	155	
	2500	For stainless steel, spring type, add				↓	133%					
	2510											
	2700	Metal frame, extruded sections, 3' x 7' door, aluminum	1 Carp	2	4	Opng.	30	88		118	175	
	2800	Bronze		2	4		68	88		156	220	
	3100	6' x 7' door, aluminum		1.20	6.670		36	145		181	280	
	3200	Bronze	↓	1.20	6.670	↓	80	145		225	325	
	3500	Threshold weatherstripping										
	3650	Door sweep, flush mounted, aluminum	1 Carp	25	.320	Ea.	8.50	7.05		15.55	21	
	3700	Vinyl		25	.320		9.80	7.05		16.85	22	
	5000	Garage door bottom weatherstrip, 12' aluminum, clear		14	.571		12.60	12.55		25.15	34	
	5010	Bronze		14	.571		50	12.55		62.55	76	
	5050	Bottom protection, 12' aluminum, clear		14	.571		17.95	12.55		30.50	40	
	5100	Bronze		14	.571	↓	62	12.55		74.55	89	
	9000	Minimum labor/equipment charge	↓	3	2.670	Job		59		59	96	

087 500 | Door/Window Acces.

			CREW	DAILY OUTPUT	MAN-HOURS	UNIT	MAT.	LABOR	EQUIP.	TOTAL	TOTAL INCL O&P	
506	0010	DOOR ACCESSORIES										506
	0020											
	1000	Knockers, brass, standard	1 Carp	16	.500	Ea.	32	11		43	53	
	1100	Deluxe		10	.800		100	17.60		117.60	140	
	4000	Security chain, standard		18	.444		5.50	9.80		15.30	22	
	4100	Deluxe		18	.444	↓	33	9.80		42.80	52	
	9000	Minimum labor/equipment charge	↓	6	1.330	Job		29		29	48	

088 | Glazing

		088 100 | Glass	CREW	DAILY OUTPUT	MAN-HOURS	UNIT	BARE COSTS MAT.	LABOR	EQUIP.	TOTAL	TOTAL INCL O&P	
112	0010	CURTAIN WALL See division 089-200										112
118	0010	FLOAT GLASS 3/16" thick, clear, plain	2 Glaz	130	.123	S.F.	3.05	2.78		5.83	7.75	118
	0200	Tempered, clear		130	.123		5.80	2.78		8.58	10.80	
	0300	Tinted		130	.123		6.95	2.78		9.73	12.05	
	0600	1/4" thick, clear, plain	↓	120	.133	↓	3.05	3.01		6.06	8.10	

For expanded coverage of these items see *Means Interior Cost Data 1991*

088 100 | Glass

			CREW	DAILY OUTPUT	MAN-HOURS	UNIT	MAT.	LABOR	EQUIP.	TOTAL	TOTAL INCL O&P	
118	0700	Tinted	2 Glaz	120	.133	S.F.	4.10	3.01		7.11	9.30	118
	0800	Tempered, clear		120	.133		5.45	3.01		8.46	10.75	
	0900	Tinted		120	.133		6.90	3.01		9.91	12.35	
	1200	5/16" thick, clear, plain		100	.160		5.90	3.61		9.51	12.20	
	1300	Tempered, clear		100	.160		9.85	3.61		13.46	16.55	
	1600	3/8" thick, clear, plain		75	.213		6.75	4.81		11.56	15.05	
	1700	Tinted		75	.213		8.65	4.81		13.46	17.15	
	1800	Tempered, clear		75	.213		14.10	4.81		18.91	23	
	1900	Tinted		75	.213		16.35	4.81		21.16	26	
	2200	1/2" thick, clear, plain		55	.291		11.05	6.55		17.60	23	
	2300	Tinted		55	.291		12.65	6.55		19.20	24	
	2400	Tempered, clear		55	.291		21.75	6.55		28.30	34	
	2500	Tinted		55	.291		24.65	6.55		31.20	38	
	2800	5/8" thick, clear, plain		45	.356		12.80	8		20.80	27	
	2900	Tempered, clear		45	.356		24.15	8		32.15	39	
	3200	3/4" thick, clear, plain		35	.457		14.75	10.30		25.05	33	
	3300	Tempered, clear		35	.457		28.95	10.30		39.25	48	
	3600	1" thick, clear, plain		30	.533		26.35	12.05		38.40	48	
	4000	For low emissivity coating, add to above					100%					
	9000	Minimum labor/equipment charge	1 Glaz	2	4	Job		90		90	145	
120	0010	FULL VISION Window system with 3/4" glass mullions, 10' high	H-2	120	.200	S.F.	17.15	4.17		21.32	26	120
	0100	10' to 20' high, minimum		100	.240		22.65	5		27.65	33	
	0150	Average		90	.267		26.50	5.55		32.05	38	
	0200	Maximum		70	.343		33.50	7.15		40.65	48	
	9000	Minimum labor/equipment charge	1 Glaz	2	4	Job		90		90	145	
128	0010	GLAZING VARIABLES										128
	0500	For high rise glazing, from exterior, add per S.F. per story				S.F.		.08		.08	.09	
	0600	For glass replacement, add				"		100%				
	0700	For gasket settings, add	ⓘ03			L.F.	2.35			2.35	2.59	
	0800	For concrete reglet settings, add				S.F.	20%	25%				
	0900	For sloped glazing, add				"		25%				
	2000	Fabrication, polished edges, 1/4" thick				Inch	.22			.22	.24	
	2100	1/2" thick					.34			.34	.37	
	2500	Mitered edges, 1/4" thick					.78			.78	.86	
	2600	1/2" thick					1.10			1.10	1.21	
132	0010	INSULATING GLASS 2 lites 1/8" float, 1/2" thk, under 15 S.F.										132
	0020	Clear	2 Glaz	95	.168	S.F.	5.90	3.80		9.70	12.50	
	0100	Tinted	"	95	.168	"	7.75	3.80		11.55	14.55	
	0110											
	0200	2 lites 3/16" float, for 5/8" thk unit, 15 to 30 S.F., clear	2 Glaz	90	.178	S.F.	7.15	4.01		11.16	14.20	
	0400	1" thick, double glazed, 1/4" float, 30 to 70 S.F., clear		75	.213		8.75	4.81		13.56	17.25	
	0500	Tinted	ⓘ03	75	.213		9.50	4.81		14.31	18.10	
	2000	Both lites, light & heat reflective		85	.188		14.50	4.24		18.74	23	
	2500	Heat reflective, film inside, 1" thick unit, clear		85	.188		12.85	4.24		17.09	21	
	2600	Tinted		85	.188		13.60	4.24		17.84	22	
	3000	Film on weatherside, clear, 1/2" thick unit		95	.168		9.70	3.80		13.50	16.70	
	3100	5/8" thick unit		90	.178		10.95	4.01		14.96	18.40	
	3200	1" thick unit		85	.188		12.60	4.24		16.84	21	
	3350	Minimum	1 Glaz	50	.160		5.50	3.61		9.11	11.75	
	3360	Maximum	"	25	.320		7.95	7.20		15.15	20	
	3370	Reflective or tinted, add					1.80			1.80	1.98	
	9000	Minimum labor/equipment charge	1 Glaz	2	4	Job		90		90	145	
160	0010	REFLECTIVE GLASS 1/4" float with fused metallic oxide, tinted	2 Glaz	115	.139	S.F.	6.35	3.14		9.49	11.95	160
	0500	1/4" float glass with reflective applied coating		115	.139		4.90	3.14		8.04	10.35	
	2000	Solar film on glass, not including glass, minimum		180	.089		1.15	2		3.15	4.44	
	2050	Maximum		225	.071		2.35	1.60		3.95	5.15	

For expanded coverage of these items see *Means Interior Cost Data 1991*

139

088 100	Glass	CREW	DAILY OUTPUT	MAN-HOURS	UNIT	BARE COSTS				TOTAL INCL O&P	
						MAT.	LABOR	EQUIP.	TOTAL		
176	0010	WINDOW GLASS Clear float, stops, putty bed, ⅛″ thick	2 Glaz	480	.033	S.F.	2.85	.75		3.60	4.33	**176**
	0500	³⁄₁₆″ thick, clear		480	.033		2.95	.75		3.70	4.44	
	0600	Tinted		480	.033		3.60	.75		4.35	5.15	
	0700	Tempered		480	.033		5.65	.75		6.40	7.40	
	2000	Replace broken window lite, ⅛″ glass (9 S.F. maximum)	1 Glaz	48	.167		2.75	3.76		6.51	9	
	2100	¼″ plate (16 S.F. maximum)	"	48	.167		3	3.76		6.76	9.25	
	9000	Minimum labor/equipment charge	2 Glaz	5	3.200	Job		72		72	115	

088 400	Plastic Glazing	CREW	DAILY OUTPUT	MAN-HOURS	UNIT	BARE COSTS				TOTAL INCL O&P	
						MAT.	LABOR	EQUIP.	TOTAL		
408	0010	POLYCARBONATE Clear, masked, cut sheets, ⅛″ thick	2 Glaz	170	.094	S.F.	3.15	2.12		5.27	6.85	**408**
	0500	³⁄₁₆″ thick		165	.097		4.25	2.19		6.44	8.15	
	1000	¼″ thick		155	.103		5.15	2.33		7.48	9.35	
	1500	⅜″ thick		150	.107		7.85	2.41		10.26	12.45	
	9000	Minimum labor/equipment charge	1 Glaz	2	4	Job		90		90	145	

089 | Glazed Curtain Wall

089 200	Glazed Curtain Wall	CREW	DAILY OUTPUT	MAN-HOURS	UNIT	BARE COSTS				TOTAL INCL O&P	
						MAT.	LABOR	EQUIP.	TOTAL		
204	0010	TUBE FRAMING For window walls and store fronts, aluminum, stock										**204**
	0020											
	0050	Plain tube frame, mill finish, 1-¾″ x 1-¾″	2 Glaz	90	.178	L.F.	2.90	4.01		6.91	9.55	
	0150	1-¾″ x 4″		90	.178		6.40	4.01		10.41	13.40	
	0200	1-¾″ x 4-½″		90	.178		5.30	4.01		9.31	12.20	
	0250	2″ x 6″		90	.178		9	4.01		13.01	16.25	
	0350	4″ x 4″		90	.178		8	4.01		12.01	15.15	
	0400	4-½″ x 4-½″		90	.178		10.35	4.01		14.36	17.75	
	0450	Glass bead		240	.067		1.75	1.50		3.25	4.31	
	1000	Flush tube frame, mill finish, ¼″ glass, 1-¾″ x 4″, open header		80	.200		4.25	4.51		8.76	11.85	
	1050	Open sill		82	.195		3.93	4.40		8.33	11.30	
	1100	Closed back header		83	.193		5.46	4.35		9.81	12.90	
	1150	Closed back sill		85	.188		5.25	4.24		9.49	12.50	
	1200	Vertical mullion, one piece		75	.213		5.40	4.81		10.21	13.55	
	1250	Two piece		73	.219		6.80	4.94		11.74	15.30	
	1300	90° or 180° vertical corner post		75	.213		10.30	4.81		15.11	18.95	
	1400	1-¾″ x 4-½″, open header		80	.200		4.63	4.51		9.14	12.25	
	1450	Open sill		82	.195		4.55	4.40		8.95	12	
	1500	Closed back header		83	.193		5.90	4.35		10.25	13.40	
	1550	Closed back sill		85	.188		6.10	4.24		10.34	13.45	
	1600	Vertical mullion, one piece		75	.213		5.90	4.81		10.71	14.10	
	1650	Two piece		73	.219		7.35	4.94		12.29	15.90	
	1700	90° or 180° vertical corner post		75	.213		12.40	4.81		17.21	21	
	2000	Flush tube frame, mill fin. for ins. glass, 2″ x 4-½″, open header		75	.213		5	4.81		9.81	13.15	
	2050	Open sill		77	.208		4.80	4.69		9.49	12.70	
	2100	Closed back header		78	.205		7.22	4.63		11.85	15.30	
	2150	Closed back sill		80	.200		6.40	4.51		10.91	14.20	
	2200	Vertical mullion, one piece		70	.229		6.35	5.15		11.50	15.15	
	2250	Two piece		68	.235		7.70	5.30		13	16.90	
	2300	90° or 180° vertical corner post		70	.229		12.70	5.15		17.85	22	
	5000	Flush tube frame, mill fin., thermal brk., 2-¼″ x 4-½″, open header		74	.216		6.30	4.88		11.18	14.65	
	5050	Open sill		75	.213		6.70	4.81		11.51	15	

For expanded coverage of these items see *Means Interior Cost Data 1991*

089 | Glazed Curtain Wall

089 200 | Glazed Curtain Wall

			CREW	DAILY OUTPUT	MAN-HOURS	UNIT	BARE COSTS MAT.	BARE COSTS LABOR	BARE COSTS EQUIP.	BARE COSTS TOTAL	TOTAL INCL O&P	
204	5100	Vertical mullion, one piece	2 Glaz	69	.232	L.F.	7.35	5.25		12.60	16.40	204
	5150	Two piece		67	.239		8.65	5.40		14.05	18.05	
	5200	90° or 180° vertical corner post		69	.232		12	5.25		17.25	21	
	5300	Door stop (snap in)	↓	380	.042	↓	1.25	.95		2.20	2.88	
	7000	For joints, 90°, clip type, add				Ea.	13.33			13.33	14.65	
	7050	Screw spline joint, add					9.33			9.33	10.25	
	7100	For joint other than 90°, add				↓	21			21	23	
	8000	For bronze finish, add				L.F.	18%					
	8020	For black finish, add					27%					
	8050	For stainless steel, add					125%					
	8100	For monumental grade, add					50%					
	8150	For steel stiffener, add	2 Glaz	200	.080	↓	5.50	1.80		7.30	8.90	
	8200	For 2 to 5 stories, add per story				Story		5%				
	9000	Minimum labor/equipment charge	2 Glaz	2	8	Job		180		180	285	
206	0010	WINDOW WALLS Aluminum, stock, including glazing, minimum	H-2	160	.150	S.F.	20	3.13		23.13	27	206
	0050	Average		140	.171		25	3.58		28.58	33	
	0100	Maximum (105)	↓	110	.218	↓	75	4.55		79.55	90	
	0200											
	0500	For translucent sandwich wall systems, see div. 074-104										
	0850	Cost of the above walls depends on material,										
	0860	finish, repetition, and size of units.										
	0870	The larger the opening, the lower the S.F. cost										
	1200	Double glazed acoustical window wall for airports,										
	1220	including 1" thick glass with 2" x 4-½" tube frame	H-2	40	.600	S.F.	39	12.50		51.50	63	

091 | Metal Support Systems

091 300 | Suspension Systems

			CREW	DAILY OUTPUT	MAN-HOURS	UNIT	BARE COSTS MAT.	BARE COSTS LABOR	BARE COSTS EQUIP.	BARE COSTS TOTAL	TOTAL INCL O&P	
304	0010	SUSPENSION SYSTEMS For boards and tile										304
	0050	Class A suspension system, T bar, 2' x 4' grid	1 Carp	800	.010	S.F.	.37	.22		.59	.77	
	0300	2' x 2' grid		650	.012		.49	.27		.76	.98	
	0400	Concealed Z bar suspension system, 12" module		520	.015		.56	.34		.90	1.17	
	0600	1-½" carrier channels, 4' O.C., add	↓	470	.017	↓	.14	.37		.51	.77	
	0700	Carrier channels for ceilings with										
	0900	recessed lighting fixtures, add				S.F.	.30			.30	.33	
	9000	Minimum labor/equipment charge	1 Carp	5	1.600	Job		35		35	57	

092 | Lath, Plaster and Gypsum Board

092 050 | Furring & Lathing

			CREW	DAILY OUTPUT	MAN-HOURS	UNIT	BARE COSTS MAT.	BARE COSTS LABOR	BARE COSTS EQUIP.	BARE COSTS TOTAL	TOTAL INCL O&P	
054	0010	FURRING Beams & columns, ¾" galvanized channels,										054
	0030	12" O.C.	1 Lath	155	.052	S.F.	.25	1.13		1.38	2.05	
	0050	16" O.C.		170	.047		.20	1.03		1.23	1.84	
	0070	24" O.C.		185	.043		.16	.95		1.11	1.66	
	0100	Ceilings, on steel, ¾" channels, galvanized, 12" O.C.		210	.038		.25	.84		1.09	1.59	
	0300	16" O.C.	↓	290	.028	↓	.20	.61		.81	1.17	

For expanded coverage of these items see *Means Interior Cost Data 1991*

092 050 | Furring & Lathing

			CREW	DAILY OUTPUT	MAN-HOURS	UNIT	BARE COSTS MAT.	LABOR	EQUIP.	TOTAL	TOTAL INCL O&P	
054	0400	24″ O.C.	1 Lath	420	.019	S.F.	.16	.42		.58	.83	054
	0600	1-½″ channels, galvanized, 12″ O.C.		190	.042		.34	.92		1.26	1.82	
	0700	16″ O.C.		260	.031		.27	.68		.95	1.36	
	0900	24″ O.C.		390	.021		.22	.45		.67	.95	
	1000	Walls, galvanized, ¾″ channels, 12″ O.C.		235	.034		.25	.75		1	1.45	
	1200	16″ O.C.		265	.030		.20	.66		.86	1.26	
	1300	24″ O.C.		350	.023		.16	.50		.66	.96	
	1500	1-½″ channels, galvanized, 12″ O.C.,		210	.038		.34	.84		1.18	1.69	
	1600	16″ O.C.		240	.033		.27	.73		1	1.44	
	1800	24″ O.C.		305	.026		.22	.58		.80	1.15	
	8000	Suspended ceilings, including carriers										
	8200	1-½″ carriers, 24″ O.C. with:										
	8300	¾″ channels, 16″ O.C.	1 Lath	165	.048	S.F.	.37	1.06		1.43	2.08	
	8320	24″ O.C.		200	.040		.33	.88		1.21	1.74	
	8400	1-½″ channels, 16″ O.C.		155	.052		.44	1.13		1.57	2.26	
	8420	24″ O.C.		190	.042		.39	.92		1.31	1.88	
	8600	2″ carriers, 24″ O.C. with:										
	8700	¾″ channels, 16″ O.C.	1 Lath	155	.052	S.F.	.39	1.13		1.52	2.21	
	8720	24″ O.C.		190	.042		.35	.92		1.27	1.83	
	8800	1-½″ channels, 16″ O.C.		145	.055		.46	1.21		1.67	2.41	
	8820	24″ O.C.		180	.044		.41	.98		1.39	1.98	
	9000	Minimum labor/equipment charge		4	2	Job		44		44	69	
056	0010	GYPSUM LATH Plain or perforated, nailed, ⅜″ thick		85	.094	S.Y.	3.27	2.07		5.34	6.85	056
	0100	½″ thick, nailed		80	.100		3.54	2.20		5.74	7.35	
	0300	Clipped to steel studs, ⅜″ thick		75	.107		3.56	2.34		5.90	7.60	
	0400	½″ thick		70	.114		3.83	2.51		6.34	8.15	
	0600	Firestop gypsum base, to steel studs, ⅜″ thick		70	.114		3.72	2.51		6.23	8.05	
	0700	½″ thick		65	.123		3.98	2.70		6.68	8.60	
	0900	Foil back, to steel studs, ⅜″ thick		75	.107		3.62	2.34		5.96	7.65	
	1000	½″ thick		70	.114		3.88	2.51		6.39	8.20	
	1500	For ceiling installations, add		2,150	.004	S.F.		.08		.08	.13	
	1600	For columns and beams, add		1,750	.005	″		.10		.10	.16	
	9000	Minimum labor/equipment charge		4.25	1.880	Job		41		41	65	
058	0011	METAL LATH										058
	0020											
	3600	2.5 lb. diamond painted, on wood framing, on walls	1 Lath	85	.094	S.Y.	1.58	2.07		3.65	4.98	
	3700	On ceilings		75	.107		1.58	2.34		3.92	5.40	
	4200	3.4 lb. diamond painted, wired to steel framing, on walls		75	.107		1.78	2.34		4.12	5.65	
	4300	On ceilings		60	.133		1.78	2.93		4.71	6.55	
	5100	Rib lath, painted, wired to steel, on walls, 2.75 lb.		75	.107		1.70	2.34		4.04	5.55	
	5200	3.4 lb.		70	.114		1.95	2.51		4.46	6.10	
	5700	Suspended ceiling system, incl. 3.4 lb. diamond lath, painted		15	.533		8.10	11.70		19.80	27	
	5800	Galvanized		15	.533		8.30	11.70		20	27	
	9000	Minimum labor/equipment charge		4.25	1.880	Job		41		41	65	

092 100 | Gypsum Plaster

			CREW	DAILY OUTPUT	MAN-HOURS	UNIT	MAT.	LABOR	EQUIP.	TOTAL	TOTAL INCL O&P	
108	0010	GYPSUM PLASTER 80# bag, less than 1 ton				Bag	12.20			12.20	13.40	108
	0020											
	0300	2 coats, no lath included, on walls	J-1 ⑪⓪	105	.381	S.Y.	3.02	7.70	.32	11.04	16	
	0400	On ceilings		92	.435		3.02	8.80	.36	12.18	17.80	
	0900	3 coats, no lath included, on walls		87	.460		4.20	9.30	.38	13.88	19.95	
	1000	On ceilings		78	.513		4.20	10.40	.43	15.03	22	
	1600	For irregular or curved surfaces, add					30%					
	1800	For columns and beams, add					50%					
	9000	Minimum labor/equipment charge	1 Plas	1	8	Job		175		175	280	

For expanded coverage of these items see *Means Interior Cost Data 1991*

092 100 | Gypsum Plaster

			CREW	DAILY OUTPUT	MAN-HOURS	UNIT	MAT.	LABOR	EQUIP.	TOTAL	TOTAL INCL O&P	
116	0010	PERLITE OR VERMICULITE PLASTER 100 lb. bags Under 200 bags				Bag	12.35			12.35	13.60	116
	0200											
	0300	2 coats, no lath included, on walls	J-1	92	.435	S.Y.	3.44	8.80	.36	12.60	18.25	
	0400	On ceilings		79	.506		3.44	10.25	.42	14.11	21	
	0900	3 coats, no lath included, on walls		74	.541		5.40	10.95	.45	16.80	24	
	1000	On ceilings		63	.635		5.40	12.85	.53	18.78	27	
	1700	For irregular or curved surfaces, add to above					30%					
	1800	For columns and beams, add to above					50%					
	1900	For soffits, add to ceiling prices						40%				
	9000	Minimum labor/equipment charge	1 Plas	1	8	Job		175		175	280	

092 150 | Veneer Plaster

			CREW	DAILY OUTPUT	MAN-HOURS	UNIT	MAT.	LABOR	EQUIP.	TOTAL	TOTAL INCL O&P	
154	0010	THIN COAT Plaster, 1 coat veneer, not incl. lath	J-1	3,600	.011	S.F.	.09	.22	.01	.32	.47	154
	1000	In 50 lb. bags				Bag	8.50			8.50	9.35	

092 300 | Aggregate Coatings

			CREW	DAILY OUTPUT	MAN-HOURS	UNIT	MAT.	LABOR	EQUIP.	TOTAL	TOTAL INCL O&P	
304	0010	STUCCO 3 coats 1" thick, float finish, with mesh, on wood frame	J-2	52	.923	S.Y.	5.80	18.95	.64	25.39	37	304
	0100	On masonry construction	J-1	55	.727		1.91	14.70	.60	17.21	26	
	0300	For trowel finish, add	1 Plas	170	.047			1.03		1.03	1.64	
	0400	For ¾" thick, on masonry, deduct	J-1	880	.045		.70	.92	.04	1.66	2.28	
	0600	For coloring and special finish, add, minimum					.30			.30	.33	
	0700	Maximum					1.10			1.10	1.21	
	0900	For soffits, add	J-2	155	.310		1.70	6.35	.21	8.26	12.25	
	0910											
	1000	Exterior stucco, with bonding agent, 3 coats, on walls	J-1	200	.200	S.Y.	2.85	4.05	.17	7.07	9.80	
	1200	Ceilings		180	.222		2.85	4.50	.18	7.53	10.55	
	1300	Beams		80	.500		2.85	10.10	.42	13.37	19.80	
	1500	Columns		100	.400		2.85	8.10	.33	11.28	16.45	
	1550	Minimum labor/equipment charge	1 Plas	1	8	Job		175		175	280	
	1560											
	1600	Mesh, painted, nailed to wood, 1.8 lb.	1 Lath	60	.133	S.Y.	2.52	2.93		5.45	7.35	
	1800	3.6 lb.		55	.145		2.89	3.19		6.08	8.20	
	1900	Wired to steel, painted, 1.8 lb.		53	.151		2.52	3.31		5.83	7.95	
	2100	3.6 lb.		50	.160		2.89	3.51		6.40	8.70	
	9000	Minimum labor/equipment charge		4	2	Job		44		44	69	

092 600 | Gypsum Board Systems

			CREW	DAILY OUTPUT	MAN-HOURS	UNIT	MAT.	LABOR	EQUIP.	TOTAL	TOTAL INCL O&P	
602	0010	BLUEBOARD For use with thin coat										602
	0100	plaster application (see 092-154)										
	1000	⅜" thick, on walls or ceilings, standard, no finish included	2 Carp	1,900	.008	S.F.	.16	.19		.35	.48	
	1100	With thin coat plaster finish		875	.018		.25	.40		.65	.93	
	1400	On beams, columns, or soffits, standard, no finish included		675	.024		.26	.52		.78	1.14	
	1450	With thin coat plaster finish		475	.034		.35	.74		1.09	1.59	
	3000	½" thick, on walls or ceilings, standard, no finish included		1,900	.008		.17	.19		.36	.49	
	3100	With thin coat plaster finish		875	.018		.26	.40		.66	.94	
	3300	Fire resistant, no finish included		1,900	.008		.21	.19		.40	.53	
	3400	With thin coat plaster finish		875	.018		.30	.40		.70	.99	
	3450	On beams, columns, or soffits, standard, no finish included		675	.024		.27	.52		.79	1.15	
	3500	With thin coat plaster finish		475	.034		.36	.74		1.10	1.61	
	3700	Fire resistant, no finish included		675	.024		.31	.52		.83	1.19	
	3800	With thin coat plaster finish		475	.034		.40	.74		1.14	1.65	
	5000	⅝" thick, on walls or ceilings, fire resistant, no finish included		1,900	.008		.21	.19		.40	.53	
	5100	With thin coat plaster finish		875	.018		.30	.40		.70	.99	
	5500	On beams, columns, or soffits, no finish included		675	.024		.31	.52		.83	1.19	
	5600	With thin coat plaster finish		475	.034		.40	.74		1.14	1.65	
	6000	For high ceilings, over 8' high, add		3,060	.005		.08	.12		.20	.28	
	6500	For over 3 stories high, add per story		6,100	.003		.04	.06		.10	.14	

For expanded coverage of these items see *Means Interior Cost Data 1991*

092 600	Gypsum Board Systems	CREW	DAILY OUTPUT	MAN-HOURS	UNIT	BARE COSTS				TOTAL INCL O&P	
						MAT.	LABOR	EQUIP.	TOTAL		
602 9000	Minimum labor/equipment charge	1 Carp	2	4	Job		88		88	145	**602**
608 0010	DRYWALL Gypsum plasterboard, nailed or screwed to studs,										**608**
0100	unless otherwise noted										
0150	⅜″ thick, on walls, standard, no finish included	2 Carp	2,000	.008	S.F.	.16	.18		.34	.46	
0200	On ceilings, standard, no finish included		1,800	.009		.16	.20		.36	.50	
0250	On beams, columns, or soffits, no finish included	↓	675	.024	↓	.26	.52		.78	1.14	
0270											
0300	½″ thick, on walls, standard, no finish included	2 Carp	2,000	.008	S.F.	.17	.18		.35	.47	
0350	Taped and finished		965	.017		.22	.36		.58	.84	
0400	Fire resistant, no finish included		2,000	.008		.21	.18		.39	.52	
0450	Taped and finished		965	.017		.26	.36		.62	.88	
0500	Water resistant, no finish included		2,000	.008		.27	.18		.45	.58	
0550	Taped and finished		965	.017		.32	.36		.68	.95	
0600	Prefinished, vinyl, clipped to studs	↓	900	.018	↓	.55	.39		.94	1.24	
0650											
1000	On ceilings, standard, no finish included	2 Carp	1,800	.009	S.F.	.17	.20		.37	.51	
1050	Taped and finished		765	.021		.22	.46		.68	.99	
1100	Fire resistant, no finish included		1,800	.009		.21	.20		.41	.55	
1150	Taped and finished		765	.021		.26	.46		.72	1.04	
1200	Water resistant, no finish included		1,800	.009		.27	.20		.47	.62	
1250	Taped and finished		765	.021		.32	.46		.78	1.10	
1500	On beams, columns, or soffits, standard, no finish included		675	.024		.27	.52		.79	1.15	
1550	Taped and finished		475	.034		.32	.74		1.06	1.56	
1600	Fire resistant, no finish included		675	.024		.31	.52		.83	1.19	
1650	Taped and finished		475	.034		.36	.74		1.10	1.61	
1700	Water resistant, no finish included		675	.024		.37	.52		.89	1.26	
1750	Taped and finished		475	.034		.42	.74		1.16	1.67	
2000	⅝″ thick, on walls, standard, no finish included		2,000	.008		.21	.18		.39	.52	
2050	Taped and finished		965	.017		.26	.36		.62	.88	
2100	Fire resistant, no finish included		2,000	.008		.23	.18		.41	.54	
2150	Taped and finished		965	.017		.28	.36		.64	.90	
2200	Water resistant, no finish included		2,000	.008		.31	.18		.49	.63	
2250	Taped and finished		965	.017		.36	.36		.72	.99	
2300	Prefinished, vinyl, clipped to studs	↓	900	.018	↓	.62	.39		1.01	1.32	
2350											
3000	On ceilings, standard, no finish included	2 Carp	1,800	.009	S.F.	.21	.20		.41	.55	
3050	Taped and finished		765	.021		.26	.46		.72	1.04	
3100	Fire resistant, no finish included		1,800	.009		.23	.20		.43	.57	
3150	Taped and finished		765	.021		.28	.46		.74	1.06	
3200	Water resistant, no finish included		1,800	.009		.31	.20		.51	.66	
3250	Taped and finished		765	.021		.36	.46		.82	1.15	
3500	On beams, columns, or soffits, standard, no finish included		675	.024		.31	.52		.83	1.19	
3550	Taped and finished		475	.034		.36	.74		1.10	1.61	
3600	Fire resistant, no finish included		675	.024		.33	.52		.85	1.21	
3650	Taped and finished		475	.034		.38	.74		1.12	1.63	
3700	Water resistant, no finish included		675	.024		.41	.52		.93	1.30	
3750	Taped and finished		475	.034		.46	.74		1.20	1.72	
4000	Fireproofing, beams or columns, 2 layers, ½″ thick, incl finish		330	.048		.52	1.07		1.59	2.31	
4050	⅝″ thick		300	.053		.56	1.17		1.73	2.53	
4100	3 layers, ½″ thick		225	.071		.78	1.56		2.34	3.41	
4150	⅝″ thick		210	.076		.84	1.68		2.52	3.66	
5050	For 1″ thick coreboard on columns	↓	480	.033		.50	.73		1.23	1.75	
5100	For foil-backed board, add					.07			.07	.08	
5200	For high ceilings, over 8′ high, add	2 Carp	3,060	.005		.08	.12		.20	.28	
5270	For textured spray, add	2 Lath	1,600	.010		.10	.22		.32	.45	
5300	For over 3 stories high, add per story	2 Carp	6,100	.003	↓	.04	.06		.10	.14	
5350	For finishing corners, inside or outside, add	"	1,100	.015	L.F.	.05	.32		.37	.58	

For expanded coverage of these items see *Means Interior Cost Data 1991*

		092 600	Gypsum Board Systems	CREW	DAILY OUTPUT	MAN-HOURS	UNIT	BARE COSTS				TOTAL INCL O&P	
								MAT.	LABOR	EQUIP.	TOTAL		
608	5500		For acoustical sealant, add per bead	1 Carp	500	.016	L.F.	.02	.35		.37	.60	608
	5550		Sealant, 1 quart tube				Ea.	4			4	4.40	
	5600		Sound deadening board, ¼″ gypsum	2 Carp	1,800	.009	S.F.	.15	.20		.35	.48	
	5650		½″ wood fiber	″	1,800	.009	″	.20	.20		.40	.54	
	9000		Minimum labor/equipment charge	1 Carp	2	4	Job		88		88	145	
612	0010		METAL STUDS, DRYWALL Partitions, 10′ high, with runners										612
	1800												
	2000		Non-load bearing, galvanized, 25 ga. 1-⅝″, 16″ O.C.	1 Carp	450	.018	S.F.	.23	.39		.62	.89	
	2100		24″ O.C.		520	.015		.19	.34		.53	.76	
	2200		2-½″ wide, 16″ O.C.		440	.018		.27	.40		.67	.95	
	2250		24″ O.C.		510	.016		.22	.35		.57	.81	
	2300		3-⅝″ wide, 16″ O.C. (108)		430	.019		.31	.41		.72	1.01	
	2350		24″ O.C.		500	.016		.26	.35		.61	.86	
	2400		4″ wide, 16″ O.C.		420	.019		.35	.42		.77	1.07	
	2450		24″ O.C.		490	.016		.29	.36		.65	.91	
	2500		6″ wide, 16″ O.C.		410	.020		.45	.43		.88	1.20	
	2550		24″ O.C.		480	.017		.37	.37		.74	1.01	
	2600		20 ga. studs, 1-⅝″ wide, 16″ O.C.		450	.018		.41	.39		.80	1.09	
	2650		24″ O.C.		520	.015		.34	.34		.68	.93	
	2700		2-½″ wide, 16″ O.C.		440	.018		.47	.40		.87	1.17	
	2750		24″ O.C.		510	.016		.39	.35		.74	.99	
	2800		3-⅝″ wide, 16″ O.C.		430	.019		.56	.41		.97	1.28	
	2850		24″ O.C.		500	.016		.46	.35		.81	1.08	
	2900		4″ wide, 16″ O.C.		420	.019		.59	.42		1.01	1.33	
	2950		24″ O.C.		490	.016		.49	.36		.85	1.13	
	3000		6″ wide, 16″ O.C.		410	.020		.74	.43		1.17	1.51	
	3050		24″ O.C.		480	.017		.61	.37		.98	1.27	
	4000		LB studs, light ga. structural, galv., 18 ga., 2-½″, 16″ O.C.		200	.040		.88	.88		1.76	2.40	
	4100		24″ O.C.		240	.033		.73	.73		1.46	2	
	4200		3-⅝″ wide, 16″ O.C. (108)		190	.042		1.01	.93		1.94	2.62	
	4250		24″ O.C.		230	.035		.84	.77		1.61	2.17	
	4300		4″ wide, 16″ O.C.		180	.044		1.07	.98		2.05	2.77	
	4350		24″ O.C.		220	.036		.88	.80		1.68	2.27	
	4400		6″ wide, 16″ O.C.		170	.047		1.29	1.04		2.33	3.11	
	4450		24″ O.C.		210	.038		1.07	.84		1.91	2.55	
	4600		16 ga. studs, 2-½″, 16″ O.C.		180	.044		1.10	.98		2.08	2.81	
	4650		24″ O.C.		220	.036		.92	.80		1.72	2.32	
	4700		3-⅝″ wide, 16″ O.C.		170	.047		1.24	1.04		2.28	3.05	
	4750		24″ O.C.		210	.038		1.03	.84		1.87	2.50	
	4800		4″ wide, 16″ O.C.		160	.050		1.33	1.10		2.43	3.26	
	4850		24″ O.C.		200	.040		1.11	.88		1.99	2.66	
	4900		6″ wide, 16″ O.C.		150	.053		1.64	1.17		2.81	3.72	
	4950		24″ O.C.		190	.042		1.36	.93		2.29	3.01	
	9000		Minimum labor/equipment charge		4	2	Job		44		44	72	
624	0010		SHAFT WALL Cavity type on 25 ga. J track & C-H studs, 24″ O.C.										624
	0030		1″ thick coreboard wall liner on shaft side										
	0040		2-hour assembly with double layer										
	0060		⅝″ fire rated gypsum board on room side	2 Carp	220	.073	S.F.	1.27	1.60		2.87	4.01	
	0100		3-hour assembly with triple layer										
	0300		⅝″ fire rated gypsum board on room side	2 Carp	180	.089	S.F.	1.50	1.96		3.46	4.84	
	0400		4-hour assembly, 1″ coreboard, ⅝″ fire rated gypsum board										
	0600		and ¾″ galv. metal furring channels, 24″ O.C., with										
	0700		Double layer ⅝″ fire rated gypsum board on room side	2 Carp	110	.145	S.F.	2.05	3.20		5.25	7.50	
	0800												
	0900		For taping & finishing, add per side	1 Carp	1,050	.008	S.F.	.05	.17		.22	.33	
	1000		For insulation, see Div. 072										

For expanded coverage of these items see *Means Interior Cost Data 1991*

092 | Lath, Plaster and Gypsum Board

092 800 | Drywall Accessories

804			CREW	DAILY OUTPUT	MAN-HOURS	UNIT	MAT.	LABOR	EQUIP.	TOTAL	TOTAL INCL O&P	804
0010	ACCESSORIES, DRYWALL Casing bead, galvanized steel		1 Carp	2.90	2.760	C.L.F.	13	61		74	115	
0100		Vinyl		2.90	2.760		19	61		80	120	
0300		Corner bead, galvanized steel, 1″ x 1″		4	2		8	44		52	81	
0400		1-¼″ x 1-¼″		3.50	2.290		10.50	50		60.50	94	
0600		Vinyl corner bead		4	2		13	44		57	86	
0700		Door casing, vinyl, for 2″ wall systems		2.50	3.200		22	70		92	140	
0900		Furring channel, galv. steel, ⅞″ deep, standard		2.60	3.080		15.50	68		83.50	130	
1000		Resilient		2.60	3.080		16.10	68		84.10	130	
1100		J trim, galvanized steel, ½″ wide		3	2.670		14.25	59		73.25	110	
1120		⅝″ wide		3	2.670		14.25	59		73.25	110	
1500		Z stud, galvanized steel, 1-½″ wide		2.60	3.080		22	68		90	135	
9000		Minimum labor/equipment charge		3	2.670	Job		59		59	96	

093 | Tile

093 100 | Ceramic Tile

102			CREW	DAILY OUTPUT	MAN-HOURS	UNIT	MAT.	LABOR	EQUIP.	TOTAL	TOTAL INCL O&P	102
0011	CERAMIC TILE											
0020												
0600		Cove base, 4-¼″ x 4-¼″ high, mud set	J-4	91	.176	L.F.	2.70	3.48		6.18	8.35	
0700		Thin set	D-7	128	.125		2.50	2.48		4.98	6.60	
0900		6″ x 4-¼″ high, mud set		100	.160		2.60	3.17		5.77	7.75	
1000		Thin set		137	.117		2.40	2.31		4.71	6.25	
1200		Sanitary cove base, 6″ x 4-¼″ high, mud set		93	.172		2.75	3.41		6.16	8.30	
1300		Thin set		124	.129		2.45	2.55		5	6.65	
1500		6″ x 6″ high, mud set		84	.190		3.30	3.77		7.07	9.50	
1600		Thin set		117	.137		3	2.71		5.71	7.50	
2400		Bullnose trim, 4-¼″ x 4-¼″, mud set		82	.195		2.70	3.86		6.56	8.95	
2500		Thin set		128	.125		2.40	2.48		4.88	6.50	
2700		6″ x 4-¼″ bullnose trim, mud set		84	.190		3.50	3.77		7.27	9.70	
2800		Thin set		124	.129		3.30	2.55		5.85	7.60	
3000		Floors, natural clay, random or uniform, thin set, color group 1		183	.087	S.F.	2.95	1.73		4.68	5.95	
3100		Color group 2		183	.087		3.15	1.73		4.88	6.15	
3300		Porcelain type, 1 color, color group 2, 1″ x 1″		183	.087		3.45	1.73		5.18	6.50	
3400		2″ x 2″ or 2″ x 1″, thin set		183	.087		3.60	1.73		5.33	6.65	
3600		For random blend, 2 colors, add					.60			.60	.66	
3700		4 colors, add					.85			.85	.94	
4300		Specialty tile, 4-¼″ x 4-¼″ x ½″, decorator finish	D-7	183	.087		5.45	1.73		7.18	8.70	
4310												
4500		Add for epoxy grout, ¹⁄₁₆″ joint, 1″ x 1″ tile	D-7	965	.017	S.F.	.45	.33		.78	1.01	
4600		2″ x 2″ tile	″	965	.017	″	.40	.33		.73	.95	
4800		Pregrouted sheets, walls, 4-¼″ x 4-¼″, 6″ x 4-¼″										
4810		and 8-½″ x 4-¼″, 4 S.F. sheets, silicone grout	D-7	240	.067	S.F.	3.55	1.32		4.87	5.95	
5100		Floors, unglazed, 2 S.F. sheets,										
5110		urethane adhesive	D-7	180	.089	S.F.	3.50	1.76		5.26	6.60	
5400		Walls, interior, thin set, 4-¼″ x 4-¼″ tile		180	.089		1.85	1.76		3.61	4.77	
5500		6″ x 4-¼″ tile		190	.084		2.15	1.67		3.82	4.96	
5700		8-½″ x 4-¼″ tile		190	.084		2.80	1.67		4.47	5.65	
5800		6″ x 6″ tile		200	.080		2.30	1.58		3.88	4.99	
6000		Decorated wall tile, 4-¼″ x 4-¼″, minimum		870	.018		2.50	.36		2.86	3.31	
6100		Maximum		580	.028		14	.55		14.55	16.25	
6600		Crystalline glazed, 4-¼″ x 4-¼″, mud set, plain		100	.160		3.20	3.17		6.37	8.45	
6700		4-¼″ x 4-¼″, scored tile		100	.160		3.30	3.17		6.47	8.55	

For expanded coverage of these items see *Means Interior Cost Data 1991*

093 100 | Ceramic Tile

		CREW	DAILY OUTPUT	MAN-HOURS	UNIT	MAT.	BARE COSTS LABOR	EQUIP.	TOTAL	TOTAL INCL O&P	
102	6900	1-⅜" squares	D-7	93	.172	S.F.	3.30	3.41		6.71	8.90	102
	7000	For epoxy grout, ⅛" joints, 4-¼" tile, add		965	.017		.45	.33		.78	1.01	
	7200	For tile set in dry mortar, add		1,735	.009			.18		.18	.28	
	7300	For tile set in portland cement mortar, add		290	.055			1.09		1.09	1.69	
	9500	Minimum labor/equipment charge		3.25	4.920	Job		97		97	150	

093 300 | Quarry Tile

		CREW	DAILY OUTPUT	MAN-HOURS	UNIT	MAT.	BARE COSTS LABOR	EQUIP.	TOTAL	TOTAL INCL O&P	
304	0010	QUARRY TILE Base, cove or sanitary, 2" or 5" high, mud set										304
	0100	½" thick	D-7	110	.145	L.F.	2.50	2.88		5.38	7.20	
	0300	Bullnose trim, red, mud set, 6" x 6" x ½" thick		120	.133		2.50	2.64		5.14	6.85	
	0400	4" x 4" x ½" thick		110	.145		2.50	2.88		5.38	7.20	
	0600	4" x 8" x ½" thick, using 8" as edge		130	.123		2.60	2.44		5.04	6.65	
	0610											
	0700	Floors, mud set, 1000 S.F. lots, red, 4" x 4" x ½" thick	D-7	120	.133	S.F.	2.90	2.64		5.54	7.30	
	0900	6" x 6" x ½" thick		140	.114		2.90	2.26		5.16	6.70	
	1000	4" x 8" x ½" thick		130	.123		3.05	2.44		5.49	7.15	
	1300	For waxed coating, add					.30			.30	.33	
	1500	For colors other than green, add					.40			.40	.44	
	1600	For abrasive surface, add					.45			.45	.50	
	1800	Brown tile, imported, 6" x 6" x ¾"	D-7	120	.133		3.50	2.64		6.14	7.95	
	1900	8" x 8" x 1"		110	.145		4.35	2.88		7.23	9.25	
	2100	For thin set mortar application, deduct		700	.023			.45		.45	.70	
	2500											
	2700	Stair tread & riser, 6" x 6" x ¾", plain	D-7	50	.320	S.F.	4.50	6.35		10.85	14.75	
	2800	Abrasive		50	.320		5.05	6.35		11.40	15.40	
	3000	Wainscot, 6" x 6" x ½", thin set, red		105	.152		2.75	3.02		5.77	7.70	
	3100	Colors other than green		105	.152		2.95	3.02		5.97	7.95	
	3300	Window sill, 6" wide, ¾" thick		90	.178	L.F.	3.60	3.52		7.12	9.40	
	3400	Corners		90	.178	Ea.	2.90	3.52		6.42	8.65	
	9000	Minimum labor/equipment charge		3.25	4.920	Job		97		97	150	

093 700 | Metal Tile

		CREW	DAILY OUTPUT	MAN-HOURS	UNIT	MAT.	BARE COSTS LABOR	EQUIP.	TOTAL	TOTAL INCL O&P	
701	0010	METAL TILE 4' x 4' sheet, 24 ga., tile pattern, nailed										701
	0200	Stainless steel	2 Carp	512	.031	S.F.	14.75	.69		15.44	17.35	
	0400	Aluminized steel	"	512	.031	"	7.80	.69		8.49	9.70	
	9000	Minimum labor/equipment charge	1 Carp	4	2	Job		44		44	72	

094 100 | Portland Cem. Terrazzo

		CREW	DAILY OUTPUT	MAN-HOURS	UNIT	MAT.	BARE COSTS LABOR	EQUIP.	TOTAL	TOTAL INCL O&P	
104	0010	TERRAZZO, CAST IN PLACE Cove base, 6" high	1 Mstz	20	.400	L.F.	1.10	8.85		9.95	14.90	104
	0100	Curb, 6" high and 6" wide		6	1.330		3.20	29		32.20	49	
	0300	Divider strip for floors, 14 ga., 1-¼" deep, zinc		375	.021		.75	.47		1.22	1.56	
	0400	Brass		375	.021		1.50	.47		1.97	2.38	
	0600	Heavy top strip ¼" thick, 1-¼" deep, zinc		300	.027		1.15	.59		1.74	2.18	
	1200	For thin set floors, 16 ga., ½" x ½", zinc		350	.023		.60	.51		1.11	1.44	
	1500	Floor, bonded to concrete, 1-¾" thick, gray cement	J-3	130	.123	S.F.	1.75	2.43	.68	4.86	6.45	
	1600	White cement		130	.123		2	2.43	.68	5.11	6.70	
	1800	Not bonded, 3" total thickness, gray cement		115	.139		2.46	2.74	.77	5.97	7.80	
	1900	White cement		115	.139		2.80	2.74	.77	6.31	8.20	

For expanded coverage of these items see *Means Interior Cost Data 1991*

094 100 | Portland Cem. Terrazzo

			CREW	DAILY OUTPUT	MAN-HOURS	UNIT	MAT.	LABOR	EQUIP.	TOTAL	TOTAL INCL O&P	
104	9000	Minimum labor/equipment charge	1 Mstz	1	8	Job		175		175	275	104
108	0010	**TILE OR TERRAZZO BASE** Scratch coat only	1 Mstz	150	.053	S.F.	.21	1.18		1.39	2.06	108
	0500	Scratch and brown coat only		75	.107	"	.42	2.36		2.78	4.12	
	9000	Minimum labor/equipment charge	↓	1	8	Job		175		175	275	

094 200 | Precast Terrazzo

			CREW	DAILY OUTPUT	MAN-HOURS	UNIT	MAT.	LABOR	EQUIP.	TOTAL	TOTAL INCL O&P	
201	0010	**TERRAZZO, PRECAST** Base, 6" high, straight	1 Mstz	35	.229	L.F.	6	5.05		11.05	14.45	201
	0100	Cove		30	.267		7.20	5.90		13.10	17.05	
	0300	8" high base, straight		30	.267		6.60	5.90		12.50	16.40	
	0400	Cove	↓	25	.320		7.80	7.05		14.85	19.55	
	0600	For white cement, add					.25			.25	.28	
	0700	For 16 ga. zinc toe strip, add					.75			.75	.83	
	0900	Curbs, 4" x 4" high	1 Mstz	19	.421		12.45	9.30		21.75	28	
	1000	8" x 8" high	"	15	.533		13.60	11.80		25.40	33	
	1200	Floor tiles, non-slip, 1" thick, 12" x 12"	D-1	29	.552	S.F.	9	11.15		20.15	28	
	1300	1-¼" thick, 12" x 12"		29	.552		10	11.15		21.15	29	
	1500	16" x 16"		23	.696		11.20	14.05		25.25	35	
	1600	1-½" thick, 16" x 16"	↓	21	.762		13.50	15.40		28.90	40	
	1800	For Venetian terrazzo, add					4			4	4.40	
	1900	For white cement, add					.36			.36	.40	
	2100	Floor tiles, 12" x 12", ³⁄₁₆" thick, ¼" to ½" chips	1 Tilf	75	.107		4.24	2.36		6.60	8.35	
	2200	¼" to 1" chips	"	75	.107		4.78	2.36		7.14	8.90	
	2400	Stair treads, 1-½" thick, non-slip, three line pattern	2 Mstz	70	.229	L.F.	21.50	5.05		26.55	31	
	2500	Nosing and two lines		70	.229		21.50	5.05		26.55	31	
	2700	2" thick treads, straight		60	.267		22.50	5.90		28.40	34	
	2800	Curved		50	.320		31.50	7.05		38.55	46	
	3000	Stair risers, 1" thick, to 6" high, straight sections		60	.267		6	5.90		11.90	15.75	
	3100	Cove		50	.320		7.50	7.05		14.55	19.20	
	3300	Curved, 1" thick, to 6" high, vertical		48	.333		9.80	7.35		17.15	22	
	3400	Cove		38	.421		10.80	9.30		20.10	26	
	3600	Stair tread and riser, single piece, straight, minimum		60	.267		28	5.90		33.90	40	
	3700	Maximum		40	.400		35	8.85		43.85	52	
	3900	Curved tread and riser, minimum		40	.400		36.50	8.85		45.35	54	
	4000	Maximum		32	.500		48.50	11.05		59.55	70	
	4200	Stair stringers, notched, 1" thick		25	.640		15.40	14.15		29.55	39	
	4300	2" thick		22	.727		18.90	16.05		34.95	46	
	4500	Stair landings, structural, non-slip, 1-½" thick		85	.188	S.F.	11.50	4.16		15.66	19.10	
	4600	3" thick	↓	14	1.140		14.80	25		39.80	55	
	4800	Wainscot, 12" x 12" x 1" tiles	1 Mstz	12	.667		9	14.75		23.75	33	
	4900	16" x 16" x 1-½" tiles	"	8	1		13.50	22		35.50	49	
	9500	Minimum labor/equipment charge	1 Tilf	2	4	Job		89		89	135	

095 | Acoustical Treatment and Wood Flooring

095 100 | Acoustical Ceilings

			CREW	DAILY OUTPUT	MAN-HOURS	UNIT	MAT.	LABOR	EQUIP.	TOTAL	TOTAL INCL O&P	
102	0010	**CEILING TILE** Stapled, cemented or installed on suspension										102
	0100	system, 12" x 12" or 12" x 24", not including furring										
	0600	Mineral fiber, vinyl coated , ⅝" thick	1 Carp	200	.040	S.F.	.48	.88		1.36	1.96	
	0700	¾" thick	"	200	.040	"	.63	.88		1.51	2.13	

For expanded coverage of these items see *Means Interior Cost Data 1991*

095 100 | Acoustical Ceilings

			DAILY OUTPUT	MAN-HOURS	UNIT	BARE COSTS				TOTAL INCL O&P	
						MAT.	LABOR	EQUIP.	TOTAL		
102	0900	Fire rated, ¾" thick, plain faced	1 Carp	200	.040	S.F.	.64	.88		1.52	2.14	102
	1000	Plastic coated face		200	.040		.69	.88		1.57	2.20	
	1200	Aluminum faced, ⅝" thick, plain		200	.040		.83	.88		1.71	2.35	
	1210											
	3000	Wood fiber tile, ½" thick	1 Carp	400	.020	S.F.	.57	.44		1.01	1.35	
	3100	¾" thick	"	400	.020		.80	.44		1.24	1.60	
	3300	For flameproofing, add					.06			.06	.07	
	3400	For sculptured 3 dimensional, add					.18			.18	.20	
	3900	For ceiling primer, add					.09			.09	.10	
	4000	For ceiling cement, add					.24			.24	.26	
	9000	Minimum labor/equipment charge	1 Carp	4	2	Job		44		44	72	
104	0010	**SUSPENDED ACOUSTIC CEILING BOARDS** Not including										104
	0100	suspension system										
	0300	Fiberglass boards, film faced, 2' x 2' or 2' x 4', ⅝" thick	1 Carp	625	.013	S.F.	.39	.28		.67	.89	
	0400	¾" thick		600	.013		.45	.29		.74	.97	
	0500	3" thick, thermal, R11		500	.016		.95	.35		1.30	1.62	
	0600	Glass cloth faced fiberglass, ¾" thick		500	.016		1.24	.35		1.59	1.94	
	0700	1" thick		500	.016		1.38	.35		1.73	2.09	
	0820	1-½" thick, nubby face		500	.016		1.68	.35		2.03	2.42	
	0900	Mineral fiber boards, ⅝" thick, aluminum faced, 24" x 24"		600	.013		1.05	.29		1.34	1.63	
	0930	24" x 48"		650	.012		1.02	.27		1.29	1.56	
	0960	Standard face		675	.012		.45	.26		.71	.92	
	1000	Plastic coated face		400	.020		.70	.44		1.14	1.49	
	1200	Mineral fiber, 2 hour rating, ⅝" thick		675	.012		.65	.26		.91	1.14	
	1300	Mirror faced panels, ¹⁵⁄₁₆" thick, 2' x 2'		500	.016		4.50	.35		4.85	5.50	
	1900	Eggcrate, acrylic, ½" x ½" x ½" cubes		500	.016		1.05	.35		1.40	1.73	
	2100	Polystyrene eggcrate, ⅜" x ⅜" x ½" cubes		500	.016		.86	.35		1.21	1.52	
	2200	½" x ½" x ½" cubes		500	.016		1.18	.35		1.53	1.87	
	2210											
	2400	Luminous panels, prismatic, acrylic	1 Carp	400	.020	S.F.	1.24	.44		1.68	2.08	
	2500	Polystyrene		400	.020		.65	.44		1.09	1.43	
	2700	Flat white acrylic		400	.020		2.20	.44		2.64	3.14	
	2800	Polystyrene		400	.020		1.50	.44		1.94	2.37	
	3000	Drop pan, white, acrylic		400	.020		3.25	.44		3.69	4.29	
	3100	Polystyrene		400	.020		2.70	.44		3.14	3.69	
	3600	Perforated aluminum sheets, .024" thick, corrugated, painted		500	.016		1.30	.35		1.65	2	
	3700	Plain		500	.016		1.16	.35		1.51	1.85	
	3720	Mineral fiber, 24" x 24" or 48", reveal edge, painted, ⅝" thick		600	.013		.70	.29		.99	1.25	
	3740	¾" thick		575	.014		1.17	.31		1.48	1.79	
	3750	Wood fiber in cementitious binder, 2' x 2' or 4', painted, 1" thick		600	.013		.89	.29		1.18	1.46	
	3760	2" thick		550	.015		1.45	.32		1.77	2.12	
	3770	2-½" thick		500	.016		1.98	.35		2.33	2.75	
	3780	3" thick		450	.018		2.28	.39		2.67	3.15	
	9000	Minimum labor/equipment charge		4	2	Job		44		44	72	
106	0010	**SUSPENDED CEILINGS, COMPLETE** Including standard										106
	0100	suspension system but not incl. 1-½" carrier channels										
	0600	Fiberglass ceiling board, 2' x 4' x ⅝", plain faced, supermarkets	1 Carp	500	.016	S.F.	.77	.35		1.12	1.42	
	0700	Offices, 2' x 4' x ¾"		380	.021		.86	.46		1.32	1.70	
	1800	Tile, Z bar suspension, ⅝" mineral fiber tile		150	.053		1.10	1.17		2.27	3.13	
	1900	¾" mineral fiber tile		150	.053		1.18	1.17		2.35	3.21	
	9000	Minimum labor/equipment charge		2	4	Job		88		88	145	

095 250 | Acoustical Space Units

			CREW	DAILY OUTPUT	MAN-HOURS	UNIT	BARE COSTS MAT.	LABOR	EQUIP.	TOTAL	TOTAL INCL O&P	
254	0010	**SOUND ABSORBING PANELS** Perforated steel facing, painted with										254
	0100	fiberglass or mineral filler, no backs, 2-¼″ thick, modular										
	0200	space units, ceiling or wall hung, white or colored	1 Carp	100	.080	S.F.	4.10	1.76		5.86	7.40	
	0300	Fiberboard sound deadening panels, ½″ thick	"	600	.013	"	.19	.29		.48	.69	
	0500	Fiberglass panels, 4′ x 8′ x 1″ thick, with										
	0600	glass cloth face for walls, cemented	1 Carp	155	.052	S.F.	1.65	1.14		2.79	3.67	
	0700	1-½″ thick, dacron covered, inner aluminum frame,										
	0710	wall mounted	1 Carp	300	.027	S.F.	5.95	.59		6.54	7.50	
	0900	Mineral fiberboard panels, fabric covered, 30″x 108″,										
	1000	¾″ thick, concealed spline, wall mounted	1 Carp	150	.053	S.F.	4.15	1.17		5.32	6.50	
	9000	Minimum labor/equipment charge	"	4	2	Job		44		44	72	

095 300 | Acoustical Insulation

			CREW	DAILY OUTPUT	MAN-HOURS	UNIT	MAT.	LABOR	EQUIP.	TOTAL	TOTAL INCL O&P	
308	0010	**SOUND ATTENUATION** Blanket, 1″ thick	1 Carp	925	.009	S.F.	.21	.19		.40	.54	308
	0500	1-½″ thick		925	.009		.23	.19		.42	.56	
	1000	2″ thick		925	.009		.28	.19		.47	.62	
	1500	3″ thick		925	.009		.45	.19		.64	.81	
	3400	Urethane plastic foam, open cell, on wall, 2″ thick	2 Carp	2,050	.008		1.95	.17		2.12	2.43	
	3500	3″ thick		1,550	.010		2.60	.23		2.83	3.23	
	3600	4″ thick		1,050	.015		3.60	.34		3.94	4.51	
	3700	On ceiling, 2″ thick		1,700	.009		1.95	.21		2.16	2.48	
	3800	3″ thick		1,300	.012		2.60	.27		2.87	3.30	
	3900	4″ thick		900	.018		3.60	.39		3.99	4.60	
	4000	Nylon matting 0.4″ thick, with carbon black spinerette										
	4010	plus polyester fabric, on floor	J-4	4,000	.004	S.F.	1.45	.08		1.53	1.72	
	4200	Fiberglass reinf. backer board underlayment, 7/16″ thick, on floor	"	800	.020	"	1.20	.40		1.60	1.93	
	9000	Minimum labor/equipment charge	1 Carp	5	1.600	Job		35		35	57	

095 600 | Wood Strip Flooring

			CREW	DAILY OUTPUT	MAN-HOURS	UNIT	MAT.	LABOR	EQUIP.	TOTAL	TOTAL INCL O&P	
604	0010	**WOOD** Fir, vertical grain, 1″ x 4″, not incl. finish, B & better	1 Carp	255	.031	S.F.	1.85	.69		2.54	3.16	604
	0100	C grade & better		255	.031		1.75	.69		2.44	3.05	
	0300	Flat grain, 1″ x 4″, not incl. finish, B & better		255	.031		2.05	.69		2.74	3.38	
	0400	C & better		255	.031		1.95	.69		2.64	3.27	
	4000	Maple, strip, 25/32″ x 2-¼″, not incl. finish, select		170	.047		2.35	1.04		3.39	4.28	
	4100	#2 & better		170	.047		2.05	1.04		3.09	3.95	
	4300	33/32″ x 3-¼″, not incl. finish, #1 grade		170	.047		2.55	1.04		3.59	4.50	
	4400	#2 & better		170	.047		2.35	1.04		3.39	4.28	
	4600	Oak, white or red, 25/32″ x 2-¼″, not incl. finish (87)										
	4700	Clear quartered	1 Carp	170	.047	S.F.	2.20	1.04		3.24	4.11	
	4900	Clear/select, 2-¼″ wide		170	.047		2.65	1.04		3.69	4.61	
	5000	#1 common		185	.043		2.45	.95		3.40	4.25	
	5200	Parquetry, standard, 5/16″ thick, not incl. finish, oak, minimum		160	.050		1.60	1.10		2.70	3.56	
	5300	Maximum		100	.080		5	1.76		6.76	8.35	
	5500	Teak, minimum		160	.050		3.50	1.10		4.60	5.65	
	5600	Maximum		100	.080		6	1.76		7.76	9.45	
	5650	13/16″ thick, select grade oak, minimum		160	.050		7.50	1.10		8.60	10.05	
	5700	Maximum		100	.080		10.85	1.76		12.61	14.80	
	5800	Custom parquetry, including finish, minimum		100	.080		10.80	1.76		12.56	14.75	
	5900	Maximum		50	.160		14.75	3.52		18.27	22	
	6100	Prefinished white oak, prime grade, 2-¼″ wide		170	.047		4.85	1.04		5.89	7.05	
	6200	3-¼″ wide		185	.043		5.85	.95		6.80	8	
	6400	Ranch plank		145	.055		5.85	1.21		7.06	8.40	
	6500	Hardwood blocks, 9″ x 9″, 25/32″ thick		160	.050		3.70	1.10		4.80	5.85	
	6700	Parquetry, 5/16″ thick, oak, minimum		160	.050		2.85	1.10		3.95	4.93	
	6800	Maximum		100	.080		8	1.76		9.76	11.65	
	7000	Walnut or teak, parquetry, minimum		160	.050		3.85	1.10		4.95	6.05	
	7100	Maximum		100	.080		6.45	1.76		8.21	9.95	

For expanded coverage of these items see *Means Interior Cost Data 1991*

095 600 | Wood Strip Flooring

			CREW	DAILY OUTPUT	MAN-HOURS	UNIT	BARE COSTS				TOTAL INCL O&P	
							MAT.	LABOR	EQUIP.	TOTAL		
604	7200	Acrylic wood parquet blocks, 12" x 12" x 5/16",										604
	7210	irradiated, set in epoxy	1 Carp	160	.050	S.F.	5.10	1.10		6.20	7.40	
	7400	Yellow pine, 3/4" x 3-1/8", T & G, C & better, not incl. finish	"	200	.040		1.70	.88		2.58	3.31	
	7500	Refinish old floors, minimum	A-1A	400	.020		.50	.35	.07	.92	1.20	
	7600	Maximum		130	.062		.75	1.08	.23	2.06	2.84	
	7800	Sanding and finishing, fill, polyurethane, wax	↓	295	.027	↓	.50	.47	.10	1.07	1.44	
	7900	Subfloor and underlayment, see division 061-164 & 168										
	9000	Minimum labor/equipment charge	1 Carp	2	4	Job		88		88	145	

095 650 | Wood Block Flooring

			CREW	DAILY OUTPUT	MAN-HOURS	UNIT	MAT.	LABOR	EQUIP.	TOTAL	TOTAL INCL O&P	
651	0010	WOOD BLOCK FLOORING End grain flooring, creosoted, 2" thick	1 Carp	295	.027	S.F.	2.10	.60		2.70	3.28	651
	0400	Natural finish, 1" thick, fir		125	.064		3.15	1.41		4.56	5.75	
	0600	1-1/2" thick, pine		125	.064		3.15	1.41		4.56	5.75	
	0700	2" thick, pine		125	.064	↓	3.40	1.41		4.81	6.05	
	9000	Minimum labor/equipment charge	↓	2	4	Job		88		88	145	

096 | Flooring and Carpet

096 150 | Marble Flooring

			CREW	DAILY OUTPUT	MAN-HOURS	UNIT	BARE COSTS				TOTAL INCL O&P	
							MAT.	LABOR	EQUIP.	TOTAL		
151	0010	MARBLE Thin gauge tile, 12" x 6", 3/8", White Carara	D-7	60	.267	S.F.	3.90	5.30		9.20	12.50	151
	0100	Filled Travertine		60	.267		3.90	5.30		9.20	12.50	
	0200	12" x 12" x 3/8", thin set, floors		60	.267		5.55	5.30		10.85	14.30	
	0300	On walls		52	.308	↓	5.55	6.10		11.65	15.55	
	9000	Minimum labor/equipment charge	↓	3	5.330	Job		105		105	165	

096 250 | Slate Flooring

			CREW	DAILY OUTPUT	MAN-HOURS	UNIT	MAT.	LABOR	EQUIP.	TOTAL	TOTAL INCL O&P	
251	0010	SLATE TILE Vermont, 6" x 6" x 1/4" thick, thin set	D-7	180	.089	S.F.	2.85	1.76		4.61	5.85	251
	9000	Minimum labor/equipment charge	"	3	5.330	Job		105		105	165	

096 350 | Brick Flooring

			CREW	DAILY OUTPUT	MAN-HOURS	UNIT	MAT.	LABOR	EQUIP.	TOTAL	TOTAL INCL O&P	
354	0010	FLOORING Acidproof shales, red, 8" x 3-3/4" x 1-1/4" thick	D-7	.43	37.210	M	495	735		1,230	1,675	354
	0050	2-1/4" thick	D-1	.40	40		525	810		1,335	1,875	
	0200	Acid proof clay brick, 8" x 3-3/4" x 2-1/4" thick	"	.40	40	↓	525	810		1,335	1,875	
	0260	Cast ceramic, pressed, 4" x 8" x 1/2", unglazed	D-7	100	.160	S.F.	4.10	3.17		7.27	9.40	
	0270	Glazed		100	.160		5.45	3.17		8.62	10.90	
	0280	Hand molded flooring, 4" x 8" x 3/4", unglazed		95	.168		5.45	3.33		8.78	11.15	
	0290	Glazed		95	.168		6.55	3.33		9.88	12.40	
	0300	8" hexagonal, 3/4" thick, unglazed		85	.188		5.85	3.73		9.58	12.20	
	0310	Glazed	↓	85	.188		10.35	3.73		14.08	17.15	
	0400	Heavy duty industrial, cement mortar bed, 2" thick, not incl. brick	D-1	80	.200		.60	4.04		4.64	7.15	
	0450	Acid proof joints, 1/4" wide	"	65	.246		.85	4.97		5.82	8.90	
	0500	Pavers, 8" x 4", 1" to 1-1/4" thick, red	D-7	95	.168		2.40	3.33		5.73	7.80	
	0510	Ironspot	"	95	.168		3.85	3.33		7.18	9.40	
	0540	1-3/8" to 1-3/4" thick, red	D-1	95	.168		2.60	3.40		6	8.30	
	0560	Ironspot		95	.168		3.60	3.40		7	9.40	
	0580	2-1/4" thick, red		90	.178		2.60	3.59		6.19	8.60	
	0590	Ironspot		90	.178		3.60	3.59		7.19	9.70	
	0600	Sidewalk or patios, on sand bed, laid flat, no mortar, 4.5 per S.F.		110	.145		2.35	2.94		5.29	7.30	
	0650	Laid on edge, 7 per S.F.	↓	70	.229		3.70	4.62		8.32	11.50	
	0800	For basket weave pattern, add						15%				

For expanded coverage of these items see *Means Interior Cost Data 1991*

096 350 | Brick Flooring

		CREW	DAILY OUTPUT	MAN-HOURS	UNIT	BARE COSTS				TOTAL INCL O&P	
						MAT.	LABOR	EQUIP.	TOTAL		
354	0850	For herringbone pattern, add				S.F.		20%				354
	0870	For epoxy joints, add	D-1	600	.027		1.75	.54		2.29	2.79	
	0880	For Furan underlayment, add	"	600	.027		1.45	.54		1.99	2.46	
	0890	For waxed surface, steam cleaned, add	D-5	1,000	.008		.09	.18	.03	.30	.43	
	9000	Minimum equipment/labor charge	1 Bric	2	4	Job		91		91	145	

096 600 | Resilient Tile Flooring

		CREW	DAILY OUTPUT	MAN-HOURS	UNIT	BARE COSTS				TOTAL INCL O&P	
						MAT.	LABOR	EQUIP.	TOTAL		
601	0010	**RESILIENT** Asphalt tile, on concrete, ⅛" thick										601
	0050	Color group B	1 Tilf	400	.020	S.F.	.80	.44		1.24	1.57	
	0100	Color group C & D	"	400	.020		.85	.44		1.29	1.62	
	0300	For wood subfloor, add to above for felt underlayment					.15			.15	.17	
	0500	For less than 500 S.F., add						.03		.03	.04	
	0600	For over 5000 S.F., deduct						.01		.01	.01	
	0800	Base, cove, rubber or vinyl, .080" thick										
	1100	Standard colors, 2-½" high	1 Tilf	315	.025	L.F.	.40	.56		.96	1.31	
	1150	4" high		315	.025		.50	.56		1.06	1.42	
	1200	6" high		315	.025		.70	.56		1.26	1.64	
	1450	⅛" thick, standard colors, 2-½" high		315	.025		.45	.56		1.01	1.37	
	1500	4" high		315	.025		.66	.56		1.22	1.60	
	1550	6" high		315	.025		.70	.56		1.26	1.64	
	1600	Corners, 2-½" high		315	.025	Ea.	.70	.56		1.26	1.64	
	1630	4" high		315	.025		.80	.56		1.36	1.75	
	1660	6" high		315	.025		.90	.56		1.46	1.86	
	1700	Conductive flooring, rubber tile, ⅛" thick		315	.025	S.F.	2.45	.56		3.01	3.57	
	1800	Homogeneous vinyl tile, ⅛" thick		315	.025		2.70	.56		3.26	3.84	
	2200	Cork tile, standard finish, ⅛" thick		315	.025		1.80	.56		2.36	2.85	
	2250	3/16" thick		315	.025		2.10	.56		2.66	3.18	
	2300	5/16" thick		315	.025		2.85	.56		3.41	4.01	
	2350	½" thick		315	.025		3.45	.56		4.01	4.67	
	2500	Urethane finish, ⅛" thick		315	.025		2.75	.56		3.31	3.90	
	2550	3/16" thick		315	.025		3.90	.56		4.46	5.15	
	2600	5/16" thick		315	.025		4.85	.56		5.41	6.20	
	2650	½" thick		315	.025		6.50	.56		7.06	8	
	3700	Polyethylene, in rolls, no base incl., landscape surfaces		275	.029		1.75	.64		2.39	2.92	
	3800	Nylon action surface, ⅛" thick		275	.029		1.90	.64		2.54	3.09	
	3900	¼" thick		275	.029		2.75	.64		3.39	4.02	
	4000	⅜" thick		275	.029		3.65	.64		4.29	5	
	5500	Polyvinyl chloride, sheet goods for gyms, ¼" thick		80	.100		2.80	2.22		5.02	6.50	
	5600	⅜" thick		60	.133		3.20	2.95		6.15	8.10	
	5900	Rubber, sheet goods, 36" wide, ⅛" thick		120	.067		2.45	1.48		3.93	4.98	
	5950	3/16" thick		100	.080		3.15	1.77		4.92	6.20	
	6000	¼" thick		90	.089		3.65	1.97		5.62	7.05	
	6050	Tile, marbleized colors, 12" x 12", ⅛" thick		400	.020		2.30	.44		2.74	3.22	
	6100	3/16" thick		400	.020		3.20	.44		3.64	4.21	
	6300	Special tile, plain colors, ⅛" thick		400	.020		2.85	.44		3.29	3.82	
	6350	3/16" thick		400	.020		3.80	.44		4.24	4.87	
	7000	Vinyl composition tile, 12" x 12", 1/16" thick		500	.016		.55	.35		.90	1.15	
	7050	Embossed		500	.016		.70	.35		1.05	1.32	
	7100	Marbleized		500	.016		.70	.35		1.05	1.32	
	7150	Solid		500	.016		.80	.35		1.15	1.43	
	7200	3/32" thick, embossed		500	.016		.75	.35		1.10	1.37	
	7250	Marbleized		500	.016		.95	.35		1.30	1.59	
	7300	Solid		500	.016		1.25	.35		1.60	1.92	
	7310											
	7350	⅛" thick, marbleized	1 Tilf	500	.016	S.F.	.85	.35		1.20	1.48	
	7400	Solid		500	.016		1.45	.35		1.80	2.14	
	7500	Vinyl tile, 12" x 12", .050" thick, minimum		500	.016		1.35	.35		1.70	2.03	

For expanded coverage of these items see *Means Interior Cost Data 1991*

096 600 | Resilient Tile Flooring

			CREW	DAILY OUTPUT	MAN-HOURS	UNIT	BARE COSTS MAT.	LABOR	EQUIP.	TOTAL	TOTAL INCL O&P	
601	7550	Maximum	1 Tilf	500	.016	S.F.	2.60	.35		2.95	3.41	601
	7600	⅛" thick, minimum		500	.016		1.75	.35		2.10	2.47	
	7650	Solid colors		500	.016		2.25	.35		2.60	3.02	
	7700	Marbleized or Travertine pattern		500	.016		2.65	.35		3	3.46	
	7750	Florentine pattern		500	.016		3.40	.35		3.75	4.29	
	7800	Maximum	↓	500	.016	↓	6.75	.35		7.10	7.95	
	7810											
	8000	Vinyl sheet goods, backed, .065" thick, minimum	1 Tilf	250	.032	S.F.	1.15	.71		1.86	2.36	
	8050	Maximum		200	.040		2.10	.89		2.99	3.68	
	8100	.080" thick, minimum		230	.035		1.30	.77		2.07	2.62	
	8150	Maximum		200	.040		2.35	.89		3.24	3.96	
	8200	.125" thick, minimum		230	.035		1.50	.77		2.27	2.84	
	8250	Maximum		200	.040		3.25	.89		4.14	4.95	
	8300	.250" thick, minimum		230	.035		2.35	.77		3.12	3.78	
	8350	Maximum	↓	200	.040	↓	4.35	.89		5.24	6.15	
	8700	Adhesive cement, 1 gallon does 200 to 300 S.F.				Gal.	11.40			11.40	12.55	
	8800	Asphalt primer, 1 gallon per 300 S.F.					7.45			7.45	8.20	
	8900	Emulsion, 1 gallon per 140 S.F.					7.45			7.45	8.20	
	8950	Latex underlayment				↓	22.75			22.75	25	
	9500	Minimum labor/equipment charge	1 Tilf	4	2	Job		44		44	69	

096 780 | Resilient Accessories

			CREW	DAILY OUTPUT	MAN-HOURS	UNIT	MAT.	LABOR	EQUIP.	TOTAL	TOTAL INCL O&P	
781	0010	STAIR TREADS AND RISERS See index for materials other										781
	0100	than rubber and vinyl										
	0300	Rubber, molded tread, 12" wide, ⁵⁄₁₆" thick, black	1 Tilf	115	.070	L.F.	4.80	1.54		6.34	7.65	
	0400	Colors		115	.070		4.95	1.54		6.49	7.85	
	0600	¼" thick, black		115	.070		4.50	1.54		6.04	7.35	
	0700	Colors		115	.070		4.85	1.54		6.39	7.70	
	0900	Grip strip safety tread, colors, ⁵⁄₁₆" thick		115	.070		6.75	1.54		8.29	9.80	
	1000	³⁄₁₆" thick		115	.070		5.50	1.54		7.04	8.45	
	1200	Landings, smooth sheet rubber, ⅛" thick		120	.067	S.F.	2.40	1.48		3.88	4.93	
	1300	³⁄₁₆" thick		120	.067	"	3.15	1.48		4.63	5.75	
	1500	Nosings, 1-½" deep, 3" wide, black		140	.057	L.F.	1.30	1.27		2.57	3.39	
	1600	Colors		140	.057		1.90	1.27		3.17	4.05	
	1800	Risers, 7" high, ⅛" thick, flat		250	.032		1.30	.71		2.01	2.53	
	1900	Coved		250	.032		1.50	.71		2.21	2.75	
	2100	Vinyl, molded tread, 12" wide, colors, ⅛" thick		115	.070		2.55	1.54		4.09	5.20	
	2200	¼" thick		115	.070	↓	3.65	1.54		5.19	6.40	
	2300	Landing material, ⅛" thick		200	.040	S.F.	2.15	.89		3.04	3.74	
	2400	Riser, 7" high, ⅛" thick, coved		175	.046	L.F.	1.25	1.01		2.26	2.94	
	2500	Tread and riser combined, ⅛" thick		80	.100	"	3.20	2.22		5.42	6.95	
	9000	Minimum labor/equipment charge	↓	3	2.670	Job		59		59	92	

096 850 | Sheet Carpet

			CREW	DAILY OUTPUT	MAN-HOURS	UNIT	MAT.	LABOR	EQUIP.	TOTAL	TOTAL INCL O&P	
852	0010	CARPET Commercial grades, direct cement										852
	0700	Nylon, level loop, 26 oz., light to medium traffic	1 Tilf	57	.140	S.Y.	13.80	3.11		16.91	20	
	0900	32 oz., medium traffic		57	.140		16.40	3.11		19.51	23	
	1100	40 oz., medium to heavy traffic		57	.140		20.80	3.11		23.91	28	
	2100	Nylon, plush, 20 oz., light traffic		57	.140		8.50	3.11		11.61	14.15	
	2800	24 oz., light to medium traffic		57	.140		9.85	3.11		12.96	15.65	
	2900	30 oz., medium traffic		57	.140		12.20	3.11		15.31	18.25	
	3000	36 oz., medium traffic		57	.140		14.90	3.11		18.01	21	
	3100	42 oz., medium to heavy traffic		50	.160		17.80	3.54		21.34	25	
	3200	46 oz., medium to heavy traffic		50	.160		21.50	3.54		25.04	29	
	3300	54 oz., heavy traffic		50	.160		24.50	3.54		28.04	32	
	3500	Olefin, 15 oz., light traffic		57	.140		4.10	3.11		7.21	9.35	
	3650	22 oz., light traffic		57	.140		5.10	3.11		8.21	10.45	
	4500	50 oz., medium to heavy traffic, level loop		57	.140		30.40	3.11		33.51	38	

For expanded coverage of these items see *Means Interior Cost Data 1991*

096 850 | Sheet Carpet

			DAILY OUTPUT	MAN-HOURS	UNIT	BARE COSTS				TOTAL INCL O&P	
						MAT.	LABOR	EQUIP.	TOTAL		
852	4700	32 oz., medium to heavy traffic, patterned	1 Tilf	49	.163	S.Y.	28.40	3.62		32.02	37	852
	4900	48 oz., heavy traffic, patterned	"	49	.163	"	38.90	3.62		42.52	48	
	4950											
	5000	For less than full roll, add					25%					
	5100	For small rooms, add						25%				
	5200	For large open areas, deduct						25%				
	5600	For bound carpet baseboard, add	1 Tilf	300	.027	L.F.	.85	.59		1.44	1.85	
	5610	For stairs, not incl. price of carpet, add	"	30	.267	Riser		5.90		5.90	9.15	
	8100											
	8950	For tackless, stretched installation, add padding to above										
	9000	Sponge rubber pad, minimum	1 Tilf	150	.053	S.Y.	2.25	1.18		3.43	4.31	
	9100	Maximum		150	.053		6	1.18		7.18	8.45	
	9200	Felt pad, minimum		150	.053		2.50	1.18		3.68	4.58	
	9300	Maximum		150	.053		4	1.18		5.18	6.25	
	9400	Bonded urethane pad, minimum		150	.053		2.75	1.18		3.93	4.86	
	9500	Maximum		150	.053		4.75	1.18		5.93	7.05	
	9600	Prime urethane pad, minimum		150	.053		1.75	1.18		2.93	3.76	
	9700	Maximum		150	.053		3	1.18		4.18	5.15	
	9900	Carpet cleaning machine, rent				Day					35	
	9910	Minimum labor/equipment charge	1 Tilf	3	2.670	Job		59		59	92	

096 900 | Carpet Tile

			DAILY OUTPUT	MAN-HOURS	UNIT	BARE COSTS				TOTAL INCL O&P	
						MAT.	LABOR	EQUIP.	TOTAL		
901	0010	**CARPET TILE**										901
	1100	Tufted, 18" x 18" or 24" x 24", 24 oz. nylon	1 Tilf	80	.100	S.Y.	18.45	2.22		20.67	24	
	1180	35 oz.		80	.100		21.30	2.22		23.52	27	
	5060	42 oz.		80	.100		27.35	2.22		29.57	34	

097 | Special Flooring and Floor Treatment

097 200 | Epoxy-Marble Flooring

			CREW	DAILY OUTPUT	MAN-HOURS	UNIT	BARE COSTS				TOTAL INCL O&P	
							MAT.	LABOR	EQUIP.	TOTAL		
201	0010	**COMPOSITION FLOORING** Acrylic, ¼" thick	C-6	520	.092	S.F.	3	1.71	.12	4.83	6.20	201
	0050											
	0600	Epoxy, with colored quartz chips, broadcast, minimum	C-6	675	.071	S.F.	1.55	1.32	.09	2.96	3.95	
	0700	Maximum		490	.098		2.45	1.81	.13	4.39	5.80	
	0900	Trowelled, minimum		560	.086		2	1.59	.11	3.70	4.89	
	1000	Maximum		480	.100		2.75	1.85	.13	4.73	6.15	
	1200	Heavy duty epoxy topping, ¼" thick,										
	1300	500 to 1,000 S.F.	C-6	420	.114	S.F.	3.70	2.12	.15	5.97	7.65	
	1500	1,000 to 2,000 S.F.		450	.107		3.10	1.98	.14	5.22	6.75	
	1600	Over 10,000 S.F.		480	.100		2.85	1.85	.13	4.83	6.30	
	1800	Epoxy terrazzo, ¼" thick, chemical resistant, minimum	J-3	200	.080		3.30	1.58	.44	5.32	6.55	
	1900	Maximum	"	150	.107		5.15	2.10	.59	7.84	9.60	

For expanded coverage of these items see *Means Interior Cost Data 1991*

098 | Special Coatings

098 150 | Glazed Coatings

			CREW	DAILY OUTPUT	MAN-HOURS	UNIT	BARE COSTS				TOTAL INCL O&P	
							MAT.	LABOR	EQUIP.	TOTAL		
150	0010	WALL COATINGS Acrylic glazed coatings, minimum	1 Pord	525	.015	S.F.	.18	.32		.50	.70	150
	0100	Maximum		305	.026		.35	.55		.90	1.25	
	0300	Epoxy coatings, minimum		525	.015		.20	.32		.52	.72	
	0400	Maximum		170	.047		.65	.98		1.63	2.27	
	2400	Sprayed perlite or vermiculite, 1/16" thick, minimum		2,935	.003		.15	.06		.21	.26	
	2500	Maximum		640	.013		.40	.26		.66	.85	
	2700	Vinyl plastic wall coating, minimum		735	.011		.20	.23		.43	.58	
	2800	Maximum		240	.033		.65	.69		1.34	1.82	
	3000	Urethane on smooth surface, 2 coats, minimum		1,135	.007		.15	.15		.30	.40	
	3100	Maximum		665	.012		.30	.25		.55	.73	

099 | Painting and Wall Coverings

099 100 | Exterior Painting

			CREW	DAILY OUTPUT	MAN-HOURS	UNIT	BARE COSTS				TOTAL INCL O&P	
							MAT.	LABOR	EQUIP.	TOTAL		
106	0010	SIDING Exterior										106
	0020	Labor cost includes protection of adjacent items not painted										
	0100	Steel siding, oil base, primer or sealer coat, brushwork	2 Pord	2,800	.006	S.F.	.07	.12		.19	.27	
	0500	Spray		5,600	.003		.06	.06		.12	.16	
	0800	Paint 2 coats, brushwork		1,850	.009		.14	.18		.32	.44	
	1000	Spray		3,600	.004		.12	.09		.21	.28	
	1200	Stucco, rough, oil base, paint 2 coats, brushwork		1,100	.015		.15	.30		.45	.65	
	1400	Roller		1,800	.009		.16	.18		.34	.47	
	1600	Spray		3,600	.004		.13	.09		.22	.29	
	1700											
	1800	Texture 1-11 or clapboard, oil base, primer coat, brushwork	2 Pord	3,000	.005	S.F.	.08	.11		.19	.26	
	2000	Spray		4,760	.003		.07	.07		.14	.19	
	2400	Paint 2 coats, brushwork		1,800	.009		.12	.18		.30	.43	
	2600	Spray		2,700	.006		.11	.12		.23	.32	
	3400	Stain 2 coats, brushwork		1,225	.013		.14	.27		.41	.59	
	4000	Spray		2,700	.006		.12	.12		.24	.33	
	4200	Wood shingles, oil base primer coat, brushwork		1,500	.011		.08	.22		.30	.44	
	4400	Spray		2,380	.007		.07	.14		.21	.30	
	5000	Paint 2 coats, brushwork		800	.020		.14	.42		.56	.82	
	5200	Spray		2,800	.006		.12	.12		.24	.32	
	6500	Stain 2 coats, brushwork		775	.021		.12	.43		.55	.82	
	7000	Spray		2,800	.006		.11	.12		.23	.31	
	8000	For latex paint, deduct					10%					

099 200 | Interior Painting

			CREW	DAILY OUTPUT	MAN-HOURS	UNIT	BARE COSTS				TOTAL INCL O&P	
							MAT.	LABOR	EQUIP.	TOTAL		
204	0010	CABINETS AND CASEWORK										204
	0020	Labor cost includes protection of adjacent items not painted										
	1000	Primer coat, oil base, brushwork	1 Pord	680	.012	S.F.	.05	.24		.29	.44	
	2000	Paint, oil base, brushwork, 1 coat		600	.013		.05	.28		.33	.50	
	3000	Stain, brushwork, wipe off		400	.020		.05	.42		.47	.72	
	4000	Shellac, 1 coat, brushwork		600	.013		.05	.28		.33	.50	
	4500	Varnish, 3 coats, brushwork, sand after 1st coat		235	.034		.16	.71		.87	1.30	
	5000	For latex paint, deduct					10%					216
216	0010	DOORS AND WINDOWS										
	0020	Labor cost includes protection of adjacent items not painted										
	0500	Flush door & frame, 3' x 7', per side, oil, primer, brushwork	1 Pord	17	.471	Ea.	1.25	9.80		11.05	16.95	
	1000	Paint, 1 coat	"	16	.500	"	1.20	10.40		11.60	17.85	

For expanded coverage of these items see *Means Interior Cost Data 1991*

		099 200	Interior Painting	CREW	DAILY OUTPUT	MAN-HOURS	UNIT	BARE COSTS				TOTAL INCL O&P	
								MAT.	LABOR	EQUIP.	TOTAL		
216	1400		Stain, brushwork, wipe off	1 Pord	18	.444	Ea.	1.10	9.25		10.35	15.90	216
	1600		Shellac, 1 coat, brushwork		25	.320		1.10	6.65		7.75	11.80	
	1800		Varnish, 3 coats, brushwork, sand after 1st coat		9	.889		3.15	18.50		21.65	33	
	1900												
	2000		Panel door & frame, 3' x 7' per side, oil, primer, brushwork	1 Pord	15	.533	Ea.	1.10	11.10		12.20	18.85	
	2200		Paint, 1 coat		14	.571		1.10	11.90		13	20	
	2600		Stain, brushwork, panel door, 3' x 7', per side, not incl. frame		16	.500		1.10	10.40		11.50	17.75	
	2800		Shellac, 1 coat, brushwork		22	.364		1.10	7.55		8.65	13.25	
	3000		Varnish, 3 coats, brushwork, sand after 1st coat		7.50	1.070		3.15	22		25.15	39	
	4400		Windows, including frame and trim, per side										
	4600		Colonial type, 6 lites, 2' x 3', oil, primer, brushwork	1 Pord	21	.381	Ea.	.65	7.90		8.55	13.30	
	5800		Paint, 1 coat		21	.381		.85	7.90		8.75	13.55	
	6200		3' x 5' opening, 6 lites, primer coat, brushwork		15	.533		1.20	11.10		12.30	18.95	
	6400		Paint, 1 coat		15	.533		1.15	11.10		12.25	18.90	
	6800		4' x 8' opening, 6 lites, primer coat, brushwork		10	.800		1.50	16.65		18.15	28	
	7000		Paint, 1 coat		10	.800		1.45	16.65		18.10	28	
	8000		Single lite type, 2' x 3', oil base, primer coat, brushwork		34	.235		.65	4.89		5.54	8.50	
	8200		Paint, 1 coat		34	.235		.85	4.89		5.74	8.70	
	8600		3' x 5' opening, primer coat, brushwork		17	.471		1.10	9.80		10.90	16.80	
	8800		Paint, 1 coat		17	.471		1.05	9.80		10.85	16.75	
	9200		4' x 8' opening, primer coat, brushwork		12	.667		1.50	13.85		15.35	24	
	9400		Paint, 1 coat		12	.667		1.45	13.85		15.30	24	
	9800		For latex paint deduct					10%					
220	0010		**MISCELLANEOUS PAINTING**										220
	0020		Labor cost includes protection of adjacent items not painted										
	0700		Fence, chain link, oil base, primer coat, brushwork	2 Pord	1,450	.011	S.F.	.05	.23		.28	.42	
	1000		Spray		4,760	.003		.06	.07		.13	.18	
	1200		Paint 1 coat, brushwork		1,450	.011		.05	.23		.28	.42	
	1400		Spray		4,760	.003		.06	.07		.13	.18	
	1600		Picket, wood, one side, primer coat, brushwork		1,700	.009		.05	.20		.25	.37	
	1800		Spray		4,760	.003		.06	.07		.13	.18	
	2000		Paint 1 coat, brushwork		1,500	.011		.05	.22		.27	.41	
	2200		Spray		4,760	.003		.06	.07		.13	.18	
	2400		Floors, conc./wood, oil base, primer/sealer coat, brushwork		5,400	.003		.04	.06		.10	.14	
	2450		Roller		3,800	.004		.05	.09		.14	.19	
	2600		Spray		6,000	.003		.04	.06		.10	.13	
	2650		Paint 1 coat, brushwork		3,200	.005		.04	.10		.14	.21	
	2800		Roller		5,200	.003		.05	.06		.11	.16	
	2850		Spray		6,000	.003		.04	.06		.10	.13	
	3000		Stain, wood floor, brushwork, 1 coat		4,760	.003		.04	.07		.11	.16	
	3200		Roller		5,400	.003		.04	.06		.10	.14	
	3250		Spray		6,000	.003		.04	.06		.10	.13	
	3400		Varnish, wood floor, brushwork		4,760	.003		.04	.07		.11	.16	
	3450		Roller		5,400	.003		.04	.06		.10	.14	
	3600		Spray		6,000	.003		.04	.06		.10	.13	
	3650		For dust proofing or anti skid, see division 033-454										
	3750												
	3800		Grilles, per side, oil base, primer coat, brushwork	1 Pord	550	.015	S.F.	.04	.30		.34	.53	
	3850		Spray		1,100	.007		.04	.15		.19	.28	
	3920		Paint 2 coats, brushwork		340	.024		.06	.49		.55	.84	
	3940		Spray		680	.012		.06	.24		.30	.46	
	4200		Gutters and downspouts, oil base, primer coat, brushwork	2 Pord	1,200	.013	L.F.	.08	.28		.36	.53	
	4300		Paint 2 coats, brushwork		800	.020		.16	.42		.58	.84	
	5000		Pipe, to 4" diameter, primer or sealer coat, oil base, brushwork		1,360	.012		.08	.24		.32	.48	
	5100		Spray		2,380	.007		.08	.14		.22	.31	
	5350		Paint 2 coats, brushwork		850	.019		.14	.39		.53	.78	
	5400		Spray		1,360	.012		.14	.24		.38	.54	

For expanded coverage of these items see *Means Interior Cost Data 1991*

099 200 | Interior Painting

		CREW	DAILY OUTPUT	MAN-HOURS	UNIT	MAT.	LABOR	EQUIP.	TOTAL	TOTAL INCL O&P	
220							**BARE COSTS**				**220**
6300	To 16" diameter, primer or sealer coat, brushwork	2 Pord	340	.047	L.F.	.16	.98		1.14	1.73	
6450	Spray		567	.028		.17	.59		.76	1.12	
6500	Paint 2 coats, brushwork		202	.079		.33	1.65		1.98	2.98	
6550	Spray		323	.050		.33	1.03		1.36	2	
7000	Trim , wood, incl. puttying, under 6" wide										
7200	Primer coat, oil base, brushwork	1 Pord	900	.009	L.F.	.03	.18		.21	.33	
7250	Paint, 1 coat, brushwork		875	.009		.03	.19		.22	.34	
7450	3 coats		370	.022		.09	.45		.54	.81	
7470											
7500	Over 6" wide, primer coat, brushwork	1 Pord	600	.013	L.F.	.03	.28		.31	.47	
7550	Paint, 1 coat, brushwork		450	.018		.04	.37		.41	.63	
7650	3 coats		190	.042		.12	.88		1	1.53	
8000	Cornice, simple design, primer coat, oil base, brushwork		550	.015	S.F.	.04	.30		.34	.53	
8250	Paint, 1 coat		500	.016		.05	.33		.38	.58	
8350	Ornate design, primer coat		300	.027		.04	.55		.59	.93	
8400	Paint, 1 coat		280	.029		.05	.59		.64	1	
8600	Balustrades, primer coat, oil base, brushwork		598	.013		.04	.28		.32	.49	
8800											
8900	Trusses and wood frames, primer coat, oil base, brushwork	1 Pord	800	.010	S.F.	.04	.21		.25	.37	
8950	Spray		1,200	.007		.04	.14		.18	.26	
9220	Paint 2 coats, brushwork		500	.016		.09	.33		.42	.63	
9240	Spray		600	.013		.09	.28		.37	.54	
9260	Stain, brushwork, wipe off		600	.013		.04	.28		.32	.49	
9280	Varnish, 3 coats, brushwork		275	.029		.15	.61		.76	1.13	
9350	For latex paint, deduct					10%					
224	**0010 WALLS AND CEILINGS**										**224**
0020	Labor cost includes protection of adjacent items not painted										
0100	Concrete, dry wall or plaster, oil base, primer or sealer coat										
0200	Smooth finish, brushwork	1 Pord	1,300	.006	S.F.	.04	.13		.17	.25	
0240	Roller		2,040	.004		.04	.08		.12	.17	
0300	Sand finish, brushwork		1,163	.007		.05	.14		.19	.28	
0340	Roller		1,700	.005		.05	.10		.15	.21	
0380	Spray		2,720	.003		.05	.06		.11	.15	
0800	Paint 2 coats, smooth finish, brushwork		680	.012		.08	.24		.32	.48	
0840	Roller		1,190	.007		.09	.14		.23	.32	
0880	Spray		1,700	.005		.11	.10		.21	.28	
0900	Sand finish, brushwork		605	.013		.10	.28		.38	.55	
0940	Roller		1,020	.008		.11	.16		.27	.38	
0980	Spray		1,700	.005		.13	.10		.23	.30	
1500											
1600	Glaze coating, 5 coats, spray, clear	1 Pord	900	.009	S.F.	.50	.18		.68	.84	
1640	Multicolor	"	900	.009		.60	.18		.78	.95	
1700	For latex paint, deduct					10%					
1800	For ceiling installations, add						25%				
1900											
2000	Masonry or concrete block, oil base, primer or sealer coat										
2100	Smooth finish, brushwork	1 Pord	1,224	.007	S.F.	.08	.14		.22	.30	
2180	Spray		2,400	.003		.09	.07		.16	.21	
2200	Sand finish, brushwork		1,089	.007		.09	.15		.24	.34	
2280	Spray		2,400	.003		.11	.07		.18	.23	
2800	Paint 2 coats, smooth finish, brushwork		756	.011		.12	.22		.34	.48	
2880	Spray		1,360	.006		.13	.12		.25	.34	
2900	Sand finish, brushwork		672	.012		.13	.25		.38	.54	
2980	Spray		1,360	.006		.14	.12		.26	.35	
3600	Glaze coating, 5 coats, spray, clear		900	.009		.50	.18		.68	.84	
3620	Multicolor		900	.009		.60	.18		.78	.95	
4000	Block filler, 1 coat, brushwork		520	.015		.09	.32		.41	.61	

For expanded coverage of these items see *Means Interior Cost Data 1991*

099 200 | Interior Painting

		CREW	DAILY OUTPUT	MAN-HOURS	UNIT	BARE COSTS MAT.	LABOR	EQUIP.	TOTAL	TOTAL INCL O&P	
224	4100	Silicone, water repellent, 2 coats, spray	1 Pord	2,000	.004	S.F.	.12	.08		.20	.26	224
	4120	For latex paint, deduct					10%					
228	0010	**VARNISH** 1 coat + sealer, on wood trim, no sanding included	1 Pord	400	.020		.07	.42		.49	.74	228
	0100	Hardwood floors, 1 coat, no sanding included, 1 brushwork		2,380	.003		.07	.07		.14	.19	
	9000	Minimum labor/equipment charge		4	2	Job		42		42	66	

099 700 | Wallpaper

		CREW	DAILY OUTPUT	MAN-HOURS	UNIT	MAT.	LABOR	EQUIP.	TOTAL	TOTAL INCL O&P	
701	0010	**WALL COVERING**										701
	0050	Aluminum foil	1 Pape	275	.029	S.F.	.62	.61		1.23	1.64	
	0100	Copper sheets, .025″ thick, vinyl backing		240	.033		3.10	.69		3.79	4.51	
	0300	Phenolic backing		240	.033		4.05	.69		4.74	5.55	
	0600	Cork tiles, light or dark, 12″ x 12″ x 3/16″		240	.033		1.80	.69		2.49	3.08	
	0700	5/16″ thick		240	.033		1.90	.69		2.59	3.19	
	0900	1/4″ basketweave		240	.033		2.85	.69		3.54	4.24	
	1000	1/2″ natural, non-directional pattern		240	.033		2.75	.69		3.44	4.13	
	1200	Granular surface, 12″ x 36″, 1/2″ thick		385	.021		.60	.43		1.03	1.35	
	1300	1″ thick		385	.021		.80	.43		1.23	1.57	
	1500	Polyurethane coated, 12″ x 12″ x 3/16″ thick		240	.033		1.78	.69		2.47	3.06	
	1600	5/16″ thick		240	.033		2.48	.69		3.17	3.83	
	1800	Cork wallpaper, paperbacked, natural		480	.017		1.05	.35		1.40	1.71	
	1900	Colors		480	.017		1.25	.35		1.60	1.93	
	2100	Flexible wood veneer, 1/32″ thick, plain woods		100	.080		1.30	1.66		2.96	4.08	
	2200	Exotic woods		100	.080		1.88	1.66		3.54	4.72	
	2400	Gypsum-based, fabric-backed, fire										
	2500	resistant for masonry walls, minimum	1 Pape	800	.010	S.F.	.58	.21		.79	.97	
	2600	Average		720	.011		.68	.23		.91	1.12	
	2700	Maximum		640	.013		.80	.26		1.06	1.29	
	2750	Acrylic, modified, semi-rigid PVC, .028″ thick	2 Carp	330	.048		.69	1.07		1.76	2.50	
	2800	.040″ thick	"	320	.050		.89	1.10		1.99	2.78	
	3000	Vinyl wall covering, fabric-backed, lightweight	1 Pape	640	.013		.62	.26		.88	1.10	
	3300	Medium weight		480	.017		.75	.35		1.10	1.38	
	3400	Heavy weight		435	.018		.98	.38		1.36	1.69	
	3600	Adhesive, 5 gal. lots				Gal.	6.95			6.95	7.65	
	3700	Wallpaper at $10.00 per double roll, average workmanship	1 Pape	640	.013	S.F.	.22	.26		.48	.66	
	3900	Paper at $20 per double roll, average workmanship		535	.015		.38	.31		.69	.91	
	4000	Paper at $44 per double roll, quality workmanship		435	.018		.76	.38		1.14	1.44	
	4010											
	4200	Grass cloths with lining paper, minimum	1 Pape	400	.020	S.F.	.55	.42		.97	1.27	
	4300	Maximum		350	.023		1.40	.48		1.88	2.30	
	6000	Wallpaper removal, 3 layer, maximum		400	.020		.05	.42		.47	.72	
	9000	Minimum labor/equipment charge		2	4	Job		83		83	130	

099 900 | Surface Preparation

		CREW	DAILY OUTPUT	MAN-HOURS	UNIT	MAT.	LABOR	EQUIP.	TOTAL	TOTAL INCL O&P	
902	0010	**REMOVAL** Existing lead paint, by chemicals,										902
	0020	refinish with 2 coats of paint										
	0050	Baseboard, to 6″ wide	1 Pord	190	.042	L.F.	.07	.88		.95	1.47	
	0070	To 12″ wide		150	.053	"	.14	1.11		1.25	1.92	
	0200	Balustrades, one side		90	.089	S.F.	.17	1.85		2.02	3.13	
	0220											
	1400	Cabinets, simple design	1 Pord	85	.094	S.F.	.14	1.96		2.10	3.27	
	1420	Ornate design		40	.200		.16	4.16		4.32	6.80	
	1600	Cornice, simple design		65	.123		.14	2.56		2.70	4.23	
	1620	Ornate design		35	.229		.16	4.75		4.91	7.75	
	2800	Doors, one side, flush		125	.064		.14	1.33		1.47	2.27	
	2820	Two panel		110	.073		.14	1.51		1.65	2.56	
	2840	Four panel		95	.084		.16	1.75		1.91	2.96	
	2880	For trim, one side, add		200	.040	L.F.	.14	.83		.97	1.48	

For expanded coverage of these items see *Means Interior Cost Data 1991*

099 900	Surface Preparation	CREW	DAILY OUTPUT	MAN-HOURS	UNIT	BARE COSTS				TOTAL INCL O&P	
						MAT.	LABOR	EQUIP.	TOTAL		
902 3000	Fence, picket, one side	1 Pord	80	.100	S.F.	.14	2.08		2.22	3.46	**902**
3020											
3200	Grilles, one side, simple design	1 Pord	95	.084	S.F.	.16	1.75		1.91	2.96	
3220	Ornate design		45	.178	"	.19	3.70		3.89	6.10	
4400	Pipes, to 4" diameter		200	.040	L.F.	.07	.83		.90	1.40	
4420	To 8" diameter		90	.089		.14	1.85		1.99	3.10	
4440	To 12" diameter		65	.123		.24	2.56		2.80	4.34	
4460	To 16" diameter		45	.178		.37	3.70		4.07	6.30	
4500	For hangers, add	↓	100	.080	Ea.	.21	1.66		1.87	2.88	
4510											
4800	Siding	1 Pord	170	.047	S.F.	.14	.98		1.12	1.71	
5000	Trusses, open		75	.107	S.F.Face	.14	2.22		2.36	3.68	
6200	Windows, one side only, double hung, ¼ light, 24" x 48" high		12	.667	Ea.	1.12	13.85		14.97	23	
6220	30" x 60" high		9	.889		1.75	18.50		20.25	31	
6240	36" x 72" high		8	1		2.52	21		23.52	36	
6280	40" x 80" high		6	1.330		3.15	28		31.15	48	
6400	Colonial window, ⅝ light, 24" x 48" high		7	1.140		1.39	24		25.39	39	
6420	30" x 60" high		5	1.600		2.17	33		35.17	55	
6440	36" x 72" high		4	2		2.95	42		44.95	69	
6480	40" x 80" high		3.50	2.290		3.63	48		51.63	80	
6600	⅝ light, 24" x 48" high		6	1.330		1.66	28		29.66	46	
6620	40" x 80" high		3	2.670		4.14	55		59.14	93	
6800	12/12 light, 24" x 48" high		5	1.600		2.12	33		35.12	55	
6820	40" x 80" high		2.50	3.200	↓	4.57	67		71.57	110	
6840	For frame & trim, add		150	.053	L.F.	.14	1.11		1.25	1.92	
9000	Minimum labor/equipment charge	↓	3	2.670	Job		55		55	88	
904 0010	SANDING And puttying interior trim, compared to										**904**
0100	painting 1 coat, on quality work				L.F.		100%				
0300	Medium work						50%				
0400	Industrial grade				↓		25%				
906 0010	SCRAPE AFTER FIRE DAMAGE										**906**
0020											
0050	Boards, 1" x 4"	1 Pord	336	.024	L.F.		.50		.50	.79	
0060	1" x 6"		260	.031			.64		.64	1.02	
0070	1" x 8"		207	.039			.80		.80	1.28	
0080	1" x 10"		174	.046			.96		.96	1.52	
0500	Framing, 2" x 4"		265	.030			.63		.63	1	
0510	2" x 6"		221	.036			.75		.75	1.20	
0520	2" x 8"		190	.042			.88		.88	1.39	
0530	2" x 10"		165	.048			1.01		1.01	1.60	
0540	2" x 12"	↓	144	.056	↓		1.16		1.16	1.84	
0550											
1000	Heavy framing, 3" x 4"	1 Pord	226	.035	L.F.		.74		.74	1.17	
1010	4" x 4"		210	.038			.79		.79	1.26	
1020	4" x 6"		191	.042			.87		.87	1.39	
1030	4" x 8"		165	.048			1.01		1.01	1.60	
1040	4" x 10"		144	.056			1.16		1.16	1.84	
1060	4" x 12"		131	.061	↓		1.27		1.27	2.02	
2900	For sealing, minimum		825	.010	S.F.	.08	.20		.28	.41	
2920	Maximum		460	.017	"	.08	.36		.44	.66	
3000	For sandblasting, see division 045-102										
3020											
9000	Minimum labor/equipment charge	1 Pord	3	2.670	Job		55		55	88	

101 100 | Chalkboards

		CREW	DAILY OUTPUT	MAN-HOURS	UNIT	BARE COSTS MAT.	BARE COSTS LABOR	BARE COSTS EQUIP.	BARE COSTS TOTAL	TOTAL INCL O&P
104	0010	**CHALKBOARDS** Porcelain enamel steel									104
	0100	Freestanding, reversible									
	0120	Economy, wood frame, 4' x 6'									
	0140	Chalkboard both sides				Ea.	231			231	255
	0160	Chalkboard one side, cork other side				"	284			284	310
	0200	Standard, lightweight satin finished aluminum, 4' x 6'									
	0220	Chalkboard both sides				Ea.	272			272	300
	0240	Chalkboard one side, cork other side				"	330			330	365
	0300	Deluxe, heavy duty extruded aluminum, 4' x 6'									
	0320	Chalkboard both sides				Ea.	854			854	940
	0340	Chalkboard one side, cork other side				"	854			854	940
	3900	Wall hung									
	4000	Aluminum frame and chalktrough									
	4300	3' x 6'	2 Carp	15	1.070	Ea.	183	23		206	240
	4600	4' x 12'	"	13	1.230	"	364	27		391	445
	4700	Wood frame and chalktrough									
	4800	3' x 4'	2 Carp	16	1	Ea.	75	22		97	120
	5300	4' x 8'	"	13	1.230	"	125	27		152	180
	5350										
	5400	Liquid chalk, white porcelain enamel, wall hung									
	5420	Deluxe units, aluminum trim and chalktrough									
	5450	4' x 4'	2 Carp	16	1	Ea.	150	22		172	200
	5550	4' x 12'	"	12	1.330	"	384	29		413	470
	5700	Wood trim and chalktrough									
	5900	4' x 4'	2 Carp	16	1	Ea.	145	22		167	195
	6200	4' x 8'		14	1.140	"	234	25		259	300
	9000	Minimum labor/equipment charge		3	5.330	Job		115		115	190

101 600 | Toilet Compartments

		CREW	DAILY OUTPUT	MAN-HOURS	UNIT	BARE COSTS MAT.	BARE COSTS LABOR	BARE COSTS EQUIP.	BARE COSTS TOTAL	TOTAL INCL O&P
602	0010	**PARTITIONS, TOILET**									602
	0100	Cubicles, ceiling hung, marble	2 Marb	2	8	Ea.	835	180		1,015	1,200
	0200	Painted metal	2 Carp	4	4		345	88		433	525
	0300	Plastic laminate on particle board		4	4		435	88		523	620
	0400	Porcelain enamel		4	4		695	88		783	910
	0500	Stainless steel		4	4		840	88		928	1,075
	0600	For handicap units, incl. 52" grab bars, add					160			160	175
	0700										
	0800	Floor & ceiling anchored, marble	2 Marb	2.50	6.400	Ea.	705	145		850	1,000
	1000	Painted metal	2 Carp	5	3.200		330	70		400	480
	1100	Plastic laminate on particle board		5	3.200		425	70		495	580
	1200	Porcelain enamel		5	3.200		685	70		755	870
	1300	Stainless steel		5	3.200		830	70		900	1,025
	1400	For handicap units, incl. 52" grab bars, add					160			160	175
	1600	Floor mounted, marble	2 Marb	3	5.330		690	120		810	950
	1700	Painted metal	2 Carp	7	2.290		305	50		355	420
	1800	Plastic laminate on particle board		7	2.290		405	50		455	530
	1900	Porcelain enamel		7	2.290		680	50		730	830
	2000	Stainless steel		7	2.290		820	50		870	985
	2100	For handicap units, incl. 52" grab bars, add					160			160	175
	2200	For juvenile units, deduct					30			30	33
	2300										
	2400	Floor mounted, headrail braced, marble	2 Marb	3	5.330	Ea.	725	120		845	990
	2500	Painted metal	2 Carp	6	2.670		305	59		364	430
	2600	Plastic laminate on particle board		6	2.670		425	59		484	565
	2700	Porcelain enamel		6	2.670		685	59		744	850
	2800	Stainless steel		6	2.670		810	59		869	985
	2900	For handicap units, incl. 52" grab bars, add					160			160	175

101 600 | Toilet Compartments

		CREW	DAILY OUTPUT	MAN-HOURS	UNIT	BARE COSTS MAT.	LABOR	EQUIP.	TOTAL	TOTAL INCL O&P	
602	3000	Wall hung partitions, painted metal	2 Carp	7	2.290	Ea.	360	50		410	480	602
	3200	Porcelain enamel		7	2.290		695	50		745	845	
	3300	Stainless steel	↓	7	2.290		810	50		860	975	
	3400	For handicap units, incl. 52" grab bars, add				↓	160			160	175	
	4000	Screens, entrance, floor mounted, 54" high										
	4100	Marble	D-1	35	.457	L.F.	225	9.25		234.25	260	
	4200	Painted metal	2 Carp	60	.267		135	5.85		140.85	160	
	4300	Plastic laminate on particle board		60	.267		165	5.85		170.85	190	
	4400	Porcelain enamel		60	.267		265	5.85		270.85	300	
	4500	Stainless steel	↓	60	.267	↓	280	5.85		285.85	320	
	4600	Urinal screen, 18" wide, ceiling braced, marble	D-1	6	2.670	Ea.	295	54		349	410	
	4700	Painted metal	2 Carp	8	2		145	44		189	230	
	4800	Plastic laminate on particle board		8	2		180	44		224	270	
	4900	Porcelain enamel		8	2		265	44		309	365	
	5000	Stainless steel	↓	8	2	↓	365	44		409	475	
	5050											
	5100	Floor mounted, head rail braced										
	5200	Marble	D-1	6	2.670	Ea.	275	54		329	390	
	5300	Painted metal	2 Carp	8	2		145	44		189	230	
	5400	Plastic laminate on particle board		8	2		185	44		229	275	
	5500	Porcelain enamel		8	2		275	44		319	375	
	5600	Stainless steel		8	2		375	44		419	485	
	5700	Pilaster, flush, marble	D-1	9	1.780		340	36		376	430	
	5800	Painted metal	2 Carp	10	1.600		190	35		225	265	
	5900	Plastic laminate on particle board		10	1.600		250	35		285	330	
	6000	Porcelain enamel		10	1.600		225	35		260	305	
	6100	Stainless steel	↓	10	1.600	↓	275	35		310	360	
	6150											
	6200	Post braced, marble	D-1	9	1.780	Ea.	325	36		361	415	
	6300	Painted metal	2 Carp	10	1.600		195	35		230	270	
	6400	Plastic laminate on particle board		10	1.600		250	35		285	330	
	6500	Porcelain enamel		10	1.600		225	35		260	305	
	6600	Stainless steel	↓	10	1.600		275	35		310	360	
	6700	Wall hung, bracket supported										
	6800	Painted metal	2 Carp	10	1.600	Ea.	195	35		230	270	
	6900	Plastic laminate on particle board		10	1.600		245	35		280	325	
	7000	Porcelain enamel		10	1.600		235	35		270	315	
	7100	Stainless steel		10	1.600	↓	240	35		275	320	
	7200											
	7400	Flange supported, painted metal	2 Carp	10	1.600	Ea.	130	35		165	200	
	7500	Plastic laminate on particle board		10	1.600		145	35		180	215	
	7600	Porcelain enamel		10	1.600		275	35		310	360	
	7700	Stainless steel		10	1.600		295	35		330	380	
	7800	Wedge type, painted metal		10	1.600		165	35		200	240	
	8000	Porcelain enamel		10	1.600		235	35		270	315	
	8100	Stainless steel	↓	10	1.600	↓	300	35		335	385	
	9000	Minimum labor/equipment charge	1 Carp	2.50	3.200	Job		70		70	115	

101 850 | Shower Compartments

		CREW	DAILY OUTPUT	MAN-HOURS	UNIT	BARE COSTS MAT.	LABOR	EQUIP.	TOTAL	TOTAL INCL O&P	
852	0010	**PARTITIONS, SHOWER** Floor mounted, no plumbing										852
	0100	Cabinet, incl. base, no door, painted steel, 1" thick walls	2 Shee	5	3.200	Ea.	450	80		530	620	
	0300	With door, fiberglass		5	3.200		340	80		420	500	
	0600	Galvanized and painted steel, 1" thick walls		5	3.200		485	80		565	660	
	0800	Stall, 1" thick wall, no base, enameled steel		5	3.200		465	80		545	635	
	1500	Circular fiberglass, cabinet 36" diameter,		4	4		375	100		475	570	
	1700	One piece, 36" diameter, less door		4	4		325	100		425	515	
	1800	With door	↓	3.50	4.570		510	115		625	740	

101 | Chalkboards, Compartments and Cubicles

101 850 | Shower Compartments

			CREW	DAILY OUTPUT	MAN-HOURS	UNIT	MAT.	LABOR	EQUIP.	TOTAL	TOTAL INCL O&P	
852	1900											852
	4100	Shower doors, economy plastic, 24" wide	1 Shee	9	.889	Ea.	70	22		92	110	
	4200	Tempered glass door, economy		8	1		120	25		145	170	
	4700	Deluxe, tempered glass, chrome on brass frame, minimum		5	1.600		265	40		305	355	
	4800	Maximum		1	8	↓	490	200		690	855	
	9990	Minimum labor/equipment charge	↓	2.50	3.200	Job		80		80	125	

102 | Louvers, Corner Protection and Access Flooring

102 100 | Metal Wall Louvers

			CREW	DAILY OUTPUT	MAN-HOURS	UNIT	MAT.	LABOR	EQUIP.	TOTAL	TOTAL INCL O&P	
104	0010	LOUVERS Aluminum with screen, residential, 8" x 8"	1 Carp	38	.211	Ea.	6	4.63		10.63	14.15	104
	0100	12" x 12"		38	.211		8.20	4.63		12.83	16.60	
	0200	12" x 18"		35	.229		10.65	5.05		15.70	19.95	
	0250	14" x 24"		30	.267		13.90	5.85		19.75	25	
	0300	18" x 24"		27	.296		16.35	6.50		22.85	29	
	0500	24" x 30"		24	.333		24	7.35		31.35	38	
	0700	Triangle, adjustable, small		20	.400		17.30	8.80		26.10	33	
	0800	Large		15	.533		82	11.75		93.75	110	
	2100	Midget, aluminum, ¾" deep, 1" diameter		85	.094		.51	2.07		2.58	3.94	
	2150	3" diameter		60	.133		1.34	2.93		4.27	6.25	
	2200	4" diameter		50	.160		1.69	3.52		5.21	7.60	
	2250	6" diameter		30	.267	↓	2.37	5.85		8.22	12.20	
	2300	Ridge vent strip, mill finish	1 Shee	155	.052	L.F.	2.51	1.29		3.80	4.79	
	2400	Under eaves vent, aluminum, mill finish, 16" x 4"	1 Carp	75	.107	Ea.	1.67	2.35		4.02	5.65	
	2500	16" x 8"		75	.107		1.89	2.35		4.24	5.90	
	7000	Vinyl gable vent, 8" x 8"		38	.211		5.25	4.63		9.88	13.35	
	7020	12" x 12"		38	.211		6.25	4.63		10.88	14.45	
	7080	12" x 18"		35	.229		8.55	5.05		13.60	17.60	
	7200	14" x 24"		30	.267	↓	11.55	5.85		17.40	22	
	9000	Minimum labor/equipment charge	↓	2.50	3.200	Job		70		70	115	

102 600 | Wall & Corner Guards

			CREW	DAILY OUTPUT	MAN-HOURS	UNIT	MAT.	LABOR	EQUIP.	TOTAL	TOTAL INCL O&P	
604	0010	CORNER GUARDS Steel angle w/anchors, 1" x 1" x ¼", 1.5#/L.F.	E-4	320	.100	L.F.	2.95	2.46	.21	5.62	7.95	604
	0100	2" x 2" x ¼" angles, 3.2#/L.F.	"	300	.107	"	4.80	2.62	.23	7.65	10.30	
	9000	Minimum labor/equipment charge	A-1	2	4	Job		70	27	97	145	

102 700 | Access Flooring

			CREW	DAILY OUTPUT	MAN-HOURS	UNIT	MAT.	LABOR	EQUIP.	TOTAL	TOTAL INCL O&P	
705	0010	PEDESTAL ACCESS FLOORS Computer room application, metal										705
	0020	Particle board or steel panels, no covering, under 6000 S. F.	2 Carp	1,000	.016	S.F.	5	.35		5.35	6.05	
	0300	Metal covered, over 6000 S.F.		1,100	.015		5	.32		5.32	6	
	0400	Aluminum, 24" panels	↓	500	.032		15	.70		15.70	17.65	
	0600	For carpet covering, add					2.35			2.35	2.59	
	0700	For vinyl floor covering, add					3.40			3.40	3.74	
	0900	For high pressure laminate covering, add					3.25			3.25	3.58	
	0910	For snap on stringer system, add	2 Carp	1,000	.016	↓	.60	.35		.95	1.23	
	0950	Office applications, to 8" high, steel panels,										
	0960	no covering, over 6000 S.F.	2 Carp	400	.040	S.F.	4.75	.88		5.63	6.65	

103 050 | Prefabricated Fireplaces

		CREW	DAILY OUTPUT	MAN-HOURS	UNIT	MAT.	LABOR	EQUIP.	TOTAL	TOTAL INCL O&P	
054	0010	**FIREPLACE, PREFABRICATED** Free standing or wall hung									054
	0100	with hood & screen, minimum	F-1	1.30	6.150	Ea.	675	135	6.15	816.15	970	
	0150	Average		1	8		885	175	8	1,068	1,275	
	0200	Maximum		.90	8.890		2,100	195	8.90	2,303	2,650	
	0500	Chimney dbl. wall, all stainless, over 8'-6", 7" diam., add		33	.242	V.L.F.	19.50	5.35	.24	25.09	30	
	0600	10" diameter, add		32	.250		32.20	5.50	.25	37.95	45	
	0700	12" diameter, add		31	.258		42.50	5.70	.26	48.46	56	
	0800	14" diameter, add		30	.267		52.60	5.85	.27	58.72	68	
	1000	Simulated brick chimney top, 4' high, 16" x 16"		10	.800	Ea.	130	17.60	.80	148.40	175	
	1100	24" x 24"		7	1.140	"	260	25	1.14	286.14	330	
	1500	Simulated logs, gas fired, 40,000 BTU, 2' long, minimum		7	1.140	Set	300	25	1.14	326.14	370	
	1600	Maximum		6	1.330		500	29	1.33	530.33	600	
	1700	Electric, 1,500 BTU, 1'-6" long, minimum		7	1.140		85	25	1.14	111.14	135	
	1800	11,500 BTU, maximum		6	1.330		190	29	1.33	220.33	260	
	2000	Fireplace, built-in, 36" hearth, radiant		1.30	6.150	Ea.	385	135	6.15	526.15	650	
	2100	Recirculating, small fan		1	8		530	175	8	713	880	
	2150	Large fan		.90	8.890		1,035	195	8.90	1,238	1,475	
	2200	42" hearth, radiant		1.20	6.670		490	145	6.65	641.65	785	
	2300	Recirculating, small fan		.90	8.890		655	195	8.90	858.90	1,050	
	2350	Large fan		.80	10		1,250	220	10	1,480	1,750	
	2400	48" hearth, radiant		1.10	7.270		930	160	7.25	1,097	1,300	
	2500	Recirculating, small fan		.80	10		1,125	220	10	1,355	1,600	
	2550	Large fan		.70	11.430		1,780	250	11.45	2,041	2,375	
	3000	See through, including doors		.80	10		1,435	220	10	1,665	1,950	
	3200	Corner (2 wall)		1	8		700	175	8	883	1,075	

103 100 | Fireplace Accessories

		CREW	DAILY OUTPUT	MAN-HOURS	UNIT	MAT.	LABOR	EQUIP.	TOTAL	TOTAL INCL O&P	
104	0010	**FIREPLACE ACCESSORIES** Chimney screens, galv., 13" x 13" flue	1 Bric	8	1	Ea.	35	23		58	75	104
	0050	Galv., 24" x 24" flue		5	1.600		95	36		131	165	
	0200	Stainless steel, 13" x 13" flue		8	1		200	23		223	255	
	0250	20" x 20" flue		5	1.600		280	36		316	365	
	0400	Cleanout doors and frames, cast iron, 8" x 8"		12	.667		12.80	15.15		27.95	38	
	0450	12" x 12"		10	.800		28.60	18.20		46.80	61	
	0500	18" x 24"		8	1		80	23		103	125	
	0550	Cast iron frame, steel door, 24" x 30"		5	1.600		175	36		211	250	
	0800	Damper, rotary control, steel, 30" opening		6	1.330		50	30		80	105	
	0850	Cast iron, 30" opening		6	1.330		55	30		85	110	
	0880	36" opening		6	1.330		62	30		92	115	
	0900	48" opening		6	1.330		80	30		110	135	
	0920	60" opening		6	1.330		155	30		185	220	
	1000	84" opening, special order		5	1.600		480	36		516	585	
	1050	96" opening, special order		4	2		495	46		541	615	
	1100											
	1200	Steel plate, poker control, 60" opening	1 Bric	8	1	Ea.	165	23		188	220	
	1250	84" opening		5	1.600		295	36		331	385	
	1400	"Universal" type, chain operated, 32" x 20" opening		8	1		125	23		148	175	
	1450	48" x 24" opening		5	1.600		195	36		231	275	
	1600	Dutch Oven door and frame, cast iron, 12" x 15" opening		13	.615		65	14		79	94	
	1650	Copper plated, 12" x 15" opening		13	.615		125	14		139	160	
	1800	Fireplace forms with registers, 25" opening		3	2.670		350	61		411	480	
	1900	34" opening		2.50	3.200		425	73		498	585	
	2000	48" opening		2	4		785	91		876	1,000	
	2100	72" opening		1.50	5.330		1,100	120		1,220	1,400	
	2400	Squirrel and bird screens, galvanized, 8" x 8" flue		16	.500		32	11.40		43.40	53	
	2450	13" x 13" flue		8	1		38	23		61	78	
	9000	Minimum labor/equipment charge		3.50	2.290	Job		52		52	83	

103 | Fireplaces, Ext. Specialties and Flagpoles

103 200 | Stoves

			CREW	DAILY OUTPUT	MAN-HOURS	UNIT	BARE COSTS MAT.	LABOR	EQUIP.	TOTAL	TOTAL INCL O&P	
201	0010	WOODBURNING STOVES Cast iron, minimum	F-2	1.30	12.310	Ea.	550	270	12.30	832.30	1,050	201
	0020	Average		1	16		800	350	16	1,166	1,475	
	0030	Maximum	↓	.80	20		2,000	440	20	2,460	2,950	
	0050	For gas log lighter, add				↓	28			28	31	

103 460 | Cupolas

			CREW	DAILY OUTPUT	MAN-HOURS	UNIT	BARE COSTS MAT.	LABOR	EQUIP.	TOTAL	TOTAL INCL O&P	
464	0010	CUPOLA Stock units, pine, painted, 18″ sq., 28″ high, alum. roof	1 Carp	4	2	Ea.	105	44		149	185	464
	0100	Copper roof		4	2		105	44		149	185	
	0300	23″ square, 33″ high, aluminum roof		3.50	2.290		167	50		217	265	
	0400	Copper roof		3.50	2.290		167	50		217	265	
	0600	30″ square, 37″ high, aluminum roof		3.50	2.290		266	50		316	375	
	0700	Copper roof		3.50	2.290		266	50		316	375	
	0900	Hexagonal, 31″ wide, 46″ high, copper roof		3	2.670		368	59		427	500	
	1000	36″ wide, 50″ high, copper roof	↓	3	2.670		431	59		490	570	
	1200	For deluxe stock units, add to above					25%					
	1400	For custom built units, add to above					50%	50%				
	1600	Fiberglass, 5′-0″ base, 63″ high minimum	F-3	6	6.670		1,000	150	63	1,213	1,400	
	1650	Maximum		4	10		2,500	225	95	2,820	3,225	
	1700	6′-0″ base, 63″ high, minimum		5	8		1,100	180	76	1,356	1,575	
	1750	Maximum	↓	3	13.330	↓	2,750	295	125	3,170	3,650	
	9000	Minimum labor/equipment charge	1 Carp	2.75	2.910	Job		64		64	105	

104 | Identifying and Pedestrian Control Devices

104 100 | Directories

			CREW	DAILY OUTPUT	MAN-HOURS	UNIT	BARE COSTS MAT.	LABOR	EQUIP.	TOTAL	TOTAL INCL O&P	
104	0010	DIRECTORY BOARDS Plastic, glass covered, 30″ x 20″	2 Carp	3	5.330	Ea.	261	115		376	480	104
	0100	36″ x 48″		2	8		539	175		714	880	
	0900	Outdoor, weatherproof, black plastic, 36″ x 24″		2	8		553	175		728	895	
	1000	36″ x 36″	↓	1.50	10.670	↓	638	235		873	1,075	
	9000	Minimum labor/equipment charge	1 Carp	1	8	Job		175		175	285	

104 150 | Bulletin Boards

			CREW	DAILY OUTPUT	MAN-HOURS	UNIT	BARE COSTS MAT.	LABOR	EQUIP.	TOTAL	TOTAL INCL O&P	
151	0011	BULLETIN BOARD										151
	0020											
	2120	Prefabricated, ¼″ cork, 3′ x 5′ with aluminum frame	2 Carp	16	1	Ea.	83	22		105	125	
	2140	Wood frame	"	16	1	"	104	22		126	150	
	2300	Glass enclosed cabinets, alum., cork panel, hinged doors										
	2600	4′ x 7′, 3 door	2 Carp	10	1.600	Ea.	1,000	35		1,035	1,150	
	9000	Minimum labor/equipment charge	"	4	4	Job		88		88	145	

104 300 | Signs

			CREW	DAILY OUTPUT	MAN-HOURS	UNIT	BARE COSTS MAT.	LABOR	EQUIP.	TOTAL	TOTAL INCL O&P	
304	0011	SIGNS, Plaques, 20″ x 30″, up to 450 letters, cast alum.	2 Carp	4	4	Ea.	620	88		708	825	304
	4000	Cast bronze		4	4		850	88		938	1,075	
	4200	30″ x 36″, up to 900 letters cast aluminum		3	5.330		1,115	115		1,230	1,425	
	4300	Cast bronze	↓	3	5.330		1,530	115		1,645	1,875	
	5100	Exit signs, 24 ga. alum., 14″ x 12″ surface mounted	1 Carp	30	.267		6	5.85		11.85	16.20	
	5200	10″ x 7″	"	20	.400		4.50	8.80		13.30	19.30	
	6400	Replacement sign faces, 6″ or 8″	1 Clab	50	.160	↓	18	2.80		20.80	24	
	9000	Minimum labor/equipment charge	1 Carp	4	2	Job		44		44	72	

105 050 | Metal Lockers

		CREW	DAILY OUTPUT	MAN-HOURS	UNIT	BARE COSTS				TOTAL INCL O&P	
						MAT.	LABOR	EQUIP.	TOTAL		
054	0010	**LOCKERS** Steel, baked enamel, 60″ or 72″, single tier, minimum	1 Shee	14	.571	Opng.	85.25	14.30		99.55	115	**054**
	0100	Maximum		12	.667		150.60	16.65		167.25	190	
	0300	2 tier, 60″ or 72″ total height, minimum ⑰		26	.308		52.30	7.70		60	70	
	0400	Maximum		20	.400		72.15	10		82.15	95	
	2400	16-person locker unit with clothing rack										
	2500	72 wide x 15″ deep x 72″ high	1 Shee	80	.100	Ea.	405	2.50		407.50	450	
	2550	18″ deep	"	8	1		425	25		450	505	
	3600	For hanger rods, add					1.25			1.25	1.38	
	9000	Minimum labor/equipment charge	1 Shee	2.50	3.200	Job		80		80	125	

105 380 | Canopies

		CREW	DAILY OUTPUT	MAN-HOURS	UNIT	BARE COSTS				TOTAL INCL O&P	
						MAT.	LABOR	EQUIP.	TOTAL		
384	0010	**CANOPIES** Wall hung, aluminum, prefinished, 8′ x 10′	K-2	1.30	18.460	Ea.	805	420	300	1,525	1,950	**384**
	0300	8′ x 20′		1.10	21.820		1,440	495	355	2,290	2,850	
	1000	12′ x 20′		1	24		2,160	545	390	3,095	3,775	
	1050											
	2300	Aluminum entrance canopies, flat soffit										
	2500	3′-6″ x 4′-0″, clear anodized	2 Carp	4	4	Ea.	510	88		598	705	
	4700	Canvas awnings, including canvas, frame & lettering										
	5000	Minimum	2 Carp	100	.160	S.F.	8	3.52		11.52	14.55	
	5300	Average		90	.178		16	3.91		19.91	24	
	5500	Maximum		80	.200		23	4.40		27.40	32	
	9000	Minimum labor/equipment charge		2	8	Job		175		175	285	
388	0010	**CANOPIES, RESIDENTIAL** Prefabricated										**388**
	0020											
	0500	Carport, free standing, baked enamel, alum., 40 psf										
	0520	16′ x 8′, 4 posts	2 Carp	3	5.330	Ea.	1,010	115		1,125	1,300	
	0600	20′ x 10′, 6 posts	"	2	8		1,350	175		1,525	1,775	
	1000	Door canopies, extruded alum., 42″ projection, 4′ wide	1 Carp	8	1		350	22		372	420	
	1020	6′ wide	"	6	1.330		420	29		449	510	
	1040	8′ wide	2 Carp	9	1.780		580	39		619	700	
	1060	10′ wide		7	2.290		660	50		710	810	
	1080	12′ wide		5	3.200		785	70		855	980	
	1100											
	1200	54″ projection, 4′ wide	1 Carp	8	1	Ea.	440	22		462	520	
	1220	6′ wide	"	6	1.330		585	29		614	690	
	1240	8′ wide	2 Carp	9	1.780		790	39		829	935	
	1260	10′ wide		7	2.290		885	50		935	1,050	
	1280	12′ wide		5	3.200		995	70		1,065	1,200	
	1300	Painted, add					20%					
	1310	Bronze anodized, add					50%					
	3000	Window awnings, aluminum, window 3′ high, 4′ wide	1 Carp	10	.800		98	17.60		115.60	135	
	3020	6′ wide	"	8	1		128	22		150	175	
	3040	9′ wide	2 Carp	9	1.780		170	39		209	250	
	3060	12′ wide	"	5	3.200		215	70		285	350	
	3100	Window, 4′ high, 4′ wide	1 Carp	10	.800		116	17.60		133.60	155	
	3120	6′ wide	"	8	1		154	22		176	205	
	3140	9′ wide	2 Carp	9	1.780		205	39		244	290	
	3160	12′ wide	"	5	3.200		260	70		330	400	
	3200	Window, 6′ high, 4′ wide	1 Carp	10	.800		180	17.60		197.60	225	
	3220	6′ wide	"	8	1		250	22		272	310	
	3240	9′ wide	2 Carp	9	1.780		310	39		349	405	
	3260	12′ wide	"	5	3.200		385	70		455	540	
	3400	Roll-up aluminum, 2′-6″ wide	1 Carp	14	.571		72	12.55		84.55	100	
	3420	3′ wide		12	.667		87	14.65		101.65	120	
	3440	4′ wide		10	.800		110	17.60		127.60	150	
	3460	6′ wide		8	1		133	22		155	180	

105 380	Canopies	CREW	DAILY OUTPUT	MAN-HOURS	UNIT	MAT.	LABOR	EQUIP.	TOTAL	TOTAL INCL O&P	
388 3480	9' wide	2 Carp	9	1.780	Ea.	185	39		224	265	388
3500	12' wide	"	5	3.200	"	240	70		310	380	
3600	Window awnings, canvas, 24" drop, 3' wide	1 Carp	30	.267	L.F.	43	5.85		48.85	57	
3620	4' wide		40	.200		40	4.40		44.40	51	
3700	30" drop, 3' wide		30	.267		49	5.85		54.85	63	
3720	4' wide		40	.200		43	4.40		47.40	54	
3740	5' wide		45	.178		40	3.91		43.91	50	
3760	6' wide		48	.167		38	3.67		41.67	48	
3780	8' wide		48	.167		31	3.67		34.67	40	
3800	10' wide		50	.160		28	3.52		31.52	37	
3900	Repair canvas, minimum		8	1	Ea.		22		22	36	
3920	Maximum		4	2	"		44		44	72	
9000	Minimum labor/equipment charge		3	2.670	Job		59		59	96	

105 510	Mail Chutes	CREW	DAILY OUTPUT	MAN-HOURS	UNIT	MAT.	LABOR	EQUIP.	TOTAL	TOTAL INCL O&P	
511 0010	MAIL CHUTES Aluminum & glass, 14-1/4" wide, 4-5/8" deep	2 Shee	4	4	Floor	530	100		630	740	511
0100	8-5/8" deep		3.80	4.210	"	585	105		690	810	
0600	Lobby collection boxes, aluminum		4.50	3.560	Ea.	1,195	89		1,284	1,450	
0700	Bronze or stainless		4.50	3.560	"	1,975	89		2,064	2,300	

105 520	Mail Boxes	CREW	DAILY OUTPUT	MAN-HOURS	UNIT	MAT.	LABOR	EQUIP.	TOTAL	TOTAL INCL O&P	
521 0010	MAIL BOXES Horiz., key lock, 5"H x 6"W x 15"D, alum., rear load	1 Carp	34	.235	Ea.	45	5.20		50.20	58	521
0100	Front loading		34	.235		49	5.20		54.20	62	
0200	Double, 5"H x 12"W x 15"D, rear loading		26	.308		82	6.75		88.75	100	
0300	Front loading		26	.308		88	6.75		94.75	110	
0500	Quadruple, 10"H x 12"W x 15"D, rear loading		20	.400		140	8.80		148.80	170	
0600	Front loading		20	.400		150	8.80		158.80	180	
1600	Vault type, horizontal, for apartments, 4" x 5"		34	.235		30	5.20		35.20	41	
1700	Alphabetical directories, 120 names		10	.800		125	17.60		142.60	165	
1900	Letter slot, residential		20	.400		26	8.80		34.80	43	
2000	Post office type		8	1		100	22		122	145	
9000	Minimum labor/equipment charge		5	1.600	Job		35		35	57	

106 150	Demountable Partitions	CREW	DAILY OUTPUT	MAN-HOURS	UNIT	MAT.	LABOR	EQUIP.	TOTAL	TOTAL INCL O&P	
152 0010	PARTITIONS, MOVABLE OFFICE Demountable, add for doors										152
0100	Do not deduct door openings from total L.F.										
0900	Demountable gypsum system on 2" to 2-1/2"										
1000	steel studs, 9' high, 3" to 3-3/4" thick										
1200	Vinyl clad gypsum	2 Carp	48	.333	L.F.	17	7.35		24.35	31	
1300	Fabric clad gypsum		44	.364		52	8		60	70	
1500	Steel clad gypsum		40	.400		56	8.80		64.80	76	
1600	1.75 system, aluminum framing, vinyl clad hardboard,										
1800	paper honeycomb core panel, 1-3/4" to 2-1/2" thick										
1900	9' high	2 Carp	48	.333	L.F.	36	7.35		43.35	52	
2100	7' high		60	.267		32	5.85		37.85	45	
2200	5' high		80	.200		28	4.40		32.40	38	
2250	Unitized gypsum system										
2300	Unitized panel, 9' high, 2" to 2-1/2" thick										

106 150 | Demountable Partitions

			CREW	DAILY OUTPUT	MAN-HOURS	UNIT	BARE COSTS MAT.	LABOR	EQUIP.	TOTAL	TOTAL INCL O&P	
152	2350	Vinyl clad gypsum	2 Carp	48	.333	L.F.	48	7.35		55.35	65	152
	2400	Fabric clad gypsum	"	44	.364	"	86	8		94	110	
	2500	Unitized mineral fiber system										
	2510	Unitized panel, 9' high, 2-¼" thick, aluminum frame										
	2550	Vinyl clad mineral fiber	2 Carp	48	.333	L.F.	60	7.35		67.35	78	
	2600	Fabric clad mineral fiber	"	44	.364	"	78	8		86	99	
	2800	Movable steel walls, modular system										
	2900	Unitized panels, 9' high, 48" wide										
	3100	Baked enamel, pre-finished	2 Carp	60	.267	L.F.	56	5.85		61.85	71	
	3200	Fabric clad steel	"	56	.286	"	87	6.30		93.30	105	
	5500	For acoustical partitions, add, minimum				S.F.	1			1	1.10	
	5550	Maximum				"	3.75			3.75	4.13	
	5700	For doors, see div. 081 & 082										
	5800	For door hardware, see div. 087										
	6000											
	6100	In-plant modular office system, w/prehung steel door										
	6200	3" thick honeycomb core panels										
	6250	12' x 12', 2 wall	2 Clab	3.80	4.210	Ea.	2,645	74		2,719	3,025	
	6300	4 wall		1.90	8.420		3,855	145		4,000	4,475	
	6350	16' x 16', 2 wall		3.60	4.440		3,745	78		3,823	4,250	
	6400	4 wall		1.80	8.890		5,370	155		5,525	6,150	
	9000	Minimum labor/equipment charge	2 Carp	3	5.330	Job		115		115	190	

106 300 | Portable Partitions

			CREW	DAILY OUTPUT	MAN-HOURS	UNIT	BARE COSTS MAT.	LABOR	EQUIP.	TOTAL	TOTAL INCL O&P	
304	0010	**PARTITIONS, PORTABLE** Divider panels, free standing, fiber core										304
	0020	Fabric face straight										
	0100	3'-0" long, 4'-0" high	2 Carp	100	.160	L.F.	75	3.52		78.52	88	
	0200	5'-0" high		90	.178		80	3.91		83.91	94	
	0500	6'-0" high		75	.213		95	4.69		99.69	110	
	0900	5'-0" long, 4'-0" high		175	.091		55	2.01		57.01	64	
	1000	5'-0" high		150	.107		60	2.35		62.35	70	
	1500	6"-0" high		125	.128		65	2.82		67.82	76	
	1600	6'-0" long, 5'-0" high		162	.099		50	2.17		52.17	59	
	3100	Curved, 3'-0" long, 5'-0" high		90	.178		110	3.91		113.91	125	
	3150	6'-0" high		75	.213		115	4.69		119.69	135	
	3200	Economical panels, fabric face, 4'-0" long, 5'-0" high		132	.121		30	2.67		32.67	37	
	3250	6'-0" high		112	.143		38	3.14		41.14	47	
	3300	5'-0" long, 5'-0" high		150	.107		28	2.35		30.35	35	
	3350	6'-0" high		125	.128		30	2.82		32.82	38	
	3380	3'-0" curved, 5'-0" high		90	.178		65	3.91		68.91	78	
	3390	6'-0" high		75	.213		70	4.69		74.69	85	
	3450	Acoustical panels, 60 to 90 NRC, 3'-0" long, 5'-0" high		90	.178		95	3.91		98.91	110	
	3550	6'-0" high		75	.213		115	4.69		119.69	135	
	3600	5'-0" long, 5'-0" high		150	.107		70	2.35		72.35	81	
	3650	6'-0" high		125	.128		79	2.82		81.82	92	
	3700	6'-0" long, 5'-0" high		162	.099		65	2.17		67.17	75	
	3750	6'-0" high		138	.116		75	2.55		77.55	87	
	3800	Economy acoustical panels, 40 NRC, 4'-0" long, 5'-0" high		132	.121		42	2.67		44.67	51	
	3850	6'-0" high		112	.143		50	3.14		53.14	60	
	3900	5'-0" long, 6'-0" high		125	.128		45	2.82		47.82	54	
	3950	6'-0" long, 5'-0" high		162	.099		32	2.17		34.17	39	
	9000	Minimum labor/equipment charge		3	5.330	Job		115		115	190	

106 520 | Panel Partitions

			CREW	DAILY OUTPUT	MAN-HOURS	UNIT	BARE COSTS MAT.	LABOR	EQUIP.	TOTAL	TOTAL INCL O&P	
522	0010	**PARTITIONS, FOLDING LEAF** Acoustic, wood										522
	0100	Vinyl faced, to 18' high, 6 psf, minimum	2 Carp	60	.267	S.F.	29	5.85		34.85	41	
	0150	Average		45	.356		35	7.80		42.80	51	
	0200	Maximum		30	.533		45	11.75		56.75	69	

106 520 | Panel Partitions

			CREW	DAILY OUTPUT	MAN-HOURS	UNIT	BARE COSTS MAT.	BARE COSTS LABOR	BARE COSTS EQUIP.	TOTAL	TOTAL INCL O&P	
522	0400	Formica or hardwood finish, minimum	2 Carp	60	.267	S.F.	30	5.85		35.85	43	522
	0500	Maximum		30	.533		32	11.75		43.75	54	
	0600	Wood, low acoustical type, 4.5 psf, to 14' high		50	.320	↓	22	7.05		29.05	36	
	9000	Minimum labor/equipment charge	↓	4	4	Job		88		88	145	

106 550 | Accordion Partitions

			CREW	DAILY OUTPUT	MAN-HOURS	UNIT	BARE COSTS MAT.	BARE COSTS LABOR	BARE COSTS EQUIP.	TOTAL	TOTAL INCL O&P	
552	0010	**PARTITIONS, FOLDING ACCORDION**										552
	0100	Vinyl covered, over 150 S.F., frame not included										
	0300	Residential, 1.25 lb. per S.F., 8' maximum height	2 Carp	300	.053	S.F.	8.40	1.17		9.57	11.15	
	0400	Commercial, 1.75 lb. per S.F., 8' maximum height		225	.071		8.85	1.56		10.41	12.30	
	0900	Acoustical, 3 lb. per S.F., 17' maximum height		100	.160		10.85	3.52		14.37	17.70	
	1200	5 lb. per S.F., 20' maximum height		95	.168		16.10	3.71		19.81	24	
	1500	Vinyl clad wood or steel, electric operation, 5.0 psf		160	.100		24.40	2.20		26.60	30	
	1900	Wood, non-acoustic, birch or mahogany, to 10' high		300	.053	↓	13.75	1.17		14.92	17.05	
	9000	Minimum labor/equipment charge	↓	4	4	Job		88		88	145	

106 750 | Storage & Shelving

			CREW	DAILY OUTPUT	MAN-HOURS	UNIT	BARE COSTS MAT.	BARE COSTS LABOR	BARE COSTS EQUIP.	TOTAL	TOTAL INCL O&P	
754	0010	**SHELVING** Metal, industrial, cross-braced, 3' wide, 12" deep	1 Sswk	175	.046	SF Shlf	5.90	1.10		7	8.50	754
	0100	24" deep		330	.024		4.25	.58		4.83	5.75	
	2200	Wide span, 1600 lb. capacity per shelf, 6' wide, 24" deep		380	.021		4.60	.51		5.11	6	
	2400	36" deep		440	.018	↓	3.75	.44		4.19	4.92	
	3000	Residential, adjustable metal, 12" deep										
	3100	23" to 37" wide	1 Carp	16	.500	Ea.	6.60	11		17.60	25	
	3200	47" to 61" wide		16	.500		9.50	11		20.50	28	
	3300	71" to 88" wide	↓	16	.500		14	11		25	33	
	3500	For closet rod, add					22%	2%				
	3600	For 14" shelves, add				↓	12%	5%				
	9000	Minimum labor/equipment charge	1 Carp	4	2	Job		44		44	72	

107 550 | Telephone Enclosures

			CREW	DAILY OUTPUT	MAN-HOURS	UNIT	BARE COSTS MAT.	BARE COSTS LABOR	BARE COSTS EQUIP.	TOTAL	TOTAL INCL O&P	
551	0010	**TELEPHONE ENCLOSURE**										551
	0020											
	0300	Shelf type, wall hung, minimum	2 Carp	5	3.200	Ea.	400	70		470	555	
	0400	Maximum		5	3.200		2,105	70		2,175	2,425	
	0600	Booth type, painted steel, indoor or outdoor, minimum		1.50	10.670		2,500	235		2,735	3,125	
	0700	Maximum (stainless steel)		1.50	10.670		8,650	235		8,885	9,900	
	1900	Outdoor, drive-up type, wall mounted		4	4		680	88		768	890	
	2000	Post mounted, stainless steel posts		3	5.330	↓	1,050	115		1,165	1,350	
	9000	Minimum labor/equipment charge	↓	4	4	Job		88		88	145	

108 200 | Bath Accessories

		CREW	DAILY OUTPUT	MAN-HOURS	UNIT	BARE COSTS				TOTAL INCL O&P	
						MAT.	LABOR	EQUIP.	TOTAL		
204	0010	**BATHROOM ACCESSORIES**									204
	0020										
	0200	Curtain rod, stainless steel, 5' long, 1" diameter	1 Carp	13	.615	Ea.	20	13.55		33.55	44
	0300	1-¼" diameter	"	13	.615	"	25	13.55		38.55	50
	0500	Dispenser units, combined soap & towel dispensers,									
	0510	mirror and shelf, flush mounted	1 Carp	10	.800	Ea.	215	17.60		232.60	265
	0600	Towel dispenser and waste receptacle,									
	0610	18 gallon capacity	1 Carp	10	.800	Ea.	335	17.60		352.60	395
	0800	Grab bar, straight, 1-¼" diameter, stainless steel, 18" long		24	.333		21	7.35		28.35	35
	1100	36" long		20	.400		27	8.80		35.80	44
	3000	Mirror with stainless steel, ¾" square frame, 18" x 24"		20	.400		70	8.80		78.80	91
	3300	72" x 24"		6	1.330		210	29		239	280
	3500	Mirror with 5" stainless steel shelf, ¾" sq. frame, 18" x 24"		20	.400		90	8.80		98.80	115
	3800	72" x 24"		6	1.330		330	29		359	410
	4200	Napkin/tampon dispenser, recessed		15	.533		236	11.75		247.75	280
	4300	Robe hook, single, regular		36	.222		9.60	4.89		14.49	18.55
	4400	Heavy duty, concealed mounting		36	.222		12	4.89		16.89	21
	4600	Soap dispenser, chrome, surface mounted, liquid		20	.400		44	8.80		52.80	63
	5600	Shelf, stainless steel, 5" wide, 18 ga., 24" long		24	.333		28	7.35		35.35	43
	6100	Toilet tissue dispenser, surface mounted, S.S., single roll		30	.267		16	5.85		21.85	27
	6200	Double roll		24	.333		25	7.35		32.35	39
	6290	Toilet seat	1 Plum	40	.200		14	5.10		19.10	23
	6400	Towel bar, stainless steel, 18" long	1 Carp	23	.348		17.60	7.65		25.25	32
	6500	30" long		21	.381		20.40	8.40		28.80	36
	6700	Towel dispenser, stainless steel, surface mounted		16	.500		41	11		52	63
	6800	Flush mounted, recessed		10	.800		72	17.60		89.60	110
	7400	Tumbler holder, tumbler only		30	.267		10.35	5.85		16.20	21
	7500	Soap, tumbler & toothbrush		30	.267		17.50	5.85		23.35	29
	7700	Wall urn ash receiver, surface mount, 11" long		12	.667		66	14.65		80.65	97
	8000	Waste receptacles, stainless steel, with top, 13 gallon		10	.800		142	17.60		159.60	185
	9000	Minimum labor/equipment charge		5	1.600	Job		35		35	57
208	0010	**MEDICINE CABINETS** With mirror, st. st. frame, 16" x 22", unlighted	1 Carp	14	.571	Ea.	57	12.55		69.55	83	208
	0100	Wood frame		14	.571		61	12.55		73.55	88
	0300	Sliding mirror doors, 20" x 16" x 4-¾", unlighted		7	1.140		56	25		81	105
	0400	24" x 19" x 8-½", lighted		5	1.600		94	35		129	160
	0600	Triple door, 30" x 32", unlighted, plywood body		7	1.140		183	25		208	240
	0700	Steel body		7	1.140		243	25		268	310
	0900	Oak door, wood body, beveled mirror, single door		7	1.140		102	25		127	155
	1000	Double door		6	1.330		200	29		229	270
	1200	Hotel cabinets, stainless, with lower shelf, unlighted		10	.800		150	17.60		167.60	195
	1300	Lighted		5	1.600		220	35		255	300
	9000	Minimum labor/equipment charge		4	2	Job		44		44	72

110 | Equipment

110 100 | Maintenance Equipment

		CREW	DAILY OUTPUT	MAN-HOURS	UNIT	BARE COSTS				TOTAL INCL O&P	
						MAT.	LABOR	EQUIP.	TOTAL		
121	0010	**VACUUM CLEANING** Central, 3 inlet, residential	1 Skwk	.90	8.890	Total	600	200		800	985	121
	0200	Commercial		.70	11.430		1,200	260		1,460	1,750
	0400	5 inlet system, residential		.50	16		690	360		1,050	1,350
	0600	7 inlet system		.40	20		780	455		1,235	1,600

110 200 | Security/Vault Equip

			CREW	DAILY OUTPUT	MAN-HOURS	UNIT	BARE COSTS MAT.	LABOR	EQUIP.	TOTAL	TOTAL INCL O&P	
261	0010	**SAFE** Office, 4 hr. rating, 30″ x 18″ x 18″ inside				Ea.	2,310			2,310	2,550	261
	0800	Money, ''B'' label, 9″ x 14″ x 14″					435			435	480	
	0900	Tool resistive, 24″ x 24″ x 20″					2,950			2,950	3,250	
	1000	Tool and torch resistive, 24″ x 24″ x 20″				↓	7,400			7,400	8,150	

110 300 | Teller And Service Equip

			CREW	DAILY OUTPUT	MAN-HOURS	UNIT	MAT.	LABOR	EQUIP.	TOTAL	INCL O&P	
301	0010	**BANK EQUIPMENT** Alarm system, police	2 Elec	1.60	10	Ea.	3,175	250		3,425	3,875	301
	0020											
	0400	Bullet resistant teller window, 44″ x 60″	1 Glaz	.60	13.330	Ea.	2,250	300		2,550	2,950	
	0500	48″ x 60″	"	.60	13.330	"	2,875	300		3,175	3,650	
	3000	Counters for banks, frontal only	2 Carp	1	16	Station	1,250	350		1,600	1,950	
	3100	Complete with steel undercounter	"	.50	32	"	2,550	705		3,255	3,950	
	5400	Partitions, bullet-resistant, 1-³⁄₁₆″ glass, 8′ high				L.F.	140			140	155	
	5600	Pass thru, bullet-resist. window, painted steel, 24″ x 36″	2 Sswk	1.60	10	Ea.	1,200	240		1,440	1,750	

110 400 | Ecclesiastical Equip

			CREW	DAILY OUTPUT	MAN-HOURS	UNIT	MAT.	LABOR	EQUIP.	TOTAL	INCL O&P	
401	0010	**CHURCH EQUIPMENT** Altar, wood, custom design, plain	1 Carp	1.40	5.710	Ea.	1,200	125		1,325	1,525	401
	0050	Deluxe	"	.20	40	"	6,625	880		7,505	8,725	
	0150	Baptistry, fiberglass, 3′-6″ deep, x 13′-7″ long,										
	0160	steps at both ends, incl. plumbing, minimum	L-8	1	20	Ea.	1,350	455		1,805	2,225	
	0200	Maximum	"	.70	28.570	↓	3,100	650		3,750	4,450	
	0250	Add for filter, heater and lights					950			950	1,050	
	1500	Pews, bench type, hardwood, minimum	1 Carp	20	.400	L.F.	46	8.80		54.80	65	
	1550	Maximum	"	15	.533	"	75	11.75		86.75	100	
	4000	Steeples, translucent fiberglass, 30″ square, 15′ high	F-3	2	20	Ea.	1,650	445	190	2,285	2,750	
	4150	25′ high	"	1.80	22.220		2,775	495	210	3,480	4,075	
	4600	Aluminum, baked finish, 14′ high, 16″ square					1,200			1,200	1,325	
	4640	35′ high, 8′ base					14,200			14,200	15,600	
	4680	152′ high, custom				↓	320,000			320,000	352,000	
	4690											
	5000	Wall cross, aluminum, extruded, 2″ x 2″ section	1 Carp	34	.235	L.F.	27	5.20		32.20	38	
	5150	4″ x 4″ section		29	.276		39	6.05		45.05	53	
	5300	Bronze, extruded, 1″ x 2″ section		31	.258		52	5.70		57.70	66	
	5350	2-½″ x 2-½″ section	↓	34	.235	↓	80	5.20		85.20	96	

110 500 | Library Equipment

			CREW	DAILY OUTPUT	MAN-HOURS	UNIT	MAT.	LABOR	EQUIP.	TOTAL	INCL O&P	
501	0010	**LIBRARY EQUIPMENT** Bookshelf, mtl, 90″ high, 10″ shelf, dbl face	1 Carp	11.50	.696	L.F.	78	15.30		93.30	110	501
	0300	Single face	"	12	.667	"	58	14.65		72.65	88	
	0600	For 8″ shelving, subtract from above					10%					
	0800	For 42″ high with countertop, subtract from above					20%					
	2500	Carrels, hardwood, 36″ x 24″, minimum	1 Carp	5	1.600	Ea.	475	35		510	580	
	2650	Maximum		4	2		675	44		719	815	
	2850	Metal, minimum		5	1.600		170	35		205	245	
	3000	Maximum	↓	4	2	↓	560	44		604	690	

110 600 | Theater/Stage Equip

			CREW	DAILY OUTPUT	MAN-HOURS	UNIT	MAT.	LABOR	EQUIP.	TOTAL	INCL O&P	
601	0010	**STAGE EQUIPMENT** Control boards with dimmers & breakers										601
	0050	Minimum	1 Elec	1	8	Ea.	2,000	200		2,200	2,500	
	0150	Maximum	"	.20	40	"	30,000	1,000		31,000	34,500	
	0160											
	0500	Curtain track, straight, light duty	2 Carp	20	.800	L.F.	11.10	17.60		28.70	41	
	0700	Curved sections		12	1.330	"	62	29		91	115	
	1000	Curtains, velour, medium weight		600	.027	S.F.	9.25	.59		9.84	11.15	
	1100	Asbestos		600	.027	"	10.25	.59		10.84	12.25	
	5000	Stages, portable with steps, folding legs, stock, 8″ high				SF Stg.	6.95			6.95	7.65	
	5100	16″ high					7.65			7.65	8.40	
	5200	32″ high					8.10			8.10	8.90	
	5300	40″ high				↓	8.70			8.70	9.55	

110 600 | Theater/Stage Equip

		CREW	DAILY OUTPUT	MAN-HOURS	UNIT	BARE COSTS				TOTAL INCL O&P	
						MAT.	LABOR	EQUIP.	TOTAL		
604	0011	MOVIE EQUIPMENT									604
	0020										
	3000	Projection screens, rigid, in wall, acrylic, ¼" thick	2 Glaz	195	.082	S.F.	30	1.85		31.85	36
	3100	½" thick	"	130	.123	"	35	2.78		37.78	43
	3700	Sound systems, incl. amplifier, single system, minimum	1 Elec	.90	8.890	Ea.	1,800	225		2,025	2,325
	3800	Dolby/Super Sound, maximum		.40	20		10,000	505		10,505	11,800
	4100	Dual system, minimum		.70	11.430		3,000	285		3,285	3,750
	4200	Dolby/Super Sound, maximum		.40	20		12,000	505		12,505	14,000
	5700	Seating, painted steel, upholstered, minimum	2 Carp	35	.457		80	10.05		90.05	105
	5800	Maximum	"	28	.571		140	12.55		152.55	175

111 | Mercantile, Commercial and Detention Equipment

111 320 | Projection Screens

		CREW	DAILY OUTPUT	MAN-HOURS	UNIT	BARE COSTS				TOTAL INCL O&P	
						MAT.	LABOR	EQUIP.	TOTAL		
321	0010	PROJECTION SCREENS Wall or ceiling hung, glass beaded									321
	0100	Manually operated, economy	2 Carp	500	.032	S.F.	3.28	.70		3.98	4.76
	0300	Intermediate		450	.036		4.76	.78		5.54	6.50
	0400	Deluxe		400	.040		5.75	.88		6.63	7.75
	9000	Minimum labor/equipment charge		3	5.330	Job		115		115	190

111 600 | Loading Dock Equipment

		CREW	DAILY OUTPUT	MAN-HOURS	UNIT	BARE COSTS				TOTAL INCL O&P	
						MAT.	LABOR	EQUIP.	TOTAL		
601	0010	LOADING DOCK Bumpers, rubber blocks 4-½" thk, 10" hg, 14" lg	1 Carp	26	.308	Ea.	47	6.75		53.75	63	601
	0200	24" long		22	.364		54	8		62	72
	0300	36" long		17	.471		62	10.35		72.35	85
	0500	12" high, 14" long		25	.320		52	7.05		59.05	69
	2200	Dock boards, heavy duty, 60" x 60", aluminum, 5000 lb. cap.					880			880	970
	4200	Platform lifter, 6' x 6', portable, 3000 lb. capacity					5,200			5,200	5,725
	6200	Shelters, fabric, for truck or train, scissor arms, minimum	1 Carp	1	8		750	175		925	1,100
	6300	Maximum	"	.50	16		1,300	350		1,650	2,000
603	0010	DOCK BUMPERS Bolts not included, 2" x 6" to 4" x 8", average	F-1	300	.027	B.F.	.43	.59	.03	1.05	1.46	603

111 700 | Waste Handling Equip

		CREW	DAILY OUTPUT	MAN-HOURS	UNIT	BARE COSTS				TOTAL INCL O&P	
						MAT.	LABOR	EQUIP.	TOTAL		
701	0010	WASTE HANDLING Compactors, 115 volt, 250#/hr., chute fed	L-4	1	24	Ea.	7,400	500		7,900	8,950	701
	0100	Hand fed		2.40	10		5,625	210		5,835	6,525
	1000	Heavy duty industrial compactor, 0.5 C.Y. capacity		1	24		4,500	500		5,000	5,750
	1050	1.0 C.Y. capacity		1	24		7,475	500		7,975	9,025

111 900 | Detention Equipment

		CREW	DAILY OUTPUT	MAN-HOURS	UNIT	BARE COSTS				TOTAL INCL O&P	
						MAT.	LABOR	EQUIP.	TOTAL		
901	0011	DETENTION EQUIPMENT										901
	0020										
	2000	Cells, prefab., 5' to 6' wide, 7' to 8' high, 7' to 8' deep,	E-4	1.50	21.330	Ea.	5,125	525	46	5,696	6,650
	2010	bar front, cot, not incl. plumbing									

114 000	Food Service Equipment	CREW	DAILY OUTPUT	MAN-HOURS	UNIT	BARE COSTS				TOTAL INCL O&P	
						MAT.	LABOR	EQUIP.	TOTAL		
002	0010	APPLIANCES Cooking range, 30" free standing, 1 oven, minimum	2 Clab	17	.941	Ea.	270	16.45		286.45	325	002
	0050	Maximum	"	10	1.600		1,200	28		1,228	1,375	
	0350	Built-in, 30" wide, 1 oven, minimum	2 Carp	16	1		430	22		452	510	
	0400	Maximum	"	8	2		990	44		1,034	1,150	
	0900	Counter top cook tops, 4 burner, standard, minimum	1 Elec	8	1		200	25		225	260	
	0950	Maximum		4	2		470	50		520	595	
	1250	Microwave oven, minimum		9	.889		100	22		122	145	
	1300	Maximum		8	1		1,525	25		1,550	1,725	
	1500	Combination range, refrigerator and sink, 30" wide, minimum	L-1	5	3.200		520	81		601	695	
	1550	Maximum	"	4	4		1,180	100		1,280	1,450	
	1640	Combination range, refrigerator, sink, microwave										
	1660	oven and ice maker	L-1	9	1.780	Ea.	3,175	45		3,220	3,550	
	1750	Compactor, residential size, 4 to 1 compaction, minimum	1 Carp	8	1		300	22		322	365	
	1800	Maximum	"	7	1.140		500	25		525	590	
	2750	Dishwasher, built-in, 2 cycles, minimum	L-1	8	2		230	51		281	330	
	2800	Maximum		8	2		390	51		441	505	
	3300	Garbage disposer, sink type, minimum		10	1.600		55	40		95	125	
	3350	Maximum		10	1.600		195	40		235	275	
	4150	Hood for range, 2 speed, vented, 30" wide, minimum	1 Elec	6	1.330		40	34		74	95	
	4200	Maximum		5	1.600		260	40		300	345	
	4220	36" wide, minimum		5	1.600		140	40		180	215	
	4250	Maximum		4	2		280	50		330	385	
	4300	42" wide, minimum		4	2		155	50		205	245	
	4350	Maximum		3	2.670		300	67		367	430	
	4400	Ventless hood, 2 speed, 30" wide, minimum		8	1		50	25		75	93	
	4450	Maximum		7	1.140		190	29		219	255	
	4510	36" wide, minimum		7	1.140		150	29		179	210	
	4550	Maximum		6	1.330		210	34		244	280	
	4580	42" wide, minimum		6	1.330		170	34		204	240	
	4600	Maximum		5	1.600		240	40		280	325	
	4650	For vented 1 speed, deduct from maximum					25			25	28	
	4700											
	5380	Oven, built in, standard	1 Elec	4	2	Ea.	350	50		400	460	
	5390	Deluxe	"	4	2		680	50		730	825	
	5500	Refrigerator, no frost, 10 C.F. to 12 C.F. minimum	2 Clab	10	1.600		300	28		328	375	
	5600	Maximum	"	7	2.290		590	40		630	715	
	6400	Sump pump cellar drainer, ⅓ H.P., minimum	1 Plum	8	1		90	25		115	140	
	6450	Maximum	"	6	1.330		300	34		334	380	
	6460	Sump pump, see also division 152-480										
	7350	Water softener, automatic, to 30 grains per gallon	2 Plum	5	3.200	Ea.	310	81		391	465	
	7400	To 100 grains per gallon	"	4	4	"	675	100		775	900	
004	0011	KITCHEN EQUIPMENT										004
	0020											
	9400	Rule of thumb: Equipment cost based										
	9410	on kitchen work area										
	9420	Office buildings, minimum	L-7	77	.364	S.F.	41	7.70		48.70	58	
	9450	Maximum		58	.483		69	10.20		79.20	92	
	9550	Public eating facilities, minimum		77	.364		54	7.70		61.70	72	
	9600	Maximum		46	.609		87	12.90		99.90	115	
	9750	Hospitals, minimum		58	.483		55	10.20		65.20	77	
	9800	Maximum		39	.718		92	15.20		107.20	125	
008	0010	WINE VAULT Redwood, air conditioned, walk-in type										008
	0020	6'-8" high, incl. racks, 2' x 4' for 156 bottles	2 Carp	2	8	Ea.	3,050	175		3,225	3,650	
	0200	4' x 6' for 614 bottles		1.50	10.670		4,975	235		5,210	5,850	
	0400	6' x 12' for 1940 bottles		1	16		9,325	350		9,675	10,800	

114 580 | Disappearing Stairs

		CREW	DAILY OUTPUT	MAN-HOURS	UNIT	BARE COSTS MAT.	BARE COSTS LABOR	BARE COSTS EQUIP.	BARE COSTS TOTAL	TOTAL INCL O&P	
581	0010	**DISAPPEARING STAIRWAY** No trim included										581
	0020	One piece, yellow pine, 8'-0" ceiling	2 Carp	4	4	Ea.	925	88		1,013	1,150	
	0030	9'-0" ceiling		4	4		950	88		1,038	1,200	
	0040	10'-0" ceiling		3	5.330		1,000	115		1,115	1,300	
	0050	11'-0" ceiling		3	5.330		1,175	115		1,290	1,475	
	0060	12'-0" ceiling		3	5.330		1,225	115		1,340	1,550	
	0080	For 90 minute mineral core fire door, add					180			180	200	
	0100	Custom grade, pine, 8'-6" ceiling, minimum	1 Carp	4	2		70	44		114	150	
	0150	Average		3.50	2.290		120	50		170	215	
	0200	Maximum		3	2.670		180	59		239	295	
	0250											
	0500	Heavy duty, pivoted, from 7'7" to 12'10" floor to floor	1 Carp	3	2.670	Ea.	195	59		254	310	
	0600	16'-0" ceiling		2	4		800	88		888	1,025	
	0800	Economy folding, pine, 8'-6" ceiling		4	2		50	44		94	125	
	0900	9'-6" ceiling		4	2		70	44		114	150	
	1000	Fire escape, galvanized steel, 8'-0" to 10'-4" ceiling	2 Carp	1	16		1,025	350		1,375	1,700	
	1010	10'-6" to 13'-6" ceiling		1	16		1,300	350		1,650	2,000	
	1100	Automatic electric, aluminum, floor to floor height, 8' to 9'		1	16		5,000	350		5,350	6,075	
	1500	11' to 12'		.90	17.780		5,350	390		5,740	6,525	
	1700	14' to 15'		.70	22.860		5,725	505		6,230	7,125	
	9000	Minimum labor/equipment charge	1 Carp	2	4	Job		88		88	145	

114 760 | Revolving Darkrm Doors

		CREW	DAILY OUTPUT	MAN-HOURS	UNIT	MAT.	LABOR	EQUIP.	TOTAL	TOTAL INCL O&P	
768	0010	**DARKROOM DOORS** Revolving, standard, 2 way, 40" diameter	2 Carp	3.50	4.570	Opng.	1,080	100		1,180	1,350	768
	0050	3 way, 50" diameter		3	5.330		1,390	115		1,505	1,725	
	1000	4 way, 50" diameter		2	8		1,490	175		1,665	1,925	
	1010											
	2000	Hinged safety, 2 way, 40" diameter	2 Carp	3.70	4.320	Opng.	1,480	95		1,575	1,775	
	2500	3 way, 50" diameter		2.90	5.520		1,690	120		1,810	2,050	
	3000	Pop out safety, 2 way, 40" diameter		3.10	5.160		1,580	115		1,695	1,925	
	4000	3 way, 50" diameter		2.80	5.710		1,890	125		2,015	2,275	

114 800 | Athletic/Recreational

		CREW	DAILY OUTPUT	MAN-HOURS	UNIT	MAT.	LABOR	EQUIP.	TOTAL	TOTAL INCL O&P	
801	0010	**HEALTH CLUB EQUIPMENT** Abdominal rack, 2 board capacity				Ea.	450			450	495	801
	0050	Abdominal board, upholstered					460			460	505	
	0200	Bicycle trainer, minimum					435			435	480	
	0300	Deluxe, electric					1,800			1,800	1,975	
	0400	Bar bell set, chrome plated steel, 25 lbs.					185			185	205	
	0420	100 lbs.					400			400	440	
	0450	200 lbs.					700			700	770	
	0500	Weight plates, cast iron, per lb.				Lb.	3			3	3.30	
	0520	Storage rack, 10 station				Ea.	750			750	825	
	0600	Circuit training apparatus, 12 machines minimum	2 Clab	1.25	12.800	Set	20,000	225		20,225	22,400	
	0700	Average		1	16		28,200	280		28,480	31,500	
	0800	Maximum		.75	21.330		35,000	375		35,375	39,100	
	0820	Dumbbell set, cast iron, with rack and 5 pair					545			545	600	
	0900	Squat racks	2 Clab	5	3.200	Ea.	320	56		376	445	
	1600	For saunas, see division 130-521										
	1640	For steam baths, see division 130-541										

116 | Laboratory, Planetarium, Observatory Equipment

		116 000	Laboratory Equipment	CREW	DAILY OUTPUT	MAN-HOURS	UNIT	BARE COSTS MAT.	BARE COSTS LABOR	BARE COSTS EQUIP.	BARE COSTS TOTAL	TOTAL INCL O&P	
001	0010		LABORATORY EQUIPMENT Cabinets, base, door units, metal	2 Carp	18	.889	L.F.	82	19.55		101.55	120	001
	0300		Drawer units		18	.889		153	19.55		172.55	200	
	0700		Tall storage cabinets, open, 7' high		20	.800		132	17.60		149.60	175	
	0900		With glazed doors		20	.800		212	17.60		229.60	260	
	1300		Wall cabinets, metal, 12-½" deep, open		20	.800		55	17.60		72.60	89	
	1500		With doors		20	.800		111	17.60		128.60	150	
	1550		Counter tops, not incl. base cabinets, acidproof, minimum		82	.195	S.F.	7.70	4.29		11.99	15.50	
	1600		Maximum		70	.229		19	5.05		24.05	29	
	1650		Stainless steel		82	.195		47	4.29		51.29	59	
	4200		Alternate pricing method: as percent of lab furniture										
	4400		Installation, not incl. plumbing & duct work				% Furn.					22%	
	4800		Plumbing, final connections, simple system									10%	
	5000		Moderately complex system									15%	
	5200		Complex system									20%	
	5400		Electrical, simple system								10%	10%	
	5600		Moderately complex system								20%	20%	
	5800		Complex system								35%	35%	
	6000		Safety equipment, eye wash, hand held				Ea.	206			206	225	
	6200		Deluge shower				"	122			122	135	

125 | Window Treatment

		125 100	Blinds	CREW	DAILY OUTPUT	MAN-HOURS	UNIT	BARE COSTS MAT.	BARE COSTS LABOR	BARE COSTS EQUIP.	BARE COSTS TOTAL	TOTAL INCL O&P	
101	0011		BLINDS, Exterior shutters										101
	0020												
	1000		Pine, louvered, primed, each 1'-2" wide, 3'-3" long	1 Carp	10	.800	Pr.	41	17.60		58.60	74	
	1100		4'-7" long		10	.800		58	17.60		75.60	93	
	1250		Each 1'-4" wide, 3'-0" long		10	.800		43	17.60		60.60	76	
	1350		5'-3" long		10	.800		52	17.60		69.60	86	
	1500		Each 1'-6" wide, 3'-3" long		10	.800		49	17.60		66.60	83	
	1600		4'-7" long		10	.800		65	17.60		82.60	100	
	1610		Door blinds, 6'-9" long 1'-3" wide		9	.889		53	19.55		72.55	90	
	1615		1'-6" wide		9	.889		60	19.55		79.55	98	
	1620		Hemlock, louvered, each 1'-2" wide, 5'-7" long		10	.800		49	17.60		66.60	83	
	1630		Each 1'-4" wide, 2'-2" long		10	.800		30	17.60		47.60	62	
	1640		3'-0" long		10	.800		30	17.60		47.60	62	
	1650		3'-3" long		10	.800		30	17.60		47.60	62	
	1660		3'-11" long		10	.800		47	17.60		64.60	80	
	1670		4'-3" long		10	.800		36	17.60		53.60	68	
	1680		5'-3" long		10	.800		42	17.60		59.60	75	
	1690		5'-11" long		10	.800		47	17.60		64.60	80	
	1700		Door blinds, 6'-9" long, 1'-3" wide		9	.889		60	19.55		79.55	98	
	1710		1'-6" wide		9	.889		62	19.55		81.55	100	
	1720		Hemlock, solid raised panel, each 1'-4" wide, 3'-0" long		9	.889		49	19.55		68.55	86	
	1730		3'-11" long		10	.800		55	17.60		72.60	89	
	1740		4'-3" long		10	.800		61	17.60		78.60	96	
	1750		4'-7" long		10	.800		64	17.60		81.60	99	
	1760		4'-11" long		10	.800		69	17.60		86.60	105	
	1770		5'-11" long		10	.800		83	17.60		100.60	120	
	1780		Door blinds, 6'-9" long, 1'-4" wide		9	.889		91	19.55		110.55	130	
	2000												

For expanded coverage of these items see *Means Interior Cost Data 1991*

125 100 | Blinds

			CREW	DAILY OUTPUT	MAN-HOURS	UNIT	MAT.	LABOR	EQUIP.	TOTAL	TOTAL INCL O&P	
101	2500	Polystyrene, solid raised panel, each 3'-3" wide, 3'-0" long	1 Carp	10	.800	Pr.	62	17.60		79.60	97	101
	2600	3'-11" long		10	.800		71	17.60		88.60	105	
	2700	4'-7" long		10	.800		78	17.60		95.60	115	
	2800	5'-3" long		10	.800		86	17.60		103.60	125	
	2900	6'-8" long	↓	9	.889	↓	115	19.55		134.55	160	
	3510											
	4500	Polystyrene, louvered, each 1'-2" wide, 3'-3" long	1 Carp	10	.800	Pair	26	17.60		43.60	57	
	4600	4'-7" long		10	.800	Pr.	33	17.60		50.60	65	
	4750	5'-3" long		10	.800		37	17.60		54.60	69	
	4850	6'-8" long		9	.889		50	19.55		69.55	87	
	6000	Vinyl, louvered, each 1'-2" x 4'-7" long		10	.800		32	17.60		49.60	64	
	6200	Each 1'-4" x 6'-8" long		9	.889	↓	49	19.55		68.55	86	
103	0010	BLINDS, INTERIOR Solid colors	↓	590	.014	S.F.	2	.30		2.30	2.69	103
	0020											
	3000	Wood folding panels with movable louvers, 7" x 20" each	1 Carp	17	.471	Pr.	18	10.35		28.35	37	
	4000	Fixed louver type, stock units, 8" x 20" each		17	.471		21	10.35		31.35	40	
	4450	18" x 40" each	↓	17	.471	↓	58	10.35		68.35	81	

125 200 | Shades

			CREW	DAILY OUTPUT	MAN-HOURS	UNIT	MAT.	LABOR	EQUIP.	TOTAL	TOTAL INCL O&P	
201	0011	SHADES Basswood roll-up, stain finish, ⅜" slats	1 Carp	300	.027	S.F.	11.80	.59		12.39	13.95	201
	0030	Double layered, heat reflective		685	.012		5.40	.26		5.66	6.35	
	0250	⅞" slats		300	.027		10.80	.59		11.39	12.85	
	0950	Mylar, single layer, non-heat reflective		685	.012		3.20	.26		3.46	3.94	
	1050	Double layered, heat reflective		685	.012		5.40	.26		5.66	6.35	
	1150	Triple layered, heat reflective		685	.012		5.40	.26		5.66	6.35	
	5000	Thermal, roll up, R-4		44	.182		6.40	4		10.40	13.55	
	5030	R-10.7	↓	44	.182	↓	7.50	4		11.50	14.80	
	5050	Magnetic clips, set of 20				Set	14			14	15.40	

126 200 | Furniture

			CREW	DAILY OUTPUT	MAN-HOURS	UNIT	MAT.	LABOR	EQUIP.	TOTAL	TOTAL INCL O&P	
205	0010	DORMITORY FURNITURE Beds, free standing, minimum				Ea.	181.30			181.30	200	205
	0100	Maximum				"	436.70			436.70	480	
	2050	Rule of thumb: Total cost for furniture, minimum				Student					1,800	
	2150	Maximum				"					3,500	
210	0010	FURNITURE, HOSPITAL Beds, manual, minimum				Ea.	594.20			594.20	655	210
	0100	Maximum				"	1,033			1,033	1,125	
	1100	Patient wall systems, not incl. plumbing, minimum				Room	594.20			594.20	655	
	1200	Maximum					1,096			1,096	1,200	
214	0010	FURNITURE, HOTEL Standard quality set, minimum					1,921			1,921	2,125	214
	0200	Maximum				↓	5,904			5,904	6,500	
222	0010	FURNITURE, OFFICE										222
	0020	Desks, 29" high, double pedestal, 30" x 60", metal, minimum				Ea.	252			252	275	
	0030	Maximum				"	702			702	770	
226	0010	FURNITURE, RESTAURANT Bars, built-in, front bar	1 Carp	5	1.600	L.F.	159	35		194	230	226
	0200	Back bar		5	1.600	"	115	35		150	185	
	0500	Booth unit, molded plastic, stub wall and 2 seats, minimum		2	4	Set	300	88		388	475	
	0600	Maximum	↓	1.50	5.330	"	648	115		763	905	

For expanded coverage of these items see *Means Interior Cost Data 1991*

126 | Furniture and Accessories

126 200	Furniture	CREW	DAILY OUTPUT	MAN-HOURS	UNIT	BARE COSTS MAT.	LABOR	EQUIP.	TOTAL	TOTAL INCL O&P	
226	0800	Booth seat, upholstered, foursome, single (end) minimum	1 Carp	5	1.600	Ea.	434	35		469	535	226
	0900	Maximum		4	2		907	44		951	1,075	
	1000	Foursome, double, minimum		4	2		648	44		692	785	
	1100	Maximum		3	2.670		2,340	59		2,399	2,675	
	1300	Circle booth, upholstered, ¼ circle, minimum		3	2.670		712	59		771	880	
	1400	Maximum		2	4		1,438	88		1,526	1,725	
	1500	¾ circle, minimum		1.50	5.330		1,282	115		1,397	1,600	
	1600	Maximum	↓	1	8	↓	3,486	175		3,661	4,125	

130 | Special Construction

130 250	Integrated Ceilings	CREW	DAILY OUTPUT	MAN-HOURS	UNIT	BARE COSTS MAT.	LABOR	EQUIP.	TOTAL	TOTAL INCL O&P	
251	0010	INTEGRATED CEILINGS Lighting, ventilating & acoustical										251
	1800	Radiant hot water system with finished acoustic ceiling,										
	1810	not including supply piping. Heating only (gross S.F.)										
	2000											
	2100	Elementary schools, minimum				S.F.					4.81	
	2200	Maximum									6.05	
	2400	High schools and colleges, minimum									4.27	
	2500	Maximum									6.05	
	2700	Libraries, minimum									4.32	
	2800	Maximum									5.70	
	3000	Hospitals, minimum									5.90	
	3100	Maximum									7.35	
	3300	Office buildings, minimum									4.21	
	3400	Maximum									5.55	
	3600	For combined heating and cooling, add, minimum					30%					
	3700	Maximum				↓	40%					

130 320 | Athletic Rooms

		CREW	DAILY OUTPUT	MAN-HOURS	UNIT	BARE COSTS MAT.	LABOR	EQUIP.	TOTAL	TOTAL INCL O&P	
321	0010	SPORT COURT Floors, No. 2 & better maple, ²⁵⁄₃₂" thick										321
	0300	Squash, regulation court in existing building, minimum				Court					11,800	
	0400	Maximum				"					21,900	
	0450	Rule of thumb for components:										
	0470	Walls	3 Carp	.15	160	Court	9,650	3,525		13,175	16,400	
	0500	Floor	"	.25	96	↓	4,300	2,100		6,400	8,175	
	0550	Lighting	1 Elec	.30	26.670	↓	1,350	670		2,020	2,500	

130 380 | Cold Storage Rooms

		CREW	DAILY OUTPUT	MAN-HOURS	UNIT	BARE COSTS MAT.	LABOR	EQUIP.	TOTAL	TOTAL INCL O&P	
381	0010	REFRIGERATORS Curbs, 12" high, 4" thick, concrete										381
	6300	Rule of thumb for complete units, not incl. doors, cooler	2 Carp	146	.110	SF Flr.	78	2.41		80.41	90	
	6400	Freezer	"	109.60	.146	"	93	3.21		96.21	110	

130 520 | Saunas

		CREW	DAILY OUTPUT	MAN-HOURS	UNIT	BARE COSTS MAT.	LABOR	EQUIP.	TOTAL	TOTAL INCL O&P	
521	0010	SAUNA Prefabricated, incl. heater & controls, 7' high, 6' x 4'	L-7	2.20	12.730	Ea.	2,750	270		3,020	3,450	521
	1700	Door only, with tempered insulated glass window	2 Carp	3.40	4.710	↓	160	105		265	345	
	1800	Prehung, incl. jambs, pulls & hardware	"	12	1.330		200	29		229	270	
	2500	Heaters only (incl. above), wall mounted, to 200 C.F.				↓	375			375	415	
	4480	For additional equipment, see div. 114-801										

130 540 | Steam Baths

			CREW	DAILY OUTPUT	MAN-HOURS	UNIT	BARE COSTS MAT.	LABOR	EQUIP.	TOTAL	TOTAL INCL O&P	
541	0010	STEAM BATH Heater, timer & head, single, to 140 C.F.	1 Plum	1.20	6.670	Ea.	790	170		960	1,125	541
	0500	To 300 C.F.	"	1.10	7.270		900	185		1,085	1,275	
	2000	Multiple baths, motels, apartment, 2 baths	Q-1	1.30	12.310		1,200	280		1,480	1,750	
	2500	4 baths	"	.70	22.860		1,475	525		2,000	2,425	

130 810 | Acoustical Enclosures

												811
811	0010	ACOUSTICAL Enclosure, 4" thick wall and ceiling panels										
	0020	8# per S.F., up to 12' span	3 Carp	72	.333	SF Surf	20	7.35		27.35	34	
	0300	Better quality panels, 10.5# per S.F.		64	.375		22	8.25		30.25	38	
	0400	Reverb-chamber, 4" thick, parallel walls		60	.400		23	8.80		31.80	40	

130 910 | Radiation Protection

			CREW	DAILY OUTPUT	MAN-HOURS	UNIT	MAT.	LABOR	EQUIP.	TOTAL	TOTAL INCL O&P	
911	0010	SHIELDING LEAD										911
	0050											
	0300	Lead lath or sheets, 1/16" thick	2 Lath	135	.119	S.F.	4.50	2.60		7.10	9.05	
	0400	1/8" thick		120	.133	"	8.50	2.93		11.43	13.95	
	0600	Lead glass, 1/4" thick, 12" x 16"		13	1.230	Ea.	167	27		194	225	
	0700	24" x 36"		8	2	"	750	44		794	895	
	1200	X-ray protection, average radiography or fluoroscopy										
	1210	room, up to 300 S.F. floor, 1/16" lead, minimum	2 Lath	.25	64	Total	3,000	1,400		4,400	5,500	
	1500	Maximum, 7'-0" walls	"	.15	107	"	4,000	2,350		6,350	8,075	
	1600	Deep therapy X-ray room, 250 KV capacity,										
	1800	up to 300 S.F. floor, 1/4" lead, minimum	2 Lath	.09	178	Total	8,500	3,900		12,400	15,500	
	1900	Maximum, 7'-0" walls	"	.07	229	"	11,500	5,025		16,525	20,500	
	2000	X-ray viewing panels, clear lead plastic										
	2010	7 mm thick, 0.3 mm LE, 2.3 lbs/S.F.	H-2	209	.115	S.F.	80	2.40		82.40	92	
	2020	12 mm thick, 0.5 mm LE, 3.9 lbs/S.F.		123	.195		98	4.07		102.07	115	
	2030	18 mm thick, 0.8mm LE, 5.9 lbs/S.F.		81	.296		106	6.20		112.20	125	
	2040	22 mm thick, 1.0 mm LE, 7.2 lbs/S.F.		66	.364		110	7.60		117.60	135	
	2050	35 mm thick, 1.5 mm LE, 11.5 lbs/S.F.		42	.571		122	11.90		133.90	155	
	2060	46 mm thick, 2.0 mm LE, 15.0 lbs/S.F.		32	.750		166	15.65		181.65	210	
	2090	For panels 12 S.F. to 48 S.F., add crating charge				Ea.					50	
	4000	X-ray barriers, modular, panels mounted within framework for										
	4002	attaching to floor, wall or ceiling, upper portion is clear lead										
	4005	plastic window panels 48"H, lower portion is opaque leaded										
	4008	steel panels 36"H, structural supports not included										
	4010	1-section barrier, 36"W x 84"H overall										
	4020	0.5 mm LE panels	H-2	9.60	2.500	Ea.	1,790	52		1,842	2,050	
	4030	0.8 mm LE panels		9.60	2.500		1,890	52		1,942	2,150	
	4040	1.0 mm LE panels		8	3		1,930	63		1,993	2,225	
	4050	1.5 mm LE panels		8	3		2,080	63		2,143	2,400	
	4060	2-section barrier, 72"W x 84"H overall										
	4070	0.5 mm LE panels	H-2	6	4	Ea.	3,860	83		3,943	4,375	
	4080	0.8 mm LE panels		6	4		4,055	83		4,138	4,600	
	4090	1.0 mm LE panels		5.30	4.530		4,140	94		4,234	4,700	
	5000	1.5 mm LE panels		4.80	5		4,445	105		4,550	5,050	
	5010	3-section barrier, 108"W x 84"H overall										
	5020	0.5 mm LE panels	H-2	4.80	5	Ea.	5,890	105		5,995	6,650	
	5030	0.8 mm LE panels		4.80	5		6,180	105		6,285	6,975	
	5040	1.0 mm LE panels		4	6		6,310	125		6,435	7,150	
	5050	1.5 mm LE panels		3.70	6.490		6,765	135		6,900	7,650	
	7000	X-ray barriers, mobile, mounted within framework w/casters on										
	7005	bottom, clear lead plastic window panels on upper portion,										
	7010	opaque on lower, 30"W x 75"H overall, incl. framework										
	7020	24"h upper w/0.5 mm LE, 48"H lower w/0.8 mm LE	1 Carp	16	.500	Ea.	1,345	11		1,356	1,500	
	7030	48"W x 75"H overall, incl. framework										
	7040	36"H upper w/0.5 mm LE, 36"H lower w/0.8 mm LE	1 Carp	16	.500	Ea.	2,220	11		2,231	2,450	
	7050	36"H upper w/1.0 mm LE, 36"H lower w/1.5 mm LE	"	16	.500	"	2,875	11		2,886	3,175	

130 | Special Construction

911	7060	72"W x 75"H overall, incl. framework										911
	7070	36"H upper w/0.5 mm LE, 36"H lower w/0.8 mm LE	1 Carp	16	.500	Ea.	2,820	11		2,831	3,125	
	7080	36"H upper w/1.0 mm LE, 36"H lower w/1.5 mm LE	"	16	.500	"	3,595	11		3,606	3,975	
	9000	Minimum labor/equipment charge	2 Lath	4.50	3.560	Job		78		78	120	

131 | Pre-Eng. Structures, Pools and Ice Rinks

131 230	Greenhouses	CREW	DAILY OUTPUT	MAN-HOURS	UNIT	BARE COSTS				TOTAL INCL O&P	
						MAT.	LABOR	EQUIP.	TOTAL		
231	0010	GREENHOUSE Shell only, stock units, not incl. 2' stub walls,										231
	0020	foundation, floors, heat or compartments										
	0300	Residential type, free standing, 8'-6" long x 7'-6" wide	2 Carp	59	.271	SF Flr.	35	5.95		40.95	48	
	0400	10'-6" wide		85	.188		27	4.14		31.14	36	
	0600	13'-6" wide		108	.148		24	3.26		27.26	32	
	0700	17'-0" wide		160	.100		27	2.20		29.20	33	
	0900	Lean-to type, 3'-10" wide		34	.471		31	10.35		41.35	51	
	1000	6'-10" wide		58	.276		24	6.05		30.05	36	
	1050	8'-0" wide	↓	60	.267	↓	22	5.85		27.85	34	
	1060											
	1100	Wall mounted, to existing window, 3' x 3'	1 Carp	4	2	Ea.	335	44		379	440	
	1120	4' x 5'	"	3	2.670	"	500	59		559	645	
	3900	For cooling, add, minimum				SF Flr.	2.10			2.10	2.31	
	4000	Maximum					5.20			5.20	5.70	
	4200	For heaters, 13.6 MBH, add					4			4	4.40	
	4300	60 MBH, add				↓	1.50			1.50	1.65	
	4500	For benches, 2' x 3'-6", add				SF Hor.	17.90			17.90	19.70	
	4600	3' x 10', add				S.F.	9.70			9.70	10.65	
	4800	For controls, add, minimum				Total	1,800			1,800	1,975	
	4900	Maximum				"	10,700			10,700	11,800	
	5100	For humidification equipment, add				M.C.F.	4.60			4.60	5.05	
	5200	For vinyl shading, add				S.F.	.97			.97	1.07	

131 240	Portable Buildings										
245	0010	KIOSKS Round, 5' diameter, 8' high, ¼" fiberglass wall				Ea.	3,750			3,750	4,125	245
	0100	1" insulated double wall, fiberglass					4,500			4,500	4,950	
	0500	Rectangular, 5' x 9', 7'-6" high, ¼" fiberglass wall					6,375			6,375	7,025	
	0600	1" insulated double wall, fiberglass				↓	7,650			7,650	8,425	

131 520	Swimming Pools										
521	0010	SWIMMING POOL ENCLOSURE Translucent, free standing,										521
	0020	not including foundations, heat or light										
	0200	Economy, minimum	2 Carp	200	.080	SF Hor.	7.25	1.76		9.01	10.85	
	0300	Maximum		100	.160		16	3.52		19.52	23	
	0400	Deluxe, minimum		100	.160		20	3.52		23.52	28	
	0600	Maximum	↓	70	.229	↓	210	5.05		215.05	240	
525	0011	SWIMMING POOLS, Outdoor, incl. equip. & houses, minimum				SF Surf					36	525
	0300	Maximum									68	
	0400	Residential, incl. equipment, permanent type, minimum									12	
	0700	Maximum									25	
	0900	Municipal, including equipment only, over 5000 S.F., minimum									25	
	1000	Maximum									51	
	1300	Motel or apt., incl. equipment only, under 5000 S.F., minimum									22	
	1400	Maximum									34	

141 | Dumbwaiters

141 100 | Manual Dumbwaiters

			CREW	DAILY OUTPUT	MAN-HOURS	UNIT	BARE COSTS				TOTAL INCL O&P	
							MAT.	LABOR	EQUIP.	TOTAL		
101	0010	DUMBWAITERS 2 stop, hand, minimum	2 Elev	.23	69.570	Ea.	2,700	1,775		4,475	5,700	101
	0100	Maximum	↓	.19	84.210	"	5,800	2,125		7,925	9,675	
	0300	For each additional stop, add	↓	.60	26.670	Stop	630	675		1,305	1,725	

141 200 | Electric Dumbwaiters

			CREW	DAILY OUTPUT	MAN-HOURS	UNIT	BARE COSTS				TOTAL INCL O&P	
							MAT.	LABOR	EQUIP.	TOTAL		
201	0010	DUMBWAITERS 2 stop, electric, minimum	2 Elev	.13	123	Ea.	4,600	3,125		7,725	9,875	201
	0100	Maximum	↓	.11	145	"	9,000	3,675		12,675	15,600	
	0600	For each additional stop, add	↓	.54	29.630	Stop	940	750		1,690	2,200	

142 | Elevators

142 010 | Elevators

			CREW	DAILY OUTPUT	MAN-HOURS	UNIT	BARE COSTS				TOTAL INCL O&P	
							MAT.	LABOR	EQUIP.	TOTAL		
011	0012	ELEVATORS										011
	5000	Passenger, pre-engineered, 5 story, hydraulic, 2,500 lb. cap.	M-1	.04	800	Ea.	27,130	19,300	1,900	48,330	61,500	
	5100	For less than 5 stops, deduct	"	.29	110	Stop	6,900	2,650	260	9,810	12,000	
	5200	For 4,000 lb. capacity, general purpose, add				Ea.	5,500			5,500	6,050	
	5800											
	7000	Residential, cab type, 1 floor, 2 stop, minimum	2 Elev	.20	80	Ea.	5,500	2,025		7,525	9,175	
	7100	Maximum		.10	160		10,000	4,050		14,050	17,300	
	7200	2 floor, 3 stop, minimum		.12	133		7,000	3,375		10,375	12,900	
	7300	Maximum		.06	267		17,000	6,750		23,750	29,100	
	7700	Stair climber (chair lift), single seat, minimum		1	16		3,850	405		4,255	4,850	
	7800	Maximum		.20	80		5,250	2,025		7,275	8,900	
	8000	Wheelchair, porch lift, minimum		1	16		3,800	405		4,205	4,800	
	8500	Maximum		.50	32		7,900	810		8,710	9,950	
	8700	Stair lift, minimum		1	16		6,375	405		6,780	7,650	
	8900	Maximum	↓	.20	80	↓	9,675	2,025		11,700	13,800	

144 | Lifts

144 010 | Lifts

			CREW	DAILY OUTPUT	MAN-HOURS	UNIT	BARE COSTS				TOTAL INCL O&P	
							MAT.	LABOR	EQUIP.	TOTAL		
011	0010	CORRESPONDENCE LIFT 1 floor 2 stop, 25 lb. capacity, electric	2 Elev	.20	80	Ea.	3,600	2,025		5,625	7,100	011
	0100	Hand, 5 lb. capacity	"	.20	80	"	1,500	2,025		3,525	4,775	

145 | Material Handling Systems

145 600	Chutes	CREW	DAILY OUTPUT	MAN-HOURS	UNIT	MAT.	LABOR	EQUIP.	TOTAL	TOTAL INCL O&P	
601	0010	CHUTES Linen or refuse, incl. sprinklers, 12' floor height										601
	0020											
	0050	Aluminized steel, 16 ga., 18" diameter	2 Shee	3.50	4.570	Floor	560	115		675	795	
	0100	24" diameter		3.20	5		695	125		820	960	
	0400	Galvanized steel, 16 ga., 18" diameter		3.50	4.570		545	115		660	780	
	0500	24" diameter		3.20	5		610	125		735	865	
	0800	Stainless steel, 18" diameter		3.50	4.570		965	115		1,080	1,250	
	0900	24" diameter		3.20	5		1,130	125		1,255	1,450	
	9000	Minimum labor/equipment charge	1 Shee	1	8	Job		200		200	315	

145 800	Tube Systems	CREW	DAILY OUTPUT	MAN-HOURS	UNIT	MAT.	LABOR	EQUIP.	TOTAL	TOTAL INCL O&P	
801	0010	PNEUMATIC TUBE SYSTEM Single tube, 2 stations,										801
	0020	100' long, stock, economy,										
	0100	3" diameter	2 Stpi	.12	133	Total	2,505	3,400		5,905	8,000	
	0300	4" diameter	"	.09	178	"	3,385	4,525		7,910	10,700	
	0400	Twin tube, two stations or more, conventional system										
	0700	3" round	2 Stpi	46	.348	L.F.	9.45	8.85		18.30	24	
	1050	Add for blower		2	8	System	2,200	205		2,405	2,725	
	1110	Plus for each round station, add		7.50	2.130	Ea.	210	54		264	315	
	1200	Alternate pricing method: base cost, minimum		.75	21.330	Total	1,700	545		2,245	2,700	
	1300	Maximum		.25	64	"	6,000	1,625		7,625	9,125	
	1500	Plus total system length, add, minimum		93.40	.171	L.F.	5.05	4.37		9.42	12.30	
	1600	Maximum		37.60	.426	"	15	10.85		25.85	33	

146 | Hoists and Cranes

146 050	Rails	CREW	DAILY OUTPUT	MAN-HOURS	UNIT	MAT.	LABOR	EQUIP.	TOTAL	TOTAL INCL O&P	
054	0010	CRANE RAIL Box beam bridge, no equipment included	E-4	3,400	.009	Lb.	.76	.23	.02	1.01	1.28	054
	0210	Running track only, 104 lb. per yard, 20' piece	"	160	.200	L.F.	22.50	4.92	.43	27.85	34	

151 | Pipe and Fittings

151 100	Miscellaneous Fittings	CREW	DAILY OUTPUT	MAN-HOURS	UNIT	MAT.	LABOR	EQUIP.	TOTAL	TOTAL INCL O&P	
105	0010	BACKFLOW PREVENTER Includes gate valves,										105
	0020	and four test cocks, corrosion resistant, automatic operation										
	4100	Threaded										
	4120	¾" pipe size	1 Plum	16	.500	Ea.	215	12.75		227.75	255	
	4140	1" pipe size		14	.571		270	14.55		284.55	320	
	4160	1-½" pipe size		10	.800		400	20		420	470	
	4500	Minimum labor/equipment charge		4	2	Job		51		51	78	
	5000	Flanged, bronze										
	5060	2-½" pipe size	Q-1	5	3.200	Ea.	1,725	73		1,798	2,000	
	5080	3" pipe size		4.50	3.560		1,900	81		1,981	2,225	
	5100	4" pipe size		3	5.330		3,450	120		3,570	3,975	
	5120	6" pipe size	Q-2	3	8		6,150	190		6,340	7,050	

For expanded coverage of these items see *Means Mechanical Cost Data* or *Means Plumbing Cost Data 1991*

151 100 | Miscellaneous Fittings

		CREW	DAILY OUTPUT	MAN-HOURS	UNIT	MAT.	LABOR	EQUIP.	TOTAL	TOTAL INCL O&P	
105	9000	Minimum labor/equipment charge	1 Plum	2	4	Job		100		100	155	105
110	0010	**CLEANOUTS**										110
	0060	Floor type										
	0080	Round or square, scoriated nickel bronze top										
	0100	2" pipe size	1 Plum	10	.800	Ea.	46	20		66	82	
	0140	4" pipe size	"	6	1.330	"	65	34		99	125	
	0980	Round top, recessed for terrazzo										
	1000	2" pipe size	1 Plum	9	.889	Ea.	78	23		101	120	
	1100	4" pipe size	"	4	2		91	51		142	180	
	1120	5" pipe size	Q-1	6	2.670		145	61		206	255	
	9000	Minimum labor/equipment charge	1 Plum	3	2.670	Job		68		68	105	
115	0010	**CLEANOUT TEE** Cast iron with countersunk plug										115
	0220	3" pipe size	1 Plum	3.60	2.220	Ea.	21	57		78	110	
	0240	4" pipe size	"	3.30	2.420		32	62		94	130	
	0500	For round smooth access cover, add					20%					
	9000	Minimum labor/equipment charge	1 Plum	2.75	2.910	Job		74		74	115	
120	0010	**CONNECTORS** Flexible, corrugated, ⅞" O.D., ½" I.D.										120
	0050	Gas, seamless brass, steel fittings										
	0200	12" long	1 Plum	36	.222	Ea.	7.25	5.65		12.90	16.70	
	0220	18" long		36	.222		9	5.65		14.65	18.60	
	0240	24" long		34	.235		10.70	6		16.70	21	
	0260	30" long		34	.235		11.50	6		17.50	22	
	9000	Minimum labor/equipment charge		4	2	Job		51		51	78	
125	0010	**DRAINS**										125
	0020											
	0140	Cornice, C.I., 45° or 90° outlet										
	0200	3" and 4" pipe size	Q-1	12	1.330	Ea.	53	31		84	105	
	0260	For galvanized body, add					10.35			10.35	11.40	
	0280	For polished bronze dome, add					9			9	9.90	
	0400	Deck, auto park, C.I., 13" top										
	0440	3", 4", 5", and 6" pipe size	Q-1	8	2	Ea.	218	46		264	310	
	0480	For galvanized body, add				"	98			98	110	
	0500											
	2000	Floor, medium duty, C.I., deep flange, 7" top										
	2040	2" and 3" pipe size	Q-1	12	1.330	Ea.	33	31		64	83	
	2080	For galvanized body, add					13.85			13.85	15.25	
	2120	For polished bronze top, add					18			18	19.80	
	2500	Heavy duty, cleanout & trap w/bucket, C.I., 15" top										
	2540	2", 3", and 4" pipe size	Q-1	6	2.670	Ea.	805	61		866	980	
	2560	For galvanized body, add					248			248	275	
	2580	For polished bronze top, add					250			250	275	
	3860	Roof, flat metal deck, C.I. body, 10" aluminum dome										
	3890	3" pipe size	Q-1	14	1.140	Ea.	115	26		141	165	
	3900	4" pipe size	"	13	1.230	"	120	28		148	175	
	3980	Precast plank deck, C.I. body, aluminum dome										
	4100	10" top, 3" pipe size	Q-1	13	1.230	Ea.	64	28		92	115	
	4120	13" top, 4" pipe size		12	1.330		83	31		114	140	
	4160	16" top, 6" pipe size		8	2		115	46		161	195	
	4220	For galvanized body, add					45			45	50	
	4620	Main, all aluminum, 12" low profile dome										
	4640	2", 3" and 4" pipe size	Q-1	14	1.140	Ea.	175	26		201	235	
	9200	Minimum labor/equipment charge	"	3.75	4.270	Job		98		98	150	
130	0010	**DIELECTRIC UNIONS** Standard gaskets for water and air										130
	0020	250 psi maximum pressure										

For expanded coverage of these items see *Means Mechanical Cost Data* or *Means Plumbing Cost Data 1991*

151 100 | Miscellaneous Fittings

			CREW	DAILY OUTPUT	MAN-HOURS	UNIT	MAT.	LABOR	EQUIP.	TOTAL	TOTAL INCL O&P	
130	0280	Female IPT to sweat, straight										**130**
	0340	¾" pipe size	1 Plum	20	.400	Ea.	2.58	10.20		12.78	18.55	
	0780	Female IPT to female IPT, straight										
	0800	½" pipe size	1 Plum	24	.333	Ea.	3.70	8.50		12.20	17.15	
	0840	¾" pipe size		20	.400	"	4.24	10.20		14.44	20	
	9000	Minimum labor/equipment charge	↓	4	2	Job		51		51	78	
141	0010	**FAUCETS/FITTINGS**										**141**
	0020											
	0150	Bath, faucets, diverter spout combination, sweat	1 Plum	8	1	Ea.	48	25		73	92	
	0200	For integral stops, IPS unions, add					21.75			21.75	24	
	0500	Drain, central lift, 1-½" IPS male	1 Plum	20	.400		29	10.20		39.20	48	
	0600	Trip lever, 1-½" IPS male	"	20	.400	↓	29	10.20		39.20	48	
	0840	Flush valves, with vacuum breaker										
	0850	Water closet										
	0860	Exposed, rear spud	1 Plum	8	1	Ea.	75	25		100	120	
	0870	Top spud		8	1		69	25		94	115	
	0880	Concealed, rear spud		8	1		84	25		109	130	
	0890	Top spud		8	1		96	25		121	145	
	0900	Wall hung	↓	8	1	↓	80	25		105	125	
	0910											
	0920	Urinal										
	0930	Exposed, stall	1 Plum	8	1	Ea.	69	25		94	115	
	0940	Wall, (washout)		8	1		76	25		101	125	
	0950	Pedestal, top spud		8	1		70	25		95	115	
	0960	Concealed, stall		8	1		76	25		101	125	
	0970	Wall (washout)		8	1		76	25		101	125	
	1000	Kitchen sink faucets, top mount, cast spout	↓	10	.800		35	20		55	70	
	1100	For spray, add					10	10%				
	2000	Laundry faucets, shelf type, IPS or copper unions	1 Plum	12	.667		30	16.95		46.95	59	
	2100	Lavatory faucet, centerset, without drain	"	10	.800		25	20		45	59	
	2200	For pop-up drain, add					11	15%				
	2800	Self-closing, center set	1 Plum	10	.800		75	20		95	115	
	3000	Service sink faucet, cast spout, pail hook, hose end		14	.571		51	14.55		65.55	79	
	4000	Shower by-pass valve with union		18	.444		37	11.30		48.30	58	
	4200	Shower thermostatic mixing valve, concealed	↓	8	1		180	25		205	235	
	4300	For inlet strainer, check, and stops, add					54	5%				
	5000	Sillcock, compact, brass, IPS or copper to hose	1 Plum	24	.333	↓	4.50	8.50		13	18.05	
	9000	Minimum labor/equipment charge	"	4	2	Job		51		51	78	
146	0010	**FLOOR RECEPTORS** For connection to 2", 3" & 4" diameter pipe										**146**
	0200	12-½" square top, 25 sq. in. open area	Q-1	10	1.600	Ea.	210	37		247	290	
	0300	For grate with 4" diam. x 3-¾" high funnel, add					31			31	34	
	0400	For grate with 6" diameter x 6" high funnel, add					46			46	51	
	0700	For acid-resisting bucket, add					44			44	48	
	0900	For stainless steel mesh bucket liner, add					35			35	39	
	2000	12-⅝" diameter top, 40 sq. in. open area	Q-1	10	1.600	↓	180	37		217	255	
	2100	For options, add same prices as square top										
	3000	8" x 4" rectangular top, 7.5 sq. in. open area	Q-1	14	1.140	Ea.	120	26		146	170	
	3100	For trap primer connections, add				"	13			13	14.30	
	9000	Minimum labor/equipment charge	Q-1	3	5.330	Job		120		120	190	
156	0010	**HYDRANTS**										**156**
	0050	Wall type, moderate climate, bronze, encased										
	0200	¾" IPS connection	1 Plum	16	.500	Ea.	120	12.75		132.75	150	
	0400	For ¾" adapter type vacuum breaker, add				"	55			55	61	
	1000	Non-freeze, bronze, exposed										
	1100	¾" IPS connection, 4" to 9" thick wall	1 Plum	14	.571	Ea.	87	14.55		101.55	120	
	1120	10" to 14" thick wall	"	12	.667		94	16.95		110.95	130	
	1280	For anti-siphon type, add	↓				15			15	16.50	

For expanded coverage of these items see *Means Mechanical Cost Data* or *Means Plumbing Cost Data* 1991

151 100 | Miscellaneous Fittings

		Description	CREW	DAILY OUTPUT	MAN-HOURS	UNIT	BARE COSTS MAT.	LABOR	EQUIP.	TOTAL	TOTAL INCL O&P	
156	9000	Minimum labor/equipment charge	1 Plum	3	2.670	Job		68		68	105	156
165												165
	0010	**SHOCK ABSORBERS**										
	0500	¾″ male I.P.S. For 1 to 11 fixtures	1 Plum	12	.667	Ea.	37	16.95		53.95	67	
	0600	1″ male I.P.S., For 12 to 32 fixtures		8	1		75	25		100	120	
	0700	For 33 to 60 fixtures		8	1		110	25		135	160	
	0800	For 61 to 113 fixtures		8	1		280	25		305	345	
	0900	For 114 to 154 fixtures		8	1		335	25		360	410	
	1000	For 155 to 330 fixtures		4	2		390	51		441	505	
	9000	Minimum labor/equipment charge		3.50	2.290	Job		58		58	90	
170												170
	0010	**SUPPORTS/CARRIERS** For plumbing fixtures										
	0020											
	3000	Lavatory, concealed arm										
	3050	Floor mounted, single										
	3100	High back fixture	1 Plum	6	1.330	Ea.	72	34		106	130	
	3200	Flat slab fixture		6	1.330		85	34		119	145	
	3220	Paraplegic		6	1.330		70	34		104	130	
	6980	Water closet, siphon jet										
	7000	Horizontal, adjustable, caulk										
	7040	Single, 4″ pipe size	1 Plum	6	1.330	Ea.	110	34		144	175	
	7060	5″ pipe size		6	1.330		145	34		179	210	
	7100	Double, 4″ pipe size		5	1.600		210	41		251	295	
	7120	5″ pipe size		5	1.600		260	41		301	350	
	8200	Water closet, residential										
	8220	Vertical centerline, floor mount										
	8240	Single, 3″ caulk, 2″ or 3″ vent	1 Plum	6	1.330	Ea.	60	34		94	120	
	8260	4″ caulk, 2″ or 4″ vent		6	1.330	″	83	34		117	145	
	9990	Minimum labor/equipment charge		3.50	2.290	Job		58		58	90	
181												181
	0010	**TRAPS**										
	0030	Cast iron, service weight										
	0050	Long P trap, 2″ pipe size	Q-1	16	1	Ea.	9.05	23		32.05	45	
	1100	12″ long		16	1		7.10	23		30.10	43	
	3000	P trap, 2″ pipe size		14	1.140		9.05	26		35.05	50	
	3040	3″ pipe size	Q-2	17	1.410		12.50	34		46.50	65	
	3800	Drum trap, 4″ x 5″, 1-½″ tapping										
	3900											
	4700	Copper, drainage, drum trap	1 Plum	16	.500	Ea.	40	12.75		52.75	64	
	4840	3″ x 6″ swivel, 1-½″ pipe size										
	5100	P trap, standard pattern	1 Plum	18	.444	Ea.	24	11.30		35.30	44	
	5200	1-¼″ pipe size		17	.471		22	12		34	43	
	5240	1-½″ pipe size		15	.533		43	13.55		56.55	68	
	5260	2″ pipe size		11	.727		55	18.50		73.50	89	
	5280	3″ pipe size		3	2.670	Job		68		68	105	
	9000	Minimum labor/equipment charge										
185												185
	0010	**VACUUM BREAKERS** Hot or cold water										
	1030	Anti-siphon, brass	1 Plum	24	.333	Ea.	14.10	8.50		22.60	29	
	1060	½″ size		20	.400		16.75	10.20		26.95	34	
	1080	¾″ size		19	.421		26	10.70		36.70	45	
	1100	1″ size		15	.533		43	13.55		56.55	68	
	1120	1-¼″ size		13	.615		51	15.65		66.65	80	
	1140	1-½″ size		11	.727		79	18.50		97.50	115	
	1160	2″ size		4	2	Job		51		51	78	
	9000	Minimum labor/equipment charge										
195												195
	0010	**VENT FLASHING**										
	1000	Aluminum with lead ring	1 Plum	17	.471	Ea.	5.70	12		17.70	25	
	1050	3″ pipe	″	16	.500	″	6.95	12.75		19.70	27	
	1060	4″ pipe										183

For expanded coverage of these items see *Means Mechanical Cost Data* or *Means Plumbing Cost Data 1991*

151 100 | Miscellaneous Fittings

			CREW	DAILY OUTPUT	MAN-HOURS	UNIT	BARE COSTS				TOTAL INCL O&P
							MAT.	LABOR	EQUIP.	TOTAL	
195	1350	Copper with neoprene ring										195
	1440	2" pipe	1 Plum	18	.444	Ea.	14.85	11.30		26.15	34
	1450	3" pipe		17	.471		17.40	12		29.40	38
	1460	4" pipe		16	.500		19.20	12.75		31.95	41
	9000	Minimum labor/equipment charge		4	2	Job		51		51	78

151 250 | Brass Pipe

			CREW	DAILY OUTPUT	MAN-HOURS	UNIT	MAT.	LABOR	EQUIP.	TOTAL	INCL O&P
251	0010	**PIPE, BRASS** Plain end,										251
	0900	Field threaded, coupling & clevis hanger 10' O.C.									
	0920	Regular weight									
	1120	½" diameter	1 Plum	48	.167	L.F.	3.03	4.24		7.27	9.85
	1140	¾" diameter		46	.174		4.13	4.43		8.56	11.35
	1160	1" diameter		43	.186		5.86	4.73		10.59	13.75
	1180	1-¼" diameter	Q-1	72	.222		8.78	5.10		13.88	17.50
	1200	1-½" diameter		65	.246		10.30	5.65		15.95	20
	1220	2" diameter		53	.302		13.57	6.90		20.47	26
	9000	Minimum labor/equipment charge	1 Plum	4	2	Job		51		51	78
258	0010	**PIPE, BRASS, FITTINGS** Rough bronze, threaded										258
	0020										
	1000	Standard wt., 90° Elbow									
	1100	½"	1 Plum	12	.667	Ea.	2.13	16.95		19.08	29
	1120	¾"		11	.727		3.13	18.50		21.63	32
	1140	1"		10	.800		4.38	20		24.38	36
	1160	1-¼"	Q-1	17	.941		7.80	22		29.80	42
	1180	1-½"	"	16	1		9.25	23		32.25	46
	1500	45° Elbow, ⅛"	1 Plum	13	.615		2.25	15.65		17.90	27
	1580	½"		12	.667		2.25	16.95		19.20	29
	1600	¾"		11	.727		3.75	18.50		22.25	33
	1620	1"		10	.800		6.50	20		26.50	39
	1640	1-¼"	Q-1	17	.941		10	22		32	44
	1660	1-½"	"	16	1		12.45	23		35.45	49
	2000	Tee, ⅛"	1 Plum	9	.889		2.25	23		25.25	37
	2080	½"		8	1		2.25	25		27.25	42
	2100	¾"		7	1.140		3.88	29		32.88	49
	2120	1"		6	1.330		6.90	34		40.90	60
	2140	1-¼"	Q-1	11	1.450		12.50	33		45.50	65
	2160	1-½"	"	9	1.780		13.50	41		54.50	78
	2500	Coupling, ⅛"	1 Plum	26	.308		2.24	7.85		10.09	14.55
	2580	½"		15	.533		2.24	13.55		15.79	23
	2600	¾"		14	.571		3.15	14.55		17.70	26
	2620	1"		13	.615		5.40	15.65		21.05	30
	2640	1-¼"	Q-1	22	.727		8.95	16.65		25.60	36
	2660	1-½"		20	.800		11.65	18.30		29.95	41
	2680	2"		18	.889		19.25	20		39.25	53
	9000	Minimum labor/equipment charge	1 Plum	4	2	Job		51		51	78

151 300 | Cast Iron Pipe

			CREW	DAILY OUTPUT	MAN-HOURS	UNIT	MAT.	LABOR	EQUIP.	TOTAL	INCL O&P
301	0010	**PIPE, CAST IRON** Soil, on hangers 5' O.C.										301
	0020	Single hub, service wt., lead & oakum joints 10' O.C.									
	2120	2" diameter	Q-1	63	.254	L.F.	2.03	5.80		7.83	11.20
	2140	3" diameter		60	.267		2.83	6.10		8.93	12.55
	2160	4" diameter		55	.291		3.67	6.65		10.32	14.30
	2180	5" diameter	Q-2	76	.316		4.95	7.50		12.45	17
	2200	6" diameter	"	73	.329		6.12	7.80		13.92	18.75
	2220	8" diameter	Q-3	59	.542		9.71	13.20		22.91	31
	4000	No hub, couplings 10' O.C.									
	4100	1-½" diameter	Q-1	71	.225	L.F.	2.58	5.15		7.73	10.80

(129)

For expanded coverage of these items see *Means Mechanical Cost Data* or *Means Plumbing Cost Data 1991*

151 300 | Cast Iron Pipe

		CREW	DAILY OUTPUT	MAN-HOURS	UNIT	MAT.	LABOR	EQUIP.	TOTAL	TOTAL INCL O&P	
301	4120	2" diameter	Q-1	67	.239	L.F.	2.71	5.45		8.16	11.40	301
	4140	3" diameter		64	.250		3.55	5.75		9.30	12.75	
	4160	4" diameter	↓	58	.276		4.57	6.30		10.87	14.75	
	4180	5" diameter	Q-2	83	.289		6.85	6.85		13.70	18.15	
	9000	Minimum labor/equipment charge	1 Plum	4	2	Job		51		51	78	

											320
320	0010	**PIPE, CAST IRON, FITTINGS, Soil**										
	0040	Hub and spigot, service weight, lead & oakum joints										
	0080	¼ Bend, 2"	Q-1	16	1	Ea.	2.87	23		25.87	38	
	0120	3"		14	1.140		5.35	26		31.35	46	
	0140	4"	↓	13	1.230		7.80	28		35.80	52	
	0160	5"	Q-2	18	1.330		10.90	32		42.90	61	
	0180	6"	"	17	1.410		13.70	34		47.70	67	
	0200	8"	Q-3	11	2.910		42	71		113	155	
	0340	⅛ Bend, 2"	Q-1	16	1		2.32	23		25.32	38	
	0350	3"		14	1.140		4.18	26		30.18	45	
	0360	4"	↓	13	1.230		6.25	28		34.25	50	
	0380	5"	Q-2	18	1.330		8.50	32		40.50	58	
	0400	6"	"	17	1.410		10.50	34		44.50	63	
	0420	8"	Q-3	11	2.910		30	71		101	140	
	0500	Sanitary Tee, 2"	Q-1	10	1.600		4.95	37		41.95	62	
	0540	3"		9	1.780		9.15	41		50.15	73	
	0620	4"	↓	8	2		11.15	46		57.15	83	
	0700	5"	Q-2	12	2		20	48		68	95	
	0800	6"	"	11	2.180		25	52		77	105	
	0880	8"	Q-3	7	4.570		65	110		175	245	
	0890	10" x 6"	"	8	4		92	97		189	250	
	5990	No hub										
	6000	Cplg. & labor required at joints not incl. in fitting										
	6010	price. Add 1 coupling per joint for installed price										
	6020	¼ Bend, 1-½"				Ea.	2.37			2.37	2.61	
	6060	2"					2.50			2.50	2.75	
	6080	3"					3.30			3.30	3.63	
	6120	4"					4.95			4.95	5.45	
	6200	⅛ Bend, 1-½"					1.90			1.90	2.09	
	6240	2"					1.95			1.95	2.15	
	6260	3"					2.70			2.70	2.97	
	6280	4"					3.60			3.60	3.96	
	6400	Sanitary Tee, 1-½"					3.15			3.15	3.47	
	6460	2"					3.50			3.50	3.85	
	6520	3"					4.20			4.20	4.62	
	6600	4"					6.55			6.55	7.20	
	8000	Coupling, standard (by CISPI Mfrs.)										
	8020	1-½"	Q-1	48	.333	Ea.	2.10	7.65		9.75	14.10	
	8040	2"		44	.364		2.10	8.35		10.45	15.15	
	8080	3"		38	.421		2.55	9.65		12.20	17.70	
	8120	4"	↓	33	.485		2.95	11.10		14.05	20	
	9000	Minimum labor/equipment charge	1 Plum	4	2	Job		51		51	78	

151 400 | Copper Pipe & Tubing

											401
401	0010	**PIPE, COPPER** Solder joints										
	0020	Type K tubing, couplings & clevis hangers 10' O.C.										
	1100	¼" diameter	1 Plum	84	.095	L.F.	.78	2.42		3.20	4.60	
	1200	1" diameter		66	.121		3.82	3.08		6.90	8.95	
	1260	2" diameter	↓	40	.200		7.53	5.10		12.63	16.15	
	2000	Type L tubing, couplings & hangers 10' O.C.										
	2100	¼" diameter	1 Plum	88	.091	L.F.	.71	2.31		3.02	4.35	
	2120	⅜" diameter	"	84	.095	"	.99	2.42		3.41	4.83	

For expanded coverage of these items see *Means Mechanical Cost Data* or *Means Plumbing Cost Data 1991*

151 400 | Copper Pipe & Tubing

		Description	CREW	DAILY OUTPUT	MAN-HOURS	UNIT	MAT.	LABOR	EQUIP.	TOTAL	TOTAL INCL O&P	
401	2140	½" diameter	1 Plum	81	.099	L.F.	1.18	2.51		3.69	5.15	401
	2160	⅝" diameter		79	.101		1.73	2.58		4.31	5.90	
	2180	¾" diameter		76	.105		1.77	2.68		4.45	6.10	
	2200	1" diameter		68	.118		2.45	2.99		5.44	7.30	
	2220	1-¼" diameter		58	.138		3.35	3.51		6.86	9.10	
	2240	1-½" diameter		52	.154		4.21	3.92		8.13	10.65	
	2260	2" diameter		42	.190		6.33	4.85		11.18	14.45	
	2280	2-½" diameter	Q-1	62	.258		9.17	5.90		15.07	19.20	
	2300	3" diameter		56	.286		12.91	6.55		19.46	24	
	2320	3-½" diameter		43	.372		17.24	8.50		25.74	32	
	2340	4" diameter		39	.410		21.06	9.40		30.46	38	
	2360	5" diameter		34	.471		53.22	10.80		64.02	75	
	2380	6" diameter	Q-2	40	.600		66.11	14.25		80.36	95	
	2410	For other than full hard temper, add					25%					
	2590	For silver solder, add						15%				
	3000	Type M tubing, couplings & hangers 10' O.C.										
	3140	½" diameter	1 Plum	84	.095	L.F.	.89	2.42		3.31	4.72	
	3180	¾" diameter		78	.103		1.35	2.61		3.96	5.50	
	3200	1" diameter		70	.114		1.88	2.91		4.79	6.55	
	4000	Type DWV tubing, couplings & hangers 10' O.C.										
	4100	1-¼" diameter	1 Plum	60	.133	L.F.	2.65	3.39		6.04	8.15	
	4120	1-½" diameter		54	.148		3.31	3.77		7.08	9.45	
	4140	2" diameter		44	.182		4.38	4.63		9.01	11.95	
	4160	3" diameter	Q-1	58	.276		7.41	6.30		13.71	17.90	
	4180	4" diameter		40	.400		13.56	9.15		22.71	29	
	4200	5" diameter		36	.444		39.22	10.20		49.42	59	
	4220	6" diameter	Q-2	42	.571		55.11	13.55		68.66	82	
	9000	Minimum labor/equipment charge	1 Plum	4	2	Job		51		51	78	
430	0010	PIPE, COPPER, FITTINGS, Wrought unless otherwise noted										430
	0040	Solder joints, copper x copper										
	0070	90° Elbow, ¼"	1 Plum	22	.364	Ea.	.73	9.25		9.98	15.05	
	0100	½"		20	.400		.29	10.20		10.49	16	
	0120	¾"		19	.421		.49	10.70		11.19	17.05	
	0130	1"		16	.500		.99	12.75		13.74	21	
	0250	45° Elbow, ¼"		22	.364		1.73	9.25		10.98	16.15	
	0270	⅜"		22	.364		1.39	9.25		10.64	15.80	
	0280	½"		20	.400		.53	10.20		10.73	16.30	
	0290	⅝"		19	.421		2.65	10.70		13.35	19.45	
	0300	¾"		19	.421		.90	10.70		11.60	17.50	
	0310	1"		16	.500		2.31	12.75		15.06	22	
	0320	1-¼"		15	.533		3.18	13.55		16.73	24	
	0330	1-½"		13	.615		3.83	15.65		19.48	28	
	0340	2"		11	.727		6.35	18.50		24.85	36	
	0350	2-½"	Q-1	13	1.230		13.60	28		41.60	58	
	0360	3"		13	1.230		20	28		48	65	
	0370	3-½"		10	1.600		20	37		57	79	
	0380	4"		9	1.780		42	41		83	110	
	0390	5" cast brass		6	2.670		155	61		216	265	
	0400	6" cast brass	Q-2	9	2.670		200	63		263	320	
	0450	Tee, ¼"	1 Plum	14	.571		2.15	14.55		16.70	25	
	0470	⅜"		14	.571		1.62	14.55		16.17	24	
	0480	½"		13	.615		.49	15.65		16.14	25	
	0490	⅝"		12	.667		3.21	16.95		20.16	30	
	0500	¾"		12	.667		1.22	16.95		18.17	28	
	0510	1"		10	.800		3.55	20		23.55	35	
	0520	1-¼"		9	.889		6.60	23		29.60	42	
	0530	1-½"		8	1		7.80	25		32.80	48	
	0540	2"		7	1.140		12.20	29		41.20	58	

For expanded coverage of these items see *Means Mechanical Cost Data* or *Means Plumbing Cost Data 1991*

151 400 | Copper Pipe & Tubing

		CREW	DAILY OUTPUT	MAN-HOURS	UNIT	BARE COSTS MAT.	LABOR	EQUIP.	TOTAL	TOTAL INCL O&P	
430 0550	2-½"	Q-1	8	2	Ea.	24	46		70	97	**430**
0560	3"		7	2.290		38	52		90	125	
0570	3-½"		6	2.670		92	61		153	195	
0580	4"		5	3.200		81	73		154	200	
0590	5" cast brass	↓	4	4		225	92		317	390	
0600	6" cast brass	Q-2	6	4		245	95		340	415	
0650	Coupling, ¼"	1 Plum	24	.333		.17	8.50		8.67	13.25	
0670	⅜"		24	.333		.32	8.50		8.82	13.45	
0680	½"		22	.364		.24	9.25		9.49	14.55	
0690	⅝"		21	.381		.73	9.70		10.43	15.75	
0700	¾"		21	.381		.49	9.70		10.19	15.50	
0710	1"		18	.444		.99	11.30		12.29	18.55	
0720	1-¼"		17	.471		1.73	12		13.73	20	
0730	1-½"		15	.533		2.31	13.55		15.86	23	
0740	2"	↓	13	.615		3.44	15.65		19.09	28	
0750	2-½"	Q-1	15	1.070		6.15	24		30.15	44	
0760	3"		13	1.230		11.30	28		39.30	56	
0770	3-½"		8	2		19.45	46		65.45	92	
0780	4"		7	2.290		22	52		74	105	
0790	5"		6	2.670		48	61		109	145	
0800	6"	Q-2	8	3	↓	76	71		147	195	
2000	DWV, solder joints, copper x copper										
2030	90° Elbow, 1-¼"	1 Plum	13	.615	Ea.	2.84	15.65		18.49	27	
2050	1-½"		12	.667		2.93	16.95		19.88	29	
2070	2"	↓	10	.800		6.50	20		26.50	39	
2090	3"	Q-1	10	1.600		13.30	37		50.30	71	
2100	4"	"	9	1.780		43	41		84	110	
2250	Tee, Sanitary, 1-¼"	1 Plum	9	.889		4.75	23		27.75	40	
2270	1-½"		8	1		4.40	25		29.40	44	
2290	2"	↓	7	1.140		8.10	29		37.10	54	
2310	3"	Q-1	7	2.290		15.90	52		67.90	98	
2330	4"	"	6	2.670		46	61		107	145	
2400	Coupling, 1-¼"	1 Plum	14	.571		1.21	14.55		15.76	24	
2420	1-½"		13	.615		1.64	15.65		17.29	26	
2440	2"	↓	11	.727		2.26	18.50		20.76	31	
2460	3"	Q-1	11	1.450		3.77	33		36.77	56	
2480	4"	"	10	1.600	↓	9.60	37		46.60	67	
9000	Minimum labor/equipment charge	1 Plum	4	2	Job		51		51	78	

151 450 | Corrosion Resistant Pipe

		CREW	DAILY OUTPUT	MAN-HOURS	UNIT	MAT.	LABOR	EQUIP.	TOTAL	TOTAL INCL O&P	
451 0010	**PIPE, CORROSION RESISTANT** No couplings or hangers										**451**
0020	Iron alloy, drain, mechanical joint										
1000	1-½" diameter	Q-1	70	.229	L.F.	17.10	5.25		22.35	27	
1100	2" diameter		66	.242		19.50	5.55		25.05	30	
1120	3" diameter		60	.267		27	6.10		33.10	39	
1140	4" diameter		52	.308	↓	35	7.05		42.05	49	
2980	Plastic, Epoxy, fiberglass filament wound										
3000	2" diameter	Q-1	62	.258	L.F.	4.68	5.90		10.58	14.25	
3100	3" diameter		51	.314		6.60	7.20		13.80	18.35	
3120	4" diameter		45	.356		8.30	8.15		16.45	22	
3160	8" diameter	Q-2	38	.632		22	15		37	47	
3200	12" diameter	"	28	.857	↓	45	20		65	81	
9800	Minimum labor/equipment charge	1 Plum	4	2	Job		51		51	78	
454 0010	**PIPE, CORROSION RESISTANT, FITTINGS**										**454**
0030	Iron alloy										
0050	Mechanical joint										
0060	¼ Bend, 1-½"	Q-1	12	1.330	Ea.	25	31		56	75	

151 450 | Corrosion Resistant Pipe

			CREW	DAILY OUTPUT	MAN-HOURS	UNIT	MAT.	LABOR	EQUIP.	TOTAL	TOTAL INCL O&P	
454	0080	2″	Q-1	10	1.600	Ea.	31	37		68	91	454
	0090	3″		9	1.780		45	41		86	110	
	0100	4″		8	2		62	46		108	140	
	0160	Tee and Y, sanitary, straight										
	0170	1-½″	Q-1	8	2	Ea.	32	46		78	105	
	0180	2″		7	2.290		44	52		96	130	
	0190	3″		6	2.670		66	61		127	165	
	0200	4″		5	3.200		115	73		188	240	
	0360	Coupling, 1-½″		14	1.140		23	26		49	66	
	0380	2″		12	1.330		27	31		58	77	
	0390	3″		11	1.450		29	33		62	83	
	0400	4″		10	1.600		31	37		68	91	
	3000	Epoxy, filament wound										
	3030	Quick-lock joint										
	3040	90° Elbow, 2″	Q-1	28	.571	Ea.	28	13.10		41.10	51	
	3060	3″		16	1		37	23		60	76	
	3070	4″		13	1.230		46	28		74	94	
	3190	Tee, 2″		19	.842		49	19.30		68.30	84	
	3200	3″		11	1.450		57	33		90	115	
	3210	4″		9	1.780		68	41		109	140	
	9000	Minimum labor/equipment charge	1 Plum	4	2	Job		51		51	78	

151 500 | Glass Pipe

			CREW	DAILY OUTPUT	MAN-HOURS	UNIT	MAT.	LABOR	EQUIP.	TOTAL	TOTAL INCL O&P	
501	0010	**PIPE, GLASS** Borosilicate, couplings & hangers 10′ O.C.										501
	0020	Drainage										
	1100	1-½″ diameter	Q-1	52	.308	L.F.	5.77	7.05		12.82	17.20	
	1120	2″ diameter		44	.364		7.72	8.35		16.07	21	
	1140	3″ diameter		39	.410		10.56	9.40		19.96	26	
	1160	4″ diameter		30	.533		18.64	12.20		30.84	39	
	1180	6″ diameter		26	.615		33.92	14.10		48.02	59	
	9000	Minimum labor/equipment charge	1 Plum	4	2	Job		51		51	78	
512	0010	**PIPE, GLASS, FITTINGS**										512
	0020	Drainage, beaded ends										
	0040	Cplg. & labor required at joints not incl. in fitting										
	0050	price. Add 1 per joint for installed price										
	0070	90° Bend or sweep, 1-½″				Ea.	12.80			12.80	14.10	
	0090	2″					16.15			16.15	17.75	
	0100	3″					27			27	30	
	0110	4″					41			41	45	
	0120	6″ (sweep only)					125			125	140	
	0350	Tee, single sanitary, 1-½″					20			20	22	
	0370	2″					20			20	22	
	0380	3″					30			30	33	
	0390	4″					54			54	59	
	0400	6″					145			145	160	
	0500	Coupling, stainless steel, TFE seal ring										
	0520	1-½″	Q-1	32	.500	Ea.	8.35	11.45		19.80	27	
	0530	2″		30	.533		10.65	12.20		22.85	31	
	0540	3″		25	.640		14.30	14.65		28.95	38	
	0550	4″		23	.696		25	15.95		40.95	52	
	0560	6″		20	.800		55	18.30		73.30	89	
	9000	Minimum labor/equipment charge	1 Plum	4	2	Job		51		51	78	

151 550 | Plastic Pipe

551	0010	**PIPE, PLASTIC** See also division 151-451										551
	0020	Fiberglass reinforced, couplings 10′ O.C., hangers 3 per 10′										

For expanded coverage of these items see *Means Mechanical Cost Data* or *Means Plumbing Cost Data 1991*

151 550	Plastic Pipe	CREW	DAILY OUTPUT	MAN-HOURS	UNIT	BARE COSTS				TOTAL INCL O&P	
						MAT.	LABOR	EQUIP.	TOTAL		
551 0240	2" diameter	Q-1	58	.276	L.F.	10.56	6.30		16.86	21	**551**
0280	4" diameter		47	.340		16.09	7.80		23.89	30	
0300	6" diameter	↓	38	.421	↓	24.35	9.65		34	42	
1800	PVC, couplings 10' O.C., hangers 3 per 10'										
1820	Schedule 40										
1860	½" diameter	1 Plum	54	.148	L.F.	.62	3.77		4.39	6.50	
1870	¾" diameter		51	.157		.69	3.99		4.68	6.90	
1880	1" diameter		46	.174		.82	4.43		5.25	7.75	
1890	1-¼" diameter		42	.190		.96	4.85		5.81	8.55	
1900	1-½" diameter	↓	36	.222		1.04	5.65		6.69	9.85	
1910	2" diameter	Q-1	59	.271		1.32	6.20		7.52	11.05	
1920	2-½" diameter		56	.286		1.90	6.55		8.45	12.20	
1930	3" diameter		53	.302		2.40	6.90		9.30	13.30	
1940	4" diameter		48	.333		3.29	7.65		10.94	15.40	
1950	5" diameter		43	.372		5.31	8.50		13.81	19	
1960	6" diameter	↓	39	.410	↓	6.08	9.40		15.48	21	
4100	DWV type, schedule 40, couplings 10' O.C., hangers 3 per 10'										
4120	ABS										
4140	1-¼" diameter	1 Plum	42	.190	L.F.	1.04	4.85		5.89	8.60	
4150	1-½" diameter	"	36	.222		1.12	5.65		6.77	9.95	
4160	2" diameter	Q-1	59	.271	↓	1.33	6.20		7.53	11.05	
4400	PVC										
4410	1-¼" diameter	1 Plum	42	.190	L.F.	.78	4.85		5.63	8.35	
4460	2" diameter	Q-1	59	.271		.94	6.20		7.14	10.60	
4470	3" diameter		53	.302		1.64	6.90		8.54	12.45	
4480	4" diameter		48	.333		2.21	7.65		9.86	14.20	
4490	6" diameter	↓	39	.410	↓	4.40	9.40		13.80	19.35	
5360	CPVC, couplings 10' O.C., hangers 3 per 10'										
5380	Schedule 40										
5460	½" diameter	1 Plum	54	.148	L.F.	1.27	3.77		5.04	7.20	
5470	¾" diameter		51	.157		1.52	3.99		5.51	7.85	
5480	1" diameter		46	.174		1.94	4.43		6.37	8.95	
5490	1-¼" diameter		42	.190		2.71	4.85		7.56	10.45	
5500	1-½" diameter	↓	36	.222		3.47	5.65		9.12	12.55	
5510	2" diameter	Q-1	59	.271		4.03	6.20		10.23	14	
5520	2-½" diameter		56	.286		5.95	6.55		12.50	16.65	
5530	3" diameter	↓	53	.302	↓	7.83	6.90		14.73	19.25	
5560											
9900	Minimum labor/equipment charge	1 Plum	4	2	Job		51		51	78	
558 0010	**PIPE, PLASTIC, FITTINGS**										**558**
0500	PVC, high impact/pressure, Schedule 40										
0530	90° Elbow, ½"	1 Plum	20	.400	Ea.	.26	10.20		10.46	16	
0550	¾"		19	.421		.30	10.70		11	16.85	
0560	1"		16	.500		.53	12.75		13.28	20	
0570	1-¼"		15	.533		.91	13.55		14.46	22	
0580	1-½"	↓	14	.571		1	14.55		15.55	24	
0590	2"	Q-1	23	.696		1.55	15.95		17.50	26	
0600	3"		16	1		6.35	23		29.35	42	
0610	4"		13	1.230		11.45	28		39.45	56	
0620	6"	↓	8	2		36	46		82	110	
0800	Tee, ½"	1 Plum	13	.615		.34	15.65		15.99	25	
0820	¾"		12	.667		.37	16.95		17.32	27	
0830	1"		11	.727		.69	18.50		19.19	29	
0840	1-¼"		10	.800		1.07	20		21.07	33	
0850	1-½"		10	.800		1.32	20		21.32	33	
0860	2"	Q-1	17	.941		1.89	22		23.89	35	
0870	3"	"	10	1.600		8.20	37		45.20	66	

		151 550 Plastic Pipe	CREW	DAILY OUTPUT	MAN-HOURS	UNIT	MAT.	BARE COSTS LABOR	EQUIP.	TOTAL	TOTAL INCL O&P	
558	0880	4"	Q-1	8	2	Ea.	14.85	46		60.85	87	558
	0890	6"	"	5	3.200	"	50	73		123	170	
	2700	PVC (white), Schedule 40, socket joints										
	2760	Elbow 90°, ½"	1 Plum	22	.364	Ea.	.17	9.25		9.42	14.45	
	2810	2"	Q-1	28	.571	"	.97	13.10		14.07	21	
	4500	DWV, ABS, non pressure, socket joints										
	4540	¼ Bend, 1-¼"	1 Plum	17	.471	Ea.	1.21	12		13.21	19.80	
	4560	1-½"	"	16	.500		.61	12.75		13.36	20	
	4570	2"	Q-1	28	.571	↓	.80	13.10		13.90	21	
	4800	Tee, sanitary										
	4820	1-¼"	1 Plum	11	.727	Ea.	2.59	18.50		21.09	31	
	4830	1-½"	"	10	.800		.96	20		20.96	32	
	4840	2"	Q-1	17	.941	↓	1.43	22		23.43	35	
	5000	PVC, Schedule 40, socket joints										
	5040	¼ Bend, 1-¼" diameter	1 Plum	17	.471	Ea.	1.31	12		13.31	19.90	
	5060	1-½"	"	16	.500		.55	12.75		13.30	20	
	5070	2"	Q-1	28	.571		.70	13.10		13.80	21	
	5080	3"		17	.941		2.15	22		24.15	36	
	5090	4"		14	1.140		4.55	26		30.55	45	
	5100	6"	↓	8	2		28	46		74	100	
	5250	Tee, sanitary 1-¼"	1 Plum	11	.727		1.64	18.50		20.14	30	
	5270	1-½"	"	10	.800		.96	20		20.96	32	
	5280	2"	Q-1	17	.941		1.39	22		23.39	35	
	5290	3"		11	1.450		3.69	33		36.69	55	
	5300	4"	↓	9	1.780	↓	7.05	41		48.05	71	
	5500	CPVC, Schedule 80, threaded joints										
	5540	90° Elbow, ¼"	1 Plum	20	.400	Ea.	4.40	10.20		14.60	21	
	5560	½"		18	.444		1.95	11.30		13.25	19.60	
	5570	¾"		17	.471		2.48	12		14.48	21	
	5580	1"		15	.533		3.94	13.55		17.49	25	
	5590	1-¼"		14	.571		8.55	14.55		23.10	32	
	5600	1-½"	↓	13	.615		9.50	15.65		25.15	35	
	5610	2"	Q-1	22	.727		14.50	16.65		31.15	42	
	5620	2-½"		18	.889		27	20		47	61	
	5630	3"	↓	14	1.140		30	26		56	73	
	6000	Coupling, ¼"	1 Plum	20	.400		5.30	10.20		15.50	22	
	6020	½"		18	.444		2.05	11.30		13.35	19.70	
	6030	¾"		17	.471		2.87	12		14.87	22	
	6040	1"		15	.533		3.87	13.55		17.42	25	
	6050	1-¼"		14	.571		5.75	14.55		20.30	29	
	6060	1-½"	↓	13	.615		7.30	15.65		22.95	32	
	6070	2"	Q-1	22	.727		8.50	16.65		25.15	35	
	6080	2-½"		20	.800		19	18.30		37.30	49	
	6090	3"	↓	19	.842	↓	21	19.30		40.30	53	
	9900	Minimum labor/equipment charge	1 Plum	4	2	Job		51		51	78	

		151 700 Steel Pipe										
701	0010	PIPE, STEEL										701
	0020	All pipe sizes are to Spec. A-53										
	0050	Schedule 40, threaded, with couplings, and clevis type										
	0060	hangers sized for covering, 10' O.C.										
	0540	Black, ¼" diameter (129)	1 Plum	66	.121	L.F.	.75	3.08		3.83	5.60	
	0570	¾" diameter		61	.131		1.06	3.34		4.40	6.30	
	0580	1" diameter	↓	53	.151		1.49	3.84		5.33	7.55	
	0590	1-¼" diameter	Q-1	89	.180		1.85	4.12		5.97	8.40	
	0600	1-½" diameter		80	.200		2.17	4.58		6.75	9.45	
	0610	2" diameter	↓	64	.250	↓	2.91	5.75		8.66	12.05	

For expanded coverage of these items see *Means Mechanical Cost Data* or *Means Plumbing Cost Data 1991*

			DAILY	MAN-		BARE COSTS				TOTAL	
151 700	**Steel Pipe**	CREW	OUTPUT	HOURS	UNIT	MAT.	LABOR	EQUIP.	TOTAL	INCL O&P	
701	0620	2-½″ diameter	Q-1	50	.320	L.F.	4.78	7.35		12.13	16.55	**701**
	0630	3″ diameter		43	.372		6.19	8.50		14.69	19.95	
	0640	3-½″ diameter		40	.400		8.61	9.15		17.76	24	
	0650	4″ diameter	↓	36	.444		9.76	10.20		19.96	26	
	0670	6″ diameter	Q-2	31	.774		21.94	18.40		40.34	52	
	0680	8″ diameter	″	27	.889		30.78	21		51.78	66	
	1290	Galvanized, ¼″ diameter	1 Plum	66	.121		1	3.08		4.08	5.85	
	1320	¾″ diameter		61	.131		1.30	3.34		4.64	6.60	
	1330	1″ diameter	↓	53	.151		1.81	3.84		5.65	7.90	
	1350	1-½″ diameter	Q-1	80	.200		2.74	4.58		7.32	10.10	
	1360	2″ diameter		64	.250		3.53	5.75		9.28	12.70	
	1370	2-½″ diameter		50	.320		5.67	7.35		13.02	17.55	
	1380	3″ diameter		43	.372		7.39	8.50		15.89	21	
	1400	4″ diameter	↓	36	.444		11.29	10.20		21.49	28	
	1420	6″ diameter	Q-2	31	.774		26.54	18.40		44.94	58	
	1430	8″ diameter	″	27	.889		38.48	21		59.48	75	
	2000	Welded, sch. 40, on yoke & roll hangers, sized for covering,										
	2040	Black, 1″ diameter	Q-15	67	.239	L.F.	1.47	5.45	.58	7.50	10.70	
	2070	2″ diameter		56	.286		2.68	6.55	.70	9.93	13.80	
	2090	3″ diameter		38	.421		4.69	9.65	1.03	15.37	21	
	2110	4″ diameter		30	.533		6.77	12.20	1.30	20.27	28	
	2120	5″ diameter	↓	24	.667		11.33	15.25	1.63	28.21	38	
	2130	6″ diameter	Q-16	31	.774	↓	13.54	18.40	1.26	33.20	45	
	9990	Minimum labor/equipment charge	1 Plum	3	2.670	Job		68		68	105	
716	0010	**PIPE, STEEL, FITTINGS, Threaded**										**716**
	0020	Cast Iron,										
	0040	Standard weight, black										
	0060	90° Elbow, straight										
	0070	¼″	1 Plum	16	.500	Ea.	1.28	12.75		14.03	21	
	0080	⅜″		16	.500		1.65	12.75		14.40	21	
	0090	½″		15	.533		.92	13.55		14.47	22	
	0100	¾″		14	.571		1.05	14.55		15.60	24	
	0110	1″	↓	13	.615		1.30	15.65		16.95	26	
	0130	1-½″	Q-1	20	.800		2.53	18.30		20.83	31	
	0140	2″		18	.889		3.92	20		23.92	36	
	0150	2-½″		14	1.140		8.10	26		34.10	49	
	0160	3″		10	1.600		13.10	37		50.10	71	
	0180	4″		6	2.670		24	61		85	120	
	0200	6″	Q-2	9	2.670	↓	55	63		118	160	
	0210	8″	″	7	3.430		115	81		196	250	
	0500	Tee, straight										
	0510	¼″	1 Plum	10	.800	Ea.	1.64	20		21.64	33	
	0520	⅜″		10	.800		1.74	20		21.74	33	
	0540	¾″		9	.889		1.69	23		24.69	37	
	0550	1″	↓	8	1		1.68	25		26.68	41	
	0570	1-½″	Q-1	13	1.230		3.70	28		31.70	48	
	0580	2″		11	1.450		5.55	33		38.55	57	
	0590	2-½″		9	1.780		11.50	41		52.50	75	
	0600	3″		6	2.670		17.40	61		78.40	115	
	0620	4″	↓	4	4		34	92		126	180	
	0640	6″	Q-2	5	4.800		77	115		192	260	
	0650	8″	″	4	6	↓	160	145		305	395	
	0700	Standard weight, galvanized										
	0720	90° Elbow, straight										
	0730	¼″	1 Plum	16	.500	Ea.	1.78	12.75		14.53	22	
	0740	⅜″		16	.500		1.78	12.75		14.53	22	
	0750	½″		15	.533		2.12	13.55		15.67	23	
	0760	¾″	↓	14	.571	↓	2.25	14.55		16.80	25	

(129)

151 700 | Steel Pipe

			CREW	DAILY OUTPUT	MAN-HOURS	UNIT	BARE COSTS				TOTAL INCL O&P	
							MAT.	LABOR	EQUIP.	TOTAL		
716	0770	1"	1 Plum	13	.615	Ea.	2.63	15.65		18.28	27	716
	0790	1-½"	Q-1	20	.800		5.55	18.30		23.85	34	
	0800	2"		18	.889		8.25	20		28.25	40	
	0810	2-½"		14	1.140		15.85	26		41.85	58	
	0820	3"		10	1.600		24	37		61	83	
	0840	4"		6	2.670		44	61		105	145	
	0860	6"	Q-2	9	2.670		96	63		159	205	
	0870	8"	"	7	3.430		200	81		281	345	
	1100	Tee, straight										
	1110	¼"	1 Plum	10	.800	Ea.	2.35	20		22.35	34	
	1120	⅜"		10	.800		2.35	20		22.35	34	
	1140	¾"		9	.889		3.05	23		26.05	38	
	1150	1"		8	1		3.44	25		28.44	43	
	1170	1-½"	Q-1	13	1.230		7.50	28		35.50	52	
	1180	2"		11	1.450		9.45	33		42.45	62	
	1190	2-½"		9	1.780		20	41		61	85	
	1200	3"		6	2.670		31	61		92	130	
	1220	4"		4	4		59	92		151	205	
	1240	6"	Q-2	5	4.800		130	115		245	320	
	1250	8"	"	4	6		270	145		415	515	
	5000	Malleable iron, 150 lb.										
	5020	Black										
	5040	90° Elbow, straight										
	5060	¼"	1 Plum	16	.500	Ea.	.74	12.75		13.49	20	
	5070	⅜"		16	.500		.74	12.75		13.49	20	
	5090	¾"		14	.571		.64	14.55		15.19	23	
	5100	1"		13	.615		1.17	15.65		16.82	25	
	5120	1-½"	Q-1	20	.800		2.53	18.30		20.83	31	
	5130	2"		18	.889		3.69	20		23.69	35	
	5140	2-½"		14	1.140		9.40	26		35.40	51	
	5150	3"		10	1.600		14	37		51	72	
	5170	4"		6	2.670		29	61		90	125	
	5190	6"	Q-2	7	3.430		80	81		161	215	
	5450	Tee, straight										
	5470	¼"	1 Plum	10	.800	Ea.	1.13	20		21.13	33	
	5480	⅜"		10	.800		1.13	20		21.13	33	
	5500	¾"		9	.889		1.03	23		24.03	36	
	5510	1"		8	1		1.82	25		26.82	41	
	5520	1-¼"	Q-1	14	1.140		2.96	26		28.96	44	
	5530	1-½"		13	1.230		3.65	28		31.65	48	
	5540	2"		11	1.450		5.35	33		38.35	57	
	5550	2-½"		9	1.780		12.45	41		53.45	77	
	5560	3"		6	2.670		18	61		79	115	
	5570	3-½"		5	3.200		37	73		110	155	
	5580	4"		4	4		43	92		135	190	
	5600	6"	Q-2	4	6		115	145		260	345	
	5650	Coupling, straight										
	5670	¼"	1 Plum	19	.421	Ea.	.85	10.70		11.55	17.45	
	5680	⅜"		19	.421		.85	10.70		11.55	17.45	
	5690	½"		19	.421		.74	10.70		11.44	17.35	
	5700	¾"		16	.500		.90	12.75		13.65	21	
	5710	1"		15	.533		1.35	13.55		14.90	22	
	5730	1-½"	Q-1	24	.667		2.25	15.25		17.50	26	
	5740	2"		21	.762		3.26	17.45		20.71	31	
	5750	2-½"		16	1		8.05	23		31.05	44	
	5760	3"		12	1.330		11.05	31		42.05	59	
	5780	4"		7	2.290		22	52		74	105	
	5800	6"	Q-2	8	3		55	71		126	170	

For expanded coverage of these items see *Means Mechanical Cost Data* or *Means Plumbing Cost Data 1991*

151 700 | Steel Pipe

		CREW	DAILY OUTPUT	MAN-HOURS	UNIT	BARE COSTS MAT.	LABOR	EQUIP.	TOTAL	TOTAL INCL O&P	
716	6000	For galvanized elbows, tees, and couplings add				Ea.	23%					**716**
	9000	Minimum labor/equipment charge	1 Plum	4	2	Job		51		51	78	
720	0010	**PIPE, STEEL, FITTINGS** Flanged, welded and special type										**720**
	3000	Weld joint, butt, carbon steel, standard weight										
	3040	90° Elbow, long										
	3050	½" pipe size	Q-15	16	1	Ea.	5.40	23	2.44	30.84	44	
	3060	¾" pipe size		16	1		5.40	23	2.44	30.84	44	
	3070	1" pipe size		16	1		5.40	23	2.44	30.84	44	
	3100	2" pipe size		10	1.600		6.10	37	3.90	47	68	
	3120	3" pipe size		7	2.290		10.20	52	5.55	67.75	98	
	3130	4" pipe size		5	3.200		16.80	73	7.80	97.60	140	
	3140	6" pipe size	Q-16	5	4.800		41	115	7.80	163.80	230	
	3350	Tee, straight										
	3360	½" pipe size	Q-15	10	1.600	Ea.	15.10	37	3.90	56	77	
	3370	¾" pipe size		10	1.600		15.10	37	3.90	56	77	
	3380	1" pipe size		10	1.600		15.10	37	3.90	56	77	
	3410	2" pipe size		6	2.670		16.10	61	6.50	83.60	120	
	3430	3" pipe size		4	4		23	92	9.75	124.75	175	
	3440	4" pipe size		3	5.330		31	120	13	164	235	
	3450	6" pipe size	Q-16	3	8		57	190	13	260	370	
	9990	Minimum labor/equipment charge	Q-15	3	5.330	Job		120	13	133	205	

151 800 | Grooved-Joint Steel Pipe

		CREW	DAILY OUTPUT	MAN-HOURS	UNIT	BARE COSTS MAT.	LABOR	EQUIP.	TOTAL	TOTAL INCL O&P	
801	0010	**PIPE, GROOVED-JOINT STEEL FITTINGS & VALVES**										**801**
	0020	Pipe includes coupling & clevis type hanger 10' O.C.										
	1000	Schedule 40, black										
	1040	¾" diameter	1 Plum	71	.113	L.F.	2.01	2.87		4.88	6.65	
	1050	1" diameter		63	.127		2.19	3.23		5.42	7.40	
	1060	1-¼" diameter		58	.138		2.71	3.51		6.22	8.40	
	1070	1-½" diameter		51	.157		3.01	3.99		7	9.45	
	1080	2" diameter		40	.200		3.60	5.10		8.70	11.80	
	1090	2-½" diameter	Q-1	57	.281		5.03	6.45		11.48	15.45	
	1100	3" diameter		50	.320		6.09	7.35		13.44	18	
	1110	4" diameter		45	.356		8.84	8.15		16.99	22	
	1120	5" diameter		37	.432		14.97	9.90		24.87	32	
	4000	Elbow, 90° or 45°, black steel										
	4030	¾" diameter	1 Plum	50	.160	Ea.	6.60	4.07		10.67	13.55	
	4040	1" diameter		50	.160		6.60	4.07		10.67	13.55	
	4050	1-¼" diameter		40	.200		8.70	5.10		13.80	17.40	
	4060	1-½" diameter		33	.242		9.33	6.15		15.48	19.80	
	4070	2" diameter		25	.320		9.33	8.15		17.48	23	
	4080	2-½" diameter	Q-1	40	.400		12.60	9.15		21.75	28	
	4100	4" diameter		25	.640		25	14.65		39.65	50	
	4110	5" diameter		20	.800		59	18.30		77.30	93	
	4250	For galvanized elbows, add					15%					
	4690	Tee, black steel										
	4700	¾" diameter	1 Plum	38	.211	Ea.	9.27	5.35		14.62	18.45	
	4740	1" diameter		33	.242		9.27	6.15		15.42	19.70	
	4750	1-¼" diameter		27	.296		10.75	7.55		18.30	23	
	4760	1-½" diameter		22	.364		11.60	9.25		20.85	27	
	4770	2" diameter		17	.471		14.40	12		26.40	34	
	4780	2-½" diameter	Q-1	27	.593		19.45	13.55		33	42	
	4800	4" diameter		17	.941		42	22		64	79	
	4810	5" diameter		13	1.230		98	28		126	150	
	4900	For galvanized tees, add					15%					
	9990	Minimum labor/equipment charge	1 Plum	4	2	Job		51		51	78	

For expanded coverage of these items see *Means Mechanical Cost Data* or *Means Plumbing Cost Data 1991*

151 950	Valves	CREW	DAILY OUTPUT	MAN-HOURS	UNIT	BARE COSTS				TOTAL INCL O&P
						MAT.	LABOR	EQUIP.	TOTAL	
955	**0010**	**VALVES, BRONZE**									**955**
1020	Angle, 150 lb., rising stem, threaded									
1070	¾" size	1 Plum	20	.400	Ea.	25	10.20		35.20	43
1080	1" size		19	.421		36	10.70		46.70	56
1100	1-½" size		13	.615		65	15.65		80.65	96
1110	2" size	↓	11	.727	↓	100	18.50		118.50	140
1380	Ball, 150 psi, threaded									
1460	¾" size	1 Plum	20	.400	Ea.	7.50	10.20		17.70	24
1750	Check, swing, class 150, regrinding disc, threaded									
1850	½" size	1 Plum	24	.333	Ea.	9.50	8.50		18	24
1860	¾" size		20	.400		11	10.20		21.20	28
1870	1" size		19	.421		14.50	10.70		25.20	32
1890	1-½" size		13	.615		25	15.65		40.65	52
1900	2" size	↓	11	.727		36	18.50		54.50	68
2850	Gate, N.R.S., soldered, 300 psi									
2920	½" size	1 Plum	24	.333	Ea.	10.90	8.50		19.40	25
2940	¾" size		20	.400		14.40	10.20		24.60	32
2950	1" size		19	.421		17.50	10.70		28.20	36
2970	1-½" size		13	.615		29	15.65		44.65	56
2980	2" size	↓	11	.727	↓	40	18.50		58.50	73
4850	Globe, class 150, rising stem, threaded									
4950	½" size	1 Plum	24	.333	Ea.	13.90	8.50		22.40	28
4960	¾" size	"	24	.333	"	16.55	8.50		25.05	31
5030										
5600	Relief, pressure & temperature, self-closing, ASME									
5640	¾" size	1 Plum	28	.286	Ea.	43	7.25		50.25	59
5650	1" size		24	.333		61	8.50		69.50	80
5660	1-¼" size		20	.400		125	10.20		135.20	155
5670	1-½" size		18	.444		245	11.30		256.30	285
5680	2" size	↓	16	.500	↓	265	12.75		277.75	310
6400	Pressure, water, ASME, threaded									
6440	¾" size	1 Plum	28	.286	Ea.	33	7.25		40.25	48
6450	1" size		24	.333		56	8.50		64.50	75
6470	1-½" size	↓	18	.444	↓	120	11.30		131.30	150
6900	Reducing, water pressure									
6940	½" size	1 Plum	24	.333	Ea.	56	8.50		64.50	75
6960	1" size	"	19	.421	"	100	10.70		110.70	125
8350	Tempering, water, sweat connections									
8400	½" size	1 Plum	24	.333	Ea.	27	8.50		35.50	43
8440	¾" size	"	20	.400	"	31	10.20		41.20	50
8650	Threaded connections									
8700	½" size	1 Plum	24	.333	Ea.	31	8.50		39.50	47
8740	¾" size	↓	20	.400	"	38	10.20		48.20	58
9000	Minimum labor/equipment charge		4	2	Job		51		51	78
960	**0010**	**VALVES, IRON BODY**									**960**
1020	Butterfly, wafer type, lever actuator									
1030	2" size	1 Plum	14	.571	Ea.	60	14.55		74.55	88
1060	4" size	Q-1	5	3.200	"	90	73		163	210
1650	Gate, 125 lb., N.R.S., threaded									
1700	2" size	1 Plum	11	.727	Ea.	140	18.50		158.50	185
1760	3" size	Q-1	13	1.230		185	28		213	245
1780	4" size	"	10	1.600	↓	370	37		407	465
2150	Flanged									
2240	2-½" size	Q-1	5	3.200	Ea.	110	73		183	235
2260	3" size		4.50	3.560		115	81		196	250
2280	4" size	↓	3	5.330	↓	190	120		310	395
3550	OS&Y, flanged									
3680	4" size	Q-1	3	5.330	Ea.	215	120		335	425

For expanded coverage of these items see *Means Mechanical Cost Data* or *Means Plumbing Cost Data 1991*

151 950	Valves	CREW	DAILY OUTPUT	MAN-HOURS	UNIT	BARE COSTS MAT.	BARE COSTS LABOR	BARE COSTS EQUIP.	BARE COSTS TOTAL	TOTAL INCL O&P	
960 3690	5" size	Q-2	3.40	7.060	Ea.	355	170		525	650	960
3700	6" size	"	3	8	"	355	190		545	685	
9000	Minimum labor/equipment charge	1 Plum	3	2.670	Job		68		68	105	
975 0010	VALVES, PLASTIC										975
0020											
1150	Ball, PVC, socket or threaded, single union	1 Plum	17	.471	Ea.	22	12		34	43	
1280	2" size										
1650	CPVC, socket or threaded, single union	1 Plum	17	.471	Ea.	55	12		67	79	
1770	2" size										
3150	Ball check, PVC, socket or threaded	1 Plum	17	.471	Ea.	45	12		57	68	
3290	2" size										
9000	Minimum labor/equipment charge	"	3.50	2.290	Job		58		58	90	

152 100	Fixtures	CREW	DAILY OUTPUT	MAN-HOURS	UNIT	BARE COSTS MAT.	BARE COSTS LABOR	BARE COSTS EQUIP.	BARE COSTS TOTAL	TOTAL INCL O&P	
104 0010	**BATHS**										104
0100	Tubs, recessed porcelain enamel on cast iron, with trim										
0180	48" x 42"	Q-1	4	4	Ea.	865	92		957	1,100	
0220	72" x 36"		3	5.330		925	120		1,045	1,200	
0300	Mat bottom, 4' long	(130)	5.50	2.910		705	67		772	880	
0380	5' long		4.40	3.640		281	83		364	440	
0560	Corner 48" x 44"		4	4		995	92		1,087	1,225	
2000	Enameled formed steel, 4'-6" long		5.80	2.760		207	63		270	325	
2200	5' long		5.50	2.910		192	67		259	315	
6000	Whirlpool, bath with vented overflow, molded fiberglass										
7000	Redwood tub system										
7050	4' diameter x 4' deep	Q-1	1	16	Ea.	690	365		1,055	1,325	
7150	6' diameter x 4' deep		.80	20		1,050	460		1,510	1,850	
7200	8' diameter x 4' deep		.80	20		1,430	460		1,890	2,275	
9000	Minimum labor/equipment charge		3	5.330	Job		120		120	190	
9600	Rough-in, supply, waste and vent, for all above tubs, add		2.07	7.730	Ea.	97.19	175		272.19	380	
112 0010	**DENTAL FOUNTAIN**										112
0020	Deck mounted, with cuspidor										
0050	Stainless steel receptor	1 Plum	4	2	Ea.	156	51		207	250	
9600	For rough-in, supply and waste, add	"	1.16	6.900	"	75.83	175		250.83	355	
116 0010	**DRINKING FOUNTAIN** For connection to cold water supply										116
1000	Wall mounted, non-recessed										
2700	Stainless steel, single bubbler, no back	1 Plum	4	2	Ea.	550	51		601	685	
2740	With back		4	2		261	51		312	365	
2780	Dual handle & wheelchair projection type		4	2		306	51		357	415	
2820	Dual level for handicapped type		3.20	2.500		620	64		684	780	
3980	For rough-in, supply and waste, add		2.21	3.620		51.01	92		143.01	200	
4000	Wall mounted, semi-recessed										
4200	Poly-marble, single bubbler	1 Plum	4	2	Ea.	339	51		390	450	
4600	Stainless steel, satin finish, single bubbler	"	4	2	"	300	51		351	410	
6000	Wall mounted, fully recessed										
6400	Poly-marble, single bubbler	1 Plum	4	2	Ea.	435	51		486	555	
6800	Stainless steel, single bubbler		4	2		354	51		405	470	
7580	For rough-in, supply and waste, add		1.83	4.370		51.01	110		161.01	230	

For expanded coverage of these items see *Means Mechanical Cost Data* or *Means Plumbing Cost Data 1991*

			DAILY OUTPUT	MAN-HOURS	UNIT	BARE COSTS				TOTAL INCL O&P
152 100	**Fixtures**	CREW				MAT.	LABOR	EQUIP.	TOTAL	
116	7590										116
	7600	Floor mounted, pedestal type								
	8600	Enameled iron, heavy duty service, 2 bubblers	1 Plum	2	4	Ea.	620	100		720	840
	8880	For freeze-proof valve system, add		2	4		174	100		274	350
	8900	For rough-in, supply and waste, add		1.83	4.370		51.01	110		161.01	230
	9000	Minimum labor/equipment charge		2	4	Job		100		100	155
124	0010	**INDUSTRIAL SAFETY FIXTURES** Rough-in not included										124
	0020										
	1000	Eye wash fountain									
	1400	Plastic bowl, pedestal mounted	Q-1	4	4	Ea.	150	92		242	305
	5000	Shower, single head, drench, ball valve, pull, freestanding		4	4		177	92		269	335
	5200	Horizontal or vertical supply		4	4		156	92		248	315
	6000	Multi-nozzle, eye/face wash combination		4	4		355	92		447	530
	6400	Multi-nozzle, 12 spray, shower only		4	4		750	92		842	965
	6600	For freeze-proof, add		6	2.670		355	61		416	485
	9000	Minimum labor/equipment charge		3	5.330	Job		120		120	190
128	0010	**INTERCEPTORS**										128
	0020										
	0150	Grease, cast iron, 4 GPM, 8 lb. fat capacity	1 Plum	4	2	Ea.	191	51		242	290
	0200	7 GPM, 14 lb. fat capacity		4	2		265	51		316	370
	1000	10 GPM, 20 lb. fat capacity		4	2		312	51		363	420
	1160	100 GPM, 200 lb. fat capacity	Q-1	2	8		2,230	185		2,415	2,725
	1240	300 GPM, 600 lb. fat capacity	"	1	16		4,930	365		5,295	6,000
	9000	Minimum labor/equipment charge	1 Plum	3	2.670	Job		68		68	105
136	0010	**LAVATORIES** With trim, white unless noted otherwise										136
	0020										
	0500	Vanity top, porcelain enamel on cast iron									
	0600	20" x 18"	Q-1	6.40	2.500	Ea.	134	57		191	235
	0640	26" x 18" oval	"	6.40	2.500	"	167	57		224	270
	0860	For color, add					25%				
	1000	Cultured marble, 19" x 17", single bowl	Q-1	6.40	2.500	Ea.	88	57		145	185
	1120	25" x 22", single bowl		6.40	2.500		110	57		167	210
	1900	Stainless steel, self-rimming, 25" x 22", single bowl, ledge		6.40	2.500		185	57		242	290
	1960	17" x 22", single bowl		6.40	2.500		93	57		150	190
	2600	Steel, enameled, 20" x 17", single bowl		6.40	2.500		81	57		138	175
	2900	Vitreous china, 20" x 16", single bowl		6.40	2.500		173	57		230	280
	2960	20" x 17", single bowl		6.40	2.500		138	57		195	240
	3580	Rough-in, supply, waste and vent for all above lavatories		2.30	6.960		88.75	160		248.75	345
	4000	Wall hung									
	4040	Porcelain enamel on cast iron, 16" x 14", single bowl	Q-1	8	2	Ea.	260	46		306	355
	4180	20" x 18", single bowl	"	8	2		110	46		156	190
	4580	For color, add					30%				
	6000	Vitreous china, 18" x 15", single bowl with backsplash	Q-1	8	2		155	46		201	240
	6960	Rough-in, supply, waste and vent for above lavatories	"	1.66	9.640		130.10	220		350.10	485
	9000	Minimum labor/equipment charge	1 Plum	3	2.670	Job		68		68	105
140	0010	**LAUNDRY SINKS** With trim										140
	0020	Porcelain enamel on cast iron, black iron frame									
	0050	24" x 20", single compartment	Q-1	6	2.670	Ea.	288	61		349	410
	0100	24" x 23", single compartment	"	6	2.670	"	315	61		376	440
	2000	Molded stone, on wall hanger or legs									
	2020	22" x 21", single compartment	Q-1	6	2.670	Ea.	143	61		204	250
	2100	45" x 21", double compartment	"	5	3.200	"	228	73		301	365
	3000	Plastic, on wall hanger or legs									
	3020	18" x 23", single compartment	Q-1	6.50	2.460	Ea.	102	56		158	200
	3100	20" x 24", single compartment		6.50	2.460		123	56		179	220
	3200	36" x 23", double compartment		5.50	2.910		135	67		202	250
	3300	40" x 24", double compartment		5.50	2.910		190	67		257	310

For expanded coverage of these items see *Means Mechanical Cost Data* or *Means Plumbing Cost Data 1991*

			CREW	DAILY OUTPUT	MAN-HOURS	UNIT	BARE COSTS				TOTAL INCL O&P	
152 100		**Fixtures**					MAT.	LABOR	EQUIP.	TOTAL		
140	5000	Stainless steel, counter top, 22" x 17" single compartment	Q-1	6	2.670	Ea.	232	61		293	350	140
	5100	19" x 22", single compartment		6	2.670		247	61		308	365	
	5200	33" x 22", double compartment		5	3.200		295	73		368	440	
	9600	Rough-in, supply, waste and vent, for all laundry sinks		2.14	7.480		95.66	170		265.66	370	
	9810	Minimum labor/equipment charge	1 Plum	3	2.670	Job		68		68	105	
148	0011	**SHOWERS**, Stall, with door and trim										148
	0020											
	0151	Baked enamel, molded stone receptor, 30" square	Q-1	2	8	Ea.	349	185		534	665	
	1520	32" square		2	8		316	185		501	630	
	1540	Terrazzo receptor, 32" square		2	8		530	185		715	865	
	1580	36" corner angle		1.80	8.890		585	205		790	960	
	3000	Fiberglass, one piece, with 3 walls, 32" x 32" square		2.40	6.670		338	155		493	605	
	3100	36" x 36" square		2.40	6.670		378	155		533	650	
	4960	Rough-in, supply, waste and vent for above showers		2.05	7.800		72.99	180		252.99	355	
	5500	Head, water economizer, 3.0 GPM	1 Plum	24	.333		46.25	8.50		54.75	64	
	9000	Minimum labor/equipment charge	Q-1	4	4	Job		92		92	140	
152	0010	**SINKS** With faucets and drain										152
	2000	Kitchen, counter top, P.E. on C.I., 24" x 21" single bowl	Q-1	5.60	2.860	Ea.	170	65		235	290	
	2100	30" x 21" single bowl		5.60	2.860		193	65		258	315	
	2200	32" x 21" double bowl		4.80	3.330		218	76		294	360	
	3000	Stainless steel, self rimming, 19" x 18" single bowl		5.60	2.860		233	65		298	355	
	3100	25" x 22" single bowl		5.60	2.860		261	65		326	390	
	3200	33" x 22" double bowl		4.80	3.330		365	76		441	520	
	3300	43" x 22" double bowl		4.80	3.330		450	76		526	615	
	4000	Steel, enameled, with ledge, 24" x 21" single bowl		5.60	2.860		104	65		169	215	
	4100	32" x 21" double bowl		4.80	3.330		115	76		191	245	
	4960	For color sinks except stainless steel, add					10%					
	4980	For rough-in, supply, waste and vent, counter top sinks	Q-1	2.14	7.480		95.66	170		265.66	370	
	5790	For rough-in, supply, waste & vent, sinks		1.85	8.650		95.66	200		295.66	410	
	6650	Service, floor, corner, P.E. on C.I., 28" x 28"		4.40	3.640		420	83		503	590	
	6790	For rough-in, supply, waste & vent, floor service sinks		1.64	9.760		152.24	225		377.24	510	
	9000	Minimum labor/equipment charge		4	4	Job		92		92	140	
168	0010	**URINALS**										168
	0020											
	3000	Wall hung, vitreous china, with hanger & self-closing valve	Q-1	3	5.330	Ea.	335	120		455	555	
	3300	Rough-in, supply, waste & vent		2.83	5.650		64.85	130		194.85	270	
	5000	Stall type, vitreous china, includes valve		2.50	6.400		410	145		555	675	
	6980	Rough-in, supply, waste and vent		1.99	8.040		107.85	185		292.85	405	
	9000	Minimum labor/equipment charge		4	4	Job		92		92	140	
176	0010	**WASH FOUNTAINS** Rigging not included										176
	1900	Group, foot control										
	2000	Precast terrazzo, circular, 36" diam., 5 or 6 persons	Q-2	3	8	Ea.	1,190	190		1,380	1,600	
	2100	54" diameter for 8 or 10 persons		2.50	9.600		1,480	230		1,710	1,975	
	2400	Semi-circular, 36" diam. for 3 persons		3	8		1,090	190		1,280	1,500	
	2500	54" diam. for 4 or 5 persons		2.50	9.600		1,325	230		1,555	1,800	
	5700	Rough-in, supply, waste and vent for above wash fountains	Q-1	1.82	8.790		120.35	200		320.35	445	
	9000	Minimum labor/equipment charge	Q-2	3	8	Job		190		190	295	
180	0010	**WATER CLOSETS**										180
	0100											
	0150	Tank type, vitreous china, incl. seat, supply pipe w/stop										
	0200	Wall hung, one piece	Q-1	5.30	3.020	Ea.	495	69		564	650	
	0400	Two piece, close coupled		5.30	3.020		339	69		408	480	
	0960	For rough-in, supply, waste, vent and carrier (130)		2.60	6.150		147.31	140		287.31	380	
	1000	Floor mounted, one piece		5.30	3.020		418	69		487	565	
	1100	Two piece, close coupled, water saver		5.30	3.020		138	69		207	260	

For expanded coverage of these items see *Means Mechanical Cost Data* or *Means Plumbing Cost Data 1991*

152 100 | Fixtures

			CREW	DAILY OUTPUT	MAN-HOURS	UNIT	BARE COSTS MAT.	LABOR	EQUIP.	TOTAL	TOTAL INCL O&P	
180	1960	For color, add				Ea.	30%					180
	1980	For rough-in, supply, waste and vent	Q-1	1.94	8.250	"	85.65	190		275.65	385	
	3000	Bowl only, with flush valve, seat										
	3100	Wall hung	Q-1	5.80	2.760	Ea.	275	63		338	400	
	3200	For rough-in, supply, waste and vent, single WC	↓	2.05	7.800	"	156.94	180		336.94	450	
	9000	Minimum labor/equipment charge		4	4	Job		92		92	140	

152 400 | Pumps

			CREW	DAILY OUTPUT	MAN-HOURS	UNIT	BARE COSTS MAT.	LABOR	EQUIP.	TOTAL	TOTAL INCL O&P	
410	0010	**PUMPS, CIRCULATING** Heated or chilled water application										410
	0100											
	0600	Bronze, sweat connections, 1/40 HP, in line										
	0640	3/4" size	Q-1	16	1	Ea.	94.60	23		117.60	140	
	1000	Flange connection, 3/4" to 1-1/2" size										
	1040	1/12 HP	Q-1	6	2.670	Ea.	230	61		291	345	
	1060	1/8 HP		6	2.670		385	61		446	520	
	1100	1/3 HP		6	2.670		424	61		485	560	
	1140	2" size, 1/8 HP		5	3.200		484	73		557	645	
	1180	2-1/2" size, 1/4 HP		5	3.200		754	73		827	940	
	1220	3" size, 1/4 HP		4	4	↓	786	92		878	1,000	
	9000	Minimum labor/equipment charge	↓	3.25	4.920	Job		115		115	175	
450	0010	**PUMPS, PRESSURE BOOSTER SYSTEM**										450
	0200	Pump system, with diaphragm tank, control, press. switch										
	0300	1 HP pump	Q-1	1.30	12.310	Ea.	2,335	280		2,615	3,000	
	0400	1-1/2 HP pump		1.25	12.800		2,365	295		2,660	3,050	
	0420	2 HP pump		1.20	13.330		2,440	305		2,745	3,150	
	0440	3 HP pump	↓	1.10	14.550		2,570	335		2,905	3,350	
	0460	5 HP pump	Q-2	1.50	16		2,725	380		3,105	3,575	
	0480	7-1/2 HP pump		1.42	16.900		3,025	400		3,425	3,950	
	0500	10 HP pump	↓	1.34	17.910	↓	3,170	425		3,595	4,150	
	1000	Pump/ energy storage system, diaphragm tank, 3 HP pump										
	1100	motor, PRV, switch, gauge, control center, flow switch			104							
	1200	125 lb. working pressure	Q-2	.70	34.290	Ea.	6,800	815		7,615	8,725	
	1300	250 lb. working pressure		.64	37.500	"	7,430	890		8,320	9,550	
	9000	Minimum labor/equipment charge	↓	.30	80	Job		1,900		1,900	2,925	
465	0010	**PUMPS, SEWAGE EJECTOR** With operating and level controls										465
	0100	Simplex system incl. tank, cover, pump 15' head										
	0500	37 gal PE tank, 12 GPM, 1/2 HP, 2" discharge	Q-1	3.20	5	Ea.	344	115		459	555	
	0510	3" discharge		3.10	5.160		397	120		517	620	
	0530	87 GPM, .7 HP, 2" discharge		3.20	5		535	115		650	765	
	0540	3" discharge		3.10	5.160		564	120		684	805	
	0600	45 gal. coated stl tank, 12 GPM, 1/2 HP, 2" discharge		3	5.330		530	120		650	770	
	0610	3" discharge		2.90	5.520		566	125		691	820	
	0630	87 GPM, .7 HP, 2" discharge		3	5.330		710	120		830	970	
	0640	3" discharge		2.90	5.520		760	125		885	1,025	
	0660	134 GPM, 1 HP, 2" discharge		2.80	5.710		820	130		950	1,100	
	0680	3" discharge		2.70	5.930		865	135		1,000	1,150	
	0700	70 gal. PE tank, 12 GPM, 1/2HP, 2" discharge		2.60	6.150		620	140		760	900	
	0710	3" discharge		2.40	6.670		662	155		817	965	
	0730	87 GPM, .7 HP, 2" discharge		2.50	6.400		825	145		970	1,125	
	0740	3" discharge		2.30	6.960		865	160		1,025	1,200	
	0760	134 GPM, 1 HP, 2" discharge		2.20	7.270		927	165		1,092	1,275	
	0770	3" discharge	↓	2	8	↓	970	185		1,155	1,350	
	9000	Minimum labor/equipment charge	↓	2.50	6.400	Job		145		145	225	
480	0010	**PUMPS, SUBMERSIBLE** Dewatering										480
	0120											
	2000	Sewage & solids mixture, 8' head, automatic										
	2020	Bronze, 1/2 HP										

For expanded coverage of these items see *Means Mechanical Cost Data or Means Plumbing Cost Data 1991*

			DAILY	MAN-		BARE COSTS				TOTAL	
152 400		**Pumps**	CREW	OUTPUT	HOURS	UNIT	MAT.	LABOR	EQUIP.	TOTAL	INCL O&P
480	2100	75 GPM, 1-¼″ or 1-½″ NPT discharge	1 Plum	8	1	Ea.	360	25		385	435	**480**
	2140	120 GPM, 2″ or 3″ discharge	"	7	1.140	"	470	29		499	560	
	7000	Sump pump, 10′ head, automatic										
	7100	Bronze, 22 GPM., ¼ HP, 1-¼″ discharge	1 Plum	6	1.330	Ea.	230	34		264	305	
	7140	68 GPM, ½ HP, 1-¼″ or 1-½″ discharge		5	1.600		375	41		416	475	
	7180	105 GPM, ½ HP, 2″ or 3″ discharge		4	2		490	51		541	615	
	7500	Cast iron, 23 GPM, ¼ HP, 1-¼″ discharge		6	1.330		109	34		143	170	
	7540	35 GPM, ⅓ HP, 1-¼″ discharge		6	1.330		123	34		157	190	
	7560	68 GPM, ½ HP, 1-¼″ or 1-½″ discharge		5	1.600		252	41		293	340	
	9000	Minimum labor/equipment charge		4	2	Job		51		51	78	

			DAILY	MAN-		BARE COSTS				TOTAL	
153 100		**Water Appliances**	CREW	OUTPUT	HOURS	UNIT	MAT.	LABOR	EQUIP.	TOTAL	INCL O&P
105	0010	**WATER COOLER**										**105**
	0100	Wall mounted, non-recessed										
	0140	4 GPH	Q-1	4	4	Ea.	291	92		383	460	
	1040	14.3 GPH		3.80	4.210		504	96		600	705	
	3300	Semi-recessed, 8.1 GPH		4	4		468	92		560	655	
	4600	Floor mounted, flush-to-wall										
	4640	4 GPH	1 Plum	3	2.670	Ea.	312	68		380	450	
	4980	For stainless steel cabinet, add				"	65.50			65.50	72	
	9000	Minimum labor/equipment charge	1 Plum	2	4	Job		100		100	155	
110	0010	**WATER HEATERS**										**110**
	0050											
	1000	Residential, electric, glass lined tank, 10 gal., single element	1 Plum	2.30	3.480	Ea.	144	89		233	295	
	1060	30 gallon, double element		2.20	3.640		202	93		295	365	
	1080	40 gallon, double element		2	4		221	100		321	400	
	1100	52 gallon, double element		2	4		244	100		344	425	
	1120	66 gallon, double element		1.80	4.440		314	115		429	520	
	1140	80 gallon, double element		1.60	5		363	125		488	595	
	2000	Gas fired, glass lined tank, vent not incl., 20 gallon		2.10	3.810		186	97		283	355	
	2040	30 gallon		2	4		189	100		289	365	
	2060	40 gallon		1.90	4.210		203	105		308	390	
	2100	75 gallon		1.50	5.330		497	135		632	755	
	3000	Oil fired, glass lined tank, vent not included, 30 gallon		2	4		673	100		773	895	
	3040	50 gallon		1.80	4.440		910	115		1,025	1,175	
	4000	Commercial, 100° rise. NOTE: for each size tank, a range of										
	4010	heaters between the ones shown are available										
	4020	Electric										
	4100	5 gal., 3 KW, 12 GPH	1 Plum	2	4	Ea.	1,070	100		1,170	1,325	
	4160	50 gal., 36 KW, 148 GPH	"	1.80	4.440		2,345	115		2,460	2,750	
	4480	400 gal., 210 KW, 860 GPH	Q-1	1	16		16,450	365		16,815	18,700	
	6000	Gas fired, flush jacket, std. controls, vent not incl.										
	6040	75 MBH input, 63 GPH	1 Plum	1.40	5.710	Ea.	807	145		952	1,100	
	6060	96 MBH input, 81 GPH		1.40	5.710		950	145		1,095	1,275	
	6180	200 MBH input, 192 GPH		.60	13.330		1,975	340		2,315	2,700	
	6340	1200 MBH input, 1150 GPH	Q-1	.40	40		9,550	915		10,465	11,900	
	8000	Oil fired, flush jacket, std. controls, vent not incl.										
	8060	103 MBH gross output, 116 GPH	1 Plum	1.10	7.270	Ea.	1,460	185		1,645	1,900	
	8080	122 MBH gross output, 141 GPH	"	1	8	"	1,490	205		1,695	1,950	

For expanded coverage of these items see *Means Mechanical Cost Data* or *Means Plumbing Cost Data 1991*

153 | Plumbing Appliances

		153 100	Water Appliances	CREW	DAILY OUTPUT	MAN-HOURS	UNIT	BARE COSTS MAT.	BARE COSTS LABOR	BARE COSTS EQUIP.	BARE COSTS TOTAL	TOTAL INCL O&P	
110	8160		225 MBH gross output, 256 GPH	Q-1	.80	20	Ea.	2,330	460		2,790	3,275	110
	8280		735 MBH gross output, 880 GPH	"	.40	40	"	7,000	915		7,915	9,125	
	9000		Minimum labor/equipment charge	1 Plum	1.75	4.570	Job		115		115	180	
160	0010	**WATER SUPPLY METERS**											160
	2000		Domestic/commercial, bronze										
	2020		Threaded										
	2060		⅝″ diameter, to 20 GPM	1 Plum	16	.500	Ea.	45	12.75		57.75	69	
	2080		¾″ diameter, to 30 GPM		14	.571		86	14.55		100.55	115	
	2100		1″ diameter, to 50 GPM	↓	12	.667	↓	104	16.95		120.95	140	
	2300		Threaded/flanged										
	2340		1-½″ diameter, to 100 GPM	1 Plum	8	1	Ea.	237	25		262	300	
	9000		Minimum labor/equipment charge	"	3.25	2.460	Job		63		63	97	

154 | Fire Extinguishing Systems

		154 100	Fire Systems	CREW	DAILY OUTPUT	MAN-HOURS	UNIT	BARE COSTS MAT.	BARE COSTS LABOR	BARE COSTS EQUIP.	BARE COSTS TOTAL	TOTAL INCL O&P	
115	0010	**FIRE EQUIPMENT CABINETS** Not equipped, 20 ga. steel box,											115
	0040		recessed, D.S. glass in door, box size given										
	1000		Portable extinguisher, single, 8″ x 12″ x 27″, alum. door & frame	Q-12	8	2	Ea.	103	48		151	185	
	1100		Steel door and frame	"	8	2	"	52.50	48		100.50	130	
	3000		Hose rack assy., 1-½″ valve & 100′ hose, 24″ x 40″ x 5-½″										
	3200		Steel door and frame	Q-12	6	2.670	Ea.	110	63		173	220	
	4000		Hose rack assy., 2-½″ x 1-½″ valve, 100′ hose, 24″ x 40″ x 8″										
	4200		Steel door and frame	Q-12	6	2.670	Ea.	116	63		179	225	
	5000		Hose rack assy., 2-½″ x 1-½″ valve, 100′ hose										
	5010		and extinguisher, 30″ x 40″ x 8″										
	5200		Steel door and frame	Q-12	5	3.200	Ea.	124	76		200	255	
	9000		Minimum labor/equipment charge	1 Plum	3	2.670	Job		68		68	105	
125	0010	**FIRE EXTINGUISHERS**											125
	0020												
	0120		CO2, portable with swivel horn, 5 lb.				Ea.	87.10			87.10	96	
	0140		With hose and "H" horn, 10 lb.				"	130			130	145	
	1000		Dry chemical, pressurized										
	1040		Standard type, portable, painted, 2-½ lb.				Ea.	18.25			18.25	20	
	1080		10 lb.					47.60			47.60	52	
	1100		20 lb.					69			69	76	
	2000		ABC all purpose type, portable, 2-½ lb.					18.25			18.25	20	
	2080		9-½ lb.				↓	47.60			47.60	52	
135	0010	**FIRE HOSE AND EQUIPMENT**											135
	2200		Hose, less couplings										
	2260		Synthetic jacket, lined, 300 lb. test, 1-½″ diameter				L.F.	1.20			1.20	1.32	
	2280		2-½″ diameter				"	1.82			1.82	2	
	2600		Hose rack, swinging, for 1-½″ diameter hose,										
	2620		Enameled steel, 50′ & 75′ lengths of hose	Q-12	20	.800	Ea.	25	19		44	57	
	3750		Hydrants, wall, w/caps, single, flush, polished brass										
	3800		2-½″ x 2-½″	Q-12	5	3.200	Ea.	77	76		153	200	
	4350		Double, projecting, polished brass										
	4400		2-½″ x 2-½″ x 4″	Q-12	5	3.200	Ea.	135	76		211	265	
	5600		Nozzles, brass										
	5620		Adjustable fog, ¾″ booster line				Ea.	68			68	75	

For expanded coverage of these items see *Means Mechanical Cost Data* or *Means Plumbing Cost Data 1991*

154 100	Fire Systems	CREW	DAILY OUTPUT	MAN-HOURS	UNIT	MAT.	LABOR	EQUIP.	TOTAL	TOTAL INCL O&P	
135 5640	1-½″ leader line				Ea.	71.40			71.40	79	**135**
5660	2-½″ direct connection				"	143			143	155	
7140	Standpipe connections, wall, w/plugs & chains										
7280	Double, flush, polished brass										
7300	2-½″ x 2-½″ x 4″	Q-12	5	3.200	Ea.	239	76		315	380	
9000	Minimum labor/equipment charge	1 Plum	2	4	Job		100		100	155	
160 0010	**FIRE VALVES**										**160**
3000	Gate, hose, wheel handle, N.R.S., rough brass, 1-½″	1 Spri	12	.667	Ea.	60	17.60		77.60	93	
3040	2-½″, 300 lb.	↓	7	1.140	"	78	30		108	130	
9990	Minimum labor/equipment charge		4	2	Job		53		53	82	
170 0010	**SPRINKLER SYSTEM COMPONENTS**										**170**
0020											
0800	Air compressor for dry pipe system, automatic, complete										
0820	280 gal. system capacity, ¾ HP	1 Spri	1.30	6.150	Ea.	560	160		720	865	
1220	Water motor, complete with gong	"	4	2	"	108	53		161	200	
1800	Firecycle system, controls, includes panel,										
1820	batteries, solenoid valves and pressure switches	Q-13	1	32	Ea.	4,860	805		5,665	6,600	
1980	Detector	1 Spri	16	.500	"	176	13.20		189.20	215	
2600	Sprinkler heads, not including supply piping										
2640	Dry, pendent, ½″ orifice, ¾″ or 1″ NPT										
2660	1″ to 4-¾″ length	1 Spri	14	.571	Ea.	27.80	15.10		42.90	54	
2700	11″ to 12-¾″ length		14	.571		30.50	15.10		45.60	57	
3600	Foam-water, pendent or upright, ½″ NPT	↓	12	.667	↓	25.10	17.60		42.70	55	
3700	Standard spray, pendent or upright, brass, 135° to 286°F										
3740	½″ NPT, ½″ orifice	1 Spri	16	.500	Ea.	3.67	13.20		16.87	24	
3790	For riser & feeder piping, see div. 151-701										
6500	Check, swing, C.I. body, brass fittings, auto. ball drip										
6520	4″ size	Q-12	3	5.330	Ea.	84.50	125		209.50	290	
8200	Dry pipe valve, incl. trim and gauges, 3″ size		2	8		700	190		890	1,075	
8220	4″ size	↓	1	16	↓	780	380		1,160	1,450	
9990	Minimum labor/equipment charge	1 Spri	3	2.670	Job		70		70	110	

155 100	Boilers	CREW	DAILY OUTPUT	MAN-HOURS	UNIT	MAT.	LABOR	EQUIP.	TOTAL	TOTAL INCL O&P	
110 0010	**BOILERS, ELECTRIC, ASME** Standard controls and trim										**110**
1000	Steam, 6 KW, 20.5 MBH	Q-19	1.20	20	Ea.	2,950	475		3,425	3,975	
1160	60 KW, 205 MBH		1	24		4,240	570		4,810	5,525	
1300	240 KW, 819 MBH	↓	.45	53.330		14,020	1,275		15,295	17,400	
1400	600 KW, 2047 MBH	Q-21	.34	94.120		19,950	2,275		22,225	25,400	
1540	1800 KW, 6141 MBH	"	.19	168		40,850	4,075		44,925	51,000	
2000	Hot water, 12 KW, 41 MBH	Q-19	1.30	18.460		2,440	435		2,875	3,350	
2040	24 KW, 82 MBH		1.20	20		2,610	475		3,085	3,600	
2060	30 KW, 103 MBH		1.20	20		2,670	475		3,145	3,675	
2070	36 KW, 123 MBH		1.20	20		2,840	475		3,315	3,850	
2080	45 KW, 154 MBH		1.10	21.820		2,860	515		3,375	3,950	
2100	60 KW, 205 MBH	↓	1.10	21.820	↓	3,400	515		3,915	4,525	
9000	Minimum labor/equipment charge	Q-20	1	20	Job		460		460	720	
115 0010	**BOILERS, GAS FIRED** Natural or propane, standard controls										**115**
1000	Cast iron, with insulated jacket										

For expanded coverage of these items see *Means Mechanical Cost Data or Means Plumbing Cost Data 1991*

155 100 | Boilers

			CREW	DAILY OUTPUT	MAN-HOURS	UNIT	MAT.	LABOR	EQUIP.	TOTAL	TOTAL INCL O&P	
115	2000	Steam, gross output, 81 MBH	Q-7	1.40	22.860	Ea.	1,055	555		1,610	2,025	115
	2060	163 MBH		.90	35.560		1,615	865		2,480	3,100	
	2200	440 MBH		.55	58.180		3,400	1,425		4,825	5,925	
	2320	1875 MBH		.28	114		11,900	2,775		14,675	17,400	
	2400	3570 MBH		.18	178		18,950	4,325		23,275	27,500	
	2540	6970 MBH		.08	400		46,750	9,750		56,500	66,500	
	2800											
	3000	Hot water, gross output, 80 MBH	Q-7	1.46	21.920	Ea.	935	535		1,470	1,850	
	3020	100 MBH		1.35	23.700		1,100	575		1,675	2,100	
	3040	122 MBH		1.10	29.090		1,215	710		1,925	2,425	
	3060	163 MBH		1	32		1,495	780		2,275	2,850	
	3200	440 MBH		.65	49.230		3,220	1,200		4,420	5,400	
	3320	2000 MBH		.38	84.210		12,920	2,050		14,970	17,400	
	3540	6970 MBH		.08	400		46,750	9,750		56,500	66,500	
	4000	Steel, insulating jacket										
	6000	Hot water, including burner & one zone valve, gross output										
	6010	51.2 MBH	Q-6	2	12	Ea.	1,090	285		1,375	1,650	
	6020	72 MBH		2	12		1,490	285		1,775	2,075	
	6040	89 MBH		1.90	12.630		1,530	300		1,830	2,150	
	6060	105 MBH		1.80	13.330		1,710	315		2,025	2,375	
	6080	132 MBH		1.70	14.120		1,960	335		2,295	2,675	
	6100	155 MBH		1.50	16		2,270	380		2,650	3,075	
	6110	186 MBH		1.40	17.140		2,740	410		3,150	3,650	
	6120	227 MBH		1.30	18.460		3,490	440		3,930	4,525	
	6140	292 MBH		1.20	20		3,940	475		4,415	5,075	
	6180	480 MBH		.70	34.290		5,715	815		6,530	7,550	
	6200	640 MBH		.60	40		6,800	950		7,750	8,950	
	6220	800 MBH		.50	48		7,910	1,150		9,060	10,500	
	6240	960 MBH		.45	53.330		9,820	1,275		11,095	12,800	
	7000	For tankless water heater on smaller gas units, add					10%					
	7050	For additional zone valves up to 312 MBH add					70			70	77	
	7060											
	9900	Minimum labor/equipment charge	Q-6	1	24	Job		570		570	880	
120	0010	**BOILERS, OIL FIRED** Standard controls, flame retention burner										120
	1000	Cast iron, with insulated flush jacket										
	2000	Steam, gross output, 109 MBH	Q-7	1.20	26.670	Ea.	1,400	650		2,050	2,550	
	2060	207 MBH		.90	35.560		1,910	865		2,775	3,425	
	2180	1084 MBH		.42	76.190		6,990	1,850		8,840	10,500	
	2200	1360 MBH		.38	84.210		8,250	2,050		10,300	12,200	
	2240	2175 MBH		.28	114		11,750	2,775		14,525	17,200	
	2340	4360 MBH		.17	188		20,650	4,575		25,225	29,800	
	2460	6970 MBH		.08	400		46,050	9,750		55,800	65,500	
	3000	Hot water, same price as steam										
	5000	Steel, insulated jacket, burner										
	7000	Hot water, gross output, 103 MBH	Q-6	1.90	12.630	Ea.	1,740	300		2,040	2,375	
	7020	122 MBH		1.80	13.330		1,770	315		2,085	2,425	
	7040	137 MBH		1.60	15		1,855	355		2,210	2,600	
	7060	168 MBH		1.50	16		2,240	380		2,620	3,050	
	7080	225 MBH		1.40	17.140		2,645	410		3,055	3,550	
	7100	315 MBH		1.10	21.820		3,910	520		4,430	5,100	
	7120	420 MBH		.80	30		4,400	715		5,115	5,950	
	9000	Minimum labor/equipment charge		1.75	13.710	Job		325		325	505	
125	0010	**BOILERS, GAS/OIL** Combination with burners and controls										125
	1000	Cast Iron with insulated jacket										
	2000	Steam, gross output, 720 MBH	Q-7	.40	80	Ea.	9,900	1,950		11,850	13,900	
	2140	2700 MBH	"	.19	168	"	19,550	4,100		23,650	27,800	

For expanded coverage of these items see *Means Mechanical Cost Data or Means Plumbing Cost Data 1991*

155 100 | Boilers

			CREW	DAILY OUTPUT	MAN-HOURS	UNIT	MAT.	LABOR	EQUIP.	TOTAL	TOTAL INCL O&P	
125	2380	6970 MBH	Q-7	.05	640	Ea.	49,150	15,600		64,750	78,000	125
	3000	Hot water, gross output, 584 MBH		.54	59.260		6,840	1,450		8,290	9,750	
	3060	1460 MBH		.45	71.110		16,750	1,725		18,475	21,100	
	3300	13,500 MBH, 403.3 BHP		.20	160		102,200	3,900		106,100	118,500	
	4000	Steel, insulated jacket, skid base, tubeless										
	4500	Steam, 150 psi gross output, 335 MBH, 10 BHP	Q-6	.65	36.920	Ea.	7,720	880		8,600	9,850	
	4640	1339 MBH, 40 BHP		.28	85.710		15,050	2,050		17,100	19,700	
	4720	2511 MBH, 75 BHP		.17	141		20,450	3,350		23,800	27,700	
130	0010	**BOILERS, SOLID FUEL**										130
	5000	Wood or coal and oil combination, circulator,										
	5050	mixing valve, controls										
	5100	Output (oil), 60 MBH, without burner	Q-5	.84	19.050	Ea.	965	435		1,400	1,725	
	5150	80 MBH, without burner		.80	20		1,130	460		1,590	1,950	
	5250	100 MBH, with burner		.73	21.920		2,020	505		2,525	3,000	
	5300	130 MBH, with burner		.69	23.190		2,270	530		2,800	3,325	

155 200 | Boiler Accessories

			CREW	DAILY OUTPUT	MAN-HOURS	UNIT	MAT.	LABOR	EQUIP.	TOTAL	TOTAL INCL O&P	
230	0010	**BURNERS**										230
	0990	Residential, conversion, gas fired, LP or natural										
	1000	Gun type, atmospheric input 35 to 180 MBH	Q-1	2.50	6.400	Ea.	162	145		307	405	
	1020	50 to 240 MBH		2	8		187	185		372	490	
	1040	200 to 400 MBH		1.70	9.410		350	215		565	715	
	3000	Flame retention oil fired assembly, input										
	3045	2.0 to 5.0 GPH	Q-1	2	8	Ea.	425	185		610	750	
240	0010	**DRAFT CONTROLS, BAROMETRIC**										240
	1000	Gas fired system only, 6" size for 5" and 6" pipes	1 Shee	20	.400	Ea.	24.70	10		34.70	43	
	1040	8" size, for 7" and 8" pipes	"	18	.444	"	32.85	11.10		43.95	54	
	2000	All fuel, oil, oil/gas, coal										
	2020	10" for 9" and 10" pipes	1 Shee	15	.533	Ea.	55.15	13.35		68.50	82	
	3260	For thermal switch for above, add	"	24	.333		25	8.35		33.35	41	
	5000	Vent damper, bi-metal, gas, 3" diameter	Q-9	24	.667		34.80	15		49.80	62	
	5010	4" diameter		24	.667		45.50	15		60.50	74	
	5020	5" diameter		23	.696		54.30	15.65		69.95	84	
	5030	6" diameter		22	.727		61.30	16.35		77.65	93	
	5040	7" diameter		21	.762		70	17.15		87.15	105	
	5050	8" diameter		20	.800		80.50	18		98.50	115	
	5101	Electric, automatic, gas, 4" diameter		24	.667		116	15		131	150	
	5110	5" diameter		23	.696		117	15.65		132.65	155	
	5121	6" diameter		22	.727		118	16.35		134.35	155	
	5130	7" diameter		21	.762		122	17.15		139.15	160	
	5140	8" diameter		20	.800		128	18		146	170	
	5150	9" diameter		20	.800		135	18		153	175	
	5160	10" diameter		19	.842		138	18.95		156.95	180	
	5170	12" diameter		19	.842		156	18.95		174.95	200	
	5180	14" diameter		18	.889		218	20		238	270	
	5190	16" diameter		17	.941		225	21		246	280	
	5200	18" diameter		16	1		237	23		260	295	
	5250	Automatic, oil, 4" diameter		24	.667		135	15		150	170	
	5260	5" diameter		23	.696		136	15.65		151.65	175	
	5270	6" diameter		22	.727		137	16.35		153.35	175	
	5280	7" diameter		21	.762		138	17.15		155.15	180	
	5290	8" diameter		20	.800		139	18		157	180	
	5300	9" diameter		20	.800		141	18		159	185	
	5310	10" diameter		19	.842		142	18.95		160.95	185	
	5320	12" diameter		19	.842		157	18.95		175.95	200	
	5330	14" diameter		18	.889		232	20		252	285	

For expanded coverage of these items see *Means Mechanical Cost Data* or *Means Plumbing Cost Data 1991*

			DAILY	MAN-		BARE COSTS				TOTAL	
		155 200 Boiler Accessories	CREW	OUTPUT	HOURS	UNIT	MAT.	LABOR	EQUIP.	TOTAL	INCL O&P	
240	5340	16" diameter	Q-9	17	.941	Ea.	237	21		258	295	240
	5350	18" diameter	"	16	1	"	248	23		271	310	
	9000	Minimum labor/equipment charge	1 Shee	4	2	Job		50		50	78	

155 400 | Warm Air Systems

			CREW	DAILY OUTPUT	MAN-HOURS	UNIT	MAT.	LABOR	EQUIP.	TOTAL	INCL O&P	
401	0010	**DUCT FURNACES** Includes burner, controls, stainless steel										401
	0020	heat exchanger. Gas fired, electric ignition										
	1000	Outdoor installation, with vent cap										
	1120	225 MBH output	Q-5	2.50	6.400	Ea.	2,400	145		2,545	2,875	
	1160	375 MBH output		1.60	10		3,715	230		3,945	4,450	
	1180	450 MBH output		1.40	11.430		3,890	260		4,150	4,675	
408	0010	**DUCT HEATERS** , Electric, 480 V, 3 ph										408
	0020	Finned tubular insert, 500°F										
	0100	8" wide x 6" high, 4.0KW	Q-20	16	1.250	Ea.	277	29		306	350	
	0120	12" high, 8.0KW		15	1.330		449	31		480	540	
	0140	18" high, 12.0KW		14	1.430		630	33		663	745	
	0160	24" high, 16.0KW		13	1.540		810	35		845	945	
	0180	30" high, 20.0KW		12	1.670		990	38		1,028	1,150	
	0300	12" wide x 6" high, 6.7KW		15	1.330		288	31		319	365	
	0320	12" high, 13.3 KW		14	1.430		465	33		498	565	
	0340	18" high, 20.0KW		13	1.540		650	35		685	770	
	8000	To obtain BTU multiply KW by 3413										
	9900	Minimum labor/equipment charge	1 Shee	4	2	Job		50		50	78	
420	0010	**FURNACES** Hot air heating, blowers, standard controls										420
	0020	not including gas, oil or flue piping.										
	1000	Electric, UL listed, heat staging, 240 volt										
	1020	30 MBH	Q-20	4	5	Ea.	290	115		405	500	
	1080	76 MBH		3.60	5.560		515	130		645	765	
	1090	85.3 MBH		3.70	5.410		575	125		700	825	
	1100	91 MBH		3.40	5.880		720	135		855	1,000	
	3000	Gas, AGA certified, direct drive models										
	3020	42 MBH output	Q-9	4	4	Ea.	400	90		490	580	
	3040	63 MBH output		3.80	4.210		420	95		515	610	
	3060	79 MBH output		3.60	4.440		475	100		575	680	
	3080	84 MBH output		3.40	4.710		515	105		620	735	
	3100	105 MBH output		3.20	5		535	115		650	765	
	3120	126 MBH output		3	5.330		690	120		810	945	
	3130	160 MBH output		2.80	5.710		1,065	130		1,195	1,375	
	3140	200 MBH output		2.60	6.150		2,650	140		2,790	3,125	
	6000	Oil, UL listed, atomizing gun type burner										
	6020	55 MBH output	Q-9	3.60	4.440	Ea.	705	100		805	935	
	6040	99 MBH output		3.40	4.710		790	105		895	1,025	
	6060	125 MBH output		3.20	5		900	115		1,015	1,175	
	6080	152 MBH output		3	5.330		1,115	120		1,235	1,425	
	6100	200 MBH output		2.60	6.150		1,765	140		1,905	2,150	
	8500	Wood, coal and oil combination complete with burner										
	8520	112 MBH output (oil)	Q-9	3	5.330	Ea.	3,495	120		3,615	4,025	
	8540	140 MBH output (oil)		2.80	5.710		3,495	130		3,625	4,050	
	8560	150 MBH output (oil)		2.60	6.150		3,495	140		3,635	4,050	
	8580	170 MBH output (oil)		2.40	6.670		3,495	150		3,645	4,075	
	9000	Minimum labor/equipment charge		2.75	5.820	Job		130		130	205	
461	0010	**MAKE-UP AIR UNIT**										461
	0020	Indoor suspension, natural/LP gas, direct fired,										
	0030	standard control. For flue see Division 155-680										
	0040	70°F temperature rise, MBH is input										
	0100	2000 CFM, 168 MBH	Q-6	3	8	Ea.	4,580	190		4,770	5,325	
	0220	12,000 CFM, 1005 MBH	"	1	24	"	7,180	570		7,750	8,775	

For expanded coverage of these items see *Means Mechanical Cost Data* or *Means Plumbing Cost Data 1991*

155 400 | Warm Air Systems

		CREW	DAILY OUTPUT	MAN-HOURS	UNIT	BARE COSTS				TOTAL INCL O&P	
						MAT.	LABOR	EQUIP.	TOTAL		
461											**461**
0300	24,000 CFM, 2007 MBH	Q-7	1	32	Ea.	9,350	780		10,130	11,500	
0400	50,000 CFM, 4180 MBH	"	.80	40	"	12,850	975		13,825	15,600	
9000	Minimum labor/equipment charge	Q-6	2.75	8.730	Job		210		210	320	
471	0010 **SOLAR ENERGY**										**471**
0020	System/Package prices, not including connecting										
0030	pipe, insulation, or special heating/plumbing fixtures										
0500	Hot water, standard package, low temperature										
0580	2 collectors, circulator, fittings, 120 gal. tank	Q-1	.40	40	Ea.	1,765	915		2,680	3,350	
1200	Water and residence heating package										
1240	10 medium temperature collectors, fittings	Q-2	.13	185	Ea.	7,115	4,375		11,490	14,600	
1300	Solar assist package, for space heating and domestic										
1340	hot water, 10 collectors, fittings	Q-2	.13	185	Ea.	7,115	4,375		11,490	14,600	
2250	Controller, liquid temperature	1 Plum	5	1.600	"	72	41		113	140	
2300	Circulators, air										
2310	Blowers										
2320	30 to 100 S.F. system, 1/20 HP	Q-9	16	1	Ea.	55	23		78	96	
2480	Shutter mounted fan, 12" diameter, 650 CFM		14	1.140		105	26		131	155	
2520	Space & DHW system, less duct work		.50	32		1,200	720		1,920	2,450	
2580	8" diameter, 150 CFM		16	1		28	23		51	66	
2660	Shutter/damper		12	1.330		27	30		57	77	
2670	Shutter motor		16	1		59	23		82	100	
2750	Circulators, liquid, 1/100 HP, 2 GPM	Q-1	14	1.140		86	26		112	135	
2770	1/100 HP, 3 GPM		14	1.140		88	26		114	135	
2800	1/25 HP, 5.3 GPM		14	1.140		100	26		126	150	
2820	1/20 HP, 17 GPM		12	1.330		77	31		108	130	
3000	Collector panels, air with aluminum absorber plate										
3010	Wall or roof mount										
3040	Flat black, plastic glazing										
3080	4' x 8'	Q-9	6	2.670	Ea.	400	60		460	535	
3300	Collector panels, liquid with copper absorber plate										
3320	Black chrome, tempered glass glazing										
3330	Alum. frame, 3' x 8', 5/32" single glazing	Q-1	9.50	1.680	Ea.	350	39		389	445	
3440	Flat black										
3450	Alum. frame, 3' x 8', 5/32" single glazing	Q-1	9	1.780	Ea.	340	41		381	435	
3500	Alum. frame, 3' x 8', 5/32" double glazing		5.50	2.910		360	67		427	500	
3600	Liquid, full wetted, plastic, alum. frame, 3' x 10'		5	3.200		153	73		226	280	
3650	Collector panel mounting, flat roof or ground rack		5	3.200		125	73		198	250	
3670	Roof clamps		70	.229		5	5.25		10.25	13.55	
3700	Roof strap, teflon	1 Plum	205	.039	L.F.	4.95	.99		5.94	7	
3900	Differential controller with two sensors										
3930	Thermostat, hard wired	1 Plum	8	1	Ea.	72	25		97	120	
4000	External adjustment		12	.667		94	16.95		110.95	130	
4050	Pool valve system, 2" pipe size (plastic)		2.50	3.200		202	81		283	350	
4150	Sensors										
4220	Freeze prevention	1 Plum	32	.250	Ea.	15.50	6.35		21.85	27	
4260											
4300	Heat exchanger										
4310	Fluid to air coil										
4315	includes coil, blower, circulator										
4316	and controller for DHW and space hot air										
4380	70 MBH	Q-1	3.50	4.570	Ea.	255	105		360	440	
4400	80 MBH		3	5.330		370	120		490	595	
4580	Fluid to fluid package includes two circulating pumps		2.50	6.400		625	145		770	915	
4620											
4650	Heat transfer fluid										
4700	Propylene glycol, inhibited anti-freeze	1 Plum	28	.286	Gal.	7.39	7.25		14.64	19.35	
4800	Solar storage tanks, knocked down										

For expanded coverage of these items see *Means Mechanical Cost Data* or *Means Plumbing Cost Data 1991*

155 400 | Warm Air Systems

		Description	CREW	DAILY OUTPUT	MAN-HOURS	UNIT	MAT.	LABOR	EQUIP.	TOTAL	TOTAL INCL O&P	
471	4860	4' high, 3' x 3', = 36 C.F./250 gallons	Q-9	2.60	6.150	Ea.	1,095	140		1,235	1,425	471
	5020	7' x 10'-6" = 459 C.F./3000 gallons	Q-10	.80	30		6,530	700		7,230	8,275	
	5210	7' x 14' = 613 C.F./4000 gallons		.60	40		8,725	935		9,660	11,100	
	5230	10'-6" x 14' = 919 C.F./6000 gallons		.40	60		11,050	1,400		12,450	14,400	
	5250	14' x 17'-6" = 1531 C.F./10,000 gallons	Q-11	.30	107		14,800	2,550		17,350	20,300	
	7000	Solar control valves and vents										
	7070	Air eliminator, automatic ¾" size	1 Plum	32	.250	Ea.	9.95	6.35		16.30	21	
	7090	Air vent, automatic, ⅛" fitting		32	.250		5.55	6.35		11.90	15.90	
	7100	Manual, ⅛" NPT		32	.250		1.43	6.35		7.78	11.40	
	7120	Backflow preventer, ½" pipe size		16	.500		44.90	12.75		57.65	69	
	7130	¾" pipe size		16	.500		44.90	12.75		57.65	69	
	7150	Balancing valve, ¾" pipe size		20	.400		14.20	10.20		24.40	31	
	7180	Draindown valve, ½" copper tube		9	.889		156	23		179	205	
	7200	Flow control valve, ½" pipe size		22	.364		26.60	9.25		35.85	44	
	7220	Expansion tank, up to 5 gal.		32	.250		38.50	6.35		44.85	52	
	7400	Pressure gauge, 2" dial		32	.250		8.35	6.35		14.70	19	
	7450	Relief valve, temp. and pressure ¾" pipe size		30	.267		10.20	6.80		17	22	
	7500	Solenoid valve, normally closed										
	7750	Vacuum relief valve, ¾" pipe size	1 Plum	32	.250	Ea.	26.25	6.35		32.60	39	
	7800	Thermometers										
	7890	In line, dial, ½" NPT	1 Plum	8	1	Ea.	17	25		42	58	
	8250	Water storage tank with heat exchanger and electric element										
	8260	66 gal. with 2" x ½ lb. density insulation	1 Plum	1.60	5	Ea.	525	125		650	775	
	8270	66 gal. with 2" x 2 lb. density insulation		1.60	5		545	125		670	795	
	8280	80 gal. with 2" x ½ lb. density insulation		1.60	5		585	125		710	840	
	8300	80 gal. with 2" x 2 lb. density insulation		1.60	5		600	125		725	855	
	8350	120 gal. with 2" x ½ lb. density insulation		1.60	5		660	125		785	920	
	8500	Water storage module, plastic										
	8600	Tubular, 12" diameter, 4' high	1 Carp	48	.167	Ea.	60	3.67		63.67	72	
	8610	12" diameter, 8' high	"	40	.200		90	4.40		94.40	105	
	8650	Cap, 12" diameter					7.95			7.95	8.75	
	9000	Minimum labor/equipment charge	1 Plum	2	4	Job		100		100	155	
480	0010	SPACE HEATERS Cabinet, grilles, fan, controls, burner,										480
	0020	thermostat, no piping. For flue see division 155-680										
	1000	Gas fired, floor mounted										
	1100	60 MBH output	Q-5	10	1.600	Ea.	420	37		457	520	
	1180	180 MBH output		6	2.670		640	61		701	800	
	2000	Suspension mounted, propeller fan, 20 MBH output		8.50	1.880		370	43		413	475	
	2040	60 MBH output		7	2.290		445	52		497	570	
	2100	130 MBH output		5	3.200		665	73		738	845	
	2240	320 MBH output		2	8		1,350	185		1,535	1,775	
	2500	For powered venter and adapter, add					195			195	215	
	5000	Wall furnace, 17.5 MBH output	Q-5	6	2.670		275	61		336	395	
	5020	24 MBH output		5	3.200		288	73		361	430	
	5040	35 MBH output		4	4		400	92		492	580	
	9000	Minimum labor/equipment charge		3.50	4.570	Job		105		105	160	

155 600 | Heating System Access.

		Description	CREW	DAILY OUTPUT	MAN-HOURS	UNIT	MAT.	LABOR	EQUIP.	TOTAL	TOTAL INCL O&P	
620	0010	HEAT RECOVERY PACKAGES										620
	0100	Air to air										
	2000	Kitchen exhaust, commercial, heat pipe exchanger										
	2040	Combined supply/exhaust air volume										
	2080	2.5 to 6.0 MCFM	Q-10	2.80	8.570	MCFM	3,318	200		3,518	3,975	
	2120	6 to 16 MCFM		5	4.800		2,037	110		2,147	2,425	
	2160	16 to 22 MCFM		6	4		1,575	93		1,668	1,875	
	9000	Minimum labor/equipment charge	1 Shee	4	2	Job		50		50	78	
630	0010	HYDRONIC HEATING Terminal units, not incl. main supply pipe										630
	1000	Radiation										

For expanded coverage of these items see *Means Mechanical Cost Data* or *Means Plumbing Cost Data 1991*

155 600 | Heating System Access.

		CREW	DAILY OUTPUT	MAN-HOURS	UNIT	BARE COSTS				TOTAL INCL O&P
						MAT.	LABOR	EQUIP.	TOTAL	
630	1150	Fin tube, wall hung, 14″ slope top cover, with damper									**630**
	1200	1-¼″ copper tube, 4-¼″ alum. fin	Q-5	38	.421	L.F.	31	9.65		40.65	49
	1250	1-¼″ steel tube, 4-¼″ steel fin		36	.444		28.10	10.20		38.30	47
	1255	2″ steel tube, 4-¼″ steel fin		32	.500		30.20	11.50		41.70	51
	1320	¾″ copper tube, alum. fin, 7″ high		58	.276		7.70	6.35		14.05	18.25
	1340	1″ copper tube, alum. fin, 8-⅞″ high		56	.286		13.80	6.55		20.35	25
	1360	1-¼″ copper tube, alum. fin, 8-⅞″ high	↓	54	.296	↓	19.42	6.80		26.22	32
	3000	Radiators, cast iron									
	3100	Free standing or wall hung, 6 tube, 25″ high	Q-5	96	.167	Section	20.16	3.83		23.99	28
	3950	Unit heaters, propeller, 230/460V 2 psi steam, 60°F entering air									
	4000	Horizontal, 12 MBH	Q-5	12	1.330	Ea.	415	31		446	505
	4060	43.9 MBH		8	2		585	46		631	715
	4240	286.9 MBH	↓	2	8		1,150	185		1,335	1,550
	4300	For vertical diffuser, add					88.50			88.50	97
	5000	Vertical flow, 40.0 MBH	Q-5	11	1.450		660	33		693	775
	5080	131.0 MBH	"	4	4		785	92		877	1,000
	5160	420 MBH	Q-6	1.80	13.330		1,560	315		1,875	2,200
	5180	500 MBH	"	1.40	17.140	↓	1,775	410		2,185	2,575
	9000	Minimum labor/equipment charge	Q-5	3.20	5	Job		115		115	175
640	0010	**HUMIDIFIERS**									**640**
	0030	Centrifugal atomizing									
	0100	10 lb. per hour	Q-5	10	1.600	Ea.	1,135	37		1,172	1,300
	0520	Steam, room or duct, filter, regulators, auto. controls, 220 V									
	0560	17 lb. per hour	Q-5	5	3.200	Ea.	1,990	73		2,063	2,300
	0580	30 lb. per hour		4	4		2,050	92		2,142	2,400
	0620	90 lb. per hour		3	5.330	↓	3,090	120		3,210	3,600
	9000	Minimum labor/equipment charge	↓	3.50	4.570	Job		105		105	160
651	0010	**INSULATION**									**651**
	0100	Rule of thumb, as a percentage of total mechanical costs				Job				10%	
	2900	Domestic water heater wrap kit									
	2920	1-½″ with vinyl jacket, 20-60 gal.	1 Plum	8	1	Ea.	23.50	25		48.50	65
	2930										
	3000	Ductwork									
	3020	Blanket type, fiberglass, flexible									
	3030	Fire resistant liner, black coating one side									
	3050	½″ thick, 2 lb. density	Q-14	380	.042	S.F.	.34	.94		1.28	1.87
	3060	1″ thick, 1-½ lb. density	"	350	.046	"	.45	1.02		1.47	2.13
	3140	FRK vapor barrier wrap, .75 lb. density									
	3160	1″ thick	Q-14	350	.046	S.F.	.23	1.02		1.25	1.88
	3170	1-½″ thick		320	.050		.30	1.11		1.41	2.11
	3180	2″ thick		300	.053		.37	1.19		1.56	2.31
	3190	3″ thick		260	.062		.56	1.37		1.93	2.81
	3200	4″ thick	↓	260	.062	↓	.73	1.37		2.10	2.99
	3490	Board type, fiberglass, 3 lb. density									
	3500	Fire resistant, black pigmented, 1 side									
	3520	1″ thick	Q-14	150	.107	S.F.	1.29	2.37		3.66	5.20
	3540	1-½″ thick	"	130	.123	"	1.50	2.74		4.24	6.05
	4000	Pipe covering									
	6120	2″ wall, ½″ iron pipe size	Q-14	145	.110	L.F.	3.01	2.45		5.46	7.25
	6210	4″ iron pipe size	"	125	.128	"	5.28	2.85		8.13	10.35
	6600	Fiberglass, with all service jacket									
	6840	1″ wall, ½″ iron pipe size	Q-14	240	.067	L.F.	.97	1.48		2.45	3.44
	6860	¾″ iron pipe size		230	.070		1.12	1.55		2.67	3.71
	6870	1″ iron pipe size		220	.073		1.15	1.62		2.77	3.86
	6880	1-¼″ iron pipe size		210	.076		1.28	1.69		2.97	4.12
	6890	1-½″ iron pipe size		210	.076		1.39	1.69		3.08	4.24
	6900	2″ iron pipe size	↓	200	.080	↓	1.51	1.78		3.29	4.51

For expanded coverage of these items see *Means Mechanical Cost Data* or *Means Plumbing Cost Data 1991*

		155 600 Heating System Access.	CREW	DAILY OUTPUT	MAN-HOURS	UNIT	BARE COSTS MAT.	LABOR	EQUIP.	TOTAL	TOTAL INCL O&P	
651	7080	1-1/2" wall, 1/2" iron pipe size	Q-14	230	.070	L.F.	1.98	1.55		3.53	4.66	651
	7140	2" iron pipe size		190	.084		2.70	1.87		4.57	5.95	
	7160	3" iron pipe size	↓	170	.094	↓	3.06	2.09		5.15	6.70	
	7879	Rubber tubing, flexible closed cell foam										
	8100	1/2" wall, 1/4" iron pipe size	1 Asbe	90	.089	L.F.	.34	2.20		2.54	3.89	
	8120	3/8" iron pipe size		90	.089		.38	2.20		2.58	3.93	
	8130	1/2" iron pipe size		90	.089		.42	2.20		2.62	3.98	
	8140	3/4" iron pipe size		90	.089		.47	2.20		2.67	4.03	
	8150	1" iron pipe size		90	.089		.52	2.20		2.72	4.09	
	8170	1-1/2" iron pipe size		90	.089		.73	2.20		2.93	4.32	
	8180	2" iron pipe size		90	.089		.89	2.20		3.09	4.50	
	8200	3" iron pipe size		90	.089		1.64	2.20		3.84	5.30	
	8220	4" iron pipe size		80	.100		1.83	2.47		4.30	5.95	
	8300	3/4" wall, 1/4" iron pipe size		90	.089		.51	2.20		2.71	4.08	
	8330	1/2" iron pipe size		90	.089		.69	2.20		2.89	4.28	
	8340	3/4" iron pipe size		90	.089		.85	2.20		3.05	4.45	
	8350	1" iron pipe size		90	.089		.97	2.20		3.17	4.58	
	8360	1-1/4" iron pipe size		90	.089		1.31	2.20		3.51	4.96	
	8370	1-1/2" iron pipe size		90	.089		1.48	2.20		3.68	5.15	
	8380	2" iron pipe size	↓	90	.089	↓	1.73	2.20		3.93	5.40	
	8800	Urethane, with ASJ, -60°F to +225°F										
	8960	1" wall, 1/2" iron pipe size	Q-14	240	.067	L.F.	1.08	1.48		2.56	3.56	
	9290	Urethane, with ultraviolet cover										
	9310	1" wall, 1/2" pipe size	Q-14	216	.074	L.F.	1.19	1.65		2.84	3.95	
	9320	3/4" pipe size	"	207	.077	"	1.19	1.72		2.91	4.06	
	9600	Minimum labor/equipment charge	1 Plum	4	2	Job		51		51	78	
671	0010	**TANKS**										671
	0020	Fiberglass, underground, U.L. listed, not including										
	0030	manway or hold-down strap										
	0100	1000 gallon capacity	Q-6	2	12	Ea.	1,570	285		1,855	2,175	
	0140	2000 gallon capacity	Q-7	2	16		2,290	390		2,680	3,125	
	0160	4000 gallon capacity		1.30	24.620		2,950	600		3,550	4,175	
	0180	6000 gallon capacity	↓	1	32		3,740	780		4,520	5,325	
	0500	For manway, fittings and hold-downs, add					20%	15%				
	2000	Steel, liquid expansion, ASME, painted, 15 gallon capacity	Q-5	17	.941		244	22		266	300	
	2040	30 gallon capacity		12	1.330		278	31		309	355	
	2080	60 gallon capacity		8	2		375	46		421	485	
	2120	100 gallon capacity		6	2.670		505	61		566	650	
	3000	Steel ASME expansion, rubber diaphragm, 19 gal. cap. accept.		12	1.330		945	31		976	1,075	
	3020	31 gallon capacity		8	2		1,050	46		1,096	1,225	
	3040	61 gallon capacity		6	2.670		1,470	61		1,531	1,700	
	3080	119 gallon capacity	↓	4	4	↓	1,675	92		1,767	1,975	
	4000	Steel, storage, above ground, including supports, coating,										
	4020	fittings, not including mat, pumps or piping										
	4040	275 gallon capacity	Q-5	5	3.200	Ea.	225	73		298	360	
	4060	550 gallon capacity	"	4	4		1,050	92		1,142	1,300	
	4080	1000 gallon capacity	Q-7	4	8		1,350	195		1,545	1,775	
	4120	2000 gallon capacity	"	3	10.670	↓	1,700	260		1,960	2,275	
	5000	Steel underground, sti-P3, set in place, incl. hold-down bars.										
	5500	Excavation, pad, pumps and piping not included										
	5520	1000 gallon capacity, 7 gauge shell	Q-7	4	8	Ea.	1,400	195		1,595	1,850	
	5540	5000 gallon capacity, 1/4" thick shell		1	32		4,000	780		4,780	5,600	
	5560	10,000 gallon capacity, 1/4" thick shell		.70	45.710		5,200	1,125		6,325	7,425	
	5600	20,000 gallon capacity, 5/16" thick shell	↓	.30	107	↓	12,000	2,600		14,600	17,200	
	9000	Minimum labor/equipment charge	Q-5	4	4	Job		92		92	140	
680	0010	**VENT CHIMNEY** Prefab metal, U.L. listed										680
	0020	Gas, double wall, galvanized steel										

For expanded coverage of these items see *Means Mechanical Cost Data or Means Plumbing Cost Data 1991*

155 600 | Heating System Access.

		CREW	DAILY OUTPUT	MAN-HOURS	UNIT	BARE COSTS				TOTAL INCL O&P
						MAT.	LABOR	EQUIP.	TOTAL	
680	0080	3" diameter	Q-9	72	.222	V.L.F.	2.47	5		7.47	10.55	**680**
	0100	4" diameter		68	.235		3.01	5.30		8.31	11.60
	0120	5" diameter		64	.250		3.56	5.65		9.21	12.75
	0140	6" diameter		60	.267		4.17	6		10.17	14
	0160	7" diameter		56	.286		6.13	6.45		12.58	16.85
	0180	8" diameter		52	.308		6.84	6.90		13.74	18.40
	0200	10" diameter		48	.333		14.15	7.50		21.65	27
	0220	12" diameter		44	.364		18.90	8.20		27.10	34
	0260	16" diameter		40	.400		43.10	9		52.10	62
	0300	20" diameter	Q-10	36	.667		65.75	15.55		81.30	97
	0340	24" diameter	"	32	.750		102	17.50		119.50	140
	5000	Vent damper bi-metal 6" flue	Q-9	16	1	Ea.	59.30	23		82.30	100
	5100	Gas, auto., electric		8	2		115	45		160	195
	5120	Oil, auto., electric		8	2		137	45		182	220
	7800	All fuel, double wall, stainless steel, 6" diameter		60	.267	V.L.F.	17.75	6		23.75	29
	7802	7" diameter		56	.286		22.90	6.45		29.35	35
	7804	8" diameter		52	.308		26.80	6.90		33.70	40
	7806	10" diameter		48	.333		39.10	7.50		46.60	55
	7808	12" diameter		44	.364		52.40	8.20		60.60	70
	7810	14" diameter		42	.381		68.80	8.55		77.35	89
	8000	All fuel, double wall, stainless steel fittings									
	8010	Roof support 6" diameter	Q-9	30	.533	Ea.	45.85	12		57.85	69
	8020	7" diameter		28	.571		51.85	12.85		64.70	77
	8030	8" diameter		26	.615		56	13.85		69.85	83
	8040	10" diameter		24	.667		73.90	15		88.90	105
	8050	12" diameter		22	.727		89	16.35		105.35	125
	8060	14" diameter		21	.762		113	17.15		130.15	150
	8100	Elbow 15°, 6" diameter		30	.533		39.90	12		51.90	63
	8120	7" diameter		28	.571		45	12.85		57.85	70
	8140	8" diameter		26	.615		51.20	13.85		65.05	78
	8160	10" diameter		24	.667		67	15		82	97
	8180	12" diameter		22	.727		81.30	16.35		97.65	115
	8200	14" diameter		21	.762		97.20	17.15		114.35	135
	8300	Insulated tee with insulated tee cap, 6" diameter		30	.533		75.35	12		87.35	100
	8340	7" diameter		28	.571		98.90	12.85		111.75	130
	8360	8" diameter		26	.615		112	13.85		125.85	145
	8380	10" diameter		24	.667		157	15		172	195
	8400	12" diameter		22	.727		220	16.35		236.35	270
	8420	14" diameter		21	.762		289	17.15		306.15	345
	8500	Joist shield, 6" diameter		30	.533		29.30	12		41.30	51
	8510	7" diameter		28	.571		31.30	12.85		44.15	55
	8520	8" diameter		26	.615		32.70	13.85		46.55	58
	8530	10" diameter		24	.667		41.40	15		56.40	69
	8540	12" diameter		22	.727		51.70	16.35		68.05	83
	8550	14" diameter		21	.762		64.20	17.15		81.35	98
	8600	Round top, 6" diameter		30	.533		25.50	12		37.50	47
	8620	7" diameter		28	.571		34.90	12.85		47.75	59
	8640	8" diameter		26	.615		46.80	13.85		60.65	73
	8660	10" diameter		24	.667		84.10	15		99.10	115
	8680	12" diameter		22	.727		120	16.35		136.35	160
	8700	14" diameter		21	.762		160	17.15		177.15	205
	8800	Adjustable roof flashing, 6" diameter		30	.533		30.40	12		42.40	52
	8820	7" diameter		28	.571		34.75	12.85		47.60	58
	8840	8" diameter		26	.615		37.75	13.85		51.60	63
	8860	10" diameter		24	.667		48.50	15		63.50	77
	8880	12" diameter		22	.727		62.50	16.35		78.85	94
	8900	14" diameter		21	.762		78	17.15		95.15	115
	9990	Minimum labor/equipment charge		3	5.330	Job		120		120	190

For expanded coverage of these items see *Means Mechanical Cost Data* or *Means Plumbing Cost Data 1991*

156 200 | Heat/Cool Piping Misc.

			CREW	DAILY OUTPUT	MAN-HOURS	UNIT	BARE COSTS				TOTAL INCL O&P
							MAT.	LABOR	EQUIP.	TOTAL	
245	0010	MIXING VALVE Automatic water tempering										245
	0050	¾" size	1 Stpi	18	.444	Ea.	284	11.35		295.35	330
	9000	Minimum labor/equipment charge	"	5	1.600	Job		41		41	63
272	0010	STEAM TRAP										272
	0020										
	0030	Cast iron body, threaded									
	0040	Inverted bucket									
	0050	½" pipe size	1 Stpi	12	.667	Ea.	69	17		86	100
	0100	1" pipe size		8	1		184	26		210	240
	0120	1-¼" pipe size	↓	8	1	↓	278	26		304	345
	1000	Float & thermostatic, 15 psig									
	1010	¾" pipe size	1 Stpi	16	.500	Ea.	68.10	12.75		80.85	95
	1020	1" pipe size		15	.533		104	13.60		117.60	135
	1040	1-½" pipe size		9	.889		185	23		208	240
	1060	2" pipe size		6	1.330	↓	336	34		370	420
	9000	Minimum labor/equipment charge	↓	4	2	Job		51		51	79

156 600 | Strainers

			CREW	DAILY OUTPUT	MAN-HOURS	UNIT	MAT.	LABOR	EQUIP.	TOTAL	TOTAL INCL O&P
608	0010	STRAINERS, Y TYPE Bronze body										608
	0020										
	0050	Screwed, 150 lb., ¼" pipe size	1 Stpi	24	.333	Ea.	18	8.50		26.50	33
	0100	½" pipe size		20	.400		21.60	10.20		31.80	39
	0120	¾" pipe size		19	.421		24.60	10.75		35.35	44
	0140	1" pipe size		17	.471		35.40	12		47.40	57
	0160	1-½" pipe size		15	.533		66	13.60		79.60	94
	0180	2" pipe size	↓	13	.615		104	15.70		119.70	140
	0220	3" pipe size	Q-5	16	1		420	23		443	495
	0240	4" pipe size	"	15	1.070		955	24		979	1,100
	1000	Flanged, 150 lb., 1-½" pipe size	1 Stpi	11	.727		270	18.55		288.55	325
	1020	2" pipe size	"	8	1		350	26		376	425
	1040	3" pipe size	Q-5	4.50	3.560		666	82		748	860
	1060	4" pipe size	"	3	5.330		1,010	120		1,130	1,300
	1100	6" pipe size	Q-6	3	8		1,930	190		2,120	2,425
	1500	For 300 lb. rating, add		40		↓	40%				
	9000	Minimum labor/equipment charge	1 Stpi	3.75	2.130	Job		54		54	84

157 100 | A.C. & Vent. Units

			CREW	DAILY OUTPUT	MAN-HOURS	UNIT	BARE COSTS				TOTAL INCL O&P
							MAT.	LABOR	EQUIP.	TOTAL	
110	0010	ABSORPTION COLD GENERATORS Water chiller										110
	3000	Gas fired, air cooled									
	3180	3 ton	Q-5	1.30	12.310	Ea.	2,170	280		2,450	2,825
	9000	Minimum labor/equipment charge	"	1	16	Job		365		365	565
130	0010	COMPUTER ROOM UNITS										130
	1000	Air cooled, includes remote condenser but not									
	1020	interconnecting tubing or refrigerant									
	1160	6 ton	Q-5	.30	53.330	Ea.	12,250	1,225		13,475	15,400
	1240	10 ton	"	.25	64		14,750	1,475		16,225	18,500
	1320	20 ton	Q-6	.29	82.760	↓	19,650	1,975		21,625	24,700

For expanded coverage of these items see *Means Mechanical Cost Data* or *Means Plumbing Cost Data 1991*

157 100	A.C. & Vent. Units	CREW	DAILY OUTPUT	MAN-HOURS	UNIT	BARE COSTS				TOTAL INCL O&P
						MAT.	LABOR	EQUIP.	TOTAL	
150	0010	**FAN COIL AIR CONDITIONING** Cabinet mounted, filters										150
	0020										
	0100	Chilled water, ½ ton cooling	Q-5	8	2	Ea.	555	46		601	680
	0120	1 ton cooling		6	2.670		630	61		691	785
	0160	2.5 ton cooling		5	3.200		1,155	73		1,228	1,375
	0180	3 ton cooling		4	4		1,265	92		1,357	1,525
	0200	10 ton cooling	Q-6	2.80	8.570		1,830	205		2,035	2,325
	0220	15 ton cooling		1.50	16		2,550	380		2,930	3,400
	0240	20 ton cooling		.80	30		3,265	715		3,980	4,700
	0260	30 ton cooling		.60	40		4,830	950		5,780	6,775
	0940	Direct expansion, air cooled condensing, 1.5 ton cooling	Q-5	5	3.200		340	73		413	485
	1000	5 ton cooling	"	3	5.330		775	120		895	1,050
	1040	10 ton cooling	Q-6	2.60	9.230		1,725	220		1,945	2,225
	1060	20 ton cooling		.70	34.290		3,210	815		4,025	4,800
	1100	40 ton cooling		.45	53.330		6,730	1,275		8,005	9,350
	9000	Minimum labor/equipment charge	Q-5	3.75	4.270	Job		98		98	150
160	0010	**HEAT PUMPS**										160
	1000	Air to air, split system, not including curbs or pads									
	1020	2 ton cooling, 8.5 MBH heat @ 0°F	Q-5	1.20	13.330	Ea.	1,600	305		1,905	2,225
	1040	3 ton cooling, 13 MBH heat @ 0°F		.80	20		2,070	460		2,530	2,975
	1080	7.5 ton cooling, 33 MBH heat @ 0°F		.30	53.330		5,040	1,225		6,265	7,425
	1120	15 ton cooling, 64 MBH heat @ 0°F	Q-6	.26	92.310		8,970	2,200		11,170	13,300
	1500	Single package, not including curbs, pads, or plenums									
	1520	2 ton cooling, 6.5 MBH heat @ 0°F	Q-5	1.50	10.670	Ea.	1,930	245		2,175	2,500
	1560	3 ton cooling, 10 MBH heat @ 0°F	"	1.20	13.330		2,325	305		2,630	3,025
	1660	15 ton cooling, 56 MBH heat @ 0°F	Q-6	.30	80		12,350	1,900		14,250	16,500
	2000	Water source to air, single package									
	2100	1 ton cooling, 13 MBH heat @ 75°F	Q-5	2	8	Ea.	860	185		1,045	1,225
	2140	2 ton cooling, 19 MBH heat @ 75°F		1.70	9.410		1,080	215		1,295	1,525
	2220	5 ton cooling, 29 MBH heat @ 75°F		.90	17.780		1,890	410		2,300	2,700
	9000	Minimum labor/equipment charge		1.75	9.140	Job		210		210	325
180	0010	**ROOF TOP AIR CONDITIONERS** Standard controls, curb, economizer										180
	0020										
	1000	Single zone, electric cool, gas heat									
	1100	3 ton cooling, 60 MBH heating	Q-5	1.30	12.310	Ea.	3,240	280		3,520	4,000
	1120	4 ton cooling, 95 MBH heating	"	1.10	14.550		3,460	335		3,795	4,325
	1160	10 ton cooling, 200 MBH heating	Q-6	.46	52.170		7,280	1,250		8,530	9,925
	1220	30 ton cooling, 540 MBH heating	Q-7	.22	145		22,350	3,550		25,900	30,000
	1240	40 ton cooling, 675 MBH heating	"	.16	200		29,100	4,875		33,975	39,500
	1260	50 ton cooling, 810 MBH heating	Q-8	.13	246		35,350	6,000	300	41,650	48,500
	1700	Gas cool, gas heat									
	1720	3 ton cooling, 90 MBH heating	Q-5	1	16	Ea.	3,080	365		3,445	3,950
	2000	Multizone, electric cool, gas heat, economizer									
	2120	20 ton cooling, 360 MBH heating	Q-7	.21	152	Ea.	33,000	3,700		36,700	42,000
	2180	30 ton cooling, 540 MBH heating		.15	213		43,800	5,200		49,000	56,000
	2220	70 ton cooling, 1500 MBH heating		.09	356		56,000	8,650		64,650	75,000
	2240	80 ton cooling, 1500 MBH heating		.08	400		64,000	9,750		73,750	85,500
	2280	105 ton cooling, 1500 MBH heating		.06	533		84,000	13,000		97,000	112,500
	2400	For hot water heat coil, deduct					5%				
	2500	For steam heat coil, deduct					2%				
	2600	For electric heat, deduct					3%	5%			
	9000	Minimum labor/equipment charge	Q-5	1.50	10.670	Job		245		245	375
185	0010	**SELF-CONTAINED SINGLE PACKAGE**										185
	0100	Air cooled, for free blow or duct, including remote condenser									
	0200	3 ton cooling	Q-5	1	16	Ea.	3,300	365		3,665	4,200
	0210	4 ton cooling	"	.83	19.280	"	3,940	440		4,380	5,025

For expanded coverage of these items see *Means Mechanical Cost Data* or *Means Plumbing Cost Data 1991*

157 100 | A.C. & Vent. Units

			CREW	DAILY OUTPUT	MAN-HOURS	UNIT	MAT.	LABOR	EQUIP.	TOTAL	TOTAL INCL O&P	
185	0240	10 ton cooling	Q-7	1	32	Ea.	7,080	780		7,860	9,000	185
	0280	30 ton cooling	"	.80	40	"	14,550	975		15,525	17,500	
	1000	Water cooled for free blow or duct, not including tower										
	1100	3 ton cooling	Q-6	1	24	Ea.	2,310	570		2,880	3,425	
	1120	5 ton cooling	"	1	24		2,750	570		3,320	3,900	
	1140	10 ton cooling	Q-7	.90	35.560		5,590	865		6,455	7,475	
	1180	30 ton cooling	"	.70	45.710		13,250	1,125		14,375	16,300	
	1240	60 ton cooling	Q-8	.30	107		24,400	2,600	130	27,130	31,000	
	9000	Minimum labor/equipment charge	Q-5	1	16	Job		365		365	565	
190	0010	**WATER CHILLERS**										190
	0500	Reciprocating, air cooled, 20 ton cooling	Q-7	.35	91.430	Ea.	13,650	2,225		15,875	18,400	
	0520	40 ton cooling		.24	133		19,750	3,250		23,000	26,700	
	0540	65 ton cooling		.14	229		31,300	5,575		36,875	43,000	
	0680	Water cooled, single compressor, semi-hermetic										
	0760	10 ton cooling	Q-6	.36	66.670	Ea.	8,250	1,575		9,825	11,500	
	0840	35 ton cooling	Q-7	.27	119	"	15,150	2,875		18,025	21,100	
	0980	Water cooled, multiple compressors, semi-hermetic										
	1100	50 ton cooling	Q-7	.21	152	Ea.	19,900	3,700		23,600	27,600	
	1160	100 ton cooling		.12	267		34,250	6,500		40,750	47,700	
	1200	140 ton cooling		.09	356		46,800	8,650		55,450	65,000	
	2000	Packaged chiller, condenserless for remote condenser										
	2040	60 ton cooling	Q-7	.21	152	Ea.	27,100	3,700		30,800	35,500	
	2100	120 ton cooling		.10	320		45,400	7,800		53,200	62,000	
	2120	160 ton cooling		.09	356		53,450	8,650		62,100	72,000	
	9000	Minimum labor/equipment charge	Q-6	1	24	Job		570		570	880	
195	0010	**WINDOW UNIT AIR CONDITIONERS**										195
	4250	Semi-permanent installation, 3 speed fan										
	4260	15 amp 125V grounded receptacle required										
	4400	High efficiency models										
	4500	10,000 BTUH, 4 way high thrust air	1 Carp	6	1.330	Ea.	520	29		549	620	
	4520	12,000 BTUH, 4 way high thrust air	L-2	8	2		575	39		614	695	
	4540	14,000 BTUH, 4 way high thrust air	"	8	2		640	39		679	770	
	9000	Minimum labor/equipment charge	1 Carp	2	4	Job		88		88	145	

157 200 | System Components

			CREW	DAILY OUTPUT	MAN-HOURS	UNIT	MAT.	LABOR	EQUIP.	TOTAL	TOTAL INCL O&P	
201	0010	**COILS, FLANGED**										201
	1500	Hot water heating, 1 row, 24" x 48"	Q-5	3	5.330	Ea.	945	120		1,065	1,225	
	9000	Minimum labor/equipment charge	1 Plum	1	8	Job		205		205	315	
210	0010	**COMPRESSORS**										210
	1000	(Ratings are ARI standard 515 group IV using R-22) 2 ton	Q-5	1	16	Ea.	2,025	365		2,390	2,800	
	1100	15 ton	Q-6	.72	33.330		4,820	795		5,615	6,525	
	1400	50 ton	"	.20	120		14,750	2,850		17,600	20,600	
	1600	130 ton	Q-7	.21	152		21,600	3,700		25,300	29,500	
230	0010	**CONDENSING UNITS**										230
	0030	Air cooled, compressor, standard controls										
	0050	1.5 ton	Q-5	2.50	6.400	Ea.	615	145		760	905	
	0200	2.5 ton		1.70	9.410		850	215		1,065	1,275	
	0300	3 ton		1.30	12.310		990	280		1,270	1,525	
	0400	4 ton		.90	17.780		1,310	410		1,720	2,075	
	0500	5 ton		.60	26.670		1,600	610		2,210	2,700	
	9000	Minimum labor/equipment charge		1.50	10.670	Job		245		245	375	
240	0010	**COOLING TOWERS** Packaged units										240
	0080	Draw thru, single flow										
	0100	Belt drive, 60 tons	Q-6	90	.267	Ton	68	6.35		74.35	85	
	0150	90 tons	"	100	.240	"	62	5.70		67.70	77	

For expanded coverage of these items see *Means Mechanical Cost Data* or *Means Plumbing Cost Data 1991*

157 200 | System Components

		CREW	DAILY OUTPUT	MAN-HOURS	UNIT	BARE COSTS MAT.	LABOR	EQUIP.	TOTAL	TOTAL INCL O&P	
240	0200	100 tons	Q-6	109	.220	Ton	60	5.25		65.25	74	240
	1000	For higher capacities, use multiples										
	1500	Induced air, double flow										
	1800	Gear drive, 125 ton	Q-6	120	.200	Ton	56	4.76		60.76	69	
	1900	150 ton		126	.190		55	4.53		59.53	67	
	2000	300 ton		129	.186		50	4.43		54.43	62	
	2100	600 ton		132	.182		38	4.33		42.33	48	
	2200	Up to 1000 tons		150	.160		30	3.81		33.81	39	
	3500	For pumps and piping, add		38	.632		34	15.05		49.05	61	
	4000	For absorption systems, add					75%	75%				
	9000	Minimum labor/equipment charge	Q-6	1	24	Job		570		570	880	
250	0010	**DUCTWORK**										250
	0020	Fabricated rectangular, includes fittings, joints, supports,										
	0100	Aluminum, alloy 3003-H14, under 100 lb.	Q-10	80	.300	Lb.	2.45	7		9.45	13.70	
	0110	100 to 500 lb.		80	.300		1.69	7		8.69	12.85	
	0120	500 to 1000 lb.		95	.253		1.50	5.90		7.40	10.90	
	0500	Galvanized steel, under 200 lb.		235	.102		.75	2.38		3.13	4.57	
	0520	200 to 500 lb.		245	.098		.65	2.29		2.94	4.31	
	0540	500 to 1000 lb.		255	.094		.55	2.20		2.75	4.06	
	0560	1000 to 2000 lb.		265	.091		.50	2.11		2.61	3.87	
	0580	Over 5,000 lb.		285	.084		.45	1.96		2.41	3.58	
	0590											
	1000	Stainless steel, type 304, under 100 lb.	Q-10	165	.145	Lb.	2.78	3.39		6.17	8.40	
	1020	100 to 500 lb.		175	.137		2	3.20		5.20	7.20	
	1030	500 to 1,000 lb.		190	.126		1.82	2.95		4.77	6.65	
	1300	Flexible, coated fiberglass fabric on corr. resist. metal helix										
	1400	pressure to 12" (WG) UL-181										
	1500	Non-insulated, 3" diameter	Q-9	400	.040	L.F.	.69	.90		1.59	2.17	
	1540	5" diameter		320	.050		.83	1.13		1.96	2.68	
	1560	6" diameter		280	.057		.96	1.29		2.25	3.08	
	1580	7" diameter		240	.067		1.13	1.50		2.63	3.60	
	1600	8" diameter		200	.080		1.27	1.80		3.07	4.23	
	1640	10" diameter		160	.100		1.60	2.25		3.85	5.30	
	1660	12" diameter		120	.133		1.87	3		4.87	6.75	
	1900	Insulated, 1" thick with ¾ lb., PE jacket, 3" diameter		380	.042		1.18	.95		2.13	2.79	
	1910	4" diameter		340	.047		1.18	1.06		2.24	2.96	
	1920	5" diameter		300	.053		1.40	1.20		2.60	3.42	
	1940	6" diameter		260	.062		1.54	1.38		2.92	3.86	
	1960	7" diameter		220	.073		1.84	1.64		3.48	4.59	
	1980	8" diameter		180	.089		1.98	2		3.98	5.30	
	2020	10" diameter		140	.114		2.50	2.57		5.07	6.80	
	2040	12" diameter		100	.160		2.92	3.60		6.52	8.85	
	2060	14" diameter		80	.200		3.49	4.50		7.99	10.90	
	9990	Minimum labor/equipment charge	1 Shee	3	2.670	Job		67		67	105	
290	0010	**FANS**										290
	0020	Air conditioning and process air handling										
	0030	Axial flow, compact, low sound, 2.5" S.P.										
	0050	3800 CFM, 5 HP	Q-20	3.40	5.880	Ea.	2,790	135		2,925	3,275	
	0120	15,600 CFM, 10 HP	"	1.60	12.500	"	4,890	290		5,180	5,825	
	0200	In-line centrifugal, supply/exhaust booster,										
	0220	aluminum wheel/hub, disconnect switch, ¼" S.P.										
	0240	500 CFM, 10" diameter connection	Q-20	3	6.670	Ea.	480	155		635	770	
	0280	1520 CFM, 16" diameter connection		2	10		750	230		980	1,175	
	0320	3480 CFM, 20" diameter connection		.80	25		1,200	575		1,775	2,225	
	2500	Ceiling fan, right angle, extra quiet, 0.10" S.P.										
	2540	210 CFM	Q-20	19	1.050	Ea.	155	24		179	210	

For expanded coverage of these items see *Means Mechanical Cost Data* or *Means Plumbing Cost Data 1991*

		157 200	System Components	CREW	DAILY OUTPUT	MAN-HOURS	UNIT	BARE COSTS				TOTAL INCL O&P	
								MAT.	LABOR	EQUIP.	TOTAL		
290	2580		885 CFM	Q-20	16	1.250	Ea.	375	29		404	455	290
	2620		2960 CFM	"	11	1.820		670	42		712	800	
	2640		For wall or roof cap, add	1 Shee	16	.500		87.50	12.50		100	115	
	4500	Corrosive fume resistant, plastic											
	4600	roof ventilators, centrifugal, V belt drive, motor											
	4620	¼" S.P., 250 CFM, ¼ HP		Q-20	6	3.330	Ea.	1,790	77		1,867	2,100	
	4640	895 CFM, ⅓ HP			5	4		1,940	92		2,032	2,275	
	4660	1630 CFM, ½ HP			4	5		2,300	115		2,415	2,700	
	4680	2240 CFM, 1 HP			3	6.670		2,390	155		2,545	2,875	
	5000	Utility set, centrifugal, V belt drive, motor											
	5020	¼" S.P., 1900 CFM, ¼ HP		Q-20	6	3.330	Ea.	3,060	77		3,137	3,475	
	5040	2170 CFM, ⅓ HP			5	4		3,090	92		3,182	3,550	
	5060	2680 CFM, ½ HP			4	5		3,090	115		3,205	3,575	
	5080	3020 CFM, ¾ HP			3	6.670		3,120	155		3,275	3,675	
	7100	Direct drive, 420 CFM, 8" sq. damper			6	3.330		415	77		492	575	
	7120	675 CFM, 12" sq. damper			6	3.330		585	77		662	765	
	8000	Ventilation, residential											
	8020	Attic, roof type											
	8030	Aluminum dome, damper & curb											
	8040	6" diameter, 300 CFM		1 Elec	16	.500	Ea.	143	12.60		155.60	175	
	8050	7" diameter, 450 CFM			15	.533		160	13.40		173.40	195	
	8060	9" diameter, 900 CFM			14	.571		285	14.35		299.35	335	
	8080	12" diameter, 1000 CFM (gravity)			10	.800		200	20		220	250	
	8090	16" diameter, 1500 CFM (gravity)			9	.889		255	22		277	315	
	8100	20" diameter, 2500 CFM (gravity)			8	1		320	25		345	390	
	8110	26" diameter, 4000 CFM (gravity)			7	1.140		390	29		419	475	
	8120	32" diameter, 6500 CFM (gravity)			6	1.330		545	34		579	650	
	8130	38" diameter, 8000 CFM (gravity)			5	1.600		770	40		810	910	
	8140	50" diameter, 13,000 CFM (gravity)			4	2		1,220	50		1,270	1,425	
	8160	Plastic, ABS dome											
	8180	1050 CFM		1 Elec	14	.571	Ea.	67	14.35		81.35	96	
	8200	1600 CFM		"	12	.667	"	101	16.75		117.75	135	
	8240	Attic, wall type, with shutter, one speed											
	8250	12" diameter, 1000 CFM		1 Elec	14	.571	Ea.	115	14.35		129.35	150	
	8260	14" diameter, 1500 CFM			12	.667		135	16.75		151.75	175	
	8270	16" diameter, 2000 CFM			9	.889		166	22		188	215	
	8290	Whole house, wall type, with shutter, one speed											
	8300	30" diameter, 4800 CFM		1 Elec	7	1.140	Ea.	290	29		319	365	
	8310	36" diameter, 7000 CFM			6	1.330		345	34		379	430	
	8320	42" diameter, 10,000 CFM			5	1.600		410	40		450	510	
	8330	48" diameter, 16,000 CFM			4	2		495	50		545	620	
	8340	For two speed, add						11			11	12.10	
	8350	Whole house, lay-down type, with shutter, one speed											
	8360	30" diameter, 4500 CFM		1 Elec	8	1	Ea.	303	25		328	370	
	8370	36" diameter, 6500 CFM			7	1.140		355	29		384	435	
	8380	42" diameter, 9000 CFM			6	1.330		420	34		454	515	
	8390	48" diameter, 12,000 CFM			5	1.600		500	40		540	610	
	8440	For two speed, add						11			11	12.10	
	8450	For 12 hour timer switch, add		1 Elec	32	.250		16.50	6.30		22.80	28	
	9000	Minimum labor/equipment charge		"	4	2	Job		50		50	77	

		157 400	Accessories										
401	0010	AIR FILTERS											401
	0020												
	0050	Activated charcoal type, full flow					MCFM	600			600	660	
	2000	Electronic air cleaner, self-contained											
	2150	500 CFM		1 Shee	2.30	3.480	Ea.	585	87		672	780	
	2200	1000 CFM		"	2.20	3.640	"	810	91		901	1,025	

For expanded coverage of these items see *Means Mechanical Cost Data* or *Means Plumbing Cost Data 1991*

157 400 | Accessories

			CREW	DAILY OUTPUT	MAN-HOURS	UNIT	BARE COSTS MAT.	LABOR	EQUIP.	TOTAL	TOTAL INCL O&P	
401	2250	1200 CFM	1 Shee	2.10	3.810	Ea.	1,190	95		1,285	1,450	401
	2300	2500 CFM	"	2	4	"	1,360	100		1,460	1,650	
	2950	Mechanical media filtration units										
	3000	High efficiency type, with frame, non-supported				MCFM	35			35	39	
	3100	Supported type					47			47	52	
	4000	Medium efficiency, extended surface					5			5	5.50	
	4500	Permanent washable					33			33	36	
	5000	Renewable disposable roll					85			85	94	
	5500	Throwaway glass or paper media type				Ea.	2.65			2.65	2.92	
	9000	Minimum labor/equipment charge	1 Shee	2.25	3.560	Job		89		89	140	
420	0010	**CONTROL COMPONENTS**										420
	5000	Thermostats										
	5030	1 set back, manual	1 Shee	8	1	Ea.	16.30	25		41.30	57	
	5040	1 set back, electric, timed		8	1		59.45	25		84.45	105	
	5050	2 set back, electric, timed		8	1		59.45	25		84.45	105	
	5200	24 hour, automatic, clock		8	1		86.90	25		111.90	135	
	6000	Valves, motorized zone										
	6100	Sweat connections, ½" C x C	1 Stpi	20	.400	Ea.	34.10	10.20		44.30	53	
	6110	¾" C x C		20	.400		34.10	10.20		44.30	53	
	6120	1" C x C		19	.421		38.15	10.75		48.90	59	
	9000	Minimum labor/equipment charge	1 Plum	4	2	Job		51		51	78	
430	0010	**CONTROL SYSTEMS, ELECTRONIC**										430
	1000	Electronic hydronic controller	1 Plum	8	1	Ea.	36.40	25		61.40	79	
	9000	Minimum labor/equipment charge	"	8	1	Job		25		25	39	
440	0010	**CURBS/PADS PREFABRICATED**										440
	6000	Pad, fiberglass reinforced concrete with polystyrene foam core										
	6050	Condenser, 2" thick, 16" x 36"	1 Shee	8	1	Ea.	8.80	25		33.80	49	
	6090	24" x 36"	"	12	.667		12.60	16.65		29.25	40	
	6280	36" x 36"	Q-9	8	2		19.60	45		64.60	92	
	6300	36" x 48"		7	2.290		26.40	51		77.40	110	
	6320	36" x 60"		7	2.290		33.50	51		84.50	120	
450	0010	**DIFFUSERS** Aluminum, opposed blade damper unless noted										450
	0100	Ceiling, linear, also for sidewall										
	0120	2" wide	1 Shee	32	.250	L.F.	18.35	6.25		24.60	30	
	0160	4" wide		26	.308		24.50	7.70		32.20	39	
	0180	6" wide		24	.333		30.30	8.35		38.65	46	
	0200	8" wide		22	.364		35.40	9.10		44.50	53	
	0500	Perforated, 24" x 24", panel size 6" x 6"		16	.500	Ea.	50	12.50		62.50	75	
	0520	8" x 8"		15	.533		51.70	13.35		65.05	78	
	0540	10" x 10"		14	.571		52.80	14.30		67.10	80	
	0560	12" x 12"		12	.667		53.90	16.65		70.55	85	
	0600	18" x 18"		10	.800		70.40	20		90.40	110	
	1000	Rectangular, 1 to 4 way blow, 6" x 6"		16	.500		29.70	12.50		42.20	52	
	1020	12" x 6"		15	.533		37.90	13.35		51.25	63	
	1040	12" x 9"		14	.571		45.60	14.30		59.90	73	
	1060	12" x 12"		12	.667		52.25	16.65		68.90	84	
	1100	24" x 12"		10	.800		89.10	20		109.10	130	
	2000	T bar mounting, 24" x 24" lay-in frame, 6" x 6"		16	.500		47.85	12.50		60.35	72	
	2020	9" x 9"		14	.571		56.10	14.30		70.40	84	
	2040	12" x 12"		12	.667		71.50	16.65		88.15	105	
	2060	15" x 15"		11	.727		92.40	18.20		110.60	130	
	2080	18" x 18"		10	.800		93.40	20		113.40	135	
	6000	For steel diffusers instead of aluminum, deduct					10%					
	9000	Minimum labor/equipment charge	1 Shee	4	2	Job		50		50	78	

For expanded coverage of these items see *Means Mechanical Cost Data* or *Means Plumbing Cost Data 1991*

157 400 | Accessories

			CREW	DAILY OUTPUT	MAN-HOURS	UNIT	BARE COSTS				TOTAL INCL O&P
							MAT.	LABOR	EQUIP.	TOTAL	
460	0010	**GRILLES**									**460**
	0020	Aluminum									
	1000	Air return, 6″ x 6″	1 Shee	26	.308	Ea.	7.15	7.70		14.85	19.95
	1020	10″ x 6″		24	.333		8.25	8.35		16.60	22
	1080	16″ x 8″		22	.364		12.10	9.10		21.20	28
	1100	12″ x 12″		22	.364		12.10	9.10		21.20	28
	1120	24″ x 12″		18	.444		22	11.10		33.10	42
	1300	48″ x 24″		12	.667		83.60	16.65		100.25	120
	9000	Minimum labor/equipment charge		4	2	Job		50		50	78
470	0010	**REGISTERS**									**470**
	0020										
	0980	Air supply									
	1000	Ceiling/wall, O.B. damper, anodized aluminum									
	1010	One or two way deflection, adj. curved face bars									
	1140	14″ x 8″	1 Shee	17	.471	Ea.	20.80	11.75		32.55	41
	3000	Baseboard, hand adj. damper, enameled steel									
	3020	10″ x 6″	1 Shee	24	.333	Ea.	6.55	8.35		14.90	20
	3040	12″ x 5″		22	.364		7.75	9.10		16.85	23
	3060	12″ x 6″		23	.348		7.15	8.70		15.85	22
	3080	12″ x 8″		22	.364		10.35	9.10		19.45	26
	3100	14″ x 6″		20	.400		7.75	10		17.75	24
	9000	Minimum labor/equipment charge		4	2	Job		50		50	78
480	0010	**DUCT ACCESSORIES**									**480**
	0020										
	0050	Air extractors, 12″ x 4″	1 Shee	24	.333	Ea.	7.70	8.35		16.05	22
	0100	8″ x 6″		22	.364		5.50	9.10		14.60	20
	0200	20″ x 8″		16	.500		15.95	12.50		28.45	37
	0240	18″ x 10″		14	.571		15.95	14.30		30.25	40
	0280	24″ x 12″		10	.800		22	20		42	56
	3000	Fire damper, curtain type, vertical, 8″ x 4″		24	.333		15.40	8.35		23.75	30
	3020	12″ x 4″		22	.364		15.40	9.10		24.50	31
	3240	16″ x 14″		18	.444		24.75	11.10		35.85	45
	3400	24″ x 20″		8	1		31.35	25		56.35	74
	5180	Mixing box, includes electric or pneumatic motor									
	5190	Recommend use with attenuator, see 157-480-9000									
	5200	Constant volume, 150 to 270 CFM	Q-9	12	1.330	Ea.	420	30		450	510
	5210	270 to 600 CFM	"	11	1.450		486	33		519	585
	6000	Multi-blade dampers, opposed blade, 12″ x 12″	1 Shee	21	.381		8	9.50		17.50	24
	6020	12″ x 18″		18	.444		11.10	11.10		22.20	30
	6080	24″ x 24″		8	1		26.70	25		51.70	69
	6180	48″ x 36″	Q-9	7	2.290		80	51		131	170
	7500	Variable volume modulating motorized damper									
	7520	12″ x 12″	1 Shee	12	.667	Ea.	107	16.65		123.65	145
	7560	24″ x 12″		8	1		128	25		153	180
	7600	30″ x 18″		4	2		161	50		211	255
	7700	For thermostat, add		8	1		42.70	25		67.70	86
	8000	Volume control, dampers									
	8100	8″ x 8″	1 Shee	24	.333	Ea.	17.10	8.35		25.45	32
	8140	16″ x 10″		20	.400		23.85	10		33.85	42
	8160	18″ x 12″		18	.444		29.60	11.10		40.70	50
	8220	28″ x 16″		10	.800		53.40	20		73.40	90
	9000	Silencers, noise control for air flow, duct				MCFM	40			40	44
	9200	Plenums, measured by panel surface				S.F.	8			8	8.80
	9900	Minimum labor/equipment charge	1 Shee	4	2	Job		50		50	78
490	0010	**VENTILATORS** Base, damper & bird screen, CFM in 5 MPH wind									**490**
	0520	8″ neck diameter, 215 CFM	Q-9	14	1.140	Ea.	41.85	26		67.85	86

For expanded coverage of these items see *Means Mechanical Cost Data* or *Means Plumbing Cost Data 1991*

157 400	Accessories	CREW	DAILY OUTPUT	MAN-HOURS	UNIT	MAT.	LABOR	EQUIP.	TOTAL	TOTAL INCL O&P	
							BARE COSTS				
490	9000 Minimum labor/equipment charge	1 Plum	2	4	Job		100		100	155	490

160 | Raceways

160 100	Cable Trays	CREW	DAILY OUTPUT	MAN-HOURS	UNIT	MAT.	LABOR	EQUIP.	TOTAL	TOTAL INCL O&P	
							BARE COSTS				
150	0010 **WIREWAY to 15' high**										150
	0100 Screw cover with fittings and supports, 2-½" x 2-½"	1 Elec	45	.178	L.F.	5.70	4.47		10.17	13.10	
	0200 4" x 4"		40	.200		6.45	5.05		11.50	14.80	
	0400 6" x 6"		30	.267		10.65	6.70		17.35	22	
	0600 8" x 8"		20	.400		14.30	10.05		24.35	31	

160 200	Conduits	CREW	DAILY OUTPUT	MAN-HOURS	UNIT	MAT.	LABOR	EQUIP.	TOTAL	TOTAL INCL O&P	
205	0010 **CONDUIT** To 15' high, includes 2 terminations, 2 elbows and										205
	0020 11 beam clamps per 100 L.F.										
	2500 Steel, intermediate conduit (IMC), ½" diameter	1 Elec	100	.080	L.F.	.87	2.01		2.88	4.03	
	2530 ¾" diameter		90	.089		1.08	2.24		3.32	4.60	
	2550 1" diameter		70	.114		1.40	2.87		4.27	5.95	
	2570 1-¼" diameter		65	.123		1.75	3.10		4.85	6.65	
	2600 1-½" diameter		60	.133		2.30	3.35		5.65	7.65	
	2630 2" diameter		50	.160		2.85	4.02		6.87	9.30	
	2650 2-½" diameter		40	.200		4.60	5.05		9.65	12.75	
	2670 3" diameter		30	.267		6.30	6.70		13	17.15	
	2700 3-½" diameter		27	.296		8.70	7.45		16.15	21	
	2730 4" diameter		25	.320		10.25	8.05		18.30	24	
	5000 Electric metallic tubing (EMT), ½" diameter		170	.047		.33	1.18		1.51	2.17	
	5020 ¾" diameter		130	.062		.47	1.55		2.02	2.88	
	5040 1" diameter		115	.070		.70	1.75		2.45	3.44	
	5060 1-¼" diameter		100	.080		1	2.01		3.01	4.17	
	5080 1-½" diameter		90	.089		1.15	2.24		3.39	4.68	
	5100 2" diameter		80	.100		1.50	2.52		4.02	5.50	
	5120 2-½" diameter		60	.133		3.30	3.35		6.65	8.75	
	5140 3" diameter		50	.160		4.10	4.02		8.12	10.65	
	5160 3-½" diameter		45	.178		5.70	4.47		10.17	13.10	
	5180 4" diameter		40	.200		6.75	5.05		11.80	15.10	
	5200 Field bends, 45° to 90°, ½" diameter		89	.090	Ea.		2.26		2.26	3.45	
	5220 ¾" diameter		80	.100			2.52		2.52	3.84	
	5240 1" diameter		73	.110			2.76		2.76	4.21	
	5260 1-¼" diameter		38	.211			5.30		5.30	8.10	
	5280 1-½" diameter		36	.222			5.60		5.60	8.55	
	5300 2" diameter		26	.308			7.75		7.75	11.80	
	5320 Offsets, ½" diameter		65	.123			3.10		3.10	4.73	
	5340 ¾" diameter		62	.129			3.25		3.25	4.95	
	5360 1" diameter		53	.151			3.80		3.80	5.80	
	5380 1-¼" diameter		30	.267			6.70		6.70	10.25	
	5400 1-½" diameter		28	.286			7.20		7.20	10.95	
	7600 EMT, "T" fittings with covers, ½" diameter, set screw		16	.500		5.50	12.60		18.10	25	
	9000 Minimum labor/equipment charge		4	2	Job		50		50	77	
230	0010 **CONDUIT IN CONCRETE SLAB** Including terminations,										230
	0020 fittings and supports										

For expanded coverage of these items see *Means Electrical Cost Data 1991*

		160 200 | Conduits	CREW	DAILY OUTPUT	MAN-HOURS	UNIT	BARE COSTS				TOTAL INCL O&P	
							MAT.	LABOR	EQUIP.	TOTAL		
230	3230	PVC, schedule 40, ½" diameter	1 Elec	270	.030	L.F.	.34	.75		1.09	1.51	**230**
	3250	¾" diameter		230	.035		.35	.87		1.22	1.72	
	3270	1" diameter		200	.040		.48	1.01		1.49	2.06	
	3300	1-¼" diameter		170	.047		.64	1.18		1.82	2.51	
	3330	1-½" diameter		140	.057		.81	1.44		2.25	3.09	
	3350	2" diameter		120	.067		1.07	1.68		2.75	3.74	
	4350	Rigid galvanized steel, ½" diameter		200	.040		1.01	1.01		2.02	2.65	
	4400	¾" diameter		170	.047		1.26	1.18		2.44	3.19	
	4450	1" diameter		130	.062		1.72	1.55		3.27	4.26	
	4500	1-¼" diameter		110	.073		2.23	1.83		4.06	5.25	
	4600	1-½" diameter		100	.080		2.75	2.01		4.76	6.10	
	4800	2" diameter		90	.089		3.55	2.24		5.79	7.30	
	9000	Minimum labor/equipment charge		4	2	Job		50		50	77	
240	0010	**CONDUIT IN TRENCH** Includes terminations and fittings										**240**
	0020	Does not include excavation or backfill, see 022-200										
	0200	Rigid galvanized steel, 2" diameter	1 Elec	151	.053	L.F.	3.40	1.33		4.73	5.75	
	0400	2-½" diameter		100	.080		5.60	2.01		7.61	9.25	
	0600	3" diameter		80	.100		7.10	2.52		9.62	11.65	
	0800	3-½" diameter		70	.114		9.30	2.87		12.17	14.60	
	1000	4" diameter		50	.160		11.50	4.02		15.52	18.80	
	1200	5" diameter		40	.200		23.20	5.05		28.25	33	
	1400	6" diameter		30	.267		33	6.70		39.70	47	
	9000	Minimum labor/equipment charge		4	2	Job		50		50	77	
260	0010	**CUTTING AND DRILLING**										**260**
	0100	Hole drilling to 10' high, concrete wall										
	0110	8" thick, ½" pipe size	1 Elec	12	.667	Ea.		16.75		16.75	26	
	0120	¾" pipe size		12	.667			16.75		16.75	26	
	0130	1" pipe size		9.50	.842			21		21	32	
	0140	1-¼" pipe size		9.50	.842			21		21	32	
	0150	1-½" pipe size		9.50	.842			21		21	32	
	0160	2" pipe size		4.40	1.820			46		46	70	
	0170	2-½" pipe size		4.40	1.820			46		46	70	
	0180	3" pipe size		4.40	1.820			46		46	70	
	0190	3-½" pipe size		3.30	2.420			61		61	93	
	0200	4" pipe size		3.30	2.420			61		61	93	
	0500	12" thick, ½" pipe size		9.40	.851			21		21	33	
	0520	¾" pipe size		9.40	.851			21		21	33	
	0540	1" pipe size		7.30	1.100			28		28	42	
	0560	1-¼" pipe size		7.30	1.100			28		28	42	
	0570	1-½" pipe size		7.30	1.100			28		28	42	
	0580	2" pipe size		3.60	2.220			56		56	85	
	0590	2-½" pipe size		3.60	2.220			56		56	85	
	0600	3" pipe size		3.60	2.220			56		56	85	
	0610	3-½" pipe size		2.80	2.860			72		72	110	
	0630	4" pipe size		2.50	3.200			80		80	125	
	0650	16" thick ½" pipe size		7.60	1.050			26		26	40	
	0670	¾" pipe size		7	1.140			29		29	44	
	0690	1" pipe size		6	1.330			34		34	51	
	0710	1-¼" pipe size		5.50	1.450			37		37	56	
	0730	1-½" pipe size		5.50	1.450			37		37	56	
	0750	2" pipe size		3	2.670			67		67	100	
	0770	2-½" pipe size		2.70	2.960			75		75	115	
	0790	3" pipe size		2.50	3.200			80		80	125	
	0810	3-½" pipe size		2.30	3.480			87		87	135	
	0830	4" pipe size		2	4			100		100	155	
	0850	20" thick ½" pipe size		6.40	1.250			31		31	48	
	0870	¾" pipe size		6	1.330			34		34	51	

For expanded coverage of these items see *Means Electrical Cost Data 1991*

			CREW	DAILY OUTPUT	MAN-HOURS	UNIT	MAT.	BARE COSTS LABOR	EQUIP.	TOTAL	TOTAL INCL O&P	
260		**160 200 | Conduits**										260
	0890	1" pipe size	1 Elec	5	1.600	Ea.		40		40	61	
	0910	1-¼" pipe size		4.80	1.670			42		42	64	
	0930	1-½" pipe size		4.60	1.740			44		44	67	
	0950	2" pipe size		2.70	2.960			75		75	115	
	0970	2-½" pipe size		2.40	3.330			84		84	130	
	0990	3" pipe size		2.20	3.640			91		91	140	
	1010	3-½" pipe size		2	4			100		100	155	
	1030	4" pipe size		1.70	4.710			120		120	180	
	1050	24" thick ½" pipe size		5.50	1.450			37		37	56	
	1070	¾" pipe size		5.10	1.570			39		39	60	
	1090	1" pipe size		4.30	1.860			47		47	71	
	1110	1-¼" pipe size		4	2			50		50	77	
	1130	1-½" pipe size		4	2			50		50	77	
	1150	2" pipe size		2.40	3.330			84		84	130	
	1170	2-½" pipe size		2.20	3.640			91		91	140	
	1190	3" pipe size		2.20	3.640			91		91	140	
	1210	3-½" pipe size		1.80	4.440			110		110	170	
	1230	4" pipe size		1.50	5.330			135		135	205	
	1500	Brick wall, 8" thick, ½" pipe size		18	.444			11.20		11.20	17.05	
	1520	¾" pipe size		18	.444			11.20		11.20	17.05	
	1540	1" pipe size		13.30	.602			15.15		15.15	23	
	1560	1-¼" pipe size		13.30	.602			15.15		15.15	23	
	1580	1-½" pipe size		13.30	.602			15.15		15.15	23	
	1600	2" pipe size		5.70	1.400			35		35	54	
	1620	2-½" pipe size		5.70	1.400			35		35	54	
	1640	3" pipe size		5.70	1.400			35		35	54	
	1660	3-½" pipe size		4.40	1.820			46		46	70	
	1680	4" pipe size		4	2			50		50	77	
	1700	12" thick, ½" pipe size		14.50	.552			13.90		13.90	21	
	1720	¾" pipe size		14.50	.552			13.90		13.90	21	
	1740	1" pipe size		11	.727			18.30		18.30	28	
	1760	1-¼" pipe size		11	.727			18.30		18.30	28	
	1780	1-½" pipe size		11	.727			18.30		18.30	28	
	1800	2" pipe size		5	1.600			40		40	61	
	1820	2-½" pipe size		5	1.600			40		40	61	
	1840	3" pipe size		5	1.600			40		40	61	
	1860	3-½" pipe size		3.80	2.110			53		53	81	
	1880	4" pipe size		3.30	2.420			61		61	93	
	1900	16" thick, ½" pipe size		12.30	.650			16.35		16.35	25	
	1920	¾" pipe size		12.30	.650			16.35		16.35	25	
	1940	1" pipe size		9.30	.860			22		22	33	
	1960	1-¼" pipe size		9.30	.860			22		22	33	
	1980	1-½" pipe size		9.30	.860			22		22	33	
	2000	2" pipe size		4.40	1.820			46		46	70	
	2010	2-½" pipe size		4.40	1.820			46		46	70	
	2030	3" pipe size		4.40	1.820			46		46	70	
	2050	3-½" pipe size		3.30	2.420			61		61	93	
	2070	4" pipe size		3	2.670			67		67	100	
	2090	20" thick, ½" pipe size		10.70	.748			18.80		18.80	29	
	2110	¾" pipe size		10.70	.748			18.80		18.80	29	
	2130	1" pipe size		8	1			25		25	38	
	2150	1-¼" pipe size		8	1			25		25	38	
	2170	1-½" pipe size		8	1			25		25	38	
	2190	2" pipe size		4	2			50		50	77	
	2210	2-½" pipe size		4	2			50		50	77	
	2230	3" pipe size		4	2			50		50	77	
	2250	3-½" pipe size		3	2.670			67		67	1	
	2270	4" pipe size		2.70	2.960			75		75		

For expanded coverage of these items see *Means Electrical Cost Data 1991*

160 200	Conduits	CREW	DAILY OUTPUT	MAN-HOURS	UNIT	BARE COSTS				TOTAL INCL O&P	
						MAT.	LABOR	EQUIP.	TOTAL		
260	2290	24" thick, ½" pipe size	1 Elec	9.40	.851	Ea.		21		21	33	260
	2310	¾" pipe size		9.40	.851			21		21	33	
	2330	1" pipe size		7.10	1.130			28		28	43	
	2350	1-¼" pipe size		7.10	1.130			28		28	43	
	2370	1-½" pipe size		7.10	1.130			28		28	43	
	2390	2" pipe size		3.60	2.220			56		56	85	
	2410	2-½" pipe size		3.60	2.220			56		56	85	
	2430	3" pipe size		3.60	2.220			56		56	85	
	2450	3-½" pipe size		2.80	2.860			72		72	110	
	2470	4" pipe size		2.50	3.200			80		80	125	
	2480											
	3000	Knockouts to 8' high, metal boxes & enclosures										
	3020	With hole saw, ½" pipe size	1 Elec	53	.151	Ea.		3.80		3.80	5.80	
	3040	¾" pipe size		47	.170			4.28		4.28	6.55	
	3050	1" pipe size		40	.200			5.05		5.05	7.70	
	3060	1-¼" pipe size		36	.222			5.60		5.60	8.55	
	3070	1-½" pipe size		32	.250			6.30		6.30	9.60	
	3080	2" pipe size		27	.296			7.45		7.45	11.40	
	3090	2-½" pipe size		20	.400			10.05		10.05	15.35	
	4010	3" pipe size		16	.500			12.60		12.60	19.20	
	4030	3-½" pipe size		13	.615			15.50		15.50	24	
	4050	4" pipe size		11	.727			18.30		18.30	28	
	4070	With hand punch set, ½" pipe size		40	.200			5.05		5.05	7.70	
	4090	¾" pipe size		32	.250			6.30		6.30	9.60	
	4110	1" pipe size		30	.267			6.70		6.70	10.25	
	4130	1-¼" pipe size		28	.286			7.20		7.20	10.95	
	4150	1-½" pipe size		26	.308			7.75		7.75	11.80	
	4170	2" pipe size		20	.400			10.05		10.05	15.35	
	4190	2-½" pipe size		17	.471			11.85		11.85	18.05	
	4200	3" pipe size		15	.533			13.40		13.40	20	
	4220	3-½" pipe size		12	.667			16.75		16.75	26	
	4240	4" pipe size		10	.800			20		20	31	
	4260	With hydraulic punch, ½" pipe size		44	.182			4.57		4.57	7	
	4280	¾" pipe size		38	.211			5.30		5.30	8.10	
	4300	1" pipe size		38	.211			5.30		5.30	8.10	
	4320	1-¼" pipe size		38	.211			5.30		5.30	8.10	
	4340	1-½" pipe size		38	.211			5.30		5.30	8.10	
	4360	2" pipe size		32	.250			6.30		6.30	9.60	
	4380	2-½" pipe size		27	.296			7.45		7.45	11.40	
	4400	3" pipe size		23	.348			8.75		8.75	13.35	
	4420	3-½" pipe size		20	.400			10.05		10.05	15.35	
275	0010	**MOTOR CONNECTIONS**										275
	0020	Flexible conduit and fittings, up to 1 HP motor, 115 volt, 1 phase	1 Elec	8	1	Ea.	2.95	25		27.95	42	
	9000	Minimum labor/equipment charge	"	4	2	Job		50		50	77	
290	0010	**WIREMOLD RACEWAY**										290
	0100	No. 500	1 Elec	100	.080	L.F.	.46	2.01		2.47	3.58	
	0400	No. 1500, small pancake		90	.089		.76	2.24		3	4.25	
	0600	No. 2000, base & cover		90	.089		.80	2.24		3.04	4.29	
	0800	No. 3000, base & cover		75	.107		1.65	2.68		4.33	5.90	
	1000	No. 4000, base & cover		65	.123		3	3.10		6.10	8.05	
	1200	No. 6000, base & cover		50	.160		4.70	4.02		8.72	11.30	
	2400	Fittings, elbows, No. 500		40	.200	Ea.	.78	5.05		5.83	8.55	
	2800	Elbow cover, No. 2000		40	.200		1.60	5.05		6.65	9.45	
	3000	Switch box, No. 500		16	.500		6.25	12.60		18.85	26	
	3400	Telephone outlet, No. 1500		16	.500		5.43	12.60		18.03	25	
	3600	Junction box, No. 1500		16	.500		3.67	12.60		16.27	23	

For expanded coverage of these items see *Means Electrical Cost Data 1991*

160 | Raceways

160 200	Conduits	CREW	DAILY OUTPUT	MAN-HOURS	UNIT	BARE COSTS				TOTAL INCL O&P
						MAT.	LABOR	EQUIP.	TOTAL	
290	3800	Plugmold wired sections, No. 2000									**290**
	4000	1 circuit, 6 outlets, 3 ft. long	1 Elec	8	1	Ea.	16.50	25		41.50	57
	4100	2 circuits, 8 outlets, 6 ft. long		5.30	1.510		21.05	38		59.05	81
	4200	Tele-power poles, aluminum, 4 outlets		2.70	2.960		98.50	75		173.50	220
	9990	Minimum labor/equipment charge		5	1.600	Job		40		40	61

161 | Conductors and Grounding

161 100	Conductors	CREW	DAILY OUTPUT	MAN-HOURS	UNIT	BARE COSTS				TOTAL INCL O&P
						MAT.	LABOR	EQUIP.	TOTAL	
105	0010	**ARMORED CABLE**									**105**
	0020									
	0050	600 volt, copper (BX), #14, 2 wire	1 Elec	2.40	3.330	C.L.F.	27.40	84		111.40	160
	0100	3 wire		2	4		32.60	100		132.60	190
	0150	#12, 2 wire		2.10	3.810		31.60	96		127.60	180
	0200	3 wire		1.80	4.440		42.40	110		152.40	215
	0250	#10, 2 wire		1.80	4.440		53.25	110		163.25	230
	0300	3 wire		1.50	5.330		66.30	135		201.30	280
	0350	#8, 3 wire		1.20	6.670		110	170		280	375
	9000	Minimum labor/equipment charge		4	2	Job		50		50	77
140	0010	**MINERAL INSULATED CABLE** 600 volt									**140**
	0100	1 conductor, #12	1 Elec	1.60	5	C.L.F.	201	125		326	415
	1500	2 conductor, #12		1.40	5.710		317	145		462	570
	1600	#10		1.20	6.670		382	170		552	675
	1800	#8		1.10	7.270		477	185		662	805
	2000	#6		1.05	7.620		626	190		816	980
	2100	#4		1	8		822	200		1,022	1,200
	2200	3 conductor, #12		1.20	6.670		371	170		541	665
	2400	#10		1.10	7.270		450	185		635	775
	2600	#8		1.05	7.620		562	190		752	910
	2800	#6		1	8		729	200		929	1,100
	3000	#4		.90	8.890		978	225		1,203	1,425
	3100	4 conductor, #12		1.20	6.670		411	170		581	710
	3200	#10		1.10	7.270		488	185		673	815
	3400	#8		1	8		625	200		825	995
	3600	#6		.90	8.890		820	225		1,045	1,250
	3620	7 conductor, #12		1.10	7.270		520	185		705	850
	3640	#10		1	8		657	200		857	1,025
	3800	M.I. terminations, 600 volt, 1 conductor, #12		8	1	Ea.	6.95	25		31.95	46
	5500	2 conductor, #12		6.70	1.190		7.50	30		37.50	54
	5600	#10		6.40	1.250		10	31		41	59
	5800	#8		6.20	1.290		10	32		42	61
	6000	#6		5.70	1.400		10	35		45	65
	6200	#4		5.30	1.510		20.55	38		58.55	81
	6400	3 conductor, #12		5.70	1.400		8.30	35		43.30	63
	6500	#10		5.50	1.450		11.35	37		48.35	68
	6600	#8		5.20	1.540		11.35	39		50.35	72
	6800	#6		4.80	1.670		11.35	42		53.35	76
	7200	#4		4.60	1.740		21.40	44		65.40	90
	7400	4 conductor, #12		4.60	1.740		11.65	44		55.65	80
	7500	#10		4.40	1.820		11.65	46		57.65	8
	7600	#8		4.20	1.900		11.65	48		59.65	

For expanded coverage of these items see *Means Electrical Cost Data 1991*

			CREW	DAILY OUTPUT	MAN-HOURS	UNIT	BARE COSTS				TOTAL INCL O&P	
	161 100	**Conductors**					MAT.	LABOR	EQUIP.	TOTAL		
140	8400	#6	1 Elec	4	2	Ea.	23.25	50		73.25	100	**140**
	8500	7 conductor, #12		3.50	2.290		14.75	57		71.75	105	
	8600	#10	↓	3	2.670		25.80	67		92.80	130	
	8800	Crimping tool, plier type					54			54	59	
	9000	Stripping tool					80			80	88	
	9200	Hand vise				↓	27			27	30	
	9500	Minimum labor/equipment charge	1 Elec	4	2	Job		50		50	77	
145	0010	**NON-METALLIC SHEATHED CABLE** 600 volt										**145**
	0100	Copper with ground wire, (Romex)										
	0150	#14, 2 wire	1 Elec	2.50	3.200	C.L.F.	16	80		96	140	
	0200	3 wire		2.30	3.480		30	87		117	165	
	0250	#12, 2 wire		2.20	3.640		25	91		116	165	
	0300	3 wire		2	4		41	100		141	200	
	0350	#10, 2 wire		2	4		41	100		141	200	
	0400	3 wire		1.40	5.710		63	145		208	290	
	0450	#8, 3 wire		1.30	6.150		134	155		289	385	
	0500	#6, 3 wire	↓	1.20	6.670	↓	187	170		357	460	
	0550	SE type SER aluminum cable, 3 RHW and										
	0600	1 bare neutral, 3 #8 & 1 #8	1 Elec	1.50	5.330	C.L.F.	50.75	135		185.75	260	
	0650	3 #6 & 1 #6		1.30	6.150		57.50	155		212.50	300	
	0700	3 #4 & 1 #6		1.10	7.270		73.30	185		258.30	360	
	0750	3 #2 & 1 #4		1	8		99.25	200		299.25	415	
	0800	3 #1/0 & 1 #2		.90	8.890		147	225		372	505	
	0850	3 #2/0 & 1 #1		.80	10		170	250		420	570	
	0900	3 #4/0 & 1 #2/0	↓	.70	11.430	↓	219	285		504	680	
	6500	Service entrance cap for copper SEU										
	6700	150 amp	1 Elec	10	.800	Ea.	9.20	20		29.20	41	
	6800	200 amp		8	1	"	12.60	25		37.60	52	
	9000	Minimum labor/equipment charge	↓	4	2	Job		50		50	77	
155	0010	**SPECIAL WIRES & FITTINGS**										**155**
	1250	Nonshielded #22-2 conductor	1 Elec	10	.800	C.L.F.	10.40	20		30.40	42	
	9000	Minimum labor/equipment charge	"	4	2	Job		50		50	77	
165	0010	**WIRE**										**165**
	0020	600 volt, type THW, copper, solid, #14	1 Elec	13	.615	C.L.F.	4.80	15.50		20.30	29	
	0030	#12		11	.727		6.70	18.30		25	35	
	0040	#10	↓	10	.800	↓	9.80	20		29.80	42	
	0051	Wire, 600 volt, stranded										
	0160	#6	1 Elec	6.50	1.230	C.L.F.	23.85	31		54.85	74	
	0180	#4		5.30	1.510		36.80	38		74.80	98	
	0200	#3		5	1.600		44.85	40		84.85	110	
	0220	#2		4.50	1.780		54.65	45		99.65	130	
	0240	#1		4	2		71.90	50		121.90	155	
	0260	1/0		3.30	2.420		84.80	61		145.80	185	
	0280	2/0		2.90	2.760		102.85	69		171.85	220	
	0300	3/0		2.50	3.200		126	80		206	260	
	0350	4/0		2.20	3.640		156	91		247	310	
	0400	250 MCM		2	4		199	100		299	375	
	0420	300 MCM		1.90	4.210		256	105		361	445	
	0450	350 MCM		1.80	4.440		270	110		380	470	
	0480	400 MCM		1.70	4.710		330	120		450	545	
	0490	500 MCM		1.60	5		372	125		497	600	
	0510	750 MCM		1.10	7.270		690	185		875	1,050	
	0530	Aluminum, stranded, #8		9	.889		7.95	22		29.95	43	
	0540	#6		8	1		9.80	25		34.80	49	
	0560	#4		6.50	1.230		13	31		44	62	
	0580	#2	↓	5.30	1.510	↓	17.45	38		55.45	77	

For expanded coverage of these items see *Means Electrical Cost Data 1991*

161 100 | Conductors

		CREW	DAILY OUTPUT	MAN-HOURS	UNIT	BARE COSTS				TOTAL INCL O&P	
						MAT.	LABOR	EQUIP.	TOTAL		
165	0600	#1	1 Elec	4.50	1.780	C.L.F.	25.30	45		70.30	96	165
	0620	1/0		4	2		28.80	50		78.80	110	
	0640	2/0		3.60	2.220		33.95	56		89.95	125	
	0680	3/0		3.30	2.420		40.70	61		101.70	140	
	0700	4/0		3.10	2.580		47.55	65		112.55	150	
	0720	250 MCM		2.90	2.760		56.45	69		125.45	170	
	0740	300 MCM		2.70	2.960		74.30	75		149.30	195	
	0760	350 MCM		2.50	3.200		80	80		160	210	
	0780	400 MCM		2.30	3.480		88	87		175	230	
	0800	500 MCM		2	4		101	100		201	265	
	0850	600 MCM		1.90	4.210		123	105		228	295	
	0880	700 MCM		1.70	4.710		143	120		263	340	
	9000	Minimum labor/equipment charge		4	2	Job		50		50	77	

161 800 | Grounding

		CREW	DAILY OUTPUT	MAN-HOURS	UNIT	BARE COSTS				TOTAL INCL O&P	
						MAT.	LABOR	EQUIP.	TOTAL		
810	0010	**GROUNDING**										810
	0030	Rod, copper clad, 8' long, ½" diameter	1 Elec	5.50	1.450	Ea.	10.15	37		47.15	67	
	0040	⅝" diameter		5.50	1.450		11.50	37		48.50	69	
	0050	¾" diameter		5.30	1.510		19	38		57	79	
	0080	10' long, ½" diameter		4.80	1.670		12.55	42		54.55	78	
	0090	⅝" diameter		4.60	1.740		15.25	44		59.25	84	
	0100	¾" diameter		4.40	1.820		24	46		70	96	
	0130	15' long, ¾" diameter		4	2		56	50		106	140	
	0260	Wire, ground, bare armored, #8-1 conductor		2	4	C.L.F.	64	100		164	225	
	0280	#4-1 conductor		1.60	5		120	125		245	325	
	0400	Bare copper, #6 wire		10	.800		23.50	20		43.50	57	
	0600	#2		5	1.600		53.50	40		93.50	120	
	0800	3/0		3.30	2.420		123.50	61		184.50	230	
	1000	4/0		2.85	2.810		153	71		224	275	
	1200	250 MCM		2.40	3.330		180	84		264	325	
	1800	Water pipe ground clamps, heavy duty										
	2000	Bronze, ½" to 1" diameter	1 Elec	8	1	Ea.	5.75	25		30.75	45	
	2100	1-¼" to 2" diameter		8	1		8.25	25		33.25	47	
	2200	2-½" to 3" diameter		6	1.330		25	34		59	79	
	2800	Brazed connections, #6 wire		12	.667		6.90	16.75		23.65	33	
	3000	#2 wire		10	.800		9.15	20		29.15	41	
	3100	3/0 wire		8	1		13.80	25		38.80	54	
	3200	4/0 wire		7	1.140		16.10	29		45.10	62	
	3400	250 MCM wire		5	1.600		18.40	40		58.40	82	
	3600	500 MCM wire		4	2		22.80	50		72.80	100	
	9000	Minimum labor/equipment charge		4	2	Job		50		50	77	

162 100 | Boxes

		CREW	DAILY OUTPUT	MAN-HOURS	UNIT	BARE COSTS				TOTAL INCL O&P	
						MAT.	LABOR	EQUIP.	TOTAL		
110	0010	**OUTLET BOXES**										110
	0020	Pressed steel, octagon, 4"	1 Elec	20	.400	Ea.	1.06	10.05		11.11	16.55	
	0100	Extension		40	.200		1.27	5.05		6.32	9.10	
	0150	Square 4"		18	.444		1.30	11.20		12.50	18.50	
	0200	Extension		40	.200		1.68	5.05		6.73	9.55	
	0250	Covers, blank		64	.125		.50	3.14		3.64		

For expanded coverage of these items see *Means Electrical Cost Data 1991*

		162 100 Boxes	CREW	DAILY OUTPUT	MAN-HOURS	UNIT	MAT.	LABOR	EQUIP.	TOTAL	TOTAL INCL O&P	
								BARE COSTS				
110	0300	Plaster rings	1 Elec	64	.125	Ea.	.77	3.14		3.91	5.65	110
	0650	Switchbox		24	.333		1.10	8.40		9.50	14	
	1100	Concrete, floor, 1 gang		4.80	1.670		42	42		84	110	
	9000	Minimum labor/equipment charge		4	2	Job		50		50	77	
120	0010	**OUTLET BOXES, PLASTIC**										120
	0050	4", round, with 2 mounting nails										
	0100	Bar hanger mounted	1 Elec	23	.348	Ea.	1.90	8.75		10.65	15.45	
	0200	Square with 2 mounting nails		23	.348		1.30	8.75		10.05	14.80	
	0300	Plaster ring		64	.125		.41	3.14		3.55	5.25	
	0400	Switch box with 2 mounting nails, 1 gang		27	.296		.68	7.45		8.13	12.15	
	0500	2 gang		23	.348		1.45	8.75		10.20	14.95	
	0600	3 gang		18	.444		2.35	11.20		13.55	19.65	
	0700	Old work box		27	.296		1.20	7.45		8.65	12.70	
	9000	Minimum labor/equipment charge		4	2	Job		50		50	77	
130	0010	**PULL BOXES & CABINETS**										130
	0100	Sheet metal, pull box, NEMA 1, type SC, 6"W x 6"H x 4"D	1 Elec	8	1	Ea.	6.10	25		31.10	45	
	0200	8"W x 8"H x 4"D		8	1		8.35	25		33.35	48	
	0300	10"W x 12"H x 6"D		5.30	1.510		14.70	38		52.70	74	
	0400	16"W x 20"H x 8"D		4	2		54.60	50		104.60	135	
	0500	20"W x 24"H x 8"D		3.20	2.500		63.35	63		126.35	165	
	0600	24"W x 36"H x 8"D		2.70	2.960		90	75		165	215	
	0650	Hinged cabinets, type A, 6"W x 6"H x 4"D		8	1		6.10	25		31.10	45	
	0800	12"W x 16"H x 6"D		4	2		19.60	50		69.60	98	
	1000	20"W x 20"H x 6"D		3.60	2.220		40.10	56		96.10	130	
	1200	20"W x 20"H x 8"D		3.20	2.500		80	63		143	185	
	1400	24"W x 36"H x 8"D		2.70	2.960		140	75		215	270	
	1600	24"W x 42"H x 8"D		2	4		212	100		312	385	
	7000	Cabinets, current transformer										
	7050	Single door, 24"H x 24"W x 10"D	1 Elec	1.60	5	Ea.	76	125		201	275	
	7100	30"H x 24"W x 10"D		1.30	6.150		91	155		246	335	
	7150	36"H x 24"W x 10"D		1.10	7.270		103	185		288	395	
	7200	30"H x 30"W x 10"D		1	8		110	200		310	430	
	7250	36"H x 30"W x 10"D		.90	8.890		135	225		360	490	
	7300	36"H x 36"W x 10"D		.80	10		140	250		390	540	
	7500	Double door, 48"H x 36"W x 10"D		.60	13.330		297	335		632	840	
	7550	24"H x 24"W x 12"D		1	8		155	200		355	480	
	9990	Minimum labor/ equipment charge		2	4	Job		100		100	155	

		162 300 Wiring Devices	CREW	DAILY OUTPUT	MAN-HOURS	UNIT	MAT.	LABOR	EQUIP.	TOTAL	TOTAL INCL O&P	
320	0010	**WIRING DEVICES**										320
	0200	Toggle switch, quiet type, single pole, 15 amp	1 Elec	40	.200	Ea.	3.20	5.05		8.25	11.20	
	0500	20 amp		27	.296		4.55	7.45		12	16.40	
	0600	3 way, 15 amp		23	.348		4.95	8.75		13.70	18.80	
	0900	4 way, 15 amp		15	.533		14.10	13.40		27.50	36	
	1650	Dimmer switch, 120 volt, incandescent, 600 watt, 1 pole		16	.500		9	12.60		21.60	29	
	2200	Receptacle, duplex, 120 V grounded, 15 amp		40	.200		1.65	5.05		6.70	9.50	
	2300	20 amp		27	.296		7.65	7.45		15.10	19.80	
	2400	Dryer, 30 amp		15	.533		9.15	13.40		22.55	31	
	2500	Range, 50 amp		11	.727		10.35	18.30		28.65	39	
	2600	Wall plates, stainless steel, 1 gang		80	.100		1.50	2.52		4.02	5.50	
	2800	2 gang		53	.151		3.75	3.80		7.55	9.90	
	3200	Lampholder, keyless		26	.308		2.50	7.75		10.25	14.55	
	3400	Pullchain with receptacle		22	.364		6.40	9.15		15.55	21	
	9000	Minimum labor/equipment charge		4	2	Job		50		50	77	

163 200 | Boards

			DAILY OUTPUT	MAN-HOURS	UNIT	MAT.	LABOR	EQUIP.	TOTAL	TOTAL INCL O&P	
			CREW			BARE COSTS					
205	0010	**CIRCUIT BREAKERS (in enclosure)**									205
	0100	Enclosed (NEMA 1), 600 volt, 3 pole, 30 amp	1 Elec	3.20	2.500	Ea.	220	63		283	340	
	0200	60 amp		2.80	2.860		220	72		292	350	
	0400	100 amp		2.30	3.480		270	87		357	430	
	0600	225 amp		1.50	5.330		578	135		713	840	
	0700	400 amp		.80	10		998	250		1,248	1,475	
	9000	Minimum labor/equipment charge		4	2	Job		50		50	77	
240	0010	**METER CENTERS AND SOCKETS**										240
	0100	Sockets, single position, 4 terminal, 100 amp	1 Elec	3.20	2.500	Ea.	23.70	63		86.70	120	
	0200	150 amp		2.30	3.480		28.50	87		115.50	165	
	0300	200 amp		1.90	4.210		37	105		142	200	
	0400	20 amp		3.20	2.500		42	63		105	140	
	0500	Double position, 4 terminal, 100 amp		2.80	2.860		70.50	72		142.50	185	
	0600	150 amp		2.10	3.810		86	96		182	240	
	0700	200 amp		1.70	4.710		115	120		235	305	
	9000	Minimum labor/equipment charge		3	2.670	Job		67		67	100	
245	0010	**PANELBOARDS (Commercial use)**										245
	0050	NQOB, w/20 amp 1 pole bolt-on circuit breakers										
	0100	3 wire, 120/240 volts, 100 amp main lugs										
	0150	10 circuits	1 Elec	1	8	Ea.	295	200		495	630	
	0200	14 circuits		.88	9.090		335	230		565	720	
	0250	18 circuits		.75	10.670		390	270		660	840	
	0300	20 circuits		.65	12.310		424	310		734	940	
	0600	4 wire, 120/208 volts, 100 amp main lugs, 12 circuits		1	8		335	200		535	675	
	0650	16 circuits		.75	10.670		385	270		655	835	
	0700	20 circuits		.65	12.310		445	310		755	960	
	0750	24 circuits		.60	13.330		485	335		820	1,050	
	0800	30 circuits		.53	15.090		575	380		955	1,200	
	0850	225 amp main lugs, 32 circuits		.45	17.780		625	445		1,070	1,375	
	0900	34 circuits		.42	19.050		640	480		1,120	1,425	
	0950	36 circuits		.40	20		665	505		1,170	1,500	
	1000	42 circuits		.34	23.530		740	590		1,330	1,725	
	1200	NEHB, w/20 amp, 1 pole bolt-on circuit breakers										
	1250	4 wire, 277/480 volts, 100 amp main lugs, 12 circuits	1 Elec	.88	9.090	Ea.	660	230		890	1,075	
	1300	20 circuits		.60	13.330		955	335		1,290	1,575	
	1350	225 amp main lugs, 24 circuits		.45	17.780		1,135	445		1,580	1,925	
	1400	30 circuits		.40	20		1,350	505		1,855	2,250	
	1450	36 circuits		.36	22.220		1,580	560		2,140	2,600	
	1600	NQOB panel, w/20 amp, 1 pole, circuit breakers										
	2000	4 wire, 120/208 volts with main circuit breaker										
	2050	100 amp main, 24 circuits	1 Elec	.47	17.020	Ea.	650	430		1,080	1,375	
	2100	30 circuits		.40	20		725	505		1,230	1,575	
	2200	225 amp main, 32 circuits		.36	22.220		1,170	560		1,730	2,150	
	2250	42 circuits		.28	28.570		1,320	720		2,040	2,550	
	2300	400 amp main, 42 circuits		.24	33.330		1,835	840		2,675	3,300	
	2350	600 amp main, 42 circuits		.20	40		2,080	1,000		3,080	3,825	
	2400	NEHB, with 20 amp, 1 pole circuit breaker										
	2450	4 wire, 227/480 volts with main circuit breaker										
	2500	100 amp main, 24 circuits	1 Elec	.42	19.050	Ea.	1,300	480		1,780	2,150	
	2550	30 circuits		.38	21.050		1,530	530		2,060	2,500	
	2600	225 amp main, 30 circuits		.36	22.220		1,995	560		2,555	3,050	
	2650	42 circuits		.28	28.570		2,375	720		3,095	3,700	
	9000	Minimum labor/equipment charge		1	8	Job		200		200	305	
250	0010	**PANELBOARD & LOAD CENTER CIRCUIT BREAKERS**										250
	0050	Bolt-on, 10,000 amp I.C., 120 volt, 1 pole										

For expanded coverage of these items see *Means Electrical Cost Data 1991*

		163 200 Boards	CREW	DAILY OUTPUT	MAN-HOURS	UNIT	BARE COSTS MAT.	LABOR	EQUIP.	TOTAL	TOTAL INCL O&P	
250	0100	15 to 50 amp	1 Elec	10	.800	Ea.	7.50	20		27.50	39	250
	0200	60 amp		8	1		7.50	25		32.50	47	
	0300	70 amp	↓	8	1	↓	15.30	25		40.30	55	
	0350	240 volt, 2 pole										
	0400	15 to 50 amp	1 Elec	8	1	Ea.	16.06	25		41.06	56	
	0500	60 amp		7.50	1.070		16.06	27		43.06	59	
	0600	80 to 100 amp		5	1.600		37.75	40		77.75	105	
	0700	3 pole, bolt-on, 15 to 60 amp		6.20	1.290		48.45	32		80.45	105	
	0800	70 amp		5	1.600		63.25	40		103.25	130	
	0900	80 to 100 amp		3.60	2.220		75	56		131	170	
	1000	22,000 amp I.C., 240 volt, 2 pole, 70 to 225 amp		2.70	2.960		219	75		294	355	
	1100	3 pole, 70 to 225 amp		2.30	3.480		337	87		424	505	
	1200	14,000 amp I.C., 277 volts, 1 pole, 15 to 30 amp	↓	8	1	↓	31	25		56	73	
	1250											
	1300	22,000 tramp I.C., 480 volts, 2 pole, 70 to 225 amp	1 Elec	2.70	2.960	Ea.	372	75		447	525	
	1400	3 pole, 70 to 225 amp		2.30	3.480	"	469	87		556	650	
	9000	Minimum labor/equipment charge	↓	3	2.670	Job		67		67	100	

		163 300 Switches										
350	0010	**RELAYS** Enclosed (NEMA 1)										350
	0100	2 pole, 12 amp	1 Elec	5	1.600	Ea.	80	40		120	150	
	0200	4 pole, 10 amp	"	4.50	1.780	"	95	45		140	175	
360	0010	**SAFETY SWITCHES**										360
	0100	General duty, 240 volt, 3 pole, fused, 30 amp	1 Elec	3.20	2.500	Ea.	38	63		101	140	
	0200	60 amp		2.30	3.480		65	87		152	205	
	0300	100 amp		1.90	4.210		114	105		219	285	
	0400	200 amp		1.30	6.150		230	155		385	490	
	0500	400 amp		.90	8.890	↓	513	225		738	905	
	9990	Minimum labor/equipment charge	↓	3	2.670	Job		67		67	100	
370	0010	**TIME SWITCHES**										370
	0100	Single pole, single throw, 24 hour dial	1 Elec	4	2	Ea.	47	50		97	130	
	0200	24 hour dial with reserve power		3.60	2.220		239	56		295	350	
	0300	Astronomic dial		3.60	2.220		80	56		136	175	
	0400	Astronomic dial with reserve power		3.30	2.420		263	61		324	380	
	0500	7 day calendar dial		3.30	2.420		75	61		136	175	
	0600	7 day calendar dial with reserve power		3.20	2.500		250	63		313	370	
	0700	Photo cell 2000 watt		8	1		17	25		42	57	
	1080	Load management device, 2 loads		4	2	↓	415	50		465	535	
	1100	Load management device, 8 loads		1	8		1,150	200		1,350	1,575	
	9000	Minimum labor/equipment charge	↓	3.50	2.290	Job		57		57	88	

164 | Transformers and Bus Ducts

		164 100 Transformers	CREW	DAILY OUTPUT	MAN-HOURS	UNIT	BARE COSTS MAT.	LABOR	EQUIP.	TOTAL	TOTAL INCL O&P	
120	0010	**DRY TYPE TRANSFORMER**										120
	0050	Single phase, 240/480 volt primary 120/240 volt secondary										
	0100	1 KVA	1 Elec	2	4	Ea.	108	100		208	270	
	0300	2 KVA	"	1.60	5	"	151	125		276	360	

For expanded coverage of these items see *Means Electrical Cost Data 1991*

164 | Transformers and Bus Ducts

164 100	Transformers	CREW	DAILY OUTPUT	MAN-HOURS	UNIT	BARE COSTS				TOTAL INCL O&P
						MAT.	LABOR	EQUIP.	TOTAL	
120										**120**
0500	3 KVA	1 Elec	1.40	5.710	Ea.	195	145		340	435
0700	5 KVA		1.20	6.670		265	170		435	550
0900	7.5 KVA		1.10	7.270		371	185		556	685
1100	10 KVA	↓	.80	10		456	250		706	885
1300	15 KVA	2 Elec	1.20	13.330	↓	615	335		950	1,200
2300	3 phase, 240/480 volt primary, 120/208 volt secondary									
2310	Ventilated, 3 KVA	1 Elec	1	8	Ea.	354	200		554	695
2700	6 KVA		.80	10		408	250		658	835
2900	9 KVA	↓	.70	11.430		525	285		810	1,025
3100	15 KVA	2 Elec	1.10	14.550		775	365		1,140	1,400
3300	30 KVA		.90	17.780		1,020	445		1,465	1,800
3500	45 KVA	↓	.80	20	↓	1,225	505		1,730	2,125
9000	Minimum labor/equipment charge	1 Elec	1	8	Job		200		200	305

165 | Power Systems and Capacitors

165 100	Power Systems	CREW	DAILY OUTPUT	MAN-HOURS	UNIT	BARE COSTS				TOTAL INCL O&P
						MAT.	LABOR	EQUIP.	TOTAL	
120										**120**
0010	**GENERATOR SET**									
0020	Gas or gasoline operated, includes battery,									
0050	charger, muffler & transfer switch									
0200	3 phase, 4 wire, 277/480 volt, 7.5 KW	R-3	.83	24.100	Ea.	5,480	600	125	6,205	7,100
0300	10 KW		.71	28.170		7,470	705	145	8,320	9,450
0400	15 KW		.63	31.750		8,860	795	165	9,820	11,100
0500	30 KW	↓	.55	36.360	↓	12,730	910	190	13,830	15,600

166 | Lighting

166 100	Lighting	CREW	DAILY OUTPUT	MAN-HOURS	UNIT	BARE COSTS				TOTAL INCL O&P
						MAT.	LABOR	EQUIP.	TOTAL	
110										**110**
0010	**EXIT AND EMERGENCY LIGHTING**									
0080	Exit light, ceiling or wall mount, incandescent, single face	1 Elec	8	1	Ea.	46	25		71	89
0100	Double face	"	6.70	1.190	"	52	30		82	105
0300	Emergency light units, battery operated									
0350	Twin sealed beam light, 25 watt, 6 volt each									
0500	Lead battery operated	1 Elec	4	2	Ea.	225	50		275	325
0700	Nickel cadmium battery operated		4	2	"	375	50		425	490
9000	Minimum labor/equipment charge	↓	4	2	Job		50		50	77
115										**115**
0010	**EXTERIOR FIXTURES** With lamps									
0400	Quartz, 500 watt	1 Elec	5.30	1.510	Ea.	80	38		118	145
0800	Wall pack, mercury vapor, 175 watt		4	2		235	50		285	335
1000	250 watt		4	2		250	50		300	350
1100	Low pressure sodium, 35 watt		4	2		166	50		216	260
1150	55 watt		4	2		250	50		300	350
1160	High pressure sodium, 70 watt		4	2		281	50		331	385
1170	150 watt	↓	4	2		300	50		350	405

For expanded coverage of these items see *Means Electrical Cost Data 1991*

			DAILY	MAN-		BARE COSTS				TOTAL	
166 100	**Lighting**	CREW	OUTPUT	HOURS	UNIT	MAT.	LABOR	EQUIP.	TOTAL	INCL O&P	
115	1180	Metal Halide, 175 watt	1 Elec	4	2	Ea.	226	50		276	325	115
	1190	250 watt	"	4	2	"	316	50		366	425	
	1200	Floodlights with ballast and lamp,										
	1400	pole mounted, pole not included										
	2250	Low pressure sodium, 55 watt	1 Elec	2.70	2.960	Ea.	435	75		510	590	
	2270	90 watt		2	4	"	485	100		585	685	
	9000	Minimum labor/equipment charge	↓	3.75	2.130	Job		54		54	82	
130	0010	**INTERIOR LIGHTING FIXTURES** Including lamps, mounting										130
	0030	hardware and connections										
	0100	Fluorescent, C.W. lamps, troffer, recess mounted in grid, RS										
	0200	Acrylic lens, 1'W x 4'L, two 40 watt	1 Elec	5.70	1.400	Ea.	45	35		80	105	
	0300	2'W x 2'L, two U40 watt		5.70	1.400		58	35		93	120	
	0600	2'W x 4'L, four 40 watt	↓	4.70	1.700	↓	63	43		106	135	
	1000	Surface mounted, RS										
	1030	Acrylic lens with hinged & latched door frame										
	1100	1'W x 4'L, two 40 watt	1 Elec	7	1.140	Ea.	50	29		79	99	
	1200	2'W x 2'L, two U40 watt		7	1.140		80	29		109	130	
	1500	2'W x 4'L, four 40 watt	↓	5.30	1.510	↓	83	38		121	150	
	2100	Strip fixture										
	2200	4' long, one 40 watt RS	1 Elec	8.50	.941	Ea.	24.90	24		48.90	64	
	2300	4' long, two 40 watt RS		8	1		26	25		51	67	
	2600	8' long, one 75 watt, SL		6.70	1.190		41	30		71	91	
	2700	8' long, two 75 watt, SL	↓	6.20	1.290	↓	47	32		79	100	
	3000	Pendent mounted, industrial, white porcelain enamel										
	3030	C.W. lamps										
	3100	4' long, two 40 watt, RS	1 Elec	5.70	1.400	Ea.	60	35		95	120	
	3200	4' long, two 60 watt, HO		5	1.600		92	40		132	165	
	3300	8' long, two 75 watt, SL	↓	4.40	1.820	↓	93	46		139	170	
	4000	High bay, aluminum reflector										
	4030	Single unit, 400 watt DX lamp	1 Elec	2.30	3.480	Ea.	245	87		332	405	
	4220	Metal halide, integral ballast, ceiling, recess mounted										
	4230	prismatic glass lens, floating door										
	4240	2'W x 2'L, 250 watt	1 Elec	3.20	2.500	Ea.	280	63		343	405	
	4250	2'W x 2'L, 400 watt		2.90	2.760		320	69		389	460	
	4260	Surface mounted, 2'W x 2'L, 250 watt		2.70	2.960		260	75		335	400	
	4270	2'W x 2'L, 400 watt	↓	2.40	3.330	↓	300	84		384	460	
	4280	High bay, aluminum reflector,										
	4290	Single unit, 400 watt	1 Elec	2.30	3.480	Ea.	280	87		367	440	
	4300	Single unit, 1000 watt		2	4		520	100		620	725	
	4310	Twin unit, 400 watt		1.60	5		560	125		685	810	
	4320	Low bay, aluminum reflector, 250W DX lamp	↓	3.20	2.500	↓	350	63		413	480	
	4340	High pressure sodium integral ballast ceiling, recess mounted										
	4350	prismatic glass lens, floating door										
	4360	2'W x 2'L, 150 watt lamp	1 Elec	3.20	2.500	Ea.	320	63		383	450	
	4370	2'W x 2'L, 400 watt lamp		2.90	2.760		345	69		414	485	
	4380	Surface mounted, 2'W x 2'L, 150 watt lamp		2.70	2.960		315	75		390	460	
	4390	2'W x 2'L, 400 watt lamp	↓	2.40	3.330	↓	340	84		424	500	
	4400	High bay, aluminum reflector,										
	4410	Single unit, 400 watt lamp	1 Elec	2.30	3.480	Ea.	435	87		522	610	
	4430	Single unit, 1000 watt lamp		2	4		650	100		750	870	
	4440	Low bay, aluminum reflector, 150 watt lamp	↓	3.20	2.500	↓	330	63		393	460	
	4450	Incandescent, high hat can, round alzak reflector, prewired										
	4470	100 watt	1 Elec	8	1	Ea.	48	25		73	91	
	4480	150 watt		8	1		50	25		75	93	
	4500	300 watt	↓	6.70	1.190	↓	55	30		85	105	
	4600	Square glass lens with metal trim, prewired										
	4630	100 watt	1 Elec	6.70	1.190	Ea.	30	30		60	79	

For expanded coverage of these items see *Means Electrical Cost Data 1991*

166 100	Lighting	CREW	DAILY OUTPUT	MAN-HOURS	UNIT	MAT.	LABOR	EQUIP.	TOTAL	TOTAL INCL O&P	
130	4700	200 watt	1 Elec	6.70	1.190	Ea.	35	30		65	84	130
	6010	Vapor tight, incandescent, ceiling mounted, 200 watt		6.20	1.290		44	32		76	98	
	6100	Fluorescent, surface mounted, 2 lamps, 4'L, RS, 40 watt		3.20	2.500		74	63		137	175	
	6850	Vandalproof, surface mounted, fluorescent, two 40 watt		3.20	2.500		105	63		168	210	
	6860	Incandescent, one 150 watt		8	1		45	25		70	88	
	6900	Mirror light, fluorescent, RS, acrylic enclosure, two 40 watt		8	1		61	25		86	105	
	6910	One 40 watt		8	1		56	25		81	100	
	6920	One 20 watt		12	.667		49	16.75		65.75	80	
	7500	Ballast replacement, by weight of ballast, to 15' high										
	7520	Indoor fluorescent, less than 2 lbs.	1 Elec	10	.800	Ea.		20		20	31	
	7540	2 40W, watt reducer, 2 to 5 lbs.		9.40	.851		18	21		39	52	
	7560	2 F96 slimline, over 5 lbs.		8	1		28	25		53	69	
	7580	Vaportite ballast, less than 2 lbs.		9.40	.851			21		21	33	
	7600	2 lbs. to 5 lbs.		8.90	.899			23		23	35	
	7620	Over 5 lbs.		7.60	1.050			26		26	40	
	9000	Minimum labor/equipment charge		3	2.670	Job		67		67	100	
140	0010	**LAMPS**										140
	0080	Fluorescent, rapid start, cool white, 2' long, 20 watt	1 Elec	1	8	C	385	200		585	730	
	0100	4' long, 40 watt		.90	8.890		224	225		449	590	
	0170	4' long, 35 watt energy saver		.90	8.890		305	225		530	675	
	0600	Mercury vapor, mogul base, deluxe white, 100 watt		.30	26.670		2,436	670		3,106	3,700	
	0800	400 watt		.30	26.670		2,570	670		3,240	3,850	
	1000	Metal halide, mogul base, 175 watt		.30	26.670		4,043	670		4,713	5,475	
	1100	250 watt		.30	26.670		4,850	670		5,520	6,350	
	1350	Sodium high pressure, 70 watt		.30	26.670		4,712	670		5,382	6,200	
	1370	150 watt		.30	26.670		5,059	670		5,729	6,600	
	1500	Low pressure, 35 watt		.30	26.670		3,993	670		4,663	5,425	
	1600	90 watt		.30	26.670		5,140	670		5,810	6,675	
	1800	Incandescent, interior, A21, 100 watt		1.60	5		183	125		308	395	
	1900	A21, 150 watt		1.60	5		218	125		343	430	
	2300	R30, 75 watt		1.30	6.150		420	155		575	700	
	2500	Exterior, PAR 38, 75 watt		1.30	6.150		605	155		760	900	
	2600	PAR 38, 150 watt		1.30	6.150		580	155		735	875	
	9000	Minimum labor/equipment charge		4	2	Job		50		50	77	
145	0010	**RESIDENTIAL FIXTURES**										145
	0200	Pendant globe with shade, 150 watt	1 Elec	20	.400	Ea.	62	10.05		72.05	84	
	0400	Fluorescent, interior, surface, circline, 32 watt & 40 watt		20	.400		49	10.05		59.05	69	
	0500	2' x 2', two U 40 watt		8	1		68	25		93	115	
	0700	Shallow under cabinet, two 20 watt		16	.500		46	12.60		58.60	70	
	0900	Wall mounted, 4'L, one 40 watt, with baffle		10	.800		42	20		62	77	
	2000	Incandescent, exterior lantern, wall mounted, 60 watt		16	.500		37	12.60		49.60	60	
	2100	Post light, 150W, with 7' post		4	2		108	50		158	195	
	2500	Lamp holder, weatherproof with 150W PAR		16	.500		17	12.60		29.60	38	
	2550	With reflector and guard		12	.667		32	16.75		48.75	61	
	2600	Interior pendent, globe with shade, 150 watt		20	.400		80	10.05		90.05	105	
	9000	Minimum labor/equipment charge		4	2	Job		50		50	77	
150	0010	**TRACK LIGHTING**										150
	0080	Track, 1 circuit, 4' section	1 Elec	6.70	1.190	Ea.	35	30		65	84	
	0100	8' section		5.30	1.510		52	38		90	115	
	0300	3 circuits, 4' section		6.70	1.190		39	30		69	89	
	0400	8' section		5.30	1.510		52	38		90	115	
	9000	Minimum labor/equipment charge		3	2.670	Job		67		67	100	

For expanded coverage of these items see *Means Electrical Cost Data 1991*

167 | Electric Utilities

		167 100	Electric Utilities	CREW	DAILY OUTPUT	MAN-HOURS	UNIT	BARE COSTS MAT.	BARE COSTS LABOR	BARE COSTS EQUIP.	BARE COSTS TOTAL	TOTAL INCL O&P	
110	0010		**ELECTRIC & TELEPHONE SITEWORK** Not including excavation,										110
	0200		backfill and cast in place concrete										
	4200		Underground duct, banks ready for concrete fill, min. of 7″										
	4400		between conduits, ctr. to ctr.(for wire & cable see div. 161)										
	4600		2 @ 2″ diameter	1 Elec	120	.067	L.F.	.72	1.68		2.40	3.35	
	4800		4 @ 2″ diameter		60	.133		1.44	3.35		4.79	6.70	
	5600		4 @ 4″ diameter		40	.200		3.24	5.05		8.29	11.25	
	6200		Rigid galvanized steel, 2 @ 2″ diameter		90	.089		7.10	2.24		9.34	11.20	
	6400		4 @ 2″ diameter		45	.178		14.20	4.47		18.67	22	
	7400		4 @ 4″ diameter		17	.471		41.60	11.85		53.45	64	
	9990		Minimum labor/equipment charge		3.50	2.290	Job		57		57	88	

168 | Special Systems

		168 100	Special Systems	CREW	DAILY OUTPUT	MAN-HOURS	UNIT	BARE COSTS MAT.	BARE COSTS LABOR	BARE COSTS EQUIP.	BARE COSTS TOTAL	TOTAL INCL O&P	
120	0010		**DETECTION SYSTEMS**										120
	0020												
	0100		Burglar alarm, battery operated, mechanical trigger	1 Elec	4	2	Ea.	190	50		240	285	
	0200		Electrical trigger	"	4	2		229	50		279	330	
	0400		For outside key control, add					54			54	59	
	0600		For remote signaling circuitry, add					85			85	94	
	0800		Card reader, flush type, standard	1 Elec	2.70	2.960		643	75		718	820	
	1000		Multi-code		2.70	2.960		824	75		899	1,025	
	1200		Door switches, hinge switch		5.30	1.510		41	38		79	105	
	1400		Magnetic switch		5.30	1.510		48	38		86	110	
	1600		Exit control locks, horn alarm		4	2		237	50		287	340	
	1800		Flashing light alarm		4	2		268	50		318	370	
	2000		Indicating panels, 1 channel		2.70	2.960		252	75		327	390	
	2200		10 channel		1.60	5		865	125		990	1,150	
	2400		20 channel		1	8		1,674	200		1,874	2,150	
	2600		40 channel		.57	14.040		3,090	355		3,445	3,950	
	2800		Ultrasonic motion detector, 12 volt		2.30	3.480		159	87		246	310	
	3000		Infrared photoelectric detector		2.30	3.480		128	87		215	275	
	3200		Passive infrared detector		2.30	3.480		195	87		282	350	
	3400		Glass break alarm switch		8	1		32	25		57	74	
	3420		Switchmats, 30″ x 5′		5.30	1.510		58	38		96	120	
	3440		25′		4	2		138	50		188	230	
	3460		Police connect panel		4	2		170	50		220	265	
	3480		Telephone dialer		5.30	1.510		263	38		301	345	
	3500		Alarm bell		4	2		53	50		103	135	
	3520		Siren		4	2		100	50		150	185	
	3540		Microwave detector, 10′ to 200′		2	4		460	100		560	660	
	3560		10′ to 350′		2	4		1,326	100		1,426	1,600	
	3600		Fire, sprinkler & standpipe alarm, control panel, 4 zone		2	4		700	100		800	925	
	3800		8 zone		1	8		975	200		1,175	1,375	
	4000		12 zone		.66	12.120		1,400	305		1,705	2,000	
	4150												
	4200		Battery and rack	1 Elec	4	2	Ea.	530	50		580	660	
	4400		Automatic charger		8	1		340	25		365	410	
	4600		Signal bell		8	1		38	25		63	80	
	4800		Trouble buzzer or manual station		8	1		28	25		53	69	

For expanded coverage of these items see *Means Electrical Cost Data 1991*

			DAILY	MAN-		BARE COSTS				TOTAL	
168 100	**Special Systems**	CREW	OUTPUT	HOURS	UNIT	MAT.	LABOR	EQUIP.	TOTAL	INCL O&P	
120	5000	Detector, rate of rise	1 Elec	8	1	Ea.	26	25		51	67	**120**
	5100	Fixed temperature		8	1		21	25		46	62	
	5200	Smoke detector, ceiling type		6.20	1.290		49	32		81	105	
	5400	Duct type		3.20	2.500		197	63		260	315	
	5600	Light and horn		5.30	1.510		81	38		119	145	
	5800	Fire alarm horn		6.70	1.190		28	30		58	77	
	6000	Door holder, electro-magnetic		4	2		59	50		109	140	
	6200	Combination holder and closer		3.20	2.500		330	63		393	460	
	6400	Code transmitter		4	2		530	50		580	660	
	6600	Drill switch		8	1		66	25		91	110	
	6800	Master box		2.70	2.960		1,600	75		1,675	1,875	
	7000	Break glass station		8	1		38	25		63	80	
	7800	Remote annunciator, 8 zone lamp		1.80	4.440		188	110		298	375	
	8000	12 zone lamp		1.30	6.150		234	155		389	495	
	8200	16 zone lamp		1.10	7.270		287	185		472	595	
	8400	Standpipe or sprinkler alarm, alarm device		8	1		100	25		125	150	
	8600	Actuating device		8	1		230	25		255	290	
	9410	Minimum labor/equipment charge		4	2	Job		50		50	77	
125	0010	**DOORBELL SYSTEM** Incl. transformer, button & signal										**125**
	0020											
	1000	Door chimes, 2 notes, minimum	1 Elec	16	.500	Ea.	21	12.60		33.60	42	
	1020	Maximum		12	.667		88	16.75		104.75	120	
	1100	Tube type, 3 tube system		12	.667		72	16.75		88.75	105	
	1180	4 tube system		10	.800		175	20		195	225	
	1900	For transformer & button, minimum add		5	1.600		21	40		61	85	
	1960	Maximum, add		4.50	1.780		45	45		90	120	
	3000	For push button only, minimum		24	.333		7.80	8.40		16.20	21	
	3100	Maximum		20	.400		19	10.05		29.05	36	
	9000	Minimum labor/equipment charge		4	2	Job		50		50	77	
130	0010	**ELECTRIC HEATING**										**130**
	1300	Baseboard heaters, 2' long, 375 watt	1 Elec	8	1	Ea.	29.70	25		54.70	71	
	1400	3' long, 500 watt		8	1		37	25		62	79	
	1600	4' long, 750 watt		6.70	1.190		45.63	30		75.63	96	
	1800	5' long, 935 watt		5.70	1.400		60.77	35		95.77	120	
	2000	6' long, 1125 watt		5	1.600		67.98	40		107.98	135	
	2400	8' long, 1500 watt		4	2		90.64	50		140.64	175	
	2600	9' long, 1680 watt		3.60	2.220		103	56		159	200	
	2800	10' long, 1875 watt		3.30	2.420		103	61		164	205	
	2950	Wall heaters with fan, 120 to 277 volt										
	3000	1500 watt	1 Elec	4	2	Ea.	82	50		132	165	
	3040	2250 watt		4	2		129	50		179	220	
	3070	4000 watt		3.50	2.290		148	57		205	250	
	3600	Thermostats, integral		16	.500		23	12.60		35.60	45	
	3800	Line voltage, 1 pole		8	1		23	25		48	64	
	3810	2 pole		8	1		23	25		48	64	
	9990	Minimum labor/equipment charge		4	2	Job		50		50	77	
150	0010	**PUBLIC ADDRESS SYSTEM**										**150**
	0100	Conventional, office	1 Elec	5.33	1.500	Speaker	67	38		105	130	
	0200	Industrial		2.70	2.960	"	132	75		207	260	
	9000	Minimum labor/equipment charge		3.50	2.290	Job		57		57	88	
155	0010	**SOUND SYSTEM**										**155**
	2020	11 station capacity	1 Elec	2	4	Ea.	474	100		574	675	
	3600	House telephone, talking station		1.60	5		221	125		346	435	
	3800	Press to talk, release to listen		5.30	1.510		52	38		90	115	
	4000	System-on button					31			31	34	
	4200	Door release	1 Elec	4	2		54	50		104	135	

For expanded coverage of these items see *Means Electrical Cost Data 1991*

168 100 | Special Systems

		Description	CREW	DAILY OUTPUT	MAN-HOURS	UNIT	BARE COSTS MAT.	LABOR	EQUIP.	TOTAL	TOTAL INCL O&P	
155	4400	Combination speaker and microphone	1 Elec	8	1	Ea.	93	25		118	140	155
	4600	Termination box		3.20	2.500		29	63		92	130	
	4800	Amplifier or power supply		5.30	1.510		337	38		375	430	
	5000	Vestibule door unit		16	.500		62	12.60		74.60	87	
	5200	Strip cabinet		27	.296		118	7.45		125.45	140	
	5400	Directory		16	.500		55	12.60		67.60	80	
	9000	Minimum labor/equipment charge		3.50	2.290	Job		57		57	88	
160	0010	**T.V. SYSTEMS**										160
	5000	T.V. Antenna only, minimum	1 Elec	6	1.330	Ea.	27	34		61	81	
	5100	Maximum		4	2	"	115	50		165	205	
	9000	Minimum labor/equipment charge		3.75	2.130	Job		54		54	82	
170	0010	**RESIDENTIAL WIRING**										170
	0020	20' avg. runs and #14/2 wiring incl. unless otherwise noted										
	1000	Service & panel, includes 24' SE-AL cable, service eye, meter,										
	1010	Socket, panel board, main bkr., ground rod, 15 or 20 amp										
	1020	1-pole circuit breakers, and misc. hardware										
	1100	100 amp, with 10 branch breakers	1 Elec	1.19	6.720	Ea.	270	170		440	555	
	1110	With PVC conduit and wire		.92	8.700		291	220		511	655	
	1120	With RGS conduit and wire		.73	10.960		384	275		659	845	
	1150	150 amp, with 14 branch breakers		1.03	7.770		447	195		642	790	
	1170	With PVC conduit and wire		.82	9.760		478	245		723	900	
	1180	With RGS conduit and wire		.67	11.940		587	300		887	1,100	
	1200	200 amp, with 18 branch breakers		.90	8.890		551	225		776	945	
	1220	With PVC conduit and wire		.73	10.960		587	275		862	1,075	
	1230	With RGS conduit and wire		.62	12.900		785	325		1,110	1,350	
	1800	Lightning surge suppressor for above services, add		32	.250		30	6.30		36.30	43	
	2000	Switch devices										
	2100	Single pole, 15 amp, Ivory, with a 1-gang box, cover plate,										
	2110	Type NM (Romex) cable	1 Elec	17.10	.468	Ea.	5.72	11.75		17.47	24	
	2120	Type MC (BX) cable		14.30	.559		10.82	14.05		24.87	33	
	2130	EMT & wire		5.71	1.400		13.78	35		48.78	69	
	2150	3-way, #14/3, type NM cable		14.55	.550		9.46	13.85		23.31	32	
	2170	Type MC cable		12.31	.650		12.79	16.35		29.14	39	
	2180	EMT & wire		5	1.600		15	40		55	78	
	2200	4-way, #14/3, type NM cable		14.55	.550		18.56	13.85		32.41	42	
	2220	Type MC cable		12.31	.650		21.79	16.35		38.14	49	
	2230	EMT & wire		5	1.600		23.40	40		63.40	87	
	2250	S.P., 20 amp, #12/2, type NM cable		13.33	.600		10.60	15.10		25.70	35	
	2270	Type MC cable		11.43	.700		14.87	17.60		32.47	43	
	2280	EMT & wire		4.85	1.650		17.78	41		58.78	83	
	2300	S.P. rotary dimmer, 600W, type NM cable		14.55	.550		13.15	13.85		27	36	
	2320	Type MC cable		12.31	.650		18.35	16.35		34.70	45	
	2330	EMT & wire		5	1.600		21.26	40		61.26	85	
	2350	3-way rotary dimmer, type NM cable		13.33	.600		19.34	15.10		34.44	44	
	2370	Type MC cable		11.43	.700		23.20	17.60		40.80	52	
	2380	EMT & wire		4.85	1.650		25.42	41		66.42	91	
	2400	Interval timer wall switch, 20 amp, 1-30 min., #12/2			.550							
	2410	Type NM cable	1 Elec	14.55	.550	Ea.	19.34	13.85		33.19	42	
	2420	Type MC cable		12.31	.650		24.28	16.35		40.63	52	
	2430	EMT & wire		5	1.600		27.56	40		67.56	92	
	2500	Decorator style										
	2510	S.P., 15 amp, type NM cable	1 Elec	17.10	.468	Ea.	8.11	11.75		19.86	27	
	2520	Type MC cable		14.30	.559		13.47	14.05		27.52	36	
	2530	EMT & wire		5.71	1.400		16.43	35		51.43	72	
	2550	3-way, #14/3, type NM cable		14.55	.550		11.28	13.85		25.13	34	
	2570	Type MC cable		12.31	.650		16.06	16.35		32.41	43	
	2580	EMT & wire		5	1.600		18.15	40		58.15	81	

For expanded coverage of these items see *Means Electrical Cost Data 1991*

168 100	Special Systems	CREW	DAILY OUTPUT	MAN-HOURS	UNIT	BARE COSTS				TOTAL INCL O&P
						MAT.	LABOR	EQUIP.	TOTAL	
170	2600	4-way, #14/3, type NM cable	1 Elec	14.55	.550	Ea.	20.80	13.85		34.65	44	170
2620	Type MC cable		12.31	.650		24.54	16.35		40.89	52
2630	EMT & wire		5	1.600		26.73	40		66.73	91
2650	S.P., 20 amp, #12/2, type NM cable		13.33	.600		14.92	15.10		30.02	39
2670	Type MC cable		11.43	.700		19	17.60		36.60	48
2680	EMT & wire		4.85	1.650		22.15	41		63.15	88
2700	S.P., slide dimmer, type NM cable		17.10	.468		18.20	11.75		29.95	38
2720	Type MC cable		14.30	.559		23.60	14.05		37.65	47
2730	EMT & wire		5.71	1.400		27	35		62	84
2770	Type MC cable		14.30	.559		30	14.05		44.05	54
2780	EMT & wire		5.71	1.400		33.28	35		68.28	90
2800	3-way touch dimmer, type NM cable		13.33	.600		35.36	15.10		50.46	62
2820	Type MC cable		11.43	.700		39.52	17.60		57.12	70
2830	EMT & wire		4.85	1.650		41.60	41		82.60	110
3100	S.P. switch/15 amp recpt., Ivory, 1-gang box, plate									
3110	Type NM cable	1 Elec	11.43	.700	Ea.	11.44	17.60		29.04	39
3120	Type MC cable		10	.800		20.43	20		40.43	53
3130	EMT & wire		4.40	1.820		23.60	46		69.60	96
3150	S.P. switch/pilot light, type NM cable		11.43	.700		11.33	17.60		28.93	39
3170	Type MC cable		10	.800		16	20		36	48
3180	EMT & wire		4.43	1.810		18.93	45		63.93	90
3200	2-S.P. switches, 2-#14/2, type NM cables		10	.800		14.71	20		34.71	47
3220	Type MC cable		8.89	.900		22.46	23		45.46	59
3230	EMT & wire		4.10	1.950		21.37	49		70.37	98
3250	3-way switch/15 amp recpt., #14/3, type NM cable		10	.800		20.33	20		40.33	53
3270	Type MC cable		8.89	.900		23.60	23		46.60	61
3280	EMT & wire		4.10	1.950		26	49		75	105
3300	2-3 way switches, 2-#14/3 type NM cables		8.89	.900		26	23		49	63
3320	Type MC cable		8	1		30.16	25		55.16	72
3330	EMT & wire		4	2		29.12	50		79.12	110
3350	S.P. switch/20 amp recpt., #12/2 type NM cable		10	.800		18.72	20		38.72	51
3370	Type MC cable		8.89	.900		23.60	23		46.60	61
3380	EMT & wire		4.10	1.950		26	49		75	105
3400	Decorator style									
3410	S.P. switch/15 amp recpt., type NM cable	1 Elec	11.43	.700	Ea.	19	17.60		36.60	48
3420	Type MC cable		10	.800		20.22	20		40.22	53
3430	EMT & wire		4.40	1.820		23.50	46		69.50	96
3450	S.P. switch/pilot light, type NM cable		11.43	.700		16.22	17.60		33.82	45
3470	Type MC cable		10	.800		20.75	20		40.75	54
3480	EMT & wire		4.40	1.820		25	46		71	97
3500	2-S.P. switches, 2-#14/2 type NM cables		10	.800		18.40	20		38.40	51
3520	Type MC cable		8.89	.900		26	23		49	63
3530	EMT & wire		4.10	1.950		25	49		74	100
3550	3-way/15 amp recpt., #14/3 type NM cable		10	.800		24	20		44	57
3580	EMT & wire		4.10	1.950		29	49		78	105
3650	2-3 way switches, 2-3 #14/3 type NM cables		8.89	.900		31.20	23		54.20	69
3670	Type MC cable		8	1		35.36	25		60.36	77
3680	EMT & wire		4	2		34.32	50		84.32	115
3700	S.P. switch/20 amp recpt., #12/2 type NM cable		10	.800		21.84	20		41.84	55
3720	Type MC cable		8.89	.900		27	23		50	64
3730	EMT & wire		4.10	1.950		29	49		78	105
4000	Receptacle devices									
4010	Duplex outlet, 15 amp recpt., Ivory, 1-gang box, plate									
4015	Type NM cable	1 Elec	12.31	.650	Ea.	5.72	16.35		22.07	31
4020	Type MC cable		12.31	.650		10.30	16.35		26.65	36
4030	EMT & wire		5.33	1.500		13.31	38		51.31	72
4050	With #12/2 type NM cable		12.31	.650		6.86	16.35		23.21	33
4070	Type MC cable		10.67	.750		12.32	18.85		31.17	42

For expanded coverage of these items see *Means Electrical Cost Data 1991*

		CREW	DAILY OUTPUT	MAN-HOURS	UNIT	BARE COSTS MAT.	LABOR	EQUIP.	TOTAL	TOTAL INCL O&P
168 100	**Special Systems**									
170	4080	EMT & wire	1 Elec	4.71	1.700	Ea.	14.20	43		57.20	81	170
4100	20 amp recpt., #12/2 type NM cable		12.31	.650		9.30	16.35		25.65	35
4120	Type MC cable		10.67	.750		13.62	18.85		32.47	44
4130	EMT & wire		4.71	1.700		16.60	43		59.60	83
4140	For GFI see line 4300 below									
4150	Decorator style, 15 amp recpt., type NM cable	1 Elec	14.55	.550	Ea.	7	13.85		20.85	29
4170	Type MC cable		12.31	.650		11.54	16.35		27.89	38
4180	EMT & wire		5.33	1.500		14.45	38		52.45	74
4200	With #12/2 type NM cable		12.31	.650		8.22	16.35		24.57	34
4220	Type MC cable		10.67	.750		12.58	18.85		31.43	43
4230	EMT & wire		4.71	1.700		15.50	43		58.50	82
4250	20 amp recpt. #12/2 type NM cable		12.31	.650		11.75	16.35		28.10	38
4270	Type MC cable		10.67	.750		15	18.85		33.85	45
4280	EMT & wire		4.71	1.700		19.45	43		62.45	87
4300	GFI, 15 amp recpt., type NM cable		12.31	.650		28	16.35		44.35	56
4320	Type MC cable		10.67	.750		32.24	18.85		51.09	64
4330	EMT & wire		4.71	1.700		35.36	43		78.36	105
4350	GFI with #12/2 type NM cable		10.67	.750		29	18.85		47.85	61
4370	Type MC cable		9.20	.870		33.28	22		55.28	70
4380	EMT & wire		4.21	1.900		36.40	48		84.40	115
4400	20 amp recpt., #12/2 type NM cable		10.67	.750		31.20	18.85		50.05	63
4420	Type MC cable		9.20	.870		35.36	22		57.36	72
4430	EMT & wire		4.21	1.900		37.44	48		85.44	115
4500	Weather-proof cover for above receptacles, add		32	.250		3.54	6.30		9.84	13.50
4550	Air conditioner outlet, 20 amp-240 volt recpt.									
4560	30' of #12/2, 2 pole circuit breaker									
4570	Type NM cable	1 Elec	10	.800	Ea.	26	20		46	59
4580	Type MC cable		9	.889		31.20	22		53.20	68
4590	EMT & wire		4	2		42.64	50		92.64	125
4600	Decorator style, type NM cable		10	.800		28	20		48	62
4620	Type MC cable		9	.889		33.28	22		55.28	71
4630	EMT & wire		4	2		43.68	50		93.68	125
4650	Dryer outlet, 30 amp-240 volt recpt., 20' of #10/3									
4660	2 pole circuit breaker									
4670	Type NM cable	1 Elec	6.41	1.250	Ea.	42.64	31		73.64	95
4680	Type MC cable		5.71	1.400		43.68	35		78.68	100
4690	EMT & wire		3.48	2.300		40.56	58		98.56	135
4700	Range outlet, 50 amp-240 volt recpt., 30' of #8/3									
4710	Type NM cable	1 Elec	4.21	1.900	Ea.	71.76	48		119.76	150
4720	Type MC cable		4	2		69.68	50		119.68	155
4730	EMT & wire		2.96	2.700		58.24	68		126.24	170
4750	Central vacuum outlet		6.40	1.250		53	31		84	105
4770	Type MC cable		5.71	1.400		37.44	35		72.44	95
4780	EMT & wire		3.48	2.300		48.90	58		106.90	140
4800	30 amp-110 volt locking recpt., #10/2 circ. bkr.									
4810	Type NM cable	1 Elec	6.20	1.290	Ea.	45.76	32		77.76	100
4820	Type MC cable		5.40	1.480		48.90	37		85.90	110
4830	EMT & wire		3.20	2.500		44.72	63		107.72	145
4900	Low voltage outlets									
4910	Telephone recpt., 20' of 4/C phone wire	1 Elec	26	.308	Ea.	5.20	7.75		12.95	17.55
4920	TV recpt., 20' of RG59U coax wire, F type connector	"	16	.500	"	7.90	12.60		20.50	28
4950	Door bell chime, transformer, 2 buttons, 60' of bellwire									
4970	Economy model	1 Elec	11.50	.696	Ea.	44.72	17.50		62.22	76
4980	Custom model		11.50	.696		91.52	17.50		109.02	125
4990	Luxury model, 3 buttons		9.50	.842		197	21		218	250
6000	Lighting outlets									
6050	Wire only (for fixture) type NM cable	1 Elec	32	.250	Ea.	3.45	6.30		9.75	13.40
6070	Type MC cable	"	24	.333	"	6.80	8.40		15.20	20

For expanded coverage of these items see *Means Electrical Cost Data 1991*

168 100	Special Systems	CREW	DAILY OUTPUT	MAN-HOURS	UNIT	BARE COSTS				TOTAL INCL O&P
						MAT.	LABOR	EQUIP.	TOTAL	
6080	EMT & wire	1 Elec	10	.800	Ea.	9.15	20		29.15	41
6100	Box (4") and wire (for fixture), type NM cable		25	.320		5.93	8.05		13.98	18.80
6120	Type MC cable		20	.400		10.60	10.05		20.65	27
6130	EMT & wire		11	.727		12.85	18.30		31.15	42
6200	Fixtures (use with lines 6050 or 6100 above)									
6210	Canopy style, economy grade	1 Elec	40	.200	Ea.	19.45	5.05		24.50	29
6220	Custom grade		40	.200		35.36	5.05		40.41	47
6250	Dining room chandelier, economy grade		19	.421		58.24	10.60		68.84	80
6270	Luxury grade		15	.533		384	13.40		397.40	445
6310	Kitchen fixture (fluorescent), economy grade		30	.267		39.52	6.70		46.22	54
6320	Custom grade		25	.320		126	8.05		134.05	150
6350	Outdoor, wall mounted, economy grade		30	.267		22.67	6.70		29.37	35
6360	Custom grade		30	.267		78	6.70		84.70	96
6370	Luxury grade		25	.320		176	8.05		184.05	205
6410	Outdoor Par floodlights, 1 lamp, 150 watt		20	.400		15.15	10.05		25.20	32
6420	2 lamp, 150 watt each		20	.400		25.22	10.05		35.27	43
6430	For infrared security sensor, add		32	.250		75	6.30		81.30	92
6450	Outdoor, quartz-halogen, 300 watt flood		20	.400		28	10.05		38.05	46
6600	Recessed downlight, round, pre-wired, 50 or 75 watt trim		30	.267		27	6.70		33.70	40
6610	With shower light trim		30	.267		32	6.70		38.70	45
6620	With wall washer trim		28	.286		40	7.20		47.20	55
6630	With eye-ball trim		28	.286		37	7.20		44.20	52
6640	For direct contact with insulation, add					1.15			1.15	1.27
6700	Porcelain lamp holder	1 Elec	40	.200		2.70	5.05		7.75	10.65
6710	With pull switch		40	.200		3	5.05		8.05	11
6750	Fluorescent strip, 1-20 watt tube, wrap around diffuser, 24"		24	.333		40	8.40		48.40	57
6760	1-40 watt tube, 48"		24	.333		57	8.40		65.40	76
6770	2-40 watt tubes, 48"		20	.400		67	10.05		77.05	89
6780	With 0° ballast		20	.400		70	10.05		80.05	92
6800	Bathroom heat lamp, 1-250 watt		28	.286		23.80	7.20		31	37
6810	2-250 watt lamps		28	.286		39.52	7.20		46.72	54
6820	For timer switch, see line 2400									
6900	Outdoor post lamp, incl. post, fixture, 35' of #14/2									
6910	Type NMC cable	1 Elec	3.50	2.290	Ea.	140	57		197	240
6920	Photo-eye, add		27	.296		22.50	7.45		29.95	36
6950	Clock dial time switch, 24 hr., w/enclosure, type NM cable		11.43	.700		39	17.60		56.60	70
6970	Type MC cable		11	.727		44	18.30		62.30	76
6980	EMT & wire		4.85	1.650		46	41		87	115
7000	Alarm systems									
7050	Smoke detectors, box, #14/3 type NM cable	1 Elec	14.55	.550	Ea.	21.63	13.85		35.48	45
7070	Type MC cable		12.31	.650		26	16.35		42.35	54
7080	EMT & wire		5	1.600		38	40		78	105
7090	For relay output to security system, add					9			9	9.90
8000	Residential equipment									
8050	Disposal hook-up, incl. switch, outlet box, 3' of flex									
8060	20 amp-1 pole circ. bkr., and 25' of #12/2									
8070	Type NM cable	1 Elec	10	.800	Ea.	17.50	20		37.50	50
8080	Type MC cable		8	1		20.95	25		45.95	61
8090	EMT & wire		5	1.600		22.67	40		62.67	86
8100	Trash compactor or dishwasher hook-up, incl. outlet box,									
8110	3' of flex, 15 amp-1 pole circ. bkr., and 25' of #14/2									
8120	Type NM cable	1 Elec	10	.800	Ea.	11.85	20		31.85	44
8130	Type MC cable		8	1		14.66	25		39.66	55
8140	EMT & wire		5	1.600		17.57	40		57.57	81
8150	Hot water sink dispensor hook-up, use line 8100									
8200	Vent/exhaust fan hook-up, type NM cable	1 Elec	32	.250	Ea.	3.43	6.30		9.73	13.35
8220	Type MC cable		24	.333		6.86	8.40		15.26	20
8230	EMT & wire		10	.800		9.15	20		29.15	41

For expanded coverage of these items see *Means Electrical Cost Data 1991*

		CREW	DAILY OUTPUT	MAN-HOURS	UNIT	BARE COSTS				TOTAL INCL O&P
168 100	**Special Systems**					MAT.	LABOR	EQUIP.	TOTAL	
8250	Bathroom vent fan, 50 CFM (use with above hook-up)									
8260	Economy model	1 Elec	15	.533	Ea.	18	13.40		31.40	40
8270	Low noise model		15	.533		23.71	13.40		37.11	47
8280	Custom model	↓	12	.667	↓	85	16.75		101.75	120
8300	Bathroom or kitchen vent fan, 110 CFM									
8310	Economy model	1 Elec	15	.533	Ea.	45.45	13.40		58.85	70
8320	Low noise model	"	15	.533	"	59.28	13.40		72.68	86
8350	Paddle fan, variable speed (w/o lights)									
8360	Economy model (AC motor)	1 Elec	10	.800	Ea.	70.72	20		90.72	110
8370	Custom model (AC motor)		10	.800		123	20		143	165
8380	Luxury model (DC motor)		8	1		251	25		276	315
8390	Remote speed switch for above, add	↓	12	.667	↓	16.22	16.75		32.97	43
8500	Whole house exhaust fan, ceiling mount, 36", variable speed									
8510	Remote switch, incl. shutters, 20 amp-1 pole circ. bkr.									
8520	30' of #12/2/ type NM cable	1 Elec	4	2	Ea.	368	50		418	480
8530	Type MC cable		3.50	2.290		376	57		433	500
8540	EMT & wire	↓	3	2.670	↓	392	67		459	535
8610	3' of flex, 20 amp-1 pole GFI circ. bkr.									
8620	30' of #12/2 type NM cable	1 Elec	10	.800	Ea.	60	20		80	97
8630	Type MC cable		8	1		63	25		88	110
8640	EMT & wire	↓	4	2	↓	65	50		115	150
8650	Hot water heater hook-up, incl. 1-2 pole circ. bkr. box;									
8660	3' of flex, 20' of #10/2 type NM cable	1 Elec	10	.800	Ea.	11.85	20		31.85	44
8670	Type MC cable		8	1		14.56	25		39.56	54
8680	EMT & wire	↓	5	1.600	↓	17.26	40		57.26	80
9000	Heating/air conditioning									
9050	Furnace/boiler hook-up, incl. firestat, local on-off switch									
9060	Emergency switch, and 40' of type NM cable	1 Elec	4	2	Ea.	29	50		79	110
9070	Type MC cable		3.50	2.290		35.15	57		92.15	125
9080	EMT & wire	↓	1.50	5.330	↓	39.52	135		174.52	250
9100	Air conditioner hook-up, incl. local 60 amp disc. switch									
9110	3' Sealtite, 40 amp, 2 pole circuit breaker									
9130	40' of #8/2 type NM cable	1 Elec	3.50	2.290	Ea.	103	57		160	200
9140	Type MC cable		3	2.670		108	67		175	220
9150	EMT & wire	↓	1.30	6.150	↓	109	155		264	355
9200	Heat pump hook-up, 1-40 & 1-100 amp 2 pole circ. bkr.									
9210	Local disconnect switch, 3' Sealtite									
9220	40' of #8/2 & 30' of #3/2									
9230	Type NM cable	1 Elec	1.30	6.150	Ea.	252	155		407	515
9240	Type MC cable		1.08	7.410		257	185		442	565
9250	EMT & wire	↓	.94	8.510	↓	247	215		462	600
9500	Thermostat hook-up, using low voltage wire									
9520	Heating only	1 Elec	24	.333	Ea.	3.43	8.40		11.83	16.55
9530	Heating/cooling	"	20	.400	"	3.75	10.05		13.80	19.50

170

For expanded coverage of these items see *Means Electrical Cost Data 1991*

ASSEMBLIES SECTION

The Assemblies Section of this book provides the costs of construction "systems" made up by combining unit prices, including overhead and profit, from the Unit Price Section.

The System Components at the head of each table show typical unit price elements that are combined to create the single total cost for each line in the table.

By choosing the assembly with characteristics nearest to those required by your job, an accurate estimate can be compiled quickly.

Assemblies Estimates are especially useful for preparing budget estimates, preparing feasibility studies, comparing the cost of optional construction methods, and checking the approximate accuracy of unit price estimates.

TABLE OF CONTENTS

HOW TO USE ASSEMBLIES COST TABLES

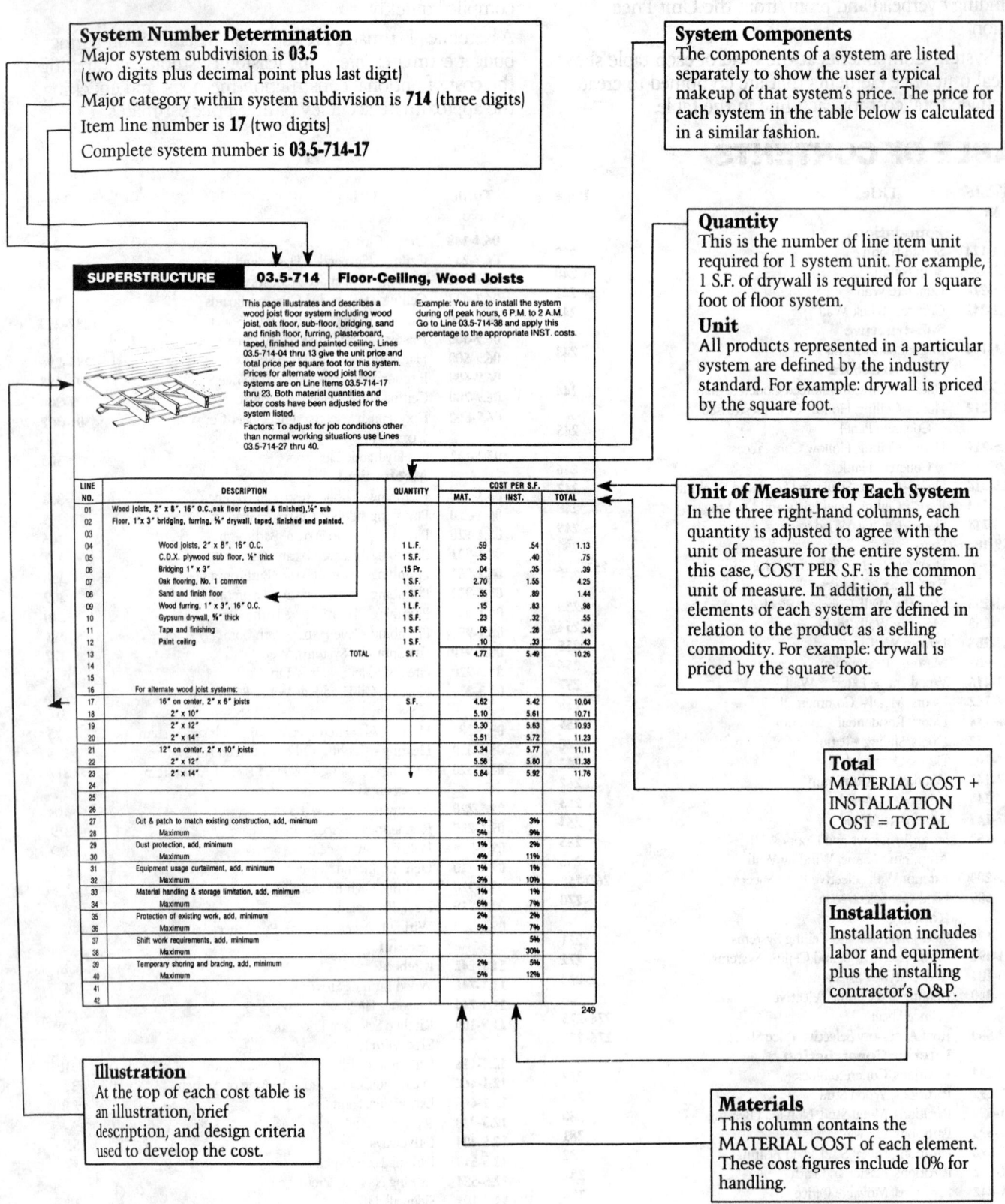

System Number Determination
Major system subdivision is **03.5**
(two digits plus decimal point plus last digit)
Major category within system subdivision is **714** (three digits)
Item line number is **17** (two digits)
Complete system number is **03.5-714-17**

System Components
The components of a system are listed separately to show the user a typical makeup of that system's price. The price for each system in the table below is calculated in a similar fashion.

Quantity
This is the number of line item unit required for 1 system unit. For example, 1 S.F. of drywall is required for 1 square foot of floor system.

Unit
All products represented in a particular system are defined by the industry standard. For example: drywall is priced by the square foot.

Unit of Measure for Each System
In the three right-hand columns, each quantity is adjusted to agree with the unit of measure for the entire system. In this case, COST PER S.F. is the common unit of measure. In addition, all the elements of each system are defined in relation to the product as a selling commodity. For example: drywall is priced by the square foot.

Total
MATERIAL COST + INSTALLATION COST = TOTAL

Installation
Installation includes labor and equipment plus the installing contractor's O&P.

Illustration
At the top of each cost table is an illustration, brief description, and design criteria used to develop the cost.

Materials
This column contains the MATERIAL COST of each element. These cost figures include 10% for handling.

SUPERSTRUCTURE	03.5-714	Floor-Ceiling, Wood Joists

This page illustrates and describes a wood joist floor system including wood joist, oak floor, sub-floor, bridging, sand and finish floor, furring, plasterboard, taped, finished and painted ceiling. Lines 03.5-714-04 thru 13 give the unit price and total price per square foot for this system. Prices for alternate wood joist floor systems are on Line Items 03.5-714-17 thru 23. Both material quantities and labor costs have been adjusted for the system listed.

Factors: To adjust for job conditions other than normal working situations use Lines 03.5-714-27 thru 40.

Example: You are to install the system during off peak hours, 6 P.M. to 2 A.M. Go to Line 03.5-714-38 and apply this percentage to the appropriate INST. costs.

LINE NO.	DESCRIPTION	QUANTITY	COST PER S.F.		
			MAT.	INST.	TOTAL
01	Wood joists, 2″ x 8″, 16″ O.C.,oak floor (sanded & finished),½″ sub				
02	Floor, 1″x 3″ bridging, furring, ⅝″ drywall, taped, finished and painted.				
03					
04	Wood joists, 2″ x 8″, 16″ O.C.	1 L.F.	.59	.54	1.13
05	C.D.X. plywood sub floor, ½″ thick	1 S.F.	.35	.40	.75
06	Bridging 1″ x 3″	.15 Pr.	.04	.35	.39
07	Oak flooring, No. 1 common	1 S.F.	2.70	1.55	4.25
08	Sand and finish floor	1 S.F.	.55	.89	1.44
09	Wood furring, 1″ x 3″, 16″ O.C.	1 L.F.	.15	.83	.98
10	Gypsum drywall, ⅝″ thick	1 S.F.	.23	.32	.55
11	Tape and finishing	1 S.F.	.06	.28	.34
12	Paint ceiling	1 S.F.	.10	.33	.43
13	TOTAL	S.F.	4.77	5.49	10.26
14					
15					
16	For alternate wood joist systems:				
17	16″ on center, 2″ x 6″ joists	S.F.	4.62	5.42	10.04
18	2″ x 10″		5.10	5.61	10.71
19	2″ x 12″		5.30	5.63	10.93
20	2″ x 14″		5.51	5.72	11.23
21	12″ on center, 2″ x 10″ joists		5.34	5.77	11.11
22	2″ x 12″		5.58	5.80	11.38
23	2″ x 14″		5.84	5.92	11.76
24					
25					
26					
27	Cut & patch to match existing construction, add, minimum		2%	3%	
28	Maximum		5%	9%	
29	Dust protection, add, minimum		1%	2%	
30	Maximum		4%	11%	
31	Equipment usage curtailment, add, minimum		1%	1%	
32	Maximum		3%	10%	
33	Material handling & storage limitation, add, minimum		1%	1%	
34	Maximum		6%	7%	
35	Protection of existing work, add, minimum		2%	2%	
36	Maximum		5%	7%	
37	Shift work requirements, add, minimum			5%	
38	Maximum			30%	
39	Temporary shoring and bracing, add, minimum		2%	5%	
40	Maximum		5%	12%	
41					
42					

249

238

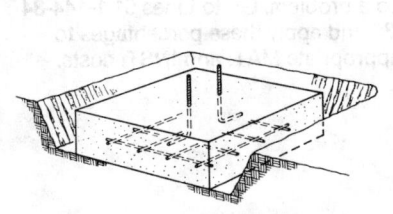

This page illustrates and describes a spread footing system including concrete, forms, reinforcing and anchor bolts. Lines 01.1-124-05 thru 10 give the unit price and total price on a cost each basis for this system. Prices for alternate spread footing systems are on Line Items 01.1-124-13 thru 23. Both material quantities and labor costs have been adjusted for the system listed.

Factors: To adjust for job conditions other than normal working situations use Lines 01.1-124-29 thru 40.

Example: You are to install the system in an existing occupied building. Access to the site and protection to the building are mandatory. Go to Lines 01.1-124-36 and 38 and apply these percentages to the appropriate MAT. and INST. costs.

LINE NO.	DESCRIPTION	QUANTITY	COST EACH		
			MAT.	INST.	TOTAL
01	Interior column footing, 3' square 1' thick, 2000 psi concrete,				
02	Including forms, reinforcing, and anchor bolts.				
03					
04					
05	Concrete, 2000 psi	.33 C.Y.	17.49		17.49
06	Placing concrete	.33 C.Y.		8.91	8.91
07	Forms, footing, 4 uses	12 SFCA	5.02	32.30	37.32
08	Reinforcing	11 Lb.	3.27	3.44	6.71
09	Anchor bolts, ¾" diameter	2 Ea.	.75	2.91	3.66
10	TOTAL	Ea.	26.53	47.56	74.09
11					
12	Above system with the following:				
13	3' square x 1' thick, 3000 psi concrete	Ea.	27.52	47.56	75.08
14	4000 psi concrete	"	28.84	47.56	76.40
15					
16	For alternate footing systems:				
17	4' square x 1' thick, 2000 psi concrete	Ea.	44.53	68.05	112.58
18	3000 psi concrete		46.30	68.05	114.35
19	4000 psi concrete	↓	48.66	68.05	116.71
20					
21	5' square x 1'-3" thick, 2000 psi concrete	Ea.	83.97	113.42	197.39
22	3000 psi concrete		87.45	113.42	200.87
23	4000 psi concrete	↓	92.09	113.42	205.51
24					
25					
26					
27					
28					
29	Cut & patch to match existing construction, add, minimum		2%	3%	
30	Maximum		5%	9%	
31	Dust protection, add, minimum		1%	2%	
32	Maximum		4%	11%	
33	Equipment usage curtailment, add, minimum		1%	1%	
34	Maximum		3%	10%	
35	Material handling & storage limitation, add, minimum		1%	1%	
36	Maximum		6%	7%	
37	Protection of existing work, add, minimum		2%	2%	
38	Maximum		5%	7%	
39	Shift work requirements, add, minimum			5%	
40	Maximum			30%	
41					
42					

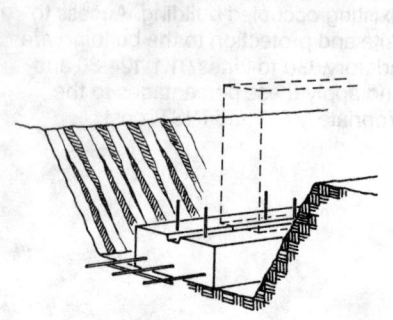

This page illustrates and describes a strip footing system including concrete, forms, reinforcing, keyway and dowels. Lines 01.1-144-04 thru 10 give the unit price and total price per linear foot for this system. Prices for alternate strip footing systems are on Line Items 01.1-144-15 thru 25. Both material quantities and labor costs have been adjusted for the system listed.

Factors: To adjust for job conditions other than normal working situations use Lines 01.1-144-29 thru 40.

Example: You are to install this footing, and due to a lack of accessibility, only hand tools can be used. Material handling is also a problem. Go to Lines 01.1-144-34 and 36 and apply these percentages to the appropriate MAT. and INST. costs.

LINE NO.	DESCRIPTION	QUANTITY	COST PER L.F.		
			MAT.	INST.	TOTAL
01	Strip footing, 2'-0" wide x 1'-0" thick, 2000 psi concrete including forms				
02	Reinforcing, keyway, and dowels.				
03					
04	Concrete, 2000 psi	.074 C.Y.	3.92		3.92
05	Placing concrete	.074 C.Y.		.93	.93
06	Forms, footing, 4 uses	2 S.F.	.77	4.63	5.40
07	Reinforcing	3.17 Lb.	.94	.99	1.93
08	Keyway, 2" x 4", 4 uses	1 L.F.	.08	.54	.62
09	Dowels, #4 bars, 2' long, 24" O.C.	.5 Ea.	.53	2.67	3.20
10	TOTAL	L.F.	6.24	9.76	16
11					
12					
13					
14	Above system with the following:				
15	2'-0" wide x 1' thick, 3000 psi concrete	L.F.	6.46	9.76	16.22
16	4000 psi concrete	"	6.76	9.76	16.52
17					
18	For alternate footing systems:				
19	2'-6" wide x 1' thick, 2000 psi concrete	L.F.	7.50	10.26	17.76
20	3000 psi concrete		7.78	10.26	18.04
21	4000 psi concrete	↓	8.15	10.26	18.41
22					
23	3'-0" wide x 1' thick, 2000 psi concrete	L.F.	8.63	10.74	19.37
24	3000 psi concrete		8.96	10.74	19.70
25	4000 psi concrete	↓	9.40	10.74	20.14
26					
27					
28					
29	Cut & patch to match existing construction, add, minimum		2%	3%	
30	Maximum		5%	9%	
31	Dust protection, add, minimum		1%	2%	
32	Maximum		4%	11%	
33	Equipment usage curtailment, add, minimum		1%	1%	
34	Maximum		3%	10%	
35	Material handling & storage limitation, add, minimum		1%	1%	
36	Maximum		6%	7%	
37	Protection of existing work, add, minimum		2%	2%	
38	Maximum		5%	7%	
39	Shift work requirements, add, minimum			5%	
40	Maximum			30%	
41					
42					

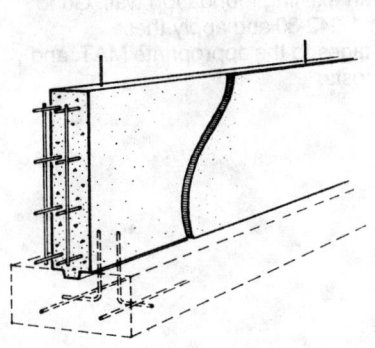

This page illustrates and describes a concrete wall system including concrete, placing concrete, forms, reinforcing, insulation, waterproofing and anchor bolts. Lines 01.1-214-04 thru 11 give the unit price and total price per linear foot for this system. Prices for alternate concrete wall systems are on Line Items 01.1-214-15 thru 26. Both material quantities and labor costs have been adjusted for the system listed.

Factors: To adjust for job conditions other than normal working situations use Lines 01.1-214-29 thru 40.

Example: You are to install this wall system where delivery of material is difficult. Go to Line 01.1-214-36 and apply these percentages to the appropriate MAT. and INST. costs.

LINE NO.	DESCRIPTION	QUANTITY	COST PER L.F.		
			MAT.	INST.	TOTAL
01	Cast in place concrete foundation wall, 8" thick, 3' high, 2500 psi				
02	Concrete including forms, reinforcing, waterproofing, and anchor bolts.				
03					
04	Concrete, 2500 psi, 8" thick, 3' high	.07 C.Y.	3.78		3.78
05	Forms, wall, 4 uses	6 S.F.	4.09	20.51	24.60
06	Reinforcing	6 Lb.	1.85	1.33	3.18
07	Placing concrete	.07 C.Y.		2.10	2.10
08	Waterproofing	3 S.F.	.53	1.72	2.25
09	Rigid insulaton, 1" polystyrene	3 S.F.	.89	1.27	2.16
10	Anchor bolts, ½" diameter, 4' O.C.	.25 Ea.	.12	.39	.51
11	TOTAL	L.F.	11.26	27.32	38.58
12					
13					
14	For alternate wall systems:				
15	8" thick, 2500 psi concrete, 4' high	L.F.	15.33	36.50	51.83
16	6' high		22.94	54.55	77.49
17	8' high		30.55	72.60	103.15
18	3500 psi concrete, 4' high		15.73	36.50	52.23
19	6' high		23.54	54.55	78.09
20	8' high		31.35	72.60	103.95
21	12" thick, 2500 psi concrete, 4' high		19.27	38.88	58.15
22	6' high		27.95	57.54	85.49
23	8' high		37.18	76.49	113.67
24	3500 psi concrete, 4' high		19.87	38.88	58.75
25	8' high		38.38	76.49	114.87
26	10' high		47.65	95.37	143.02
27					
28					
29	Cut & patch to match existing construction, add, minimum		2%	3%	
30	Maximum		5%	9%	
31	Dust protection, add, minimum		1%	2%	
32	Maximum		4%	11%	
33	Equipment usage curtailment, add, minimum		1%	1%	
34	Maximum		3%	10%	
35	Material handling & storage limitation, add, minimum		1%	1%	
36	Maximum		6%	7%	
37	Protection of existing work, add, minimum		2%	2%	
38	Maximum		5%	7%	
39	Shift work requirements, add, minimum			5%	
40	Maximum			30%	
41					
42					

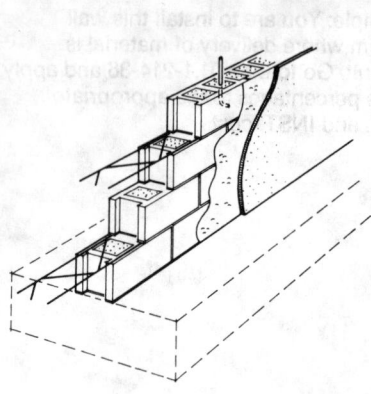

This page illustrates and describes a concrete block wall system including concrete block, masonry reinforcing, parging, waterproofing insulation and anchor bolts. Lines 01.1-242-04 thru 10 give the unit price and total price per linear foot for this system. Prices for alternate concrete block wall systems are on Line Items 01.1-242-13 thru 26. Both material quantities and labor costs have been adjusted for the system listed.

Factors: To adjust for job conditions other than normal working situations use Lines 01.1-242-29 thru 40.

Example: You are to install the system to match an existing foundation wall. Go to Line 01.1-242-30 and apply these percentages to the appropriate MAT. and INST. costs.

LINE NO.	DESCRIPTION	QUANTITY	COST PER L.F.		
			MAT.	INST.	TOTAL
01	**Concrete block, 8″ thick, masonry reinforcing, parged and**				
02	**Waterproofed, insulation and anchor bolts, wall 2′-8″ high.**				
03					
04	Concrete block, 8″ x 8″ x 16″	2.67 S.F.	4.70	8.25	12.95
05	Masonry reinforcing	2 L.F.	.40	.20	.60
06	Parging	.193 S.Y.	.61	1.28	1.89
07	Waterproofing	2.67 S.F.	.46	1.54	2
08	Insulation, 1″ rigid polystyrene	2.67 S.F.	.79	1.13	1.92
09	Anchor bolts, ½″ diameter, 4′ O.C.	.25 Ea.	.12	.39	.51
10	TOTAL	L.F.	7.08	12.79	19.87
11					
12	For alternate wall systems:				
13	8″ thick block, 4′ high	L.F.	10.55	19.01	29.56
14	6′ high		15.69	28.24	43.93
15	8′ high		21.02	37.59	58.61
16	Grouted solid, 4′ high		14.25	24.39	38.64
17	6′ high		21.23	36.32	57.55
18	8′ high		28.41	48.36	76.77
19					
20					
21	12″ thick block, 4′ high	L.F.	13.10	22.61	35.71
22	6′ high		19.50	33.63	53.13
23	8′ high		26.10	44.81	70.91
24	Grouted solid, 4′ high		19.17	28.34	47.51
25	6′ high		28.61	42.22	70.83
26	8′ high		38.24	56.27	94.51
27					
28					
29	Cut & patch to match existing construction, add, minimum		2%	3%	
30	Maximum		5%	9%	
31	Dust protection, add, minimum		1%	2%	
32	Maximum		4%	11%	
33	Equipment usage curtailment, add, minimum		1%	1%	
34	Maximum		3%	10%	
35	Material handling & storage limitation, add, minimum		1%	1%	
36	Maximum		6%	7%	
37	Protection of existing work, add, minimum		2%	2%	
38	Maximum		5%	7%	
39	Shift work requirements, add, minimum			5%	
40	Maximum			30%	
41					
42					

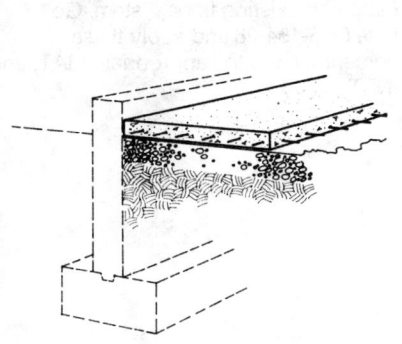

This page illustrates and describes a slab on grade system including slab, bank run gravel, bulkhead forms, placing concrete, welded wire fabric, vapor barrier, steel trowel finish and curing paper. Lines 02.1-104-04 thru 11 give the unit price and total price per square foot for this system. Prices for alternate slab on grade systems are on Line Items 02.1-104-15 thru 26. Both material quantities and labor costs have been adjusted for the system listed.

Factors: To adjust for job conditions other than normal working situations use Line 02.1-104-29 thru 40.

Example: You are to install the system at a site where protection of the existing building is required. Go to Line 02.1-104-38 and apply these percentages to the appropriate MAT. and INST. costs.

LINE NO.	DESCRIPTION	QUANTITY	COST PER S.F.		
			MAT.	INST.	TOTAL
01	**Ground slab, 4″ thick, 3000 psi concrete, 4″ granular base, vapor barrier**				
02	**Welded wire fabric, screed and steel trowel finish.**				
03					
04	Concrete, 4″ thick, 3000 psi concrete	.012 C.Y.	.67		.67
05	Bank run gravel, 4″ deep	.074 C.Y.	.12	.04	.16
06	Polyethylene vapor barrier, 10 mil.	.011 C.S.F.	.03	.10	.13
07	Bulkhead forms, expansion material	.1 L.F.	.02	.19	.21
08	Welded wire fabric, 6 x 6 - #10/10	.011 C.S.F.	.08	.22	.30
09	Place concrete	.012 C.Y.		.16	.16
10	Screed & steel trowel finish	1 S.F.		.56	.56
11	TOTAL	S.F.	.92	1.27	2.19
12					
13					
14	Above system with the following:				
15	4″ thick slab, 3000 psi concrete, 6″ deep bank run gravel	S.F.	.98	1.29	2.27
16	12″ deep bank run gravel	"	1.16	1.33	2.49
17					
18					
19					
20	For alternate slab systems:				
21	5″ thick slab, 3000 psi concrete, 6″ deep bank run gravel	S.F.	1.15	1.34	2.49
22	12″ deep bank run gravel	"	1.33	1.38	2.71
23					
24					
25	6″ thick slab, 3000 psi concrete, 6″ deep bank run gravel	S.F.	1.37	1.39	2.76
26	12″ deep bank run gravel	"	1.55	1.43	2.98
27					
28					
29	Cut & patch to match existing construction, add, minimum		2%	3%	
30	Maximum		5%	9%	
31	Dust protection, add, minimum		1%	2%	
32	Maximum		4%	11%	
33	Equipment usage curtailment, add, minimum		1%	1%	
34	Maximum		3%	10%	
35	Material handling & storage limitation, add, minimum		1%	1%	
36	Maximum		6%	7%	
37	Protection of existing work, add, minimum		2%	2%	
38	Maximum		5%	7%	
39	Shift work requirements, add, minimum			5%	
40	Maximum			30%	
41					
42					

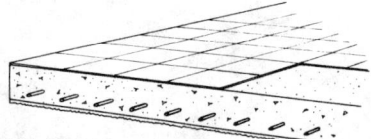

This page illustrates and describes a reinforced concrete slab system including concrete, placing concrete, formwork, reinforcing, steel trowel finish, V.A. floor tile and acoustical spray ceiling finish. Lines 03.5-154-04 thru 12 give the unit price per square foot for this system. Prices for alternate reinforced concrete slab systems are on Line Items 03.5-154-15 thru 18. Both material quantities and labor costs have been adjusted for the system listed.

Factors: To adjust for job conditions other than normal working situations use Lines 03.5-154-27 thru 40.

Example: You are to install the system to match an existing floor system. Go to Line 03.5-154-28 and apply these percentages to the appropriate MAT. and INST. costs.

LINE NO.	DESCRIPTION	QUANTITY	COST PER S.F. MAT.	INST.	TOTAL
01	**Flat slab system, reinforced concrete with vinyl tile floor, sprayed**				
02	**Acoustical ceiling finish, not including columns.**				
03					
04	Concrete, 4000 psi, 6″ thick	.02 C.Y.	1.20		1.20
05	Placing concrete	.02 C.Y.		.48	.48
06	Formwork, 4 use	1 S.F.	.64	3.09	3.73
07	Edge form	.12 S.F.	.06	.60	.66
08	Reinforcing steel	3 Lb.	.89	.94	1.83
09	Steel trowel finish	1 S.F.		.49	.49
10	Vinyl asbestos floor tile	1 S.F.	1.93	.54	2.47
11	Acoustical spray ceiling finish	1 S.F.	.17	.09	.26
12	TOTAL	S.F.	4.89	6.23	11.12
13					
14	For alternate slab systems:				
15	Concrete, 4000 psi, 7″ thick	S.F.	5.13	6.51	11.64
16	8″ thick		5.51	6.86	12.37
17	9″ thick		5.81	7.16	12.97
18	10″ thick	↓	6.19	7.51	13.70
19					
20					
21					
22					
23					
24					
25					
26					
27	Cut & patch to match existing construction, add, minimum		2%	3%	
28	Maximum		5%	9%	
29	Dust protection, add, minimum		1%	2%	
30	Maximum		4%	11%	
31	Equipment usage curtailment, add, minimum		1%	1%	
32	Maximum		3%	10%	
33	Material handling & storage limitation, add, minimum		1%	1%	
34	Maximum		6%	7%	
35	Protection of existing work, add, minimum		2%	2%	
36	Maximum		5%	7%	
37	Shift work requirements, add, minimum			5%	
38	Maximum			30%	
39	Temporary shoring and bracing, add, minimum		2%	5%	
40	Maximum		5%	12%	
41					
42					

This page illustrates and describes a hollow core prestressed concrete panel system including hollow core slab, grout, carpet, carpet padding, and sprayed ceiling. Lines 03.5-212-04 thru 08 give the unit price and total price per square foot for this system. Prices for alternate hollow core prestressed concrete panel systems are on Line Items 03.5-212-13 thru 16. Both material and labor costs have been adjusted for the system listed.

Factors: To adjust for job conditions other than normal working situations use Lines 03.5-212-27 thru 38.

Example: You are to install the system where dust control is a major concern. Go to Line 03.5-212-28 and apply these percentages to the appropriate MAT. and INST. costs.

LINE NO.	DESCRIPTION	QUANTITY	COST PER S.F.		
			MAT.	INST.	TOTAL
01	**Precast hollow core plank with carpeted floors, padding**				
02	**And sprayed textured ceiling.**				
03					
04	Hollow core plank, 4″ thick with grout topping	1 S.F.	3.47	.93	4.40
05	Nylon carpet, 26 oz. medium traffic	.11 S.Y.	1.98	.55	2.53
06	Carpet padding, minimum quality	.11 S.Y.	.26	.21	.47
07	Sprayed texture ceiling	1 S.F.	.11	.34	.45
08	TOTAL	S.F.	5.82	2.03	7.85
09					
10					
11					
12	For alternate floor systems:				
13	Hollow core concrete plank, with grout, 6″ thick	S.F.	5.98	1.88	7.86
14	8″ thick		6.11	1.73	7.84
15	10″ thick		6.61	1.62	8.23
16	12″ thick		6.90	1.60	8.50
17					
18					
19					
20					
21					
22					
23					
24					
25					
26					
27	Dust protection, add, minimum		1%	2%	
28	Maximum		4%	11%	
29	Equipment usage curtailment, add, minimum		1%	1%	
30	Maximum		3%	10%	
31	Material handling & storage limitation, add, minimum		1%	1%	
32	Maximum		6%	7%	
33	Protection of existing work, add, minimum		2%	2%	
34	Maximum		5%	7%	
35	Shift work requirements, add, minimum			5%	
36	Maximum			30%	
37	Temporary shoring and bracing, add, minimum		2%	5%	
38	Maximum		5%	12%	
39					
40					
41					
42					

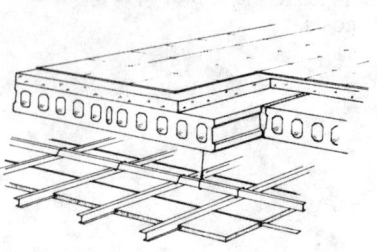

This page illustrates and describes a hollow core plank system including precast hollow core plank, concrete topping, V.A. tile, concealed suspension system and ceiling tile. Lines 03.5-214-04 thru 09 give the unit price and total price per square foot for this system. Prices for alternate hollow core plank systems are on Line Items 03.5-214-15 thru 18. Both material quantities and labor costs have been adjusted for the system listed.

Factors: To adjust for job conditions other than normal working situations use Lines 03.5-214-29 thru 40.

Example: You are to install the system where equipment usage is a problem. Go to Line 03.5-214-32 and apply these percentages to the appropriate MAT. and INST. costs.

LINE NO.	DESCRIPTION	QUANTITY	COST PER S.F.		
			MAT.	INST.	TOTAL
01	Precast hollow core plank with 2″ topping, vinyl composition floor tile				
02	And suspended acoustical tile ceiling.				
03					
04	Hollow core concrete plank, 4″ thick	1 S.F.	3.47	.93	4.40
05	Concrete topping, 2″ thick	1 S.F.	.61	1.69	2.30
06	Vinyl composition tile, .08″ thick	1 S.F.	.83	1.69	2.30
07	Concealed Z bar suspension system, 12″ module	1 S.F.	.62	.54	1.37
08	Ceiling tile, mineral fiber, ¾″ thick	1 S.F.	.69	1.44	2.13
09	TOTAL	S.F.	6.22	5.15	11.37
10					
11					
12					
13					
14	For alternate floor systems:				
15	Hollow core concrete plank, 6″ thick	S.F.	6.38	5	11.38
16	8″ thick		6.51	4.85	11.36
17	10″ thick		7.01	4.74	11.75
18	12″ thick		7.30	4.72	12.02
19					
20					
21					
22					
23					
24					
25					
26					
27					
28					
29	Dust protection, add, minimum		1%	2%	
30	Maximum		4%	11%	
31	Equipment usage curtailment, add, minimum		1%	1%	
32	Maximum		3%	10%	
33	Material handling & storage limitation, add, minimum		1%	1%	
34	Maximum		6%	7%	
35	Protection of existing work, add, minimum		2%	2%	
36	Maximum		5%	7%	
37	Shift work requirements, add, minimum			5%	
38	Maximum			30%	
39	Temporary shoring and bracing, add, minimum		2%	5%	
40	Maximum		5%	12%	
41					
42					

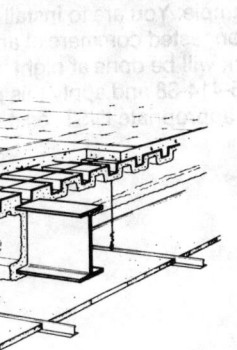

This page illustrates and describes a structural steel w/metal decking and concrete system including steel beams, steel decking, shear studs, concrete, placing concrete, edge form, steel trowel finish, curing, wire fabric, fireproofing, beams and decking, tile floor, and suspended ceiling. Lines 03.5-314-05 thru 16 give the unit price and total price per square foot for this system. Prices for alternate structural steel w/metal decking and concrete systems are on Line Items 03.5-314-19 thru 24. Both material quantities and labor costs have been adjusted for the system listed.

Factors: To adjust for job conditions other than normal working situations use Lines 03.5-314-29 thru 40.

Example: You are to install the system where material handling and storage are a problem. Go to Line 03.5-314-34 and apply these percentages to the appropriate MAT. and INST. costs.

LINE NO.	DESCRIPTION	QUANTITY	COST PER S.F.		
			MAT.	INST.	TOTAL
01	**Composite structural beams, 20 ga. 3″ deep steel decking, shear studs,**				
02	**3000 psi, concrete, placing concrete, edge forms, steel trowel finish,**				
03	**Welded wire fabric fireproofing, tile floor and suspended ceiling.**				
04					
05	Steel deck, 20 gage, 3″ deep, galvanized	1 S.F.	1.11	.50	1.61
06	Structural steel framing	.004 Ton	4.62	1.48	6.10
07	Shear studs, ¾″	.25 Ea.	.11	.29	.40
08	Concrete, 3000 psi, 5″ thick	.015 C.Y.	.84		.84
09	Place concrete	.015 C.Y.		.36	.36
10	Edge form	.14 L.F.	.03	.33	.36
11	Steel trowel finish	1 S.F.		.49	.49
12	Welded wire fabric 6 x 6 - #10/10	.011 C.S.F.	.08	.22	.30
13	Fireproofing, sprayed	1.88 S.F.	.72	1.25	1.97
14	Vinyl composition floor tile	1 S.F.	.61	.54	1.15
15	Suspended acoustical ceiling	1 S.F.	.95	.75	1.70
16	TOTAL	S.F.	9.07	6.21	15.28
17					
18	For alternate floor systems:				
19	Composite deck, galvanized, 3″ deep, 22 ga.	S.F.	8.88	6.18	15.06
20	18 ga.		9.41	6.24	15.65
21	16 ga.		9.79	6.27	16.06
22	Composite deck, galvanized, 2″ deep, 22 ga.		8.81	6.10	14.91
23	20 ga.		8.98	6.13	15.11
24	16 ga.		9.64	6.18	15.82
25					
26					
27					
28					
29	Dust protection, add, minimum		1%	2%	
30	Maximum		4%	11%	
31	Equipment usage curtailment, add, minimum		1%	1%	
32	Maximum		3%	10%	
33	Material handling & storage limitation, add, minimum		1%	1%	
34	Maximum		6%	7%	
35	Protection of existing work, add, minimum		2%	2%	
36	Maximum		5%	7%	
37	Shift work requirements, add, minimum			5%	
38	Maximum			30%	
39	Temporary shoring and bracing, add, minimum		2%	5%	
40	Maximum		5%	12%	
41					
42					

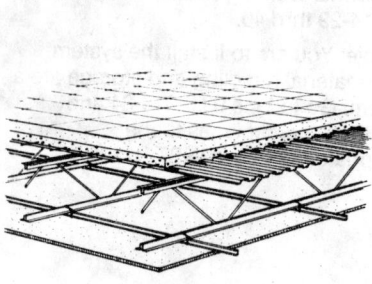

This page illustrates and describes an open web joist and steel slab-form system including open web steel joist, slab form, concrete, placing concrete, wire fabric, steel trowel finish, tile floor and plasterboard. Lines 03.5-414-05 thru 15 give the unit price and total price per square foot for this system. Prices for alternate open web joists and steel slab-form systems are on Line Items 03.5-414-19 thru 22. Both material quantities and labor costs have been adjusted for the systems listed.

Factors: To adjust for job conditions other than normal working situations use Lines 03.5-414-29 thru 40.

Example: You are to install the system in a congested commercial area and most work will be done at night. Go to Line 03.5-414-38 and apply this percentage to the appropriate INST. cost.

LINE NO.	DESCRIPTION	QUANTITY	COST PER S.F. MAT.	COST PER S.F. INST.	COST PER S.F. TOTAL
01	Open web steel joists, 24" O.C., slab form, 3000 psi concrete, welded				
02	Wire fabric, steel trowel finish, floor tile, ⅝" drywall ceiling.				
03					
04					
05	Open web steel joists, 12" deep, 5.2#/L.F., 24" O.C.	2.63 Lb.	.95	.63	1.58
06	Slab form, 28 gage, ⁹⁄₁₆" deep, galvanized	1 S.F.	.43	.27	.70
07	Concrete, 3000 psi, 2-½" thick	.008 C.Y.	.45		.45
08	Placing concrete	.008 C.Y.		.19	.19
09	Welded wire fabric 6 x 6 - #10/10	.011 C.S.F.	.08	.22	.30
10	Steel trowel finish	1 S.F.		.49	.49
11	Vinyl composition floor tile	1 S.F.	.61	.54	1.15
12	Ceiling furring, ¾" channels, 24" O.C.	1 S.F.	.18	.65	.83
13	Gypsum drywall, ⅝" thick, finished	1 S.F.	.29	.75	1.04
14	Paint ceiling	1 S.F.	.10	.33	.43
15	**TOTAL**	S.F.			
16			3.09	4.07	7.16
17					
18	For alternate floor systems:				
19	Open web joists, 16" deep, 6.6#/L.F.	S.F.	3.33	4.23	7.56
20	20" deep, 8.4# L.F.		3.66	4.44	8.10
21	24" deep, 11.5# L.F.		4.22	4.81	9.03
22	26" deep, 12.8# L.F.		4.46	4.96	9.42
23					
24					
25					
26					
27					
28					
29	Dust protection, add, minimum		1%	2%	
30	Maximum		4%	11%	
31	Equipment usage curtailment, add, minimum		1%	1%	
32	Maximum		3%	10%	
33	Material handling & storage limitation, add, minimum		1%	1%	
34	Maximum		6%	7%	
35	Protection of existing work, add, minimum		2%	2%	
36	Maximum		5%	7%	
37	Shift work requirements, add, minimum			5%	
38	Maximum			30%	
39	Temporary shoring and bracing, add, minimum		2%	5%	
40	Maximum		5%	12%	
41					
42					

This page illustrates and describes a wood joist floor system including wood joist, oak floor, sub-floor, bridging, sand and finish floor, furring, plasterboard, taped, finished and painted ceiling. Lines 03.5-714-04 thru 13 give the unit price and total price per square foot for this system. Prices for alternate wood joist floor systems are on Line Items 03.5-714-17 thru 23. Both material quantities and labor costs have been adjusted for the system listed.

Factors: To adjust for job conditions other than normal working situations use Lines 03.5-714-27 thru 40.

Example: You are to install the system during off peak hours, 6 P.M. to 2 A.M. Go to Line 03.5-714-38 and apply this percentage to the appropriate INST. costs.

LINE NO.	DESCRIPTION	QUANTITY	COST PER S.F.		
			MAT.	INST.	TOTAL
01	Wood joists, 2″ x 8″, 16″ O.C.,oak floor (sanded & finished),½″ sub				
02	Floor, 1″x 3″ bridging, furring, ⅝″ drywall, taped, finished and painted.				
03					
04	Wood joists, 2″ x 8″, 16″ O.C.	1 L.F.	.59	.54	1.13
05	C.D.X. plywood sub floor, ½″ thick	1 S.F.	.35	.40	.75
06	Bridging 1″ x 3″	.15 Pr.	.04	.35	.39
07	Oak flooring, No. 1 common	1 S.F.	2.70	1.55	4.25
08	Sand and finish floor	1 S.F.	.55	.89	1.44
09	Wood furring, 1″ x 3″, 16″ O.C.	1 L.F.	.15	.83	.98
10	Gypsum drywall, ⅝″ thick	1 S.F.	.23	.32	.55
11	Tape and finishing	1 S.F.	.06	.28	.34
12	Paint ceiling	1 S.F.	.10	.33	.43
13	TOTAL	S.F.	4.77	5.49	10.26
14					
15					
16	For alternate wood joist systems:				
17	16″ on center, 2″ x 6″ joists	S.F.	4.62	5.42	10.04
18	2″ x 10″		5.10	5.61	10.71
19	2″ x 12″		5.30	5.63	10.93
20	2″ x 14″		5.51	5.72	11.23
21	12″ on center, 2″ x 10″ joists		5.34	5.77	11.11
22	2″ x 12″		5.58	5.80	11.38
23	2″ x 14″		5.84	5.92	11.76
24					
25					
26					
27	Cut & patch to match existing construction, add, minimum		2%	3%	
28	Maximum		5%	9%	
29	Dust protection, add, minimum		1%	2%	
30	Maximum		4%	11%	
31	Equipment usage curtailment, add, minimum		1%	1%	
32	Maximum		3%	10%	
33	Material handling & storage limitation, add, minimum		1%	1%	
34	Maximum		6%	7%	
35	Protection of existing work, add, minimum		2%	2%	
36	Maximum		5%	7%	
37	Shift work requirements, add, minimum			5%	
38	Maximum			30%	
39	Temporary shoring and bracing, add, minimum		2%	5%	
40	Maximum		5%	12%	
41					
42					

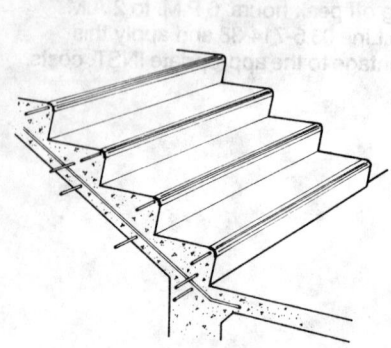

This page illustrates and describes a stair system based on a cost per flight price. Prices for various stair systems are on Line Items 03.9-104-07 thru 32. Both material quantities and labor costs have been adjusted for the system listed.

Factors: To adjust for job conditions other than normal working situations use Lines 03.9-104-35 thru 42.

Example: You are to install the system during evenings only. Go to Line 03.9-104-42 and apply this percentage to the appropriate MAT. and INST. costs.

LINE NO.	DESCRIPTION	QUANTITY	COST PER FLIGHT		
			MAT.	INST.	TOTAL
01	Below are various stair systems based on cost per flight of stairs, no side				
02	Walls. Stairs are 4'-0" wide, railings are included unless otherwise noted				
03					
04					
05					
06					
07	Concrete, cast in place, no nosings, no railings, 12 risers	Flight	256.08	991.92	1248
08	24 risers		512.16	1983.84	2496
09	Add for 1 intermediate landing		70.40	281.60	352
10	Concrete, cast in place, with nosings, no railings, 12 risers		601.92	1164.48	1766.40
11	24 risers		1203.84	2328.96	3532.80
12	Add for 1 intermediate landing		99.22	295.98	395.20
13	Steel, grating tread, safety nosing, 12 risers		1016.40	567.60	1584
14	24 risers		2032.80	1135.20	3168
15	Add for intermediate landing		528	144	672
16	Steel, cement fill pan tread, 12 risers		871.20	580.80	1452
17	24 risers		1742.40	1161.60	2904
18	Add for intermediate landing		528	144	672
19	Spiral, industrial, 4' - 6" diameter, 12 risers		1056	384	1440
20	24 risers		2112	768	2880
21	Wood, box stairs, oak treads, 12 risers		1459.70	431.30	1891
22	24 risers		2919.40	862.60	3782
23	Add for 1 intermediate landing		116.56	74.31	190.87
24	Wood, basement stairs, no risers, 12 steps		108.90	160.60	269.50
25	24 steps		217.80	321.20	539
26	Add for 1 intermediate landing		13.95	14.65	28.60
27	Wood, open, rough sawn cedar, 12 steps		1609.30	260.70	1870
28	24 steps		3218.60	521.40	3740
29	Add for 1 intermediate landing		12.52	15.20	27.72
30	Wood, residential, oak treads, 12 risers		1190.64	1449.36	2640
31	24 risers		2381.28	2898.72	5280
32	Add for 1 intermediate landing		109.58	67.04	176.62
33					
34					
35	Dust protection, add, minimum		1%	2%	
36	Maximum		4%	11%	
37	Material handling & storage limitation, add, minimum		1%	1%	
38	Maximum		6%	7%	
39	Protection of existing work, add, minimum		2%	2%	
40	Maximum		5%	7%	
41	Shift work requirements, add, minimum			5%	
42	Maximum			30%	

LINE NO.	DESCRIPTION	COST PER S.F.		
		MAT.	INST.	TOTAL
01	Flooring, carpet, nylon, level loop, 26 oz. light traffic	1.68	.54	2.22
02	40 oz. heavy traffic	2.53	.58	3.11
03	Nylon, plush, 20 oz. light traffic	1.03	.54	1.57
04	24 oz. medium traffic	1.82	.51	2.33
05	26 oz. heavy traffic	2.62	.60	3.22
06	28 oz. heavy traffic	2.98	.57	3.55
07	Tile, foamed back, needle punch	2.20	.42	2.62
08	Tufted loop	1.93	.42	2.35
09	Wool, 36 oz. medium traffic, level loop	3.47	.64	4.11
10	48 oz. heavy traffic, patterned	4.74	.59	5.33
11	Composition, epoxy, with colored chips, minimum	1.71	2.24	3.95
12	Maximum	2.70	3.10	5.80
13	Trowelled, minimum	2.20	2.69	4.89
14	Maximum	3.03	3.12	6.15
15	Terrazzo, ¼" thick, chemical resistant, minimum	3.63	2.92	6.55
16	Maximum	5.67	3.93	9.60
17	Resilient, asphalt tile, ⅛" thick	.88	.69	1.57
18	Conductive flrg, rubber, ⅛" thick	2.70	.87	3.57
19	Cork tile, ⅛" thick, standard finish	1.98	.87	2.85
20	Urethane finish	3.03	.87	3.90
21	PVC sheet goods for gyms, ¼" thick	3.08	3.42	6.50
22	⅜" thick	3.52	4.58	8.10
23	Vinyl composition 12" x 12" tile, plain, 1/16" thick	.61	.54	1.15
24	⅛" thick	1.60	.54	2.14
25	Vinyl tile, 12" x 12" x ⅛" thick, minimum	1.93	.54	2.47
26	Maximum	7.43	.52	7.95
27	Vinyl sheet goods, backed, .093" thick	1.43	1.19	2.62
28	.250" thick	2.59	1.19	3.78
29	Wood, maple strip 25/32" x 2-¼", finished, select	3.14	2.58	5.72
30	2nd and better	2.81	2.58	5.39
31	Oak, 25/32" x 2-¼" finished, clear	3.47	2.58	6.05
32	No. 1 common	3.25	2.44	5.69
33	Parquet, standard 5/16" thick finished, minimum	2.31	2.69	5
34	Maximum	6.05	3.74	9.79
35	Custom finished, minimum	11.88	2.87	14.75
36	Maximum	16.23	5.77	22
37	Prefinished, oak, 2-¼" wide	5.34	1.71	7.05
38	Ranch plank	6.44	1.96	8.40
39	Sleepers on concrete, treated, 24" O.C., 1" x 2"	.08	.20	.28
40	1" x 3"	.11	.24	.35
41	2" x 4"	.29	.32	.61
42	2" x 6"	.42	.37	.79
43	Ceiling, plaster, gypsum, 2 coats	.36	1.62	1.98
44	3 coats	.51	1.93	2.44
45	Perlite or vermiculite, 2 coats	.42	1.91	2.33
46	3 coats	.65	2.35	3
47	Gypsum lath, plain, ⅜" thick	.40	.36	.76
48	½" thick	.43	.39	.82
49	Firestop, ⅜" thick	.45	.44	.89
50	½" thick	.48	.47	.95
51	Metal lath, rib, 2.75 lb.	.20	.42	.62
52	3.40 lb.	.23	.45	.68
53	Diamond, 2.50 lb.	.19	.41	.60
54	3.40 lb.	.21	.52	.73
55	Drywall, taped and finished, standard, ½" thick	.24	.75	.99
56	⅝" thick	.29	.75	1.04
57	Fire resistant, ½" thick	.29	.75	1.04
58	⅝" thick	.31	.75	1.06

251

LINE NO.	DESCRIPTION	COST PER S.F.		
		MAT.	INST.	TOTAL
59	Water resistant, ½" thick			
60	⅝" thick	.35	.75	1.10
61	Finish, instead of taping	.40	.59	.99
62	For thin coat plaster, add			
63	Finish, textured spray, add	.10	.37	.47
64	Drywall, no finish included, see system 069-700	.11	.34	.45
65	Tile, stapled or glued, plastic coated min. fiber, ⅝" thick			
66	¾" thick	.53	1.43	1.96
67	Wood fiber, ½" thick	.69	1.44	2.13
68	¾" thick	.63	.72	1.35
69	Suspended, film faced fiberglass boards, ⅝" thick	.88	.72	1.60
70	3" thick	.43	.46	.89
71	⅝" thick min. fiber boards, standard face	1.05	.57	1.62
72	Aluminum faced	.50	.42	.92
73	Reveal edge wood fiber, 1" thick	1.16	.47	1.63
74	3" thick	.98	.48	1.46
75	Ceiling suspension systems, for tile, class "A", "T" bar, 2' x 4' grid	2.51	.64	3.15
76	2' x 2' grid	.41	.36	.77
77	Concealed "Z" bar, 12" module	.54	.44	.98
78		.62	.55	1.17
79	For plaster or drywall, ¾" channels, steel furring 16" O.C.			
80	24" O. C.	.22	.95	1.17
81	1-½" channels, 16" O. C.	.18	.65	.83
82	24" O. C.	.30	1.06	1.36
83	Ceiling framing, 2" x 4" studs, 16" O.C.	.24	.71	.95
84	24" O.C.	.22	.50	.72
		.14	.34	.48

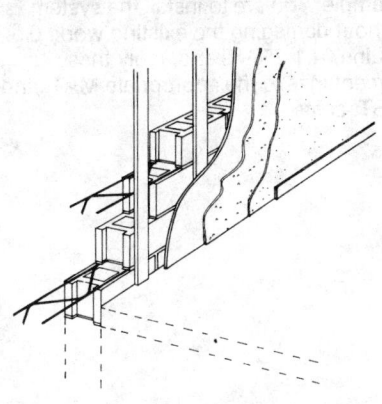

This page illustrates and describes a masonry concrete block system including concrete block wall, pointed, reinforcing, waterproofing, gypsum plaster on gypsum lath, and furring. Lines 04.1-208-04 thru 12 give the unit price and total price per square foot for this system. Prices for alternate masonry concrete block wall systems are on Line Items 04.1-208-15 thru 22. Both material quantities and labor costs have been adjusted for the system listed.

Factors: To adjust for job conditions other than normal working situations use Lines 04.1-208-27 thru 40.

Example: You are to install the system and match existing construction at several locations. Go to Line 04.1-208-28 and apply these percentages to the appropriate MAT. and INST. costs.

LINE NO.	DESCRIPTION	QUANTITY	COST PER S.F.		
			MAT.	INST.	TOTAL
01	Concrete block, 8" thick, reinforced every 2 courses, waterproofing gypsum				
02	Plaster over gypsum lath on 1" x 3" furring, interior painting & baseboard.				
03					
04	Concrete block, 8" x 8" x 16", reinforced	1 S.F.	1.87	3.13	5
05	Silicone waterproofing, 2 coats	1 S.F.	.20	.14	.34
06	Bituminous coating, ⅛" thick	1 S.F.	.15	.58	.73
07	Furring, 1" x 3", 16" O.C.	1 L.F.	.17	1.10	1.27
08	Gypsum lath, ⅜" thick	.11 S.Y.	.39	.36	.75
09	Gypsum plaster, 2 coats	.11 S.Y.	.36	1.40	1.76
10	Painting, 2 coats	1 S.F.	.07	.22	.29
11	Baseboard wood, ⁹⁄₁₆" x 2-⅝"	.1 L.F.	.09	.15	.24
12	TOTAL	S.F.	3.30	7.08	10.38
13					
14	For alternate exterior wall systems:				
15	8" thick block, fluted 2 sides	S.F.	5.10	8.38	13.48
16	Deep grooved		4.27	8.36	12.63
17	Slump block		4.41	7.27	11.68
18	Split rib		4.71	8.32	13.03
19					
20	12" thick block, regular	S.F.	3.47	8.51	11.98
21	Slump block		6.41	7.92	14.33
22	Split rib		5.40	9.63	15.03
23					
24					
25					
26					
27	Cut & patch to match existing construction, add, minimum		2%	3%	
28	Maximum		5%	9%	
29	Dust protection, add, minimum		1%	2%	
30	Maximum		4%	11%	
31	Equipment usage curtailment, add, minimum		1%	1%	
32	Maximum		3%	10%	
33	Material handling & storage limitation, add, minimum		1%	1%	
34	Maximum		6%	7%	
35	Protection of existing work, add, minimum		2%	2%	
36	Maximum		5%	7%	
37	Shift work requirements, add, minimum			5%	
38	Maximum			30%	
39	Temporary shoring and bracing, add, minimum		2%	5%	
40	Maximum		5%	12%	
41					
42					

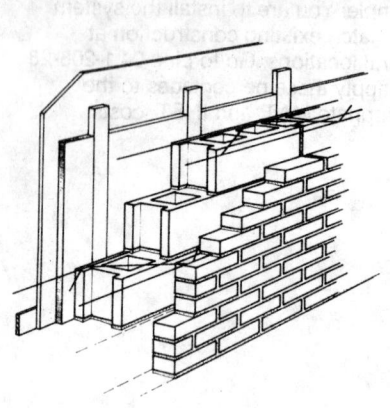

This page illustrates and describes a masonry wall, brick-stone system including brick, concrete block, durawall, insulation, plasterboard, taped and finished, furring, baseboard and painting interior. Lines 04.1-258-04 thru 12 give the unit price and total price per square foot for this system. Prices for alternate masonry wall, brick-stone systems are on Line Item 04.1-258-15 thru 25. Both material quantities and labor costs have been adjusted for the system listed.

Factors: To adjust for job conditions other than normal working situations use Lines 04.1-258-31 thru 42.

Example: You are to install the system without damaging the existing work. Go to Line 04.1-258-39 and apply these percentages to the appropriate MAT. and INST. costs.

LINE NO.	DESCRIPTION	QUANTITY	COST PER S.F.		
			MAT.	INST.	TOTAL
01	**Face brick, 4″thick, concrete block back-up, reinforce every second course,**				
02	**¾″insulation, furring, ½″drywall, taped, finish, and painted, baseboard**				
03					
04	Face brick, 4″ brick $275 per M	1 S.F.	2.34	6.06	8.40
05	Concrete back-up block, reinforced 8″ thick	1 S.F.	1.63	3.36	4.99
06	¾″ rigid polystyrene insulation	1 S.F.	.45	.36	.81
07	Furring, 1″ x 3″, wood, 16″ O.C.	1 L.F.	.17	.58	.75
08	Drywall, ½″ thick	1 S.F.	.19	.28	.47
09	Taping & finishing	1 S.F.	.06	.28	.34
10	Painting, 2 coats	1 S.F.	.10	.33	.43
11	Baseboard, wood, ⁹⁄₁₆″ x 2-⅝″	.1 L.F.	.09	.15	.24
12	TOTAL	S.F.	5.03	11.40	16.43
13					
14	For alternate exterior wall systems:				
15	Face brick, Norman, 4″ x 2-⅔″ x 12″ (4.5 per S.F.) $510 per M	S.F.	5.68	9.50	15.18
16	Roman, 4″ x 2″ x 12″ (6.0 per S.F.) $565 per M		6.85	10.68	17.53
17	Engineer, 4″ x 3-⅛″ x 8″ (5.63 per S.F.) $385 per M		5.39	10.44	15.83
18	S.C.R., 6″ x 2-⅔″ x 12″ (4.5 per S.F.) $685 per M		6.51	9.62	16.13
19	Jumbo, 6″ x 4″ x 12″ (3.0 per S.F.) $985 per M		6.25	8.38	14.63
20	Norwegian, 6″ x 3-⅛″ x 12″ (3.75 per S.F.) $585 per M		5.40	8.88	14.28
21					
22					
23	Stone, veneer, fieldstone, 6″ thick	S.F.	8.05	9.66	17.71
24	Marble, 2″ thick		33.49	16.54	50.03
25	Limestone, 2″ thick		15.62	19.41	35.03
26					
27					
28					
29					
30					
31	Cut & patch to match existing construction, add, minimum		2%	3%	
32	Maximum		5%	9%	
33	Dust protection, add, minimum		1%	2%	
34	Maximum		4%	11%	
35	Equipment usage curtailment, add, minimum		1%	1%	
36	Maximum		3%	10%	
37	Material handling & storage limitation, add, minimum		1%	1%	
38	Maximum		6%	7%	
39	Protection of existing work, add, minimum		2%	2%	
40	Maximum		5%	7%	
41	Shift work requirements, add, minimum			5%	
42	Maximum			30%	

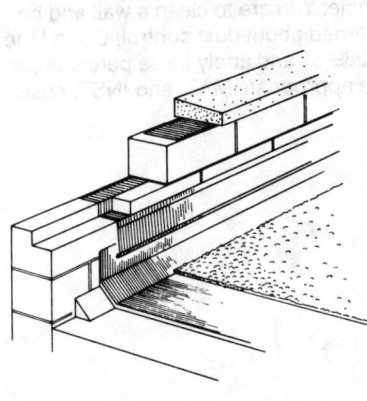

This page illustrates and describes a parapet wall system including wall, coping, flashing and cant strip. Lines 04.1-288-04 thru 10 give the unit price and cost per lineal foot for this system. Prices for alternate parapet wall systems are on Lines 04.1-288-13 thru 24. Both material quantities and labor costs have been adjusted for the system listed.

Factors: To adjust for job conditions other than normal working situations, use Lines 04.1-288-29 thru 40.

Example: You are to install the system without damaging the adjacent property. Go to Line 04.1-288-36 and apply these percentages to the appropriate MAT. and INST. costs.

LINE NO.	DESCRIPTION	QUANTITY	COST PER L.F.		
			MAT.	INST.	TOTAL
01	Concrete block parapet, incl. reinf., coping, flashing & cant strip, 2'high				
02					
03					
04	8" concrete block	2 S.F.	3.74	6.26	10
05	Masonry reinforcing	1 L.F.	.20	.10	.30
06	Coping, precast	1 L.F.	18.04	5.96	24
07	Roof cant	1 L.F.	.57	.88	1.45
08	Through wall flashing	1 L.F.	2.09	1.28	3.37
09	Cap flashing	1 L.F.	2.86	2.74	5.60
10	TOTAL	L.F.	27.50	17.22	44.72
11					
12	For alternate systems:				
13	Concrete block				
14	12" thick	L.F.	35.80	19.97	55.77
15	Split rib block, 8" thick		30.32	19.70	50.02
16	12" thick		38.70	23.37	62.07
17	Brick, common brick @ $216/M, 8" wall		31.28	30.54	61.82
18	12" wall		35.33	38.89	74.22
19	4" brick, 4" backup block		29.74	29.38	59.12
20	8" backup block		30.80	30	60.80
21	Stucco on masonry		27.73	19.85	47.58
22	On wood frame		10.20	20.99	31.19
23	Wood, T 1-11 siding		14.09	20.78	34.87
24	Boards, 1" x 6" cedar		15.30	23.36	38.66
25					
26					
27					
28					
29	Dust protection, add, minimum		1%	2%	
30	Maximum		4%	11%	
31	Equipment usage curtailment, add, minimum		1%	1%	
32	Maximum		3%	10%	
33	Material handling & storage limitation, add, minimum		1%	1%	
34	Maximum		6%	7%	
35	Protection of existing work, add, minimum		2%	2%	
36	Maximum		5%	7%	
37	Shift work requirements, add, minimum			5%	
38	Maximum			30%	
39	Temporary shoring and bracing, add, minimum		2%	5%	
40	Maximum		5%	12%	
41					
42					

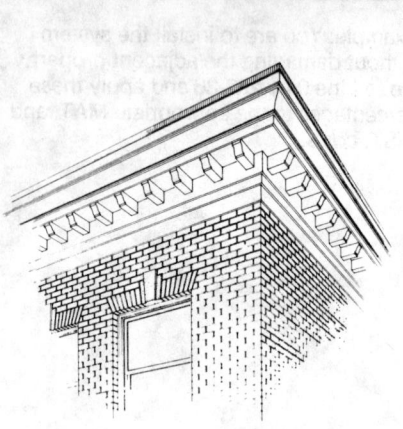

This page illustrates and describes a masonry cleaning and restoration system, including staging, cleaning, repointing. Lines 04.1-289-04 thru 08 give the unit price and cost per square foot for this system. Prices for alternate systems are on Lines 04.1-289-13 thru 25. Both material quantities and labor costs have been adjusted for the system listed.

Factors: To adjust for conditions other than normal working situations, use Lines 04.1-289-29 thru 42.

Example: You are to clean a wall and be concerned about dust control. Go to Line 04.1-289-31 and apply these percentages to the appropriate MAT. and INST. costs.

LINE NO.	DESCRIPTION	QUANTITY	COST PER S.F.		
			MAT.	INST.	TOTAL
01	Repoint existing building, brick, running bond, high pressure cleaning,				
02	Water only, soft old mortar.				
03					
04	Scaffold, building exterior	.01 C.S.F.	.14	.52	.66
05	Cleaning, high pressure water only	1 S.F.		.87	.87
06	Repoint, brick running bond	1 S.F.	.22	3.01	3.23
07					
08	TOTAL	S.F.	.36	4.40	4.76
09					
10					
11					
12	For alternate masonry surfaces:				
13	Brick, common bond	S.F.	.36	4.52	4.88
14	Flemish bond		.37	4.72	5.09
15	English bond		.37	5.04	5.41
16	Add for wire cut face brick		.28		.28
17	Stone, 2' x 2' blocks		.45	3.30	3.75
18	2' x 4' blocks		.37	2.85	3.22
19	For hard mortar, brick or stone, add	↓		30%	
20	Add to above prices for alternate cleaning systems:				
21	Chemical brush and wash	S.F.	.04	.66	.70
22	High pressure chemical and water		.03	.33	.36
23	Sandblasting, wet system		.21	.90	1.11
24	Dry system		.17	.52	.69
25	Steam cleaning	↓		.26	.26
26					
27					
28					
29	Cut & patch to match existing construction, add, minimum		2%	3%	
30	Maximum		5%	9%	
31	Dust protection, add, minimum		1%	2%	
32	Maximum		4%	11%	
33	Equipment usage curtailment, add, minimum		1%	1%	
34	Maximum		3%	10%	
35	Material handling & storage limitation, add, minimum		1%	1%	
36	Maximum		6%	7%	
37	Pedestrian protection, add				
38					
39	Protection of existing work, add, minimum		2%	2%	
40	Maximum		5%	7%	
41	Shift work requirements, add, minimum			5%	
42	Maximum			30%	

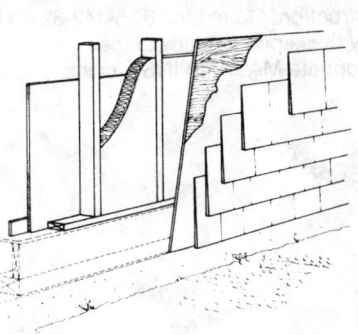

This page illustrates and describes a wood frame exterior wall system including wood studs, sheathing, felt, insulation, plasterboard, taped and finished, baseboard and painted interior. Lines 04.1-416-05 thru 13 give the unit price and total price per square foot for this system. Prices for alternate wood frame exterior wall systems are on Line Items 04.1-416-17 thru 27. Both material quantities and labor costs have been adjusted for the system listed.

Factors: To adjust for job conditions other than normal working situations use Lines 04.1-416-31 thru 42.

Example: You are to install the system with need for complete temporary bracing. Go to Line 04.1-416-42 and apply these percentages to the appropriate MAT. and INST. costs.

LINE NO.	DESCRIPTION	QUANTITY	COST PER S.F.		
			MAT.	INST.	TOTAL
01	Wood stud wall, cedar shingle siding, building paper, plywood sheathing,				
02	Insulation, ⅝″ drywall, taped, finished and painted, baseboard.				
03					
04					
05	2″ x 4″ wood studs, 16″ O.C.	.1 L.F.	.31	.60	.91
06	½″ CDX sheathing	1 S.F.	.35	.42	.77
07	18″ No. 1 red cedar shingles, 7-½″ exposure	.008 C.S.F.	.70	1.02	1.72
08	15# felt paper	.01 C.S.F.	.02	.09	.11
09	3-½″ fiberglass insulation	1 S.F.	.30	.18	.48
10	⅝″ drywall, taped and finished	1 S.F.	.29	.57	.86
11	Baseboard trim, stock pine, 9⁄16″ x 3-½″, painted	.1 L.F.	.09	.15	.24
12	Paint, 2 coats, interior	1 S.F.	.10	.33	.43
13	TOTAL	S.F.	2.16	3.36	5.52
14					
15					
16	For alternate exterior wall systems:				
17	Aluminum siding, horizontal clapboard	S.F.	2.56	3.47	6.03
18	Cedar bevel siding, ½″ x 6″, vertical , painted		3.04	3.49	6.53
19	Redwood siding 1″ x 4″ to 1″ x 6″ vertical, T & G		2.85	3.77	6.62
20	Board and batten		2.67	3.45	6.12
21	Ship lap siding		2.66	3.49	6.15
22	Plywood, grooved (T1-11) fir		2.03	3.22	5.25
23	Redwood		2.53	3.22	5.75
24	Southern yellow pine		2.09	3.22	5.31
25	Masonry on stud wall, stucco, wire and plaster		2.15	5.72	7.87
26	Stone veneer		6.82	6.66	13.48
27	Brick veneer, brick $275 per M		3.80	8.40	12.20
28					
29					
30					
31	Cut & patch to match existing construction, add, minimum		2%	3%	
32	Maximum		5%	9%	
33	Dust protection, add, minimum		1%	2%	
34	Maximum		4%	11%	
35	Material handling & storage limitation, add, minimum		1%	1%	
36	Maximum		6%	7%	
37	Protection of existing work, add, minimum		2%	2%	
38	Maximum		5%	7%	
39	Shift work requirements, add, minimum			5%	
40	Maximum			30%	
41	Temporary shoring and bracing, add, minimum		2%	5%	
42	Maximum		5%	12%	

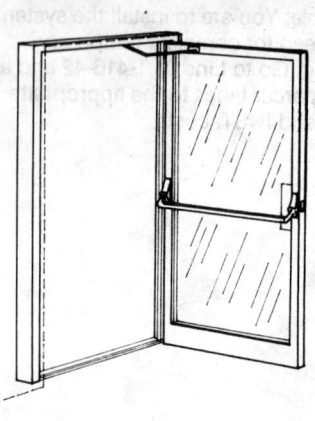

This page illustrates and describes a commercial metal door system, including a single aluminum and glass door, narrow stiles, jamb, hardware weatherstripping, panic hardware and closer. Lines 04.6-142-04 thru 10 give the unit price and total price on a cost each basis for this system. Prices for alternate commercial metal door systems are on Line Items 04.6-142-13 thru 25. Both material quantities and labor costs have been adjusted for the system listed.

Factors: To adjust for job conditions other than normal working situations, use Lines 04.6-142-31 thru 40.

Example: You are to install the system and cut and patch to match existing construction. Go to Line 04.6-142-32 and apply these percentages to the appropriate MAT. and INST. costs.

LINE NO.	DESCRIPTION	QUANTITY	COST EACH		
			MAT.	INST.	TOTAL
01	Single aluminum and glass door, 3'-0"x7'-0", with narrow stiles, ext. jamb.				
02	Weatherstripping, ½" tempered insul. glass, panic hardware, and closer.				
03					
04	Aluminum door, 3'-0" x 7'-0" x 1-¾", narrow stiles	1 Ea.	292.05	175.45	467.50
05	Exterior jamb and trim	1 Set	175.23	105.27	280.50
06	Hardware	1 Set	116.82	70.18	187
07	Tempered insulating glass, ½" thick	20 S.F.	478.50	201.50	680
08	Panic hardware	1 Set	258.50	96.50	355
09	Automatic closer	1 Ea.	66	44	110
10	TOTAL	Ea.	1387.10	692.90	2080
11					
12	For alternate door systems:				
13	Single aluminum and glass with transom, 3'-0" x 10'-0"	Ea.	1670.90	821.10	2492
14	Anodized aluminum and glass, 3'-0" x 7'-0"		1591.54	815.71	2407.25
15	With transom, 3'-0" x 10'-0"		1907.68	960.57	2868.25
16	Steel, deluxe, hollow metal 3'-0" x 7'-0"		777.26	295.74	1073
17	With transom 3'-0" x 10'-0"		905.96	329.94	1235.90
18	Fire door, "A" label, 3'-0" x 7'-0"		805.09	295.41	1100.50
19	Double, aluminum and glass, 6'-0" x 7'-0"		2321	1174	3495
20	With transom, 6'-0" x 10'-0"		2813.80	1425.20	4239
21	Anodized aluminum and glass 6'-0" x 7'-0"		2648.25	1363	4011.25
22	With transom, 6'-0" x 10'-0"		3179.55	1645.70	4825.25
23	Steel, deluxe, hollow metal 6'-0" x 7'-0"		1288.54	499.46	1788
24	With transom, 6'-0" x 10'-0"		1545.94	567.86	2113.80
25	Fire door, "A" label, 6'-0" x 7'-0"		1299.54	494.46	1794
26					
27					
28					
29					
30					
31	Cut & patch to match existing construction, add, minimum		2%	3%	
32	Maximum		5%	9%	
33	Dust protection, add, minimum		1%	2%	
34	Maximum		4%	11%	
35	Material handling & storage limitation, add, minimum		1%	1%	
36	Maximum		6%	7%	
37	Protection of existing work, add, minimum		2%	2%	
38	Maximum		5%	7%	
39	Shift work requirements, add, minimum			5%	
40	Maximum			30%	
41					
42					

This page illustrates and describes residential door systems including a door, frame, trim, hardware, weatherstripping, stained and finished. Lines 04.6-144-04 thru 11 give the unit price and total price on a cost each basis for this system. Prices for alternate residential door systems are on Line Items 04.6-144-15 thru 22. Both material quantities and labor costs have been adjusted for the system listed.

Factors: To adjust for job conditions other than normal working situations use Lines 04.6-144-31 thru 40.

Example: You are to install the system with a material handling and storage limitation. Go to Line 04.6-144-36 and apply these percentages to the appropriate MAT. and INST. costs.

LINE NO.	DESCRIPTION	QUANTITY	COST EACH		
			MAT.	INST.	TOTAL
01	Single solid wood colonial door, 3' x 6'-8" with wood frame and trim,				
02	Stained and finished, including hardware and weatherstripping.				
03					
04	Solid wood colonial door, fir 3' x 6'-8" x 1-¾", hinges	1 Ea.	318	38	356
05	Exterior frame, with trim	1 Set	66.39	27.11	93.50
06	Interior trim	1 Set	9.35	48.65	58
07	Lockset	1 Set	24.20	20.80	45
08	Sill, oak 8" deep	3.5 Ea.	33.88	20.72	54.60
09	Weatherstripping	1 Set	7.59	37.41	45
10	Stained and finished	1 Ea.	5.89	33.11	39
11	TOTAL	Ea.	465.30	225.80	691.10
12					
13					
14	For alternate exterior door system:				
15	Single doors				
16	Hollow metal exterior door, plain	Ea.	325.84	220.26	546.10
17	Solid core wood door, plain	"	259.84	221.26	481.10
18					
19	Double doors, 6' x 6'-8"				
20	Solid colonial double doors, fir	Ea.	804.92	296.08	1101
21	Hollow metal exterior doors, plain		527.96	293.04	821
22	Hollow core wood doors, plain		395.96	295.04	691
23					
24					
25					
26					
27					
28					
29					
30					
31	Cut & patch to match existing construction, add, minimum		2%	3%	
32	Maximum		5%	9%	
33	Dust protection, add, minimum		1%	2%	
34	Maximum		4%	11%	
35	Material handling & storage limitation, add, minimum		1%	1%	
36	Maximum		6%	7%	
37	Protection of existing work, add, minimum		2%	2%	
38	Maximum		5%	7%	
39	Shift work requirements, add, minimum			5%	
40	Maximum			30%	
41					
42					

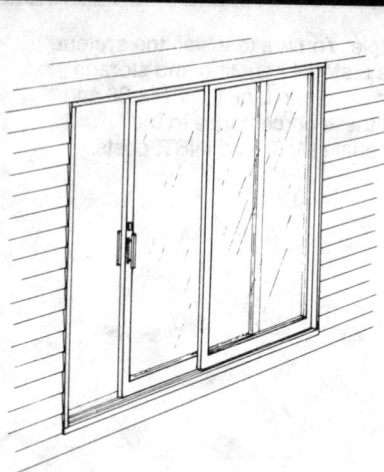

This page illustrates and describes sliding door systems including a sliding door, frame, interior and exterior trim with exterior staining. Lines 04.6-152-04 thru 07 give the unit price and total price on a cost each basis for this system. Prices for alternate sliding door systems are on Line Items 04.6-152-11 thru 22. Both material quantities and labor costs have been adjusted for the system listed.

Factors: To adjust for job conditions other than normal working situations use Lines 04.6-152-27 thru 40.

Example: You are to install the system with temporary shoring and bracing. Go to Line 04.6-152-39 and apply these percentages to the appropriate MAT. and INST. costs.

LINE NO.	DESCRIPTION	QUANTITY	COST EACH		
			MAT.	INST.	TOTAL
01	**Sliding wood door, 6′-0″ x 6′-8″, with wood frame, interior and exterior**				
02	**Trim and exterior staining.**				
03					
04	Sliding wood door, standard, 6′-0″ x 6′-8″, insulated glass	1 Ea.	517	143	660
05	Interior & exterior trim	1 Set	12.98	24.02	37
06	Stain door & trim	1 Ea.	4.06	21.44	25.50
07	TOTAL	Ea.	534.04	188.46	722.50
08					
09					
10	For alternate sliding door systems:				
11	Wood, standard, 8′-0″ x 6′-8″, insulated glass	Ea.	678.47	238.35	916.82
12	12′-0″ x 6′-8″		1078.69	281.42	1360.11
13	Vinyl coated, 6′-0″ x 6′-8″		869.54	187.96	1057.50
14	8′-0″ x 6′-8″		1118.47	248.35	1366.82
15	12′-0″ x 6′-8″		1661.69	298.42	1960.11
16					
17	Aluminum, standard, 6′-0″ x 6′-8″, insulated glass	Ea.	339.24	203.56	542.80
18	8′-0″ x 6′-8″		429.26	256.32	685.58
19	12′-0″ x 6′-8″		536.71	304.43	841.14
20	Anodized, 6′-0″ x 6′-8″		504.24	203.56	707.80
21	8′-0″ x 6′-8″		528.26	257.32	785.58
22	12′-0″ x 6′-8″		642.31	303.83	946.14
23					
24	Deduct for single glazing	Ea.	55		55
25					
26					
27	Cut & patch to match existing construction, add, minimum		2%	3%	
28	Maximum		5%	9%	
29	Dust protection, add, minimum		1%	2%	
30	Maximum		4%	11%	
31	Equipment usage curtailment, add, minimum		1%	1%	
32	Maximum		3%	10%	
33	Material handling & storage limitation, add, minimum		1%	1%	
34	Maximum		6%	7%	
35	Protection of existing, work, add, minimum		2%	2%	
36	Maximum		5%	7%	
37	Shift work requirements, add, minimum			5%	
38	Maximum			30%	
39	Temporary shoring and bracing, add, minimum		2%	5%	
40	Maximum		5%	12%	
41					
42					

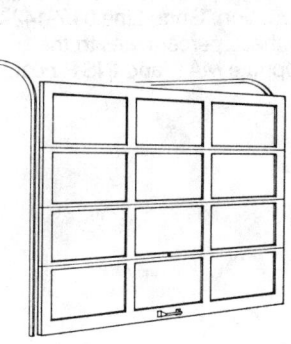

This page illustrates and describes overhead door systems including an overhead door, track, hardware, trim, and electric door opener. Lines 04.6-702-04 thru 08 give the unit price and total price on a cost each basis for this system. Prices for alternate overhead door systems are on Line Items 04.6-702-13 thru 23. Both material quantities and labor costs have been adjusted for the system listed.

Factors: To adjust for job conditions other than normal working situations use Lines 04.6-702-31 thru 40.

Example: You are to install the system and match existing construction. Go to Line 04.6-702-32 and apply these percentages to the appropriate MAT. and INST. costs.

LINE NO.	DESCRIPTION	QUANTITY	COST EACH MAT.	INST.	TOTAL
01	Wood overhead door, commercial sectional, including track, door, hardware				
02	And trim, electrically operated.				
03					
04	Commercial, heavy duty wood door, 8' x 8' x 1-¾" thick	1 Ea.	443.30	286.70	730
05	Wood frames and trim	1 Set	35.90	66.34	102.24
06	Painting, two coats	1 Ea.	11.09	105.55	116.64
07	Electric trolley operator	1 Ea.	396	144	540
08	TOTAL	Ea.	886.29	602.59	1488.88
09					
10					
11					
12	For alternate sectional overhead door systems:				
13	Commercial, wood, 1-¾" thick, 12' x 12'	Ea.	1381.38	864.70	2246.08
14	14' x 14'		2026.77	1007.35	3034.12
15	Fiberglass & aluminum 12' x 12'		1630.73	884.79	2515.52
16	20' x 20'		4570.94	2215.06	6786
17	Residential, wood, 9' x 7'		336.29	236.40	572.69
18	16' x 7'		645.95	352.74	998.69
19	Hardboard faced, 9' x 7'		325.29	232.40	557.69
20	16' x 7'		557.95	350.74	908.69
21	Fiberglass & aluminum, 9' x 7'		435.29	232.40	667.69
22	16' x 7'		717.45	351.24	1068.69
23	For residential electric opener, add		330	35	365
24					
25					
26					
27					
28					
29					
30					
31	Cut & patch to match existing construction, add, minumum		2%	3%	
32	Maximum		5%	9%	
33	Dust protection, add, minimum		1%	2%	
34	Maximum		4%	11%	
35	Material handling & storage limitation, add, minimum		1%	1%	
36	Maximum		6%	7%	
37	Protection of existing work, add, minimum		2%	2%	
38	Maximum		5%	7%	
39	Shift work requirements, add, minimum			5%	
40	Maximum			30%	
41					
42					

This page illustrates and describes an aluminum window system including double hung aluminum window, exterior and interior trim, hardware and insulating glass. Lines 04.7-142-04 thru 08 give the unit price and total price on a cost each basis for this system. Prices for alternate aluminum window systems are on Line Items 04.7-142-11 thru 24. Both material quantities and labor costs have been adjusted for the system listed.

Factors: To adjust for job conditions other than normal working situations use Lines 04.7-142-31 thru 40.

Example: You are to install the system and cut and patch to match existing construction. Go to Line 04.7-142-32 and apply these percentages to the appropriate MAT. and INST. costs.

LINE NO.	DESCRIPTION	QUANTITY	COST EACH		
			MAT.	INST.	TOTAL
01	Double hung aluminum window, 2'-4" x 2'-6" exterior and interior trim,				
02	Hardware glazed with insulating glass.				
03					
04	Double hung aluminum window, 2'-4" x 2' 6"	1 Ea.	62.84	20.53	83.37
05	Exterior and interior trim	1 Set	11.55	22.45	34
06	Hardware	1 Set	1.71	11.99	13.70
07	Insulating glass	5.83 S.F.	46.49	16.77	63.26
08	TOTAL	Ea.	122.59	71.74	194.33
09					
10	For alternate window systems:				
11	Aluminum, double hung, 2'-8" x 4'-6"	Ea.	243.27	117.23	360.50
12	3'- 4" x 5'-6"		367.48	177.22	544.70
13	Casement, 3'-6" x 2'-4"		230.03	85.73	315.76
14	4'-6" x 2'-4"		296.88	106.42	403.30
15	5'-6" x 2'-4"		364.26	140.58	504.84
16	Projected window, 2'-1" x 3'-0"		172.41	74.35	246.76
17	3'-5" x 3'-0"		279.22	105.94	385.16
18	4'-0" x 4'-0"		431.15	162.15	593.30
19	Horizontal sliding 3'-0" x 3'-0"		189.98	91.72	281.70
20	3'-6" x 4'-0"		293.10	129.60	422.70
21	5'-0" x 6'-0"		612.76	250.94	863.70
22	Picture window, 3'-8" x 3'-1"		200	106.58	306.58
23	4'-4" x 4'-5"		334.73	162.77	497.50
24	5'-8" x 4'-9"		468.87	231.96	700.83
25					
26					
27					
28					
29					
30					
31	Cut & patch to match existing construction, add, minimum		2%	3%	
32	Maximum		5%	9%	
33	Dust protection, add, minimum		1%	2%	
34	Maximum		4%	11%	
35	Material handling & storage limitation, add, minimum		1%	1%	
36	Maximum		6%	7%	
37	Protection of existing work, add, minimum		2%	2%	
38	Maximum		5%	7%	
39	Shift work requirements, add, minimum			5%	
40	Maximum			30%	
41					
42					

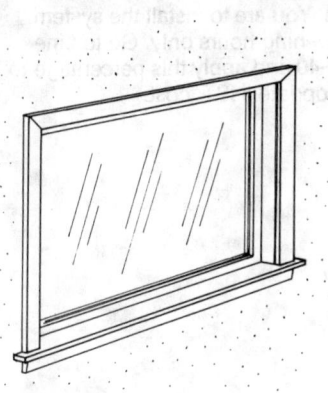

This page illustrates and describes a wood window system including wood picture window, exterior and interior trim, hardware and insulating glass. Lines 04.7-144-04 thru 07 give the unit price and total price on a cost each basis for this system. Prices for alternate wood window systems are on Line Items 04.7-144-13 thru 18. Both material quantities and labor costs have been adjusted for the system listed.

Factors: To adjust for job conditions other than normal working situations use Lines 04.7-144-31 thru 40.

Example: You are to install the above system where dust control is vital. Go to Line 04.7-144-34 and apply these percentages to the appropriate MAT. and INST. costs.

LINE NO.	DESCRIPTION	QUANTITY	COST EACH		
			MAT.	INST.	TOTAL
01	Wood picture window 4'-0" x 4'-6", exterior and interior trim, hardware,				
02	Glazed with insulating glass.				
03					
04	4'-0" x 4'-6" wood picture window with insulating glass	1 Ea.	242	53	295
05	Exterior and interior trim	1 Set	22	48	70
06	Hardware	1 Set	1.71	11.99	13.70
07	TOTAL	Ea.	265.71	112.99	378.70
08					
09					
10					
11					
12	For alternate window systems:				
13	Picture window 5'-0" x 4'-0"	Ea.	284.41	114.29	398.70
14	6'-0" x 4'-6"		317.41	116.29	433.70
15	Bow, bay window, 8'-0" x 5'-0", standard		1184.21	124.49	1308.70
16	Deluxe		980.71	127.99	1108.70
17	Bow, bay window, 12'-0" x 6'-0" standard		1464.71	143.99	1608.70
18	Deluxe		2680.21	153.49	2833.70
19					
20					
21					
22					
23					
24					
25					
26					
27					
28					
29					
30					
31	Cut & patch to match existing construction, add, minimum		2%	3%	
32	Maximum		5%	9%	
33	Dust protection, add, minimum		1%	2%	
34	Maximum		4%	11%	
35	Material handling & storage limitation, add, minimum		1%	1%	
36	Maximum		6%	7%	
37	Protection of existing work, add, minimum		2%	2%	
38	Maximum		5%	7%	
39	Shift work requirements, add, minimum			5%	
40	Maximum			30%	
41					
42					

This page illustrates and describes a wood window system including double hung wood window, exterior and interior trim, hardware and insulating glass. Lines 04.7-145-04 thru 07 give the unit price and total price on a cost each basis for this system. Prices for alternate wood window systems are on Line Items 04.7-145-13 thru 23. Both material quantities and labor costs have been adjusted for the system listed.

Factors: To adjust for job conditions other than normal working situations use Lines 04.7-145-31 thru 40.

Example: You are to install the system during evening hours only. Go to Line 04.7-145-40 and apply this percentage to the appropriate INST. cost.

LINE NO.	DESCRIPTION	QUANTITY	COST EACH		
			MAT.	INST.	TOTAL
01	Double hung wood window 2'-0" x 3'-0", exterior and interior trim,				
02	Hardware, glazed with insulating glass.				
03					
04	2'-0" x 3'-0" double hung wood window, with insulating glass	1 Ea.	134.20	30.80	165
05	Exterior and interior trim	1 Set	11.55	22.45	34
06	Hardware	1 Set	1.71	11.99	13.70
07	TOTAL	Ea.	147.46	65.24	212.70
08					
09					
10					
11					
12	For alternate window systems:				
13	Double hung, 3'-0" x 4'-0"	Ea.	173.31	70.39	243.70
14	4'-0" x 4'-6"		210.71	97.99	308.70
15	Casement 2'-0" x 3'-0"		198.06	64.64	262.70
16	2 leaf, 4'-0" x 4'-0"		495.61	103.09	598.70
17	3 leaf, 6'-0" x 6'-0"		1046.71	161.99	1208.70
18	Awning, 2'-10" x 1'-10"		178.26	64.44	242.70
19	3'-6" x 2'-4"		216.21	72.49	288.70
20	4'-0" x 3'-0"		261.31	97.39	358.70
21	Horizontal siding 3'-0" x 2'-0"		157.36	65.34	222.70
22	4'-0" x 3'-6"		223.91	74.79	298.70
23	6'-0" x 5'-0"		359.21	94.49	453.70
24					
25					
26					
27					
28					
29					
30					
31	Cut & patch to match existing construction, add, minimum		2%	3%	
32	Maximum		5%	9%	
33	Dust protection, add, minimum		1%	2%	
34	Maximum		4%	11%	
35	Material handling & storage limitation, add, minimum		1%	1%	
36	Maximum		6%	7%	
37	Protection of existing work, add, minimum		2%	2%	
38	Maximum		5%	7%	
39	Shift work requirements, add, minimum			5%	
40	Maximum			30%	
41					
42					

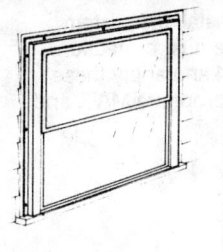

This page illustrates and describes storm window and door systems based on a cost each price. Prices for alternate storm window and door systems are on Line Items 04.7-152-04 thru 24. Both material quantities and labor costs have been adjusted for the system listed.

Factors: To adjust for job conditions other than normal working situations use Lines 04.7-152-31 thru 40 and apply these percentages to the appropriate MAT and INST. costs.

Example: You are to install the system and protect all existing construction. Go to Line 04.7-152-38 and apply these percentages to the appropriate MAT. and INST. costs.

LINE NO.	DESCRIPTION	QUANTITY	COST EACH		
			MAT.	INST.	TOTAL
01	**Storm window and door systems, single glazed**				
02					
03					
04	Window, custom aluminum anodized, 2'-0" x 3'-5"	Ea.	77	20	97
05	2'-6" x 5'-0"		116.60	23.40	140
06	4'-0" x 6'-0"		209	26	235
07	White painted aluminum, 2'-0" x 3'-5"		72.60	19.40	92
08	2'-6" x 5'-0"		116.60	23.40	140
09	4'-0" x 6'-0"		209	26	235
10	Average quality aluminum, anodized, 2'-0" x 3'-5"		62.70	19.30	82
11	2'-6" x 5'-0"		79.20	20.80	100
12	4'-0" x 6'-0"		88	22	110
13	White painted aluminum, 2'-0" x 3'-5"		59.40	19.60	79
14	2'-6" x 5'-0"		66	21	87
15	4'-0" x 6'-0"		71.50	23.50	95
16	Mill finish, 2'-0" x 3'-5"		55	20	75
17	2'-6" x 5'-0"		61.60	21.40	83
18	4'-0" x 6'-0"		68.20	23.80	92
19					
20	Door, aluminum anodized 3'-0" x 6'-8"	Ea.	209	41	250
21	White painted aluminum, 3'-0" x 6'-8"		220	40	260
22	Mill finish, 3'-0" x 6'-8"		200.20	44.80	245
23					
24	Wood, storm and screen, painted	Ea.	195.80	64.20	260
25					
26					
27					
28					
29					
30					
31	Cut & patch to match existing construction, add, minimum		2%	3%	
32	Maximum		5%	9%	
33	Dust protection, add, minimum		1%	2%	
34	Maximum		4%	11%	
35	Material handling & storage limitation, add, minimum		1%	1%	
36	Maximum		6%	7%	
37	Protection of existing work, add, minimum		2%	2%	
38	Maximum		5%	7%	
39	Shift work requirements, add, minimum			5%	
40	Maximum			30%	
41					
42					

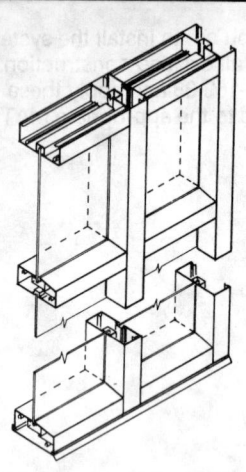

This page illustrates and describes a window wall system including aluminum tube framing, caulking, and glass. Lines 04.7-700-04 thru 10 give the unit price and total price per square foot for this system. Prices for alternate window wall systems are on Line Items 04.7-700-14 thru 26. Both material quantities and labor costs have been adjusted for the system listed.

Factors: To adjust for job conditions other than normal working situations use Lines 04.7-700-31 thru 40.

Example: You are to install the system with need for complete dust protection. Go to Line 04.7-700-34 and apply these percentages to the appropriate MAT. and INST. costs.

LINE NO.	DESCRIPTION	QUANTITY	COST PER S.F.		
			MAT.	INST.	TOTAL
01	Window wall, including aluminum header, sill, mullions, caulking				
02	And glass				
03					
04	Header, mill finish, 1-¾" x 4-½" deep	.167 L.F.	1.08	1.16	2.24
05	Sill, mill finish, 1-¾" x 4-½" deep	.167 L.F.	1.11	1.14	2.25
06	Vertical mullion, 1-¾" x 4-½" deep, 6' O.C.	.191 L.F.	1.54	1.50	3.04
07	Horizontal mullion, 1-¾" x 4-½" deep	.167 L.F.	1.34	1.32	2.66
08	Caulking	.381 L.F.	.02	.44	.46
09	Glass, ¼" plate	1 S.F.	3.36	4.74	8.10
10	TOTAL	S.F.	8.45	10.30	18.75
11					
12	For alternate systems:				
13					
14	Mill finish, 2" x 4-½" deep, insulating glass	S.F.	12.03	11.91	23.94
15	Thermo-break, 2-¼" x 4-½" deep, insulating glass		12.29	12.09	24.38
16	Bronze finish, 1-¾" x 4-½" deep, ¼" plate glass		9.38	11.18	20.56
17	2" x 4-½" deep, insulating glass		13.02	12.88	25.90
18	Thermo-break, 2-¼" x 4-½" deep, insulating glass	↓	13.31	13.12	26.43
19					
20	Black finish, 1-¾" x 4-½" deep, ¼" plate glass	S.F.	9.84	11.63	21.47
21	2" x 4-½" deep, insulating glass		13.52	13.36	26.88
22	Thermo-break, 2-¼" x 4-½" deep, insulating glass	↓	13.84	13.61	27.45
23					
24	Stainless steel, 1-¾" x 4-½" deep, ¼" plate glass	S.F.	12.28	14.07	26.35
25	2" x 4-½" deep, insulating glass		16.16	15.99	32.15
26	Thermo-break, 2-¼" x 4-½" deep, insulating glass	↓	16.61	16.31	32.92
27					
28					
29					
30					
31	Cut & patch to match existing construction, add, minimum		2%	3%	
32	Maximum		5%	9%	
33	Dust protection, add, minimum		1%	2%	
34	Maximum		4%	11%	
35	Material handling, & storage limitation, add, minimum		1%	1%	
36	Maximum		6%	7%	
37	Protection of existing work, add, minimum		2%	2%	
38	Maximum		5%	7%	
39	Shift work requirements, add, minimum			5%	
40	Maximum			30%	
41					
42					

LINE NO.	DESCRIPTION	COST PER S.F.		
		MAT.	INST.	TOTAL
01	Exterior surface, masonry, concrete block, standard 4″ thick	.91	3.09	4
02	6″ thick	1.12	3.32	4.44
03	8″ thick	1.43	3.54	4.97
04	12″ thick	2.04	4.56	6.60
05	Split rib, 4″ thick	2.06	3.84	5.90
06	8″ thick	3.28	4.37	7.65
07	Brick running bond, standard size, 6.75/S.F.	2.34	6.06	8.40
08	Buff, 6.75/S.F.	2.94	6.06	9
09	Stucco, on frame	.70	3.41	4.11
10	On masonry	.23	2.66	2.89
11	Metal, aluminum, horizontal, plain	1.10	1.13	2.23
12	Insulated	1.25	1.13	2.38
13	Vertical, plain	1.43	1.13	2.56
14	Insulated	1.49	1.18	2.67
15	Wood, beveled siding, ''A'' grade cedar, ½″ x 6″	1.58	1.15	2.73
16	½″ x 8″	1.97	1.04	3.01
17	Shingles, 16″ #1 red, 7-½″ exposure	.84	1.41	2.25
18	18″ perfections, 7-½″ exposure	.68	1.17	1.85
19	Handsplit, 10″ exposure	1.01	1.14	2.15
20	White cedar, 7-½″ exposure	.84	1.41	2.25
21	Vertical, board & batten, redwood	1.39	1.43	2.82
22	White pine	.57	1.05	1.62
23	T. & G. boards, redwood, 1″ x 4″	2.52	1.98	4.50
24	1′ x 8″	2.27	1.58	3.85
25				
26	Interior surface, drywall, taped & finished, standard, ½″	.24	.60	.84
27	⅝″ thick	.29	.59	.88
28	Fire resistant, ½″ thick	.29	.59	.88
29	⅝″ thick	.31	.59	.90
30	Moisture resistant, ½″ thick	.35	.60	.95
31	⅝″ thick	.40	.59	.99
32	Core board, 1″ thick	.55	1.20	1.75
33	Plaster, gypsum, 2 coats	.36	1.42	1.78
34	3 coats	.51	1.70	2.21
35	Perlite or vermiculite, 2 coats	.42	1.61	2.03
36	3 coats	.65	2.01	2.66
37	Gypsum lath, standard, ⅜″ thick	.40	.36	.76
38	½″ thick	.43	.39	.82
39	Fire resistant, ⅜″ thick	.45	.44	.89
40	½″ thick	.48	.47	.95
41	Metal lath, diamond, 2.5 lb.	.19	.36	.55
42	Rib, 3.4 lb.	.23	.45	.68
43	Framing metal studs including top and bottom			
44	Runners, walls 10′ high			
45	24″ O.C., non load bearing 20 gauge, 2-½″ wide	.43	.56	.99
46	3-⅝″ wide	.51	.57	1.08
47	4″ wide	.54	.59	1.13
48	6″ wide	.67	.60	1.27
49	Load bearing 18 gauge, 2-½″ wide	.80	1.20	2
50	3-⅝″ wide	.92	1.25	2.17
51	4″ wide	.97	1.30	2.27
52	6″ wide	1.18	1.37	2.55
53	16″ O.C., non load bearing 20 gauge, 2-½″ wide	.52	.65	1.17
54	3-⅝″ wide	.62	.66	1.28
55	4″ wide	.65	.68	1.33
56	6″ wide	.81	.70	1.51
57	Load bearing 18 gauge, 2-½″ wide	.97	1.43	2.40
58	3-⅝″ wide	1.11	1.51	2.62

LINE NO.	DESCRIPTION	COST PER S.F.		
		MAT.	INST.	TOTAL
59	4" wide	1.18	1.59	2.77
60	6" wide	1.42	1.69	3.11
61	Framing wood studs incl. double top plate and			
62	Single bottom plate, walls 10' high			
63	24" O.C., 2" x 4"	.24	.48	.72
64	2" x 6"	.35	.53	.88
65	16" O.C., 2" x 4"	.31	.60	.91
66	2" x 6"	.46	.67	1.13
67	Sheathing, boards, 1" x 6"	.74	.91	1.65
68	1" x 8"	.74	.77	1.51
69	Plywood, ⅜" thick	.29	.49	.78
70	½" thick	.35	.53	.88
71	⅝" thick	.44	.56	1
72	¾" thick	.51	.61	1.12
73	Wood fiber, ⅝" thick	.51	.49	1
74	Gypsum weatherproof, ½" thick	.36	.56	.92
75	Insulation, fiberglass batts, 3-½" thick, R11	.24	.41	.65
76	6" thick, R19	.41	.48	.89
77	Poured 4" thick, fiberglass wool, R4/inch	.35	1.44	1.79
78	Mineral wool, R3/inch	.88	1.44	2.32
79	Polystyrene, R4/inch	1.80	1.44	3.24
80	Perlite or vermiculite, R2.7/inch	1.95	1.43	3.38
81	Rigid, fiberglass, R4.3/inch, 1" thick	.44	.29	.73
82	R8.7/inch, 2" thick	.97	.32	1.29
83	Urethane, R5.8/inch, 1" thick	.55	.36	.91
84	R11.7/inch, 2" thick	1.10	.39	1.49

LINE NO.	DESCRIPTION	COST EACH		
		MAT.	INST.	TOTAL
01	Door closer, rack and pinion	74.80	45.20	120
02	Backcheck and adjustable power	80.30	49.70	130
03	Regular, hinge face mount, all sizes, regular arm	73.70	46.30	120
04	Hold open arm	83.60	46.40	130
05	Top jamb mount, all sizes, regular arm	71.50	48.50	120
06	Hold open arm	82.50	47.50	130
07	Stop face mount, all sizes, regular arm	73.70	46.30	120
08	Hold open arm	84.70	45.30	130
09	Fusible link, hinge face mount, all sizes, regular arm	83.60	46.30	129.90
10	Hold open arm	93.50	46.40	139.90
11	Top jamb mount, all sizes, regular arm	81.40	48.50	129.90
12	Hold open arm	92.40	47.50	139.90
13	Stop face mount, all sizes, regular arm	83.60	46.30	129.90
14	Hold open arm	94.60	45.30	139.90
15				
16				
17	Door stops			
18				
19	Holder & bumper, floor or wall	11	12	23
20	Wall bumper	4.68	11.97	16.65
21	Floor bumper	3.85	11.95	15.80
22	Plunger type, door mounted	26.40	11.60	38
23	Hinges, full mortise, material only, per pair			
24	Low frequency, 4-½″ x 4-½″, steel base, USP	11.95		11.95
25	Brass base, US10	39		39
26	Stainless steel base, US32	58		58
27	Average frequency, 4-½″ x 4-½″, steel base, USP	20		20
28	Brass base, US10	50		50
29	Stainless steel base, US32	66		66
30	High frequency, 4-½″ x 4-½″, steel base, USP	50		50
31	Brass base, US10	77		77
32	Stainless steel base, US32	99		99
33	Kick plate			
34				
35	6″ high, for 3′-0″ door, aluminum	18.70	19.30	38
36	Bronze	38.50	19.50	58
37	Panic device			
38				
39	For rim locks, single door, exit	258.50	96.50	355
40	Outside key and pull	313.50	56.50	370
41	Bar and vertical rod, exit only	396	59	455
42	Outside key and pull	473	72	545
43	Lockset			
44				
45	Heavy duty, cylindrical, passage doors	95.70	24.30	120
46	Classroom	203.50	36.50	240
47	Bedroom, bathroom, and inner office doors	123.20	21.80	145
48	Apartment, office, and corridor doors	165	30	195
49	Standard duty, cylindrical, exit doors	63.80	29.20	93
50	Inner office doors	29.15	23.85	53
51	Passage doors	39.60	24.40	64
52	Public restroom, classroom, & office doors	79.20	35.80	115
53	Deadlock, mortise, heavy duty	93.50	31.50	125
54	Double cylinder	104.50	30.50	135
55	Entrance lock, cylinder, deadlocking latch	96.80	33.20	130
56	Deadbolt	115.50	34.50	150
57	Commercial, mortise, wrought knob, keyed, minimum	115.50	34.50	150
58	Maximum	225.50	39.50	265

LINE NO.	DESCRIPTION	COST EACH		
		MAT.	INST.	TOTAL
59	Cast knob, keyed, minimum	159.50	30.50	190
60	Maximum	330	30	360
61	Push-pull			
62				
63	Aluminum	52.80	24.20	77
64	Bronze	66	24	90
65	Door pull, designer style, minimum	55	24	79
66	Maximum	236.50	33.50	270
67	Threshold			
68				
69	3'-0" long door saddles, aluminum, minimum	24.20	14.80	39
70	Maximum	74.80	24.20	99
71	Bronze, minimum	48.40	14.60	63
72	Maximum	132	23	155
73	Rubber, ½" thick, 5-½" wide	29.70	14.30	44
74	2-¾" wide	13.75	14.25	28
75	Weatherstripping, per set			
76				
77	Doors, wood frame, interlocking for 3' x 7' door, zinc	11	94	105
78	Bronze	16.50	93.50	110
79	Wood frame, spring type for 3' x 7' door, bronze	7.59	37.41	45
80	Metal frame, spring type for 3' x 7' door, bronze	33	97	130
81	For stainless steel, spring type, add	133%		
82				
83	Metal frame, extruded sections, 3' x 7' door, aluminum	33	142	175
84	Bronze	74.80	145.20	220

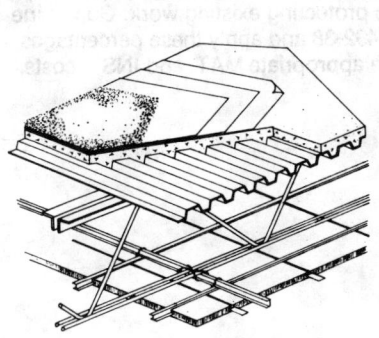

This page illustrates and describes a flat roof bar joist system including asphalt and gravel roof, lightweight concrete, metal decking, open web joists, and suspended acoustic ceiling. Lines 05.1-192-04 thru 09 give the unit price and total price per square foot for this system. Prices for alternate flat roof bar joist systems are on Line Items 05.1-192-13 thru 16. Both material quantities and labor costs have been adjusted for the system listed.

Factors: To adjust for job conditions other than normal working situations use Lines 05.1-192-27 thru 40.

Example: You are to install the system where material handling and storage present some problem. Go to Line 05.1-192-33 and apply these percentages to the appropriate MAT. and INST. costs.

LINE NO.	DESCRIPTION	QUANTITY	COST PER S.F.		
			MAT.	INST.	TOTAL
01	**Three ply asphalt and gravel roof on lightweight concrete, metal decking on**				
02	**Web joist 4' O.C., with suspended acoustic ceiling.**				
03					
04	Open web joist, 10" deep, 4' O.C.	1.25 Lb.	.45	.30	.75
05	Metal decking, 1-½" deep, 22 Ga., galvanized	1 S.F.	.78	.34	1.12
06	Gravel roofing, 3 ply asphalt	.01 C.S.F.	.41	.89	1.30
07	Perlite roof fill, 3" thick	1 S.F.	.75	.30	1.05
08	Suspended ceiling, ¾" mineral fiber on "Z" bar suspension	1 S.F.	1.30	1.91	3.21
09	TOTAL	S.F.	3.69	3.74	7.43
10					
11					
12	For alternate roof systems:				
13	Open web bar joist, 16" deep, 5#/L.F.	S.F.	3.83	3.84	7.67
14	18" deep, 6.6#/L.F.		3.97	3.91	7.88
15	20" deep, 9.6#/L.F.		4.11	4.01	8.12
16	24" deep, 11.52#/L.F.	↓	4.29	4.12	8.41
17					
18					
19					
20					
21					
22					
23					
24					
25					
26					
27	Cut & patch to match existing construction, add, minimum		2%	3%	
28	Maximum		5%	9%	
29	Dust protection, add, minimum		1%	2%	
30	Maximum		4%	11%	
31	Equipment usage curtailment, add, minimum		1%	1%	
32	Maximum		3%	10%	
33	Material handling & storage limitation, add, minimum		1%	1%	
34	Maximum		6%	7%	
35	Protection of existing work, add, minimum		2%	2%	
36	Maximum		5%	7%	
37	Shift work requirements, add, minimum			5%	
38	Maximum			30%	
39	Temporary shoring and bracing, add, minimum		2%	5%	
40	Maximum		5%	12%	
41					
42					

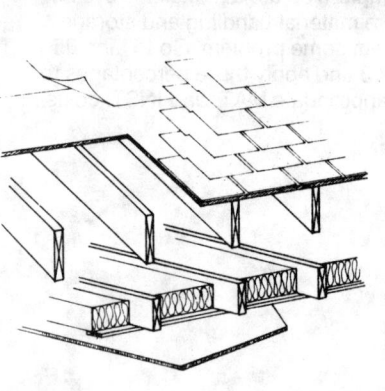

This page illustrates and describes a wood frame roof system including rafters, ceiling joists, sheathing, building paper, asphalt shingles, roof trim, furring, insulation, plaster and paint. Lines 05.1-492-04 thru 15 give the unit price and total price per square foot for this system. Prices for alternate wood frame roof systems are on Line Items 05.1-492-19 thru 27. Both material quantities and labor costs have been adjusted for the system listed.

Factors: To adjust for job conditions other than normal working situations use Lines 05.1-492-33 thru 40.

Example: You are to install the system while protecting existing work. Go to Line 05.1-492-38 and apply these percentages to the appropriate MAT. and INST. costs.

LINE NO.	DESCRIPTION	QUANTITY	COST PER S.F.		
			MAT.	INST.	TOTAL
01	Wood frame roof system, 4 in 12 pitch, including rafters, sheathing,				
02	Shingles, insulation, drywall, thin coat plaster, and painting.				
03					
04	Rafters, 2″ x 6″, 16″ O.C., 4 in 12 pitch	1.08 L.F.	.46	.64	1.10
05	Ceiling joists, 2″ x 6″, 16 O.C.	1 L.F.	.44	.47	.91
06	Sheathing, ½″ CDX	1.08 S.F.	.37	.46	.83
07	Building paper, 15# felt	.011 C.S.F.	.03	.09	.12
08	Asphalt shingles, 240#	.011 C.S.F.	.39	.63	1.02
09	Roof trim	.1 L.F.	.03	.14	.17
10	Furring, 1″ x 3″, 16″ O.C.	1 L.F.	.15	.83	.98
11	Fiberglass insulation, 6″ batts	1 S.F.	.43	.21	.64
12	Gypsum board, ½″ thick	1 S.F.	.19	.32	.51
13	Thin coat plaster	1 S.F.	.10	.37	.47
14	Paint, roller, 2 coats	1 S.F.	.10	.33	.43
15	TOTAL	S.F.	2.69	4.49	7.18
16					
17					
18	For alternate roof systems:				
19	Rafters 16″ O.C., 2″ x 8″	S.F.	2.88	4.53	7.41
20	2″ x 10″		3.20	4.87	8.07
21	2″ x 12″		3.44	4.96	8.40
22	Rafters 24″ O.C., 2″ x 6″		2.46	4.21	6.67
23	2″ x 8″		2.60	4.23	6.83
24	2″ x 10″		2.84	4.48	7.32
25	2″ x 12″		3.01	4.56	7.57
26	Roof pitch, 6 in 12, add		3%	10%	
27	8 in 12, add		5%	12%	
28					
29					
30					
31					
32					
33	Cut & patch to match existing construction, add, minimum		2%	3%	
34	Maximum		5%	9%	
35	Material handling & storage limitation, add, minimum		1%	1%	
36	Maximum		6%	7%	
37	Protection of existing work, add, minimum		2%	2%	
38	Maximum		5%	7%	
39	Shift work requirements, add, minimum			5%	
40	Maximum			30%	
41					
42					

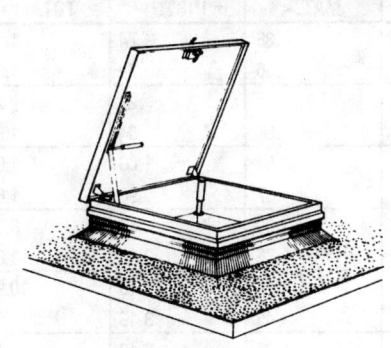

This page illustrates and describes a roof hatch system. Lines 05.8-104-04 thru 09 give the unit price and total cost each for this system. Prices for alternate systems are on Lines 05.8-104-15 thru 24. Both material quantities and labor costs have been adjusted for the system listed.

Factors: To adjust for job conditions other than normal working situations use Lines 05.8-104-29 thru 40.

Example: You are to install the system and cut & patch to match existing construction. Use line 05.8-104-30 and apply these percentages to the appropriate MAT. and INST. Costs.

LINE NO.	DESCRIPTION	QUANTITY	COST EACH		
			MAT.	INST.	TOTAL
01	Roof hatch, 2'-6" x 3'-0", aluminum, curb and cover included.				
02	Through steel construction.				
03					
04	Roof hatch, 2'-6" x 3'-0", aluminum	1 Ea.	302.50	112.50	415
05	Cutout decking	11 L.F.	4.84	45.32	50.16
06	Frame opening	44 L.F.	36.30	37.18	73.48
07	Flashing, 16 oz. copper	16 S.F.	45.76	43.84	89.60
08	Cant strip, 4 x 4, foamglass	12 L.F.	6.86	10.54	17.40
09	TOTAL	Ea.	396.26	249.38	645.64
10					
11					
12					
13					
14	For alternate systems:				
15	Roof hatch, aluminum, curb and cover, 2'-6" x 4'-6"	Ea.	574.64	296.47	871.11
16	2'-6" x 8'-0"		916.34	417.80	1334.14
17	Skylight, plexiglass dome, with curb mounting, 30" x 32"		423.76	231.88	655.64
18	30" x 45"		503.14	282.97	786.11
19	40" x 45"		586.34	362.80	949.14
20	Smoke hatch, unlabeled, 2'-6" x 3'-0"		471.89	277.50	749.39
21	2'-6" x 4'-6"		688.77	327.34	1016.11
22	2'-6" x 8'-0"		991.14	434.50	1425.64
23					
24	For steel ladder, add per vertical linear foot	V.L.F.	24.20	17.80	42
25					
26					
27					
28					
29	Cut & patch to match existing construction, add, minimum		2%	3%	
30	Maximum		5%	9%	
31	Dust protection, add, minimum		1%	2%	
32	Maximum		4%	11%	
33	Equipment usage curtailment, add, minimum		1%	1%	
34	Maximum		3%	10%	
35	Material handling & storage limitation, add, minimum		1%	1%	
36	Maximum		6%	7%	
37	Protection of existing work, add, minimum		2%	2%	
38	Maximum		5%	7%	
39	Shift work requirements, add, minimum			5%	
40	Maximum			30%	
41					
42					

LINE NO.	DESCRIPTION	COST PER S.F.		
		MAT.	INST.	TOTAL
01	Roofing, built-up, asphalt roll roof, 3 ply organic/mineral surface	.36	.74	1.10
02	3 plies glass fiber felt type iv, 1 ply mineral surface	.56	.79	1.35
03	Cold applied, 3 ply		.28	.28
04	Coal tar pitch, 4 ply asbestos felt	.66	.94	1.60
05	Mopped, 3 ply glass fiber	.54	1.06	1.60
06	4 ply organic felt	.67	.98	1.65
07	Elastomeric, hypalon, neoprene unreinforced	1.60	1.76	3.36
08	Polyester reinforced	1.76	2.08	3.84
09	Neoprene, 5 coats 60 mils	4.79	6.16	10.95
10	Over 10,000 S.F.	4.46	3.19	7.65
11	PVC, traffic deck sprayed	1.27	3.19	4.46
12	With neoprene	1.38	1.29	2.67
13	Shingles, asbestos, strip, 14" x 30", 325#/sq.	1.42	.73	2.15
14	12' x 24", 167#/sq.	.72	.83	1.55
15	Shake, 9.35" x 16" 500#/sq.	2.26	1.29	3.55
16				
17	Asphalt, strip, 210-235#/sq.	.35	.52	.87
18	235-240#/sq.	.35	.58	.93
19	Class A laminated	.57	.63	1.20
20	Class C laminated	.52	.73	1.25
21	Slate, buckingham, 3/16" thick	3.85	1.65	5.50
22	Black, 1/4" thick	3.85	1.65	5.50
23	Wood, shingles, 16" no. 1, 5" exp.	1.16	1.14	2.30
24	Red cedar, 18" perfections	1.22	1.03	2.25
25	Shakes, 24", 10" exposure	1.01	1.14	2.15
26	18", 8-1/2" exposure	.88	1.42	2.30
27	Insulation, ceiling batts, fiberglass, 3-1/2" thick, R11	.28	.17	.45
28	6" thick, R19	.43	.21	.64
29	9" thick, R30	.66	.25	.91
30	12" thick, R38	.86	.29	1.15
31	Mineral, 3-1/2" thick, R13	.31	.18	.49
32	Fiber, 6" thick, R19	.48	.18	.66
33	Roof deck, fiberboard, 1" thick, R2.78	.30	.35	.65
34	Mineral, 2" thick, R5.26	.67	.36	1.03
35	Perlite boards, 3/4" thick, R2.08	.35	.36	.71
36	2" thick, R5.26	.83	.40	1.23
37	Polystyrene extruded, R5.26, 1" thick,	.55	.19	.74
38	2" thick	1.03	.23	1.26
39	Urethane paperbacked, 1" thick, R6.7	.62	.28	.90
40	3" thick, R25	1.40	.35	1.75
41	Foam glass sheets, 1-1/2" thick, R3.95	1.79	.36	2.15
42	2" thick, R5.26	2.27	.35	2.62
43	Ceiling, plaster, gypsum, 2 coats	.36	1.62	1.98
44	3 coats	.51	1.93	2.44
45	Perlite or vermiculite, 2 coats	.42	1.91	2.33
46	3 coats	.65	2.35	3
47	Gypsum lath, plain 3/8" thick	.40	.36	.76
48	1/2" thick	.43	.39	.82
49	Firestop, 3/8" thick	.45	.44	.89
50	1/2" thick	.48	.47	.95
51	Metal lath, rib, 2.75 lb.	.20	.42	.62
52	3.40 lb.	.23	.45	.68
53	Diamond, 2.50 lb.	.19	.41	.60
54	3.40 lb.	.21	.52	.73
55	Drywall, taped and finished standard, 1/2" thick	.24	.75	.99
56	5/8" thick	.29	.75	1.04
57	Fire resistant, 1/2" thick	.29	.75	1.04
58	5/8" thick	.31	.75	1.06

LINE NO.	DESCRIPTION	COST PER S.F.		
		MAT.	INST.	TOTAL
59	Water resist., ½" thick	.35	.75	1.10
60	⅝" thick	.40	.75	1.15
61	Finish, instead of taping			
62	For thin coat plaster, add	.10	.37	.47
63	Finish textured spray, add	.11	.34	.45
64	Drywall, no finish included, see system 069-700			
65	Tile, stapled glued, mineral fiber plastic coated, ⅝" thick	.53	1.43	1.96
66	¾" thick	.69	1.44	2.13
67	Wood fiber, ½" thick	.63	.72	1.35
68	¾" thick	.88	.72	1.60
69	Suspended, fiberglass film faced, ⅝" thick	.43	.46	.89
70	3" thick	1.05	.57	1.62
71	Mineral fiber ⅝" thick, standard face	.50	.42	.92
72	Aluminum faced	1.16	.47	1.63
73	Wood fiber reveal edge, 1" thick	.98	.48	1.46
74	3" thick	2.51	.64	3.15
75	Ceiling suspension systems, for tile, "T" bar, class "A", 2' x 4' grid	.41	.36	.77
76	2' x 2' grid	.54	.44	.98
77	Concealed "Z" bar, 12" module	.62	.55	1.17
78				
79	Plaster/drywall, ¾" channels, steel furring, 16" O.C.	.22	.95	1.17
80	24" O.C.	.18	.65	.83
81	1-½" channels, 16" O.C.	.30	1.06	1.36
82	24" O.C.	.24	.71	.95
83	Ceiling framing, 2" x 4" studs, 16" O.C.	.22	.50	.72
84	24" O.C.	.14	.34	.48

LINE NO.	DESCRIPTION	COST PER L.F.		
		MAT.	INST.	TOTAL
01	Downspouts per L.F., aluminum, enameled .024″ thick, 2″ x 3″	.58	1.65	2.23
02	3″ x 4″	.94	2.24	3.18
03	Round .025″ thick, 3″ diam.	.72	1.65	2.37
04	4″ diam.	1.05	2.24	3.29
05	Copper, round 16 oz. stock, 2″ diam.	3.47	1.63	5.10
06	3″ diam.	4.68	1.67	6.35
07	4″ diam.	5.94	2.16	8.10
08	5″ diam.	6.11	2.39	8.50
09	Rectangular, 2″ x 3″	5.23	1.67	6.90
10	3″ x 4″	6.38	2.17	8.55
11	Lead coated copper, round, 2″ diam.	4.51	1.64	6.15
12	3″ diam.	5.45	1.65	7.10
13	4″ diam.	6.71	2.14	8.85
14	5″ diam.	8.58	2.42	11
15	Rectangular, 2″ x 3″	5.94	1.66	7.60
16	3″ x 4″	7.70	2.15	9.85
17	Steel galvanized, round 28 gauge, 3″ diam.	.50	1.65	2.15
18	4″ diam.	.63	2.16	2.79
19	5″ diam.	.77	2.41	3.18
20	6″ diam.	1.16	2.98	4.14
21	Rectangular, 2″ x 3″	.50	1.65	2.15
22	3″ x 4″	1.38	2.16	3.54
23	Elbows, aluminum, round, 3″ diam.	.83	3.13	3.96
24	4″ diam.	.96	3.13	4.09
25	Rectangular, 2″ x 3″	.56	3.14	3.70
26	3″ x 4″	.88	3.14	4.02
27	Copper, round 16 oz., 2″ diam.	4.68	3.12	7.80
28	3″ diam.	5.83	3.12	8.95
29	4″ diam.	10.34	3.16	13.50
30	5″ diam.	12.82	3.13	15.95
31	Rectangular, 2″ x 3″	9.90	3.15	13.05
32	3″ x 4″	13.97	3.13	17.10
33	Drip edge per L.F., aluminum, 5″ girth	.14	.72	.86
34	8″ girth	.30	.72	1.02
35	28″ girth	1.30	2.87	4.17
36				
37	Steel galvanized, 5″ girth	.23	.72	.95
38	8″ girth	.29	.71	1
39				
40				
41				
42				
43	Flashing 12″ wide per S.F., aluminum, mill finish, .013″ thick	.29	2.16	2.45
44	.019″ thick	.72	2.16	2.88
45	.040″ thick	1.45	2.17	3.62
46	.050″ thick	1.76	2.16	3.92
47	Copper, mill finish, 16 oz.	2.86	2.74	5.60
48	20 oz.	3.58	2.87	6.45
49	24 oz.	4.29	3.01	7.30
50	32 oz.	5.56	3.14	8.70
51	Lead, 2.5 lb./S.F., 12″ wide	3.47	2.13	5.60
52	Over 12″ wide	3.47	2.13	5.60
53	Lead-coated copper, fabric backed, 2 oz.	1.38	.95	2.33
54	5 oz.	1.60	.95	2.55
55	Mastic backed, 2 oz.	.96	.95	1.91
56	5 oz.	1.54	.95	2.49
57	Paper backed, 2 oz.	.88	.95	1.83
58	3 oz.	1.08	.95	2.03

LINE NO.	DESCRIPTION	COST PER L.F.		
		MAT.	INST.	TOTAL
59	Polyvinyl chloride, black, .010″ thick	.11	1	1.11
60	.020″ thick	.17	1	1.17
61	.030″ thick	.28	1	1.28
62	.056″ thick	.50	1	1.50
63	Steel, galvanized, 20 gauge	.61	2.41	3.02
64	30 gauge	.24	1.96	2.20
65	Stainless, 32 gauge, .010″ thick	2.04	2.02	4.06
66	28 gauge, .015″ thick	2.53	2.02	4.55
67	26 gauge, .018″ thick	3.14	2.01	5.15
68	24 gauge, .025″ thick	3.80	2	5.80
69	Gutters per L.F., aluminum, 5″ box, .027″ thick	.88	2.61	3.49
70	.032″ thick	1.10	2.61	3.71
71	Copper, half round, 4″ wide	4.46	2.59	7.05
72	6″ wide	5.01	2.74	7.75
73	Steel, 26 gauge galvanized, 5″ wide	.83	2.61	3.44
74	6″ wide	1.05	2.61	3.66
75	Wood, treated hem-fir, 3″ x 4″	4.24	2.86	7.10
76	4″ x 5″	5.34	2.86	8.20
77	Reglet per L.F., aluminum, .025″ thick	.99	1.28	2.27
78	Copper, 10 oz.	2.09	1.28	3.37
79	Steel, galvanized, 24 gauge	.67	1.28	1.95
80	Stainless, .020″ thick	1.27	1.27	2.54
81	Counter flashing 12″ wide per L.F., aluminum, .032″ thick	1.43	2.09	3.52
82	Copper, 10 oz.	2.37	2.09	4.46
83	Steel, galvanized, 24 gauge	.58	2.09	2.67
84	Stainless, .020″ thick	1.71	2.09	3.80

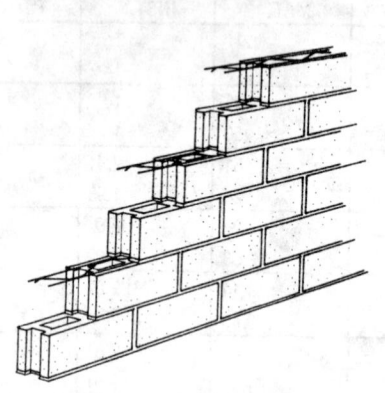

This page illustrates and describes a concrete block wall system including concrete block, horizontal reinforcing alternate courses, mortar, and tooled joints both sides. Lines 06.1-222-04 thru 06 give the unit price and total price per square foot for this system. Prices for alternate concrete block wall systems are on Line Items 06.1-222-09 thru 21. Both material quantities and labor costs have been adjusted for the system listed.

Factors: To adjust for job conditions other than normal working situations use Lines 06.1-222-29 thru 40.

Example: You are to install the system and protect all existing work. Go to Line 06.1-222-36 and apply these percentages to the appropriate MAT. and INST. costs.

LINE NO.	DESCRIPTION	QUANTITY	COST PER S.F.		
			MAT.	INST.	TOTAL
01	Concrete block partition, including horizontal reinforcing every second				
02	Course, mortar, tooled joints both sides.				
03					
04	8" x 16" concrete block, normal weight, 4" thick	1 S.F.	.91	3.09	4
05	Horizontal reinforcing every second course	.75 L.F.	.14	.09	.23
06	TOTAL	S.F.			
07			1.05	3.18	4.23
08	For alternate block partition systems:				
09	8" x 16" concrete block, normal weight, 6" thick	S.F.	1.26	3.41	4.67
10	8" thick		1.57	3.63	5.20
11	10" thick		2.12	3.76	5.88
12	12" thick		2.18	4.65	6.83
13	8" x 16" concrete block, lightweight, 4" thick		1.15	3.11	4.26
14	6" thick		1.49	3.33	4.82
15	8" thick		1.91	3.52	5.43
16	10" thick		2.77	3.66	6.43
17	12" thick		2.82	4.51	7.33
18	8" x 16" glazed concrete block, 4" thick		6.08	3.95	10.03
19	8" thick		7.13	4.40	11.53
20	Structural facing tile, 6T series, glazed 2 sides, 4" thick		7.73	6.90	14.63
21	6" thick		9.77	7.26	17.03
22					
23					
24					
25					
26					
27					
28					
29	Cut & patch to match existing construction, add, minimum		2%	3%	
30	Maximum		5%	9%	
31	Dust protection, add, minimum		1%	2%	
32	Maximum		4%	11%	
33	Material handling & storage limitation, add, minimum		1%	1%	
34	Maximum		6%	7%	
35	Protection of existing work, add, minimum		2%	2%	
36	Maximum		5%	7%	
37	Shift work requirements, add, minimum			5%	
38	Maximum			30%	
39	Temporary shoring and bracing, add, minimum		2%	5%	
40	Maximum		5%	12%	
41					
42					

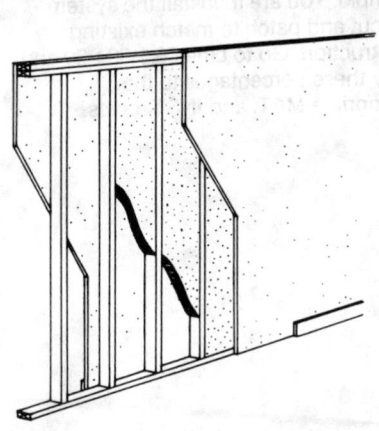

This page illustrates and describes a wood stud partition system including wood studs with plates, gypsum plasterboard - taped and finished, insulation, baseboard and painting. Lines 06.1-592-04 thru 10 give the unit price and total price per square foot for this system. Prices for alternate wood stud partition systems are on Line Items 06.1-592-13 thru 27. Both material quantities and labor costs have been adjusted for the system listed.

Factors: To adjust for job conditions other than normal working situations use Lines 06.1-592-29 thru 40.

Example: You are to install the system where material handling and storage present a serious problem. Go to Line 06.1-592-34 and apply these percentages to the appropriate MAT. and INST. costs.

LINE NO.	DESCRIPTION	QUANTITY	COST PER S.F. MAT.	COST PER S.F. INST.	COST PER S.F. TOTAL
01	Wood stud wall, 2"x4",16"O.C., dbl. top plate, sngl bot. plate, ⅝" dwl.				
02	Taped, finished and painted on 2 faces, insulation, baseboard, wall 8' high				
03					
04	Wood studs, 2" x 4", 16" O.C., 8' high	1 S.F.	.34	.74	1.08
05	Gypsum drywall, ⅝" thick	2 S.F.	.46	.58	1.04
06	Taping and finishing	2 S.F.	.11	.57	.68
07	Insulation, 3-½" fiberglass batts	1 S.F.	.30	.18	.48
08	Baseboard, painted	.2 L.F.	.24	.50	.74
09	Painting, roller, 2 coats	2 S.F.	.20	.66	.86
10	TOTAL	S.F.	1.65	3.23	4.88
11					
12	For alternate wood stud systems:				
13	2" x 3" studs, 8' high, 16" O.C.	S.F.	1.63	3.19	4.82
14	24" O.C.		1.55	3.06	4.61
15	10' high, 16" O.C.		1.61	3.05	4.66
16	24" O.C.		1.53	2.95	4.48
17	2" x 4" studs, 8' high, 24" O.C.		1.57	3.09	4.66
18	10' high, 16" O.C.		1.62	3.09	4.71
19	24" O.C.		1.55	2.97	4.52
20	12' high, 16" O.C.		1.60	3.08	4.68
21	24" O.C.		1.52	2.97	4.49
22	2" x 6" studs, 8' high, 16" O.C.		1.81	3.32	5.13
23	24" O.C.		1.70	3.14	4.84
24	10' high, 16" O.C.		1.77	3.16	4.93
25	24" O.C.		1.66	3.02	4.68
26	12' high, 16" O.C.		1.74	3.17	4.91
27	24" O.C.		1.63	3.02	4.65
28					
29	Cut & patch to match existing construction, add, minimum		2%	3%	
30	Maximum		5%	9%	
31	Dust protection, add, minimum		1%	2%	
32	Maximum		4%	11%	
33	Material handling & storage limitation, add, minimum		1%	1%	
34	Maximum		6%	7%	
35	Protection of existing work, add, minimum		2%	2%	
36	Maximum		5%	7%	
37	Shift work requirements, add, minimum			5%	
38	Maximum			30%	
39	Temporary shoring and bracing, add, minimum		2%	5%	
40	Maximum		5%	12%	
41					
42					

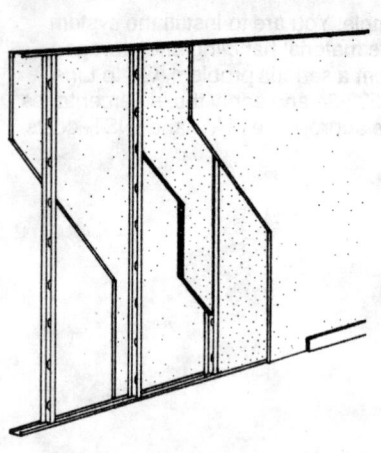

This page illustrates and describes a non-load bearing metal stud partition system including metal studs with runners, gypsum plasterboard, taped and finished, insulation, baseboard and painting. Lines 06.1-594-04 thru 10 give the unit price and total price per square foot for this system. Prices for alternate non-load bearing metal stud partition systems are on Line Items 06.1-594-13 thru 23. Both material quantities and labor costs have been adjusted for the system listed.

Factors: To adjust for job conditions other than normal working situations use Lines 06.1-594-29 thru 40.

Example: You are to install the system and cut and patch to match existing construction. Go to Line 06.1-594-30 and apply these percentages to the appropriate MAT. and INST. costs.

LINE NO.	DESCRIPTION	QUANTITY	COST PER S.F. MAT.	COST PER S.F. INST.	COST PER S.F. TOTAL
01	Non-load bearing metal studs, including top & bottom runners, ⅝" drywall,				
02	Taped, finished and painted 2 faces, insulation, painted baseboard.				
03					
04	Metal studs, 25 ga., 3-⅝" wide, 24" O.C.	1 S.F.	.29	.57	.86
05	Gypsum drywall, ⅝" thick	2 S.F.	.46	.58	1.04
06	Taping & finishing	2 S.F.	.11	.57	.68
07	Insulation, 3-½" fiberglass batts	1 S.F.	.30	.18	.48
08	Baseboard, painted	.2 L.F.	.20	.40	.60
09	Painting, roller work, 2 coats	2 S.F.	.20	.66	.86
10	TOTAL	S.F.	1.56	2.96	4.52
11					
12	For alternate metal stud systems:				
13	Non-load bearing, 25 ga., 24" O.C., 2-½" wide	S.F.	1.51	2.96	4.47
14	6" wide		1.68	2.99	4.67
15	16" O.C., 2-½" wide		1.57	3.10	4.67
16	3-⅝" wide		1.62	3.12	4.74
17	6" wide		1.78	3.14	4.92
18	20 ga., 24" O.C., 2-½" wide		1.70	2.95	4.65
19	3-⅝" wide		1.81	2.98	4.79
20	6" wide		1.94	2.99	4.93
21	16" O.C., 2-½" wide		1.80	3.10	4.90
22	3-⅝" wide		1.94	3.13	5.07
23	6" wide		2.11	3.14	5.25
24					
25					
26					
27					
28					
29	Cut & patch to match existing construction, add, minimum		2%	3%	
30	Maximum		5%	9%	
31	Dust protection, add, minimum		1%	2%	
32	Maximum		4%	11%	
33	Material handling & storage limitation, add, minimum		1%	1%	
34	Maximum		6%	7%	
35	Protection of existing work, add, minimum		2%	2%	
36	Maximum		5%	7%	
37	Shift work requirements, add, minimum			5%	
38	Maximum			30%	
39	Temporary shoring and bracing, add, minimum		2%	5%	
40	Maximum		5%	12%	
41					
42					

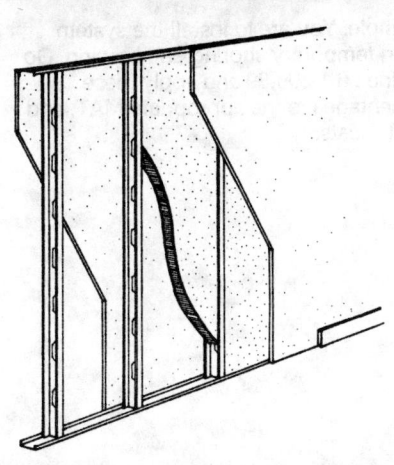

This page illustrates and describes a drywall system including gypsum plasterboard, taped and finished, metal studs with runners, insulation, baseboard and painting. Lines 06.1-595-04 thru 10 give the unit price and total price per square foot for this system. Prices for alternate drywall systems are on Line Items 06.1-595-13 thru 19. Both material quantities and labor costs have been adjusted for the system listed.

Factors: To adjust for job conditions other than normal working situations use Lines 06.1-595-29 thru 40.

Example: You are to install the system and control dust in the work area. Go to Line 06.1-595-31 and apply these percentages to the appropriate MAT. and INST. costs.

LINE NO.	DESCRIPTION	QUANTITY	COST PER S.F.		
			MAT.	INST.	TOTAL
01	**Gypsum drywall, taped, finished and painted 2 faces, galvanized metal studs**				
02	**Including top & bottom runners, insulation, painted baseboard, wall 10′ high**				
03					
04	Gypsum drywall, ⅝″ thick, standard	2 S.F.	.46	.58	1.04
05	Taping and finishing	2 S.F.	.11	.57	.68
06	Metal studs, 20 ga., 3-⅝″ wide, 24″ O.C.	1 S.F.	.29	.57	.86
07	Insulation, 3-½″ fiberglass batts	1 S.F.	.30	.18	.48
08	Baseboard, painted	.2 L.F.	.19	.29	.48
09	Painting, roller 2 coats	2 S.F.	.20	.66	.86
10	TOTAL	S.F.	1.55	2.85	4.40
11					
12	For alternate drywall systems:				
13	Gypsum drywall, ⅝″ thick, fire resistant	S.F.	1.60	2.84	4.44
14	Water resistant		1.77	2.85	4.62
15	½″ thick, standard		1.46	2.84	4.30
16	Fire resistant		1.55	2.85	4.40
17	Water resistant		1.68	2.84	4.52
18	⅝″ thick, vinyl faced, standard		2.30	3.54	5.84
19	⅝″ thick, vinyl faced, fire resistant		2.45	3.55	6
20					
21					
22					
23					
24					
25					
26					
27					
28					
29	Cut & patch to match existing construction, add, minimum		2%	3%	
30	Maximum		5%	9%	
31	Dust protection, add, minimum		1%	2%	
32	Maximum		4%	11%	
33	Material handling & storage limitation, add, minimum		1%	1%	
34	Maximum		6%	7%	
35	Protection of existing work, add, minimum		2%	2%	
36	Maximum		5%	7%	
37	Shift work requirements, add, minimum			5%	
38	Maximum			30%	
39	Temporary shoring and bracing, add, minimum		2%	5%	
40	Maximum		5%	12%	
41					
42					

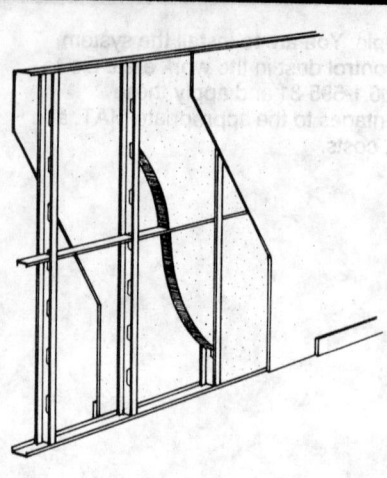

This page illustrates and describes a load bearing metal stud wall system including metal studs, sheetrock-taped and finished, insulation, baseboard and painting. Lines 06.1-596-06 thru 12 give the unit price and total price per square foot for this system. Prices for alternate load bearing metal stud wall systems are on Line Items 06.1-596-15 thru 25. Both material quantities and labor costs have been adjusted for the system listed.

Factors: To adjust for job conditions other than normal working situations use Lines 06.1-596-29 thru 40.

Example: You are to install the system using temporary shoring and bracing. Go to Line 06.1-596-39 and apply these percentages to the appropriate MAT. and INST. costs.

LINE NO.	DESCRIPTION	QUANTITY	COST PER S.F.		
			MAT.	**INST.**	**TOTAL**
01	Load bearing, 18 ga., 3-⅝", galvanized metal studs, 24" O.C., including				
02	Top and bottom runners, ½" drywall, taped, finished and painted 2				
03	Faces, 3" insulation, and painted baseboard, wall 10' high.				
04					
05					
06	Metal studs, 24" O.C., 18 ga., 3-⅝" wide, galvanized	1 S.F.	.92	1.25	2.17
07	Gypsum drywall ½" thick	2 S.F.	.37	.57	.94
08	Taping and finishing	2 S.F.	.11	.57	.68
09	Insulation, 3-½" fiberglass batts	1 S.F.	.30	.18	.48
10	Baseboard, painted	.2 L.F.	.20	.40	.60
11	Painting, roller 2 coats	2 S.F.	.20	.66	.86
12	TOTAL	S.F.	2.10	3.63	5.73
13					
14	For alternate metal stud systems:				
15	Load bearing, 18 ga., 24" O.C., 2-½" wide	S.F.	1.98	3.58	5.56
16	6" wide		2.36	3.75	6.11
17	16" O.C. 2-½" wide		2.18	3.88	6.06
18	3-⅝" wide		2.34	3.93	6.27
19	6" wide		2.64	4.11	6.75
20	16 ga., 24" O.C., 2-½" wide		2.19	3.69	5.88
21	3-⅝" wide		2.31	3.75	6.06
22	6" wide		2.68	3.89	6.57
23	16" O.C., 2-½" wide		2.45	4.01	6.46
24	3-⅝" wide		2.59	4.10	6.69
25	6" wide		3.05	4.27	7.32
26					
27					
28					
29	Cut & patch to match existing construction, add, minimum		2%	3%	
30	Maximum		5%	9%	
31	Dust protection, add, minumum		1%	2%	
32	Maximum		4%	11%	
33	Material handling & storage limitation, add, minimum		1%	1%	
34	Maximum		6%	7%	
35	Protection of existing work, add, minimum		2%	2%	
36	Maximum		5%	7%	
37	Shift work requirements, add, minimum			5%	
38	Maximum			30%	
39	Temporary shoring and bracing, add, minimum		2%	5%	
40	Maximum		5%	12%	
41					
42					

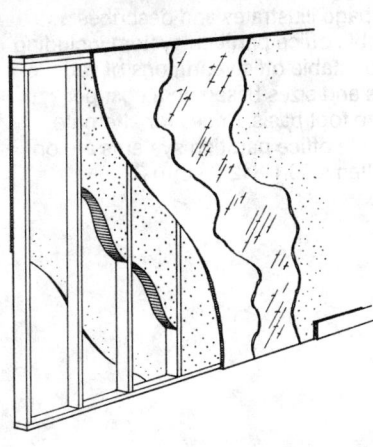

This page illustrates and describes a plaster and lath system including gypsum plaster, gypsum lath, wood studs with plates, insulation, baseboard and painting. Lines 06.1-692-06 thru 12 give the unit price and price per square foot for this system. Prices for alternate plaster and lath systems are on Line Items 06.1-692-15 thru 25. Both material quantities and labor costs have been adjusted for the system listed.

Factors: To adjust for job conditions other than normal working situations use Lines 06.1-692-29 thru 40.

Example: You are to install the system during evening hours only. Go to Line 06.1-692-38 and apply this percentage to the appropriate INST. costs.

LINE NO.	DESCRIPTION	QUANTITY	COST PER S.F.		
			MAT.	INST.	TOTAL
01	Gypsum plaster, 2 coats, over ⅜″ lath, 2 faces, 2″ x 4″ wood				
02	Stud partition, 24″O.C. incl. double top plate,single bottom plate,3-½″				
03	Insulation, baseboard and painting.				
04					
05					
06	Gypsum plaster, 2 coats	.22 S.Y.	.73	2.79	3.52
07	Lath, gypsum, ⅜″ thick	.22 S.Y.	.78	.73	1.51
08	Wood studs, 2″ x 4″, 24″ O.C.	1 S.F.	.24	.48	.72
09	Insulation, 3-½″ fiberglass batts	1 S.F.	.30	.18	.48
10	Baseboard, ⁹⁄₁₆″ x 3-½″, painted	.2 L.F.	.19	.29	.48
11	Paint, 2 coats	2 S.F.	.20	.66	.86
12	TOTAL	S.F.	2.44	5.13	7.57
13					
14	For alternate plaster systems:				
15	Gypsum plaster, 3 coats	S.F.	2.72	5.72	8.44
16	Perlite plaster, 2 coats		2.54	5.53	8.07
17	3 coats	↓	3.01	6.32	9.33
18					
19					
20	For alternate lath systems:				
21	Gypsum lath, ½″ thick	S.F.	2.61	5.57	8.18
22	Foil back, ⅜″ thick		2.63	5.61	8.24
23	½″ thick		2.70	5.66	8.36
24	Metal lath, 2.5 Lb. diamond		2.13	5.53	7.66
25	3.4 Lb. diamond	↓	2.19	5.61	7.80
26					
27					
28					
29	Cut & patch to match existing construction, add, minimum		2%	3%	
30	Maximum		5%	9%	
31	Dust protection, add, minimum		1%	2%	
32	Maximum		4%	11%	
33	Material handling & storage limitation, add, minimum		1%	1%	
34	Maximum		6%	7%	
35	Protection of existing work, add, minimum		2%	2%	
36	Maximum		5%	7%	
37	Shift work requirements, add, minimum			5%	
38	Maximum			30%	
39	Temporary shoring and bracing, add, minimum		2%	5%	
40	Maximum		5%	12%	
41					
42					

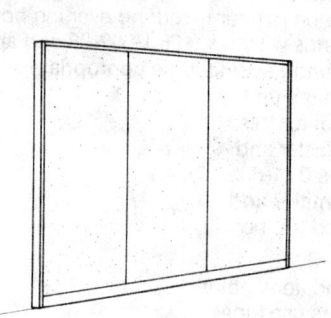

This page illustrates and describes a movable office partition system including demountable office partitions of various styles and sizes based on a cost per square foot basis. Prices for alternate movable office partition systems are on Line Items 06.1-842-03 thru 37.

LINE NO.	DESCRIPTION	QUANTITY	COST PER S.F.		
			MAT.	INST.	TOTAL
01	Office partition, demountable, no deduction for door opening, add for doors				
02					
03	Air wall, cork finish, semi acoustic, 1-⅝" thick, minimum	S.F.	14.30	1.75	16.05
04	Maximum		17.88	3.12	21
05	Acoustic, 2" thick, minimum		15.40	1.90	17.30
06	Maximum	↓	22	3	25
07					
08					
09	In-plant modular office system, w/prehung steel door				
10	3" thick honeycomb core panels				
11	12' x 12', 2 wall	S.F.	5.82	.23	6.05
12	4 wall	"	12.72	.71	13.43
13					
14					
15	Gypsum, demountable, 3" to 3-¾" thick x 9' high, vinyl clad	S.F.	2.07	1.37	3.44
16	Fabric clad	"	6.35	1.42	7.77
17	1.75 system, vinyl clad hardboard, paper honeycomb core panel				
18	1-¾" to 2-½" thick x 9' high	S.F.	4.39	1.38	5.77
19	Unitized gypsum panel, 2" to 2-½" thick x 9' high, vinyl clad	↓	5.85	1.37	7.22
20	Fabric clad		10.49	1.72	12.21
21					
22					
23	Unitized mineral fiber panel system, 2-¼" thick x 9' high				
24	Vinyl clad mineral fiber	S.F.	7.33	1.33	8.66
25	Fabric clad mineral fiber	"	9.52	1.47	10.99
26					
27	Movable steel walls, modular system				
28	Unitized panels, 48" wide x 9' high				
29	Baked enamel, pre-finished	S.F.	6.83	1.05	7.88
30	Fabric clad		10.62	1.04	11.66
31	For acoustical partitions, add, minimum		1.10		1.10
32	Maximum	↓	4.13		4.13
33					
34					
35	Note: For door prices, see divisions 081 & 082				
36	For door hardware prices, see division 087				
37					
38					
39					
40					
41					
42					

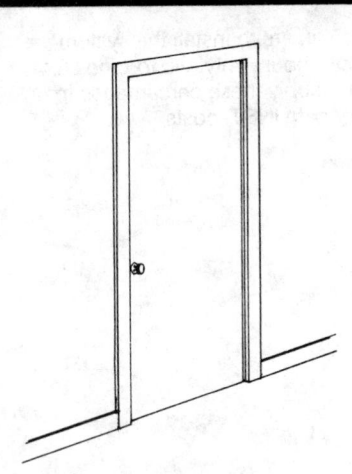

This page illustrates and describes flush interior door systems including hollow core door, jamb, header and trim with hardware. Lines 06.4-142-04 thru 08 give the unit price and total price on a cost each basis for this system. Prices for alternate flush interior door systems are on Line items 06.4-142-11 thru 24. Both material quantities and labor costs have been adjusted for the system listed.

Factors: To adjust for job conditions other than normal working situations use Lines 06.1-142-27 thru 38.

Example: You are to install the system in an area where dust protection is vital. Go to Line 06.4-142-30 and apply these percentages to the appropriate MAT. and INST. costs.

LINE NO.	DESCRIPTION	QUANTITY	COST EACH MAT.	COST EACH INST.	COST EACH TOTAL
01	Single hollow core door, include jamb, header, trim and hardware, painted.				
02					
03					
04	Hollow core Lauan, 1-⅜" thick, 2'-0" x 6'-8", painted	1 Ea.	33.54	49.76	83.30
05	Wood jamb, 4-⁹⁄₁₆" deep	1 Set	24.64	25.28	49.92
06	Trim, casing	1 Set	20.77	38.43	59.20
07	Hardware, hinges, lockset	1 Set	23.23	17.72	40.95
08	TOTAL	Ea.	102.18	131.19	233.37
09					
10	For alternate door systems:				
11	Lauan (Mahogany) hollow core, 1-⅜" x 2'-6" x 6'-8"	Ea.	104.70	133.08	237.78
12	2'-8" x 6'-8"		105.74	134.67	240.41
13	3'-0" x 6'-8"	↓	109.61	139.28	248.89
14					
15	Birch, hollow core, 1-⅜" x 2'-0" x 6'-8"	Ea.	105.48	130.89	236.37
16	2'-6" x 6'-8"		111.30	132.48	243.78
17	2'-8" x 6'-8"		112.34	134.07	246.41
18	3'-0" x 6'-8"		117.31	139.58	256.89
19	Solid core, pre-hung, 1-⅜" x 2'-6" x 6'-8"		232.30	130.48	362.78
20	2'-8" x 6'-8"		236.64	128.77	365.41
21	3'-0" x 6'-8"	↓	243.81	136.08	379.89
22					
23					
24	For metal frame instead of wood, add	Ea.	50%	20%	
25					
26					
27	Cut & patch to match existing construction, add, minimum		2%	3%	
28	Maximum		5%	9%	
29	Dust protection, add, minimum		1%	2%	
30	Maximum		4%	11%	
31	Equipment usage curtailment, add, minimum		1%	1%	
32	Maximum		3%	10%	
33	Material handling & storage limitation, add, minimum		1%	1%	
34	Maximum		6%	7%	
35	Protection of existing work, add, minimum		2%	2%	
36	Maximum		5%	7%	
37	Shift work requirements, add, minimum			5%	
38	Maximum			30%	
39					
40					
41					
42					

285

This page illustrates and describes interior, solid and louvered door systems including a pine panel door, wood jambs, header, and trim with hardware. Lines 06.4-144-04 thru 08 give the unit price and total price on a cost each basis for this system. Prices for alternate interior, solid and louvered systems are on Line Items 06.4-144-12 thru 24. Both material quantities and labor costs have been adjusted for the system listed.

Factors: To adjust for job conditions other than normal working situations use Lines 06.4-144-29 thru 40.

Example: You are to install the system during night hours only. Go to Line 06.4-144-40 and apply these percentages to the appropriate INST. costs.

LINE NO.	DESCRIPTION	QUANTITY	COST EACH		
			MAT.	INST.	TOTAL
01	**Single interior door, including jamb, header, trim and hardware, painted.**				
02					
03					
04	Solid pine panel door, painted 1-⅜" thick, 2'-0" x 6'-8"	1 Ea.	100.64	48.66	149.30
05	Wooden jamb, 4-⅝" deep	1 Set	24.64	25.28	49.92
06	Trim, casing	1 Set	20.77	38.43	59.20
07	Hardware, hinges, lockset	1 Set	23.23	17.72	40.95
08	TOTAL	Ea.	169.28	130.09	299.37
09					
10					
11	For alternate door systems:				
12	Solid pine, painted raised panel, 1-⅜" x 2'-6" x 6'-8"	Ea.	183.90	133.88	317.78
13	2'-8" x 6'-8"		191.54	133.87	325.41
14	3'-0" x 6'-8"	↓	202.01	137.88	339.89
15					
16	Louvered pine, painted 1'-6" x 6'-8"	Ea.	153.88	130.49	284.37
17	2'-0" x 6'-8"		160.80	131.98	292.78
18	2'-6" x 6'-8"		172.84	132.57	305.41
19	3'-0" x 6'-8"	↓	196.51	138.38	334.89
20					
21					
22	For prehung door, deduct	Ea.	5%	30%	
23					
24	For metal frame instead of wood, add	Ea.	50%	20%	
25					
26					
27					
28					
29	Cut & patch to match existing construction, add, minimum		2%	3%	
30	Maximum		5%	9%	
31	Dust protection, add, minimum		1%	2%	
32	Maximum		4%	11%	
33	Equipment usage curtailment, add, minimum		1%	1%	
34	Maximum		3%	10%	
35	Material handling & storage limitation, add, minimum		1%	1%	
36	Maximum		6%	7%	
37	Protection of existing work, add, minimum		2%	2%	
38	Maximum		5%	7%	
39	Shift work requirements, add, minimum			5%	
40	Maximum			30%	
41					
42					

This page illustrates and describes interior metal door systems including a metal door, metal frame and hardware. Lines 06.4-146-04 thru 07 give the unit price and total price on a cost each for this system. Prices for alternate interior metal door systems are on Line Items 06.4-146-11 thru 21. Both material quantities and labor costs have been adjusted for the system listed.

Factors: To adjust for job conditions other than normal working situations use Lines 06.5-146-29 thru 40.

Example: You are to install the system while protecting existing construction. Go to Line 06.4-146-37 and apply these percentages to the appropriate MAT. and INST. costs.

LINE NO.	DESCRIPTION	QUANTITY	COST EACH		
			MAT.	INST.	TOTAL
01	**Single metal door, including frame and hardware.**				
02					
03					
04	Hollow metal door, 1-⅜" thick, 2'-6" x 6'-8", painted	1 Ea.	226.04	58.26	284.30
05	Metal frame, 5-¾" deep	1 Set	73.70	36.30	110
06	Hinges and passage lockset	1 Set	66.28	17.72	84
07	TOTAL	Ea.	366.02	112.28	478.30
08					
09					
10	For alternate systems:				
11	Hollow metal doors, 1-⅜" thick, 2'-8" x 6'-8"	Ea.	370.42	112.88	483.30
12	3'-0" x 7'-0"	"	373.72	119.58	493.30
13					
14	Interior fire door, 1-⅜" thick, 2'-6" x 6'-8"	Ea.	404.52	108.78	513.30
15	2'-8" x 6'-8"		407.82	110.48	518.30
16	3'-0" x 7'-0"	↓	423.22	115.08	538.30
17					
18	Add to fire doors:				
19	Baked enamel finish	Ea.	30%	60%	
20	Galvanizing	↓	15%		
21	Porcelain finish	↓	125%	100%	
22					
23					
24					
25					
26					
27					
28					
29	Cut & patch to match existing construction, add, minimum		2%	3%	
30	Maximum		5%	9%	
31	Dust protection, add, minimum		1%	2%	
32	Maximum		4%	11%	
33	Equipment usage curtailment, add, minimum		1%	1%	
34	Maximum		3%	10%	
35	Material handling & storage limitation, add, minimum		1%	1%	
36	Maximum		6%	7%	
37	Protection of existing work, add, minimum		2%	2%	
38	Maximum		5%	7%	
39	Shift work requirements, add, minimum			5%	
40	Maximum			30%	
41					
42					

287

This page illustrates and describes an interior closet door system including an interior closet door, painted, with trim and hardware. Prices for alternate interior closet door system are on Line Items 06.4-148-05 thru 22. Both material quantities and labor costs have been adjusted for the system listed.

Factors: To adjust for job conditions other than normal working situations use Lines 0.64-148-29 thru 40.

Example: You are to install the system and match the existing construction. Go to Line 06.4-148-29 and apply these percentages to the appropriate MAT. and INST. costs.

LINE NO.	DESCRIPTION	QUANTITY	COST PER SET		
			MAT.	INST.	TOTAL
01	Interior closet door painted, including frame, trim and hardware, prehung.				
02					
03					
04	Bi-fold doors				
05	Pine paneled, 3'-0" x 6'-8"	Set	161.78	107.42	269.20
06	6'-0" x 6'-8"		285.76	146.84	432.60
07	Birch, hollow core, 3'-0" x 6'-8"		107.88	117.32	225.20
08	6'-0" x 6'-8"		182.36	160.24	342.60
09	Lauan, hollow core, 3'-0" x 6'-8"		95.78	108.42	204.20
10	6'-0" x 6'-8"		159.26	148.34	307.60
11	Louvered pine, 3'-0" x 6'-8"		118.88	110.32	229.20
12	6'-0" x 6'-8"		208.76	148.84	357.60
13					
14	Sliding, bi-passing closet doors				
15	Pine paneled, 4'-0" x 6'-8"	Set	238.49	115.41	353.90
16	6'-0" x 6'-8"		296.76	145.84	442.60
17	Birch, hollow core, 4'-0" x 6'-8"		130.69	118.21	248.90
18	6'-0" x 6'-8"		162.56	150.04	312.60
19	Lauan, hollow core, 4'-0" x 6'-8"		119.69	114.21	233.90
20	6'-0" x 6'-8"		142.76	149.84	292.60
21	Louvered pine, 4'-0" x 6'-8"		149.39	114.51	263.90
22	6'-0" x 6'-8"		183.46	149.14	332.60
23					
24					
25					
26					
27					
28					
29	Cut & patch to match existing construction, add, minimum		2%	3%	
30	Maximum		5%	9%	
31	Dust protection, add, minimum		1%	2%	
32	Maximum		4%	11%	
33	Equipment usage curtailment, add, minimum		1%	1%	
34	Maximum		3%	10%	
35	Material handling & storage limitation, add, minimum		1%	1%	
36	Maximum		6%	7%	
37	Protection of existing work, add, minimum		2%	2%	
38	Maximum		5%	7%	
39	Shift work requirements, add, minimum			5%	
40	Maximum			30%	
41					
42					

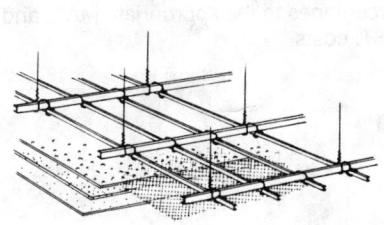

This page illustrates and describes suspended plaster and lath systems including gypsum plaster, lath, furring and runners with ceiling painted. Lines 06.7-242-04 thru 09 give the unit price and total price per square foot for this system. Prices for alternate suspended plaster and lath systems are on Line Items 06.7-242-13 thru 23. Both material quantities and labor costs have been adjusted for the system listed.

Factors: To adjust for job conditions other than normal working situations use Lines 06.7-242-29 thru 40.

Examples: You are to install the system to match existing construction. Go to Line 06.7-242-29 and apply these percentages to the appropriate MAT. and INST. costs.

LINE NO.	DESCRIPTION	QUANTITY	COST PER S.F.		
			MAT.	INST.	TOTAL
01	Gypsum plaster, 3 coats, on 3.4# rib lath, on ¾″ C.R.C. furring on				
02	1-½″ main runners, ceiling painted.				
03					
04	Gypsum plaster, 3 coats	.11 S.Y.	.51	1.91	2.42
05	3.4# rib lath	.11 S.Y.	.23	.44	.67
06	Main runners, 1-½″ C.R.C. 24″ O.C.	.333 S.F.	.24	.71	.95
07	Furring ¾″ C.R.C. 16″ O.C.	1 S.F.	.22	.95	1.17
08	Painting, 2 coats, roller work	1 S.F.	.10	.33	.43
09	TOTAL	S.F.	1.30	4.34	5.64
10					
11					
12	For alternate plaster ceiling systems:				
13	Gypsum plaster, 3 coats, on 2.5# diamond lath	S.F.	1.26	4.30	5.56
14	On ⅜″ gypsum lath		1.50	4.31	5.81
15	2 coats, on 3.4# rib lath		1.53	4.34	5.87
16	On 2.5# diamond lath		1.11	3.99	5.10
17	On ⅜″ gypsum lath		1.35	4	5.35
18	Perlite plaster, 3 coats, on 3.4# rib lath		1.44	4.75	6.19
19	On 2.5# diamond lath		1.40	4.71	6.11
20	On ⅜″ gypsum lath		1.64	4.72	6.36
21	2 coats, on 3.4# rib lath		1.20	4.33	5.53
22	On 2.5# diamond lath		1.16	4.29	5.45
23	On ⅜″ gypsum lath		1.40	4.30	5.70
24					
25					
26					
27					
28					
29	Cut & patch to match existing construction, add, minimum		2%	3%	
30	Maximum		5%	9%	
31	Dust protection, add, minimum		1%	2%	
32	Maximum		4%	11%	
33	Equipment usage curtailment, add, minimum		1%	1%	
34	Maximum		3%	10%	
35	Material handling & storage limitation, add, minimum		1%	1%	
36	Maximum		6%	7%	
37	Protection of existing work, add, minimum		2%	2%	
38	Maximum		5%	7%	
39	Shift work requirements, add, minimum			5%	
40	Maximum			30%	
41					
42					

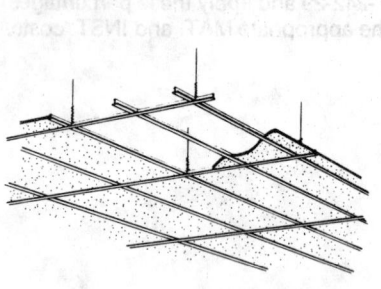

This page illustrates suspended acoustical board systems including acoustic ceiling board, hangers, and T bar suspension. Lines 06.7-342-04 thru 07 give the unit price and total price per square foot for this system. Prices for alternate suspended acoustical board systems are on Line Items 06.7-342-11 thru 24. Both material quantitites and labor costs have been adjusted for the system listed.

Factors: To adjust for job conditions other than normal working situations use Lines 06.7-342-29 thru 40.

Example: You are to install the system and protect existing construction. Go to Line 06.7-342-38 and apply these percentages to the appropriate MAT. and INST. costs.

LINE NO.	DESCRIPTION	QUANTITY	COST PER S.F.		
			MAT.	INST.	TOTAL
01	**Suspended acoustical ceiling board installed on exposed grid system.**				
02					
03					
04	Fiberglass boards, film faced, 2' x 4', ⅝" thick	1 S.F.	.43	.46	.89
05	Hangers, #12 wire	1 S.F.	.01	.05	.06
06	T bar suspension system, 2' x 4' grid	1 S.F.	.41	.36	.77
07	TOTAL	S.F.	.85	.87	1.72
08					
09					
10	For alternate suspended ceiling systems:				
11	2' x 4' grid, mineral fiber board, aluminum faced, ⅝" thick	S.F.	1.54	.85	2.39
12	Standard faced		.92	.83	1.75
13	Plastic faced		1.19	1.13	2.32
14	Fiberglass, film faced, 3" thick, R11		1.47	.98	2.45
15	Grass cloth faced, ¾" thick		1.78	.99	2.77
16	1" thick		1.94	.98	2.92
17	1-½" thick, nubby face		2.27	.98	3.25
18	Wood fiber, reveal edge, painted, 1" thick		1.40	.89	2.29
19	2" thick		2.02	.93	2.95
20	2-½" thick		2.60	.98	3.58
21	3" thick		2.93	1.05	3.98
22					
23					
24	Add for 2' x 2' grid system	S.F.	.10	.10	.20
25					
26					
27					
28					
29	Cut & patch to match existing construction, add, minimum		2%	3%	
30	Maximum		5%	9%	
31	Dust protection, add, minimum		1%	2%	
32	Maximum		4%	11%	
33	Equipment usage curtailment, add, minimum		1%	1%	
34	Maximum		3%	10%	
35	Material handling & storage limitation, add, minimum		1%	1%	
36	Maximum		6%	7%	
37	Protection of existing work, add, minimum		2%	2%	
38	Maximum		5%	7%	
39	Shift work requirements, add, minimum			5%	
40	Maximum			30%	
41					
42					

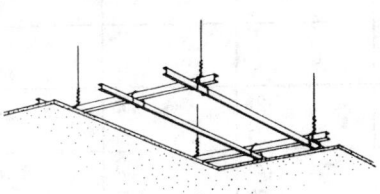

This page illustrates and describes suspended gypsum board systems including gypsum board, metal furring, taping, finished and painted. Lines 06.7-442-05 thru 10 give the unit price and total price per square foot for this system. Prices for alternate suspended gypsum board systems are on Line Items 06.7-442-15 thru 17. Both material quantities and labor costs have been adjusted for the system listed.

Factors: To adjust for job conditions other than normal working situations use Lines 06.7-442-29 thru 40.

Example: You are to install the system and control dust in the working area. Go to Line 06.7-442-32 and apply these percentages to the appropriate MAT. and INST. costs.

LINE NO.	DESCRIPTION	QUANTITY	COST PER S.F.		
			MAT.	INST.	TOTAL
01	Suspended ceiling gypsum board, 4' x 8' x ⅝" thick,				
02	On metal furring, taped, finished and painted.				
03					
04					
05	Gypsum drywall, 4' x 8', ⅝" thick, screwed	1 S.F.	.23	.32	.55
06	Main runners, 1-½" C.R.C., 4' O.C.	.5 S.F.	.12	.36	.48
07	25 ga., channels, 2' O.C.	1 S.F.	.18	.65	.83
08	Taped and finished	1 S.F.	.06	.28	.34
09	Paint, 2 coats, roller work	1 S.F.	.10	.33	.43
10	TOTAL	S.F.	.69	1.94	2.63
11					
12					
13					
14	For alternate drywall ceiling systems:				
15	Thin coat plaster, 2 coats paint	S.F.	.63	1.70	2.33
16	Spray-on sand finish, no paint		.64	1.67	2.31
17	12" x 12" x ¾" acoustical wood fiber tile	↓	1.41	2.05	3.46
18					
19					
20					
21					
22					
23					
24					
25					
26					
27					
28					
29	Cut & patch to match existing construction, add, minimum		2%	3%	
30	Maximum		5%	9%	
31	Dust protection, add, minimum		1%	2%	
32	Maximum		4%	11%	
33	Equipment usage curtailment, add, minimum		1%	1%	
34	Maximum		3%	10%	
35	Material handling & storage limitation, add, minimum		1%	1%	
36	Maximum		6%	7%	
37	Protection of existing work, add, minimum		2%	2%	
38	Maximum		5%	7%	
39	Shift work requirements, add, minimum			5%	
40	Maximum			30%	
41					
42					

LINE NO.	DESCRIPTION	COST PER S.F.		
		MAT.	INST.	TOTAL
01	Studs			
02				
03	24″ O.C. metal, 10′ high wall, including			
04	Top and bottom runners			
05	Non load bearing, galvanized 25 Ga., 1-⅝″ wide	.21	.55	.76
06	2-½″ wide	.24	.57	.81
07	3-⅝″ wide	.29	.57	.86
08	4″ wide	.32	.59	.91
09	6″ wide	.41	.60	1.01
10				
11	Galvanized 20 Ga., 2-½″ wide	.43	.56	.99
12	3-⅝″ wide	.51	.57	1.08
13	4″ wide	.54	.59	1.13
14	6″ wide	.67	.60	1.27
15	Load bearing, painted 18 Ga., 2-½″ wide	.80	1.20	2
16	3-⅝″ wide	.92	1.25	2.17
17	4″ wide	.97	1.30	2.27
18	6″ wide	1.18	1.37	2.55
19	Galvanized 18 Ga., 2-½″ wide	.80	1.20	2
20	3-⅝″ wide	.92	1.25	2.17
21	4″ wide	.97	1.30	2.27
22	6″ wide	1.18	1.37	2.55
23	Galvanized 16 Ga., 2-½″ wide	1.01	1.31	2.32
24	3-⅝″ wide	1.13	1.37	2.50
25	4″ wide	1.22	1.44	2.66
26	6″ wide	1.50	1.51	3.01
27				
28				
29	24″ O.C. wood, 10′ high wall, including			
30	Double top plate and shoe			
31	2″ x 4″	.28	.56	.84
32	2″ x 6″	.41	.61	1.02
33				
34				
35	Furring, 24″ O.C. 10′ high wall, metal, ¾″ channels	.18	.65	.83
36	1-½″ channels	.24	.71	.95
37	Wood, on wood, 1″ x 2″ strips	.06	.26	.32
38	1″ x 3″ strips	.08	.26	.34
39	On masonry, 1″ x 2″ strips	.06	.29	.35
40	1″ x 3″ strips	.08	.30	.38
41	On concrete, 1″ x 2″ strips	.06	.56	.62
42	1″ x 3″ strips	.08	.56	.64
43	Studs			
44				
45	16″ O.C., metal, 10′ high wall, including			
46	Top and bottom runners			
47	Non load bearing, galvanized 25 Ga., 1-⅝″ wide	.25	.64	.89
48	2-½″ wide	.30	.65	.95
49	3-⅝″ wide	.34	.67	1.01
50	4″ wide	.39	.68	1.07
51	6″ wide	.50	.70	1.20
52				
53	Galvanized 20 Ga., 2-½″ wide	.52	.65	1.17
54	3-⅝″ wide	.62	.66	1.28
55	4″ wide	.65	.68	1.33
56	6″ wide	.81	.70	1.51
57	Load bearing, painted 18 Ga., 2-½″ wide	.97	1.43	2.40
58	3-⅝″ wide	1.11	1.51	2.62

LINE NO.	DESCRIPTION	COST PER S.F.		
		MAT.	INST.	TOTAL
59	4″ wide	1.18	1.59	2.77
60	6″ wide	1.42	1.69	3.11
61	Galvanized 18 Ga., 2-½″ wide	.97	1.43	2.40
62	3-⅝″ wide	1.11	1.51	2.62
63	4″ wide	1.18	1.59	2.77
64	6″ wide	1.42	1.69	3.11
65	Galvanized 16 Ga., 2-½″ wide	1.21	1.60	2.81
66	3-⅝″ wide	1.36	1.69	3.05
67	4″ wide	1.46	1.80	3.26
68	6″ wide	1.80	1.92	3.72
69				
70				
71	16″ O.C., wood, 10′ high wall, including			
72	Double top plate and shoe			
73	2″ x 4″	.34	.65	.99
74	2″ x 6″	.41	.58	.99
75				
76				
77	Furring, 16″ O.C. 10′ high wall, metal, ¾″ channels	.22	.95	1.17
78	1-½″ channels	.30	1.06	1.36
79	Wood, on wood, 1″ x 2″ strips	.08	.39	.47
80	1″ x 3″ strips	.11	.40	.51
81	On masonry, 1″ x 2″ strips	.09	.44	.53
82	1″ x 3″ strips	.12	.44	.56
83	On concrete, 1″ x 2″ strips	.09	.83	.92
84	1″ x 3″ strips	.12	.83	.95

LINE NO.	DESCRIPTION	COST PER S.F.		
		MAT.	INST.	TOTAL
01	Lath, gypsum perforated			
02				
03	Regular, ⅜" thick	.40	.36	.76
04	½" thick	.46	.44	.90
05	Fire resistant, ⅜" thick	.45	.44	.89
06	½" thick	.48	.47	.95
07	Foil back, ⅜" thick	.44	.41	.85
08	½" thick	.47	.44	.91
09				
10	Metal lath			
11	Diamond painted, 2.5 lb.	.19	.36	.55
12	3.4 lb.	.21	.42	.63
13	Rib painted, 2.75 lb	.20	.42	.62
14	3.40 lb	.23	.45	.68
15				
16				
17	Plaster, gypsum, 2 coats	.36	1.42	1.78
18	3 coats	.51	1.70	2.21
19	Perlite/vermiculite, 2 coats	.42	1.61	2.03
20	3 coats	.65	2.01	2.66
21	Bondcrete, 1 coat	.34	.75	1.09
22				
25				
26				
27	Drywall, standard, ⅜" thick, no finish included	.18	.28	.46
28	½" thick, no finish included	.19	.28	.47
29	Taped and finished	.24	.60	.84
30	⅝" thick, no finish included	.23	.29	.52
31	Taped and finished	.29	.59	.88
32	Fire resistant, ½" thick, no finish included	.23	.29	.52
33	Taped and finished	.29	.59	.88
34	⅝" thick, no finish included	.25	.29	.54
35	Taped and finished	.31	.59	.90
36	Water resistant, ½" thick, no finish included	.30	.28	.58
37	Taped and finished	.35	.60	.95
38	⅝" thick, no finish included	.34	.29	.63
39	Taped and finished	.40	.59	.99
40	Finish, instead of taping			
41	For thin coat plaster, add	.10	.37	.47
42	Finish, textured spray, add	.11	.34	.45

LINE NO.	DESCRIPTION	COST EACH		
		MAT.	INST.	TOTAL
01	Door closer, rack and pinion	74.80	45.20	120
02	Backcheck and adjustable power	80.30	49.70	130
03	Regular, hinge face mount, all sizes, regular arm	73.70	46.30	120
04	Hold open arm	83.60	46.40	130
05	Top jamb mount, all sizes, regular arm	71.50	48.50	120
06	Hold open arm	82.50	47.50	130
07	Stop face mount, all sizes, regular arm	73.70	46.30	120
08	Hold open arm	84.70	45.30	130
09	Fusible link, hinge face mount, all sizes, regular arm	83.60	46.30	129.90
10	Hold open arm	93.50	46.40	139.90
11	Top jamb mount, all sizes, regular arm	81.40	48.50	129.90
12	Hold open arm	92.40	47.50	139.90
13	Stop face mount, all sizes, regular arm	83.60	46.30	129.90
14	Hold open arm	94.60	45.30	139.90
15				
16				
17	Door stops			
18				
19	Holder & bumper, floor or wall	11	12	23
20	Wall bumper	4.68	11.97	16.65
21	Floor bumper	3.85	11.95	15.80
22	Plunger type, door mounted	26.40	11.60	38
23	Hinges, full mortise, material only, per pair			
24	Low frequency, 4-½″ x 4-½″, steel base, USP	11.95		11.95
25	Brass base. US10	39		39
26	Stainless steel base, US32	58		58
27	Average frequency, 4-½″ x 4-½″, steel base, USP	20		20
28	Brass base, US10	50		50
29	Stainless steel base, US32	66		66
30	High frequency, 4-½″ x 4-½″, steel base, USP	50		50
31	Brass base, US10	77		77
32	Stainless steel base, US32	99		99
33	Kick plate			
34				
35	6″ high, for 3′-0″ door, aluminum	18.70	19.30	38
36	Bronze	38.50	19.50	58
37	Panic device			
38				
39	For rim locks, single door, exit	258.50	96.50	355
40	Outside key and pull	313.50	56.50	370
41	Bar and vertical rod, exit only	396	59	455
42	For touch bar	473	72	545
43	Lockset			
44				
45	Heavy duty, cylindrical, passage doors	95.70	24.30	120
46	Classroom	203.50	36.50	240
47	Bedroom, bathroom, and inner office doors	123.20	21.80	145
48	Apartment, office, and corridor doors	165	30	195
49	Standard duty, cylindrical, exit doors	63.80	29.20	93
50	Inner office doors	29.15	23.85	53
51	Passage doors	39.60	24.40	64
52	Public restroom, classroom, & office doors	79.20	35.80	115
53	Deadlock, mortise, heavy duty	93.50	31.50	125
54	Double cylinder	104.50	30.50	135
55	Entrance lock, cylinder, deadlocking latch	96.80	33.20	130
56	Deadbolt	115.50	34.50	150
57	Commercial, mortise, wrought knob, keyed, minimum	115.50	34.50	150
58	Maximum	225.50	39.50	265

LINE NO.	DESCRIPTION	COST EACH		
		MAT.	INST.	TOTAL
59	Cast knob, keyed, minimum	159.50	30.50	190
60	Maximum	330	30	360
61	Push-pull			
62				
63	Aluminum	52.80	24.20	77
64	Bronze	66	24	90
65	Door pull, designer style, minimum	55	24	79
66	Maximum	236.50	33.50	270
67	Threshold			
68				
69	3'-0" long door saddles, aluminum, minimum	24.20	14.80	39
70	Maximum	74.80	24.20	99
71	Bronze, minimum	48.40	14.60	63
72	Maximum	132	23	155
73	Rubber, ½" thick, 5-½" wide	29.70	14.30	44
74	2-¾" wide	13.75	14.25	28
75	Weatherstripping, per set			
76				
77	Doors, wood frame, interlocking for 3' x 7' door, zinc	11	94	105
78	Bronze	16.50	93.50	110
79	Wood frame, spring type for 3' x 7' door, bronze	7.59	37.41	45
80	Metal frame, spring type for 3' x 7' door, bronze	33	97	130
81	For stainless steel, spring type, add	133%		
82				
83	Metal frame, extruded sections, 3' x 7' door, aluminum	33	142	175
84	Bronze	74.80	145.20	220

LINE NO.	DESCRIPTION	COST PER S.F.		
		MAT.	INST.	TOTAL
01	Painting, on plaster or drywall, brushwork, primer and 1 ct.	.09	.39	.48
02	Primer and 2 ct.	.13	.60	.73
03	Rollerwork, primer and 1 ct.	.10	.22	.32
04	Primer and 2 ct.	.14	.35	.49
05	Woodwork incl. puttying, brushwork, primer and 1 ct.	.09	.59	.68
06	Primer and 2 ct.	.11	.78	.89
07	Wood trim to 6″ wide, enamel, primer and 1 ct.	.07	.24	.31
08	Primer and 2 ct.	.09	.33	.42
09	Cabinets and casework, enamel, primer and 1 ct.	.08	.61	.69
10	Primer and 2 ct.	.12	.78	.90
11	On masonry or concrete, latex, brushwork, primer and 1 ct.	.13	.53	.66
12	Primer and 2 ct.	.15	.76	.91
13	For block filler, add	.09	.14	.23
14				
15	Varnish, wood trim, sealer 1 ct., sanding, puttying, quality work	.13	1.20	1.33
16	Meduim work	.10	.94	1.04
17	Without sanding	.08	.11	.19
18				
19	Wall coverings, wall paper, at $9.70 per double roll, average workmanship	.24	.42	.66
20	At $20.00 per double roll, average workmanship	.42	.49	.91
21	At $44.00 per double roll, quality workmanship	.84	.60	1.44
22				
23	Grass cloths with lining paper, minimum	.61	.66	1.27
24	Maximum	1.54	.76	2.30
25	Vinyl, fabric backed, light weight	.68	.42	1.10
26	Medium weight	.83	.55	1.38
27	Heavy weight	1.08	.61	1.69
28				
29	Cork tiles, 12″ x 12″, ³⁄₁₆″ thick	1.98	1.10	3.08
30	⁵⁄₁₆″ thick	2.09	1.10	3.19
31	Granular surface, 12″ x 36″, ½″ thick	.66	.69	1.35
32	1″ thick	.88	.69	1.57
33	Aluminum foil	.68	.96	1.64
34				
35	Tile, ceramic, adhesive set, 4-¼″ x 4-¼″	2.04	2.73	4.77
36	6″ x 6″	2.53	2.46	4.99
37	Decorated, 4-¼″ x 4-¼″, minimum	2.75	.56	3.31
38	Maximum	15.40	.85	16.25
39	For epoxy grout, add	.50	.51	1.01
40	Pregrouted sheets	3.91	2.04	5.95
41	Glass mosaics, ¾″ tile on 12″ sheets, color groups 1 & 2, minimum		10.75	10.75
42	Maximum	11.33	6.72	18.05
43	Metal, tile pattern, 4′ x 4′ sheet, 24 ga., nailed	2.86	3.49	6.35
44	Stainless steel	16.23	1.12	17.35
45	Aluminized steel	8.58	1.12	9.70
46		9.35	3.50	12.85
47	Brick, interior veneer, 4″ face brick, running bond, minimum	1.89	6.16	8.05
48	Maximum	1.89	6.16	8.05
49	Simulated, urethane pieces, set in mastic	4.35	1.90	6.25
50	Fiberglass panels	2.04	1.43	3.47
51	Wall coating, on drywall, thin coat, plain	.10	.37	.47
52	Stipple	.10	.37	.47
53	Textured spray	.11	.34	.45
54				
55	Paneling not incl. furring or trim, hardboard, tempered, ⅛″ thick	.30	1.19	1.49
56	¼″ thick	.36	1.19	1.55
57	Plastic faced, ⅛″ thick	.48	1.19	1.67
58	¼″ thick	.66	1.19	1.85

LINE NO.	DESCRIPTION	COST PER S.F.		
		MAT.	INST.	TOTAL
59	Woodgrained, ¼" thick, minimum	.44	1.19	1.63
60	Maximum	.83	1.39	2.22
61	Plywood, 4' x 8' shts. ¼" thick, prefin., birch faced, min.	.66	1.19	1.85
62	Maximum	1.38	1.69	3.07
63	Walnut, minimum	2.26	1.19	3.45
64	Maximum	4.24	1.46	5.70
65	Mahogany, african	1.71	1.48	3.19
66	Philippine	.55	1.19	1.74
67	Chestnut	3.96	1.59	5.55
68	Pecan	1.71	1.48	3.19
69	Rosewood	10.45	1.85	12.30
70	Teak	2.64	1.48	4.12
71	Aromatic cedar, plywood	1.60	1.48	3.08
72	Particle board	.77	1.48	2.25
73	Wood board, ¾" thick, knotty pine	.97	1.98	2.95
74	Rough sawn cedar	1.34	1.98	3.32
75	Redwood, clear	3.08	1.97	5.05
76	Aromatic cedar	2.09	2.16	4.25

LINE NO.	DESCRIPTION	COST PER S.F.		
		MAT.	INST.	TOTAL
01	Ceiling, plaster, gypsum, 2 coats	.36	1.62	1.98
02	3 coats	.51	1.93	2.44
03	Perlite or vermiculite, 2 coats	.42	1.91	2.33
04	3 coats	.65	2.35	3
05	Gypsum lath, plain, ⅜″ thick	.40	.36	.76
06	½″ thick	.43	.39	.82
07	Firestop, ⅜″ thick	.45	.44	.89
08	½″ thick	.48	.47	.95
09	Metal lath, rib, 2.75 lb.	.20	.42	.62
10	3.40 lb.	.23	.45	.68
11	Diamond, 2.50 lb.	.19	.41	.60
12	3.40 lb.	.21	.52	.73
13				
14				
15	Drywall, standard, ½″ thick, no finish included	.19	.32	.51
16	Taped and finished	.24	.75	.99
17	⅝″ thick, no finish included	.23	.32	.55
18	Taped and finished	.29	.75	1.04
19	Fire resistant, ½″ thick, no finish included	.23	.32	.55
20	Taped and finished	.29	.75	1.04
21	⅝″ thick, no finish included	.25	.32	.57
22	Taped and finished	.31	.75	1.06
23	Water resistant, ½″ thick, no finish included	.30	.32	.62
24	Taped and finished	.35	.75	1.10
25	⅝″ thick, no finish included	.34	.32	.66
26	Taped and finished	.40	.75	1.15
27	Finish, instead of taping			
28	For thin coat plaster, add	.10	.37	.47
29	Finish, textured spray, add	.11	.34	.45
30				
31				
32				
33	Tile, stapled or glued, mineral fiber plastic coated, ⅝″ thick	.53	1.43	1.96
34	¾″ thick	.69	1.44	2.13
35	Wood fiber, ½″ thick	.63	.72	1.35
36	¾″ thick	.88	.72	1.60
37	Suspended, fiberglass, film faced, ⅝″ thick	.43	.46	.89
38	3″ thick	1.05	.57	1.62
39	Mineral fiber, ⅝″ thick, standard	.50	.42	.92
40	Aluminum	1.16	.47	1.63
41	Wood fiber, reveal edge, 1″ thick	.98	.48	1.46
42	3″ thick	2.51	.64	3.15
43	Framing, metal furring, ¾″ channels, 12″ O.C.	.28	1.31	1.59
44	16″ O.C.	.22	.95	1.17
45	24″ O.C.	.18	.65	.83
46				
47	1-½″ channels, 12″ O.C.	.37	1.45	1.82
48	16″ O.C.	.30	1.06	1.36
49	24″ O.C.	.24	.71	.95
50				
51				
52				
53	Ceiling suspension systems, for tile,			
54	Concealed ''Z'' bar, 12″ module	.62	.55	1.17
55	Class A, ''T'' bar 2'-0″ x 4'-0″ grid	.41	.36	.77
56	2'-0″ x 2'-0″ grid	.54	.44	.98
57	Carrier channels for lighting fixtures, add	.33		.33
58				

LINE NO.	DESCRIPTION	COST PER S.F.		
		MAT.	INST.	TOTAL
59				
60				
61	Wood, furring 1″ x 3″, on wood, 12″ O.C.	.15	.83	.98
62	16″ O.C..	.11	.63	.74
63	24″ O.C.	.08	.41	.49
64				
65	On concrete, 12″ O.C.	.17	1.36	1.53
66	16″ O.C.	.12	1.03	1.15
67	24″ O.C.	.08	.69	.77
68				
69	Joists, 2″ x 4″, 12″ O.C.	.37	.62	.99
70	16″ O.C.	.32	.51	.83
71	24″ O.C.	.25	.42	.67
72	32″ O.C.	.19	.32	.51
73	2″ x 6″, 12″ O.C.	.57	.62	1.19
74	16″ O.C.	.47	.52	.99
75	24″ O.C.	.37	.41	.78
76	32″ O.C.	.28	.30	.58

LINE NO.	DESCRIPTION	COST PER S.F.		
		MAT.	INST.	TOTAL
01	Flooring, carpet, acrylic, 26 oz. light traffic	1.68	.54	2.22
02	35 oz. heavy traffic	2.53	.58	3.11
03	Nylon anti-static, 15 oz. light traffic	1.03	.54	1.57
04	22 oz. medium traffic	1.82	.51	2.33
05	26 oz. heavy traffic	2.62	.60	3.22
06	28 oz. heavy traffic	2.98	.57	3.55
07	Tile, foamed back, needle punch	2.20	.42	2.62
08	Tufted loop	1.93	.42	2.35
09	Wool, 36 oz. medium traffic	3.47	.64	4.11
10	42 oz. heavy traffic	4.74	.59	5.33
11	Composition, epoxy, with colored chips, minimum	1.71	2.24	3.95
12	Maximum	2.70	3.10	5.80
13	Trowelled, minimum	2.20	2.69	4.89
14	Maximum	3.03	3.12	6.15
15	Terrazzo, ¼" thick, chemical resistant, minimum	3.63	2.92	6.55
16	Maximum	5.67	3.93	9.60
17	Resilient, asphalt tile, ⅛" thick	.88	.69	1.57
18	Conductive flooring, rubber, ⅛" thick	2.70	.87	3.57
19	Cork tile ⅛" thick, standard finish	1.98	.87	2.85
20	Urethane finish	3.03	.87	3.90
21	PVC sheet goods for gyms, ¼" thick	3.08	3.42	6.50
22	⅜" thick	3.52	4.58	8.10
23	Vinyl composition 12" x 12" tile, plain, ¹⁄₁₆" thick	.61	.54	1.15
24	⅛" thick	1.60	.54	2.14
25	Vinyl tile, 12" x 12" x ⅛" thick, minimum	1.93	.54	2.47
26	Maximum	7.43	.52	7.95
27	Vinyl sheet goods, backed, .093" thick	1.43	1.19	2.62
28	.250" thick	2.59	1.19	3.78
29	Slate, random rectangular, ¼" thick	3.88	3.47	7.35
30	½" thick	3.64	4.96	8.60
31	Natural cleft, irregular. ¾" thick	1.63	5.62	7.25
32	For sand rubbed finish, add	2.50		2.50
33	Terrazzo, cast in place, bonded 1-¾" thick, gray cement	1.93	4.52	6.45
34	White cement	2.20	4.50	6.70
35	Not bonded 3" thick, gray cement	2.71	5.09	7.80
36	White cement	3.08	5.12	8.20
37	Precast, 12" x 12" x 1" thick	9.90	18.10	28
38	1-¼" thick	11	18	29
39	16" x 16" x 1-¼" thick	12.32	22.68	35
40	1-½" thick	14.85	25.15	40
41	Marble travertine, standard, 12" x 12" x ¾" thick	4.29	8.21	12.50
42				
43	Tile, ceramic, natural clay, thin set	3.25	2.70	5.95
44	Porcelain, thin set	3.96	2.69	6.65
45	Specialty, decorator finish	6	2.70	8.70
46				
47	Quarry, red, mud set, 4" x 4" x ½" thick	3.19	4.11	7.30
48	6" x 6" x ½" thick	3.19	3.51	6.70
49	Brown, imported, 6" x 6" x ¾" thick	3.85	4.10	7.95
50	8" x 8" x 1" thick	4.79	4.46	9.25
51	Slate, vermont, thin set, 6" x 6" x ¼" thick	3.14	2.71	5.85
52				
53	Wood, maple strip, ²⁵⁄₃₂" x 2-¼", finished, select	2.67	1.80	4.47
54	2nd and better	2.34	1.80	4.14
55	Oak, ²⁵⁄₃₂" x 2-¼" finished, clear	2.50	1.80	4.30
56	No. 1 common	2.78	1.66	4.44
57	Parquet, standard, ⁵⁄₁₆", finished, minimum	1.84	1.91	3.75
58	Maximum	5.58	2.96	8.54

LINE NO.	DESCRIPTION	COST PER S.F.		
		MAT.	INST.	TOTAL
59	Custom, finished, minimum	11.88	2.87	14.75
60	Maximum	16.23	5.77	22
61	Prefinished, oak, 2-¼" wide	5.34	1.71	7.05
62	Ranch plank	6.44	1.96	8.40
63	Sleepers on concrete, treated, 24" O.C., 1" x 2"	.10	.20	.30
64	1" x 3"	.18	.25	.43
65	2" x 4"	.35	.36	.71
66	2" x 6"	.53	.41	.94
67	Refinish old floors, minimum	.55	.65	1.20
68	Maximum	.83	2.01	2.84
69	Subflooring, plywood, CDX, ½" thick	.35	.40	.75
70	⅝" thick	.43	.44	.87
71	¾" thick	.51	.47	.98
72				
73	1" x 10" boards, S4S, laid regular	.75	.54	1.29
74	Laid diagonal	.77	.66	1.43
75	1" x 8" boards, S4S, laid regular	.74	.59	1.33
76	Laid diagonal	.76	.70	1.46
77	Underlayment, plywood, underlayment grade, ⅜" thick	.35	.40	.75
78	½" thick	.40	.41	.81
79	⅝" thick	.48	.42	.90
80	¾" thick	.56	.46	1.02
81	Particle board, ⅜" thick	.29	.40	.69
82	½" thick	.31	.41	.72
83	⅝" thick	.34	.42	.76
84	¾" thick	.39	.46	.85

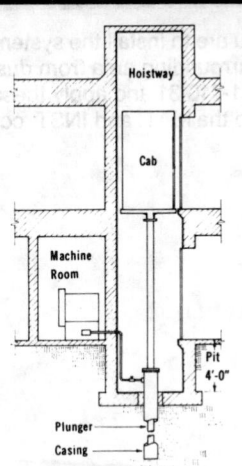

This page illustrates and describes oil hydraulic elevator systems. Prices for various oil hydraulic elevator systems are on Line Items 07.1-142-04 thru 20. Both material quantities and labor costs have been adjusted for the system listed.

Factors: To adjust for job conditions other than normal working situations use Lines 07.1-142-31 thru 40.

Example: You are to install the system with limited equipment usage. Go to Line 07.1-142-34 and apply this percentage to the appropriate TOTAL costs.

LINE NO.	DESCRIPTION	QUANTITY	COST EACH		
			MAT.	INST.	TOTAL
01	**Oil-hydraulic elevator systems**				
02	**Including piston and piston shaft.**				
03					
04	1500 Lb. passenger, 2 floors	Ea.	25,300	14,200	39,500
05	3 floors		32,411.50	20,178.50	52,590
06	4 floors		45,218.25	26,782	72,000.25
07	5 floors		52,818.75	32,511.50	85,330.25
08	6 floors		59,369.25	38,241	97,610.25
09					
10	2500 Lb. passenger, 2 floors	Ea.	29,575	14,200	43,775
11	3 floors		36,686.50	20,178.50	56,865
12	4 floors		48,068.25	26,782	74,850.25
13	5 floors		55,668.75	32,511.50	88,180.25
14	6 floors		63,644.25	38,241	101,885.25
15					
16	4000 Lb. passenger, 2 floors	Ea.	31,700	14,200	45,900
17	3 floors		39,346.50	20,178.50	59,525
18	4 floors		51,618.25	26,782	78,400.25
19	5 floors		59,218.75	32,511.50	91,730.25
20	6 floors		65,769.25	38,241	104,010.25
21					
22					
23					
24					
25					
26					
27					
28					
29					
30					
31	Dust protection, add, minimum		1%	2%	
32	Maximum		4%	11%	
33	Equipment usage curtailment, add, minimum		1%	1%	
34	Maximum		3%	10%	
35	Material handling & storage limitation, add, minimum		1%	1%	
36	Maximum		6%	7%	
37	Protection of existing work, add, minimum		2%	2%	
38	Maximum		5%	7%	
39	Shift work requirements, add, minimum			5%	
40	Maximum			30%	
41					
42					

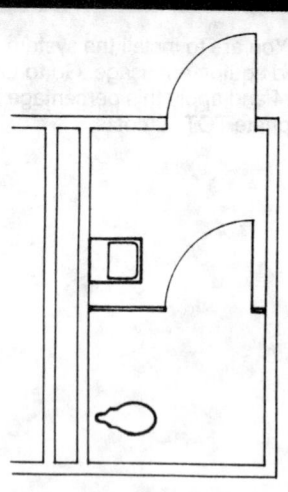

This page illustrates and describes a women's public restroom system including a water closet, lavatory, accessories, and service piping. Lines 08.1-710-04 thru 10 give the unit price and total price on a cost each basis for this system. Prices for alternate women's public restroom systems are on Line Items 08.1-710-14, thru 17. Both material quantities and labor costs have been adjusted for the system listed.

Factors: To adjust for job conditions other than normal working situations use Lines 08.1-710-29 thru 40.

Examples: You are to install the system and protect surrounding area from dust. Go to Line 08.1-710-31 and apply these percentages to the MAT. and INST. costs.

LINE NO.	DESCRIPTION	QUANTITY	COST EACH		
			MAT.	INST.	TOTAL
01	Public women's restroom incl. water closet, lavatory, accessories and				
02	Necessary service piping to install this system in one wall.				
03					
04	Water closet, wall mounted, one piece	1 Ea.	302.50	97.50	400
05	Rough-in waste and vent for water closet	1 Set	172.63	277.37	450
06	Lavatory, 20″ x 18″ P.E. cast iron with accessories	1 Ea.	121	69	190
07	Rough-in waste and vent for lavatory	1 Set	143.11	341.89	485
08	Partition, painted metal between walls, floor mounted, access.	1 Ea.	569.90	123.10	693
09	Accessories	1 Set	469.70	78.30	548
10	TOTAL	System	1778.84	987.16	2766
11					
12					
13	For alternate size restrooms:				
14	Two water closets, two lavatories	System	3018.58	1887.42	4906
15					
16	For each additional water closet over 2, add	System	828.23	468.77	1297
17	For each additional lavatory over 2, add	"	411.51	441.49	853
18					
19					
20					
21					
22					
23					
24	Note: **PLUMBING APPROXIMATIONS**				
25	WATER CONTROL: water meter, backflow preventer,				
26	Shock absorbers, vacuum breakers, mixer....10 to 15% of fixtures				
27	PIPE AND FITTINGS: .... 30 to 60% of fixtures				
28					
29	Cut & patch to match existing construction, add, minimum		2%	3%	
30	Maximum		5%	9%	
31	Dust protection, add, minimum		1%	2%	
32	Maximum		4%	11%	
33	Equipment usage curtailment, add, minimum		1%	1%	
34	Maximum		3%	10%	
35	Material handling & storage limitation, add, minimum		1%	1%	
36	Maximum		6%	7%	
37	Protection of existing work, add, minimum		2%	2%	
38	Maximum		5%	7%	
39	Shift work requirements, add, minimum			5%	
40	Maximum			30%	
41					
42					

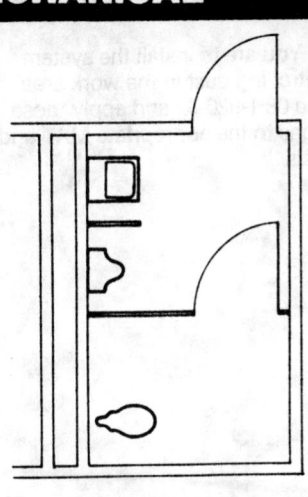

This page illustrates and describes a men's public restroom system including a water closet, urinal, lavatory, accessories and service piping. Lines 08.1-720-04 thru 13 give the unit price and total price on a cost each basis for this system. Prices for alternate men's public restroom systems are on Line Items 08.1-720-18 thru 22. Both material quantities and labor costs have been adjusted for the system listed.

Factors: To adjust for job conditions other than normal working situations use Lines 08.1-720-29 thru 40.

Example: You are to install the system and match existing construction. Go to Line 08.1-720-30 and apply these percentages to the appropriate MAT. and INST. costs.

LINE NO.	DESCRIPTION	QUANTITY	COST EACH		
			MAT.	INST.	TOTAL
01	Public men's restroom incl. water closet, urinal, lavatory, accessories,				
02	And necessary service piping to install this system in one wall.				
03					
04	Water closet, wall mounted, one piece	1 Ea.	302.50	97.50	400
05	Rough-in waste & vent for water closet	1 Set	172.63	277.37	450
06	Urinal, wall hung	1 Ea.	368.50	186.50	555
07	Rough-in waste & vent for urinal	1 Set	71.34	198.66	270
08	Lavatory, 20″ x 18″, P.E. cast iron with accessories	1 Ea.	121	69	190
09	Rough-in waste & vent for lavatory	1 Set	143.11	341.89	485
10	Partition, painted mtl., between walls, floor mntd., accessories	1 Ea.	569.90	123.10	693
11	Urinal screen, painted metal, wall mounted	1 Ea.	214.50	55.50	270
12	Accessories	1 Set	210.10	57.90	268
13	TOTAL	System	2173.58	1407.42	3581
14					
15					
16					
17	For alternate size restrooms:				
18	Two water closets, two urinals, two lavatories	System	4067.66	2748.34	6816
19					
20	For each additional water closet over 2, add	System	828.23	468.77	1297
21	For each additional urinal over 2, add		654.34	440.66	1095
22	For each additional lavatory over 2, add		411.51	441.49	853
23					
24	NOTE: PLUMBING APPROXIMATIONS				
25	WATER CONTROL: water meter, backflow preventer,				
26	Shock absorbers, vacuum breakers, mixer....10 to 15% of fixtures				
27	PIPE AND FITTINGS: .... 30 to 60% of fixtures				
28					
29	Cut & patch to match existing construction, add, minimum		2%	3%	
30	Maximum		5%	9%	
31	Dust protection, add, minimum		1%	2%	
32	Maximum		4%	11%	
33	Equipment usage curtailment, add, minimum		1%	1%	
34	Maximum		3%	10%	
35	Material handling, & storage limitation, add, minimum		1%	1%	
36	Maximum		6%	7%	
37	Protection of existing work, add, minimum		2%	2%	
38	Maximum		5%	7%	
39	Shift work requirements, add, minimum			5%	
40	Maximum			30%	
41					
42					

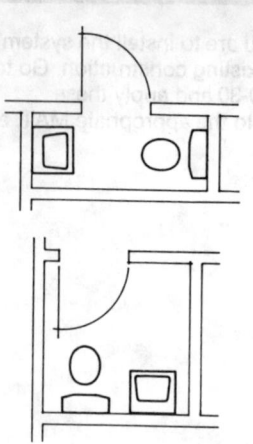

This page illustrates and describes a two fixture lavatory system including a water closet, lavatory, accessories and all service piping. Lines 08.1-920-04 thru 16 give the unit price and total price on a cost each basis for this system. Prices for an alternate two fixture lavatory system are on Line Item 08.1-920-19. Both material quantities and labor costs have been adjusted for the system listed.

Factors: To adjust for job conditions other than normal working situations use Lines 08.1-920-29 thru 40.

Example: You are to install the system while controlling dust in the work area. Go to Line 08.1-920-32 and apply these percentages to the appropriate MAT. and INST. costs.

LINE NO.	DESCRIPTION	QUANTITY	COST EACH MAT.	COST EACH INST.	TOTAL
01	Two fixture bathroom incl. water closet, lavatory, accessories and				
02	Necessary service piping to install this system in 2 walls.				
03					
04	Water closet, floor mounted, 2 piece, close coupled	1 Ea.	151.80	108.20	260
05	Rough in waste & vent for water closet	1 Set	94.22	290.78	385
06	Lavatory, 20″ x 18″, P.E. cast iron with accessories	1 Ea.	121	69	190
07	Rough in waste & vent for lavatory	1 Set	143.11	341.89	485
08	Additional service piping				
09	½″ copper pipe with sweat solder joints	10 L.F.	12.98	38.52	51.50
10	2″ black schedule 40 steel pipe with threaded couplings	12 L.F.	38.41	106.19	144.60
11	4″ cast iron soil pipe with lead and oakum joints	7 L.F.	28.26	71.84	100.10
12	Accessories				
13	Toilet tissue dispenser, chrome, single roll	1 Ea.	17.60	9.40	27
14	18″ long stainless steel towel bar	1 Ea.	19.36	12.64	32
15	Medicine cabinet with mirror, 20″ x 16″, unlighted	1 Ea.	62.70	20.30	83
16	TOTAL	System	689.44	1068.76	1758.20
17					
18					
19	Above system installed in 1 wall with all necessary service piping	System	644.45	941.75	1586.20
20					
21					
22					
23					
24	**NOTE: PLUMBING APPROXIMATIONS**				
25	WATER CONTROL: water meter, backflow preventer,				
26	Shock absorbers, vacuum breakers, mixer....10 to 15% of fixtures				
27	PIPE AND FITTINGS: .... 30 to 60% of fixtures				
28					
29	Cut & patch to match existing construction, add, minimum		2%	3%	
30	Maximum		5%	9%	
31	Dust protection, add, minimum		1%	2%	
32	Maximum		4%	11%	
33	Equipment usage curtailment, add, minimum		1%	1%	
34	Maximum		3%	10%	
35	Material handling & storage limitation, add, minimum		1%	1%	
36	Maximum		6%	7%	
37	Protection of existing work, add, minimum		2%	2%	
38	Maximum		5%	7%	
39	Shift work requirements, add, minimum			5%	
40	Maximum			30%	
41					
42					

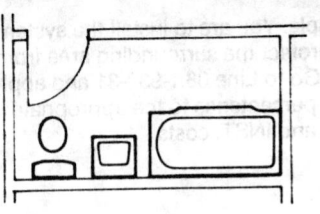

This page illustrates and describes a three fixture bathroom system including a water closet, tub, lavatory, accessories and service piping. Lines 08.1-931-04 thru 14 give the unit price and total price on a cost each basis for this system. Prices for an alternate three fixture bathroom system are on Line Item 08.1-931-17. Both material quantities and labor costs have been adjusted for the system listed.

Factors: To adjust for job conditions other than normal working situations use Lines 08.1-931-29 thru 40.

Example: You are to install the system and protect all existing work. Go to Line 08.1-931-38 and apply these percentages to the appropriate MAT. and INST. costs.

LINE NO.	DESCRIPTION	QUANTITY	COST EACH		
			MAT.	INST.	TOTAL
01	Three fixture bathroom incl. water closet, bathtub, lavatory, accessories,				
02	And necessary service piping to install this system in 2 walls.				
03					
04	Water closet, floor mounted, 2 piece, close coupled	1 Ea.	151.80	108.20	260
05	Rough-in waste & vent for water closet	1 Set	87.62	270.43	358.05
06	Bathtub, P.E. cast iron 5' long with accessories	1 Ea.	309.10	130.90	440
07	Rough-in waste & vent for bathtub	1 Set	101.56	259.44	361
08	Lavatory, 20" x 18" P.E. cast iron with accessories	1 Ea.	121	69	190
09	Rough-in waste & vent for lavatory	1 Set	143.11	341.89	485
10	Accessories				
11	Toilet tissue dispenser, chrome, single roll	1 Ea.	17.60	9.40	27
12	18" long stainless steel towel bar	2 Ea.	38.72	25.28	64
13	Medicine cabinet with mirror, 20" x 16", unlighted	1 Ea.	62.70	20.30	83
14	TOTAL	System	1033.21	1234.84	2268.05
15					
16					
17	Above system installed in one wall with all necessary service piping	System	1022.52	1207.53	2230.05
18					
19					
20					
21					
22					
23					
24	NOTE: PLUMBING APPROXIMATIONS				
25	WATER CONTROL: water meter, backflow preventer,				
26	Shock absorbers, vacuum breakers, mixer....10 to 15% of fixtures				
27	PIPE AND FITTINGS: .... 30 to 60% of fixtures				
28					
29	Cut & patch to match existing construction, add, minimum		2%	3%	
30	Maximum		5%	9%	
31	Dust protection, add, minimum		1%	2%	
32	Maximum		4%	11%	
33	Equipment usage curtailment, add, minimum		1%	1%	
34	Maximum		3%	10%	
35	Material handling & storage limitation, add, minimum		1%	1%	
36	Maximum		6%	7%	
37	Protection of existing work, add, minimum		2%	2%	
38	Maximum		5%	7%	
39	Shift work requirements, add, minimum			5%	
40	Maximum			30%	
41					
42					

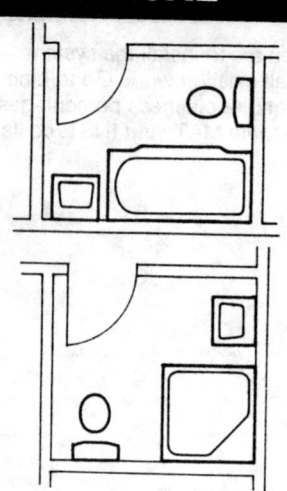

This page illustrates and describes a three fixture bathroom system including a water closet, tub, lavatory, accessories, and service piping. Lines 08.1-932-04 thru 17, give the unit price and total price on a cost each basis for this system. Prices for an alternate three fixture bathroom system are on Line Item 08.1-932-20. Both material quantities and labor costs have been adjusted for the system listed.

Factors: To adjust for job conditions other than normal working situations use Lines 08.1-932-29 thru 40.

Example: You are to install the system and protect the surrounding area from dust. Go to Line 08.1-932-31 and apply these percentages to the appropriate MAT. and INST. costs.

LINE NO.	DESCRIPTION	QUANTITY	COST EACH		
			MAT.	INST.	TOTAL
01	Three fixture bathroom incl. water closet, bathtub, lavatory, accessories,				
02	And necessary service piping to install this system in 2 walls.				
03					
04	Water closet, floor mounted, 2 piece, close coupled	1 Ea.	151.80	108.20	260
05	Rough-in waste & vent for water closet	1 Set	94.22	290.78	385
06	Bathtub, P.E. cast iron, 5' long with accessories	1 Ea.	309.10	130.90	440
07	Rough-in waste & vent for bathtub	1 Set	106.91	273.09	380
08	Lavatory, 20" x 18" P.E. cast iron with accessories	1 Ea.	121	69	190
09	Rough-in waste & vent for lavatory	1 Set	143.11	341.89	485
10	Additional service piping				
11	1-¼" copper DWV type tubing, sweat solder joints	6 L.F.	17.49	31.41	48.90
12	2" black schedule 40 steel pipe with threaded couplings	12 L.F.	38.41	106.19	144.60
13	Accessories				
14	Toilet tissue dispenser, chrome, single roll	1 Ea.	17.60	9.40	27
15	18" long stainless steel towel bar	2 Ea.	38.72	25.28	64
16	Medicine cabinet with mirror, 20" x 16", unlighted	1 Ea.	62.70	20.30	83
17	TOTAL	System			
18			1101.06	1406.44	2507.50
19					
20	Above system with corner contour tub, P.E. cast iron	System	1886.46	1406.04	3292.50
21					
22					
23					
24	NOTE: PLUMBING APPROXIMATIONS				
25	WATER CONTROL: water meter, backflow preventer,				
26	Shock absorbers, vacuum breakers, mixer....10 to 15% of fixtures				
27	PIPE AND FITTINGS: .... 30 to 60% of fixtures				
28					
29	Cut & patch to match existing construction, add, minimum		2%	3%	
30	Maximum		5%	9%	
31	Dust protection, add, minimum		1%	2%	
32	Maximum		4%	11%	
33	Equipment usage curtailment, add, minimum		1%	1%	
34	Maximum		3%	10%	
35	Material handling & storage limitation, add, minimum		1%	1%	
36	Maximum		6%	7%	
37	Protection of existing work, add, minimum		2%	2%	
38	Maximum		5%	7%	
39	Shift work requirements, add, minimum			5%	
40	Maximum			30%	
41					
42					

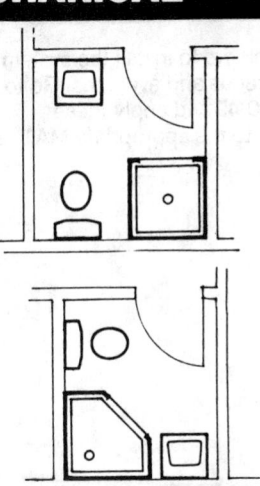

This page illustrates and describes a three fixture bathroom system including a water closet, shower, lavatory, accessories, and service piping, Lines 08.1-933-04 thru 18, give the unit price and total price on a cost each basis for this system. Prices for an alternate three fixture bathroom system are on Line Item 08.1-933-22. Both material quantities and labor costs have been adjusted for the system listed.

Factors: To adjust for job conditions other than normal working situations use Lines 08.1-933-29 thru 40.

Example: You are to install the system and protect existing construction. Go to Line 08.1-933-37 and apply these percentages to the appropriate MAT. and INST. costs.

LINE NO.	DESCRIPTION	QUANTITY	COST EACH		
			MAT.	INST.	TOTAL
01	Three fixture bathroom incl. water closet, shower, lavatory, accessories,				
02	And necessary service piping to install this system in 2 walls.				
03					
04	Water closet, floor mounted, 2 piece, close coupled	1 Ea.	151.80	108.20	260
05	Rough-in waste & vent for water closet	1 Set	94.22	290.78	385
06	32″ shower, enameled stall, molded stone receptor	1 Ea.	347.60	282.40	630
07	Rough-in waste & vent for shower	1 Set	80.29	274.71	355
08	Lavatory, 20″ x 18″ P.E. cast iron with accessories	1 Ea.	121	69	190
09	Rough-in waste & vent for lavatory	1 Set	143.11	341.89	485
10	Additional service piping				
11	1-¼″ Copper DWV type tubing, sweat solder joints	4 L.F.	11.66	20.94	32.60
12	2″ black schedule 40 steel pipe with threaded couplings	6 L.F.	19.21	53.09	72.30
13	4″ cast iron soil with lead and oakum joints	7 L.F.	28.26	71.84	100.10
14	Accessories				
15	Toilet tissue dispenser, chrome, single roll	1 Ea.	17.60	9.40	27
16	18″ long stainless steel towel bar	2 Ea.	38.72	25.28	64
17	Medicine cabinet with mirror, 20″ x 16″, unlighted	1 Ea.	62.70	20.30	83
18	TOTAL	System	1116.17	1567.83	2684
19					
20					
21	Above system installed with 36″ corner angle shower, enameled steel,				
22	Molded stone receptor	System	1412.07	1601.93	3014
23					
24	NOTE: PLUMBING APPROXIMATIONS				
25	WATER CONTROL: water meter, backflow preventer,				
26	Shock absorbers, vacuum breakers, mixer....10 to 15% of fixtures				
27	PIPE AND FITTINGS: .... 30 to 60% of fixtures				
28					
29	Cut & patch to match existing construction, add, minimum		2%	3%	
30	Maximum		5%	9%	
31	Dust protection, add, minimum		1%	2%	
32	Maximum		4%	11%	
33	Equipment usage curtailment, add, minimum		1%	1%	
34	Maximum		3%	10%	
35	Material handling & storage limitation, add, minimum		1%	1%	
36	Maximum		6%	7%	
37	Protection of existing work, add, minimum		2%	2%	
38	Maximum		5%	7%	
39	Shift work requirements, add, minimum			5%	
40	Maximum			30%	
41					
42					

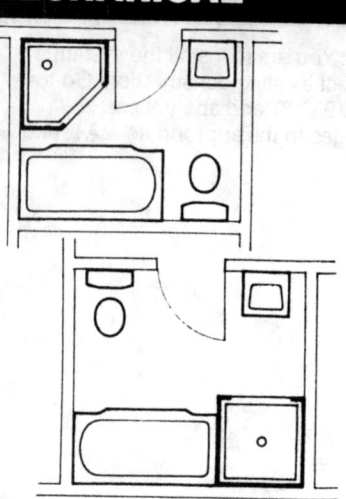

This page illustrates and describes a four fixture bathroom system including a water closet, shower, bathtub, lavatory, accessories, and service piping, Lines 08.1-940-04 thru 16 give the unit price and total price on a cost each basis for this system. Prices for an alternate four fixture bathroom system are on Line Item 08.1-940-19. Both material quantities and labor costs have been adjusted for the system listed.

Factors: To adjust for job conditions other than normal working situations use Lines 08.1-940-29 thru 40.

Example: You are to install the system during weekends and evenings. Go to Line 08.1-940-40 and apply these percentages to the appropriate MAT. and INST. costs.

LINE NO.	DESCRIPTION	QUANTITY	COST EACH		
			MAT.	INST.	TOTAL
01	Four fixture bathroom incl. water closet, shower, bathtub, lavatory				
02	Accessories and necessary service piping to install this system in 2 walls.				
03					
04	Water closet, floor mounted, 2 piece, close coupled	1 Ea.	151.80	108.20	260
05	Rough-in waste & vent for water closet	1 Set	94.22	290.78	385
06	32" shower, enameled steel stall, molded stone receptor	1 Ea.	347.60	282.40	630
07	Rough-in waste & vent for shower	1 Set	80.29	274.71	355
08	Bathtub, P.E. cast iron, 5' long with accessories	1 Ea.	309.10	130.90	440
09	Rough-in waste & vent for bathtub	1 Set	106.91	273.09	380
10	Lavatory, 20" x 18" P.E. cast iron with accessories	1 Ea.	121	69	190
11	Rough-in waste & vent for lavatory	1 Set	143.11	341.89	485
12	Accessories				
13	Toilet tissue dispenser, chrome, single roll	1 Ea.	17.60	9.40	27
14	18" long stainless steel towel bar	2 Ea.	38.72	25.28	64
15	Medicine cabinet with mirror, 20" x 16" unlighted	1 Ea.	62.70	20.30	83
16	TOTAL	System	1473.05	1825.95	3299
17					
18					
19	Above system with 36" corner angle shower, plumbing in 3 walls	System	1820.34	2004.76	3825.10
20					
21					
22					
23					
24	NOTE: PLUMBING APPROXIMATIONS				
25	WATER CONTROL: water meter, backflow preventer,				
26	Shock absorbers, vacuum breakers, mixer....10 to 15% of fixtures				
27	PIPE AND FITTINGS: .... 30 to 60% of fixtures				
28					
29	Cut & patch to match existing construction, add, minimum		2%	3%	
30	Maximum		5%	9%	
31	Dust protection, add, minimum		1%	2%	
32	Maximum		4%	11%	
33	Equipment usage curtailment, add, minimum		1%	1%	
34	Maximum		3%	10%	
35	Material handling & storage limitation, add, minimum		1%	1%	
36	Maximum		6%	7%	
37	Protection of existing work, add, minimum		2%	2%	
38	Maximum		5%	7%	
39	Shift work requirements, add, minimum			5%	
40	Maximum			30%	
41					
42					

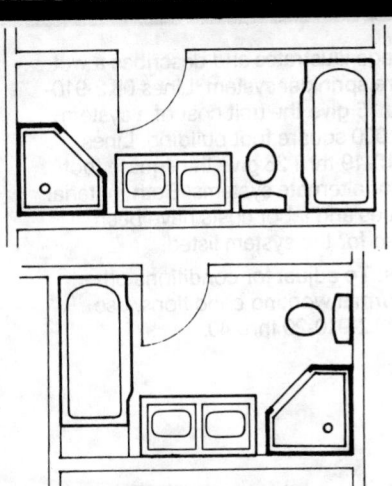

This page illustrates and describes a five fixture bathroom system including a water closet, shower, bathtub, lavatories, accessories, and service piping. Lines 08.1-950-04 thru 16 give the unit price and total price on a cost each basis for this system. Prices for an alternate five fixture bathroom system are on Line Item 08.1-950-19. Both material quantities and labor costs have been adjusted for the system listed.

Factors: To adjust for job conditions other than normal working situations use Lines 08.1-950-29 thru 40

Example: You are to install and match any existing construction. Go to Line 08.1-950-29 and apply these percentages to the appropriate MAT. and INST. costs.

LINE NO.	DESCRIPTION	QUANTITY	COST EACH		
			MAT.	INST.	TOTAL
01	Five fixture bathroom incl. water closet, shower, bathtub, 2 lavatories,				
02	Accessories and necessary service piping to install this system in 1 wall.				
03					
04	Water closet, floor mounted, 2 piece, close coupled	1 Ea.	151.80	108.20	260
05	Rough-in waste & vent for water closet	1 Set	94.22	290.78	385
06	36" corner angle shower, enameled steel stall, molded stone receptor	1 Ea.	643.50	316.50	960
07	Rough-in waste & vent for shower	1 Set	80.29	274.71	355
08	Bathtub, P.E. cast iron, 5' long with accessories	1 Ea.	309.10	130.90	440
09	Rough-in waste & vent for bathtub	1 Set	106.91	273.09	380
10	Lavatories, 20" x 18" cabinet mntd, PECI with access. & cabinet	2 Ea.	561	267	828
11	Rough-in waste & vent for lavatories	1.6 Set	156.20	395.80	552
12	Accessories				
13	Toilet tissue dispenser, chrome, single roll	1 Ea.	17.60	9.40	27
14	18" long stainless steel towel bars	2 Ea.	38.72	25.28	64
15	Medicine cabinet with mirror, 20" x 16", unlighted	2 Ea.	125.40	40.60	166
16	TOTAL	System	2284.74	2132.26	4417
17					
18					
19	Above system installed in 2 walls with all necessary service piping	System	2307.24	2195.76	4503
20					
21					
22					
23					
24	NOTE: PLUMBING APPROXIMATIONS				
25	WATER CONTROL: water meter, backflow preventer,				
26	Shock absorbers, vacuum breakers, mixer....10 to 15% of fixtures				
27	PIPE AND FITTINGS: .... 30 to 60% of fixtures				
28					
29	Cut & patch to match existing construction, add, minimum		2%	3%	
30	Maximum		5%	9%	
31	Dust protection, add, minimum		1%	2%	
32	Maximum		4%	11%	
33	Equipment usage curtailment, add, minimum		1%	1%	
34	Maximum		3%	10%	
35	Material handling & storage limitation, add, minimum		1%	1%	
36	Maximum		6%	7%	
37	Protection of existing work, add, minimum		2%	2%	
38	Maximum		5%	7%	
39	Shift work requirements, add, minimum			5%	
40	Maximum			30%	
41					
42					

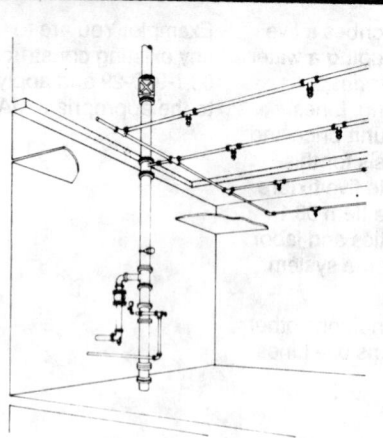

This page illustrates and describes a wet type fire sprinkler system. Lines 08.2-910-04 thru 15 give the unit cost of a system for a 2,000 square foot building. Lines 08.2-910-19 thru 26 give the square foot costs for alternate systems. Both material quantities and labor costs have been adjusted for the system listed.

Factors: To adjust for conditions other than normal working conditions, use Lines 08.2-910-29 thru 40.

LINE NO.	DESCRIPTION	QUANTITY	COST EACH		
			MAT.	INST.	TOTAL
01	Wet pipe fire sprinkler system, ordinary hazard, open area to 2000 S.F.				
02	On one floor.				
03					
04	4" OS & Y valve	1 Ea.	236.50	188.50	425
05	Wet pipe alarm valve	1 Ea.	555.50	294.50	850
06	Water motor alarm	1 Ea.	118.80	81.20	200
07	3" check valve	1 Ea.	92.95	197.05	290
08	Pipe riser, 4" diameter	10 L.F.	74.47	205.53	280
09	Water gauges and trim	1 Set	858	592	1450
10	Electric fire horn	1 Ea.	30.80	46.20	77
11	Sprinkler head supply piping	168 L.F.	454.02	1260.88	1714.90
12	Pipe fittings	1 L.S.	197.52	1202.48	1400
13	Sprinkler heads	16 Ea.	64.59	319.41	384
14	Fire department connection	1 Ea.	262.90	117.10	380
15	TOTAL	System	2946.05	4504.85	7450.90
16					

LINE NO.	DESCRIPTION	QUANTITY	COST PER S.F.		
17					
18			MAT.	INST.	TOTAL
19	Ordinary hazard, one floor, area to 2000 S.F./floor	S.F.	1.47	2.25	3.72
20	For each additional floor, add per floor		.40	1.49	1.89
21	Area to 3200 S.F./floor		1.08	2.02	3.10
22	For each additional floor, add per floor		.40	1.55	1.95
23	Area to 5000 S.F./floor		.93	1.94	2.87
24	For each additional floor, add per floor		.50	1.64	2.14
25	Area to 8000 S.F./floor		.68	1.59	2.27
26	For each additional floor, add per floor		.41	1.40	1.81
27					
28					
29	Cut & patch to match existing construction, add, minimum		2%	3%	
30	Maximum		5%	9%	
31	Dust protection, add, minimum		1%	2%	
32	Maximum		4%	11%	
33	Equipment usage curtailment, add, minimum		1%	1%	
34	Maximum		3%	10%	
35	Material handling & storage limitation, add, minimum		1%	1%	
36	Maximum		6%	7%	
37	Protection of existing work, add, minimum		2%	2%	
38	Maximum		5%	7%	
39	Shift work requirements, add, minimum			5%	
40	Maximum			30%	
41					
42					

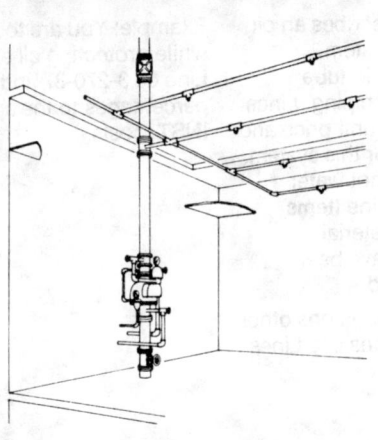

This page illustrates and describes a dry type fire sprinkler system. Lines 08.2-920-04 thru 13 give the unit cost of a system for a 2,000 square foot building. Lines 08.2-920-17 thru 24 give the square foot costs for alternate systems. Both material quantities and labor costs have been adjusted for the system listed.

Factors: To adjust for conditions other than normal working conditions, use Lines 08.2-920-29 thru 40.

LINE NO.	DESCRIPTION	QUANTITY	COST EACH		
			MAT.	INST.	TOTAL
01	Pre-action fire sprinkler system, ordinary hazard, open area to 2000 S.F.				
02	On one floor.				
03					
04					
05	Air compressor	1 Ea.	616	249	865
06	4" OS & Y valve	1 Ea.	236.50	188.50	425
07	Pipe riser 4" diameter	10 L.F.	74.47	205.53	280
08	Sprinkler head supply piping	163 L.F.	434.66	1222.49	1657.15
09	Pipe fittings	1 L.S.	191.45	1176.55	1368
10	Detectors	2 Ea.	387.20	42.80	430
11	Sprinkler heads	16 Ea.	64.59	319.41	384
12	Fire department connection	1 Ea.	262.90	117.10	380
13	TOTAL	System	2267.77	3521.38	5789.15
14					

LINE NO.	DESCRIPTION	QUANTITY	COST PER S.F.		
15			MAT.	INST.	TOTAL
16					
17	Ordinary hazard, one floor, area to 2000 S.F./floor	S.F.	1.13	1.76	2.89
18	For each additional floor, add per floor		.38	1.46	1.84
19	Area to 3200 S.F./ floor		.87	1.73	2.60
20	For each additional floor, add per floor		.40	1.55	1.95
21	Area to 5000 S.F./ floor		.88	1.77	2.65
22	For each additional floor, add per floor		.50	1.64	2.14
23	Area to 8000 S.F./ floor		.67	1.48	2.15
24	For each additional floor, add per floor		.41	1.40	1.81
25					
26					
27					
28					
29	Cut & patch to match existing construction, add, minimum		2%	3%	
30	Maximum		5%	9%	
31	Dust protection, add, minimum		1%	2%	
32	Maximum		4%	11%	
33	Equipment usage curtailment, add, minimum		1%	1%	
34	Maximum		3%	10%	
35	Material handling & storage limitation, add, minimum		1%	1%	
36	Maximum		6%	7%	
37	Protection of existing work, add, minimum		2%	2%	
38	Maximum		5%	7%	
39	Shift work requirements, add, minimum			5%	
40	Maximum			30%	
41					
42					

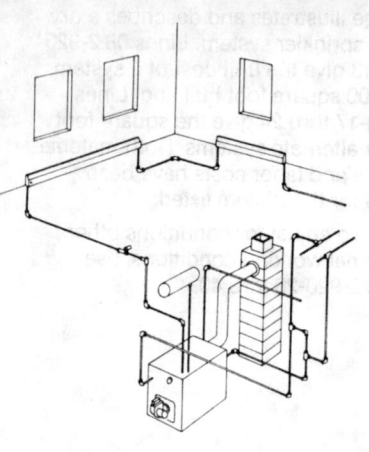

This page illustrates and describes an oil fired hot water baseboard system including an oil fired boiler, fin tube radiation and all fittings and piping. Lines 08.3-270-04 thru 13 give the unit price and total price per square foot for this system. Prices for alternate oil fired hot water baseboard systems are on Line Items 08.3-270-17 thru 21. Both material quantities and labor costs have been adjusted for the system listed.

Factors: To adjust for job conditions other than normal working situations use Lines 08.3-270-29 thru 40.

Example: You are to install the system while protecting all existing work. Go to Line 08.3-270-37 and apply these percentages to the appropriate MAT. and INST. costs.

LINE NO.	DESCRIPTION	QUANTITY	COST PER S.F.		
			MAT.	INST.	TOTAL
01	Oil fired hot water baseboard system including boiler, fin tube radiation				
02	And all necessary fittings and piping.				
03					
04	Area to 800 S.F.				
05	Boiler, cast iron, 97 MBH	1 Ea.	1914	461	2375
06	Oil piping	.25 L.S.	478.50	115.25	593.75
07	Copper piping	130 L.F.	479.05	703.95	1183
08	Fin tube radiation	68 L.F.	575.96	665.04	1241
09	Circulator	1 Ea.	253	92	345
10	Oil tank	1 Ea.	247.50	112.50	360
11	Expansion tank, ASME	1 Ea.	1039.50	35.50	1075
12	TOTAL	System	4987.51	2185.24	7172.75
13		S.F.	6.23	2.73	8.96
14					
15					
16	For alternate hot water systems:				
17	Cast iron boiler, area to 1000 S.F.	S.F.	5.19	2.43	7.62
18	To 1200 S.F.		4.47	2.21	6.68
19	To 1600 S.F.		3.93	2.22	6.15
20	To 2000 S.F.		3.31	1.97	5.28
21	To 3000 S.F.		2.49	1.67	4.16
22					
23					
24					
25					
26					
27					
28					
29	Cut & patch to match existing construction, add, minimum		2%	3%	
30	Maximum		5%	9%	
31	Dust protection, add, minimum		1%	2%	
32	Maximum		4%	11%	
33	Equipment usage curtailment, add, minimum		1%	1%	
34	Maximum		3%	10%	
35	Material handling & storage limitation, add, minimum		1%	1%	
36	Maximum		6%	7%	
37	Protection of existing work, add, minimum		2%	2%	
38	Maximum		5%	7%	
39	Shift work requirements, add, minimum			5%	
40	Maximum			30%	
41					
42					

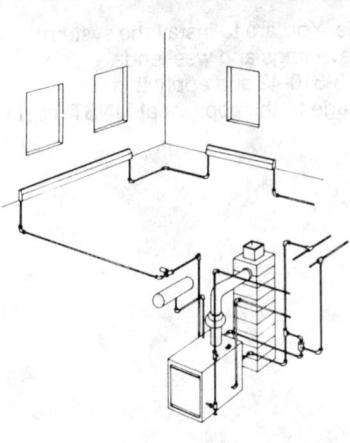

This page illustrates and describes a gas fired hot water baseboard system including a gas fired boiler, fin tube radiation, fittings and piping. Lines 08.3-280-04 thru 12 give the unit price and total price per square foot for this system. Prices for alternate gas fired hot water baseboard systems are on Line Items 08.3-280-15 thru 19. Both material quantities and labor costs have been adjusted for the system listed.

Factors: To adjust for job conditions other than normal working situations use Lines 08.3-280-29 thru 40.

Example: You are to install the system with minimal equipment usage. Go to Line 08.3-280-34 and apply these percentages to the appropriate MAT. and INST. costs.

LINE NO.	DESCRIPTION	QUANTITY	COST PER S.F.		
			MAT.	INST.	TOTAL
01	Gas fired hot water baseboard system including boiler, fin tube radiation,				
02	All necessary fittings and piping.				
03					
04	Area to 800 S.F.				
05	Cast iron boiler, insulating jacket, 80 MBH	Ea.	1028.50	821.50	1850
06	Gas piping	1 L.S.	257.13	205.37	462.50
07	Copper piping	130 L.F.	479.05	703.95	1183
08	Fin tube radiation	68 L.F.	575.96	665.04	1241
09	Circulator, flange connection	1 Ea.	253	92	345
10	Expansion tank, ASME	1 Ea.	1039.50	35.50	1075
11	TOTAL	System	3633.14	2523.36	6156.50
12		S.F.	4.54	3.15	7.69
13					
14	For alternate hot water systems:				
15	Cast iron boiler, area to 1000 S.F.	S.F.	3.83	2.77	6.60
16	To 1200 S.F.		3.34	2.49	5.83
17	To 1600 S.F.		3.08	2.43	5.51
18	To 2000 S.F.		2.72	2.18	4.90
19	To 3000 S.F.		2.24	1.91	4.15
20					
21					
22					
23					
24					
25					
26					
27					
28					
29	Cut & patch to match existing construction, add, minimum		2%	3%	
30	Maximum		5%	9%	
31	Dust protection, add, minimum		1%	2%	
32	Maximum		4%	11%	
33	Equipment usage curtailment, add, minimum		1%	1%	
34	Maximum		3%	10%	
35	Material handling & storage limitation, add, minimum		1%	1%	
36	Maximum		6%	7%	
37	Protection of existing work, add, minimum		2%	2%	
38	Maximum		5%	7%	
39	Shift work requirements, add, minimum			5%	
40	Maximum			30%	
41					
42					

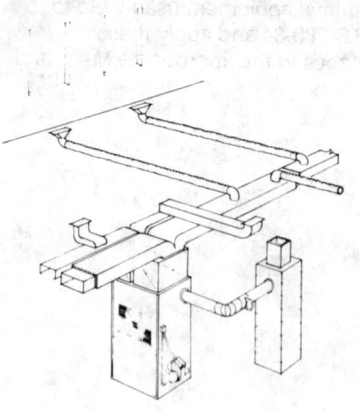

This page illustrates and describes an oil fired forced air system including an oil fired furnace, ductwork registers and hookups. Lines 08.3-310-04 thru 14 give the unit price and total price per square foot for this system. Prices for alternate oil fired forced air systems are on Line Items 08.3-310-17 thru 28. Both material quantities and labor costs have been adjusted for the system listed.

Factors: To adjust for job conditions other than normal working situations use Lines 08.3-310-31 thru 42.

Example: You are to install the system during evenings and weekends. Go to Line 08.3-310-42 and apply this percentage to the appropriate INST. cost.

LINE NO.	DESCRIPTION	QUANTITY	COST PER S.F. MAT.	COST PER S.F. INST.	COST PER S.F. TOTAL
01	Oil fired hot air heating system including furnace, ductwork, registers				
02	And all necessary hookups.				
03					
04	Area to 800 S.F., heat only				
05	Furnace, oil, atomizing gun type burner	1 Ea.	775.50	159.50	935
06	Oil piping	1 L.S.	193.88	39.87	233.75
07	Oil tank, steel, 275 gallon	1 Ea.	247.50	112.50	360
08	Duct, galvanized, steel	312 Lb.	257.40	1168.44	1425.84
09	Insulation, blanket type, duct work	270 S.F.	133.65	441.45	575.10
10	Flexible duct, 6" diameter, insulated	100 L.F.	169.40	216.60	386
11	Registers, baseboard, gravity, 12" x 6"	8 Ea.	62.92	113.08	176
12	Return, damper, 36" x 18"	1 Ea.	78.10	31.90	110
13	TOTAL	System	1918.35	2283.34	4201.69
14		S.F.	2.40	2.85	5.25
15					
16	For alternate heating systems:				
17	Oil fired, area to 1000 S.F.	S.F.	1.93	2.31	4.24
18	To 1200 S.F.		1.70	2.04	3.74
19	To 1600 S.F.		1.40	1.70	3.10
20	To 2000 S.F.		1.46	2.18	3.64
21	To 3000 S.F.		1.21	1.75	2.96
22	For combined heating and cooling systems:				
23	Oil fired, heating and cooling, area to 800 S.F.	S.F.	4.05	3.38	7.43
24	To 1000 S.F.		3.32	2.76	6.08
25	To 1200 S.F.		2.85	2.41	5.26
26	To 1600 S.F.		2.38	2	4.38
27	To 2000 S.F.		2.31	2.44	4.75
28	To 3000 S.F.		1.97	1.93	3.90
29					
30					
31	Cut & patch to match existing construction, add, minimum		2%	3%	
32	Maximum		5%	9%	
33	Dust protection, add, minimum		1%	2%	
34	Maximum		4%	11%	
35	Equipment usage curtailment, add, minimum		1%	1%	
36	Maximum		3%	10%	
37	Material handling & storage limitation, add, minimum		1%	1%	
38	Maximum		6%	7%	
39	Protection of existing work, add, minimum		2%	2%	
40	Maximum		5%	7%	
41	Shift work requirements, add, minimum			5%	
42	Maximum			30%	

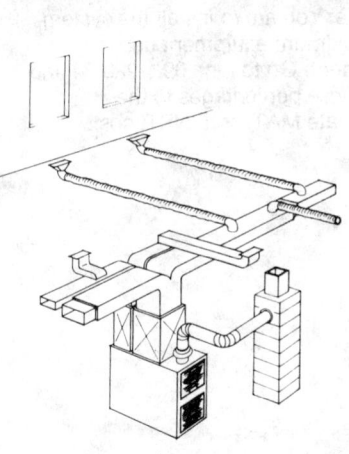

This page illustrates and describes a gas fired forced air system including a gas fired furnace, ductwork, registers and hookups. Lines 08.3-320-04 thru 13 give the unit price and total price per square foot for this system. Prices for alternate gas fired forced air systems are on Line Items 08.3-320-15 thru 26. Both material quantities and labor costs have been adjusted for the system listed.

Factors: To adjust for job conditions other than normal working situations use Lines 08.3-320-29 thru 40.

Example: You are to install the system with material handling and storage limitations. Go to Line 08.3-320-35 and apply these percentages to the appropriate MAT. and INST. costs.

LINE NO.	DESCRIPTION	QUANTITY	COST PER S.F.		
			MAT.	INST.	TOTAL
01	Gas fired hot air heating system including furnace, ductwork, registers				
02	And all necessary hookups.				
03					
04	Area to 800 S.F., heat only				
05	Furnace, gas, AGA certified, direct drive, 44 MBH	1 Ea.	440	140	580
06	Gas piping	1 L.S.	110	35	145
07	Duct, galvanized steel	312 Lb.	257.40	1168.44	1425.84
08	Insulation, blanket type, ductwork	270 S.F.	133.65	441.45	575.10
09	Flexible duct, 6″ diameter, insulated	100 L.F.	169.40	216.60	386
10	Registers, baseboard, gravity, 12″ x 6″	8 Ea.	62.92	113.08	176
11	Return, damper, 36″ x 18″	1 Ea.	78.10	31.90	110
12	TOTAL	System	1251.47	2146.47	3397.94
13		S.F.	1.56	2.68	4.24
14	For alternate heating systems:				
15	Gas fired, area to 1000 S.F.	S.F.	1.29	2.18	3.47
16	To 1200 S.F.		1.16	1.93	3.09
17	To 1600 S.F.		1.12	1.81	2.93
18	To 2000 S.F.		1.14	2.13	3.27
19	To 3000 S.F.	↓	.94	1.70	2.64
20	For combined heating and cooling systems:				
21	Gas fired, heating and cooling, area to 800 S.F.	S.F.	3.21	3.21	6.42
22	To 1000 S.F.		2.68	2.63	5.31
23	To 1200 S.F.		2.31	2.30	4.61
24	To 1600 S.F.		2.10	2.10	4.20
25	To 2000 S.F.		1.99	2.38	4.37
26	To 3000 S.F.	↓	1.70	1.89	3.59
27					
28					
29	Cut & patch to match existing construction, add, minimum		2%	3%	
30	Maximum		5%	9%	
31	Dust protection, add, minimum		1%	2%	
32	Maximum		4%	11%	
33	Equipment usage curtailment, add, minimum		1%	1%	
34	Maximum		3%	10%	
35	Material handling & storage limitation, add, minimum		1%	1%	
36	Maximum		6%	7%	
37	Protection of existing work, add, minimum		2%	2%	
38	Maximum		5%	7%	
39	Shift work requirements, add, minimum			5%	
40	Maximum			30%	
41					
42					

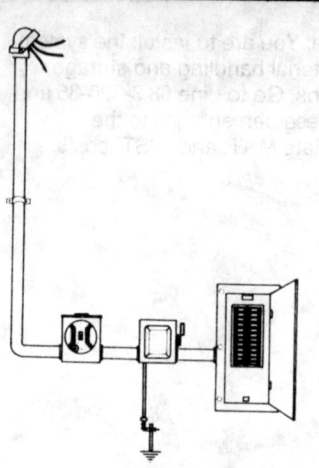

This page illustrates and describes commercial service systems including a meter socket, service head and cable, entrance switch, steel conduit, copper wire, panel board, ground rod, wire, and conduit. Lines 09.1-220-04 thru 12 give the unit price and total price on a cost each basis for this system. Prices for alternate commercial service systems are on Line Items 09.1-220-15 thru 17. Both material quantities and labor costs have been adjusted for the system listed.

Factors: To adjust for job conditions other than normal working situations use Lines 09.1-220-29 thru 40.

Example: You are to install the system with maximum equipment usage curtailment. Go to Line 09.1-220-34 and apply these percentages to the appropriate MAT. and INST. costs.

LINE NO.	DESCRIPTION	QUANTITY	COST EACH MAT.	COST EACH INST.	COST EACH TOTAL
01	Commercial electric service including service breakers, metering				
02	120/208 Volt, 3 phase, 4 wire, feeder, and panel board.				
03					
04	100 Amp Service				
05	Meter socket	1 Ea.	26.07	93.93	120
06	Service head and cable	1 EA	26.52	86.48	113
07	Service entrance switch	1 Ea.	125.40	159.60	285
08	Rigid steel conduit	20 L.F.	38.50	94.50	133
09	600 volt copper wire #3	1 C.L.F.	49.34	60.66	110
10	Panel board, 24 circuits, 20 Amp breakers	1 Ea.	533.50	516.50	1050
11	Ground rod plus wire and conduit	1 Ea.	24.37	75.13	99.50
12	TOTAL	Ea.	823.70	1086.80	1910.50
13					
14	For alternate size services:				
15	120/208 Volt, 3 phase, 4 wire service, 60 Amp	Ea.	562.62	805.88	1368.50
16	200 Amp		1492.13	1839.87	3332
17	400 Amp		3028.54	2307.50	5336.04
18					
19					
20					
21					
22					
23					
24					
25					
26					
27					
28					
29	Cut & patch to match existing construction, add, minimum		2%	3%	
30	Maximum		5%	9%	
31	Dust protection, add, minimum		1%	2%	
32	Maximum		4%	11%	
33	Equipment usage curtailment, add, minimum		1%	1%	
34	Maximum		3%	10%	
35	Material handling & storage limitation, add, minimum		1%	1%	
36	Maximum		6%	7%	
37	Protection of existing work, add, minimum		2%	2%	
38	Maximum		5%	7%	
39	Shift work requirements, add, minimum			5%	
40	Maximum			30%	
41					
42					

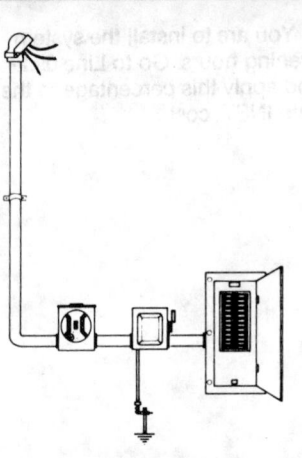

This page illustrates and describes a residential, single phase system including a weather cap, service entrance cable, meter socket, entrance switch, ground rod, ground cable, EMT, and panelboard. Lines 09.1-230-03 thru 10 give the unit price and total price on a cost each basis for this system. Prices for an alternate residential, single phase system are on Line Item 09.1-230-25. Both material quantities and labor costs have been adjusted for the system listed.

Factors: To adjust for job conditions other than normal working situations use Lines 09.1-230-29 thru 40.

Example: You are to install the system with a minimum equipment usage curtailment. Go to Line 09.1-230-33 and apply these percentages to the appropriate MAT. and INST. costs.

LINE NO.	DESCRIPTION	QUANTITY	COST EACH		
			MAT.	INST.	TOTAL
01	**100 Amp Service, single phase**				
02					
03	Weathercap	1 Ea.	4.68	25.32	30
04	Service entrance cable	.2 C.L.F.	21.84	61.16	83
05	Meter socket	1 Ea.	26.07	93.93	120
06	Entrance disconnect switch	1 Ea.	125.40	159.60	285
07	Ground rod, with clamp	1 Ea.	13.81	64.19	78
08	Ground cable	.1 C.L.F.	13.20	19.30	32.50
09	Panelboard, 12 circuit	1 Ea.	110	255	365
10	TOTAL	Ea.	315	678.50	993.50
11					
12					

LINE NO.	DESCRIPTION	QUANTITY	COST EACH		
			MAT.	INST.	TOTAL
13					
14					
15	200 Amp Service , single phase				
16					
17	Weathercap	1 Ea.	13.86	38.14	52
18	Service entrance cable	.2 C.L.F.	48.18	87.82	136
19	Meter socket	1 Ea.	40.70	159.30	200
20	Entrance disconnect switch	1 Ea.	253	237	490
21	Ground rod, with clamp	1 Ea.	26.40	69.60	96
22	Ground cable	.1 C.L.F.	11.31	10.69	22
23	¾" EMT	10 L.F.	5.17	23.63	28.80
24	Panelboard, 24 circuit	1 Ea.	270.60	474.40	745
25	TOTAL	Ea.	669.22	1100.58	1769.80
26					
27					
28					
29	Cut & patch to match existing construction, add, minimum		2%	3%	
30	Maximum		5%	9%	
31	Dust protection, add, minimum		1%	2%	
32	Maximum		4%	11%	
33	Equipment usage curtailment, add, minimum		1%	1%	
34	Maximum		3%	10%	
35	Material handling & storage limitation, add, minimum		1%	1%	
36	Maximum		6%	7%	
37	Protection of existing work, add, minimum		2%	2%	
38	Maximum		5%	7%	
39	Shift work requirements, add, minimum			5%	
40	Maximum			30%	
41					
42					

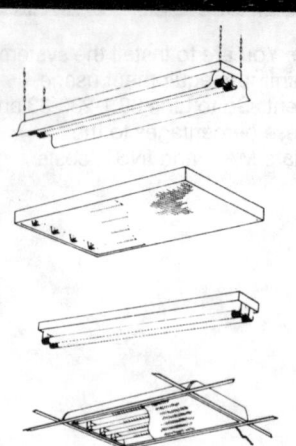

This page illustrates and describes fluorescent lighting systems including a fixture, lamp, outlet box and wiring. Lines 09.2-900-04 thru 08 give the unit price and total price on a cost each basis for this system. Prices for alternate fluorescent lighting systems are on Line Items 09.2-900-13 thru 15. Both material quantities and labor costs have been adjusted for the system listed.

Factors: To adjust for job conditions other than normal working situations use Lines 09.2-900-29 thru 40.

Example: You are to install the system during evening hours. Go to Line 09.2-900-39 and apply this percentage to the appropriate INST. cost.

LINE NO.	DESCRIPTION	QUANTITY	COST EACH		
			MAT.	INST.	TOTAL
01	Fluorescent lighting, including fixture, lamp, outlet box and wiring.				
02					
03					
04	Recessed lighting fixture, on suspended system	1 Ea.	69.30	65.70	135
05	Outlet box	1 Ea.	1.98	21.87	23.85
06	#12 wire	.66 C.L.F.	4.09	18.35	22.44
07	Conduit, EMT, ½" conduit	20 L.F.	7.26	36.14	43.40
08	TOTAL	Ea.	82.63	142.06	224.69
09					
11					
12	For alternate lighting fixtures:				
13	Surface mounted, 2' x 4', acrylic prismatic diffuser	Ea.	106.04	144.21	250.25
14	Strip fixture, 8' long, two 8' lamps		66.44	133.81	200.25
15	Pendant mounted, industrial, 8' long, with reflectors	↓	117.04	153.21	270.25
16					
17					
18					
19					
20					
21					
22					
23					
24					
25					
26					
27					
28					
29	Cut & patch to match existing construction, add, minimum		2%	3%	
30	Maximum		5%	9%	
31	Dust protection, add, minimum		1%	2%	
32	Maximum		4%	11%	
33	Equipment usage curtailment, add, minimum		1%	1%	
34	Maximum		3%	10%	
35	Material handling & storage limitation, add, minimum		1%	1%	
36	Maximum		6%	7%	
37	Protection of existing work, add, minimum		2%	2%	
38	Maximum		5%	7%	
39	Shift work requirements, add, minimum			5%	
40	Maximum			30%	
41					
42					

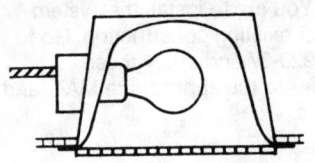

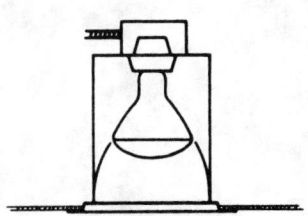

This page illustrates and describes incandescent lighting systems including a fixture, lamp, outlet box, conduit and wiring. Lines 09.2-910-04 thru 08 give the unit price and total price on a cost each basis for this system. Prices for an alternate incandescent lighting system are on Line Item 09.2-910-18. Both material quantities and labor costs have been adjusted for the system listed.

Factors: To adjust for conditions other than normal working situations use Lines 09.2-910-29 thru 40.

Example: You are to install the system and cut and match existing construction. Go to Line 09.2-910-30 and apply this percentage to the appropriate INST. costs.

LINE NO.	DESCRIPTION	QUANTITY	COST EACH		
			MAT.	INST.	TOTAL
01	Incandescent light fixture, including lamp, outlet box, conduit and wiring.				
02					
03					
04	Recessed wide reflector with flat glass lens	1 Ea.	38.50	45.50	84
05	Outlet box	1 Ea.	1.98	21.87	23.85
06	Armored cable, 3 wire	.2 C.L.F.	7.17	30.83	38
07					
08	TOTAL	Ea.	47.65	98.20	145.85
09					
10					
11					
12					
13	Recessed, R-40 flood lamp with reflector skirt	1 Ea.	55	38	93
14	150 watt R-40 flood lamp	.01 Ea.	5.06	2.34	7.40
15	Outlet box	1 Ea.	1.98	21.87	23.85
16	Romex, 12-2 with ground	.2 C.L.F.	5.50	27.50	33
17	Conduit, ½" EMT	20 L.F.	7.26	36.14	43.40
18	TOTAL	Ea.	74.80	125.85	200.65
19					
20					
21					
22					
23					
24					
25					
26					
27					
28					
29	Cut & patch to match existing construction, add, minimum		2%	3%	
30	Maximum		5%	9%	
31	Dust protection, add, minimum		1%	2%	
32	Maximum		4%	11%	
33	Equipment usage curtailment, add, minimum		1%	1%	
34	Maximum		3%	10%	
35	Material handling & storage limitation, add, minimum		1%	1%	
36	Maximum		6%	7%	
37	Protection of existing work, add, minimum		2%	2%	
38	Maximum		5%	7%	
39	Shift work requirements, add, minimum			5%	
40	Maximum			30%	
41					
42					

 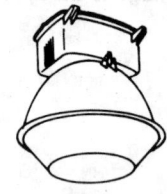

This page illustrates and describes high intensity lighting systems including a lamp, EMT conduit, EMT "T" fitting with cover, and wire. Lines 09.2-920-04 thru 08 give the unit price and total price on a cost each basis for this system. Prices for alternate high intensity lighting systems are on Line Items 09.2-920-11 thru 18. Both material quantities and labor costs have been adjusted for the system listed.

Factors: To adjust for job conditions other than normal working situations use Lines 09.2-920-29 thru 40.

Example: You are to install the system and protect existing construction. Go to Line 09.2-920-37 and apply these percentages to the appropriate MAT. and INST. costs.

LINE NO.	DESCRIPTION	QUANTITY	COST EACH		
			MAT.	INST.	TOTAL
01	High intensity lighting system consisting of 400 watt mercury vapor fixture				
02	And lamp with ½" EMT conduit and fittings using #12 wire, high bay.				
03					
04	400 watt mercury vapor fixture and lamp, high bay	1 Ea.	269.50	135.50	405
05	½" EMT conduit	30 L.F.	10.89	54.21	65.10
06	½" EMT "T" fitting with cover	1 Ea.	6.05	18.95	25
07	#12 wire	.6 L.F.	3.73	16.67	20.40
08	TOTAL	Ea.	290.17	225.33	515.50
09					
10	For alternate high intensity systems:				
11	High bay: 400 watt, metal halide fixture and lamp	Ea.	328.67	221.83	550.50
12	High pressure sodium		499.17	221.33	720.50
13	1000 watt, mercury vapor		482.67	242.83	725.50
14	Metal halide		592.67	242.83	835.50
15	High pressure sodium		735.67	244.83	980.50
16	Low bay: 250 watt, mercury vapor		290.17	225.33	515.50
17	Metal halide		405.67	184.83	590.50
18	150 watt, high pressure sodium		383.67	186.83	570.50
19					
20					
21					
22					
23					
24					
25					
26					
27					
28					
29	Cut & patch to match existing construction, add, minimum		2%	3%	
30	Maximum		5%	9%	
31	Dust protection, add, minimum		1%	2%	
32	Maximum		4%	11%	
33	Equipment usage curtailment, add, minimum		1%	1%	
34	Maximum		3%	10%	
35	Material handling & storage limitation, add, minimum		1%	1%	
36	Maximum		6%	7%	
37	Protection of existing work, add, minimum		2%	2%	
38	Maximum		5%	7%	
39	Shift work requirements, add, minimum			5%	
40	Maximum			30%	
41					
42					

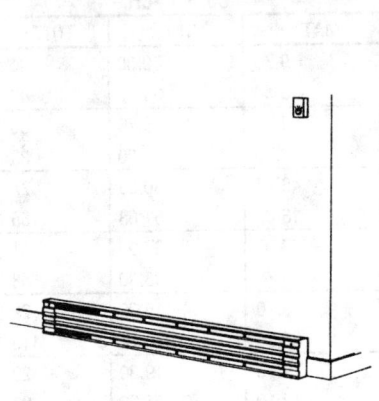

This page illustrates and describes baseboard heat systems including a thermostat, outlet box, breaker, and feed. Lines 09.4-910-04 thru 09 give the unit price and total price on a cost each basis for this system. Prices for alternate baseboard heat systems are on Line Items 09.4-910-15 thru 18. Both material quantities and labor costs have been adjusted for the system listed.

Factors: To adjust for job conditions other than normal working situations use Lines 09.4-910-29 thru 40.

Example: You are to install the system during evenings and weekends only. Go to Line 09.4-910-40 and apply this percentage to the appropriate INST. cost.

LINE NO.	DESCRIPTION	QUANTITY	COST EACH		
			MAT.	INST.	TOTAL
01	Baseboard heat including thermostat, outlet box, breaker and feed.				
02					
03					
04	Electric baseboard heater, 4' long	1 Ea.	50.19	45.81	96
05	Thermostat, integral	1 Ea.	25.30	19.70	45
06	Romex, 12-3 with ground	.4 C.L.F.	18.04	61.96	80
07	Panel board breaker, 20 Amp	1 Ea.	17.67	38.33	56
08	Outlet box	1 Ea.	1.43	17.07	18.50
09	TOTAL	Ea.	112.63	182.87	295.50
10					
11					
12					
13					
14	For alternate baseboard heating systems:				
15	Electric baseboard, 2' long	Ea.	95.11	175.39	270.50
16	6' long		137.22	197.28	334.50
17	8' long		162.14	212.36	374.50
18	10' long		175.74	228.76	404.50
19					
20					
21					
22					
23					
24					
25					
26					
27					
28					
29	Cut & patch to match existing construction, add, minimum		2%	3%	
30	Maximum		5%	9%	
31	Dust protection, add, minimum		1%	2%	
32	Maximum		4%	11%	
33	Equipment usage curtailment, add, minimum		1%	1%	
34	Maximum		3%	10%	
35	Material handling & storage limitation, add, minimum		1%	1%	
36	Maximum		6%	7%	
37	Protection of existing work, add, minimum		2%	2%	
38	Maximum		5%	7%	
39	Shift work requirements, add, minimum			5%	
40	Maximum			30%	
41					
42					

LINE NO.	DESCRIPTION	COST EACH		
		MAT.	INST.	TOTAL
01	Using non-metallic sheathed, cable, air conditioning receptacle	9.70	30.30	40
02	Disposal wiring	8.50	34.50	43
03	Dryer circuit	24.24	55.76	80
04	Duplex receptacle	9.70	23.30	33
05	Fire alarm or smoke detector	46.45	30.55	77
06	Furnace circuit & switch	13.37	51.63	65
07	Ground fault receptacle	43.05	37.95	81
08	Heater circuit	9.70	38.30	48
09	Lighting wiring	9.70	19.30	29
10	Range circuit	36.26	78.74	115
11	Switches single pole	9.70	19.30	29
12	3-way	12.10	25.90	38
13	Water heater circuit	12.10	61.90	74
14	Weatherproof receptacle	69.30	50.70	120
15	Using BX cable, air conditioning receptacle	14.56	37.44	52
16	Disposal wiring	12.10	40.90	53
17	Dryer circuit	30.36	66.64	97
18	Duplex receptacle	14.56	28.44	43
19	Fire alarm or smoke detector	46.45	30.55	77
20	Furnace circuit & switch	17	61	78
21	Ground fault receptacle	48.72	46.28	95
22	Heater circuit	13.37	46.63	60
23	Lighting wiring	14.56	23.44	38
24	Range circuit	49.50	95.50	145
25	Switches, single pole	14.56	23.44	38
26	3-way	18.13	30.87	49
27	Water heater circuit	17	73	90
28	Weatherproof receptacle	74.80	60.20	135
29	Using EMT conduit, air conditioning receptacle	17	46	63
30	Disposal wiring	15.75	51.25	67
31	Dryer circuit	33.99	81.01	115
32	Duplex receptacle	17	35	52
33	Fire alarm or smoke detector	52.12	45.88	98
34	Furnace circuit & switch	21.87	77.13	99
35	Ground fault receptacle	52.12	57.88	110
36	Heater circuit	17	57	74
37	Lighting wiring	18.13	28.87	47
38	Range circuit	57.20	112.80	170
39	Switches, single pole	17	29	46
40	3-way	20.63	38.37	59
41	Water heater circuit	20.63	89.37	110
42	Weatherproof receptacle	79.20	75.80	155
43	Using aluminum conduit, air conditioning receptacle	20.90	61.10	82
44	Disposal wiring	19.26	68.74	88
45	Dryer circuit	36.26	108.74	145
46	Duplex receptacle	20.39	47.61	68
47	Fire alarm or smoke detector	61.18	63.82	125
48	Furnace circuit & switch	28.33	101.67	130
49	Ground fault receptacle	56.65	78.35	135
50	Heater circuit	21.64	76.36	98
51	Lighting wiring	22.66	38.34	61
52	Range circuit	64.90	155.10	220
53	Switches, single pole	27.19	38.81	66
54	3-way	29.46	51.54	81
55	Water heater circuit	29.46	120.54	150
56	Weatherproof receptacle	88	102	190
57	Using galvanized steel conduit	19.94	65.06	85
58	Disposal wiring	18.02	72.98	91

LINE NO.	DESCRIPTION	COST EACH		
		MAT.	INST.	TOTAL
59	Dryer circuit	33.99	116.01	150
60	Duplex receptacle	19.26	50.74	70
61	Fire alarm or smoke detector	58.92	66.08	125
62	Furnace circuit & switch	27.19	107.81	135
63	Ground fault receptacle	55.52	79.48	135
64	Heater circuit	20.39	79.61	100
65	Lighting wiring	21.53	40.47	62
66	Range circuit	61.60	163.40	225
67	Switches, single pole	23.79	41.21	65
68	3-way	26.06	52.94	79
69	Water heater circuit	26.06	128.94	155
70	Weatherproof receptacle	83.60	111.40	195
71				
72				
73				
74				
75				
76				
77				
78				
79				
80				
81				
82				
83				
84				

This page illustrates and describes kitchen systems including top and bottom cabinets, custom laminated plastic top, single bowl sink, and appliances. Lines 11.1-242-04 thru 11 give the unit price and total price on a cost each basis for this system. Prices for alternate kitchen systems are on Line Items 11.1-242-15 and 16. Both material quantities and labor costs have been adjusted for the system listed.

Factors: To adjust for job conditions other than normal working situations use Lines 11.1-242-29 thru 40.

Example: You are to install the system and protect the work area from dust. Go to Line 11.1-242-32 and apply these percentages to the appropriate MAT. and INST. costs.

LINE NO.	DESCRIPTION	QUANTITY	COST EACH		
			MAT.	INST.	TOTAL
01	**Kitchen cabinets including wall and base cabinets, custom laminated**				
02	**Plastic top,sink & appliances,no plumbing or electrical rough-in included.**				
03					
04	Prefinished wood cabinets, average quality, wall and base	20 L.F.	1320	380	1700
05	Custom laminated plastic counter top	20 L.F.	114.40	191.60	306
06	Stainless steel sink, 22″ x 25″	1 Ea.	287.10	102.90	390
07	Faucet, top mount	1 Ea.	38.50	31.50	70
08	Dishwasher, built-in	1 Ea.	253	77	330
09	Compactor, built-in	1 Ea.	330	35	365
10	Range hood, 30″, ductless	1 Ea.	99	89	188
11	TOTAL	Ea.	2442	907	3349
12					
13					
14	For alternate kitchen systems:				
15	Prefinished wood cabinets, high quality	Ea.	4219.60	1003.40	5223
16	Custom cabinets, built in place, high quality	″	5237.10	1185.90	6423
17					
18					
19					
20					
21					
22	NOTE: No plumbing or electric rough-ins are included in the above				
23	Prices, for plumbing see Division 15, for electric see Division 16.				
24					
25					
26					
27					
28					
29	Cut & patch to match existing construction, add, minimum		2%	3%	
30	Maximum		5%	9%	
31	Dust protection, add, minimum		1%	2%	
32	Maximum		4%	11%	
33	Equipment usage curtailment, add, minimum		1%	1%	
34	Maximum		3%	10%	
35	Material handling & storage limitation, add, minimum		1%	1%	
36	Maximum		6%	7%	
37	Protection of existing work, add, minimum		2%	2%	
38	Maximum		5%	7%	
39	Shift work requirements, add, minimum			5%	
40	Maximum			30%	
41					
42					

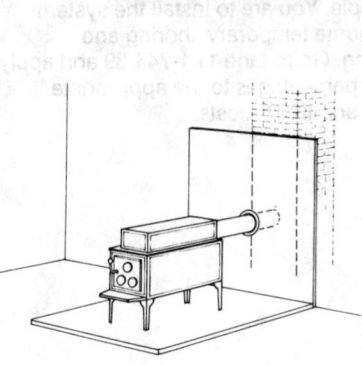

This page illustrates and describes a wood burning stove system including a free standing stove, preformed hearth, masonry chimney, and necessary piping and fittings. Lines 11.1-742-06 thru 11 give the unit price and total price on a cost each basis for this system. Prices for alternate wood burning stove systems are on Line Items 11.1-742-16 thru 20. Both material quantities and labor costs have been adjusted for the system listed.

Factors: To adjust for job conditions other than normal working situations use Lines 11.1-742-29 thru 40.

Example: You are to install the system and protect existing construction. Go to Line 11.1-742-35 and apply these percentages to the appropriate MAT., and INST. costs.

LINE NO.	DESCRIPTION	QUANTITY	COST EACH		
			MAT.	INST.	TOTAL
01	Cast iron, free standing, wood burning stove with preformed hearth, masonry				
02	Chimney, and all necessary piping and fittings to install in chimney.				
03					
04					
05					
06	Cast iron wood burning stove, stove pipe	1 Ea.	605	445	1050
07	Preformed hearth	1 Ea.	192.50	42.50	235
08	Non-combustible wall panel	1 Ea.	173.25	38.25	211.50
09	16" x 16" brick chimney, 8" x 8" flue	20 V.L.F.	321.75	703.25	1025
10	Foundation	.5 C.Y.	52.80	59.70	112.50
11	TOTAL	Ea.	1345.30	1288.70	2634
12					
13					
14	For alternate wood burning systems:				
15					
16	Installed in existing fireplace	Ea.	797.50	487.50	1285
17					
18	Installed with insulated metal chimney				
19	System including ceiling package,				
20	Metal chimney to 10 L.F.	Ea.	1297.18	680.82	1978
21					
22					
23					
24					
25					
26					
27					
28					
29	Dust protection, add, minimum		1%	2%	
30	Maximum		4%	11%	
31	Equipment usage curtailment, add, minimum		1%	1%	
32	Maximum		3%	10%	
33	Material handling & storage limitation, add, minimum		1%	1%	
34	Maximum		6%	7%	
35	Protection of existing work, add, minimum		2%	2%	
36	Maximum		5%	7%	
37	Shift work requirements, add, minimum			5%	
38	Maximum			30%	
39	Temporary shoring and bracing, add, minimum		2%	5%	
40	Maximum		5%	12%	
41					
42					

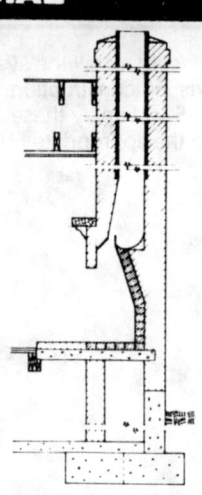

This page illustrates and describes masonry fireplace systems including a brick fireplace, footing, foundation, hearth, firebox, chimney and flue. Lines 11.1-744-04 thru 10 give the unit price and total price on a cost each basis for this system. Prices for alternate masonry fireplace systems are on Line Items 11.1-744-13 thru 15. Both material quantities and labor costs have been adjusted for the system listed.

Factors: To adjust for job conditions other than normal working situations use Lines 11.1-744-27 thru 40.

Example: You are to install the system with some temporary shoring and bracing. Go to Line 11.1-744-39 and apply these percentages to the appropriate MAT. and INST. costs.

LINE NO.	DESCRIPTION	QUANTITY	COST EACH MAT.	COST EACH INST.	COST EACH TOTAL
01	Brick masonry fireplace, including footing, foundation, hearth, firebox,				
02	Chimney and flue, chimney 12' above firebox.				
03					
04	Footing, 4' x 7' x 12" thick, 3000 psi concrete	1.04 C.Y.	96.10	96.30	192.40
05	Foundation, 12" concrete block	180 S.F.	429.66	713.34	1143
06	Fireplace, brick faced, 6'-0" wide x 5'-0" high	1 Ea.	363	1287	1650
07	Hearth	1 Ea.	110	260	370
08	Chimney, 20" x 20", one 12" x 12" flue, 12 V.L.F.	12 V.L.F.	225.72	458.28	684
09	Mantle, wood	7 L.F.	28.11	55.89	84
10	TOTAL	Ea.	1252.59	2870.81	4123.40
11					
12	Above system with the following:				
13	Chimney, 20" x 20", one 12" x 12" flue, 18 V.L.F.	Ea.	1365.45	3099.95	4465.40
14	24 V.L.F.		1478.31	3329.09	4807.40
15	Fieldstone face instead of brick	↓	1530.09	2870.81	4400.90
16					
17					
18					
19					
20					
21					
22					
23					
24					
25					
26					
27	Cut & patch to match existing construction, add, minimum		2%	3%	
28	Maximum		5%	9%	
29	Dust protection, add, minimum		1%	2%	
30	Maximum		4%	11%	
31	Equipment usage curtailment, add, minimum		1%	1%	
32	Maximum		3%	10%	
33	Material handling & storage limitation, add, minimum		1%	1%	
34	Maximum		6%	7%	
35	Protection of existing work, add, minimum		2%	2%	
36	Maximum		5%	7%	
37	Shift work requirements, add, minimum			5%	
38	Maximum			30%	
39	Temporary shoring and bracing, add, minimum		2%	5%	
40	Maximum		5%	12%	
41					
42					

LINE NO.	DESCRIPTION	COST EACH		
		MAT.	INST.	TOTAL
01	Cabinets standard wood, base, one drawer one door, 12″ wide	104.50	25.50	130
02	15″ wide	110	25	135
03	18″ wide	115.50	24.50	140
04	21″ wide	126.50	23.50	150
05	24″ wide	137.50	27.50	165
06				
07	Two drawers two doors, 27″ wide	176	24	200
08	30″ wide	181.50	28.50	210
09	33″ wide	192.50	27.50	220
10	36″ wide	198	27	225
11	42″ wide	214.50	30.50	245
12	48″ wide	225.50	29.50	255
13	Drawer base (4 drawers), 12″ wide	115.50	24.50	140
14	15″ wide	126.50	23.50	150
15	18″ wide	132	23	155
16	24″ wide	154	26	180
17	Sink or range base, 30″ wide	115.50	24.50	140
18	33″ wide	137.50	27.50	165
19	36″ wide	148.50	26.50	175
20	42″ wide	159.50	30.50	190
21	Corner base, 36″ wide	148.50	31.50	180
22	Lazy susan with revolving door	203.50	36.50	240
23	Cabinets standard wood, wall two doors, 12″ high, 30″ wide	88	22	110
24	36″ wide	99	26	125
25	15″ high, 30″ high	93.50	21.50	115
26	36″ wide	110	25	135
27	24″ high, 30″ wide	110	25	135
28	36″ wide	132	23	155
29	30″ high, 30″ wide	132	28	160
30	36″ wide	137.50	32.50	170
31	42″ wide	148.50	31.50	180
32	48″ wide	165	30	195
33	One door, 30″ high, 12″ wide	88	27	115
34	15″ wide	93.50	26.50	120
35	18″ wide	104.50	25.50	130
36	24″ wide	110	30	140
37	Corner, 30″ high, 24″ wide	132	33	165
38	36″ wide	170.50	34.50	205
39	Broom, 84″ high, 24″ deep, 18″ wide	154	56	210
40	Oven, 84″ high, 24″ deep, 27″ wide	176	74	250
41	Valance board, 4′ long	22	5.80	27.80
42	6″ long	33	8.70	41.70
43	Counter tops, laminated plastic, stock 25″ wide w/backsplash, min.	5.72	9.58	15.30
44	Maximum	17.60	11.40	29
45	Custom, ⅞″ thick, no splasn	16.78	9.22	26
46	Cove splash	21.89	9.11	31
47	1-¼″ thick, no splash	19.75	10.25	30
48	Square splash	24.75	10.25	35
49	Post formed	8.86	9.59	18.45
50				
51	Maple laminated 1-½″ thick, no splash	33	10	43
52	Square splash	37.40	10.60	48
53				
54				
55	Appliances, range, free standing, minimum	297	28	325
56	Maximum	1320	55	1375
57	Built-in, minimum	473	37	510
58	Maximum	1089	61	1150

LINE NO.	DESCRIPTION	COST EACH		
		MAT.	INST.	TOTAL
59	Counter top range 4 burner. maximum	220	40	260
60	Maximum	517	78	595
61	Compactor, built-in, minimum	330	35	365
62	Maximum	550	40	590
63	Dishwasher, built-in, minimum	253	77	330
64	Maximum	429	76	505
65	Garbage disposer, sink-pipe, minimum	60.50	64.50	125
66	Maximum	214.50	60.50	275
67	Range hood, 30″ wide, 2 speed, minimum	44	51	95
68	Maximum	286	59	345
69	Refrigerator, no frost, 12 cu. ft.	330	45	375
70	20 cu. ft.	649	66	715
71	Plumb. not incl. rough-ins, sinks porc. C.I., single bowl, 21″ x 24″	187	103	290
72	21″ x 30″	212.30	102.70	315
73	Double bowl, 20″ x 32″	239.80	120.20	360
74				
75	Stainless steel, single bowl, 19″ x 18″	256.30	98.70	355
76	22″ x 25″	287.10	102.90	390
77				
78				
79				
80				
81				
82				
83				
84				

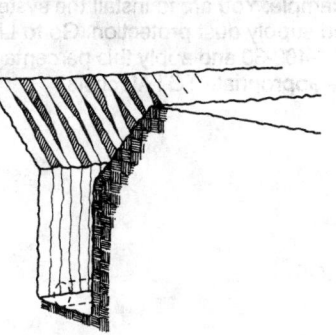

This page illustrates and describes utility trench excavation with back fill systems including a trench, back fill, excavated material, utility pipe or ductwork not included. Lines 12.1-116-06 thru 09 give the unit price and total price on a linear foot basis for this system. Prices for alternate utility trench excavation with back fill systems are on Line Items 12.1-116-11 thru 25. Both material quantities and labor costs have been adjusted for the system listed.

Factors: To adjust for job conditions other than normal working situations use Lines 12.1-116-31 thru 40.

Example: You are to install the system and protect existing construction. Go to Line 12.1-116-36 and apply these percentages to the appropriate TOTAL cost.

LINE NO.	DESCRIPTION	QUANTITY	COST PER L.F.		
			EQUIP.	LABOR	TOTAL
01	Cont. utility trench excav. with ⅜ C.Y.wheel mtd. backhoe,				
02	Backfilling with excav. mat. and comp. in 12″ lifts. Trench is 2′ wide by				
03	4′ dp. Cost of utility piping or ductwork is not incl. Cost based on an				
04	Excavation production rate of 150 C.Y. daily. no hauling included.				
05					
06	Machine excavate trench, 2′ wide by 4′ deep	.296 C.Y.	.41	1.01	1.42
07	Dozer backfill with excavated material	.296 C.Y.	.08	.10	.18
08	Compact in 12″ lifts, vibrating plate	.296 C.Y.	.19	.76	.95
09	TOTAL	L.F.	.68	1.87	2.55
10	Alternate size trenches:				
11	2′ wide x 2′ deep	L.F.	.32	.95	1.27
12	3′ deep		.51	1.39	1.90
13	With sloping sides, 2′ wide x 5′ deep		1.92	5.27	7.19
14	6′ deep		2.58	7.01	9.59
15	7′ deep		3.32	8.99	12.31
16	8′ deep		4.13	11.24	15.37
17	9′ deep		5.05	13.68	18.73
18	10′ deep		6.03	16.39	22.42
19			COST PER L.F.		
20			MAT.	INST.	TOTAL
21	Gravel, 2′ wide by 2′ deep, add	L.F.	1.79	1.91	3.70
22	4′ deep, add		3.60	3.82	7.42
23	6′ deep, add		13.50	14.33	27.83
24	8′ deep, add		21.60	22.94	44.54
25	10′ deep, add		31.52	33.43	64.95
26					
27					
28	For shallow, hand excavated trenches, no backfill, to 6′ dp, light soil	C.Y.		29	29
29	Heavy soil	"		57	57
30					
31	Equipment usage curtailment, add, minimum		1%	1%	
32	Maximum		3%	10%	
33	Material handling & storage limitation, add, minimum		1%	1%	
34	Maximum		6%	7%	
35	Protection of existing work, add, minimum		2%	2%	
36	Maximum		5%	7%	
37	Shift work requirements, add, minimum			5%	
38	Maximum			30%	
39	Temporary shoring and bracing, add, minimum		2%	5%	
40	Maximum		5%	12%	
41					
42					

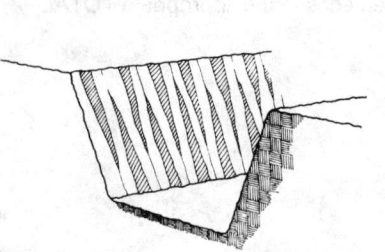

This page illustrates and describes continuous footing and trench excavation systems including a wheel mounted backhoe, operator, equipment rental, fuel, oil, mobilization, no hauling or backfill. Lines 12.1-462-06 thru 11 give the unit price and total price per linear foot for this system. Prices for alternate continuous footing and trench excavation systems are on Line Items 12.1-462-13 thru 27. Both material quantities and labor costs have been adjusted for the system listed.

Factors: To adjust for job conditions other than normal working situations use Lines 12.1-462-29 thru 40.

Example: You are to install the system and supply dust protection. Go to Line 12.1-462-30 and apply this percentage to the appropriate TOTAL costs.

LINE NO.	DESCRIPTION	QUANTITY	COST PER L.F. EQUIP.	LABOR	TOTAL
01	Continuous footing or trench excav. w/ ¾ C.Y. wheel mntd backhoe				
02	Including operator, equipment rental, fuel, oil and mobilization. Trench				
03	Is 4′ wide at bottom, 3′ deep w/sides sloped 1 to 2 and 200′ long. Costs				
04	Are based on a production rate of 240 C.Y./day. No hauling/backfill incl.				
05					
06	Equipment operator	4 Hr.		142.50	142.50
07	¾ C.Y. backhoe, wheel mounted	.5 Day	72.50		72.50
08	Operating expense (fuel, oil)	4 Hr.	32.20		32.20
09	Mobilization	1 Ea.		280	280
10	TOTAL	120 C.Y.	104.70	422.50	527.20
11		L.F.	.52	2.11	2.63
12	For alternate trench sizes, 4′ bottom with sloped sides:				
13	2′ deep, 50′ long	L.F.	1.57	8.45	10.02
14	100′ long		.85	4.23	5.08
15	300′ long		.34	1.41	1.75
16	4′ deep, 50′ long		1.69	8.45	10.14
17	100′ long		.97	4.23	5.20
18	300′ long		.73	1.94	2.67
19	6′ deep, 50′ long		1.88	8.45	10.33
20	100′ long		1.16	4.76	5.92
21	300′ long		1.40	2.83	4.23
22	8′ deep, 50′ long		2.09	8.45	10.54
23	100′ long		2.09	5.65	7.74
24	300′ long		.66	6.82	7.48
25	10′ deep, 50′ long		3.14	9.88	13.02
26	100′ long		3.01	6.90	9.91
27	300′ long	↓	2.93	4.91	7.84
28					
29	Dust protection, add, minimum		1%	2%	
30	Maximum		4%	11%	
31	Equipment usage curtailment, add, minimum		1%	1%	
32	Maximum		3%	10%	
33	Material handling & storage limitation, add, minimum		1%	1%	
34	Maximum		6%	7%	
35	Protection of existing work, add, minimum		2%	2%	
36	Maximum		5%	7%	
37	Shift work requirements, add, minimum			5%	
38	Maximum			30%	
39	Temporary shoring and bracing, add, minimum		2%	5%	
40	Maximum		5%	12%	
41					
42					

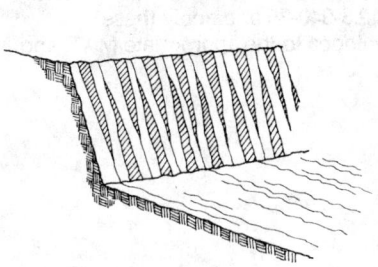

This page illustrates and describes foundation excavation systems including a backhoe-loader, operator, equipment rental, fuel, oil, mobilization, hauling material, no back filling. Lines 12.1-464-06 thru 12 give the unit price and total price per cubic yard for this system. Prices for alternate foundation excavation systems are on Line Items 12.1-464-16 thru 22. Both material quantities and labor costs have been adjusted for the system listed.

Factors: To adjust for job conditions other than normal working situations use Lines 12.1-464-29 thru 40.

Example: You are to install the system with the use of temporary shoring and bracing. Go to Line 12.1-464-39 and apply these percentages to the appropriate TOTAL costs.

LINE NO.	DESCRIPTION	QUANTITY	COST PER C.Y.		
			EQUIP.	LABOR	TOTAL
01	Foundation excav w/ ¾ C.Y. backhoe-loader, incl. operator, equip				
02	Rental, fuel, oil, and mobilization. Hauling of excavated material is				
03	Included. Prices based on one day production of 360 C.Y. in medium soil				
04	Without backfilling.				
05					
06	Equipment operator	8 Hr.		285	285
07	Backhoe-loader, ¾ C.Y.	1 Day	145		145
08	Operating expense (fuel, oil)	8 Hr.	64.40		64.40
09	Hauling, 12 C.Y. trucks, 1 mile round trip	360 C.Y.		882	882
10	Mobilization	1 Ea.		280	280
11	TOTAL	360 C.Y.	209.40	1447	1656.40
12		C.Y.	.58	4.02	4.60
13					
14	For alternate size excavations:				
15					
16	100 C.Y.	C.Y.	.19	6.85	7.04
17	200 C.Y.		.59	4.65	5.24
18	300 C.Y.		.60	4.18	4.78
19	400 C.Y.		.58	3.93	4.51
20	500 C.Y.		.59	3.81	4.40
21	600 C.Y.		.57	3.71	4.28
22	700 C.Y.		.57	3.62	4.19
23					
24					
25					
26					
27					
28					
29	Dust protection, add, minimum		1%	2%	
30	Maximum		4%	11%	
31	Equipment usage curtailment, add, minimum		1%	1%	
32	Maximum		3%	10%	
33	Material handling & storage limitation, add, minimum		1%	1%	
34	Maximum		6%	7%	
35	Protection of existing work, add, minimum		2%	2%	
36	Maximum		5%	7%	
37	Shift work requirements, add, minimum			5%	
38	Maximum			30%	
39	Temporary shoring and bracing, add, minimum		2%	5%	
40	Maximum		5%	12%	
41					
42					

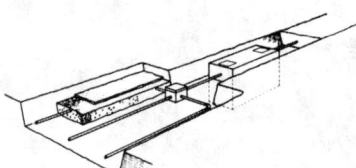

This page illustrates and describes drainage and utilities — septic system including a tank, distribution box, excavation, piping, crushed stone and backfill. Line 12.3-940-04 thru 13 give the unit price and total price on a cost each basis for this system. Prices for alternate drainage and utilities — septic systems are on Line Items 12.4-940-17 thru 20. Both material quantities and labor costs have been adjusted for the system listed.

Factors: To adjust for job conditions other than normal working situations use Lines 12.3-940-31 thru 40.

Example: You are to install the system with a material handling limitation. Go to Line 12.3-940-35 and apply these percentages to the appropriate MAT. and INST. costs.

LINE NO.	DESCRIPTION	QUANTITY	COST EACH		
			MAT.	INST.	TOTAL
01	Septic system including tank, distribution box, excavation, piping, crushed				
02	Stone and backfill for a 1000 S.F. leaching field.				
03					
04	Precast concrete septic tank, 1000 gal. capacity	1 Ea.	473	132	605
05	Concrete distribution box	1 Ea.	56.10	28.90	85
06	4" bituminous fiber pipe	25 L.F.	40.70	30.05	70.75
07	4" bituminous fiber perforated pipe	145 L.F.	239.25	147.90	387.15
08	4" pipe fittings	8 Ea.	22.88	521.12	544
09	Trench and tank excavation	120 C.Y.		441.80	441.80
10	Crushed stone backfill	76 C.Y.	907.06	476.14	1383.20
11	Backfill with excavated material	26 C.Y.		27.04	27.04
12	Building paper	6 C.S.F.	15.51	46.59	62.10
13	**TOTAL**	System	1754.50	1851.54	3606.04
14					
15					
16	For alternate septic systems:				
17	1000 gal. tank with 2000 S.F. field	System	2597.99	2583.62	5181.61
18	With leaching pits		1801.20	573.35	2374.55
19	2000 gal. tank with 2000 S.F. field		3015.99	2660.62	5676.61
20	With leaching pits		2300.60	834.65	3135.25
21					
22					
23					
24					
25					
26					
27					
28					
29					
30					
31	Dust protection, add, minimum		1%	2%	
32	Maximum		4%	11%	
33	Equipment usage curtailment, add, minimum		1%	1%	
34	Maximum		3%	10%	
35	Material handling & storage limitation, add, minimum		1%	1%	
36	Maximum		6%	7%	
37	Protection of existing work, add, minimum		2%	2%	
38	Maximum		5%	7%	
39	Shift work requirements, add, minimum			5%	
40	Maximum			30%	
41					
42					

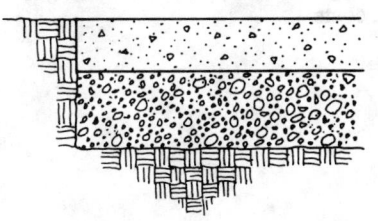

This page illustrates and describes driveway systems including concrete slab, gravel base, broom finish, compaction, and joists. Lines 12.5-404-04 thru 10 give the unit price and total price on a cost each basis for this system. Prices for alternate driveway systems are on Line Items 12.5-404-13 thru 27. Both material quantities and labor costs have been adjusted for the system listed.

Factors: To adjust for job conditions other than normal working situations use Lines 12.5-404-29.

Example: You are to install the system and match existing construction. Go to Line 12.5-404-30 and apply these percentages to the appropriate MAT. and INST. costs.

LINE NO.	DESCRIPTION	QUANTITY	COST EACH		
			MAT.	**INST.**	**TOTAL**
01	Complete driveway, 10' x 30', 4" concrete slab, 3000 psi, on 6" compacted				
02	Gravel base, concrete broom finished and cured with 10' x 10' joints.				
03					
04	Grade and compact subgrade	33 S.Y.		38.28	38.28
05	Place and compact 6" crushed stone base	33 S.Y.	156.09	20.46	176.55
06	Place and remove edgeforms (4 use)	80 L.F.	20.24	148.56	168.80
07	Place and broom finish 4" concrete slab, 3000 psi	4 C.Y.	224	192.80	416.80
08	Spray on membrane curing compound	3 S.F.	6.47	14.38	20.85
09	Saw cut 3" deep joints	60 L.F.	17.16	48.84	66
10	TOTAL	Ea.	423.96	463.32	887.28
11					
12	For alternate driveway systems:				
13	Concrete: 10' x 30' with 6" concrete on 6" crushed stone	Ea.	535.96	490.72	1026.68
14	20' x 30' with 4" concrete		863.21	890.26	1753.47
15	With 6" concrete		1087.21	945.06	2032.27
16	10' wide, for each additional 10' length over 30', 4" thick, add		138.30	149.25	287.55
17	6" thick, add		180.30	159.52	339.82
18	20' wide, for each additional 10' length over 30', 4" thick, add		280.12	285.75	565.87
19	6" thick, add		350.12	302.88	653
20	Asphalt: 10' x 30' with 2" binder, 1" topping on 6" cr. st., sealed		327.43	407.31	734.74
21	3" binder, 1" topping		380.07	414.40	794.47
22	20' x 30' with 2" binder, 1" topping		654.86	534.62	1189.48
23	3" binding, 1" topping		760.13	548.81	1308.94
24	10' wide for each add'l. 10' length over 30', 2" b + 1" t, add		109.15	42.43	151.58
25	3" binder, 1" topping		126.70	44.79	171.49
26	20' wide for each add'l. 10' length over 30', 2" b + 1" t, add		218.28	84.88	303.16
27	3" binder, 1" topping		253.37	89.61	342.98
28					
29	Cut & patch to match existing construction, add, minimum		2%	3%	
30	Maximum		5%	9%	
31	Dust protection, add, minimum		1%	2%	
32	Maximum		4%	11%	
33	Equipment usage curtailment, add, minimum		1%	1%	
34	Maximum		3%	10%	
35	Material handling & storage limitation, add, minimum		1%	1%	
36	Maximum		6%	7%	
37	Protection of existing work, add, minimum		2%	2%	
38	Maximum		5%	7%	
39	Shift work requirements, add, minimum			5%	
40	Maximum			30%	
41					
42					

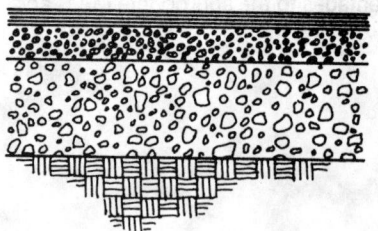

This page illustrates and describes asphalt parking lot systems including asphalt binder, topping, crushed stone base, painted parking stripes and concrete parking blocks. Lines 12.5-514-04 thru 11 give the unit price and total price per square yard for this system. Prices for alternate asphalt parking lot systems are on Line Items 12.5-514-15 thru 25. Both material quantities and labor costs have been adjusted for the system listed.

Factors: To adjust for job conditions other than normal working situations use Lines 12.5-514-29 thru 40.

Example: You are to install the system and match existing construction. Go to Line 12.5-514-29 and apply these percentages to the appropriate MAT. and INST. costs.

LINE NO.	DESCRIPTION	QUANTITY	COST PER S.Y.		
			MAT.	INST.	TOTAL
01	Parking lot consisting of 2″ asphalt binder and 1″ topping on 6″				
02	Crushed stone base with painted parking stripes and concrete parking blocks				
03					
04	Fine grade and compact subgrade	1 S.Y.		1.16	1.16
05	6″ crushed stone base, stone	.32 Ton	4.73	.62	5.35
06	2″ asphalt binder	1 S.Y.	3.20	.69	3.89
07	1″ asphalt topping	1 S.Y.	1.71	.46	2.17
08	Paint parking stripes	.5 L.F.	.01	.05	.06
09	6″ x 10″ x 6′ precast concrete parking blocks	.02 Ea.	.48	.20	.68
10	Mobilization of equipment	.005 EA		1.40	1.40
11	TOTAL	S.Y.	10.13	4.58	14.71
12					
13					
14	For alternate parking lot systems:				
15	Above system on 9″ crushed stone	S.Y.	12.53	4.78	17.31
16	12″ crushed stone		14.81	5	19.81
17	On bank run gravel, 6″ deep		7.09	4.41	11.50
18	9″ deep		7.94	4.02	11.96
19	12″ deep		8.79	4.71	13.50
20	3″ binder plus 1″ topping on 6″ crushed stone		11.73	4.79	16.52
21	9″ deep crushed stone		14.13	4.99	19.12
22	12″ deep crushed stone		16.41	5.21	21.62
23	On bank run gravel, 6″ deep		8.69	4.62	13.31
24	9″ deep		9.54	4.23	13.77
25	12″ deep		10.39	4.92	15.31
26					
27					
28					
29	Cut & patch to match existing construction, add, minimum		2%	3%	
30	Maximum		5%	9%	
31	Dust protection, add, minimum		1%	2%	
32	Maximum		4%	11%	
33	Equipment usage curtailment, add, minimum		1%	1%	
34	Maximum		3%	10%	
35	Material handling & storage limitation, add, minimum		1%	1%	
36	Maximum		6%	7%	
37	Protection of existing work, add, minimum		2%	2%	
38	Maximum		5%	7%	
39	Shift work requirements, add, minimum			5%	
40	Maximum			30%	
41					
42					

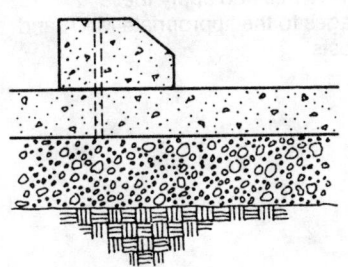

This page illustrates and describes concrete parking lot systems including a concrete slab, unreinforced, compacted gravel base, broom finished, cured, joints, painted parking stripes, and parking blocks. Lines 12.5-524-04 thru 12 give the unit price and total price per square yard for this system. Prices for alternate concrete parking lot systems are on Line Items 12.5-524-15 thru 25. All materials have been adjusted according to the system listed.

Factors: To adjust for job conditions other than normal working situations use Line 12.5-524-29 thru 40.

Example: You are to install the system and protect existing construction. Go to Line 12.5-524-38 and apply these percentages to the appropriate MAT. and INST. costs.

LINE NO.	DESCRIPTION	QUANTITY	COST PER S.Y.		
			MAT.	INST.	TOTAL
01	Parking lot, 4″ slab, 3000 psi concrete on 6″ compacted gravel base				
02	With 12′ x 12′ joints, painted parking strips, conc. parking blocks.				
03					
04	Grade and compact subgrade	1 S.Y.		1.16	1.16
05	6″ crushed stone base, stone	.32 Ton	4.73	.62	5.35
06	Place and remove edge forms, 4 use	.25 L.F.	.06	.47	.53
07	Place and broom finish 4″ slab, 3000 psi concrete	1 S.F.	6.16	1.51	7.67
08	Spray on membrane curing compound	1 S.F.	.18	.30	.48
09	Saw cut 3″ deep joints	2.25 L.F.	.64	1.84	2.48
10	Paint parking stripes	.5 L.F.	.01	.05	.06
11	6″ x 10″ x 6′ precast concrete parking blocks	.02 Ea.	.48	.20	.68
12	TOTAL	S.Y.	12.26	6.15	18.41
13					
14	For alternate parking lot systems:				
15	Above system on 3″ crushed stone	S.Y.	7.53	5.53	13.06
16	9″ crushed stone		14.66	6.35	21.01
17	On bank run gravel, 6″ deep		9.22	5.98	15.20
18	9″ deep		10.07	5.59	15.66
19	On compacted subgrade only		7.53	5.53	13.06
20	6″ concrete slab, 3000 psi on 3″ crushed stone		10.50	6.25	16.75
21	6″ crushed stone		15.23	6.87	22.10
22	9″ crushed stone		17.63	7.07	24.70
23	On bank run gravel, 6″ deep		12.19	6.70	18.89
24	9″ deep		13.04	6.31	19.35
25	On compacted subgrade only		10.50	6.25	16.75
26					
27					
28					
29	Cut & patch to match existing construction, add, minimum		2%	3%	
30	Maximum		5%	9%	
31	Dust protection, add, minimum		1%	2%	
32	Maximum		4%	11%	
33	Equipment usage curtailment, add, minimum		1%	1%	
34	Maximum		3%	10%	
35	Material handling & storage limitation, add, minimum		1%	1%	
36	Maximum		6%	7%	
37	Protection of existing work, add, minimum		2%	2%	
38	Maximum		5%	7%	
39	Shift work requirements, add, minimum			5%	
40	Maximum			30%	
41					
42					

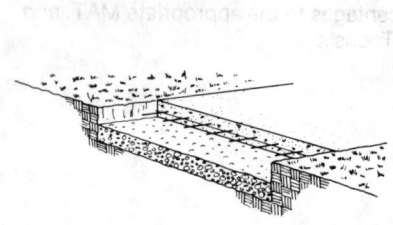

This page illustrates and describes sidewalk systems including concrete, welded wire and broom finish. Lines 12.7-104-05 thru 13 give unit price and total price per square foot for this system. Prices for alternate sidewalk systems are on Line Items 12.7-104-17 thru 19. Both material quantities and labor costs have been adjusted for the system listed.

Factors: To adjust for job conditions other than normal working situations use Lines 12.7-104-29 thru 40.

Example: You are to install the system and match existing construction. Go to Line 12.7-104-29 and apply these percentages to the appropriate MAT. and INST. costs.

LINE NO.	DESCRIPTION	QUANTITY	COST PER S.F.		
			MAT.	INST.	TOTAL
01	4″ thick concrete sidewalk with welded wire fabric				
02	3000 psi air entrained concrete, broom finish.				
03					
04					
05	Gravel fill, 4″ deep	.012 C.Y.	.04	.06	.10
06	Compact fill	.012 C.Y.		.02	.02
07	Hand grade	1 S.F.		.12	.12
08	Edge form	.25 L.F.	.06	.47	.53
09	Welded wire fabric	.011 S.F.	.08	.22	.30
10	Concrete, 3000 psi air entrained	.012 C.Y.	.67		.67
11	Place concrete	.012 C.Y.		.16	.16
12	Broom finish	1 S.F.		.46	.46
13	TOTAL	S.F.	.85	1.51	2.36
14					
15					
16	For alternate sidewalk systems:				
17	Asphalt (bituminous), 2″ thick	S.F.	.41	.45	.86
18	Brick, on sand, bed, 4.5 brick per S.F.		2.70	4.70	7.40
19	Flagstone, slate, 1″ thick, rectangular	↓	2.54	5.61	8.15
20					
21					
22					
23					
24					
25					
26					
27					
28					
29	Cut & patch to match existing construction, add, minimum		2%	3%	
30	Maximum		5%	9%	
31	Dust protection, add, minimum		1%	2%	
32	Maximum		4%	11%	
33	Equipment usage curtailment, add, minimum		1%	1%	
34	Maximum		3%	10%	
35	Material handling & storage limitation, add, minimum		1%	1%	
36	Maximum		6%	7%	
37	Protection of existing work, add, minimum		2%	2%	
38	Maximum		5%	7%	
39	Shift work requirements, add, minimum			5%	
40	Maximum			30%	
41					
42					

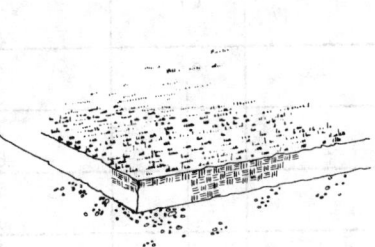

This page describes landscaping — lawn establishment systems including loam, lime, fertilizer, side and top mulching. Lines 12.7-604-04 thru 08 give the unit price and total price per square yard for this system. Prices for alternate landscaping — lawn establishment systems are on Line Items 12.7-604-11 and 12. Both material quantities and labor costs have been adjusted for the system listed.

Factors: To adjust for job conditions other than normal working situations use Lines 12.7-604-29 thru 40.

Example: You are to install the system and provide dust protection. Go to Line 12.7-604-32 and apply these percentages to the appropriate MAT. and INST. costs.

LINE NO.	DESCRIPTION	QUANTITY	COST PER S.Y.		
			MAT.	INST.	TOTAL
01	**Establishing lawns with loam, lime, fertilizer, seed and top mulching**				
02	**On rough graded areas.**				
03					
04	Furnish and place loam 4″ deep	.11 C.Y.	2.06	.57	2.63
05	Fine grade, lime, fertilize and seed	1 S.Y.	.20	1.65	1.85
06	Hay mulch, 1 bale/M.S.F.	1 S.Y.	.03	.28	.31
07	Rolling with hand roller	1 S.Y.		.53	.53
08	TOTAL	S.Y.	2.29	3.03	5.32
09					
10	For alternate lawn systems:				
11	Above system with jute mesh in place of hay mulch	S.Y.	2.94	3.03	5.97
12	Above system with sod in place of seed	″	3.79	2.25	6.04
13					
14					
15					
16					
17					
18					
19					
20					
21					
22					
23					
24					
25					
26					
27					
28					
29	Cut & patch to match existing construction, add, minimum		2%	3%	
30	Maximum		5%	9%	
31	Dust protection, add, minimum		1%	2%	
32	Maximum		4%	11%	
33	Equipment usage curtailment, add, minimum		1%	1%	
34	Maximum		3%	10%	
35	Material handling & storage limitation, add, minimum		1%	1%	
36	Maximum		6%	7%	
37	Protection of existing work, add, minimum		2%	2%	
38	Maximum		5%	7%	
39	Shift work requirements, add, minimum			5%	
40	Maximum			30%	
41					
42					

LINE NO.	DESCRIPTION	UNIT	TOTAL COST
01	Cabinets, base	L.F.	5.70
02	Wall	"	5.70
03	Carpet, bonded	S.F.	.23
04	Tackless		.05
05	Ceiling, tile, adhesive bonded		.51
06	On suspension system		.60
07	Sheetrock, on furring		.57
08	On suspension system		.63
09	Plaster, on wire lath		.80
10	On suspension system	↓	.80
11	Chimney, brick, 16″ x 16″	V.L.F.	16.05
12	20″ x 20″	"	29
13	Concrete, footing, 1′ thick, 2′ wide	L.F.	10.20
14	2′ thick, 3′ wide	"	17.50
15	Slab, 6″ thick, plain	S.F.	3.49
16	Mesh reinforced		3.85
17	Wall, interior, 6″ thick		8.20
18	12″ thick	↓	13.10
19	Door and frame, wood	EA.	35.30
20	Hollow metal	"	50.30
21	Ducts, small size, 4″ x 8″	L.F.	1.14
22	Large size, 30″ x 72″		4.57
23	Fascia, to 6″ wide		.46
24	To 10″ wide	↓	.57
25	Flooring, brick	S.F.	.96
26	Ceramic		.82
27	Linoleum		.33
28	Resilient tile		.46
29	Terrazzo, cast in place	↓	1.65
30			
31	Wood, strip	S.F.	.89
32	Block		.72
33	Subflooring, tongue and groove boards		.89
34	Plywood	↓	.48
35	Framing, steel girders, 10″ x 12″	L.F.	9.95
36			
37	Wood, studs, 2″ x 4″	L.F.	.23
38	Rafters, 2″ x 8″	"	.63
39			
40			
41			
42			
43	Gutters, attached	L.F.	.95
44	Built in	"	2.29
45	Masonry, veneer, by hand, brick to 4″ thick	S.F.	2.06
46	Marble to 2″ thick		1.60
47	Granite to 4″ thick		1.70
48	Stone to 8″ thick		1.65
49	Walls, brick, 4″ thick		3.69
50	8″ thick		4.88
51	12″ thick		5.95
52	16″ thick		7.26
53	Block, 4″ thick		.81
54	6″ thick		1.07
55	8″ thick		1.62
56	12″ thick		1.98
57	Paneling, plywood		.23
58	Woodboards, tongue and groove	↓	.65

LINE NO.	DESCRIPTION	UNIT	TOTAL COST
59	Piping, to 2" diameter	L.F.	1.57
60	To 4" diameter		2.09
61	To 8" diameter		6.30
62	To 16" diameter	↓	10.45
63			
64			
65	Roofing, built up, 5 ply	S.F.	.73
66	Shingles, asphalt strip	"	.33
67	Stairs, wood, minimum	RISER	11.45
68	Maximum	"	17.60
69	Toilet fixtures, bathtub	EA.	78
70	Sink		39
71	Shower		63
72	Toilet		45
73	Urinal		45
74	Vanity	↓	31
75	Walls, partitions, studs and sheet rock	S.F.	1.37
76	Studs and plaster	"	2.74
77	Windows, wood, to 12 S.F.	EA.	10.40
78	To 50 S.F.		17.60
79	Metal, to 12 S.F.		20
80	To 50 S.F.	↓	65

LINE NO.	DESCRIPTION	COST PER UNIT		
		MAT.	INST.	TOTAL
01	Shrubs and trees, Evergreen, in prepared beds			
02				
03	Arborvitae pyramidal, 4'-5'	28.60	50.40	79
04	Globe, 12"-15"	9.52	7.43	16.95
05	Cedar, blue, 8'-10'	115.50	84.50	200
06	Hemlock, Canadian, 2-½'-3'	20.19	19.81	40
07	Juniper, andora, 18"-24"	10.12	8.93	19.05
08	Wiltoni, 15"-18"	14.96	9.04	24
09	Skyrocket, 4½'-5'	36.30	27.70	64
10	Blue pfitzer, 2'-2-½'	15.73	16.27	32
11	Ketleerie, 2-½'-3'	24.20	13.80	38
12	Pine, black, 2-½'-3'	27.50	14.50	42
13	Mugo, 18"-24"	30.80	12.20	43
14	White, 4'-5'	34.10	19.90	54
15	Spruce, blue, 18"-24"	22	12	34
16	Norway, 4'-5'	42.90	20.10	63
17	Yew, denisforma, 12"-15"	11.39	11.61	23
18	Capitata, 18"-24"	16.83	24.17	41
19	Hicksi, 2'-2-½'	17.99	24.01	42
20				
21	Trees, Deciduous, in prepared beds			
22				
23	Beech, 5'-6'	42.90	30.10	73
24	Dogwood, 4'-5'	39.60	37.40	77
25	Elm, 8'-10'	66	74	140
26	Magnolia, 4'-5'	50.60	74.40	125
27	Maple, red, 8'-10', 1-½" caliper	49.50	150.50	200
28	Oak, 2-½"-3" caliper	187	498	685
29	Willow, 6'-8', 1" caliper	27.50	77.50	105
30				
31	Shrubs, Broadleaf Evergreen, in prepared beds			
32				
33	Andromeda, 15"-18", container	18.10	7.90	26
34	Azalea, 15"-18", container	17.93	7.07	25
35	Barberry, 9"-12", container	6.38	5.47	11.85
36	Boxwood, 15"-18", B & B	24.20	7.80	32
37	Euonymus, emerald gaiety, 12"-15", container	9.24	6.21	15.45
38	Holly, 15"-18", B & B	18.26	7.74	26
39	Mount laurel, 18"-24", B & B	20.63	9.37	30
40	Privet, 18"-24", B & B	2.48	5.47	7.95
41	Rhodendron, 18"-24", container	20.35	14.65	35
42	Rosemary, 1 gal. container	3.96	1.19	5.15
43	Deciduous, amalanchier, 2'-3', B & B	16.12	12.88	29
44	Azalea, 15"-18", B & B	15.02	6.98	22
45	Bayberry, 2'-3', B & B	19.31	12.69	32
46	Cotoneaster, 15"-18", B & B	9.08	8.92	18
47	Dogwood, 3'-4', B & B	15.73	37.27	53
48	Euonymus, alatus compacta, 15"-18", container	10.12	8.93	19.05
49	Forsythia, 2'-3', container	10.67	12.33	23
50	Honeysuckle, 3'-4', B & B	27.50	11.50	39
51	Hydrangea, 2'-3', B & B	11.22	12.78	24
52	Lilac, 3'-4', B & B	22	38	60
53	Quince, 2'-3', B & B	18.03	12.97	31
54				
55	Ground Cover			
56				
57	Plants, Pachysandra, prepared beds, per hundred	13.97	71.03	85
58	Vinca minor or English ivy, per hundred	51.70	73.30	125

LINE NO.	DESCRIPTION	COST PER UNIT		
		MAT.	INST.	TOTAL
59	Plant bed prep. 18″ dp., mach., per square foot	1.45	.07	1.52
60	By hand, per square foot	1.45	1.48	2.93
61	Stone chips, Georgia marble, 50# bags, per bag	3.03	1.37	4.40
62	Onyx gemstone, per bag	11.55	2.75	14.30
63	Quartz, per bag	5.15	2.75	7.90
64	Pea gravel, truck load lots, per cubic yard	18.37	25.63	44
65	Mulch, polyethylene mulch, per square yard	.15	.23	.38
66	Wood chips, 2″ deep, per square yard	.91	1.04	1.95
67	Peat moss, 1″ deep, per square yard	.90	.26	1.16
68	Erosion control per square yard, Jute mesh stapled	.68	.28	.96
69	Plastic netting, stapled	.36	.28	.64
70	Polypropylene mesh, stapled	1.79	.28	2.07
71	Tobacco netting, stapled	.03	.28	.31
72	Lawns per square yard, seeding incl. fine grade, limestone, fert. and seed	.20	1.65	1.85
73	Sodding, incl. fine grade, level ground	1.66	1.65	3.31
74	On slope	1.77	2.06	3.83
75	Edging, per linear foot, Redwood, untreated, 1″ x 4″	.96	1.19	2.15
76	2″ x 4″	.89	1.80	2.69
77	Stl. edge strips, ¼″ x 5″ inc. stakes	2.75	1.80	4.55
78	$\frac{3}{16}$″ x 4″	2.05	1.80	3.85
79	Brick edging, set on edge	1.67	3.83	5.50
80	Set flat	.85	1.40	2.25
81				
82				
83				
84				

REFERENCE SECTION

Following items in the cost pages, there are frequently found large numbers in circles. These "circle numbers" refer to reference tables, explanations and estimating information which explain the derivation of the unit price data. Also included are alternate pricing methods, technical data and information that can be used for material lists, design and economy in construction.

CIRCLE REFERENCE NUMBERS

② Builder's Risk Insurance (Div. 010-040)

Builder's Risk Insurance is insurance on a building during construction. Premiums are paid by the owner or the contractor. Blasting, collapse and underground insurance would raise total insurance costs above those listed. Floater policy for materials delivered to the job runs $.75 to $1.25 per $100 value. Contractor equipment insurance runs $.50 to $1.50 per $100 value.

Tabulated below are New England Builder's Risk insurance rates in dollars per $100 value for $1,000 deductible. For $25,000 deductible, rates can be reduced 13% to 34%. On contracts over $1,000,000, rates may be lower than those tabulated. Policies are written annually for the total completed value in place. For "all risk" insurance (excluding flood, earthquake and certain other perils) add $.025 to total rates below.

Coverage	Frame Construction (Class 1)		Brick Construction (Class 4)		Fire Resistive (Class 6)	
	Range	Average	Range	Average	Range	Average
Fire Insurance	$.300 to .420	$.394	$.132 to .189	$.174	$.052 to .080	$.070
Extended Coverage	.115 to .150	.144	.080 to .105	.101	.081 to .105	.100
Vandalism	.012 to .016	.015	.008 to .011	.011	.008 to .011	.010
Total Annual Rate	$.427 to .586	$.553	$.220 to .305	$.286	$.141 to .196	$.180

④ General Contractor's Overhead (Div. 010)

The table below shows a contractor's overhead as a percentage of direct cost in two ways. The figures on the right are for the overhead, markup based on both material and labor. The figures on the left are based on the entire overhead applied only to the labor. This figure would be used if the owner supplied the materials or if a contract is for labor only.

Items of General Contractor's Indirect Costs	% of Direct Costs	
	As a Markup of Labor Only	As a Markup of Both Material and Labor
Field Supervision	6.0%	3.2%
Main Office Expense (see details below)	14.7	7.7
Tools and Minor Equipment	0.8	0.4
Workers' Compensation & Employers' Liability. See ⑦	15.1	7.9
Field Office, Sheds, Photos, Etc.	1.5	0.8
Performance and Payment Bond, 0.5% to 0.9%.	0.7	0.7
Unemployment Tax See ⑤ (Combined Federal and State)	6.2	3.3
Social Security and Medicare (7.65% of first $51,300)	7.7	4.0
Sales Tax — add if applicable 38/80 x % as markup of total direct costs including both material and labor. See ⑧		
Sub Total	52.7%	28.0%
*Builder's Risk Insurance ranges from 0.141% to 0.586%. See ②	0.3	0.3
*Public Liability Insurance	1.5	1.5
Grand Total	54.5%	29.8%

*Paid by Owner or Contractor

346

CIRCLE REFERENCE NUMBERS

④ Main Office Expense

A General Contractor's main office expense consists of many items not detailed in the front portion of the book. The percentage of main office expense declines with increased annual volume of the contractor. Typical main office expense ranges from 2% to 20% with the median about 7.2% of total volume. This equals about 7.7% of direct costs. The following are approximate percentages of total overhead for different items usually included in a General Contractor's main office overhead. With different accounting procedures, these percentages may vary.

Item	Typical Range	Average
Managers', clerical and estimators' salaries	40% to 55%	48%
Profit sharing, pension and bonus plans	2 to 20	12
Insurance	5 to 8	6
Estimating and project management (not including salaries)	5 to 9	7
Legal, accounting and data processing	0.5 to 5	3
Automobile and light truck expense	2 to 8	5
Depreciation of overhead capital expenditures	2 to 6	4
Maintenance of office equipment	0.1 to 1.5	1
Office rental	3 to 5	4
Utilities including phone and light	1 to 3	2
Miscellaneous	5 to 15	8
Total		100%

⑤ Unemployment Taxes and Social Security Taxes (Div. 010–086)

Mass. State Unemployment tax ranges from 1.8% to 6.0% plus an experience rating assessment the following year, on the first $7,000 of wages. Federal Unemployment tax is 5.4% of the first $7,000 of wages. This is reduced by a credit for payment to the state. The minimum Federal Unemployment tax is .8% after all credits.

Combined rates in Mass. thus vary from 2.6% to 6.8% of the first $7,000 of wages. Combined average U.S. rate is about 6.2% of the first $7,000. Contractors with permanent workers will pay less since the average annual wages for skilled workers is $22.65 x 2,000 hours or about $45,300 per year. The average combined rate for U.S. would thus be 6.2% x $7,000 ÷ $45,300 = .96% of total wages for permanent employees.

Rates not only vary from state to state but also with the experience rating of the contractor.

Social Security (FICA) for 1991 is estimated at time of publication to be 7.65% of wages up to $51,300.

⑥ Contractor's Overhead & Profit (Div. 010)

Listed below in the last two columns are **average** billing rates for the installing contractor's labor.

The Base Rates are averages for the building construction industry and include the usual negotiated fringe benefits. Workers' Compensation is a national average of state rates established for each trade. Average Fixed Overhead is a total of average rates for U.S. and State Unemployment, 6.2%; Social Security (FICA), 7.65%; Builders' Risk, 0.34%; and Public Liability, 1.55%. These are analyzed in ② and ⑤ . All the rates except

Social Security vary from state to state as well as from company to company. The installing contractor's overhead presumes annual billing of $500,000 and up. Overhead percentages may increase with smaller annual billing.

Overhead varies greatly within each trade. Some controlling factors are annual volume, job type, job size, location, local economic conditions, engineering and logistical support staff and equipment requirements. All factors should be examined carefully for each job.

Abbr.	Trade	Base Rate Incl. Fringes		Workers' Comp. Ins.	Average Fixed Overhead	Overhead	Profit	Total Overhead & Profit		Rate with O & P	
		Hourly	Daily					%	Amount	Hourly	Daily
Skwk	Skilled Workers Average (35 trades)	$22.65	$181.20	15.1%	15.7%	16.0%	15.00%	61.8%	$14.00	$36.65	$293.20
	Helpers Average (5 trades)	17.10	136.80	16.2				62.9	10.75	27.85	222.80
	Foremen Average, Inside (50¢ over trade)	23.15	185.20	15.1				61.8	14.30	37.45	299.60
	Foremen Average, Outside ($2.00 over trade)	24.65	197.20	15.1				61.8	15.20	39.85	318.80
Clab	Common Building Laborers	17.50	140.00	16.6				63.3	11.10	28.60	228.80
Asbe	Asbestos Workers	24.70	197.60	13.5				60.2	14.85	39.55	316.40
Boil	Boilermakers	25.05	200.40	9.3				56.0	14.05	39.10	312.80
Bric	Bricklayers	22.75	182.00	13.7				60.4	13.75	36.50	292.00
Brhe	Bricklayer Helpers	17.65	141.20	13.7				60.4	10.65	28.30	226.40
Carp	Carpenters	22.00	176.00	16.6				63.3	13.95	35.95	287.60
Cefi	Cement Finishers	21.65	173.20	9.6				56.3	12.20	33.85	270.80
Elec	Electricians	25.15	201.20	6.0				52.7	13.25	38.40	307.20
Elev	Elevator Constructors	25.35	202.80	7.7				54.4	13.80	39.15	313.20
Eqhv	Equipment Operators, Crane or Shovel	23.30	186.40	10.4				57.1	13.30	36.60	292.80
Eqmd	Equipment Operators, Medium Equipment	22.50	180.00	10.4				57.1	12.85	35.35	282.80
Eqlt	Equipment Operators, Light Equipment	21.40	171.20	10.4				57.1	12.20	33.60	268.80
Eqol	Equipment Operators, Oilers	19.20	153.60	10.4				57.1	10.95	30.15	241.20
Eqmm	Equipment Operators, Master Mechanics	24.00	192.00	10.4				57.1	13.70	37.70	301.60
Glaz	Glaziers	22.55	180.40	11.9				58.6	13.20	35.75	286.00
Lath	Lathers	21.95	175.60	10.2				56.9	12.50	34.45	275.60
Marb	Marble Setters	22.55	180.40	13.7				60.4	13.60	36.15	289.20
Mill	Millwrights	22.95	183.60	9.9				56.6	13.00	35.95	287.60
Mstz	Mosaic and Terrazzo Workers	22.10	176.80	8.3				55.0	12.15	34.25	274.00
Pord	Painters, Ordinary	20.80	166.40	12.4				59.1	12.30	33.10	264.80
Psst	Painters, Structural Steel	21.45	171.60	42.9				89.6	19.20	40.65	325.20
Pape	Paper Hangers	20.80	166.40	12.4				59.1	12.30	33.10	264.80
Pile	Pile Drivers	22.20	177.60	25.4				72.1	16.00	38.20	305.60
Plas	Plasterers	21.80	174.40	13.4				60.1	13.10	34.90	279.20
Plah	Plasterer Helpers	17.90	143.20	13.4				60.1	10.75	28.65	229.20
Plum	Plumbers	25.45	203.60	7.5				54.2	13.80	39.25	314.00
Rodm	Rodmen (Reinforcing)	23.90	191.20	27.6				74.3	17.75	41.65	333.20
Rofc	Roofers, Composition	20.35	162.80	28.9				75.6	15.40	35.75	286.00
Rots	Roofers, Tile & Slate	20.45	163.60	28.9				75.6	15.45	35.90	287.20
Rohe	Roofer Helpers (Composition)	14.90	119.20	28.9				75.6	11.25	26.15	209.20
Shee	Sheet Metal Workers	25.00	200.00	10.2				56.9	14.25	39.25	314.00
Spri	Sprinkler Installers	26.40	211.20	7.7				54.4	14.35	40.75	326.00
Stpi	Steamfitters or Pipefitters	25.50	204.00	7.5				54.2	13.80	39.30	314.40
Ston	Stone Masons	22.65	181.20	13.7				60.4	13.70	36.35	290.80
Sswk	Structural Steel Workers	24.10	192.80	35.4				82.1	19.80	43.90	351.20
Tilf	Tile Layers (Floor)	22.15	177.20	8.3				55.0	12.20	34.35	274.80
Tilh	Tile Layer Helpers	17.45	139.60	8.3				55.0	9.60	27.05	216.40
Trlt	Truck Drivers, Light	18.10	144.80	13.5				60.2	10.90	29.00	232.00
Trhv	Truck Drivers, Heavy	18.40	147.20	13.5				60.2	11.10	29.50	236.00
Sswl	Welders, Structural Steel	24.10	192.80	35.4				82.1	19.80	43.90	351.20
Wrck	*Wrecking	17.50	140.00	35.4				82.1	14.35	31.85	254.80

*Not included in Averages.

CIRCLE REFERENCE NUMBERS

(7) **Workers' Compensation** (Div. 010)

The table below tabulates the national averages for Workers' Compensation insurance rates by trade and type of building. The average "Insurance Rate" is multiplied by the "% of Building Cost" for each trade. This produces the "Workers' Compensation Cost" by % of total labor cost, to be added for each trade by building type to determine the weighted average Workers' Compensation rate for the building types analyzed.

Trade	Insurance Rate (% of Labor Cost)		% of Building Cost			Workers' Compensation Cost		
	Range	Average	Office Bldgs.	Schools & Apts.	Mfg.	Office Bldgs.	Schools & Apts.	Mfg.
Excavation, Grading, etc.	4.7% to 27.2%	10.4%	4.8%	4.9%	4.5%	.50%	.51%	.47%
Piles & Foundations	5.0 to 54.8	25.4	7.1	5.2	8.7	1.80	1.32	2.21
Concrete	5.0 to 44.3	15.3	5.0	14.8	3.7	.77	2.26	.57
Masonry	4.0 to 44.2	13.7	6.9	7.5	1.9	.95	1.03	.26
Structural Steel	5.0 to 162.5	35.4	10.7	3.9	17.6	3.79	1.38	6.23
Miscellaneous & Ornamental Metals	3.9 to 22.4	10.8	2.8	4.0	3.6	.30	.43	.39
Carpentry & Millwork	5.0 to 43.4	16.6	3.7	4.0	0.5	.61	.66	.08
Metal or Composition Siding	5.0 to 34.2	13.7	2.3	0.3	4.3	.32	.04	.59
Roofing	5.0 to 62.5	28.9	2.3	2.6	3.1	.66	.75	.90
Doors & Hardware	4.7 to 20.9	9.6	0.9	1.4	0.4	.09	.13	.04
Sash & Glazing	4.1 to 23.4	11.9	3.5	4.0	1.0	.42	.48	.12
Lath & Plaster	5.0 to 38.5	13.4	3.3	6.9	0.8	.44	.92	.11
Tile, Marble & Floors	3.2 to 23.1	8.3	2.6	3.0	0.5	.22	.25	.04
Acoustical Ceilings	3.7 to 26.9	10.2	2.4	0.2	0.3	.24	.02	.03
Painting	4.6 to 44.2	12.4	1.5	1.6	1.6	.19	.20	.20
Interior Partitions	5.0 to 43.4	16.6	3.9	4.3	4.4	.65	.71	.73
Miscellaneous Items	2.2 to 139.7	15.1	5.2	3.7	9.7	.79	.56	1.46
Elevators	2.5 to 16.2	7.7	2.1	1.1	2.2	.16	.08	.17
Sprinklers	2.2 to 15.1	7.7	0.5	—	2.0	.04	—	.15
Plumbing	2.7 to 16.0	7.5	4.9	7.2	5.2	.37	.54	.39
Heat., Vent., Air Conditioning	4.1 to 25.5	10.2	13.5	11.0	12.9	1.38	1.12	1.32
Electrical	2.4 to 12.9	6.0	10.1	8.4	11.1	.61	.50	.67
Total	2.2% to 162.5%	—	100.0%	100.0%	100.0%	15.30%	13.89%	17.13%
Overall Weighted Average							15.44%	

The table below lists the weighted average Workers' Compensation base rate for each state with a factor comparing this with the national average of 15.1%.

State	Weighted Average	Factor	State	Weighted Average	Factor	State	Weighted Average	Factor
Alabama	13.2%	87	Kentucky	13.2%	87	North Dakota	13.5%	89
Alaska	22.2	147	Louisiana	12.8	85	Ohio	13.4	89
Arizona	15.8	105	Maine	24.1	160	Oklahoma	12.8	85
Arkansas	11.7	77	Maryland	15.4	102	Oregon	27.9	185
California	16.5	109	Massachusetts	25.3	168	Pennsylvania	15.2	101
Colorado	22.1	146	Michigan	15.1	100	Rhode Island	19.2	127
Connecticut	23.5	156	Minnesota	25.2	167	South Carolina	10.6	70
Delaware	12.1	80	Mississippi	10.0	66	South Dakota	10.3	68
District of Columbia	21.0	139	Missouri	8.2	54	Tennessee	10.5	70
Florida	30.4	201	Montana	37.5	248	Texas	19.2	127
Georgia	13.9	92	Nebraska	9.5	63	Utah	9.7	64
Hawaii	17.6	117	Nevada	16.2	107	Vermont	10.7	71
Idaho	12.8	85	New Hampshire	19.0	126	Virginia	9.1	60
Illinois	22.6	150	New Jersey	7.9	52	Washington	13.3	88
Indiana	6.5	43	New Mexico	18.5	123	West Virginia	10.1	67
Iowa	13.1	87	New York	11.7	77	Wisconsin	13.9	92
Kansas	9.5	63	North Carolina	7.3	48	Wyoming	5.5	36
Weighted Average for U.S. is 15.4% of payroll = 100								

Rates in the following table are the base or manual costs per $100 of payroll for Workers' Compensation in each state. Rates are usually applied to straight time wages only and not to premium time wages and bonuses.

The weighted average skilled worker rate for 35 trades is 15.1%. For bidding purposes, apply the full value of Workers' Compensation directly to total labor costs, or if labor is 32%, materials 48% and overhead and profit 20% of total cost, carry 32/80 x 15.1% = 6.0% of cost (before overhead and profit) into overhead. Rates vary not only from state to state but also with the experience rating of the contractor.

Rates are the most current available at the time of publication.

⑦ Workers' Compensation (cont.)

STATE	CARPENTRY — 3 stories or less	CARPENTRY — interior cab. work	CARPENTRY — general	CONCRETE WORK—NOC	CONCRETE WORK — flat (flr. sdwk.)	ELECTRICAL WIRING — inside	EXCAVATION — earth NOC	EXCAVATION — rock	GLAZIERS	INSULATION WORK	LATHING	MASONRY	PAINTING & DECORATING	PILE DRIVING	PLASTERING	PLUMBING	ROOFING	SHEET METAL WORK (HVAC)	STEEL ERECTION — door & sash	STEEL ERECTION — inter. ornam.	STEEL ERECTION — structure	STEEL ERECTION — NOC	TILE WORK — (interior ceramic)	WATERPROOFING	WRECKING
	5651	5437	5403	5213	5221	5190	6217	6217	5462	5479	5443	5022	5474	6003	5480	5183	5551	5538	5102	5102	5040	5057	5348	9014	5701
AL	14.97	8.55	14.18	11.06	7.25	5.86	8.68	8.68	11.20	11.17	8.35	10.59	11.54	32.21	9.38	6.53	25.13	13.57	7.78	7.78	23.21	20.90	8.16	3.60	20.90
AK	16.10	11.21	14.48	18.19	10.46	9.07	12.87	12.87	18.59	19.83	13.80	12.49	10.31	41.05	21.55	12.61	29.27	14.16	19.77	19.77	74.36	56.94	10.21	6.51	74.36
AZ	19.74	7.60	23.31	12.41	10.23	9.30	7.07	7.07	13.56	17.90	8.95	16.91	11.59	27.00	19.08	6.60	24.77	10.19	12.74	12.74	35.69	17.97	6.70	7.24	17.97
AR	11.85	6.30	15.05	12.20	6.23	4.71	8.50	8.50	8.80	8.61	8.14	10.02	10.68	16.64	10.01	4.49	13.70	7.62	6.18	6.18	44.38	16.97	5.54	5.34	44.38
CA	21.96	8.01	21.96	9.33	9.33	7.82	7.55	7.55	12.94	22.76	8.13	14.05	14.90	19.03	15.92	10.36	36.32	13.11	12.18	12.18	28.27	24.76	7.68	14.90	28.35
CO	22.81	13.43	19.38	20.39	14.20	6.56	13.75	13.75	11.58	20.41	10.43	26.22	17.00	36.16	38.54	11.54	48.70	12.88	12.23	12.23	52.23	30.77	10.81	9.24	30.77
CT	18.58	20.34	25.91	26.78	15.09	8.87	12.75	12.75	21.83	25.42	15.47	28.35	17.77	42.59	21.19	10.09	51.57	15.49	16.81	16.81	50.82	26.21	9.81	4.61	50.82
DE	11.44	13.73	11.44	9.20	6.25	5.71	8.49	8.49	11.11	11.44	10.45	9.77	13.30	11.50	10.45	5.32	22.88	10.11	10.09	10.09	26.31	10.09	7.46	9.77	25.47
DC	10.32	9.62	14.81	27.57	12.88	11.76	14.68	14.68	14.11	13.79	10.72	21.41	11.84	37.37	12.32	14.82	32.92	11.25	20.59	20.59	47.36	46.44	23.24	5.77	47.36
FL	25.21	18.82	30.26	44.32	18.17	12.89	20.27	20.27	23.43	25.66	26.91	26.88	30.18	46.97	31.60	16.01	55.77	19.12	22.40	22.40	48.03	58.10	12.70	10.32	48.03
GA	15.40	10.28	17.30	11.49	9.47	5.67	11.86	11.86	9.06	12.63	13.51	11.69	10.65	27.66	11.35	6.07	24.90	9.60	8.21	8.21	18.44	26.83	5.90	7.36	26.83
HI	11.66	9.32	39.81	13.93	9.03	8.89	10.81	10.81	14.08	20.18	9.51	17.35	7.68	28.43	19.16	5.36	40.64	8.32	14.10	14.10	27.94	26.82	7.72	10.50	27.94
ID	13.17	7.20	18.71	12.78	5.84	4.67	10.20	10.20	11.13	13.14	7.93	16.25	12.51	20.62	10.67	4.53	27.80	8.95	7.45	7.45	18.09	16.83	6.59	7.26	18.09
IL	16.54	11.14	17.49	26.51	12.88	8.75	10.19	10.19	21.11	17.72	13.22	18.01	15.60	31.52	13.46	12.20	31.86	15.94	13.72	13.72	61.21	84.88	12.92	5.22	84.88
IN	9.00	4.97	6.19	6.08	3.94	2.64	5.07	5.07	6.83	5.21	3.67	5.54	4.65	13.11	5.37	2.67	10.84	3.97	3.90	3.90	8.73	14.57	3.58	3.14	14.57
IA	9.23	7.01	10.64	17.71	4.89	5.38	9.85	9.85	8.25	12.47	6.76	10.58	8.30	17.54	8.92	7.78	18.49	8.41	9.80	9.80	38.15	37.61	6.25	4.27	37.61
KS	10.35	5.61	7.94	9.82	5.58	3.66	4.70	4.70	6.25	15.84	7.81	9.75	6.57	15.96	7.75	4.59	19.70	6.82	4.74	4.74	11.50	23.34	5.08	4.75	23.34
KY	14.27	6.19	14.56	11.42	7.61	4.67	7.40	7.40	8.65	10.16	10.07	13.10	12.37	28.02	9.53	5.21	23.87	9.76	9.85	9.85	30.22	23.44	7.65	4.39	30.22
LA	16.33	10.25	16.12	8.59	7.84	5.59	9.61	9.61	12.34	8.20	7.25	7.74	13.24	37.62	8.86	6.29	20.21	7.76	7.91	7.91	24.13	13.43	6.64	6.06	24.13
ME	12.13	10.90	38.36	24.84	10.99	8.38	16.53	16.53	18.35	19.96	13.34	19.36	19.26	43.24	19.49	12.27	44.12	13.43	17.46	17.46	52.06	57.53	13.46	7.99	52.06
MD	12.59	12.33	10.70	17.28	9.77	7.64	12.85	12.85	18.78	12.60	8.29	12.40	9.07	19.45	10.26	9.40	32.69	13.58	11.27	11.27	26.18	32.54	9.59	4.30	37.63
MA	14.16	12.04	30.31	33.73	15.36	6.21	9.34	9.34	21.21	16.86	15.19	23.65	14.05	26.22	15.99	9.21	79.41	13.85	16.23	16.23	83.51	48.66	13.21	9.81	65.27
MI	11.40	7.35	12.74	12.71	9.27	5.75	12.42	12.42	14.56	15.48	9.32	16.14	13.24	26.48	14.23	7.23	26.63	8.70	11.42	11.42	29.17	20.06	8.13	NA	29.17
MN	20.86	20.86	36.23	21.03	14.74	6.95	21.20	21.20	17.38	22.76	22.29	17.84	17.92	35.90	22.29	11.77	45.20	14.86	16.41	16.41	51.17	54.54	12.08	9.34	51.17
MS	10.19	6.87	12.79	8.41	4.70	4.93	8.18	8.18	8.45	6.92	7.85	6.29	6.64	22.56	9.94	3.36	15.30	7.80	6.98	6.98	22.27	13.33	6.72	3.86	22.27
MO	7.88	5.54	7.19	6.30	5.92	3.52	6.07	6.07	5.38	9.22	5.76	7.22	4.95	16.62	6.26	3.33	17.56	4.75	5.91	5.91	15.91	13.15	4.58	3.66	13.15
MT	23.27	13.56	43.40	38.99	21.88	8.30	27.19	27.19	19.61	25.49	23.34	44.23	44.16	54.83	28.08	15.32	62.45	16.29	17.94	17.94	162.26	55.63	14.05	16.25	162.26
NE	8.22	5.45	8.84	9.53	7.35	3.72	7.11	7.11	6.84	9.78	6.05	8.01	7.17	12.60	7.58	4.53	18.79	7.89	5.42	5.42	13.54	27.21	4.76	4.92	27.21
NV	14.25	14.25	14.25	11.31	11.31	7.21	10.61	10.61	11.46	13.98	13.68	11.93	12.97	10.90	13.68	8.97	23.39	25.48	13.01	13.01	33.81	33.81	9.98	10.90	N.A.
NH	15.51	9.63	18.51	26.18	13.21	5.16	12.90	12.90	10.28	14.57	11.62	14.47	13.26	52.73	18.31	9.06	60.55	11.19	11.94	11.94	31.55	16.57	8.96	6.26	31.55
NJ	6.83	5.66	6.83	5.78	4.90	2.43	6.26	6.26	4.91	6.54	6.43	7.98	7.58	10.12	6.43	3.18	15.67	4.20	7.66	7.66	22.64	10.49	3.16	3.51	27.74
NM	14.96	11.73	19.31	24.57	9.63	5.95	10.09	10.09	16.15	14.70	11.32	16.70	11.44	49.12	16.45	10.09	37.80	12.36	18.46	18.46	21.61	25.84	10.07	7.75	21.61
NY	9.63	4.67	9.88	12.46	9.73	5.00	10.17	10.17	8.89	8.52	8.23	12.36	9.56	17.49	9.18	7.53	22.73	9.75	7.68	7.68	17.83	24.06	7.13	5.44	19.97
NC	6.06	5.70	9.29	8.46	3.63	4.32	5.94	5.94	5.09	6.14	4.15	4.01	5.05	11.79	8.46	4.46	12.70	5.31	4.73	4.73	19.37	8.00	3.27	2.59	19.37
ND	12.86	12.86	12.86	9.98	9.98	4.54	8.28	8.28	14.60	7.64	5.87	6.34	9.54	27.52	5.87	10.27	15.98	10.27	12.86	12.86	27.52	27.52	5.69	15.98	NA
OH	9.93	9.93	9.93	10.32	10.32	3.87	10.32	10.32	11.36	8.83	8.83	10.48	11.36	10.32	8.83	5.69	20.00	12.71	N.A.	N.A.	58.93	N.A.	5.03	10.32	10.32
OK	14.20	7.38	11.62	10.27	8.38	3.78	9.39	9.39	8.02	9.81	7.80	9.85	8.30	24.54	10.73	5.06	20.50	8.02	7.74	7.74	37.18	29.25	5.53	6.60	37.18
OR	34.23	14.01	33.37	26.01	19.95	8.49	20.99	20.99	15.10	28.80	13.90	24.25	24.82	53.64	26.62	11.10	60.89	14.90	16.80	16.80	48.90	36.45	23.07	17.85	48.90
PA	12.32	11.41	12.32	18.08	8.25	5.50	9.57	9.57	10.58	12.32	13.37	12.19	12.85	17.01	13.37	7.30	27.01	9.87	16.25	16.25	41.51	16.25	8.23	12.19	64.42
RI	15.52	7.13	12.32	14.83	16.49	6.22	12.89	12.89	18.05	15.23	11.26	14.94	17.87	30.71	15.05	4.69	31.27	6.80	10.13	10.13	78.01	40.30	9.28	8.00	78.01
SC	14.28	7.39	15.32	7.64	6.55	6.30	7.43	7.43	11.77	11.46	6.85	11.01	9.36	14.13	9.58	3.44	17.21	8.84	4.23	4.23	11.28	24.72	8.80	4.44	11.28
SD	8.52	5.88	13.80	10.26	5.60	4.22	7.72	7.72	8.18	11.76	6.24	7.27	8.01	17.08	8.45	7.03	27.14	6.16	6.92	6.92	17.97	13.96	5.02	3.91	17.97
TN	11.89	6.11	11.70	9.84	5.61	4.17	7.59	7.59	7.91	9.44	6.15	8.76	8.99	21.90	7.96	5.32	18.61	8.59	8.03	8.03	20.93	15.57	4.73	4.33	20.93
TX	22.09	15.58	22.09	18.49	14.60	9.81	14.91	14.91	13.54	20.74	9.35	16.27	14.06	30.47	14.57	12.00	38.41	16.85	11.50	11.50	39.93	21.52	8.82	7.96	38.04
UT	NA	NA	8.79	9.82	4.93	5.05	6.25	6.25	6.84	6.73	8.06	13.07	9.18	15.75	8.37	5.67	20.31	5.72	5.90	5.90	NA	24.80	4.77	3.07	24.80
VT	7.62	5.25	12.12	12.62	5.63	3.63	6.85	6.85	8.04	8.22	7.31	12.12	8.01	19.05	9.67	5.05	19.24	6.31	6.66	6.66	20.53	26.40	5.90	5.02	20.53
VA	7.28	5.90	9.08	9.23	6.29	3.45	5.52	5.52	8.03	8.84	6.87	7.69	8.54	9.24	5.64	4.72	25.24	7.11	5.10	5.10	21.59	15.51	3.78	3.34	21.59
WA	12.37	12.37	12.37	10.52	8.78	3.12	8.18	8.18	13.15	10.10	12.37	12.99	10.69	21.35	13.05	4.47	11.99	5.92	10.52	10.52	29.26	29.26	7.92	11.94	14.80
WV	10.80	10.80	10.80	14.90	14.90	4.00	8.48	8.48	4.13	4.13	15.81	7.03	15.81	8.40	15.81	3.52	12.91	4.13	9.94	9.93	7.96	9.93	7.03	2.73	7.96
WI	10.26	7.57	17.21	11.12	6.31	5.41	8.58	8.58	10.62	12.79	8.28	11.77	11.34	28.37	12.80	7.34	28.44	8.70	9.19	9.19	29.73	24.92	10.03	5.38	63.36
WY	5.00	5.00	5.00	5.00	5.00	5.00	5.00	5.00	5.00	5.00	5.00	5.00	5.00	5.00	5.00	5.00	5.00	5.00	5.00	5.00	5.00	5.00	5.00	5.00	5.00
AVG.	13.72	9.61	16.64	15.29	9.55	5.97	10.37	10.37	11.90	13.48	10.22	13.71	12.36	25.40	13.39	7.45	28.91	10.24	10.79	10.79	35.36	27.59	8.28	7.09	35.54

⑦ Workers' Compensation (cont.) (Canada in Canadian dollars)

PROVINCE		Alberta	British Columbia	Manitoba	Ontario	New Brunswick	Newfndld. & Labrador	Northwest Territories	Nova Scotia	Prince Edward Island	Quebec	Saskatchewan	Yukon
CARPENTRY—3 stories or less	Rate	5.33	2.99	5.65	3.98	3.66	4.75	6.50	2.90	5.65	8.21	5.25	2.00
	Code	8-04	060412	401	062-08	403	403	4-41	4013	401	40010	B12-02	4-042
CARPENTRY—interior cab. work	Rate	5.33	2.99	5.65	3.98	3.66	4.75	6.50	2.90	5.65	8.21	4.50	2.00
	Code	8-04	060412	401	062-08	403	403	4-41	4013	401	40010	B11-25	4-042
CARPENTRY—general	Rate	5.33	2.99	5.65	3.98	3.66	4.75	6.50	2.90	5.65	8.21	5.25	2.00
	Code	8-04	060412	401	062-08	403	403	4-41	4013	401	40010	B12-02	4-042
CONCRETE WORK—NOC	Rate	6.48	5.36	5.65	8.23	3.66	4.75	6.50	2.90	5.65	11.36	8.65	2.50
	Code	6-01	070604	401	744-09	403	403	4-41	4222	401	40080	B14-04	2-032
CONCRETE WORK—flat (flr., sidewalk)	Rate	6.48	5.36	5.65	8.23	3.66	4.75	6.50	2.90	5.65	11.36	8.65	2.50
	Code	6-01	070604	401	744-09	403	403	4-41	4222	401	40080	B14-04	2-032
ELECTRICAL Wiring—inside	Rate	2.70	2.46	2.97	5.99	3.66	2.75	4.50	1.33	3.05	5.75	4.50	2.00
	Code	6-06	071100	402	864-07	403	400	4-46	4261	402	40150	B11-05	4-041
EXCAVATION—earth NOC	Rate	4.88	3.63	7.11	12.25	3.66	4.75	7.00	2.73	5.65	8.14	5.90	2.50
	Code	6-07	072607	407	753-13	403	403	4-43	4214	401	40021	R11-06	2-016
EXCAVATION—rock	Rate	4.88	3.63	7.11	12.25	3.66	4.75	7.00	2.73	5.65	8.14	5.90	2.50
	Code	6-07	072607	407	753-13	403	403	4-43	4214	401	40021	R11-06	2-016
GLAZIERS	Rate	3.65	1.52	5.65	8.64	3.66	2.60	6.50	2.90	3.05	6.83	7.35	2.00
	Code	6-03	060236	401	873-11	403	402	4-41	4233	402	40110	B13-04	4-042
INSULATION WORK	Rate	6.64	6.39	5.65	8.64	3.66	2.60	6.50	2.90	5.65	9.03	5.25	2.50
	Code	6-03	070504	401	873-11	403	402	4-41	4234	401	40170	B12-07	2-035
LATHING	Rate	6.53	6.39	5.65	9.81	3.66	2.60	6.50	2.90	3.05	9.03	7.35	2.50
	Code	6-03	070500	401	854-12	403	402	4-41	4271	402	40170	B13-02	2-036
MASONRY	Rate	7.56	5.36	5.65	9.81	3.66	4.75	6.50	2.90	5.65	11.36	11.50	2.50
	Code	6-04	070602	401	854-12	403	403	4-41	4231	401	40080	B15-01	2-032
PAINTING & DECORATING	Rate	4.91	6.39	5.65	8.64	3.66	2.60	4.50	2.90	3.05	9.03	5.25	2.50
	Code	6-03	070501	401	873-11	403	402	4-49	4275	402	40170	B12-01	2-036
PILE DRIVING	Rate	6.48	15.33	7.11	10.69	3.66	5.50	7.00	3.03	5.65	8.14	5.25	2.50
	Code	6-01	072502	407	836-13	403	404	4-43	4221	401	40021	B12-10	2-030
PLASTERING	Rate	6.53	6.39	5.65	9.68	3.66	2.60	6.50	2.90	3.05	9.03	7.35	2.50
	Code	6-03	070502	401	854-12	403	402	4-41	4271	402	40170	B13-02	2-036
PLUMBING	Rate	3.16	2.99	2.97	5.99	3.66	3.75	4.50	1.97	3.05	6.95	4.50	2.00
	Code	6-02	070712	402	864-07	403	401	4-46	4241	402	40130	B11-01	4-039
ROOFING	Rate	11.07	5.36	7.94	9.68	3.66	4.75	6.50	2.90	5.65	12.52	11.50	2.50
	Code	6-05	070600	404	854-12	403	403	4-41	4235	401	40121	B15-02	2-031
SHEET METAL WORK (HVAC)	Rate	2.20	2.99	7.94	5.99	3.66	3.75	4.50	2.90	3.05	6.95	4.50	2.00
	Code	6-02	070714	404	864-07	403	401	4-46	4236	402	40130	B11-07	4-040
STEEL ERECTION—door & sash	Rate	4.12	15.33	9.51	10.10	3.66	5.50	6.50	2.90	5.65	16.57	11.50	2.50
	Code	8-03	072509	405	827-09	403	404	4-41	4223	401	40100	B15-03	2-012
STEEL ERECTION—inter., ornam.	Rate	4.48	15.33	9.51	10.10	3.66	4.75	6.50	2.90	5.65	16.57	11.50	2.50
	Code	6-01	072509	405	827-09	403	403	4-41	4223	401	40100	B15-03	2-012
STEEL ERECTION—structure	Rate	11.40	15.33	9.51	23.26	3.66	5.50	4.25	5.48	5.65	16.57	11.50	2.50
	Code	6-08	072509	405	809-14	403	404	4-44	4227	401	40100	B15-04	2-012
STEEL ERECTION—NOC	Rate	11.40	15.33	9.51	10.10	3.66	5.50	6.50	5.48	5.65	16.57	11.50	2.50
	Code	6-08	072509	405	827-09	403	404	4-41	4227	401	40100	B15-03	2-012
TILE WORK—inter. (ceramic)	Rate	6.19	6.39	5.65	5.99	3.66	2.60	4.50	2.90	3.05	9.03	7.35	2.50
	Code	6-03	070506	401	864-07	403	402	4-49	4276	402	40170	B13-01	2-034
WATERPROOFING	Rate	5.39	1.52	7.11	8.64	3.66	4.75	6.50	2.90	3.05	11.38	4.50	2.50
	Code	6-02	060237	407	873-11	403	403	4-41	4239	402	40122	B11-17	2-030
WRECKING	Rate	17.18	5.36	5.65	23.28	3.66	4.75	7.00	2.90	5.65	11.36	8.65	2.50
	Code	6-08	070600	401	859-15	403	403	4-43	4211	401	40080	B14-07	2-030

CIRCLE REFERENCE NUMBERS

⑧ Sales Tax (Div. 010-086)

State sales tax on materials is tabulated below (5 states have no sales tax). Many states allow local jurisdictions, such as a county or city, to levy additional sales tax.

Some projects may be sales tax exempt, particularly those constructed with public funds.

State	Tax	State	Tax	State	Tax	State	Tax
Alabama	4%	Illinois	5%	Montana	0%	Rhode Island	6%
Alaska	0	Indiana	5	Nebraska	4	South Carolina	5
Arizona	5	Iowa	4	Nevada	3.5	South Dakota	4
Arkansas	4	Kansas	4.25	New Hampshire	0	Tennessee	5.5
California	6	Kentucky	5	New Jersey	6	Texas	6
Colorado	3	Louisiana	4	New Mexico	4.75	Utah	6
Connecticut	7.5	Maine	5	New York	4	Vermont	4
Delaware	0	Maryland	5	North Carolina	5	Virginia	4.5
District of Columbia	6	Massachusetts	5	North Dakota	6	Washington	6.5
Florida	6	Michigan	4	Ohio	5.5	West Virginia	6
Georgia	3	Minnesota	6	Oklahoma	4	Wisconsin	5
Hawaii	4	Mississippi	6	Oregon	0	Wyoming	3
Idaho	5	Missouri	4.225	Pennsylvania	6	Average	4.44%

⑩ Architectural Fees (Div. 010-004)

Tabulated below are typical percentage fees by project size, for good professional architectural service. Fees may vary from those listed depending upon degree of design difficulty and economic conditions in any particular area.

Rates can be interpolated horizontally and vertically. Various portions of the same project requiring different rates should be

adjusted proportionately. For alterations, add 50% to the fee for the first $500,000 of project cost and add 25% to the fee for project cost over $500,000.

Architectural fees tabulated below include Engineering Fees.

Building Type	Total Project Size in Thousands of Dollars						
	100	250	500	1,000	5,000	10,000	50,000
Factories, garages, warehouses repetitive housing	9.0%	8.0%	7.0%	6.2%	5.3%	4.9%	4.5%
Apartments, banks, schools, libraries, offices, municipal buildings	11.7	10.8	8.5	7.3	6.4	6.0	5.6
Churches, hospitals, homes, laboratories, museums, research	14.0	12.8	11.9	10.9	8.5	7.8	7.2
Memorials, monumental work, decorative furnishings	—	16.0	14.5	13.1	10.0	9.0	8.3

CIRCLE REFERENCE NUMBERS

⑪ Engineering Fees (Div. 010-028)

Typical **Structural Engineering Fees** based on type of construction and total project size. These fees are included in Architectural Fees.

Type of Construction	Total Project Size (in thousands of dollars)			
	$500	$500-$1,000	$1,000-$5,000	Over $5,000
Industrial buildings, factories & warehouses	Technical payroll times 2.0 to 2.5	1.60%	1.25%	1.00%
Hotels, apartments, offices, dormitories, hospitals, public buildings, food stores		2.00%	1.70%	1.20%
Museums, banks, churches and cathedrals		2.00%	1.75%	1.25%
Thin shells, prestressed concrete, earthquake resistive		2.00%	1.75%	1.50%
Parking ramps, auditoriums, stadiums, convention halls, hangars & boiler houses		2.50%	2.00%	1.75%
Special buildings, major alterations, underpinning & future expansion		Add to above 0.5%	Add to above 0.5%	Add to above 0.5%

For complex reinforced concrete or unusually complicated structures, add 20% to 50%.

Typical **Mechanical and Electrical Engineering Fees** based on the size of the subcontract. These fees are included in Architectural Fees.

Type of Construction	Subcontract Size							
	$25,000	$50,000	$100,000	$225,000	$350,000	$500,000	$750,000	$1,000,000
Simple structures	6.4%	5.7%	4.8%	4.5%	4.4%	4.3%	4.2%	4.1%
Intermediate structures	8.0	7.3	6.5	5.6	5.1	5.0	4.9	4.8
Complex structures	12.0	9.0	9.0	8.0	7.5	7.5	7.0	7.0

For renovations, add 15% to 25% to applicable fee.

(13) Contractor Equipment (Div. 016)

Rental Rates shown in the front of the book pertain to late model high quality machines in excellent working condition, rented from equipment dealers. Rental rates from contractors may be substantially lower than the rental rates from equipment dealers depending upon economic conditions. For older, less productive machines, reduce rates by a maximum of 15%. Any overtime must be added to the base rates. For shift work, rates are lower. Usual rule of thumb is 150% of one shift rate for two shifts; 200% for three shifts.

For periods of less than one week, operated equipment is usually more economical to rent than renting bare equipment and hiring an operator.

Equipment moving and mobilization costs must be added to rental rates where applicable. A large crane, for instance, may take two days to erect and two days to dismantle.

Rental rates vary throughout the country with larger cities generally having lower rates. Lease plans for new equipment are available for periods in excess of six months with a percentage of payments applying toward purchase.

Monthly rental rates vary from 2% to 5% of the cost of the equipment depending on the anticipated life of the equipment and its wearing parts. Weekly rates are about 1/3 the monthly rates and daily rental rates about 1/3 the weekly rate.

The hourly operating costs for each piece of equipment include costs to the user such as fuel, oil, lubrication, normal expendables for the equipment, and a percentage of mechanic's wages chargeable to maintenance. The hourly operating costs listed do not include the operator's wages.

The daily cost for equipment used in the standard crews (foreword) is figured by dividing the weekly rate by five, then adding eight times the hourly operating cost to give the total daily equipment cost, not including the operator. This figure is in the right hand column of Division 016 under Crew Equip. Cost.

Pile Driving rates shown for pile hammer and extractor do not include leads, crane, boiler or compressor. Vibratory pile driving requires an added field specialist at $310 per day during set-up and pile driving operation for the electric model. The hydraulic model requires a field specialist for set-up only. Up to 125 reuses of sheet piling are possible using vibratory drivers. For normal conditions, crane capacity for hammer type and size are as follows.

Crane Capacity	Hammer Type and Size		
	Air or Steam	Diesel	Vibratory
25 ton	to 8,750 ft.-lb.		70 H.P.
40 ton	15,000 ft.-lb.	to 32,000 ft.-lb.	170 H.P.
60 ton	25,000 ft.-lb.		300 H.P.
100 ton		112,000 ft.-lb.	

Cranes should be specified for the job by size, building and site characteristics, availability, performance characteristics, and duration of time required.

Backhoes & Shovels rent for about the same as equivalent size cranes but maintenance and operating expense is higher. Crane operators rate must be adjusted for high boom heights. Average adjustments: for 150' boom add 55¢ per hour; over 185', add $1.05 per hour; over 210', add $1.30 per hour; over 250', add $2.00 per hour and over 295', add $2.80 per hour.

Tower Cranes

Capacity in Kip-Feet	Typical Jib Length in Feet	Speed at Maximum Reach and Load	Purchase Price (New)		Monthly Rental, to 6 mo.	
			Crane & 80' Mast	Mast Sections	Crane & 80' Mast	Mast Sections
725	100	350 FPM	$204,000	$505/L.F.	$ 6,275	$14.30/L.F.
900	100	500	245,000	585	5,800	15.30
*1100	130	1000	335,000	670	8,425	19.40
1450	150	1000	467,000	935	12,100	23
2150	200	1000	589,000	1,150	15,200	27
3000	200	1000	827,000	1,225	19,900	32

*Most widely used.

Tower Cranes of the climbing or static type have jibs from 50' to 200' and capacities at maximum reach range from 4,000 to 14,000 pounds. Lifting capacities increase up to maximum load as the hook radius decreases.

Typical rental rates, based on purchase price are about 2% to 3% per month.

Erection and dismantling runs between $12,000 and $71,000. Climbing operation takes three men three hours per 20' climb. Crane dead time is about five hours per 40' climb. If crane is bolted to side of the building add cost of ties and extra mast sections. Mast sections cost $510 to $1,225 per vertical foot or can be rented at 2% to 3% of purchase price per month. Contractors using climbers claim savings of $1.50 per C.Y. of concrete placed, plus 12¢ per S.F. of formwork. Climbing cranes have from 80' to 180' of mast while static cranes have 80' to 800' of mast.

Truck Cranes can be converted to tower cranes by using tower attachments. Mast heights over 400' have been used. See Division 016-460 for rental rates of high boom cranes.

A single 100' high material **Hoist and Tower** can be erected and dismantled for about $13,250; a double 100' high hoist and tower for about $19,400. Erection costs for additional heights are $92 and $112 per vertical foot respectively up to 150' and $92 to $148 per vertical foot over 150' high. A 40' high portable Buck hoist costs about $4,700 to erect and dismantle. Additional heights run $72 per vertical foot to 80' and $100 per vertical foot for the next 100'. Most material hoists do not meet local code requirements for carrying personnel.

A 150' high **Personnel Hoist** requires about 500 to 800 man-hours to erect and dismantle with costs ranging from $11,625 to $25,500. Budget erection cost is $130 per vertical foot for all trades. Local code requirements or labor scarcity requiring overtime can add up to 50% to any of the above erection costs.

Earthmoving Equipment: The selection of earthmoving equipment depends upon the type and quantity of material, moisture content, haul distance, haul road, time available, and equipment available. Short haul cut and fill operations may require dozers only, while another operation may require excavators, a fleet of trucks, and spreading and compaction equipment. Stockpiled material and granular material are easily excavated with front end loaders. Scrapers are most economically used with hauls between 300' and 1-1/2 miles if adequate haul roads can be maintained. Shovels are often used for blasted rock and any material where a vertical face of 8' or more can be excavated. Special conditions may dictate the use of draglines, clamshells, or backhoes. Spreading and compaction equipment must be matched to the soil characteristics, the compaction required and the rate the fill is being supplied.

(14) Steel Tubular Scaffolding (Div. 015-254)

On new construction, tubular scaffolding is efficient up to 60' high or five stories. Above this it is usually better to use a hung scaffolding if construction permits.

In repairing or cleaning the front of an existing building the cost of tubular scaffolding per S.F. of building front increases as the height increases above the first tier. The first tier cost is relatively high due to leveling and alignment. Swing scaffolding operations may interfere with tenants. In this case the tubular is more practical at all heights.

The minimum efficient crew for erection is three men. For heights over 50', a four-man crew is more efficient. Use two or more on top and two at the bottom for handing up or hoisting. Four men can erect and dismantle about nine frames per hour up to five stories. From five to eight stories they will average six frames per hour. With 7' horizontal spacing this will run about 300 S.F. and 200 S.F. of wall surface, respectively. Time for placing planks must be added to the above. On heights above 50', five planks can be placed per man-hour.

The cost per 1,000 S.F. of building front in the table below was developed by pricing the materials required for a typical tubular scaffolding system eleven frames long and two frames high. Planks were figured five wide for standing plus two wide for materials.

Frames are 2', 4' and 5' wide and usually spaced 7' O.C. horizontally. Sidewalk frames are 6' wide. Rental rates will be lower for jobs over three months duration.

For jobs under twenty-five frames, figure rental at $6.00 per frame. For jobs over one hundred frames, rental can go as low as $2.65 per frame. These figures do not include accessories which are listed separately below. Large quantities for long periods can reduce rental rates by 20%.

Item	Unit	Purchase, Each		Monthly Rent, Each		Per 1,000 S.F. of Building Front	
		Regular	Heavy Duty	Regular	Heavy Duty	No. of Frames	Rental per Mo.
5' Wide Frames, 3' High	Ea.	$55	$ —	$3.65	$ —	—	—
*5'-0" High		70	—	3.65	—	—	—
*6'-6" High		85	—	3.65	—	24	$ 87.60
2' & 4' Wide, 5' High		—	75	—	3.75	—	—
6'-0" High		—	85	—	3.75	—	—
6' Wide Frame, 7'-6" High		130	155	7.95	10	—	—
Sidewalk Bracket, 20"		20	—	1.60	—	12	19.20
Guardrail Post		15	—	1.10	—	12	13.20
Guardrail, 7' section		7	—	.80	—	11	8.80
Cross Braces		15	17	.75	.75	44	33.00
Screw Jacks & Plates		20	30	2.00	2.50	24	48.00
8" Casters		50	—	5.75	—	—	—
16' Plank, 2" x 10"		22	—	5.10	—	35	178.50
8' Plank, 2" x 10"		11	—	3.75	—	7	26.25
1' to 6' Extension Tube		—	70	—	2.50	—	—
Shoring Stringers, steel, 10' to 12' long	L.F.	—	7	—	.40	—	—
Aluminum, 12' to 16' long		—	16	—	.60	—	—
Aluminum joists with nailers, 10' to 22' long		—	12.50	—	.50	—	—
Flying Truss System, Aluminum	S.F.C.A.	—	10	—	.60	—	—
						Total	$414.55
						2 Use/Mo.	$207.28

*Most commonly used

Scaffolding is often used as falsework over 15' high during construction of cast-in-place concrete beams and slabs. Two ft. wide scaffolding is generally used for heavy beam construction. The span between frames depends upon the load to be carried with a maximum span of 5'.

Heavy duty scaffolding with a capacity of 10,000#/leg can be spaced up to 10' O.C. depending upon form support design and loading.

Scaffolding used as horizontal shoring requires less than half the material required with conventional shoring.

On new construction, erection is done by carpenters.

Rolling towers supporting horizontal shores can reduce labor and speed the job. For maintenance work, catwalks with spans up to 70' can be supported by the rolling towers.

(15) Concrete Pipe (Div. 027-162)

Prices given are for inside 20 mile delivery zone. Add $1.45 per ton of pipe for each additional 10 miles. Minimum truckload is 10 tons. The non-reinforced pipe listed in the front of the book is designation ASTM C14-59 extra strength. The reinforced pipe listed is ASTM C76-65T class 3, no gaskets. The installation cost given includes shaping bottom of the trench, placing the pipe, and backfilling and tamping to the top of the pipe only.

(16) Excavating (Div. 022-200)

The selection of equipment used for structural excavation and bulk excavation or for grading is determined by the following factors.

1. Quantity of material.
2. Type of material.
3. Depth or height of cut.
4. Length of haul.
5. Condition of haul road.
6. Accessibility of site.
7. Moisture content and dewatering requirements.
8. Availability of excavating and hauling equipment.

Some additional costs must be allowed for hand trimming the sides and bottom of concrete pours and other excavation below the general excavation.

When planning excavation and fill, the following should also be considered.

1. Swell factor.
2. Compaction factor.
3. Moisture content.
4. Density requirements.

A typical example for scheduling and estimating the cost of excavation of a 15' deep basement on a dry site when the material must be hauled off the site, is outlined below.

Assumptions:

1. Swell factor, 18%.
2. No mobilization or demobilization.
3. Allowance included for idle time and moving on job.
4. No dewatering, sheeting, or bracing.
5. No truck spotter or hand trimming.

Number of B.C.Y. per truck $= 1.5$ C.Y. bucket x 8 passes $=12$ loose C.Y.

$$= 12 \times \frac{100}{118} = 10.2 \text{ B.C.Y. per truck}$$

Truck haul cycle:

Load truck 8 passes	=	4 minutes
Haul distance 1 mile	=	9 minutes
Dump time	=	2 minutes
Return 1 mile	=	7 minutes
Spot under machine	=	1 minute
		23 minute cycle

Fleet Haul Production per day in B.C.Y.

$$4 \text{ trucks} \times \frac{50 \text{ min. hr.}}{23 \text{ min. haul cycle}} \times 8 \text{ hrs.} \times 10.2 \text{ B.C.Y.}$$

$$= 4 \times 2.2 \times 8 \times 10.2 = 718 \text{ B.C.Y./day}$$

Excavating Cost with a 1-1/2 Hydraulic Excavator 15' Deep, 2 Mile Round Trip Haul	M.H./Day	Hourly Cost	Daily Cost	Subtotal	Unit Price
1 Equipment Operator	8	$23.30	$ 186.40		$1.29
1 Oiler	8	19.20	153.60		
4 Truck Drivers	32	18.40	588.80	$ 928.80	
1 Hydraulic Excavator			679.60		
4 Dump Trucks			1,460.80	2,140.40	2.97
Total for 720 BCY			$3,069.20	$3,069.20	$4.26

Description	1-1/2 C.Y. Hyd. Backhoe 15' Deep		1-1/2 C.Y. Power Shovel 7' Bank		1-1/2 C.Y. Dragline 7' Deep		2-1/2 C.Y. Trackloader Stockpile	
Operator (and Oiler, if required)		$ 340.00		$ 340.00		$ 340.00		$ 180.00
Truck Drivers	3 Ea.	441.60	4 Ea.	588.80	3 Ea.	441.60	4 Ea.	588.80
Equipment Rental		679.60		796.80		701.20		807.80
20 C.Y. Trailer Dump Trucks	3 Ea.	1,134.00	4 Ea.	1,512.00	3 Ea.	1,134.00	4 Ea.	1,512.00
Total Cost per Day		$2,595.20		$3,237.60		$2,616.80		$3,088.60
Daily Production, C.Y. Bank Measure		720		960		640		1000
Cost per C.Y.		$3.60		$3.37		$4.09		$3.09

Add the mobilization and demobilization costs to the total excavation costs. When equipment is rented for more than three days, there is often no mobilization charge by the equipment dealer. On larger jobs outside of urban areas, scrapers can move earth economically provided a dump site or fill area and adequate haul roads are available. Excavation within sheeting bracing, or cofferdam bracing is usually done with a clamshell and production is low, since the clamshell may have to be guided by hand between the bracing. When excavating or filling an area enclosed with a wellpoint system, add 10% to 15% to the cost to allow for restricted access. When estimating earth excavation quantities for structures, allow work space outside the building footprint for construction of the foundation, and a slope of 1:1 unless sheeting is used.

CIRCLE REFERENCE NUMBERS

⑰ Excavating Equipment (Div. 022-238)

The table below lists THEORETICAL hourly production in C.Y./hr. bank measure for some typical excavation equipment. Figures assume 50 minute hours, 83% job efficiency, 100% operator efficiency, 90° swing and properly sized hauling units, which must be modified for adverse digging and loading conditions. Actual production costs in the front of the book average about 50% of the theoretical values listed here.

Equipment	Soil Type	B.C.Y. Weight	% Swell	1 C.Y.	1½ C.Y.	2 C.Y.	2½ C.Y.	3 C.Y.	3½ C.Y.	4 C.Y.
Hydraulic Excavator "Backhoe" 15' deep cut	Moist loam, sandy clay	3400 lb.	40%	85	125	175	220	275	330	380
	Sand and gravel	3100	18	80	120	160	205	260	310	365
	Common earth	2800	30	70	105	150	190	240	280	330
	Clay, hard, dense	3000	33	65	100	130	170	210	255	300
Power Shovel Optimum Cut (Ft.)	Moist loam, sandy clay	3400	40	170 (6.0)	245 (7.0)	295 (7.8)	335 (8.4)	385 (8.8)	435 (9.1)	475 (9.4)
	Sand and gravel	3100	18	165 (6.0)	225 (7.0)	275 (7.8)	325 (8.4)	375 (8.8)	420 (9.1)	460 (9.4)
	Common earth	2800	30	145 (7.8)	200 (9.2)	250 (10.2)	295 (11.2)	335 (12.1)	375 (13.0)	425 (13.8)
	Clay, hard, dense	3000	33	120 (9.0)	175 (10.7)	220 (12.2)	255 (13.3)	300 (14.2)	335 (15.1)	375 (16.0)
Drag line Optimum Cut (Ft.)	Moist loam, sandy clay	3400	40	130 (6.6)	180 (7.4)	220 (8.0)	250 (8.5)	290 (9.0)	325 (9.5)	385 (10.0)
	Sand and gravel	3100	18	130 (6.6)	175 (7.4)	210 (8.0)	245 (8.5)	280 (9.0)	315 (9.5)	375 (10.0)
	Common earth	2800	30	110 (8.0)	160 (9.0)	190 (9.9)	220 (10.5)	250 (11.0)	280 (11.5)	310 (12.0)
	Clay, hard, dense	3000	33	90 (9.3)	130 (10.7)	160 (11.8)	190 (12.3)	225 (12.8)	250 (13.3)	280 (12.0)

				Wheel Loaders				Track Loaders		
				3 C.Y.	4 C.Y.	6 C.Y.	8 C.Y.	2¼ C.Y.	3 C.Y.	4 C.Y.
Loading Tractors	Moist loam, sandy clay	3400	40	260	340	510	690	135	180	250
	Sand and gravel	3100	18	245	320	480	650	130	170	235
	Common earth	2800	30	230	300	460	620	120	155	220
	Clay, hard, dense	3000	33	200	270	415	560	110	145	200
	Rock, well blasted	4000	50	180	245	380	520	100	130	180

⑱ Compacting Backfill (Div. 022-222)

Compaction of fill in embankments, around structures, in trenches, and under slabs is important to control settlement. Factors affecting compaction are:

1. Soil gradation
2. Moisture content
3. Equipment used
4. Depth of fill per lift
5. Density required

The costs for testing and soil analyses are listed in Division 014-108. Also, see Division 022 for further backfill, borrow, and compaction costs.

Example:
Compact granular fill around a building foundation using a 21" wide x 24" vibratory plate in 8" lifts. Operator moves at 50 FPM working a 50 minute hour to develop 95% Modified Proctor Density with 4 passes.

Production Rate:
$$\frac{1.75' \text{ plate width} \times 50 \text{ F.P.M.} \times 50 \text{ min./hr.} \times .67' \text{ lift}}{27 \text{ C.F. per C.Y.}} = 108.5 \text{ C.Y./hr.}$$

Production Rate for 4 Passes:
$$\frac{108.5 \text{ C.Y.}}{4 \text{ passes}} = 27 \text{ C.Y./hr.} \times 8 \text{ hrs.} = 216 \text{ C.Y./day}$$

Compacting 216 C.Y. with 21" Wide Vibratory Plate	M.H./Day	Hourly Cost	Daily Cost	Unit Price
1 Laborer	8	$17.50	$140.00	$ .65
1 Vibratory Plate Compactor			$ 43.40	$ .20
Total for 216 C.Y./day			$183.40/day	$ .85/C.Y.

CIRCLE REFERENCE NUMBERS

⑲ Caissons (Div. 023-804)

The three principal types of caissons are (1) Belled Caisson which, except for shallow depths and poor soil conditions, are generally recommended. They provide more bearing than shaft area. Because of its conical shape, no horizontal reinforcement of the bell is required.

(2) Straight Shaft Caissons are used where relatively light loads are to be supported by a caisson that rests on high value bearing strata. While the shaft is larger in diameter than for belled types this is more than offset by the saving in time and labor. (3) Keyed Caissons are used when extremely heavy loads are to be carried. A keyed or socketed caisson transfers its load into rock by a combination of end-bearing and shear reinforcing of the shaft. The most economical shaft often consists of a steel casing, a steel wide flange core and concrete. Allowable compressive stresses of 0.225 f'c for concrete, 16,000 psi for the wide flange core, and 9,000 psi for the steel casing are commonly used. The usual range of shaft diameter is from 18" to 84". The number of sizes specified for any one project should be limited due to the problems of casing and auger storage. When hand work is to be performed, shaft diameters should not be less than 32". When inspection of borings is required a minimum shaft diameter of 30" is recommended. Concrete caissons are intended to be poured against earth excavation so permanent forms which add to cost should not be used if the excavation is clean and the earth sufficiently impervious to prevent excessive loss of concrete.

Soil Conditions for Belling

Good	Requires Handwork	Not Recommended
Clay	Hard Shale	Silt
Sandy Clay	Limestone	Sand
Silty Clay	Sandstone	Gravel
Clayey Silt	Weathered Mica	Igneous Rock
Hard-pan		
Soft Shale		
Decomposed Rock		

⑳ Wood Sheet Piling (Div. 021-614)

Wood sheet piling may be used for depths to 20' where there is no ground water. If moderate ground water is encountered Tongue & Groove sheeting will help to keep it out. When considerable ground water is present, steel sheeting must be used.

For estimating purposes on trench excavation, sizes are as follows:

Depth	Sheeting	Wales	Braces	B.F. per S.F.
To 8'	3 x 12's	6 x 8's, 2 line	6 x 8's, @ 10'	4.0 @ 8'
8' x 12'	3 x 12's	10 x 10's, 2 line	10 x 10's, @ 9'	5.0 average
12' to 20'	3 x 12's	12 x 12's, 3 line	12 x 12's, @ 8'	7.0 average

Sheeting to be toed in at least 2' depending upon soil conditions. A five man crew with an air compressor and sheeting driver can drive and brace 440 SF/day at 8' deep, 360 SF/day at 12' deep, and 320 SF/day at 16' deep. For normal soils, piling can be pulled in 1/3 the time to install. Pulling difficulty increases with the time in the ground. Production can be increased by high pressure jetting. Figures below assume 50% of lumber is salvaged and includes pulling costs. Some jurisdictions require an equipment operator in addition to Crew B-31.

Sheeting Pulled

Daily Cost Crew B-31	M.H./Day	Hourly Cost	Daily Cost	8' Depth, 440 S.F./Day		16' Depth, 320 S.F./Day	
				To Drive (1 Day)	To Pull (1/3 Day)	To Drive (1 Day)	To Pull (1/3 Day)
1 Foreman	8	$19.50	$156.00	$ 156.00	$ 52.00	$ 156.00	$ 52.00
3 Laborers	24	17.50	420.00	420.00	140.00	420.00	140.00
1 Carpenter	8	22.00	176.00	176.00	58.67	176.00	58.67
1 Air Compressor			88.80	88.80	29.60	88.80	29.60
1 Sheeting Driver			9.20	9.20	3.07	9.20	3.07
2-50 Ft. Air Hoses, 1-1/2" Diam.			10.80	10.80	3.60	10.80	3.60
Lumber (50% salvage)				1.76 MBF 404.80		2.24 MBF 515.20	
Total			$860.80	$1,265.60	$286.94	$1,376.00	$286.94
Total/S.F.				$ 2.88	$ .65	$ 4.30	$ .90
Total (Drive and Pull)/S.F.				$ 3.53		$ 5.20	

Sheeting Left in Place

Daily Cost	8' Depth 440 S.F./Day		10' Depth 400 S.F./Day		12' Depth 360 S.F./Day		16' Depth 320 S.F./Day		18' Depth 305 S.F./Day		20' Depth 280 S.F./Day	
Crew B-31		$ 860.80		$ 860.80		$ 860.80		$ 860.80		$ 860.80		$ 860.80
Lumber	1.76M	800.80	1.8M	819.00	1.8M	819.00	2.24M	1,019.20	2.1M	955.50	1.9M	864.50
Total in Place		$1,661.60		$1,679.80		$1,679.80		$1,880.00		$1,816.30		$1,725.30
Total/S.F.		$ 3.78		$ 4.20		$ 4.67		$ 5.88		$ 5.96		$ 6.16
Total/M.B.F.		$ 944.09		$ 933.22		$ 933.22		$ 839.29		$ 864.90		$ 908.05

㉒ Wood Bearing Piles (Div. 023–612)

Untreated Southern Yellow Pine is most generally used for pile foundations cut off below the low water line. These are driven with the bark on. All piles cut off above the low water line should be treated with a preservative or encased in concrete for the section above water.

Item	Unit Cost	Units	Quantity	Total Cost	Cost/Pile	Cost/L.F.
50' Treated Piles, 13" Butt, 7" Tip	$ 8.45	L.F.	200	$ 84,500.00	$422.50	$ 8.45
Installation, Crew B-19	2,697.40	Day	13	35,066.20	175.33	3.51
Mobilization & Demobilization, Crew B-19	2,697.40	Day	3	8,092.20	40.46	.81
Transportation of Eq. One Way	848.40	Day	1	848.40	4.24	.08
Totals				$128,506.80	$642.53	$12.85

The above figures are based on driving 800 L.F. daily which can be considered average. Time is included for moving rig, cutoff & ordinary delays. A general observation is that the cost of a pile in place complete, is about two times the cost of pile only. See also equipment rental division 016–408 and ⑬ for equipment capacities.

㉖ Bituminous Paving (Div. 025–100)

City	Bituminous Asphalt per Ton*	Pavement (3") 6.13 S.Y./ton				Sidewalks (2") 9.2 S.Y./ton			
		Cost per S.Y.			per Ton	Cost per S.Y.			per Ton
		Material*	Installation	Total	Total	Material*	Installation	Total	Total
Atlanta	$23.45	$3.83	$ .60	$4.43	$27.16	$2.55	$ .96	$3.51	$32.29
Baltimore	31.20	5.09	.66	5.75	35.25	3.39	1.18	4.57	42.04
Boston	31.50	5.14	.81	5.95	36.47	3.42	1.63	5.05	46.46
Buffalo	33.60	5.48	.77	6.25	38.31	3.65	1.52	5.17	47.56
Chicago	31.05	5.07	.79	5.86	35.92	3.38	1.59	4.97	45.72
Cincinnati	27.75	4.53	.74	5.27	32.31	3.02	1.49	4.51	41.49
Cleveland	27.15	4.43	.81	5.24	32.12	2.95	1.72	4.67	42.96
Columbus	24.60	4.01	.71	4.72	28.93	2.67	1.36	4.03	37.08
Dallas	30.90	5.04	.59	5.63	34.51	3.36	.90	4.26	39.19
Denver	24.75	4.04	.63	4.67	28.63	2.69	1.07	3.76	34.59
Detroit	27.65	4.51	.78	5.29	32.43	3.01	1.57	4.58	42.14
Houston	26.50	4.32	.62	4.94	30.28	2.88	1.03	3.91	35.97
Indianapolis	27.55	4.49	.70	5.19	31.81	2.99	1.29	4.28	39.38
Kansas City	25.00	4.08	.72	4.80	29.42	2.72	1.39	4.11	37.81
Los Angeles	27.15	4.43	.88	5.31	32.55	2.95	1.92	4.87	44.80
Memphis	26.20	4.27	.59	4.86	29.79	2.85	.98	3.83	35.24
Milwaukee	23.80	3.88	.74	4.62	28.32	2.59	1.47	4.06	37.35
Minneapolis	23.65	3.86	.75	4.61	28.26	2.57	1.49	4.06	37.35
Nashville	22.60	3.69	.58	4.27	26.18	2.46	.91	3.37	31.00
New Orleans	28.80	4.70	.62	5.32	32.61	3.13	1.07	4.20	38.64
New York City	38.35	6.26	.91	7.17	43.95	4.17	2.02	6.19	56.95
Philadelphia	34.15	5.57	.79	6.36	38.99	3.71	1.66	5.37	49.40
Phoenix	26.80	4.37	.70	5.07	31.08	2.91	1.32	4.23	38.92
Pittsburgh	34.25	5.59	.71	6.30	38.62	3.72	1.32	5.04	46.37
St. Louis	24.65	4.02	.79	4.81	29.49	2.68	1.61	4.29	39.47
San Antonio	31.65	5.16	.58	5.74	35.19	3.44	.90	4.34	39.93
San Diego	29.25	4.77	.88	5.65	34.63	3.18	1.92	5.10	46.92
San Francisco	38.25	6.24	.88	7.12	43.65	4.16	1.89	6.05	55.66
Seattle	28.25	4.61	.76	5.37	32.92	3.07	1.55	4.62	42.50
Washington, D.C.	34.25	5.59	.67	6.26	38.37	3.72	1.19	4.91	45.17
Average	$28.82	$4.70	$ .73	$5.43	$33.27	$3.13	$1.40	$4.53	$41.68

Assumed density is 145 lb. per C.F.
* Includes delivery within 20 miles

Table below shows quantities and bare costs for 1000 S.Y. of Bituminous Paving.

Item	Roads and Parking Areas, 3" Thick (025-104-0460)		Sidewalks, 2" Thick (025-128-0010)	
	Quantities	Cost	Quantities	Cost
Bituminous asphalt	163 tons @ $28.82 per ton	$4,697.66	109 tons @ $28.82 per ton	$3,141.38
Installation using	Crew B-25B @ $3,553.80/4900 SY/day x 1000	725.27	Crew B-37 @ $1,005.60/720 SY/day x 1000	1,396.67
Total per 1000 S.Y.		$5,422.93		$4,538.05
Total per S.Y.		$ 5.42		$ 4.54
Total per Ton		$ 33.27		$ 41.63

CIRCLE REFERENCE NUMBERS

(29) Seeding (Div. 029-300)

The type of grass is determined by light, shade and moisture content of soil plus intended use. Fertilizer should be disked 4″ before seeding. For steep slopes disk five tons of mulch and lay two tons of hay or straw on surface per acre after seeding. Surface mulch can be staked, lightly disked or tar emulsion sprayed. Material for mulch can be wood chips, peat moss, partially rotted hay or straw, wood fibers and sprayed emulsions. Hemp seed blankets with fertilizer are also available. For spring seeding, watering is necessary. Late fall seeding may have to be reseeded in the spring. Hydraulic seeding, power mulching, and aerial seeding can be used on large areas.

(30) Cost of Trees (Deciduous) (Div. 029-536)

Tree Diameter	Normal Height	Catalog List Price of Tree	Guying Material	Equipment Charge	Installation Labor	Total
2 to 3 inch	14 feet	$ 88	$12	$46	$ 59	$ 205
3 to 4 inch	16 feet	170	29	77	98	374
4 to 5 inch	18 feet	245	46	93	117	501
6 to 7 inch	22 feet	655	80	116	147	998
8 to 9 inch	26 feet	1300	98	155	196	1749

Installation Time & Cost for Planting Trees, Bare Costs

Ball Size Diam. x Depth	Soil in Ball	Weight of Ball	Hole Diam. Req'd.	Hole Exca-vation	Amount of Soil Displ.	Topsoil Handled	Time Required in Man-Hours						Cost	
							Dig & Lace	Handle Ball	Dig Hole	Plant & Prune	Water & Guy	Total M.H.	Crew	Total per Tree
Inches	C.F.	Lbs.	Feet	C.F.	C.F.	C.F.								
12x12	.7	56	2	4	3	11	.25	.17	.33	.25	.07	1.1	1 Clab	$ 19.25
18x16	2	160	2.5	8	6	21	.50	.33	.47	.35	.08	1.7	2 Clab	29.75
24x18	4	320	3	13	9	38	1.00	.67	1.08	.82	.20	3.8	3 Clab	66.50
30x21	7.5	600	4	27	19.5	76	.82	.71	.79	1.22	.26	3.8		$102.00
36x24	12.5	980	4.5	38	25.5	114	1.08	.95	1.11	1.32	.30	4.76		127.00
42x27	19	1520	5.5	64	45	185	1.90	1.27	1.87	1.43	.34	6.8	B-6	182.00
48x30	28	2040	6	85	57	254	2.41	1.60	2.06	1.55	.39	8.0	@	214.00
54x33	38.5	3060	7	127	88.5	370	2.86	1.90	2.39	1.76	.45	9.4	$26.73	251.00
60x36	52	4160	7.5	159	107	474	3.26	2.17	2.73	2.00	.51	10.7	per	286.00
66x39	68	5440	8	196	128	596	3.61	2.41	3.07	2.26	.58	11.9	man-	318.00
72x42	87	7160	9	267	180	785	3.90	2.60	3.71	2.78	.70	13.7	hour	366.00

CIRCLE REFERENCE NUMBERS

(31) Formwork Labor Hours (Div. 031)

Item	Unit	Hours Required Fabricate	Hours Required Erect & Strip	Hours Required Clean & Move	Total Hours 1 Use	Multiple Use 2 Use	Multiple Use 3 Use	Multiple Use 4 Use
Beam and Girder, interior beams, 12" wide	100 S.F.	6.4	8.3	1.3	16.0	13.3	12.4	12.0
Hung from steel beams		5.8	7.7	1.3	14.8	12.4	11.6	11.2
Beam sides only, 36" high		5.8	7.2	1.3	14.3	11.9	11.1	10.7
Beam bottoms only, 24" wide		6.6	13.0	1.3	20.9	18.1	17.2	16.7
Box out for openings		9.9	10.0	1.1	21.0	16.6	15.1	14.3
Buttress forms, to 8' high		6.0	6.5	1.2	13.7	11.2	10.4	10.0
Centering, steel, 3/4" rib lath			1.0		1.0			
3/8" rib lath or slab form			0.9		0.9			
Chamfer strip or keyway	100 L.F.		1.5		1.5	1.5	1.5	1.5
Columns, fiber tube 8" diameter			20.6		20.6			
12"			21.3		21.3			
16"			22.9		22.9			
20"			23.7		23.7			
24"			24.6		24.6			
30"			25.6		25.6			
Round Steel, 12" diameter			22.0		22.0	22.0	22.0	22.0
16"			25.6		25.6	25.6	25.6	25.6
20"			30.5		30.5	30.5	30.5	30.5
24"			37.7		37.7	37.7	37.7	37.7
Plywood 8" x 8"	100 S.F.	7.0	11.0	1.2	19.2	16.2	15.2	14.7
12" x 12"		6.0	10.5	1.2	17.7	15.2	14.4	14.0
16" x 16"		5.9	10.0	1.2	17.1	14.7	13.8	13.4
24" x 24"		5.8	9.8	1.2	16.8	14.4	13.6	13.2
Steel framed plywood 8" x 8"			10.0	1.0	11.0	11.0	11.0	11.0
12" x 12"			9.3	1.0	10.3	10.3	10.3	10.3
16" x 16"			8.5	1.0	9.5	9.5	9.5	9.5
24" x 24"			7.8	1.0	8.8	8.8	8.8	8.8
Drop head forms, plywood		9.0	12.5	1.5	23.0	19.0	17.7	17.0
Coping forms		8.5	15.0	1.5	25.0	21.3	20.0	19.4
Culvert, box			14.5	4.3	18.8	18.8	18.8	18.8
Curb forms, 6" to 12" high, on grade		5.0	8.5	1.2	14.7	12.7	12.1	11.7
On elevated slabs		6.0	10.8	1.2	18.0	15.5	14.7	14.3
Edge forms to 6" high, on grade	100 L.F.	2.0	3.5	0.6	6.1	5.6	5.4	5.3
7" to 12" high	100 S.F.	2.5	5.0	1.0	8.5	7.8	7.5	7.4
Equipment foundations		10.0	18.0	2.0	30.0	25.5	24.0	23.3
Flat slabs, including drops		3.5	6.0	1.2	10.7	9.5	9.0	8.8
Hung from steel		3.0	5.5	1.2	9.7	8.7	8.4	8.2
Closed deck for domes		3.0	5.8	1.2	10.0	9.0	8.7	8.5
Open deck for pans		2.2	5.3	1.0	8.5	7.9	7.7	7.6
Footings, continuous, 12" high		3.5	3.5	1.5	8.5	7.3	6.8	6.6
Spread, 12" high		4.7	4.2	1.6	10.5	8.7	8.0	7.7
Pile caps, square or rectangular		4.5	5.0	1.5	11.0	9.3	8.7	8.4
Grade beams, 24" deep		2.5	5.3	1.2	9.0	8.3	8.0	7.9
Lintel or Sill forms		8.0	17.0	2.0	27.0	23.5	22.3	21.8
Spandrel beams, 12" wide		9.0	11.2	1.3	21.5	17.5	16.2	15.5
Stairs			25.0	4.0	29.0	29.0	29.0	29.0
Trench forms in floor		4.5	14.0	1.5	20.0	18.3	17.7	17.4
Walls, Plywood, at grade, to 8' high		5.0	6.5	1.5	13.0	11.0	9.7	9.5
8' to 16'		7.5	8.0	1.5	17.0	13.8	12.7	12.1
16' to 20'		9.0	10.0	1.5	20.5	16.5	15.2	14.5
Foundation walls, to 8' high		4.5	6.5	1.0	12.0	10.3	9.7	9.4
8' to 16' high		5.5	7.5	1.0	14.0	11.8	11.0	10.6
Retaining wall to 12' high, battered		6.0	8.5	1.5	16.0	13.5	12.7	12.3
Radial walls to 12' high, smooth		8.0	9.5	2.0	19.5	16.0	14.8	14.3
But in 2' chords		7.0	8.0	1.5	16.5	13.5	12.5	12.0
Prefabricated modular, to 8' high		—	4.3	1.0	5.3	5.3	5.3	5.3
Steel, to 8' high		—	6.8	1.2	8.0	8.0	8.0	8.0
8' to 16' high		—	9.1	1.5	10.6	10.3	10.2	10.2
Steel framed plywood to 8' high		—	6.8	1.2	8.0	7.5	7.3	7.2
8' to 16' high		—	9.3	1.2	10.5	9.5	9.2	9.0

(32) Forms for Reinforced Concrete (Div. 031)

Design Economy

Avoid many sizes in proportioning beams and columns.

From story to story avoid changing column dimensions. Gain strength by adding steel or using a richer mix. If a change in size of column is necessary, vary one dimension only to minimize form alterations. Keep beams and columns the same width.

From floor to floor in a multi-story building vary beam depth, not width, as that will leave slab panel form unchanged. It is cheaper to vary the strength of a beam from floor to floor by means of steel area than by 2″ changes in either width or depth.

Cost Factors

Material includes the cost of lumber, cost of rent or metal pans or forms if used, nails, form ties, form oil, bolts and accessories.

Labor includes the cost of carpenters to make up, erect, remove and repair, plus common labor to clean and move. Having carpenters remove forms minimizes repairs.

Improper alignment and condition of forms will increase finishing cost. When forms are heavily oiled, concrete surfaces must be neutralized before finishing. Special curing compounds will cause spillages to spall off in first frost. Gang forming methods will reduce costs on large projects.

Materials Used

Boards are seldom used unless their architectural finish is required. Generally, steel, fiberglass and plywood are used for contact surfaces. Labor on plywood is 10% less than with boards. The plywood is backed up with 2 x 4's at 12″ to 32″ O.C. Walers are generally 2 - 2 x 4's. Column forms are held together with steel yokes or bands. Shoring is with adjustable shoring or scaffolding for high ceilings.

Reuse

Floor and column forms can be reused four or possibly five times without excessive repair. Remember to allow for 10% waste on each reuse.

When modular sized wall forms are made, up to twenty uses can be expected with exterior plyform.

When forms are reused, the cost to erect, strip, clean and move will not be affected. 10% replacement of lumber should be included and about one hour of carpenter time for repairs on each reuse per 100 S.F.

The reuse cost for certain accessory items normally rented on a monthly basis will be lower than the cost for the first use.

After fifth use, new material required plus time needed for repair prevent form cost from dropping further and it may go up. Much depends on care in stripping, the number of special bays, changes in beam or column sizes and other factors.

1. Costs for multiple use of formwork may be developed as follows:

2 Uses	3 Uses	4 Uses
$\dfrac{\text{1st Use + Reuse}}{2}$ = avg. cost/2 uses	$\dfrac{\text{1st Use + 2 Reuse}}{3}$ = avg. cost/3 uses	$\dfrac{\text{1st Use + 3 Reuse}}{4}$ = avg. cost/4 uses

(33) Forms In Place (Div. 031)

This section assumes that all cuts are made with power saws, that adjustable shores are employed and that maximum use is made of commercial form ties and accessories. Bare costs are used in the table below.

BEAM AND GIRDER, INTERIOR, 12″ Wide (Line 138-2000)	First Use			Reuse		
	Quantities	Material	Installation	Quantities	Material	Installation
5/8″ exterior plyform at $695 per M.S.F.	115 S.F.	$ 79.90		11.5 S.F.	$ 8.00	
Lumber at $366 per M.B.F.	200 B.F.	73.20		20.0 B.F.	7.30	
Accessories, incl. adjustable shores	Allow	21.40		Allow	21.40	
Make up, crew C-2 at $22.24 per man-hour	6.4 M.H.		$142.35	1.0 M.H.		$ 22.25
Erect and strip	8.3 M.H.		184.60	8.3 M.H.		184.60
Clean and move	1.3 M.H.		28.90	1.3 M.H.		28.90
Total per 100 S.F.C.A.	16.0 M.H.	$174.50	$355.85	10.6 M.H.	$36.70	$235.75

For structural steel frame with beams encased, subtract 1.2 man-hours, and 50 B.F. lumber or about $45 per 100 S.F.C.A. for the first use and $29 for each reuse.

BOX CULVERT, 5′ to 8′ Square or Rectangular (Line 146-0010)	First Use			Reuse		
	Quantities	Material	Installation	Quantities	Material	Installation
3/4″ exterior plyform at $750 per M.S.F.	110 S.F.	$ 82.50		11.0 S.F.	$ 8.25	
Lumber at $366 per M.B.F.	170 B.F.	62.20		17.0 B.F.	6.20	
Accessories	Allow	16.85		Allow	16.85	
Build in place, crew C-1 at $21.62 per man-hour	14.5 M.H.		$313.50	14.5 M.H.		$313.50
Strip and salvage	4.3 M.H.		92.95	4.3 M.H.		92.95
Total per 100 S.F.C.A.	18.8 M.H.	$161.55	$406.45	18.8 M.H.	$31.30	$406.45

CIRCLE REFERENCE NUMBERS

(33) Forms in Place (cont.)

COLUMNS, 24" x 24" (Line 142-6500)	First Use			Reuse		
	Quantities	Material	Installation	Quantities	Material	Installation
5/8" exterior plyform @ $695 per M.S.F.	120 S.F.	$ 83.40		12.0 S.F.	$ 8.35	
Lumber @ $366 per M.B.F.	125 B.F.	45.75		12.5 B.F.	4.60	
Clamps, chamfer strips and accessories	Allow	21.55		Allow	21.55	
Make up, crew C-1 at $21.62 per man-hour	5.8 M.H.		$125.40	1.0 M.H.		$ 21.60
Erect and strip	9.8 M.H.		211.90	9.8 M.H.		211.90
Clean and move	1.2 M.H.		25.95	1.2 M.H.		25.95
Total per 100 S.F.C.A.	16.8 M.H.	$150.70	$363.25	12.0 M.H.	$34.50	$259.45

FLAT SLAB WITH DROP PANELS (Line 150-2000)	First Use			Reuse		
	Quantities	Material	Installation	Quantities	Material	Installation
5/8" exterior plyform @ $695 per M.S.F.	115 S.F.	$ 79.90		11.5 S.F.	$ 7.80	
Lumber @ $366 per M.B.F.	210 B.F.	76.85		21 B.F.	7.70	
Accessories, incl. adjustable shores	Allow	23.20		Allow	23.20	
Make up, crew C-2 at $22.24 per man-hour	3.5 M.H.		$ 77.85	1.0 M.H.		$ 22.25
Erect and strip	6.0 M.H.		133.45	6.0 M.H.		133.45
Clean and move	1.2 M.H.		26.70	1.2 M.H.		26.70
Total per 100 S.F.C.A.	10.7 M.H.	$179.95	$238.00	8.2 M.H.	$38.70	$182.40

Drop panels included but column caps figure with columns.

FOOTINGS, SPREAD Line (158-5000)	First Use			Reuse		
	Quantities	Material	Installation	Quantities	Material	Installation
Lumber @ $366 per M.B.F.	260 B.F.	$ 95.15		26 B.F.	$ 9.50	
Accessories	Allow	7.05		Allow	7.05	
Make up, crew C-1 at $21.62 per man-hour	4.7 M.H.		$101.60	1.0 M.H.		$ 21.60
Erect and strip	4.2 M.H.		90.80	4.2 M.H.		90.80
Clean and move	1.6 M.H.		34.60	1.6 M.H.		34.60
Total per 100 S.F.C.A.	10.5 M.H.	$102.20	$227.00	6.8 M.H.	$16.55	$147.00

FOUNDATION WALL, 8' High (Line 182-2000)	First Use			Reuse		
	Quantities	Material	Installation	Quantities	Material	Installation
5/8" exterior plyform @ $695 per M.S.F.	110 S.F.	$ 76.45		11.0 S.F.	$ 7.65	
Lumber @ $366 per M.B.F.	140 B.F.	51.25		14.0 B.F.	5.10	
Accessories	Allow	22.30		Allow	22.30	
Make up, crew C-2 at $22.24 per man-hour	5.0 M.H.		$111.20	1.0 M.H.		$ 22.25
Erect and strip	6.5 M.H.		144.55	6.5 M.H.		144.55
Clean and move	1.5 M.H.		33.35	1.5 M.H.		33.35
Total per 100 S.F.C.A.	13.0 M.H.	$150.00	$289.10	9.0 M.H.	$35.05	$200.15

PILE CAPS, Square or Rectangular (Line 158-3000)	First Use			Reuse		
	Quantities	Material	Installation	Quantities	Material	Installation
5/8" exterior plyform @ $695 per M.S.F.	110 S.F.	$ 76.45		11.0 S.F.	$ 7.65	
Lumber @ $366 per M.B.F.	160 B.F.	58.55		16.0 B.F.	5.85	
Accessories	Allow	6.75		Allow	6.75	
Make up, crew C-1 at $21.62 per man-hour	4.5 M.H.		$ 97.30	1.0 M.H.		$ 21.60
Erect and strip	5.0 M.H.		108.10	5.0 M.H.		108.10
Clean and move	1.5 M.H.		32.45	1.5 M.H.		32.45
Total per 100 S.F.C.A.	11.0 M.H.	$141.75	$237.85	7.5 M.H.	$20.25	$162.15

STAIRS, Average Run (Inclined Length x Width) (Line 174-0010)	First Use			Reuse		
	Quantities	Material	Installation	Quantities	Material	Installation
5/8" exterior plyform @ $695 per M.S.F.	110 S.F.	$ 76.45		11.0 S.F.	$ 7.65	
Lumber @ $366 per M.B.F.	425 B.F.	155.55		42.5 B.F.	15.55	
Accessories	Allow	17.35		Allow	17.35	
Build in place, crew C-2 at $22.24 per man-hour	25.0 M.H.		$556.00	25.0 M.H.		$556.00
Strip and salvage	4.0 M.H.		88.95	4.0 M.H.		88.95
Total per 100 S.F.	29.0 M.H.	$249.35	$644.95	29.0 M.H.	$40.55	$644.95

(34) Wall Form Materials (Div. 031-182)

Aluminum Forms

Approximate weight is 3 lbs. per S.F.C.A. Standard widths are available from 4" to 36" with 36" most common. Standard lengths of 2', 4', 6' to 8' are available. Forms are lightweight and fewer ties are needed with the wider widths. The form face is either smooth or textured.

Cost of bare forms per S.F.C.A. with different facing surface is listed below. Typical material cost including usual accessories but not including form ties is $15.80 per S.F.C.A.

Forms may also be rented.

	Purchase Cost Per S.F.							
Finish	3' x 8'	2' x 8'	12" x 8'	6" x 8'	3' x 4'	2' x 4'	12" x 4'	6" x 4'
Smooth Aluminum (.096" Face Sheet)	$10.90	$13.50	$18.25	$30.25	$11.70	$16.10	$21.10	$33.00
Textured Brick Aluminum	12.70	16.20	21.85	34.85	14.60	19.35	24.75	39.35

Metal Framed Plywood Forms

Manufacturers claim over 75 reuses of plywood and over 300 reuses of steel frames. Sale price for steel framed forms is $8.45 per S.F. for 2' x 8' to $11.25 per S.F. for 2' x 3'. Narrower forms range between $11.75 per S.F. to $19.75 per S.F. for 1' x 3'. Many specials such as corners, fillers, pilasters, etc. are available. Monthly rental is generally about 15% of purchase price for first month and 9% per month thereafter with 90% of rental applied to purchase for the first month and decreasing percentages thereafter. Aluminum framed forms cost 25% to 30% more than steel framed.

Rule of thumb purchase price including corners, specials, etc.; steel framed $11.50 per S.F.; aluminum framed $15.00 per S.F.

Reconditioned steel framed forms are rented for an average of $1.05 per S.F per month. Aluminum framed forms are rented for $1.95 per S.F. for the first month and $1.25 per S.F. per month thereafter.

After the first month, extra days may be prorated from the monthly charge. Rental rates do not include ties, accessories, cleaning, loss of hardware or freight in and out. Approximate weight is 5 lbs. per S.F. for steel; 3 lbs. per S.F. for aluminum.

Forms can be rented with option to buy.

Plywood Forms, Job Fabricated

There are two types of plywood used for concrete forms.

1. Exterior plyform which is completely waterproof. This is face oiled to facilitate stripping. Ten reuses can be expected with this type with 25 reuses possible.

2. An overlaid type consists of a resin fiber fused to exterior plyform. No oiling is required except to facilitate cleaning. This is available in both high density (HDO) and medium density overlaid (MDO). Using HDO, 50 reuses can be expected with 200 possible.

Plyform is available in 5/8" and 3/4" thickness. High density overlaid is available in 3/8", 1/2", 5/8" and 3/4" thickness.

5/8" thick is sufficient for most building forms, while 3/4" is best on heavy construction.

For prices on plywood and framing lumber see (83) and (86).

Plywood Forms, Modular, Prefabricated

There are many plywood forming systems without frames. Most of these are manufactured from 1-1/8" (HDO) plywood and have some hardware attached. These are used principally for foundation walls 8' or less high. With care and maintenance, 100 reuses can be attained with decreasing quality of surface finish. Sale price of 2' x 8' panels is $5.30 with 4' x 8' fillers costing $21.25 per S.F. Typical forms and accessories for 1000 S.F. of form area cost $6,650 not including form ties.

Steel Forms

Approximate weight is 6-1/2 lbs. per S.F.C.A. including accessories. Standard widths are available from 2" to 24", with 24" most common. Standard lengths are from 2' to 8', with 4' the most common. Forms are easily ganged into modular units.

Sale price for typical hand set job is $13.00 per S.F.C.A. including all usual accessories, specials, corners, etc., but not including ties or freight. Cost of bare forms runs from $10.75 for wide forms to over $22 per S.F.C.A. for narrow widths and/or short lengths. Forms are usually leased for 15% of the purchase price per month prorated daily over 30 days. Standard 6000 lb. wall ties for 12" walls cost $77 per hundred and 24" long ties cost $108 per hundred.

Rental may be applied to sale price and usually rental forms are bought. With careful handling and cleaning 200 to 400 reuses are possible.

Straight wall gang forms up to 12' x 20' or 8' x 30' can be fabricated. These crane handled forms usually cost from $18 to $26 per S.F.C.A. or can be leased for 8.7% per month. Straight wall gang forms utilizing 40,000 lb. and 60,000 lb. ties with sizes from 24' x 10' to 24' x 24' cost from $20 to $28 per S.F. including all accessories and taper ties. Rental rates on these gang forms run from 30¢ to 55¢ per week per S.F.C.A.

Individual job analysis is available from the manufacturer at no charge.

(35) Concrete for Entire Building (Div. 033-130)

These figures give an average cost per C.Y. for all concrete in a typical multi-storied reinforced concrete building. It is assumed that ready mixed concrete is used and that forms will be used four times. Rubbing is not included; 5% concrete waste is included. When forms and reinforcing quantity ratios are available compute the total C.Y. costs from the unit prices in front of the book. See (33) for handling equipment.

Item	Material	Labor	Hoist & Equipment	Total
Concrete in place	$ 54.44	$ 13.52	$ 4.32	$ 72.28
Forms	30.61	92.86	3.02	126.49
Reinforcing steel	31.69	17.90	1.35	50.94
Total per C.Y.	$116.74	$124.28	$ 8.69	$249.71

CIRCLE REFERENCE NUMBERS

(36) Floor Pans and Domes (Div. 031-150)

For 8' to 15' Ceiling Heights Using Crew C-2 at $22.24 per Man-Hour	20" Pans				19" Domes			
	(Line 3500)	1 Use	(Line 3650)	4 Use	(Line 4000)	1 Use	(Line 4150)	4 Use
Rent 20" pans or 19" domes		$ 66.50		$ 21.60		$ 66.50		$ 21.60
Adjustable shores and accessories		34.00		18.00		34.00		18.00
110 S.F. 3/4" plyform at $750 per M.S.F.		82.50		26.80		82.50		26.80
210 B.F. supporting joists, girts and braces at $366 per MBF		76.85		25.00		76.85		25.00
Labor handle, place, strip, oil and clean pans and domes	1.5 hr.	33.35	1.0 hr.	22.25	1.8 hr.	40.05	1.2 hr.	26.70
Make up, erect, remove wood decking	10.0 hr.	222.40	8.5 hr.	189.05	10.0 hr.	222.40	8.5 hr.	189.05
Total per 100 S.F. of floor area		$515.60		$302.70		$522.30		$307.15

The figures above are for closed deck forming. For open deck forming deduct $23 per 100 S.F. for four uses. For pan rental, figure 67¢ per S.F. of total area for one use; 37¢ for two uses; 27¢ for three uses; and 21¢ for four uses, which is about the most that can be expected for any one project. The purchase price of fiberglass long pans is $5.05 per S.F.C.A. or an average $10.10 per S.F. of floor area.*

For two-way grid system, 19" steel domes are leased for 21¢ to 62¢ and 30" fiberglass domes are leased for 46¢ to $2.05 per S.F. of floor area. The purchase price of custom size fiberglass forms runs from $5.50 to $15.00 per S.F.C.A.*

For slab height from 15' to 20' add $43 per 100 S.F. and for 20' to 35' add $61 per 100 S.F., both for four uses.

*It is necessary to divide the purchase cost by the number of expected uses to determine the cost per use.

(41) Materials for One C.Y. of Concrete (Div. 033)

This is an approximate method of figuring quantities of cement, sand and gravel for a field mix with waste allowance included.

With gravel as coarse aggregate for barrels of cement required, divide 10 by total mix; that is, for 1:2:4 mix, 10 divided by 7 = 1-3/7 barrels.

For tons of sand multiply barrels of cement by parts of sand and then by 0.2; that is, for the 1:2:4 mix, as above, 1-3/7 x 2 x .2 = .57 tons.

Tons of gravel are in the same ratio to tons of sand as parts in mix, or 4/2 x .57 = 1.14 tons.

If course aggregate is crushed stone, use 10-1/2 instead of 10 as given for gravel.

1 bag cement = 94#	1 C.Y. sand or gravel = 2700#	1 C.Y. crushed stone = 2575#
4 bags = 1 barrel	1 ton = 20 C.F.	1 ton = 21 C.F.

Average carload of cement is 692 bags; of sand or gravel is 56 tons.

Do not stack stored cement over 10 bags high.

42 Concrete Material Net Prices (Div. 033)

Costs below are C.Y. of concrete delivered; per ton of bulk cement; per bag cement delivered T.L.L.; per ton for stone and sand aggregates loaded at plant (no trucking included) and per 4 C.F. bag for perlite or vermiculite aggregate delivered T.L.L.

City	Ready Mix Concrete Regular Weight		Cement T.L. Lots		Aggregates per Ton Crushed Stone		Sand	Vermiculite or Perlite 4 C.F. Bag
	3000 psi	5000 psi	Bulk per Ton	Bags per Bag	1-1/2"	3/4"		
Atlanta	$44.40	$52.20	$67.75	$5.95	$ 8.40	$ 8.45	$10.55	$6.50
Baltimore	52.80	58.50	67.20	6.70	7.10	7.05	8.35	5.75
Boston	56.40	63.20	68.55	6.95	9.40	10.40	9.45	6.35
Buffalo	59.00	66.00	72.00	6.65	8.00	8.25	10.30	6.65
Chicago	50.60	58.40	70.65	6.50	8.10	8.45	8.10	5.70
Cincinnati	49.55	54.90	66.15	6.40	8.80	9.30	6.05	6.75
Cleveland	49.20	55.10	68.00	6.70	8.90	9.00	9.70	6.70
Columbus	51.75	57.10	73.55	6.70	7.65	8.00	8.15	6.65
Dallas	47.15	52.50	63.70	5.25	9.60	9.90	8.10	6.30
Denver	59.65	68.30	76.30	7.10	9.35	9.50	7.70	6.55
Detroit	55.60	62.80	67.50	6.25	8.40	8.60	7.40	6.75
Houston	48.25	52.50	66.00	5.75	8.45	8.20	8.40	6.00
Indianapolis	50.40	56.50	68.80	6.95	7.90	8.45	8.70	6.60
Kansas City	49.05	54.45	76.25	7.05	9.15	9.60	7.65	6.75
Los Angeles	51.00	58.00	76.70	7.40	9.80	10.30	10.10	7.20
Memphis	49.20	55.40	67.10	7.00	8.10	8.50	8.10	6.40
Milwaukee	50.00	54.10	71.45	6.75	8.40	9.10	7.40	6.45
Minneapolis	49.35	56.00	76.80	6.05	11.40	11.75	8.00	6.40
Nashville	49.60	57.15	69.05	6.60	8.85	9.10	9.00	6.20
New Orleans	43.40	47.60	64.10	6.15	10.00	10.80	10.25	6.60
New York City	71.00	76.60	72.10	7.40	13.25	14.80	14.50	6.65
Philadelphia	48.00	55.60	69.30	6.30	10.00	10.60	8.90	6.00
Phoenix	48.65	56.75	68.90	6.05	7.75	8.25	8.40	6.35
Pittsburgh	48.80	55.60	69.35	6.50	13.00	13.30	11.00	6.30
St. Louis	49.20	54.95	63.10	6.00	9.15	10.40	10.60	6.60
San Antonio	41.35	47.00	63.35	5.85	7.45	7.80	6.90	6.10
San Diego	52.25	60.00	71.45	6.85	9.00	9.75	9.85	6.00
San Francisco	56.00	61.00	73.50	6.95	9.80	11.25	13.75	6.20
Seattle	45.45	54.65	69.40	6.70	9.75	10.00	9.00	6.70
Washington, D.C.	55.00	62.50	68.60	6.75	11.25	12.20	10.60	6.75
Average	$51.05	$57.50	$69.60	$6.55	$ 9.20	$ 9.70	$ 9.20	$6.45

CIRCLE REFERENCE NUMBERS

㊸ Ready Mix Material Prices (Div. 033)

Table below lists national average prices per C.Y. of concrete. Prices in the key cities for different strengths can be closely estimated by factoring against the 3000 psi or 5000 psi price from ㊷ or from the City Cost Indexes, (in the Appendix) cast-in-place concrete material factor.

Strength in psi	Design Mix Nominal Mix	Bags per C.Y.	Heavy Weight Regular	Heavy Weight High Early	Light Weight 110# per C.F.	Light Weight All Light Weight	Admixtures and Special Items Add to Each C.Y. of Concrete for the Following Items:		
2,000	1:3:5	4.5	$47.80	$51.00	$62.65	$71.50	Calcium chloride, 1%	$1.45	per C.Y.
2,500	1:2½:4½	5	49.45	52.65	64.80	73.90	" " 2%	2.90	per C.Y.
3,000	1:2:4	5.5	51.05	54.25	66.90	76.25	Water reducing agent	.40	per bag
3,500	1:2:3½	6	52.70	55.85	69.10	78.80	Set retarder	.45	per bag
3,750	1:2:3¼	6.3	53.50	56.75	70.10	79.90	High early cement	.85	per bag
4,000	1:2:3	6.5	54.30	57.50	71.20	81.15	White cement	9.00	per bag
4,500	1:2:2½	7	55.90	59.05	73.25	83.50	Pump aid	1.85	per C.Y.
5,000	1:1½:2½	7.5	57.50	60.75	75.40	85.95	Winter concrete	3.00	per C.Y.
	1:1:2	8	59.10	62.30	—	Perlite	Fibermesh	9.00	per C.Y.
	1:4 topping	7.5	58.25	61.45	—	1:6=			
	1:3 topping	8.5	60.75	63.95	—	$72.70			
	1:2 topping	11	68.80	72.00	—				

㊻ Placing Ready Mixed Concrete (Div. 033-172)

For ground pours allow for 5% waste when figuring quantities.

Prices in the front of the book assume normal deliveries. If deliveries are made before 8 A.M. or after 5 P.M. or on Saturday afternoons add $10 per C.Y. Large volume discounts are not included in prices in front of the book.

For the lower floors without truck access, concrete may be wheeled in rubber tired buggies, conveyer handled, crane handled or pumped. Pumping is economical if there is top steel. Conveyers are more efficient for thick slabs. Concrete pump with an operator can be rented from $520 per day for up to 25 C.Y. to $1,200 per day for a 400 C.Y. pour. Figures include travel time if done at straight time. Pumping lightweight concrete costs an extra $1.85 per C.Y.

At higher floors the rubber tired buggies may be hoisted by a hoisting tower and then wheeled to location. Placement by a conveyer is limited to three floors and is best for high volume pours. Pumped concrete is best when building has no crane access. Concrete may be pumped directly as high as thirty-six stories using special pumping techniques. Normal maximum height is about fifteen stories.

Best pumping aggregate is screened and graded bank gravel rather than crushed stone.

Pumping downward is more difficult than pumping upwards. Horizontal distance from pump to pour may increase preparation time prior to pour. Placing by cranes, either mobile, climbing or tower types continues as the most efficient method for high rise concrete buildings.

Item	10 C.F. Walking Cart Hourly Cost	10 C.F. Walking Cart Wheeled up to 50 ft.	10 C.F. Walking Cart Wheeled up to 150 ft.	10 C.F. Walking Cart Wheeled up to 250 ft.	18 C.F. Riding Cart Hourly Cost	18 C.F. Riding Cart Wheeled up to 50 ft.	18 C.F. Riding Cart Wheeled up to 150 ft.	18 C.F. Riding Cart Wheeled up to 250 ft.
Laborer	$17.50	$4.38	$5.83	$ 7.88	$17.50	$1.75	$2.34	$3.12
.125 Labor foreman	2.44	.61	.82	1.10	2.44	.24	.33	.44
Concrete cart	5.50	1.38	1.82	2.48	13.25	1.33	1.77	2.36
Total Cost/C.Y.	—	$6.37	$8.47	$11.46	—	$3.32	$4.44	$5.92
Hourly production		4 C.Y.				10 C.Y.		

Cost per C.Y. for Wheeled Concrete, Dumped Only (Add to appropriate placing cost)

(47) Proportionate Quantities (Div. 033)

The tables below show both quantities per S.F. of floor areas as well as form and reinforcing quantities per C.Y. Unusual structural requirements would increase the ratios below. High strength reinforcing would reduce the steel weights. Figures are for 3000 psi concrete and 60,000 psi reinforcing unless specified otherwise.

Type of Construction	Live Load	Span	Per S.F. of Floor Area				Per C.Y. of Concrete		
			Concrete	Forms	Reinf.	Pans	Forms	Reinf.	Pans
Flat Plate	50 psf	15 Ft.	.46 C.F.	1.06 S.F.	1.71 lb.		62 S.F.	101 lb.	
		20	.63	1.02	2.4		44	104	
		25	.79	1.02	3.03		35	104	
	100	15	.46	1.04	2.14		61	126	
		20	.71	1.02	2.72		39	104	
		25	.83	1.01	3.47		33	113	
Flat Plate (waffle construction) 20" domes	50	20	.43	1.0	2.1	.84 S.F.	63	135	53 S.F.
		25	.52	1.0	2.9	.89	52	150	46
		30	.64	1.0	3.7	.87	42	155	37
	100	20	.51	1.0	2.3	.84	53	125	45
		25	.64	1.0	3.2	.83	42	135	35
		30	.76	1.0	4.4	.81	36	160	29
Waffle Construction 30" domes	50	25	.69	1.06	1.83	.68	42	72	40
		30	.74	1.06	2.39	.69	39	87	39
		35	.86	1.05	2.71	.69	33	85	39
		40	.78	1.0	4.8	.68	35	165	40
Flat Slab (two way with drop panels)	50	20	.62	1.03	2.34		45	102	
		25	.77	1.03	2.99		36	105	
		30	.95	1.03	4.09		29	116	
	100	20	.64	1.03	2.83		43	119	
		25	.79	1.03	3.88		35	133	
		30	.96	1.03	4.66		29	131	
	200	20	.73	1.03	3.03		38	112	
		25	.86	1.03	4.23		32	133	
		30	1.06	1.03	5.3		26	135	
One Way Joists 20" pans	50	15	.36	1.04	1.4	.93	78	105	70
		20	.42	1.05	1.8	.94	67	120	60
		25	.47	1.05	2.6	.94	60	150	54
	100	15	.38	1.07	1.9	.93	77	140	66
		20	.44	1.08	2.4	.94	67	150	58
		25	.52	1.07	3.5	.94	55	185	49
One Way Joists 8" x 16" filler blocks	50	15	.34	1.06	1.8	.81 Ea.	84	145	64 Ea.
		20	.40	1.08	2.2	.82	73	145	55
		25	.46	1.07	3.2	.83	63	190	49
	100	15	.39	1.07	1.9	.81	74	130	56
		20	.46	1.09	2.8	.82	64	160	48
		25	.53	1.10	3.6	.83	56	190	42
One Way Beam & Slab	50	15	.42	1.30	1.73		84	111	
		20	.51	1.28	2.61		68	138	
		25	.64	1.25	2.78		53	117	
	100	15	.42	1.30	1.9		84	122	
		20	.54	1.35	2.69		68	154	
		25	.69	1.37	3.93		54	154	
	200	15	.44	1.31	2.24		80	137	
		20	.58	1.40	3.30		65	163	
		25	.69	1.42	4.89		53	183	
Two Way Beam & Slab	100	15	.47	1.20	2.26		69	130	
		20	.63	1.29	3.06		55	131	
		25	.83	1.33	3.79		43	123	
	200	15	.49	1.25	2.70		41	149	
		20	.66	1.32	4.04		54	165	
		25	.88	1.32	6.08		41	187	

47 **Proportionate Quantities** (cont.)

4000 psi Concrete and 60,000 psi Reinforcing — Form and Reinforcing Quantities per C.Y.

Item	Size	Forms	Reinforcing	Minimum	Maximum
Columns (square tied)	10" x 10"	130 S.F.C.A.	#5 to #11	220 lbs.	875 lbs.
	12" x 12"	108	#6 to #14	200	955
	14" x 14"	92	#7 to #14	190	900
	16" x 16"	81	#6 to #14	187	1082
	18" x 18"	72	#6 to #14	170	906
	20" x 20"	65	#7 to #18	150	1080
	22" x 22"	59	#8 to #18	153	902
	24" x 24"	54	#8 to #18	164	884
	26" x 26"	50	#9 to #18	169	994
	28" x 28"	46	#9 & #18	147	864
	30" x 30"	43	#10 to #18	146	983
	32" x 32"	40	#10 to #18	175	866
	34" x 34"	38	#10 to #18	157	772
	36" x 36"	36	#10 to #18	175	852
	38" x 38"	34	#10 to #18	158	765
	40" x 40"	32	#10 to #18	143	692

Item	Size	Forms	Spirals	Reinforcing	Minimum	Maximum
Columns (spirally reinforced)	12" diameter	34.5 L.F.	190 lb.	#4 to #11	165 lb.	1505 lb.
		34.5	190	#14 & #18	—	1100
	14"	25	170	#4 to #11	150	970
		25	170	#14 & #18	800	1000
	16"	19	160	#4 to #11	160	950
		19	160	#14 & #18	605	1080
	18"	15	150	#4 to #11	160	915
		15	150	#14 & #18	480	1075
	20"	12	130	#4 to #11	155	865
		12	130	#14 & #18	385	1020
	22"	10	125	#4 to #11	165	775
		10	125	#14 & #18	320	995
	24"	9	120	#4 to #11	195	800
		9	120	#14 & #18	290	1150
	26"	7.3	100	#4 to #11	200	729
		7.3	100	#14 & #18	235	1035
	28"	6.3	95	#4 to #11	175	700
		6.3	95	#14 & #18	200	1075
	30"	5.5	90	#4 to #11	180	670
		5.5	90	#14 & #18	175	1015
	32"	4.8	85	#4 to #11	185	615
		4.8	85	#14 & #18	155	955
	34"	4.3	80	#4 to #11	180	600
		4.3	80	#14 & #18	170	855
	36"	3.8	75	#4 to #11	165	570
		3.8	75	#14 & #18	155	865
	40"	3.0	70	#4 to #11	165	500
		3.0	70	#14 & #18	145	765

(47) Proportionate Quantities (cont.)

3000 psi Concrete and 60,000 psi Reinforcing — Form and Reinforcing Quantities per C.Y.

Item	Type	Loading	Height	C.Y./L.F.	Forms/C.Y.	Reinf./C.Y.
Retaining Walls	Cantilever	Level Backfill	4 Ft.	0.2 C.Y.	49 S.F.	35 lb.
			8	0.5	42	45
			12	0.8	35	70
			16	1.1	32	85
			20	1.6	28	105
		Highway Surcharge	4	0.3	41	35
			8	0.5	36	55
			12	0.8	33	90
			16	1.2	30	120
			20	1.7	27	155
		Railroad Surcharge	4	0.4	28	45
			8	0.8	25	65
			12	1.3	22	90
			16	1.9	20	100
			20	2.6	18	120
	Gravity, with Vertical Face	Level Backfill	4	0.4	37	None
			7	0.6	27	↓
			10	1.2	20	
		Sloping Surcharge	4	0.3	31	
			7	0.8	21	
			10	1.6	15	

		Live Load in Kips per Linear Foot						
	Span	Under 1 Kip		2 to 3 Kips		4 to 5 Kips		6 to 7 Kips	
		Forms	Reinf.	Forms	Reinf.	Forms	Reinf.	Forms	Reinf.
Beams	10 Ft.	—	—	90 S.F.	170#	85 S.F.	175#	75 S.F.	185#
	16	130 S.F.	165#	85	180	75	180	65	225
	20	110	170	75	185	62	200	51	200
	26	90	170	65	215	62	215	—	—
	30	85	175	60	200	—	—	—	—

Item	Size	Type	Forms per C.Y.	Reinforcing per C.Y.
Spread Footings	Under 1 C.Y.	1,000 psf soil	24 S.F.	44 lb.
		5,000	24	42
		10,000	24	52
	1 C.Y. to 5 C.Y.	1,000	14	49
		5,000	14	50
		10,000	14	50
	Over 5 C.Y.	1,000	9	54
		5,000	9	52
		10,000	9	56
Pile Caps (30 Ton Concrete Piles)	Under 5 C.Y.	shallow caps	20	65
		medium	20	50
		deep	20	40
	5 C.Y. to 10 C.Y.	shallow	14	55
		medium	15	45
		deep	15	40
	10 C.Y. to 20 C.Y.	shallow	11	60
		medium	11	45
		deep	12	35
	Over 20 C.Y.	shallow	9	60
		medium	9	45
		deep	10	40

(47) Proportionate Quantities (cont.)

			3000 psi Concrete and 60,000 psi Reinforcing — Form and Reinforcing Quantities per C.Y.			
Item	Size	Pile Spacing	50 T Pile	100 T Pile	50 T Pile	100 T Pile
Pile Caps (Steel H Piles)	Under 5 C.Y.	24″ O.C.	24 S.F.	24 S.F.	75 lb.	90 lb.
		30″	25	25	80	100
		36″	24	24	80	110
	5 C.Y. to 10 C.Y.	24″	15	15	80	110
		30″	15	15	85	110
		36″	15	15	75	90
	Over 10 C.Y.	24″	13	13	85	90
		30″	11	11	85	95
		36″	10	10	85	90

		8″ Thick		10″ Thick		12″ Thick		15″ Thick	
	Height	Forms	Reinf.	Forms	Reinf.	Forms	Reinf.	Forms	Reinf.
Basement Walls	7 Ft.	81 S.F.	44 lb.	65 S.F.	45 lb.	54 S.F.	44 lb.	41 S.F.	43 lb.
	8		44		45		44		43
	9		46		45		44		43
	10		57		45		44		43
	12		83		50		52		43
	14		116		65		64		51
	16				86		90		65
	18						106		70

(48) Lift Slabs (Div. 033-130)

The cost advantage of the lift slab method is due to placing all concrete, reinforcing steel, inserts and electrical conduit at ground level and in reduction of formwork. Minimum economical project size is about 30,000 S.F. Slabs may be tilted for parking garage ramps.

It is now used in all types of buildings and has gone up to 22 stories high in apartment buildings. Current trend is to use post-tensioned flat plate slabs with spans from 22′ to 35′. Cylindrical void forms are used when deep slabs are required. One pound of prestressing steel is about equal to seven pounds of conventional reinforcing.

To be considered cured for stressing and lifting, a slab must have attained 75% of design strength. Seven days are usually sufficient with four to five days possible if high early strength cement is used. Slabs can be stacked using two coats of a non-bonding agent to insure that slabs do not stick to each other. Lifting is done by companies specializing in this work. Lift rate is 5′ to 15′ per hour with an average of 10′ per hour. Total areas up to 33,000 S.F. have been lifted at one time. 24 to 36 jacking columns are common. Most economical bay sizes are 24′ to 28′ with four to fourteen stories most efficient. Continuous design reduces reinforcing steel cost. Use of post-tensioned slabs allows larger bay sizes. Supplementary reinforcing, post-tensioned tendons, accessories and field labor cost about $2.10 per pound of tendon.

Table below shows the usual range and average S.F. cost of a typical lift slab project that has been designed to take advantage of lift slab techniques. Figures include Subs Overhead and Profit.

	Typical Cost per S.F. for Lift Slabs Project Incl. Subs O & P				
Item	Description	Typical Range		Average	Sub Total
Concrete Slabs	Edge and bulkhead forms	$ .05 to $ .36		$ .21	$4.46
	Reinforcing and post-tensioning steel	1.40 to 2.47		1.94	
	Reinforcing and accessories	.06 to .14		.10	
	Concrete cast at ground level	1.29 to 2.20		1.75	
	Float or trowel finish	.29 to .63		.46	
Lifting Slabs	Separating and curing compound	.03 to .07		.05	2.37
	Lifting collars	.23 to .48		.36	
	Lifting and welding	1.13 to 2.78		1.96	
Columns	Fabricated columns & accessories	.65 to 1.91		1.28	1.64
	Set initial stage	.05 to .20		.13	
	Column splicing	.12 to .33		.23	
Miscellaneous	Grout plates, patching, bolts, etc.	.05 to .20		.13	.13
Total in place per S.F.		$5.35 to $11.57		$8.60	$8.60

(49) Average C.Y. of Concrete (Div. 033-130)
Rubbing and floor finish not included — 4 uses of forms assumed, 5 story building.

Item	Line Number	Description	Strength in psi	Cost for 1 C.Y. of Concrete				
				Ready Mix	Place	Forms	Reinforcing	Total
Beams, 5 kip/L.F.	0300	10' span	3000	$51.05	$44.60	$238.75	$153.70	$488.10
	0350	25'		51.05	44.60	218.00	90.10	403.75
Beam & Slab, 1 way	2700	15'	3000	51.05	20.15	301.55	51.25	424.00
100 psf L.L.	2750	25'		51.05	20.15	196.55	64.65	332.40
Beam & Slab, 2 way	2900	15'	3000	51.05	20.15	227.00	54.60	352.80
100 psf L.L.	2950	25'		51.05	20.15	145.75	51.70	268.65
Columns, square	0820	16" x 16"	4000	54.30	36.35	287.55	285.30	663.50
tied	0920	24" x 24"		54.30	28.60	189.55	235.80	508.25
Columns, Tied	1220	16" diameter	4000	54.30	35.40	198.95	338.00	626.65
reinforced	1420	24" diameter		54.30	28.65	140.95	286.55	510.45
Flat Plate,	2100	15' span	4000	54.30	18.25	162.25	51.70	286.50
100 psf L.L.	2150	25'		54.30	18.30	87.80	46.35	206.75
Flat Slab,	1900	20'	4000	54.30	18.25	118.25	48.80	239.60
100 psf L.L.	1950	30'		54.30	18.30	79.75	53.75	206.10
Grade Wall,	4200	8" thick	3000	51.05	10.65*	232.50	18.45	312.65
8' high	4300	15" thick		51.05	9.15*	117.70	18.05	195.95
Metal Pan Joists,	2500	15' span	4000	54.30	18.25	157.90	19.25	249.70
100 psf L.L.	2550	25' span		54.30	18.25	129.50	33.25	235.30
Pile Caps	5900	under 5 C.Y.	3000	51.05	10.70*	46.60	21.85	130.20
	5950	over 10 C.Y.		51.05	4.50*	23.30	20.15	99.00
Slab on Grade	4650	4" thick	3500	52.70	8.70*	12.40	14.60	88.40
	4700	6" thick		52.70	5.85*	8.15	10.30	77.00
Spread Footings	3800	under 1 C.Y.	3000	51.05	17.40*	60.60	21.65	150.70
	3850	over 5 C.Y.		51.05	8.70*	18.45	21.05	99.25
Strip Footings	3900	9" x 18" plain	3000	51.05	7.95*	66.60	—	125.60
	3950	12" x 36" reinforced		51.05	7.95*	33.30	18.80	111.10
Waffle 30" Domes	2300	20' span	4000	54.30	18.30	129.70	55.65	257.95
100 psf L.L.	2350	30' span		54.30	18.30	103.25	53.05	228.90

*Direct chuting assumed. For equipment handling add 40% to 50% to placing costs.

CIRCLE REFERENCE NUMBERS

50 Granolithic Finish and Base (Div. 033)

Description	Granolithic Topping, Mix 1:1:1-1/2				Granolithic Base	
	(Line 454-0850)	1" Topping	(Line 454-0950)	2" Topping	(Line 130-0200) 1" x 5" Straight	
Cement @ $6.55 per bag	3.6 bags	$ 23.58	7.2 bags	$ 47.16	1.7 bags	$ 11.14
Sand @ $12.40 per C.Y.	3.6 C.F.	1.65	7.2 C.F.	3.31	1.7 C.F.	.78
Peastone @ $15.00 per C.Y.	5.4 C.F.	3.00	10.8 C.F.	6.00	1.85 C.F.	1.03
Clean, mix, place forms and finish. Crew C-10 @ $23.16/M.H.	4.07 M.H.	94.26	4.80 M.H.	111.17	13.70 M.H.	317.29
Total	100 S.F.	$122.49	100 S.F.	$167.64	100 L.F.	$330.24

51 Lightweight Concrete (Div. 035-212)

Vermiculite or Perlite comes in bags of 4 C.F. under various trade names. Weight is about 8 lbs. per C.F. For insulating roof fill use 1:6 mix. For structural deck use 1:4 mix over gypsum boards, steeltex, steel centering, etc. supported by closely spaced joists or bulb tees. For structural slabs use 1:3:2 vermiculite sand concrete over steeltex, metal lath, steel centering, etc. on joists spaced 2'-0" O.C. for maximum L.L. of 80#/S.F. Use same mix for slab base fill over steel flooring or regular reinforced concrete slab when tile, terrazzo or other finish is to be laid over.

For slabs on grade use 1:3:2 mix when tile, etc. finish is to be laid over. If radiant heating units are installed use a 1:6 mix for a base. After coils are in place, cover with a regular granolithic finish (mix 1:3:2) to a minimum depth of 1-1/2" over top of units.

Reinforce all slabs with 6 x 6 or 10 x 10 welded wire mesh.

Vermiculite concrete can be purchased ready mixed but the following breakdown is included for field mix. Prices given below are for a one story building and assume 50 C.Y. or more. For less than 50 C.Y. add 10%. For over one story add $2.60 per C.Y. Screed finish cost is included below. Ready mix 1:6 costs $72.70 per C.Y. delivered in truckload lots.

See 43 for prices of ready mix lightweight concrete.

Quantities per C.Y., Field Mix	(Line 0110) 1:6 Mix Insulating Roof Fill		1:3:2 Mix Lightweight Structural Concrete	
Portland cement @ $6.55 per bag	5.0 bags	$ 32.75	6.2 bags	$ 40.61
Vermiculite or Perlite @ $6.40 per bag	7.5 bags	48.00		
treated type @ $7.75 per bag			4.7 bags	36.43
Sand @ $12.40 per C.Y.			12.5 C.F.	5.74
Plant and water				
Labor, machine mix, hoist and place, Crew C-8 at $28.86/M.H.	1.12 M.H.	32.32	.91 M.H.	26.26
Total in place, per C.Y.		$113.07		$109.04

54 Integral Floor Finish (Div. 033-454)

Mix 1:1:2	(Line 0450) 1/2" Thick		(Line 0600) 1" Thick	
Cement at $6.55 per bag	1.7 bag	$11.14	3.4 bag	$22.27
Sand at $12.40 per C.Y.	0.8 C.F.	.37	1.6 C.F.	.73
Gravel at $13.10 per C.Y.	3.2 C.F.	1.55	6.4 C.F.	3.11
Mix, place and finish using Crew C-10 @ $69.50 per hr.	.84 hrs.	58.38	1.07 hrs.	74.37
Total for 100 S.F.		$71.44		$100.48

(55) Prestressed Precast Concrete Structural Units (Div. 034)

Type	Location	Depth	Span in Ft.	Live Load Lb. per S.F.	Cost per S.F. Incl. Subs O & P	
					Delivered	Erected
Double Tee	Floor	28" to 34"	60 to 80	50 to 80	$6.10 to 8.15	$6.75 to 8.90
(8' to 10')	Roof	12" to 24"	30 to 50	40	$4.05 to 6.10	$4.70 to 7.25
	Wall	Width 8'	Up to 55' high	Wind	$4.50 to $8.20	$5.25 to $9.65
Multiple Tee	Roof	8" to 12"	15 to 40	40	$ 3.65	$ 4.25
(8')	Floor	8" to 12"	15 to 30	100	$ 3.85	$ 4.45
Plank	Roof or Floor	Roof / Floor 4" 13/12, 6" 22/18, 8" 26/25, 10" 33/29, 12" 42/32		40 for Roof / 100 for Floor	$ 3.50 / 3.65 / 3.80 / 4.30 / 4.60	$ 4.45 / 4.55 / 4.60 / 4.95 / 5.20
Single Tee (8' to 10')	Roof	28" / 32" / 36" / 48"	40 / 80 / 100 / 120	40	$ 6.40* / 7.60* / 10.95* / 11.10*	$ 6.70* / 8.45* / 11.75* / 11.90*
AASHO Girder	Bridges	Type 4 / 5 / 6	100 / 110 / 125	Highway	$ 99/L.F. / 133 / 160	$129/L.F. / 158 / 186
Box Beam (4')	Bridges	15" / 27" / 33"	40 to 100	Highway	$101/L.F. / 144 / 167	$132/L.F. / 165 / 188

*Costs are for 10' wide members; for 8' wide members add $.55 per S.F.

The costs above are based on a project of 10,000 to 20,000 S.F. with a haul distance of 25 to 50 miles.

The majority of precast projects today utilize double tees rather than single tees because of speed and ease of installation. As a result casting beds at manufacturing plants are normally formed for double tees. Single tee projects will therefore require an initial set up charge of approximately $6600 to be spread over the individual single tee costs. The prices above for single tees includes this cost based on a project size of 15,000 square feet.

For floors, a 2" to 3" topping is field cast over the shapes. For roofs, insulating concrete or rigid insulation is placed over the shapes. Topping is not included in the above costs.

Hauling costs (incl. above) run from 45¢ per S.F. for short haul to 80¢ for 50 mile haul. Member lengths up to 40' are standard haul, 40' to 60' require special permits and lengths over 60' must be escorted. Over width and/or over length can add up to 100% on hauling costs.

Multiple tee erection runs between 55¢ to $1.25 per S.F. and single tee erection runs between 70¢ to $1.50 per S.F. Large heavy members may require two cranes for lifting which would increase these erection costs by about 45%. An eight man erection crew and crane will run about $3700 to $4200 per day

depending on location and size of crane. The crew can install 12 to 20 double tees, or 45 to 70 quad tees or planks per day.

The cost of supporting beams must be added to the above costs. Simple support beams run from $45 to $105 per L.F. delivered. Inverted tee beams run from $75 to $265 per L.F. delivered. Standard sized columns run $21 to $135 per L.F. delivered.

Grouting of connections must be added to the above. Typical costs are about $21 per connection but can go as high as $47 per connection.

Grouting planks run between 40¢ to 60¢ per S.F.

Single story buildings, including double tee roof members, supporting columns and girders, but no foundations, cost from $9.00 to $20.00 per S.F. of floor area. Parking garages run from $11.00 to $17.00 per S.F. above the foundations. Optimum parking garage design runs .02 C.Y. per S.F with overall costs for concrete in place running between $540 to $750 per C.Y.

Several system buildings utilizing precast members are available. Heights can go to 22 stories for apartment buildings with costs ranging from $12.00 to $29.00 per S.F. depending on the system, its components, its location and the degree of interior finish supplied. Optimum design ratio is 3 S.F. of surface to 1 S.F. of floor area.

(59) Brick, Block & Mortar Quantities (Div. 4)

	Running Bond				For Other Bonds Standard Size Add to S.F. Quantities in Table to Left		
Number of Brick per S.F. of Wall - Single Wythe with 3/8" Joints				**C.F. of Mortar per M Bricks, Waste Included**			
Type Brick	Nominal Size (incl. mortar) L H W	Modular Coursing	Number of Brick per S.F.	3/8" Joint	1/2" Joint	Bond Type	Description	Factor
Standard	8 x 2-2/3 x 4	3C=8"	6.75	10.3	12.9	Common	full header every fifth course	+20%
Economy	8 x 4 x 4	1C=4"	4.50	11.4	14.6		full header every sixth course	+16.7%
Engineer	8 x 3-1/5 x 4	5C=16"	5.63	10.6	13.6	English	full header every second course	+50%
Fire	9 x 2-1/2 x 4-1/2	2C=5"	6.40	550 # Fireclay	—	Flemish	alternate headers every course	+33.3%
Jumbo	12 x 4 x 6 or 8	1C=4"	3.00	23.8	30.8		every sixth course	+5.6%
Norman	12 x 2-2/3 x 4	3C=8"	4.50	14.0	17.9	Header = W x H exposed		+100%
Norwegian	12 x 3-1/5 x 4	5C=16"	3.75	14.6	18.6	Rowlock = H x W exposed		+100%
Roman	12 x 2 x 4	2C=4"	6.00	13.4	17.0	Rowlock stretcher = L x W exposed		+33.3%
SCR	12 x 2-2/3 x 6	3C=8"	4.50	21.8	28.0	Soldier = H x L exposed		—
Utility	12 x 4 x 4	1C=4"	3.00	15.4	19.6	Sailor = W x L exposed		–33.3%

Concrete Blocks Nominal Size	Approximate Weight per S.F.		Blocks per 100 S.F.	Mortar per M block	
	Standard	Lightweight		Partitions	Back up
2" x 8" x 16"	20 PSF	15 PSF	113	16 C.F.	36 C.F.
4"	30	20		31	51
6"	42	30		46	66
8"	55	38		62	82
10"	70	47		77	97
12"	85	55		92	112

(61) Common and Face Brick Prices (Div. 042)

Prices are based on truckload lot purchases for Common Brick and carload lots for Face Brick. Prices are per M (thousand) brick.

City	Material			Installation				Total			
	Brick per M Delivered		Mortar 3/8" Joint	Common in 8" Wall		Face Brick, 4" Veneer		Common in 8" Wall		Face Brick, 4" Veneer	
	Common	Face		Bare Costs	Incl. O&P	Bare Costs	Incl. O&P	Bare Costs	Incl. O&P	Bare Costs	Incl. O&P
Atlanta	$170	$210	$35 for 8" Wall and $29 for 4" Wall	$318	$ 475	$381	$ 570	$ 523	$ 698	$ 620	$ 829
Baltimore	200	260		432	646	519	775	667	902	808	1,089
Boston	255	325		601	897	721	1,077	891	1,214	1,075	1,463
Buffalo	265	295		516	771	619	925	816	1,098	943	1,278
Chicago	210	320		508	758	609	910	753	1,026	958	1,291
Cincinnati	185	225		455	679	546	815	675	919	800	1,091
Cleveland	190	245		550	821	659	985	775	1,066	933	1,283
Columbus	175	250		438	654	526	785	648	883	805	1,089
Dallas	195	230		308	460	370	552	538	711	629	834
Denver	160	215		374	558	448	670	569	771	692	935
Detroit	215	250		517	772	620	927	767	1,045	899	1,230
Houston	205	245		350	522	420	627	590	785	694	925
Indianapolis	215	285		430	642	516	770	680	915	830	1,112
Kansas City	220	260		443	662	532	795	698	941	821	1,110
Los Angeles	255	325		641	958	770	1,150	931	1,275	1,124	1,536
Memphis	195	235		318	475	381	569	548	725	645	857
Milwaukee	210	280		466	697	559	836	711	964	868	1,172
Minneapolis	215	290		476	712	572	854	726	984	891	1,202
Nashville	170	230		312	466	374	559	517	689	633	841
New Orleans	175	215		354	529	425	634	564	757	669	900
New York City	300	330		677	1,011	812	1,214	1,012	1,378	1,171	1,605
Philadelphia	210	270		554	827	664	993	799	1,095	963	1,318
Phoenix	245	350		378	565	454	679	658	871	833	1,092
Pittsburgh	235	285		458	684	549	821	728	979	863	1,163
St. Louis	195	260		515	769	618	923	745	1,020	907	1,237
San Antonio	220	270		301	450	362	540	556	729	661	866
San Diego	285	375		526	785	631	942	846	1,135	1,035	1,383
San Francisco	290	390		675	1,009	810	1,211	1,000	1,364	1,229	1,668
Seattle	285	350		505	755	606	906	825	1,105	985	1,319
Washington, D.C.	245	340		417	623	500	747	697	929	869	1,150
Average	$220	$280	▼	$460	$ 688	$552	$ 825	$ 715	$ 966	$ 862	$1,162

In figuring above installation costs, a D-8 Crew (with a daily output of 1.5M) was used for the 4" veneer. A D-8 Crew (with a daily output of 1.8M) was used for the 8" solid wall.

In figuring the total cost including overhead and profit, an allowance of 10% was added to the sum of the cost of the brick and mortar. Also, 3% breakage was included for both the bare costs and the costs with overhead and profit. If bricks are delivered palletized with 280 to 300 per pallet, or packaged, allow only 1-1/2% for breakage. Then add $10 per M to the cost of brick and deduct two hours helper time. The net result is a savings of $30 to $40 per M in place. Packaged or palletized delivery is practical when a job is big enough to have a crane or other equipment available to handle a package of brick. This is so on all industrial work but not always true on small commercial buildings.

There are many types of red face brick. The prices above are for the most usual type used in commercial, apartment house or industrial construction. If it is possible to obtain the price of the actual brick to be used, it should be done and substituted in the table. The use of buff and gray face is increasing, and there is a continuing trend to the Norman, Roman, Jumbo and SCR brick.

See (59) for brick quantities per S.F. and mortar quantities per M brick. (Average prices for the various sizes are listed in Division 4)

Common red clay brick for backup is not used that often. Concrete block is the most usual backup material with occasional use of sand lime or cement brick. Sand lime cost about $15 per M less than red clay and cement brick are about $5 per M less than red clay. These figures may be substituted in the common brick breakdown for the cost of these items in place, as labor is about the same. Occasionally common brick is being used in solid walls for strength and as a fire stop.

Several recent jobs have brick panels built on the ground floor and then crane erected to the upper floors. This allows the work to be done under cover and without scaffolding.

CIRCLE REFERENCE NUMBERS

(65) Brick Chimneys (Div. 042-054)

Quantities	16" x 16"		20" x 20"		20" x 24"		20" x 32"	
Brick at $250 per M	28 brick	$ 7.00	37 brick	$ 9.25	42 brick	$10.50	51 brick	$12.75
Type M mortar at $3.52 per C.F.	.5 C.F.	1.76	.6 C.F.	2.11	1.0 C.F.	3.52	1.3 C.F.	4.58
Flue tile (square)	8" x 8"	2.95	12" x 12"	5.75	2 @ 8" x 12"	9.20	2 @ 12" x 12"	11.50
Install tile & brick, crew D-1	.055 day	17.78	.073 day	23.59	.083 day	26.83	.10 day	32.32
Total per L.F. high		$29.49		$40.71		$50.05		$61.15

Labor costs are bare costs and do not include contractor's O&P. Labor for chimney brick using D-1 crew is 31 hours per thousand brick or about $626 per thousand brick. An 8" x 12" flue takes 33 brick and two 8" x 8" flues take 37 brick.

(70) Interlocking Grout Block Walls (Div. 042)

Cost per 100 S.F. of 8" x 16" Interlocking Grout Block, "Open End" Type						
8" x 16" Sand Aggregate	8" Thick Block		12" Thick Block		16" Thick Block	
113 block delivered	$1.06 ea.	$119.78	$1.55 ea.	$175.15	$2.25 ea.	$254.25
Type M mortar @ $3.52 per C.F.	7.0 C.F.	24.64	10.4 C.F.	36.61	13.8 C.F.	48.58
Minimum reinf. req. @ $.30 per lb.						
8" block #4 @ 24" horiz., 1 face	33.4 lb.	10.02				
32" vert., 1 face	40.0 lb	12.00				
12" to 16" block 8" horiz., 1 face			99.7 lb.	29.91	99.7 lb.	29.91
16" vert., 1 face			50.2 lb.	15.06	50.2 lb.	15.06
Grout type PM @ $3.00 C.F.	25.8 C.F.	77.40	42.2 C.F.	126.60	56.1 C.F.	168.30
Installation Crew D-4 @ $755.00 per day	.408 days	308.04	.455 days	343.53	.541 days	408.46
Total per 100 S.F.		$551.88		$726.85		$924.55
1 S.F.		5.52		7.27		9.25

CIRCLE REFERENCE NUMBERS

 Structural Steel Extras (Div. 051)

Principal Extras in Dollars Per Ton

Item quantity — using 5 tons per size as base price. Under 5 tons to 3 tons inclusive add $5 per ton. Under 3 tons to 2 tons inclusive add $10 per ton. Under 2 tons to 1 ton inclusive add $15 per ton. Under 1 ton to 1/2 ton add $50 per ton. Under 1/2 ton add $100 per ton.

Size Extras:

Wide Flange Shapes: W 36 x 359–135, W 33 x 354–118, W 30 x 235–99, W 27 x 217–84, W 24 x 176–104; W 21 x 166–101, W 18 x 143–76, W 16 x 100–67, W 14 x 426–145, W 12 x 190–136, add $82; W 24 x 103–55, W 21 x 93–44, W 18 x 71–35, W 16 x 57–26, W 14 x 132–22, W 12 x 50–14, W 10 x 112–12, W 8 x 67–10, W 6 x 25–9, W 5 x 19–16, and W 4 x 13 add $42; W 14 x 730–455 add $102.

Miscellaneous Shapes: M 14 x 18 to M 5 x 18.9 add $42. Standard Beams: S 24 x 121–80, S 20 x 96–66 add $82; S 18 x 70 & 54.7, S 15 x 50 & 42.9, S 12 x 50 & 40.8 add $62; S 12 x 35 & 31.8, S 10 x 35 & 25.4, S 8 x 23 & 18.4, S 7 x 20 & 15.3 and S 6 x 17.25 & 12.5 add $42.

Standard Channels: C 15 x 50–33.9 add $62; C 12 x 30–20.7, C 10 x 30–15.3, C 9 x 20–13.4, C 8 x 18.75–11.5, C 7 x 14.75–9.8 add $42.

Miscellaneous Channels: MC 18 x 58 to MC 6 x 12 add $82.

Car Building Sections: add $62.

Bulb Angles: add $82.

Angles Equal & Unequal Leg: add $62.

Cambering: Channels, standard beams, tees & wide flange shapes to 50 lbs./ft. add $60; 50 lbs./ft. to 300 lbs./ft. add $40.

Galvanizing under 1 ton, $550; over 20 tons, $440 per ton. For color coating of galvanizing add 40% to prices.

Government Specifications: MIL-5-20166 B, class U, type 1, grade M–medium add $24; grade HT–high tensile add $77. American Association of State Highway & Transportation Officials (AASHTO): Without Charpy Impacts, Grade 188 add $66; Grade M 222 add $141; M 223, Gr. 42 add $50; M 223, Gr. 50 add $56.

Cut Lengths: 10' to 20' add $40; 20' to 30' add $20; 30' to 40' add $5; over 40' to 50' add $3; over 50' to 60' add $4; over 60' to 65' add $4; over 65' to 80' add $21; over 80' to 90' add $24; over 90' to 100' add $26; over 100' add $26 + $4 for each additional 5' or fraction of 5'.

Milling One or Two Ends: 10 lbs./ft. to 50 lbs./ft., 10'–25' add $80; over 25' add $70; over 50 lbs./ft. to 100 lbs./ft., 10'–25' add $70; over 25' add $60; over 100 lbs./ft., 10'–25' add $60; over 25' add $50.

Handling & Loading: Under 5 tons to 3 tons add $2; 3 tons to 2 tons add $3; under 2 tons add $4. Banding or Wiring: over 1 ton add $1; under 1 ton add $2.

Special Testing: Band Tests, $7, Charpy Impact Testing: one set of tests — impact strength values only, $5; one set of tests — impact strength values, lateral expansion, percent shear fracture, $7.

Special Straightening: For tolerances not more restrictive than 50% of standard camber or sweep, $5.

Splitting Beams to Produce Tees: over 15 lbs./ft.–30 lbs./ft., $60, over 30 lbs./ft.–150 lbs./ft. $40, over 150 lbs./ft., $35.

Chemistry: Ladle analysis limits for carbon, manganese, silicon & copper tests run from $1 to $36 per ton.

(80) Welded Structural Steel (Div. 050-575)

Usual weight reductions with welded design run 10% to 20% compared with bolted or riveted connections. This amounts to about the same total cost compared with bolted structures since field welding is more expensive than bolts. For normal spans of 18' to 24' figure 6 to 7 connections per ton.

Trusses — For welded trusses add 4% to weight of main members for connections. Up to 15% less steel can be expected in a welded truss compared to one that is shop bolted. Cost of erection is the same whether shop bolted or welded.

General — Typical electrodes for structural steel welding are E6010, E6011, E60T and E70T. Typical buildings vary between

2# to 8# of weld rod per ton of steel. Buildings utilizing continuous design require about three times as much welding as conventional welded structures. In estimating field erection by welding, it is best to use the average linear feet of weld per ton to arrive at the welding cost per ton. The type, size and position of the weld will have a direct bearing on the cost per linear foot. A typical field welder will deposit 1.8# to 2# of weld rod per hour manually. Using semiautomatic methods can increase production by as much as 50% to 75%. Below is the cost per hour for manual welding.

Welded Structural Steel in Field			
Item	No Operating Engr.	1/2 Operating Engr.	1 Operating Engr.
3 lb. weld rod $.85 per lb.	$ 2.55	$ 2.55	$ 2.55
Equipment (for welding only) $160/40 hrs.	4.00	4.00	4.00
Operating cost at $4.40 per hour	4.40	4.40	4.40
Welder 1 hr. @ $24.10 per hour	24.10	24.10	24.10
Operating engineer @ $21.40 per hour	—	10.70	21.40
Total per Welder Hour (Bare Costs)	$35.05	$45.75	$56.45

The welding costs per ton of structural steel will vary from $22 to $182 per ton depending on Union requirements, crew size, design and inspection required.

(83) Plywood (Div. 061)

There are two types of plywood used in construction: interior, which is moisture resistant but not waterproofed, and exterior, which is waterproofed.

The grade of the exterior surface of the plywood sheets is designated by the first letter: A, for smooth surface with patches allowed; B, for solid surface with patches and plugs allowed; C, which may be surface plugged or may have knot holes up to 1″ wide; and D, which is used only for interior type plywood and may have knot holes up to 2-1/2″ wide. "Structural Grade" is specifically designed for engineered applications such as box beams. All CC & DD grades have roof and floor spans marked on them.

Underlayment grade plywood runs from 1/4″ to 1-1/4″ thick. Thicknesses 5/8″ and over have optional tongue and groove joints which eliminates the need for blocking the edges. Underlayment 19/32″ and over may be referred to as Sturd-i-Floor.

The price of plywood can fluctuate widely due to geographic and economic conditions. When one or two local prices are known, the relative prices for other types and sizes may be found by direct factoring of the prices in the table below.

Typical uses for various plywood grades are as follows:
AA-AD Interior — cupboards, shelving, paneling, furniture
B-B Plyform — concrete form plywood
CDX — wall and roof sheathing
Structural — box beams, girders, stressed skin panels
AA-AC Exterior — fences, signs, siding, soffits, etc.
Underlayment — base for resilient floor coverings
Overlaid HDO — high density for concrete forms & highway signs
Overlaid MDO — medium density for painting, siding, soffits & signs
303 siding — exterior siding, textured, striated, embossed, etc.

Grade	National Average Price in Lots of 10 MSF, per MSF-January 1991					
	Type	4′x8′	Type		4′x8′	4′x10′
Sanded Grade	1/4″ Interior AD	$330	1/4″ Exterior AC		$380	$390
	3/8″	385	3/8″		440	450
	1/2″	480	1/2″		500	510
	5/8″	525	5/8″		560	570
	3/4″	565	3/4″		630	640
	1″	675	1″		720	730
	1-1/4″	810	Exterior AA, add		105	100
	Interior AA, add	105	Exterior AB, add		90	90
			CD Structural 1		Underlayment	
Un-sanded Grade 4′x8′ Sheets	5/16″ CDX	$225			3/8″, 4′x8′ sheets	$290
	3/8″	240	5/16″, 4′x8′ sheets	$230	1/2″	325
	1/2″	290	3/8″	260	5/8″	400
	5/8″	360	1/2″	325	3/4″	460
	3/4″	420	5/8″	375	1-1/8″ 2-4-1	675
	3/4″ T&G	460	3/4″	425		
Form Plywood	5/8″ Exterior, oiled BB, plyform	$ 620	5/8″ HDO (overlay 2 sides)			$1,325
	3/4″ Exterior, oiled BB, plyform	710	3/4″ HDO (overlay 2 sides)			1,700
Overlaid 4′x8′ Sheets	Overlay 2 Sides MDO		Overlay 1 Side MDO			
	3/8″ thick	$ 740	3/8″ thick			$ 590
	1/2″	860	1/2″			670
	5/8″	950	5/8″			810
	3/4″	1,030	3/4″			870
303 Siding	Fir, rough sawn, natural finish, 3/8″ thick	$ 490	Texture 1-11	5/8″ thick, Fir		$ 630
	Redwood	1,350		Redwood		1,785
	Cedar	1,350		Cedar		1,785
	Southern Yellow Pine	340		Southern Yellow		460
Waferboard/ O.S.B.	1/4″ sheathing	$ 175	19/32″ T&G			$ 300
	7/16″ sheathing	225	23/32″ T&G			320

For 2 MSF to 10 MSF, add 10%. For less than 2 MSF, add 15%.

CIRCLE REFERENCE NUMBERS

84 Wood Stair, Residential (Div. 064-308)

One Flight with 8'-6" Story Height, 3'-6" Wide Oak Treads Open One Side, Built in Place

Item	Quantity	Unit Cost	Bare Costs	Costs Incl. Subs O & P
Treads 10-1/2" x 1-1/16" thick	11 Ea.	$ 20.00	$ 220.00	$ 242.00
Landing tread nosing, rabbeted	1 Ea.	7.00	7.00	7.70
Risers 3/4" thick	12 Ea.	10.10	121.20	133.32
Single end starting step (range $90 to $155)	1 Ea.	122.50	122.50	134.75
Balusters (range $3.50 to $9.00)	22 Ea.	6.25	137.50	151.25
Newels, starting & landing (range $36 to $85)	2 Ea.	60.50	121.00	133.10
Rail starter (range $29 to $80)	1 Ea.	54.50	54.50	59.95
Handrail (range $3.50 to $7.50)	26 L.F.	5.50	143.00	157.30
Cove trim	50 L.F.	.37	18.50	20.35
Rough stringers three - 2 x 12's, 14' long	84 B.F.	.46	38.64	42.50
Carpenter's installation: Bare Cost	36 Hrs.	$ 22.00	$ 792.00	
Cost incl. Subs O & P		33.75		$1,215.00
Total per Flight			$1,775.84	$2,297.22

Add for rail return on second floor and for varnishing or other finish. Adjoining walls or landings must be figured separately.

85 Wood Roof Trusses (Div. 061-908)

Loading figures represent live load. An additional load of 10 psf on the top chord and 10 psf on the bottom chord is included in the truss design. Spacing is 24" O.C.

Span in Feet	Flat	4 in 12 Pitch		5 in 12 Pitch		8 in 12 Pitch
	40 psf	30 psf	40 psf	30 psf	40 psf	30 psf
20	$ 49.00	$32.00	$37.00	$31.00	$36.00	$35.00
22	54.00	34.00	40.00	34.00	39.00	38.00
24	58.00	36.00	44.00	37.00	42.00	41.00
26	63.00	42.00	47.00	40.00	46.00	45.00
28	67.00	44.00	51.00	42.00	49.00	48.00
30	72.00	46.00	54.00	45.00	53.00	52.00
32	87.00	51.00	58.00	55.00	56.00	55.00
34	93.00	52.00	62.00	55.00	60.00	67.00
36	99.00	58.00	66.00	57.00	64.00	71.00
38	113.00	59.00	70.00	61.00	71.00	84.00
40	120.00	69.00	83.00	64.00	80.00	102.00

CIRCLE REFERENCE NUMBERS

⑧⑥ Thirty City Lumber Prices (Jan. 1st, 1991) (Div. 061)

Prices for boards are for #2 or better or sterling, whichever is in best supply. Dimension lumber is "Standard or Better" either Southern Yellow Pine (S.Y.P.), Spruce-Pine-Fir (S.P.F.), Hem-Fir (H.F.) or Douglas Fir (D.F.). The species of lumber used in a geographic area is listed by city. Rough sawn lumber is Douglas Fir, Hem-Fir, or a variety of hardwood, sheathing or lagging grade. Plyform is 3/4" BB oil sealed fir or S.Y.P. whichever prevails locally, 5/8" CDX is S.Y.P. or Fir.

These are prices at the time of publication and should be checked against the current market price. Relative differences between cities will stay approximately constant.

City	Species	Contractor Purchases per M.B.F.								Contractor Purchases per M.S.F.	
		S4S					Rough Sawn Lumber			3/4" Ext. Plyform	5/8" Thick CDX
		Dimensions			Boards		3"x12"	6"x12"	12"x12"		
		2"x4"	2"x6"	2"x10"	1"x6"	1"x12"					
Atlanta	S.Y.P.	$395	$370	$425	$515	$665	$575	$670	$720	$740	$375
Baltimore	S.P.F.	450	390	490	540	700	605	705	760	730	355
Boston	S.P.F.	350	350	420	490	635	600	700	755	730	350
Buffalo	S.P.F.	355	350	450	485	625	540	630	680	670	330
Chicago	S.P.F.	390	380	525	525	680	585	680	735	760	410
Cincinnati	S.Y.P.	345	325	430	450	585	510	595	640	700	355
Cleveland	S.P.F.	475	470	575	650	830	705	750	805	800	460
Columbus	S.P.F.	330	330	425	460	600	520	610	660	760	405
Dallas	S.Y.P.	430	400	510	560	725	625	725	780	790	405
Denver	H.F.	330	320	410	445	570	490	570	615	750	360
Detroit	S.P.F.	340	320	420	445	575	495	575	620	700	310
Houston	S.Y.P.	325	305	410	420	540	465	540	585	620	350
Indianapolis	S.P.F.	340	340	435	470	605	520	605	650	600	275
Kansas City	D.F.	385	380	460	520	670	575	670	720	650	360
Los Angeles	D.F.	410	405	500	560	725	625	725	780	770	375
Memphis	S.Y.P.	380	365	455	500	645	555	645	695	720	340
Milwaukee	S.P.F.	380	380	520	525	680	585	680	735	640	315
Minneapolis	S.P.F.	365	350	445	480	620	535	625	675	630	360
Nashville	S.Y.P.	305	305	395	420	545	470	545	590	680	345
New Orleans	S.Y.P.	355	350	450	485	630	545	635	685	600	295
New York City	H.F.	485	485	560	660	850	735	855	910	700	425
Philadelphia	H.F.	410	405	500	560	725	625	730	785	720	365
Phoenix	S.Y.P.	365	355	455	490	635	550	640	690	810	395
Pittsburgh	S.P.F.	390	385	485	530	690	595	695	750	640	305
St. Louis	S.Y.P.	345	340	380	470	610	525	610	660	800	330
San Antonio	S.Y.P.	350	340	445	470	610	530	615	665	600	310
San Diego	D.F.	330	325	405	450	590	545	640	690	750	375
San Francisco	D.F.	370	370	475	510	660	570	660	710	790	390
Seattle	D.F.	280	280	390	385	500	435	505	545	670	365
Washington, DC	H.F.	380	380	470	525	680	585	680	735	800	420
Average		$370	$360	$460	$500	$650	$560	$650	$700	$710	$360

To convert square feet of surface to board feet, 4% waste included

S4S Size	Multiply S.F. by	T & G Size	Multiply S.F. by	Flooring Size	Multiply S.F. by
				25/32" x 2-1/4"	1.37
1 x 4	1.18	1 x 4	1.27	25/32" x 3-1/4"	1.29
1 x 6	1.13	1 x 6	1.18	15/32" x 1-1/2"	1.54
1 x 8	1.11	1 x 8	1.14	1" x 3"	1.28
1 x 10	1.09	2 x 6	2.36	1" x 4"	1.24

(87) Lumber Product Material Prices (Div. 061, 062, 073 & 096)

The price of forest products fluctuates widely from location to location and from season to season depending upon economic conditions. The table below indicates National Average material prices in effect Jan. 1, 1991. The table shows relative differences between various sizes, grades and species. These percentage differentials remain fairly constant even though lumber prices in general may change significantly during the year.

Availability of certain items depends upon geographic location and must be checked prior to firm price bidding.

National Average Contractor Price, Quantity Purchase											
Dimension Lumber, S4S, #2 & Better, KD							**Heavy Timbers, Fir**				
	Species	2"x4"	2"x6"	2"x8"	2"x10"	2"x12"					
Framing Lumber per MBF	Douglas Fir	$355	$350	$340	$450	$465		3"x4" thru 3"x12"			$560
	Spruce	335	370	365	470	475		4"x4" thru 4"x12"			560
	Southern Yellow Pine	360	345	345	435	450		6"x6" thru 6"x12"			650
	Hem-Fir	400	400	410	485	490		8"x8" thru 8"x12"			650
	Redwood	940	940	940	995	995		10"x10" and 10"x12"			700

	S4S "D" Quality or Clear, KD					**S4S #2 & Better or Sterling, KD**					
	Species	1"x4"	1"x6"	1"x8"	1"x10"	1"x12"	Species	1"x4"	1"x6"	1"x8"	1"x10"	1"x12"
Boards per MBF *see also Cedar Siding	Sugar Pine	$1,375	$1,375	$1,200	$1,300	$1,740	Sugar Pine	—	$440	$430	$430	$650
	Idaho Pine	695	1,100	1,000	1,435	1,650	Idaho Pine	$700	700	700	700	710
	Engleman Spruce	600	975	975	1,000	1,400	Engleman Spruce	400	475	405	400	630
	So. Yellow Pine	450	575	560	530	670	So. Yellow Pine	310	320	310	305	375
	Ponderosa Pine	740	1,040	870	1,450	1,710	Ponderosa Pine	445	440	425	410	650
	Redwood, CVG	1,990	2,300	2,350	2,800	3,000						
	Finger Jt. Heart	1,530	1,800	1,990	2,140	2,140						

Flooring per MSF	1"x4" Vertical grain, Fir "B" & better	$1,800
	2-1/4"x25/32", Oak, clear	2,300
	Select	2,200
	#1 common	1,850
	Oak, prefinished, standard & better	2,650
	Standard	2,250
	2-1/4"x25/32" Maple, select	$2,400
	#2 & better	2,000
	2-1/4"x33/32" Maple, #2 & better	2,400
	3-1/4"x33/32" Maple, #2 & better	2,300
	Parquet, unfinished, 5/16", minimum	1,450
	Maximum	4,600

Siding per MSF	Clapboard, Cedar, beveled	
	1/2" x6" thru 1/2"x8", clear	$1,100
	"A" grade	930
	"B" grade	700
	3/4"x10" "clear"	1,820
	"A" grade	1,680
	Redwood, beveled	
	1/2"x6" thru 1/2"x8", vertical grain, clear	910
	3/4"x10" vertical grain, clear	1,720
	*Rough sawn, Cedar, T&G, "A" grade 1"x4"	$1,750
	1"x6"	2,150
	"STK" grade, 1"x6"	1,200
	Board, "STK" grade, 1"x8"	1,150
	Board & Batten 1" x 12"	1,200
	Cedar channel siding 1"x8", #3 & better	$1,150
	Factory stained	1,250
	White Pine siding, T&G, rough sawn	$390
	Factory stained	430

Shingles per CFS	Red Cedar	
	5X—16" long #1 regular	$90
	#2	60
	18" long perfections #1	95
	#2	60
	Resquared & Rebutted #1	70
	#2	65
	Handsplit shakes, resawn	
	24" long, 1/2" to 3/4"	90
	18" long, 1/2" to 3/4"	95
	White Cedar shingles	
	16" long, extra grade	$100
	Clear, 1st grade	90
	Fire retardant Red Cedar	
	5X — 16" long	$160
	18" long perfections	175
	Handsplit & resawn	
	24" long, 1/2" to 3/4"	175
	3/4" to 5/4"	185

CIRCLE REFERENCE NUMBERS

(90) Metal and Fiberglass Sandwich Panels (Div. 072-116 & 074-602)

Aluminum facing panels may be field erected with corrugated sheets or perforated acoustical type on the inside face and either corrugated, V-beam or ribbed exterior sheets. The most usual gauges are .032″ for the exterior and .024″ for the interior sheets. The V-beam type is available in .032″, .040″ and .050″. The insulation is generally 1″ thick. Individual sheets can be as large as 30′ x 2′.

The cost of the particular sandwich can be figured by adding the appropriate siding costs (Div. 074-602) to the insulation costs (Div. 072-116). Then add 15% per S.F. for multi-story construction and 25% per S.F. additional for jobs less than 2000 S.F. These figures do not include any supporting framework or flashings; add 2% to 4% for the usual flashings.

(91) Roof Decks — Comparisons (Div. 074-106)

Poured gypsum is cheaper than concrete plank but should not be used where interior humidity is excessive as in laundries, paper mills, swimming pools, etc. A 2″ thickness of gypsum is standard. This can be poured on insulating board, 1″ fiberglass, acoustical board or various combinations.

Precast lightweight concrete plank is not affected by moisture or temperature but costs more than poured gypsum. With metal edging it is suitable for spans up to 8′. Plank can be used on a

sloped roof. It makes a good nailing base for slate and is relatively permanent, requiring no upkeep.

Steel roof deck is lighter, hence lighter joists, beams and columns. Varying insulation characteristics.

Wood fiber planks cost more than gypsum, but have good acoustical and insulation characteristics.

(93) Mineral Fiber Roofing Shingles (Div. 073-103)

Quantities per Square	Strip Shingles 14″ x 30″ x 5/32″		Shakes 9.35″ x 16″ x 1/4″	
Colored Shingles	325 lb.	$118.00	500 lb.	$194.00
Felt, aluminum nails, special pieces		11.45		11.45
Carpenter at $22.00 per hour	2.0 M.H.	44.00	3.64 M.H.	80.10
Total Bare Cost per Square		$173.45		$285.55

(94) Front Door, Residential (Div. 082-078)

Figures below do not include Subs O & P.

Size 3′-0″ x 6′-8″ x 1-3/4″	Item Location	Deluxe Colonial Design Pine		Modern Flush Solid Core Birch	
Door, glazed with small panels	082-078		$260.00		$120.00
Frame and sill, stock unit	082-054		66.00		66.00
Entrance and interior trim (Colonial, range $140 to $1,150)	082-054		255.00		140.00
Hardware (with brass cylinder set)	087-116 & 120		100.00		100.00
Install door, carpenter @ $22.00 per hour		1.0 hr.	22.00	1.0 hr.	22.00
Install hardware		1.0 hr.	22.00	1.0 hr.	22.00
Install entrance, frame and trim		3.0 hrs.	66.00	1.5 hrs.	33.00
Complete in place			$791.00		$503.00

(95) Hollow Metal Doors (Div. 081-103 & 110)

Table below lists material prices only, not including hardware or labor.

Door Thickness and Size		Full Flush Doors, 18 Ga.				Flush Fire Doors			
		Hollow Core		Composite Core		Hollow Core "B"		Composite Core "A"	
		Plain	Glazed	Plain	Glazed	20 Ga.	18 Ga.	16 Ga.	18 Ga.
1-3/4″	3′-0″ x 6′-8″	$181	$230	$198	$254	$175	$198	—	$200
	3′-0″ x 7′-0″	188	238	205	265	184	205	$235	210
	3′-6″ x 7′-0″	209	250	225	285	—	225	266	227
	3′-0″ x 8′-0″	233	300	260	320	—	250	282	253
	4′-0″ x 8′-0″	267	330	290	345	—	285	—	285
1-3/8″	2′-0″ x 6′-8″	122*	163*	—	—	156	—	—	—
	2′-6″ x 6′-8″	130*	173*	—	—	172	—	—	—
	3′-0″ x 7′-0″	146*	184*	—	—	185	—	—	—

*Indicates 20 gauge doors.

CIRCLE REFERENCE NUMBERS

97 Wood Doors (Div. 082)

Table below lists price per door only, not including frame, hardware or labor. For pre-hung exterior door units up to 3' x 7', add $133 per door for wood frame and hardware for types not listed under pre-hung. Pricing is for ten or more doors. Doors are factory trimmed for butts and locksets.

Door Thickness and Size		Flush Type Doors					Architectural				Pre-hung	
		Hollow Core		Solid Particle Core			Pine		Fir		Pine Panel	Flush Birch Solid Core
		Lauan	Birch	Lauan	Birch	Oak	Panel	Glazed	Panel	Glazed		
1-3/4"	2'-6" x 6'-8"	$32	$36	$46	$49*	$58	$218	$208	$145	$170	$320	$177
	3'-0" x 6'-8"	34	39	49	53*	64	235	224	150	176	355	190
	3'-0" x 7'-0"	41	47	54	58*	69	247	248	160	177	371	198
	3'-6" x 7'-0"	—	—	—	68*	82	—	—	—	—	—	—
	4'-0" x 7'-0"	—	—	—	73*	88	—	—	—	—	—	—
1-3/8"	2'-0" x 6'-8"	28	31	41	43	51	98	—	147	—	197	155
	2'-6" x 6'-8"	29	35	43	46	55	102	—	149	—	202	157
	3'-0" x 6'-8"	31	38	46	50	61	120	—	155	—	220	165
	3'-0" x 7'-0"	38	46	51	55	66	—	—	162	—	—	—

*Add to the above for the following birch face door types:

Solid wood core, add $35
3/4 hour label door, add $60
1 hour label door, add $70

1-1/2 hour label door, add 100%
8' high door, add 30%
Acoustical door, add 220%

Static shielded door, add 500%
Lead lined door, add 450%
Vinyl laminated door, add 100%

98 Wood Interior Door, Residential (Div. 082-078)

Figures below do not include Subs Overhead & Profit.

Item	Item Location	6 Panel Pine	Flush Birch Hollow Core	Louvered Pine
Door 3'-0" x 6'-8" x 1-3/8" thick	082-078	$120.00	$ 38.00	$124.00
Frame 1-3/8" x 4-5/8", stock pine	082-054	35.00	35.00	35.00
Trim interior	062-212	24.48	24.48	24.48
Hardware incl. hinge & lockset	087-116 & 120	50.00	50.00	50.00
Install door, 2 Carp. @ $22.00 per hr./each	.9 hrs.	39.60	39.60	39.60
Install hardware, 1 Carp.	.7 hrs.	15.40	15.40	15.40
Install frame and trim, 1 Carp.	2.0 hrs.	44.00	44.00	44.00
Total in Place		$328.48	$246.48	$332.48

100 Steel Sash (Div. 085-102 & 104)

Ironworker crew will erect 25 S.F. or 1.3 sash unit per hour, whichever is less.

Mechanic will point 30 L.F. per hour.

Painter will paint 90 S.F. per coat per hour.
Glazier production depends on light size.
Allow 1 lb. special steel sash putty per 16" x 20" light.

CIRCLE REFERENCE NUMBERS

(101) Hinges (Div. 087-116)

All closer equipped doors should have ball bearing hinges. Lead lined or extremely heavy doors require special strength hinges.

Usually 1-1/2 pair of hinges are used per door up to 7'-6" high openings. Table below shows typical hinge requirements.

Use Frequency	Type Hinge Required	Type of Opening	Type of Structure
High	Heavy weight ball bearing	Entrances Toilet Rooms	Banks, Office buildings, Schools, Stores & Theaters Office buildings and Schools
Average	Standard weight ball bearing	Entrances Corridors Toilet Rooms	Dwellings Office buildings and Schools Stores
Low	Plain bearing	Interior	Dwellings

Door Thickness	Weight of Doors in Pounds per Square Foot				
	White Pine	Oak	Hollow Core	Solid Core	Hollow Metal
1-3/8"	3 psf	6 psf	1-1/2 psf	3-1/2 — 4 psf	6-1/2 psf
1-3/4"	3-1/2	7	2	4-1/2 — 5-1/4	6-1/2
2-1/4"	4-1/2	9	—	5-1/2 — 6-3/4	6-1/2

(103) Glazing Labor (Div. 088)

Glass sizes are estimated by the "united inch" (height + width). Table below shows the number of lights glazed in an eight hour period by the crew size indicated, for glass up to 1/4" thick. Square or nearly square lights are more economical on a S.F. basis. Long slender lights will have a high S.F. installation cost.

For insulated glass reduce production by 33%. For 1/2" plate glass reduce production by 50%. Production time for glazing with two glaziers per day averages; 1/4" plate glass 120 S.F.; 1/2" plate glass 55 S.F.; 1/2" insulated glass 95 S.F.; insulated glass 75 S.F.

Glazing Method	United Inches per Light							
	40"	60"	80"	100"	135"	165"	200"	240"
Number of Men in Crew	1	1	1	1	2	3	3	4
Industrial sash, putty	60	45	24	15	18	—	—	—
With stops, putty bed	50	36	21	12	16	8	4	3
Wood stops, rubber	40	27	15	9	11	6	3	2
Metal stops, rubber	30	24	14	9	9	6	3	2
Structural glass	10	7	4	3	—	—	—	—
Corrugated glass	12	9	7	4	4	4	3	—
Store fronts	16	15	13	11	7	6	4	4
Skylights, puttyglass	60	36	21	12	16	—	—	—
Thiokol set	15	15	11	9	9	6	3	2
Vinyl set, snap on	18	18	13	12	12	7	5	4
Maximum area per light	2.8 S.F.	6.3 S.F.	11.1 S.F.	17.4 S.F.	31.6 S.F.	47 S.F.	69 S.F.	100 S.F.
Daily Crew Cost	$180.40	$180.40	$180.40	$180.40	$360.80	$541.20	$541.20	$721.60

(105) Window Walls (Div. 089-206)

The table below shows the S.F. costs for 1-3/4" x 4-1/2" clear anodized tubular aluminum framing, flush glazed with fixed

1/4" clear polished float glass for jobs over 250 S.F. and includes overhead & profit.

Description	Total Window Wall Height								
	3'-0"			6'-0"			10'-0"		
Mullion Spacing	3'	5'	7'	3'	5'	7'	3'	5'	7'
Mullions only	$26.00	$22.00	$21.00	$21.00	$18.00	—	$18.00	—	—
1 Intermediate horizontal member	—	28.00	25.00	26.00	22.00	$21.00	22.00	$21.00	—
2	—	—	—	29.00	25.00	22.00	25.00	22.00	$21.00
3	—	—	—	30.00	28.00	25.00	28.00	24.00	22.00

Add to the above erected costs for the following:
1" insulating glass, plus 2" x 4-1/2" clear tube frame, add $12.00 per S.F.

Bronze anodized aluminum tubing, add 15% to 20%.
For screw applied square stops, add 15%.
For operating sash, add cost of each sash to total job.

(108) Studs, Joists and Track (Div. 092-612)

Material prices per 1000 L.F., for galvanized studs, joists and tracks. Panhead, framing screws, 7/16″ long are $8.45 per thousand; 1-5/8″ long are $13.60 per thousand.

Non-load bearing, 20 ga. stud and track are primarily used for curtain wall. (C.W.)

| Size | Non-Load Bearing | | | | Load Bearing 1-5/8″ Flange—Light Gauge Structural | | | | | |
| | 25 Ga. | | 20 Ga. (C.W.) | | 18 Ga. | | 16 Ga. | | 14 Ga. | |
	Stud	Track	Stud	Track	Stud	Track	Stud	Track	Stud	Track
1-5/8″	$195	$190	$345	$340						
2-1/2″	225	220	390	380	$ 755	$ 630	$ 945	$ 795	$1,140	$ 965
3-5/8″	260	250	465	455	860	745	1,060	920	1,270	1,130
4″	295	290	490	485	910	780	1,140	970	1,320	1,180
6″	375	365	615	600	1,100	960	1,390	1,260	1,540	1,490
8″					1,210	1,190	1,650	1,520	1,830	1,760

| Size | Non-Load Bearing | | | | Load Bearing-Extra Wide Flange - 2″ Flange | | | | | | | |
| | 25 Ga. | | 22 Ga. | | 18 Ga. | | 16 Ga. | | 14 Ga. | | 12 Ga. | |
	C-H Stud	J-Track	C-H Stud	J-Track	Joist	Track	Joist	Track	Joist	Track	Joist	Track
2-1/2″	$600	$480	$ 855	$670	—	$ 600	—	$ 755	—	$ 925	—	—
4″	725	585	1,200	860	$ 945	760	$1,150	930	$1,420	1,145	$2,060	$1,675
6″					1,140	970	1,390	1,180	1,715	1,450	2,400	2,115
8″					1,350	1,160	1,650	1,430	2,030	1,745	2,820	2,545
10″							1,910	1,670	2,370	2,060	3,425	2,960
12″							2,160	1,905	2,675	2,370	3,850	3,285

(110) Gypsum Plaster (Div. 092-108)

| Quantities for 100 S.Y. | 2 Coat, 5/8″ Thick | | 3 Coat, 3/4″ Thick | | |
| | Base | Finish | Scratch | Brown | Finish |
	1:3 Mix	2:1 Mix	1:2 Mix	1:3 Mix	2:1 Mix
Gypsum plaster	1300 lb.		1350 lb.	650 lb.	
Sand	2.6 C.Y.		1.85 C.Y.	1.35 C.Y.	
Finish hydrated lime		340 lb.			340 lb.
Gauging plaster		170 lb.			170 lb.

| Total, in Place for 100 S.Y. on Walls | 2 Coat, 5/8″ Thick | | | 3 Coat, 3/4″ Thick | | |
	Quantities	Bare Cost	Incl. O & P	Quantities	Bare Cost	Incl. O & P
Gypsum plaster @ $12.20 per 80 lb. bag	1300 lb.	$ 198.25	$ 218.10	2000 lb.	$ 305.00	$ 335.50
Finish hydrated lime @ $4.35 per 50 lb. bag	340 lb.	29.60	32.55	340 lb.	29.60	32.55
Gauging plaster @ $16.05 per 100 lb. bag	170 lb.	27.30	30.05	170 lb.	27.30	30.05
Sand @ $18.00 per C.Y.	2.6 C.Y.	46.80	51.50	3.2 C.Y.	57.60	63.35
J-1 crew @ $842.80 & $1,250.90 per day	.87 days	733.25	1,088.30	1.05 days	884.95	1,313.45
Cleaning, staging, handling, patching	.09 days	75.85	112.60	.10 days	84.30	125.10
Total per 100 S.Y. in place		$1,111.05	$1,533.10		$1,388.75	$1,900.00

CIRCLE REFERENCE NUMBERS

(117) Steel Lockers (Div. 105-054)

Single Tier	1-Wide	3-Wide	Double Tier	1-Wide	3-Wide
12″ x 12″ x 60″	$102.00	$ 85.25	12″ x 12″ x 30″	$61.15	$52.30
12″ x 15″ x 60″	108.15	89.00	12″ x 12″ x 36″	64.25	53.30
12″ x 18″ x 60″	115.70	93.90	12″ x 15″ x 36″	66.50	56.15
12″ x 12″ x 72″	112.55	94.90	12″ x 18″ x 36″	70.90	58.40
12″ x 15″ x 72″	119.10	99.35	15″ x 15″ x 36″	72.15	62.80
12″ x 18″ x 72″	123.45	102.20	**Multiple Tier**	**1-Wide**	**3-Wide**
15″ x 18″ x 72″	133.95	114.45	5-High 12″ x 12″ x 12″	$27.65	$23.65
18″ x 18″ x 72″	141.60	119.20	12″ x 15″ x 12″	29.70	24.50
18″ x 21″ x 72″	150.60	125.75	15″ x 15″ x 12″	32.45	27.95
			6-High 12″ x 12″ x 12″	26.55	22.45
			12″ x 15″ x 12″	27.95	23.25

Price per Opening, Material Only, Based on 100 Openings, Not Including Locks

(129) Pipe Material Costs and Considerations (Div. 026, 027 & 151)

1. Malleable fittings should be used for gas service.
2. Malleable fittings are used where there are stresses/strains due to expansion and vibration.
3. Cast fittings may be broken as an aid to disassembling of heating lines frozen by long use, temperature and minerals.
4. Cast iron pipe is extensively used for underground and submerged service.
5. Type M (light wall) copper tubing is available in hard temper only and is used for nonpressure and less severe applications than K and L.
6. Type L (medium wall) copper tubing, available hard or soft for interior service.
7. Type K (heavy wall) copper tubing, available in hard or soft temper for use where conditions are severe. For underground and interior service.
8. Hard drawn tubing requires fewer hangers or supports but should not be bent. Silver brazed fittings are recommended, however soft solder is normally used.
9. Type DMV (very light wall) copper tubing designed for drainage, waste and vent plus other non-critical pressure services.

Domestic/Imported Pipe and Fittings Cost

The prices shown in this publication for steel/cast iron pipe and steel, cast iron, malleable iron fittings are based on domestic production sold at the normal trade discounts. The above listed items of foreign manufacture may be available at prices of 1/3 to 1/2 those shown. Some imported items after minor machining or finishing operations are being sold as domestic to further complicate the system.

Caution: Most pipe prices in this book also include a coupling and pipe hangers which for the larger sizes can add significantly to the per foot cost and should be taken into account when comparing "book cost" with quoted supplier's cost.

CIRCLE REFERENCE NUMBERS

(130) Plumbing Fixtures (Div. 152)

Total labor hours to install fixtures.

Item	Rough-In	Set	Total Hours	Item	Rough-In	Set	Total Hours
Bath tub	5	5	10	Shower head only	2	1	3
Bath tub and shower, cast iron	6	6	12	Shower drain	3	1	4
Fire hose reel and cabinet	4	2	6	Shower stall, slate		15	15
Floor drain to 4 inch diameter	3	1	4	Slop sink	5	3	8
Grease trap, single, cast iron	5	3	8	Test 6 fixtures			14
Kitchen gas range		4	4	Urinal, wall	6	2	8
Kitchen sink, single	4	4	8	Urinal, pedestal or floor	6	4	10
Kitchen sink, double	6	6	12	Water closet and tank	4	3	7
Laundry tubs	4	2	6	Water closet and tank, wall hung	5	3	8
Lavatory wall hung	5	3	8	Water heater, 45 gals. gas, automatic	5	2	7
Lavatory pedestal	5	3	8	Water heaters, 65 gals. gas, automatic	5	2	7
Shower and stall	6	4	10	Water heaters, electric, plumbing only	4	2	6

Fixture prices in front of book are based on the cost per fixture set in place. The rough-in cost, which must be added for each fixture, includes carrier, if required, some supply, waste and vent pipe connecting fittings and stops. The lengths of rough-in pipe are nominal runs which would connect to the larger runs and stacks. The supply runs and DWV runs and stacks must be accounted for in separate entries. In the eastern half of the United States it is common for the plumber to carry these to a point 5' outside the building.

(137) Ductwork (Div. 157-250)

Duct weight in pounds per L.F., straight runs

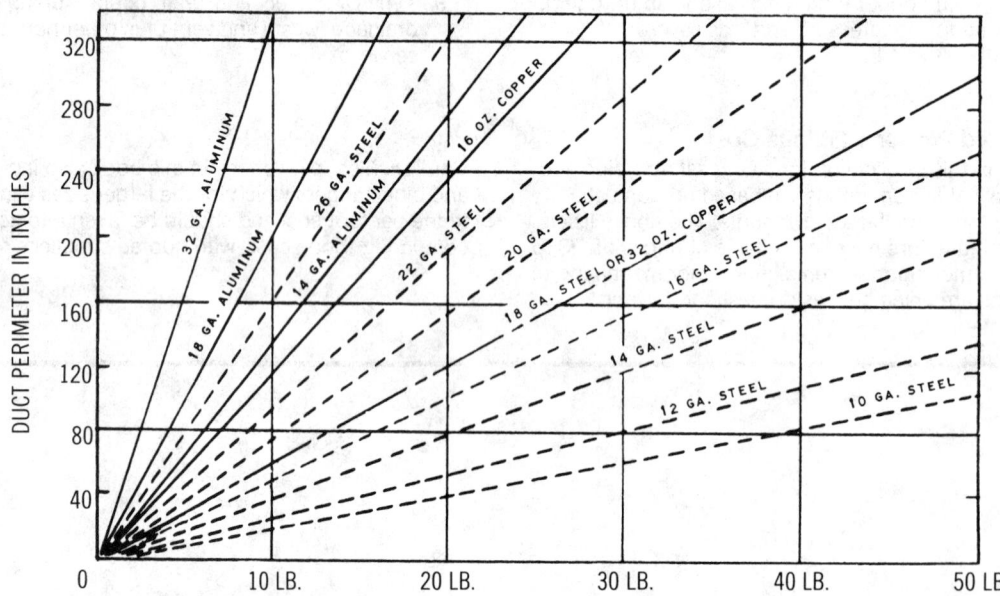

Add to the above for fittings; 90° elbow is 3 L.F.; 45° elbow is 2.5 L.F.; offset is 4 L.F.; transition offset is 6 L.F.; square-to-round transition is 4 L.F.; 90° reducing elbow is 5 L.F. For bracing and waste, add 20% to aluminum and copper, 15% to steel.

(139) Conductors in Conduit (Div. 160)

Table below lists maximum number of conductors for various sized conduit using THW, TW or THWN insulations.

Copper Wire Size	1/2"			3/4"			1"			1-1/4"			1-1/2"			2"			2-1/2"			3"		3-1/2"		4"	
	TW	THW	THWN	TW	THW	THWN	TW	THW	THWN	TW	THW	THWN	TW	THW	THWN	TW	THW	THWN	TW	THW	THWN	THW	THWN	THW	THWN	THW	THWN
#14	9	6	13	15	10	24	25	16	39	44	29	69	60	40	94	99	65	154	142	93		143		192			
#12	7	4	10	12	8	18	19	13	29	35	24	51	47	32	70	78	53	114	111	76	164	117		157			
#10	5	4	6	9	6	11	15	11	18	26	19	32	36	26	44	60	43	73	85	61	104	95	160	127		163	
#8	2	1	3	4	3	5	7	5	9	12	10	16	17	13	22	28	22	36	40	32	51	49	79	66	106	85	136
#6		1	1		2	4		4	6		7	11		10	15		16	26		23	37	36	57	48	76	62	98
#4		1	1		1	2		3	4		5	7		7	9		12	16		17	22	27	35	36	47	47	60
#3		1	1		1	1		2	3		4	6		6	8		10	13		15	19	23	29	31	39	40	51
#2		1	1		1	1		2	3		4	5		5	7		9	11		13	16	20	25	27	33	34	43
#1					1	1		1	1		3	3		4	5		6	8		9	12	14	18	19	25	25	32
1/0					1	1		1	1		2	3		3	4		5	7		8	10	12	15	16	21	21	27
2/0					1	1		1	1		1	2		3	3		5	6		7	8	10	13	14	17	18	22
3/0					1	1		1	1		1	1		2	3		4	5		6	7	9	11	12	14	15	18
4/0						1		1	1		1	1		1	2		3	4		5	6	7	9	10	12	13	15
250MCM								1	1		1	1		1	1		2	3		4	4	6	7	8	10	10	12
300								1	1		1	1		1	1		2	3		3	4	5	6	7	8	9	11
350									1		1	1		1	1		1	2		3	3	4	5	6	7	8	9
400											1	1		1	1		1	1		2	3	4	5	5	6	7	8
500											1	1		1	1		1	1		1	2	3	4	4	5	6	7
600															1		1	1		1	1	3	3	4	4	5	5
700																	1	1		1	1	2	3	3	3	5	5
750																	1	1		1	1	2	2	3	3	4	4

(140) Cable Cost Comparisons (Div. 161)

Table below lists material prices per C.L.F. for copper conductor cables. Aluminum wiring generally requires larger conductor sizes than copper wiring and is subject to very close tolerance in torque tightening of connections. Size of conduit must allow for the increased size of the aluminum conductors.

600V Wire Capacity		Size	Cost of Single Conductor, per C.L.F., Material Only, Copper Wire					
Aluminum THW	Copper THW		TW	THW	THWN/THHN	Bare Copper	5KV CLP Shielded	15KV CLP Shielded
	15 amp	#14	$ 4.00*	$ 4.80*	$ 4.10*	$ 4.05*	—	—
	20	#12	5.60*	6.70*	5.65*	5.60*	—	—
	30	#10	8.70*	9.80*	9.40*	9.35*	—	—
40 amp	45	#8	16.95	18.90	18.35	16.60*	—	—
50	65	#6	22.30	23.85	25.10	23.50*	—	—
65	85	#4	35.65	36.80	39.70	36.20*	$115.00	—
90	115	#2	53.40	54.65	70.95	53.50	139.00	—
100	130	#1	70.35	71.90	92.65	70.45	155.00	$183.00
120	150	1/0	—	84.80	110.00	83.40	178.00	227.00
135	175	2/0	—	102.85	127.00	100.80	217.00	257.00
155	200	3/0	—	126.00	156.00	123.50	—	—
180	230	4/0	—	156.00	192.00	153.00	291.00	338.00
205	255	250 MCM	—	199.00	226.00	180.00	320.00	353.00
230	285	300	—	256.00	269.00	235.00	—	—
250	310	350	—	270.00	298.00	250.00	435.00	480.00
310	380	400	—	330.00	356.00	310.00	—	—
		500	—	372.00	422.00	340.00	555.00	595.00
		600	—	565.00	—	443.00	—	—
		750	—	690.00	—	556.00	—	—
		1,000	—	1,005.00	—	810.00	—	—

* Solid conductor wire

CIRCLE REFERENCE NUMBERS

(143) Repair and Remodeling

Cost figures in MEANS REPAIR AND REMODELING COST DATA are based on new construction utilizing the most cost-effective combination of labor, equipment and material with the work scheduled in proper sequence to allow the various trades to accomplish their work in an efficient manner.

The costs for repair and remodeling work must be modified due to the following factors that may be present in any given repair and remodeling project:

1. Equipment usage curtailment due to the physical limitations of the project, with only hand-operated equipment being used.
2. Increased requirement for shoring and bracing to hold up the building while structural changes are being made and to allow for temporary storage of construction materials on above-grade floors.
3. Material handling becomes more costly due to having to move within the confines of an enclosed building. For multi-story construction, low capacity elevators and stairwells may be the only access to the upper floors.
4. Large amount of cutting and patching and attempting to match the existing construction is required. It is often more economical to remove entire walls rather than create many new door and window openings. This sort of trade-off has to be carefully analyzed.
5. Cost of protection of completed work is increased since the usual sequence of construction usually can not be accomplished.
6. Economies of scale usually associated with new construction may not be present. If small quantities of components must be custom fabricated due to job requirements, unit costs will naturally increase. Also, if only small work areas are available

at a given time, job scheduling between trades becomes difficult and subcontractor quotations may reflect the excessive start-up and shut-down phases of the job.

7. Work may have to be done on other than normal shifts and may have to be done around an existing production facility which has to stay in production during the course of the repair and remodeling.
8. Dust and noise protection of adjoining non-construction areas can involve substantial special protection and alter usual construction methods.
9. Job may be delayed due to unexpected conditions discovered during demolition or removal. These delays ultimately increase construction costs.
10. Piping and ductwork runs may not be as simple as for new construction. Wiring may have to be snaked through walls and floors.
11. Matching "existing construction" may be impossible because materials may no longer be manufactured. Substitutions may be expensive.
12. Weather protection of existing structure requires additional temporary structures to protect building at opening.
13. On small projects, because of local conditions, it may be necessary to pay a tradesman for a minimum of four hours for a task that is completed in one hour.

All of the above areas can contribute to increased costs for a repair and remodeling project. Each of the above factors should be considered in the planning, bidding and construction stage in order to minimize the increased costs associated with repair and remodeling jobs.

APPENDIX

TABLE OF CONTENTS

Historical Cost Indexes

The table below lists both the Means City Cost Index based on Jan. 1, 1975 = 100 as well as the computed value of an index based on January 1, 1991 costs. Since the Jan. 1, 1991 figure is estimated, space is left to write in the actual index figures as they become available thru either the quarterly "Means Construction Cost Indexes" or as printed in the "Engineering News-Record". To compute the actual index based on Jan. 1, 1991 = 100, divide the Quarterly City Cost Index for a particular year by the actual Jan. 1, 1991 Quarterly City Cost Index. Space has been left to advance the index figures as the year progresses.

Year	"Quarterly City Cost Index" Jan. 1, 1975 = 100 Est.	Actual	Current Index Based on Jan. 1, 1991 = 100 Est.	Actual	Year	"Quarterly City Cost Index" Jan. 1, 1975 = 100 Actual	Current Index Based on Jan. 1, 1991 = 100 Est.	Actual	Year	"Quarterly City Cost Index" Jan. 1, 1975 = 100 Actual	Current Index Based on Jan. 1, 1991 = 100 Est.	Actual
Oct. 1991					July 1978	122.4	56.0		July 1962	46.2	21.1	
July 1991					1977	113.3	51.9		1961	45.4	20.8	
April 1991					1976	107.3	49.1		1960	45.0	20.6	
Jan. 1991	218.5		100.0	100.0	1975	102.6	47.0		1959	44.2	20.2	
July 1990		215.9	98.8		1974	94.7	43.3		1958	43.0	19.7	
1989		210.9	96.5		1973	86.3	39.5		1957	42.2	19.3	
1988		205.7	94.1		1972	79.7	36.5		1956	40.4	18.5	
1987		200.7	91.9		1971	73.5	33.6		1955	38.1	17.4	
1986		192.8	88.2		1970	65.8	30.1		1954	36.7	16.8	
1985		189.1	86.5		1969	61.6	28.2		1953	36.2	16.6	
1984		187.6	85.9		1968	56.9	26.0		1952	35.3	16.2	
1983		183.5	84.0		1967	53.9	24.7		1951	34.4	15.7	
1982		174.3	79.8		1966	51.9	23.8		1950	31.4	14.4	
1981		160.2	73.3		1965	49.7	22.7		1949	30.4	13.9	
1980		144.0	65.9		1964	48.6	22.2		1948	30.4	13.9	
1979		132.3	60.5		1963	47.3	21.6		1947	27.6	12.6	

City Cost Indexes

Tabulated on the following pages are average construction cost indexes for 162 major U.S. and Canadian cities. Index figures for both material and installation are based on the 30 major city average of 100 and represent the cost relationship as of July 1, 1990. The index for each division is computed from representative material and labor quantities for that division. The weighted average for each city is a weighted total of the components listed above it, but does not include relative productivity between trades or cities.

The material index for the weighted average includes about 100 basic construction materials with appropriate quantities of each material to represent typical "average" building construction projects.

The installation index for the weighted average includes the contribution of about 30 construction trades with their representative man-days in proportion to the material items installed. Also included in the installation costs are the representative equipment costs for those items requiring equipment.

Since each division of the book contains many different items, any particular item multiplied by the particular city index may give incorrect results. However, when all the book costs for a particular division are summarized and then factored, the result should be very close to the actual costs for that particular division for that city.

If a project has a preponderance of materials from any particular division (say structural steel), then the weighted average index should be adjusted in proportion to the value of the factor for that division.

Adjustments to Costs

Time Adjustment using the Historical Cost Indexes:

$$\frac{\text{Index for Year A}}{\text{Index for Year B}} \times \text{Cost in Year B} = \text{Cost in Year A}$$

Location Adjustment using the City Cost Indexes:

$$\frac{\text{Index for City A}}{\text{Index for City B}} \times \text{Cost in City B} = \text{Cost in City A}$$

Adjustment from the National Average:

$$\text{National Average Cost} \times \frac{\text{Index for City A}}{100} = \text{Cost in City A}$$

Note: The City Cost Indexes for Canada can be used to convert U.S. national averages to local costs in Canadian dollars.

CITY COST INDEXES

	DIVISION	ALABAMA												ALASKA			ARIZONA		
		BIRMINGHAM			HUNTSVILLE			MOBILE			MONTGOMERY			ANCHORAGE			PHOENIX		
		MAT.	INST.	TOTAL	MAT.	INST.	TOTAL	MAT.	INST.	TOTAL	MAT.	INST.	TOTAL	MAT.	INST.	TOTAL	MAT.	INST.	TOTAL
2	SITE WORK	100.0	89.2	95.2	119.5	87.0	105.0	122.5	86.2	106.4	91.9	85.4	89.0	158.9	126.5	144.5	92.4	93.9	93.1
3.1	FORMWORK	97.7	70.2	76.4	103.3	61.0	70.5	106.6	72.0	79.8	112.5	64.2	75.1	124.4	136.2	133.6	108.6	85.3	90.5
3.2	REINFORCING	94.5	69.8	84.5	95.7	62.8	82.4	82.9	69.9	77.6	82.9	69.8	77.6	117.7	130.6	123.0	110.4	89.8	102.1
3.3	CAST IN PLACE CONC.	89.2	91.2	90.4	101.8	89.5	94.3	99.9	92.5	95.4	101.1	88.9	93.7	225.7	111.2	155.8	105.4	92.2	97.4
3	CONCRETE	92.1	81.1	85.1	100.8	76.0	85.0	97.4	82.5	87.9	99.3	77.6	85.5	181.3	122.7	144.1	107.2	89.3	95.8
4	MASONRY	81.6	75.2	76.7	88.4	62.1	68.3	93.8	75.4	79.7	86.5	50.9	59.2	149.9	133.3	137.2	93.5	77.1	81.0
5	METALS	95.6	76.9	89.0	100.0	72.1	90.2	93.4	77.7	87.9	95.7	76.9	89.1	116.3	122.4	118.4	99.1	90.5	96.1
6	WOOD & PLASTICS	92.1	71.5	80.6	107.5	64.0	83.3	91.9	74.2	82.0	101.7	68.0	83.0	117.9	132.6	126.1	99.3	83.8	90.7
7	MOISTURE PROTECTION	84.5	59.0	76.4	92.1	57.2	81.0	87.3	60.6	78.8	88.6	57.4	78.7	102.6	137.0	113.5	92.6	83.7	89.8
8	DOORS, WINDOWS, GLASS	90.7	70.6	80.3	100.9	58.9	79.2	98.4	71.5	84.5	98.0	65.3	81.1	128.5	125.8	127.1	103.0	81.5	91.9
9.1	LATH & PLASTER	95.8	67.9	74.7	91.1	66.2	72.3	91.8	77.8	81.2	108.3	66.7	76.7	120.3	137.8	133.5	93.4	89.1	90.2
9.2	DRYWALL	100.8	70.7	86.8	108.6	63.3	87.5	92.8	74.4	84.2	101.0	68.0	85.6	121.9	134.5	127.8	90.6	83.9	87.5
9.5	ACOUSTICAL WORK	97.7	70.7	83.1	100.1	63.0	80.0	93.1	73.2	82.3	93.1	66.8	78.9	123.9	133.8	129.2	103.6	82.7	92.3
9.6	FLOORING	112.0	71.4	101.1	97.5	62.1	88.0	114.0	76.8	104.1	100.9	45.1	86.0	117.3	130.2	120.7	93.2	86.1	91.3
9.8	PAINTING	104.2	66.7	74.5	110.6	65.4	74.7	121.4	75.7	85.2	119.6	74.4	83.8	123.1	137.8	134.8	96.3	79.1	82.6
9	FINISHES	103.3	69.2	85.1	105.3	64.1	83.3	100.4	75.2	87.0	102.4	68.4	84.3	121.1	135.5	128.8	92.8	82.6	87.4
10-14	TOTAL DIV. 10-14	100.0	76.3	93.1	100.0	75.5	92.9	100.0	80.3	94.3	100.0	74.9	92.7	100.0	133.8	109.8	100.0	89.8	97.0
15	MECHANICAL	96.5	69.6	83.1	99.3	70.0	84.7	97.3	72.2	84.8	99.0	67.4	83.3	107.2	123.6	115.4	98.4	84.6	91.5
16	ELECTRICAL	94.9	70.0	77.8	92.6	69.1	76.4	90.5	73.6	78.9	91.6	59.5	69.5	107.5	133.0	125.1	105.3	76.8	85.7
1-16	WEIGHTED AVERAGE	95.0	74.2	83.8	100.2	69.4	83.8	97.8	76.2	86.3	96.9	67.4	81.1	125.4	128.4	127.0	99.1	84.2	91.1

	DIVISION	ARIZONA			ARKANSAS						CALIFORNIA								
		TUCSON			FORT SMITH			LITTLE ROCK			ANAHEIM			BAKERSFIELD			FRESNO		
		MAT.	INST.	TOTAL	MAT.	INST.	TOTAL	MAT.	INST.	TOTAL	MAT.	INST.	TOTAL	MAT.	INST.	TOTAL	MAT.	INST.	TOTAL
2	SITE WORK	110.2	96.1	103.9	99.9	89.4	95.2	106.8	91.6	100.0	104.6	112.6	108.2	97.0	111.3	103.4	94.7	120.2	106.0
3.1	FORMWORK	109.1	85.1	90.5	111.2	63.8	74.4	103.4	63.6	72.6	104.6	122.6	118.6	124.8	122.6	123.1	110.1	122.3	119.5
3.2	REINFORCING	95.0	89.8	92.9	124.5	63.9	99.9	117.7	59.1	94.0	99.3	129.3	111.4	96.0	129.3	109.5	106.5	129.3	115.7
3.3	CAST IN PLACE CONC.	105.5	96.8	100.2	90.4	90.0	90.1	98.4	90.4	93.5	109.3	109.8	109.6	103.2	109.9	107.3	92.9	107.9	102.1
3	CONCRETE	103.8	91.6	96.1	102.2	77.4	86.5	103.7	77.2	86.9	106.1	116.5	112.7	105.8	116.6	112.7	99.4	115.4	109.6
4	MASONRY	92.3	77.1	80.7	95.3	70.3	76.1	88.8	70.3	74.6	108.8	129.4	124.6	100.9	112.7	109.9	119.7	110.4	112.6
5	METALS	90.8	92.1	91.3	96.5	73.2	88.3	106.2	70.2	93.5	99.1	121.6	107.0	99.3	121.8	107.2	94.9	123.1	104.8
6	WOOD & PLASTICS	106.2	83.3	93.5	107.2	65.0	83.7	94.9	65.0	78.3	96.2	117.6	108.1	95.2	117.6	107.7	96.8	119.4	109.4
7	MOISTURE PROTECTION	105.5	73.5	95.3	84.8	61.8	77.5	84.3	61.8	77.2	107.9	129.2	114.6	84.9	114.7	94.4	107.5	109.7	108.2
8	DOORS, WINDOWS, GLASS	88.1	81.5	84.7	92.7	58.0	74.7	95.1	58.1	76.0	93.3	122.1	108.2	99.9	116.1	108.3	101.1	118.0	109.9
9.1	LATH & PLASTER	109.0	86.0	91.5	92.9	70.4	75.8	98.4	70.4	77.2	97.2	131.2	122.9	92.2	99.4	97.7	102.1	114.0	111.1
9.2	DRYWALL	82.1	83.9	82.9	95.0	63.7	80.4	114.8	63.7	91.0	97.4	122.4	109.0	97.9	112.0	104.5	98.9	118.8	108.2
9.5	ACOUSTICAL WORK	113.6	82.7	96.9	83.7	63.7	72.9	83.7	63.7	72.9	81.4	118.2	101.3	93.2	118.2	106.8	96.7	120.2	109.4
9.6	FLOORING	110.0	82.1	102.5	89.5	70.9	84.5	88.7	70.9	84.0	116.9	137.6	122.4	111.7	100.3	108.7	88.7	99.1	91.5
9.8	PAINTING	98.5	78.1	82.3	111.0	49.1	61.9	104.7	62.0	70.8	108.3	117.9	115.9	120.1	119.7	119.8	107.9	98.0	100.0
9	FINISHES	93.1	81.8	87.1	94.5	59.6	75.9	105.1	64.0	83.2	101.6	122.1	112.5	102.8	113.6	108.5	97.5	110.0	104.2
10-14	TOTAL DIV. 10-14	100.0	89.0	96.8	100.0	72.4	92.0	100.0	73.0	92.1	100.0	126.6	107.7	100.0	124.1	107.0	100.0	143.9	112.7
15	MECHANICAL	98.7	88.7	93.8	97.2	62.9	80.2	96.8	66.7	81.8	96.8	120.5	108.6	95.0	95.3	95.2	92.8	111.7	102.2
16	ELECTRICAL	103.2	81.4	88.2	100.1	68.6	78.4	94.2	71.9	78.8	99.5	117.1	111.6	107.1	100.0	102.2	110.6	94.7	99.6
1-16	WEIGHTED AVERAGE	99.0	85.6	91.9	97.2	69.3	82.3	98.9	70.7	83.8	101.0	120.9	111.7	99.1	110.1	105.0	99.6	112.9	106.7

	DIVISION	CALIFORNIA																	
		LOS ANGELES			OXNARD			RIVERSIDE			SACRAMENTO			SAN DIEGO			SAN FRANCISCO		
		MAT.	INST.	TOTAL	MAT.	INST.	TOTAL	MAT.	INST.	TOTAL	MAT.	INST.	TOTAL	MAT.	INST.	TOTAL	MAT.	INST.	TOTAL
2	SITE WORK	97.8	116.0	105.9	101.9	104.6	103.1	98.7	111.6	104.5	86.4	105.1	94.7	95.3	108.3	101.1	102.1	116.4	108.5
3.1	FORMWORK	111.4	123.0	120.4	98.5	123.1	117.6	114.0	122.6	120.7	110.1	125.9	122.3	105.2	123.1	119.1	103.2	136.2	128.8
3.2	REINFORCING	86.6	129.3	103.9	99.3	129.3	111.4	124.5	129.3	126.4	99.3	129.3	111.4	117.6	129.3	122.3	123.1	129.3	125.6
3.3	CAST IN PLACE CONC.	96.7	112.9	106.6	102.3	110.5	107.3	102.3	110.2	107.1	115.8	107.1	110.5	99.4	104.7	102.6	100.3	117.6	110.8
3	CONCRETE	97.4	118.3	110.7	100.9	117.1	111.2	109.6	116.7	114.1	111.0	116.4	114.4	104.6	114.1	110.6	106.0	125.9	118.6
4	MASONRY	108.7	129.4	124.5	100.8	123.6	118.3	105.2	114.9	112.7	103.3	112.0	109.9	110.4	109.6	109.8	126.4	149.9	144.4
5	METALS	101.6	122.5	109.0	105.2	121.8	111.1	99.1	121.6	107.0	111.1	123.0	115.3	99.1	120.6	106.6	103.9	126.1	111.7
6	WOOD & PLASTICS	99.6	118.6	110.2	92.7	118.5	107.1	94.7	117.6	107.4	78.3	124.0	103.7	96.1	118.2	108.4	93.0	135.1	116.4
7	MOISTURE PROTECTION	103.9	131.7	112.7	90.2	128.8	102.4	90.7	124.7	101.5	85.3	119.6	96.2	94.5	110.0	99.4	100.4	134.8	111.3
8	DOORS, WINDOWS, GLASS	102.6	122.1	112.7	102.6	122.1	112.7	103.2	122.1	112.9	91.8	118.5	105.7	107.4	122.2	115.1	113.5	132.7	123.5
9.1	LATH & PLASTER	96.3	131.2	122.7	97.6	121.0	115.4	97.6	123.6	117.3	99.1	125.6	119.2	101.9	109.6	107.7	101.5	143.9	133.6
9.2	DRYWALL	89.1	122.4	104.6	98.7	119.7	108.5	94.8	122.4	107.6	97.4	124.3	109.9	99.9	118.8	108.7	81.0	138.0	107.6
9.5	ACOUSTICAL WORK	98.9	118.2	109.4	87.5	118.2	104.1	87.5	118.2	104.1	85.6	124.8	106.9	100.9	118.9	110.7	100.9	136.8	120.3
9.6	FLOORING	96.2	137.6	107.3	95.9	137.6	107.1	95.9	137.6	107.1	86.0	126.8	96.9	98.2	128.7	106.4	107.0	136.1	114.8
9.8	PAINTING	83.9	127.0	118.1	92.1	112.9	108.6	100.7	117.9	114.4	112.2	128.0	124.8	91.4	126.4	119.2	102.1	143.7	135.2
9	FINISHES	91.1	125.3	109.3	96.5	118.6	108.3	95.1	121.6	109.2	95.5	125.9	111.7	98.8	121.6	110.9	91.0	140.1	117.1
10-14	TOTAL DIV. 10-14	100.0	126.9	107.8	100.0	126.5	107.7	100.0	126.4	107.6	100.0	145.6	113.2	100.0	124.6	107.1	100.0	151.6	114.9
15	MECHANICAL	97.6	124.0	110.7	98.5	120.1	109.3	96.6	123.3	109.9	97.9	117.0	107.4	102.8	121.4	112.1	101.0	172.3	136.5
16	ELECTRICAL	101.9	126.4	118.8	99.5	112.7	108.6	99.0	120.9	114.1	110.6	90.8	97.0	105.7	103.1	103.9	108.0	148.6	136.0
1-16	WEIGHTED AVERAGE	99.3	123.7	112.3	99.4	118.9	109.8	99.6	119.7	110.3	99.8	115.1	108.0	101.7	115.0	108.8	103.3	143.7	124.9

393

	DIVISION	CALIFORNIA									COLORADO						CONNECTICUT		
		SANTA BARBARA			STOCKTON			VALLEJO			COLO SPRINGS			DENVER			BRIDGEPORT		
		MAT.	INST.	TOTAL	MAT.	INST.	TOTAL	MAT.	INST.	TOTAL	MAT.	INST.	TOTAL	MAT.	INST.	TOTAL	MAT.	INST.	TOTAL
2	SITE WORK	125.2	112.3	119.5	121.0	113.5	117.6	106.9	114.9	110.5	98.7	92.2	95.8	105.6	99.1	102.7	121.1	97.5	110.6
3.1	FORMWORK	115.0	122.8	121.0	106.7	122.2	118.7	113.6	135.2	130.3	103.4	69.5	77.1	94.4	78.9	82.4	122.5	85.9	94.1
3.2	REINFORCING	99.3	129.3	111.4	83.3	129.3	101.9	99.3	129.3	111.4	96.0	84.9	91.5	107.4	84.9	98.3	112.7	110.7	111.9
3.3	CAST IN PLACE CONC.	125.5	122.0	123.4	102.9	107.5	105.7	102.9	107.1	105.5	113.8	95.0	102.3	123.0	91.9	104.0	102.8	99.4	100.8
3	CONCRETE	117.5	122.9	121.0	99.2	115.2	109.3	104.2	120.1	114.3	107.8	84.1	92.7	113.8	86.2	96.3	108.9	95.1	100.2
4	MASONRY	118.5	121.0	120.4	112.3	112.0	112.1	109.8	137.0	130.6	105.9	81.4	87.1	104.0	81.8	87.0	106.3	94.6	97.4
5	METALS	96.1	125.5	106.4	91.6	122.8	102.6	88.8	123.0	100.8	91.4	87.5	90.1	95.7	86.6	92.5	91.8	105.6	96.6
6	WOOD & PLASTICS	108.6	118.2	113.9	86.1	119.5	104.7	99.2	135.5	119.4	86.3	69.0	76.7	91.4	80.5	85.3	107.0	82.9	93.6
7	MOISTURE PROTECTION	89.9	106.5	95.2	89.3	111.6	96.4	87.6	126.2	99.9	85.5	77.0	82.8	116.4	78.9	104.5	100.1	113.3	104.3
8	DOORS, WINDOWS, GLASS	103.9	122.1	113.3	95.3	116.5	106.3	99.1	132.7	116.5	98.3	77.3	87.4	90.8	80.9	85.7	102.1	96.3	99.1
9.1	LATH & PLASTER	104.4	112.7	110.7	100.2	116.5	112.5	100.2	122.4	117.0	101.9	89.5	92.5	88.0	92.8	91.6	106.0	91.4	95.0
9.2	DRYWALL	122.7	120.0	121.4	107.2	118.9	112.7	107.9	131.4	118.8	93.2	76.2	85.3	88.3	86.7	87.5	113.0	81.1	98.1
9.5	ACOUSTICAL WORK	87.5	118.2	104.1	85.6	120.2	104.4	88.5	136.8	114.6	94.7	67.9	80.2	96.5	80.1	87.6	105.9	82.8	93.4
9.6	FLOORING	101.9	123.2	107.6	84.9	104.6	90.2	84.3	136.1	98.2	108.0	76.0	99.4	99.5	97.9	99.1	85.1	95.1	87.8
9.8	PAINTING	118.9	112.9	114.1	102.9	101.4	101.7	105.9	128.0	123.5	117.5	74.4	83.3	104.1	87.4	90.9	121.4	80.6	89.0
9	FINISHES	114.5	117.1	115.9	100.0	111.8	106.3	100.7	130.4	116.6	99.3	75.7	86.7	93.0	87.6	90.1	106.9	82.7	94.0
10-14	TOTAL DIV. 10-14	100.0	124.8	107.2	100.0	144.1	112.7	100.0	148.4	114.0	100.0	88.9	96.8	100.0	90.5	97.2	100.0	110.7	103.1
15	MECHANICAL	98.6	119.9	109.2	96.9	111.3	104.1	95.8	134.8	115.2	97.8	86.3	92.1	97.1	85.8	91.4	103.6	96.7	100.1
16	ELECTRICAL	98.8	112.9	108.5	101.2	110.5	107.6	110.8	123.1	119.3	98.8	78.7	84.9	96.3	80.3	85.2	103.8	90.1	94.3
1-16	WEIGHTED AVERAGE	105.3	119.4	112.8	99.2	114.8	107.5	99.4	129.0	115.2	98.6	82.5	90.0	101.1	85.0	92.5	104.0	95.2	99.3

	DIVISION	CONNECTICUT												DELAWARE			D.C.		
		HARTFORD			NEW HAVEN			STAMFORD			WATERBURY			WILMINGTON			WASHINGTON		
		MAT.	INST.	TOTAL	MAT.	INST.	TOTAL	MAT.	INST.	TOTAL	MAT.	INST.	TOTAL	MAT.	INST.	TOTAL	MAT.	INST.	TOTAL
2	SITE WORK	100.2	97.8	99.1	116.9	95.7	107.5	124.8	99.3	113.4	109.1	96.8	103.6	115.1	101.6	109.1	89.9	94.1	91.8
3.1	FORMWORK	111.3	88.7	93.8	110.7	87.6	92.8	112.1	83.5	89.9	101.8	87.5	90.7	105.1	101.3	102.2	106.7	88.3	92.5
3.2	REINFORCING	115.2	110.7	113.4	112.7	110.7	111.9	130.7	110.7	122.6	115.2	110.7	113.4	112.7	100.7	107.8	108.6	87.7	100.1
3.3	CAST IN PLACE CONC.	98.3	99.4	99.0	95.1	98.9	97.4	117.8	99.8	106.8	114.9	99.5	105.5	99.0	113.5	107.8	102.7	91.0	95.6
3	CONCRETE	104.7	96.2	99.3	102.1	95.5	97.9	119.5	94.4	103.6	112.4	95.8	101.8	103.3	107.6	106.0	104.8	89.7	95.2
4	MASONRY	99.6	94.8	96.0	122.7	94.7	101.3	122.8	95.9	102.2	109.7	94.7	98.2	101.7	91.1	93.5	92.5	91.9	92.0
5	METALS	92.5	105.6	97.1	85.2	105.6	92.4	85.7	105.6	92.7	88.1	105.6	94.2	85.7	104.7	92.3	103.0	91.8	99.1
6	WOOD & PLASTICS	114.4	86.4	98.8	115.3	85.4	98.7	111.2	80.1	93.9	107.3	85.1	95.0	103.6	101.4	102.4	105.4	89.3	96.5
7	MOISTURE PROTECTION	101.2	96.0	99.5	88.2	110.6	95.3	87.9	113.3	96.0	88.7	95.8	90.9	89.1	120.4	99.0	106.7	94.1	102.7
8	DOORS, WINDOWS, GLASS	92.4	98.3	95.5	98.8	97.1	97.9	95.0	94.8	94.9	88.8	97.1	93.1	86.8	101.0	94.1	99.7	92.3	95.9
9.1	LATH & PLASTER	113.5	93.9	98.6	121.3	87.7	95.8	98.6	94.6	95.6	113.7	93.7	98.5	94.1	91.0	91.8	98.8	102.6	101.7
9.2	DRYWALL	108.2	89.6	97.6	113.0	81.0	98.1	117.9	80.9	100.7	113.4	84.6	100.0	102.5	101.3	101.9	121.1	89.2	106.2
9.5	ACOUSTICAL WORK	105.9	86.4	95.3	105.9	84.6	94.4	105.1	79.4	91.2	86.5	84.6	85.5	90.7	101.5	96.5	107.1	89.3	97.5
9.6	FLOORING	94.6	95.7	94.9	97.1	95.4	96.6	96.1	95.7	96.0	102.1	95.4	100.3	85.2	98.5	88.8	94.5	102.8	96.8
9.8	PAINTING	107.3	97.0	99.1	121.2	96.4	101.5	121.2	110.6	112.8	114.4	76.5	84.3	100.6	94.5	95.8	98.2	102.1	101.3
9	FINISHES	105.0	90.8	97.4	109.9	88.0	98.3	112.0	92.9	101.8	108.9	83.1	95.2	97.3	98.1	97.8	111.2	95.5	102.8
10-14	TOTAL DIV. 10-14	100.0	111.5	103.3	100.0	111.1	103.2	100.0	111.4	103.3	100.0	110.0	102.9	100.0	101.9	100.5	100.0	91.9	97.6
15	MECHANICAL	101.4	93.8	97.6	101.9	96.2	99.1	101.4	103.5	102.4	100.4	91.6	96.0	100.1	96.9	98.5	101.8	90.7	96.3
16	ELECTRICAL	100.2	91.7	94.3	93.8	89.7	90.9	95.6	118.6	111.4	92.4	75.1	80.5	106.3	119.1	115.1	97.3	92.5	94.0
1-16	WEIGHTED AVERAGE	100.6	95.6	97.9	101.4	95.6	98.3	104.3	101.0	102.5	100.8	92.1	96.2	98.8	103.0	101.1	101.8	91.7	96.4

	DIVISION	FLORIDA															GEORGIA		
		FT LAUDERDALE			JACKSONVILLE			MIAMI			ORLANDO			TAMPA			ATLANTA		
		MAT.	INST.	TOTAL	MAT.	INST.	TOTAL	MAT.	INST.	TOTAL	MAT.	INST.	TOTAL	MAT.	INST.	TOTAL	MAT.	INST.	TOTAL
2	SITE WORK	108.2	85.3	98.1	117.9	82.4	102.1	97.2	82.2	90.5	97.1	87.9	93.0	109.9	89.7	100.9	103.8	91.9	98.5
3.1	FORMWORK	107.3	73.0	80.7	104.5	69.9	77.7	109.0	72.8	80.9	103.8	70.5	78.0	99.8	73.9	79.8	85.7	74.9	77.4
3.2	REINFORCING	100.1	81.7	92.7	87.3	68.5	79.7	100.1	81.7	92.7	100.1	70.4	88.1	100.1	80.7	92.2	85.1	78.2	82.3
3.3	CAST IN PLACE CONC.	91.5	93.5	92.7	96.9	89.6	92.5	88.6	97.1	93.8	94.3	91.0	92.3	99.1	108.8	105.0	88.4	95.0	92.4
3	CONCRETE	96.6	84.5	88.9	96.2	80.0	85.9	95.2	86.2	89.5	97.5	81.2	87.1	99.5	92.7	95.2	87.2	85.6	86.2
4	MASONRY	98.9	86.1	89.1	90.5	60.1	67.2	92.9	72.8	77.5	93.7	60.7	68.5	96.9	76.1	80.9	89.2	74.0	77.6
5	METALS	87.0	84.8	86.2	94.8	76.9	88.5	86.7	87.7	87.0	86.7	77.9	83.6	98.0	90.8	95.4	110.2	83.6	100.9
6	WOOD & PLASTICS	107.1	78.4	91.1	102.8	73.2	86.3	107.8	77.6	91.0	102.7	71.5	85.4	102.3	77.7	88.6	89.4	78.2	83.2
7	MOISTURE PROTECTION	88.1	76.4	84.4	87.4	68.8	81.5	86.4	77.2	83.4	87.5	63.2	81.3	104.5	63.2	91.4	98.0	71.6	89.7
8	DOORS, WINDOWS, GLASS	86.6	74.7	80.4	89.7	68.0	78.5	93.2	74.7	83.6	88.3	67.5	77.6	96.5	65.2	80.3	92.4	75.4	83.6
9.1	LATH & PLASTER	100.8	85.8	89.4	102.9	65.6	74.6	104.0	76.8	83.4	102.2	63.5	72.9	100.8	65.8	74.3	112.0	77.4	85.8
9.2	DRYWALL	103.2	76.7	90.9	107.2	72.1	90.8	102.8	76.6	90.6	103.2	67.3	86.4	96.8	76.8	87.5	115.5	76.7	97.4
9.5	ACOUSTICAL WORK	91.0	76.8	83.3	91.0	72.2	80.8	100.2	76.8	87.6	91.0	70.5	79.9	92.9	76.9	84.2	92.0	77.3	84.0
9.6	FLOORING	104.1	86.7	99.4	104.1	63.0	93.1	105.6	74.9	97.4	103.0	63.4	92.4	100.2	77.3	94.1	101.2	80.8	95.8
9.8	PAINTING	113.1	67.4	76.8	102.7	65.9	73.5	111.8	67.4	76.5	100.1	70.5	76.6	108.6	63.7	73.0	95.0	82.6	85.2
9	FINISHES	103.4	74.8	87.3	104.7	68.9	85.6	104.2	73.3	87.7	101.8	68.2	83.9	98.6	71.6	84.2	108.3	79.1	92.8
10-14	TOTAL DIV. 10-14	100.0	87.2	96.3	100.0	74.9	92.7	100.0	86.8	96.1	100.0	80.0	94.2	100.0	80.9	94.4	100.0	78.8	93.8
15	MECHANICAL	100.7	80.2	90.5	99.9	74.9	87.5	97.5	88.9	93.2	96.8	76.9	86.9	97.2	80.1	88.7	102.7	79.6	91.2
16	ELECTRICAL	99.1	80.1	86.0	100.3	69.1	78.8	100.7	89.6	93.1	93.5	70.3	77.5	93.5	76.0	81.4	95.6	83.7	87.4
1-16	WEIGHTED AVERAGE	97.6	81.8	89.2	98.6	72.4	84.6	96.2	82.8	89.1	95.5	73.6	83.8	99.2	80.5	89.2	99.1	80.7	89.3

GEORGIA / HAWAII / IDAHO / ILLINOIS

	DIVISION	COLUMBUS MAT.	COLUMBUS INST.	COLUMBUS TOTAL	MACON MAT.	MACON INST.	MACON TOTAL	SAVANNAH MAT.	SAVANNAH INST.	SAVANNAH TOTAL	HONOLULU MAT.	HONOLULU INST.	HONOLULU TOTAL	BOISE MAT.	BOISE INST.	BOISE TOTAL	CHICAGO MAT.	CHICAGO INST.	CHICAGO TOTAL
2	SITE WORK	122.7	85.1	106.0	115.9	88.1	103.5	117.8	87.4	104.3	132.6	103.7	119.7	92.8	97.6	94.9	104.6	105.8	105.2
3.1	FORMWORK	105.2	57.1	67.9	98.7	64.6	72.3	101.2	68.5	75.9	132.2	107.5	113.1	110.3	92.1	96.2	80.9	112.6	105.5
3.2	REINFORCING	97.1	78.2	89.4	86.2	78.2	83.0	107.1	69.2	91.7	117.7	97.3	109.4	117.7	80.8	102.8	105.7	119.6	111.3
3.3	CAST IN PLACE CONC.	108.2	86.3	94.8	102.3	99.1	100.4	94.8	101.5	98.9	113.6	101.3	106.1	96.5	98.6	97.8	101.7	99.6	100.5
3	CONCRETE	105.1	74.1	85.4	98.0	83.7	88.9	98.8	85.7	90.5	118.2	103.4	108.8	104.0	94.5	98.0	98.5	106.5	103.6
4	MASONRY	92.6	47.2	57.8	84.3	54.6	61.5	92.1	68.9	74.4	125.6	109.4	113.1	101.7	81.7	86.4	97.3	108.1	105.6
5	METALS	95.6	82.9	91.1	90.6	87.1	89.4	100.5	82.2	94.1	110.2	98.8	106.2	93.4	86.9	91.1	89.7	111.5	97.4
6	WOOD & PLASTICS	103.5	59.3	78.9	92.5	66.5	78.0	102.6	70.4	84.7	121.3	109.2	114.6	92.3	90.8	91.5	93.6	111.5	103.6
7	MOISTURE PROTECTION	92.8	63.4	83.4	92.6	65.0	83.8	90.7	60.2	81.1	107.6	111.0	108.7	105.6	85.4	99.2	97.8	114.5	103.1
8	DOORS, WINDOWS, GLASS	89.7	61.1	74.9	93.3	66.4	79.4	91.4	63.1	76.7	114.0	103.1	108.3	104.9	80.5	92.2	110.1	113.1	111.6
9.1	LATH & PLASTER	103.3	53.0	65.2	102.9	59.5	70.0	97.6	57.6	67.3	116.8	115.7	116.0	114.3	90.5	96.3	107.0	108.2	107.9
9.2	DRYWALL	91.9	58.2	76.2	91.9	64.6	79.2	111.8	68.5	91.6	151.7	110.8	132.7	94.8	91.5	93.3	99.0	111.6	104.9
9.5	ACOUSTICAL WORK	101.5	57.9	77.9	101.5	65.3	81.9	101.6	69.4	84.2	118.7	109.0	113.4	112.9	90.5	100.7	98.6	111.7	105.7
9.6	FLOORING	82.2	49.7	73.5	88.1	57.2	79.8	95.7	70.5	88.9	132.7	112.1	127.2	94.0	88.5	92.5	98.9	110.2	101.9
9.8	PAINTING	117.0	55.4	68.1	116.5	67.7	77.7	116.8	68.6	78.5	124.1	117.4	118.8	100.4	67.6	74.3	92.8	111.4	107.5
9	FINISHES	93.3	56.3	73.6	94.6	64.9	78.8	107.6	68.1	86.6	141.3	113.3	126.4	97.1	82.9	89.5	98.5	111.2	105.3
10-14	TOTAL DIV. 10-14	100.0	73.9	92.4	100.0	75.5	92.9	100.0	74.1	92.5	100.0	120.7	106.0	100.0	84.2	95.4	100.0	109.8	102.8
15	MECHANICAL	98.4	63.7	81.1	98.7	68.5	83.7	98.4	70.6	84.6	111.2	104.3	107.8	96.5	92.9	94.7	96.6	104.7	100.6
16	ELECTRICAL	97.8	57.9	70.3	108.4	67.6	80.3	101.6	69.7	79.6	106.0	99.1	101.3	96.7	75.0	81.8	94.1	111.6	106.1
1-16	WEIGHTED AVERAGE	99.1	64.1	80.4	97.7	71.0	83.5	99.9	73.9	86.0	115.2	105.8	110.2	99.0	87.0	92.6	98.1	108.5	103.7

ILLINOIS / INDIANA

	DIVISION	PEORIA MAT.	PEORIA INST.	PEORIA TOTAL	ROCKFORD MAT.	ROCKFORD INST.	ROCKFORD TOTAL	SPRINGFIELD MAT.	SPRINGFIELD INST.	SPRINGFIELD TOTAL	EVANSVILLE MAT.	EVANSVILLE INST.	EVANSVILLE TOTAL	FORT WAYNE MAT.	FORT WAYNE INST.	FORT WAYNE TOTAL	GARY MAT.	GARY INST.	GARY TOTAL
2	SITE WORK	114.3	96.8	106.5	112.3	101.6	107.6	108.9	95.7	103.0	103.5	97.1	100.7	97.6	96.9	97.3	111.3	94.5	103.9
3.1	FORMWORK	117.2	94.3	99.5	125.1	101.3	106.6	104.3	93.1	95.7	105.7	91.7	94.8	119.1	90.3	96.7	113.7	100.1	103.1
3.2	REINFORCING	112.7	89.5	103.3	112.9	106.1	110.1	92.7	80.6	87.8	93.8	98.8	95.8	93.8	95.6	94.5	93.8	97.2	95.2
3.3	CAST IN PLACE CONC.	86.7	81.9	83.8	97.4	101.7	100.1	101.7	97.2	98.9	113.7	95.3	102.5	91.7	94.6	93.5	97.2	111.6	106.0
3	CONCRETE	98.6	87.5	91.5	106.4	101.9	103.6	100.2	94.2	96.4	107.6	94.2	99.1	97.6	93.0	94.7	99.7	105.8	103.6
4	MASONRY	99.2	87.9	90.6	90.1	98.5	96.5	107.1	83.2	88.8	88.1	94.1	92.7	95.4	83.1	86.0	95.5	104.5	102.4
5	METALS	87.7	85.0	86.8	108.0	104.3	106.7	92.9	85.9	90.4	93.8	95.3	94.3	99.8	96.2	98.5	89.5	101.7	93.8
6	WOOD & PLASTICS	113.6	94.7	103.1	107.2	99.2	102.7	103.1	93.1	97.5	93.0	90.9	91.9	102.9	89.9	95.7	113.0	102.0	106.9
7	MOISTURE PROTECTION	93.6	98.6	95.2	107.0	104.4	106.2	99.0	90.3	96.2	88.0	93.5	89.8	96.3	86.2	93.1	93.7	105.0	97.3
8	DOORS, WINDOWS, GLASS	93.8	88.5	91.1	100.5	97.2	98.6	105.4	84.6	94.7	105.4	84.4	94.7	98.1	85.7	91.7	98.6	95.7	97.1
9.1	LATH & PLASTER	103.9	94.3	96.6	105.2	96.4	98.6	103.8	87.3	91.3	101.2	95.0	96.5	101.9	75.7	82.0	117.3	94.4	100.0
9.2	DRYWALL	110.1	94.4	102.8	94.8	99.2	96.9	116.2	92.7	105.3	113.6	90.5	102.8	100.8	89.4	95.5	100.2	101.9	101.0
9.5	ACOUSTICAL WORK	84.0	94.5	89.7	83.9	99.4	92.3	83.9	92.8	88.7	103.8	90.6	96.6	84.9	89.6	87.4	90.2	102.0	96.6
9.6	FLOORING	105.1	90.7	101.2	123.3	94.8	115.7	115.2	86.3	107.5	97.9	90.5	95.9	98.9	85.8	95.4	100.9	106.4	102.4
9.8	PAINTING	115.8	91.3	96.3	96.9	98.1	97.9	90.8	80.9	82.9	103.5	90.2	92.9	95.2	88.0	89.5	92.4	93.8	93.5
9	FINISHES	107.4	93.0	99.7	100.7	98.4	99.5	110.6	87.9	98.5	108.0	90.7	98.8	98.6	87.8	92.9	99.2	99.0	99.1
10-14	TOTAL DIV. 10-14	100.0	91.8	97.6	100.0	100.9	100.2	100.0	89.6	97.0	100.0	95.3	98.6	100.0	98.7	99.6	100.0	107.6	102.2
15	MECHANICAL	96.1	88.9	92.5	101.6	94.9	98.3	97.1	82.1	89.6	97.1	82.1	89.6	98.3	83.3	90.8	97.6	93.3	95.5
16	ELECTRICAL	94.8	94.6	94.6	93.6	92.5	92.8	95.4	78.7	83.9	95.1	93.3	93.9	97.4	77.8	83.9	101.5	94.0	96.3
1-16	WEIGHTED AVERAGE	98.5	90.2	94.1	102.5	98.4	100.3	100.7	86.6	93.2	99.6	93.1	96.1	98.3	87.3	92.4	98.6	100.0	99.4

INDIANA / IOWA / KANSAS

	DIVISION	INDIANAPOLIS MAT.	INDIANAPOLIS INST.	INDIANAPOLIS TOTAL	SOUTH BEND MAT.	SOUTH BEND INST.	SOUTH BEND TOTAL	TERRE HAUTE MAT.	TERRE HAUTE INST.	TERRE HAUTE TOTAL	DAVENPORT MAT.	DAVENPORT INST.	DAVENPORT TOTAL	DES MOINES MAT.	DES MOINES INST.	DES MOINES TOTAL	TOPEKA MAT.	TOPEKA INST.	TOPEKA TOTAL
2	SITE WORK	101.3	94.5	98.3	101.3	97.8	99.7	92.3	86.1	89.5	107.8	88.4	99.2	103.5	91.0	97.9	109.9	84.2	98.5
3.1	FORMWORK	114.0	96.7	100.6	109.0	90.5	94.7	115.5	89.6	95.4	104.0	92.0	94.7	112.9	77.8	85.7	101.4	68.5	75.9
3.2	REINFORCING	109.8	98.8	105.4	105.9	90.5	99.7	93.8	92.9	93.5	97.7	86.4	93.1	101.8	76.6	91.6	87.6	95.1	90.6
3.3	CAST IN PLACE CONC.	83.5	95.0	90.5	90.0	103.3	98.1	92.8	98.2	96.1	91.6	96.4	94.5	115.3	92.6	101.4	91.8	94.5	93.5
3	CONCRETE	95.4	96.0	95.8	97.4	97.2	97.2	97.5	94.3	95.5	95.4	93.8	94.4	111.8	85.4	95.0	92.8	84.4	87.4
4	MASONRY	100.3	94.4	95.7	94.2	90.3	91.2	86.8	96.5	94.2	115.4	83.2	90.7	101.2	78.4	83.7	100.0	74.0	80.1
5	METALS	98.9	97.5	98.4	101.2	94.9	99.0	91.5	93.7	92.3	88.8	89.1	88.9	88.7	81.1	86.0	101.5	94.9	99.2
6	WOOD & PLASTICS	100.5	97.9	99.1	107.0	90.5	97.8	109.3	90.8	99.0	105.4	93.9	99.0	104.7	76.9	89.3	94.9	68.1	80.0
7	MOISTURE PROTECTION	92.6	95.7	93.6	90.1	90.9	90.3	93.3	89.3	92.0	90.5	89.2	90.1	87.4	74.1	83.2	96.6	73.9	89.4
8	DOORS, WINDOWS, GLASS	109.7	97.3	103.3	104.9	82.7	93.5	99.0	90.0	94.3	92.1	82.6	87.2	88.9	79.0	83.8	98.5	74.6	86.1
9.1	LATH & PLASTER	96.7	96.2	96.4	105.9	88.2	92.5	107.5	94.8	97.9	100.3	84.4	88.3	106.1	72.3	80.5	87.1	72.5	76.0
9.2	DRYWALL	89.6	97.2	93.2	101.1	89.5	95.7	100.9	90.4	96.0	102.6	90.4	96.9	115.9	75.1	96.9	91.2	66.9	79.9
9.5	ACOUSTICAL WORK	101.1	97.3	99.0	88.3	89.6	89.0	91.1	90.5	90.8	91.2	90.5	90.8	104.2	76.1	89.0	94.0	67.0	79.4
9.6	FLOORING	108.0	96.4	104.9	103.7	81.6	97.8	109.7	96.3	104.8	107.9	77.1	99.7	93.3	68.7	86.7	107.5	73.2	98.3
9.8	PAINTING	91.3	94.4	93.8	97.5	73.9	78.8	120.4	81.9	89.8	103.2	89.1	92.0	104.2	78.6	83.9	89.7	74.9	77.9
9	FINISHES	94.9	96.1	95.6	100.4	83.5	91.4	103.8	88.2	95.5	102.9	88.6	95.3	108.5	75.8	91.1	94.8	70.4	81.8
10-14	TOTAL DIV. 10-14	100.0	94.9	98.5	100.0	91.3	97.4	100.0	93.8	98.2	100.0	84.1	95.4	100.0	82.9	95.0	100.0	86.9	96.2
15	MECHANICAL	100.7	93.2	97.0	98.2	85.5	91.8	98.0	90.0	94.0	99.7	79.1	89.5	96.5	82.1	89.3	99.2	74.9	87.2
16	ELECTRICAL	99.8	97.2	98.0	97.9	85.0	89.0	109.1	92.5	97.6	98.0	87.0	90.4	99.4	81.9	87.3	98.7	74.5	82.0
1-16	WEIGHTED AVERAGE	99.1	95.5	97.1	98.7	89.8	93.9	97.9	92.2	94.8	98.6	86.6	92.2	99.5	81.5	89.9	98.5	78.0	87.6

DIVISION		KANSAS			KENTUCKY						LOUISIANA								
		WICHITA			LEXINGTON			LOUISVILLE			BATON ROUGE			NEW ORLEANS			SHREVEPORT		
		MAT.	INST.	TOTAL	MAT.	INST.	TOTAL	MAT.	INST.	TOTAL	MAT.	INST.	TOTAL	MAT.	INST.	TOTAL	MAT.	INST.	TOTAL
2	SITE WORK	114.1	91.7	104.1	92.4	89.4	91.1	106.8	92.0	100.2	98.5	76.9	88.9	101.8	85.7	94.6	106.2	78.8	94.0
3.1	FORMWORK	102.1	64.0	72.5	107.4	71.9	79.9	100.3	81.3	85.6	122.6	68.8	80.9	94.0	78.1	81.7	102.8	62.1	71.3
3.2	REINFORCING	87.6	98.8	92.1	101.3	92.2	97.6	117.7	92.2	107.4	100.4	63.1	85.3	94.1	73.0	85.5	100.4	61.9	84.8
3.3	CAST IN PLACE CONC.	94.4	99.8	97.7	83.7	92.0	88.8	77.9	102.7	93.0	97.2	85.8	90.2	90.6	87.6	88.8	86.4	88.5	87.7
3	CONCRETE	94.4	85.7	88.9	92.4	84.1	87.1	91.3	93.4	92.6	102.9	77.1	86.6	92.1	82.6	86.0	92.8	75.8	82.0
4	MASONRY	96.6	68.2	74.8	89.3	75.4	78.6	92.1	69.7	74.9	105.3	84.7	89.6	103.2	78.3	84.1	114.4	61.9	74.2
5	METALS	98.1	98.9	98.4	83.6	92.2	86.6	79.9	95.1	85.2	104.8	70.4	92.7	94.7	77.0	88.5	92.3	71.6	85.0
6	WOOD & PLASTICS	94.1	65.0	77.9	105.7	71.7	86.8	102.2	83.0	91.5	104.3	71.2	85.9	95.8	79.7	86.9	87.7	65.6	75.4
7	MOISTURE PROTECTION	96.9	64.3	86.6	84.8	70.9	80.4	84.9	73.5	81.3	111.2	69.3	97.9	111.4	69.3	98.0	87.1	62.3	79.2
8	DOORS, WINDOWS, GLASS	93.1	72.9	82.6	103.1	73.0	87.5	99.1	80.0	89.2	89.4	63.9	76.2	97.7	78.7	87.9	102.7	60.6	80.9
9.1	LATH & PLASTER	94.5	65.6	72.6	87.9	78.3	80.6	95.7	81.7	85.0	88.1	74.3	77.7	89.3	77.5	80.4	103.1	67.7	76.3
9.2	DRYWALL	90.7	65.0	78.7	92.0	75.3	84.2	90.3	82.3	86.6	100.3	70.1	86.2	88.3	78.4	83.7	91.0	64.3	78.6
9.5	ACOUSTICAL WORK	85.9	63.9	74.0	85.1	70.6	77.3	89.8	82.3	85.7	80.0	70.2	74.7	105.5	78.8	91.1	100.4	63.5	80.4
9.6	FLOORING	98.0	69.1	90.2	104.2	68.3	94.6	113.0	72.6	102.2	94.8	84.6	92.0	97.2	70.2	89.9	98.7	63.4	89.2
9.8	PAINTING	104.6	66.5	74.4	120.4	65.1	76.5	110.6	69.1	77.6	115.4	59.6	71.1	112.2	68.0	77.1	107.1	54.7	65.5
9	FINISHES	93.4	65.7	78.7	97.0	71.1	83.2	97.5	77.0	86.9	98.7	67.8	82.2	94.1	74.2	83.5	95.4	61.1	77.1
10-14	TOTAL DIV. 10-14	100.0	76.4	93.1	100.0	84.4	95.4	100.0	85.9	95.9	100.0	67.4	90.5	100.0	79.5	94.0	100.0	71.1	91.6
15	MECHANICAL	96.9	71.9	84.5	100.3	76.6	88.5	101.0	84.8	93.0	96.7	67.2	82.0	99.0	76.1	87.6	96.0	76.1	86.1
16	ELECTRICAL	103.2	71.2	81.1	103.1	76.3	84.6	101.0	76.1	83.8	98.0	79.6	85.4	98.0	78.8	84.8	95.4	73.5	80.3
1-16	WEIGHTED AVERAGE	97.9	75.6	86.0	95.6	78.6	86.5	95.9	82.8	88.9	100.3	74.0	86.2	98.4	78.6	87.8	96.8	70.5	82.7

DIVISION		MAINE						MARYLAND			MASSACHUSETTS								
		LEWISTON			PORTLAND			BALTIMORE			BOSTON			LAWRENCE			LOWELL		
		MAT.	INST.	TOTAL	MAT.	INST.	TOTAL	MAT.	INST.	TOTAL	MAT.	INST.	TOTAL	MAT.	INST.	TOTAL	MAT.	INST.	TOTAL
2	SITE WORK	98.0	95.1	96.7	101.1	105.6	103.1	100.6	87.0	94.5	108.6	104.1	106.6	107.4	102.4	105.2	102.9	100.4	101.8
3.1	FORMWORK	107.2	82.6	88.1	107.2	82.6	88.1	111.8	89.4	94.4	105.6	127.8	122.8	119.1	115.4	116.2	111.5	115.6	114.7
3.2	REINFORCING	117.7	81.2	102.9	117.7	81.2	102.9	104.3	99.4	102.3	117.8	122.2	119.5	98.2	110.4	103.2	116.2	119.7	117.6
3.3	CAST IN PLACE CONC.	92.7	92.6	92.7	92.7	93.1	92.9	123.5	93.4	105.1	114.5	121.8	119.0	111.5	106.9	108.7	108.3	107.5	107.8
3	CONCRETE	101.2	87.7	92.6	101.2	87.9	92.8	116.9	92.3	101.3	113.5	124.2	120.3	110.0	110.5	110.3	110.7	111.8	111.4
4	MASONRY	94.1	63.9	71.0	94.3	63.9	71.0	85.5	97.2	94.4	110.5	131.7	126.7	105.8	124.4	120.1	110.3	124.4	121.1
5	METALS	94.0	88.1	91.9	107.3	88.1	100.5	97.3	96.1	96.9	110.7	120.4	114.1	96.1	108.6	100.5	93.6	114.2	100.9
6	WOOD & PLASTICS	108.5	83.3	94.5	105.1	83.3	93.0	101.9	91.1	95.9	105.0	129.0	118.4	112.7	113.7	113.2	110.5	113.7	112.2
7	MOISTURE PROTECTION	88.3	62.3	80.0	86.8	62.3	79.0	92.2	84.7	89.8	106.9	135.4	115.9	98.7	132.1	109.3	99.3	132.7	109.9
8	DOORS, WINDOWS, GLASS	107.1	71.5	88.7	95.2	71.5	82.9	96.9	97.7	97.3	105.4	126.8	116.5	107.0	108.1	107.6	101.5	109.7	105.8
9.1	LATH & PLASTER	107.3	77.9	85.0	108.2	78.3	85.5	114.4	91.3	96.9	129.6	120.2	122.5	99.2	119.4	114.5	97.0	108.3	105.6
9.2	DRYWALL	106.0	89.8	98.5	106.0	89.8	98.5	111.8	90.5	101.9	123.4	119.0	121.4	105.7	109.9	107.7	106.7	109.9	108.2
9.5	ACOUSTICAL WORK	94.8	83.5	88.7	94.8	83.5	88.7	100.6	90.6	95.2	87.6	130.2	110.6	105.9	115.0	110.8	105.9	115.0	110.8
9.6	FLOORING	104.1	68.0	94.5	101.9	68.0	92.8	103.4	98.4	102.1	118.8	128.3	121.3	97.3	125.4	104.8	103.4	125.4	109.3
9.8	PAINTING	98.3	48.9	59.1	102.2	48.9	59.9	94.6	87.1	88.7	115.2	140.9	135.6	101.1	133.1	126.5	96.6	133.1	125.6
9	FINISHES	104.0	72.9	87.4	103.9	72.9	87.4	107.4	90.0	98.1	118.9	128.2	123.8	103.2	120.0	112.2	104.7	119.4	112.5
10-14	TOTAL DIV. 10-14	100.0	83.8	95.3	100.0	83.8	95.3	100.0	92.0	97.6	100.0	122.5	106.5	100.0	120.6	105.9	100.0	121.5	106.2
15	MECHANICAL	96.6	92.8	94.7	96.6	92.8	94.7	100.4	92.4	96.4	104.0	118.5	111.2	98.8	111.7	105.2	98.4	101.3	99.8
16	ELECTRICAL	94.4	70.6	78.0	97.8	70.6	79.0	101.2	94.5	96.6	98.9	134.1	123.2	101.1	88.8	92.6	101.1	88.7	92.5
1-16	WEIGHTED AVERAGE	98.2	80.1	88.5	99.0	80.6	89.1	101.6	93.1	97.0	107.4	125.3	117.0	102.5	111.5	107.3	102.0	110.1	106.4

DIVISION		MASSACHUSETTS						MICHIGAN											
		SPRINGFIELD			WORCESTER			ANN ARBOR			DETROIT			FLINT			GRAND RAPIDS		
		MAT.	INST.	TOTAL	MAT.	INST.	TOTAL	MAT.	INST.	TOTAL	MAT.	INST.	TOTAL	MAT.	INST.	TOTAL	MAT.	INST.	TOTAL
2	SITE WORK	96.2	103.5	99.4	105.5	103.8	104.8	98.6	102.6	100.4	92.5	104.3	97.7	97.9	97.0	97.5	81.7	87.6	84.3
3.1	FORMWORK	114.0	107.8	109.2	114.7	115.6	115.4	110.4	91.5	95.7	96.2	112.9	109.2	113.0	80.8	88.0	120.7	79.3	88.6
3.2	REINFORCING	98.2	114.2	104.7	98.2	119.7	106.9	109.2	109.2	109.2	88.0	109.2	96.6	109.2	109.2	109.2	117.7	79.2	102.1
3.3	CAST IN PLACE CONC.	99.4	106.7	103.8	106.2	103.8	104.7	88.9	104.4	98.3	110.9	113.6	112.6	103.7	95.5	98.7	96.2	93.2	94.4
3	CONCRETE	102.0	107.8	105.7	106.1	109.8	108.5	97.7	99.8	99.0	102.8	113.0	109.3	106.8	90.9	96.7	105.9	86.5	93.6
4	MASONRY	112.8	106.0	107.6	102.9	130.8	124.3	98.5	112.1	109.0	102.2	112.4	110.0	107.9	98.7	100.8	94.6	69.2	75.2
5	METALS	96.1	110.6	101.2	95.1	114.3	101.8	91.6	111.0	98.5	105.5	114.6	108.7	91.1	107.9	97.0	104.5	85.7	97.9
6	WOOD & PLASTICS	98.2	106.8	103.0	114.5	113.7	114.0	119.3	86.2	100.9	93.6	112.6	104.2	111.5	79.0	93.4	115.7	79.6	95.6
7	MOISTURE PROTECTION	97.8	111.0	102.0	99.9	122.6	107.1	90.6	102.4	94.4	94.8	116.7	101.8	85.1	84.4	84.9	111.5	69.8	98.3
8	DOORS, WINDOWS, GLASS	98.5	101.1	99.9	109.5	117.1	113.5	104.0	89.0	96.2	103.7	107.8	105.8	107.2	81.8	94.1	103.8	71.4	87.1
9.1	LATH & PLASTER	96.1	87.5	89.6	93.5	103.3	100.9	87.0	115.2	108.4	97.0	115.4	111.0	90.1	78.5	81.4	91.2	75.4	79.2
9.2	DRYWALL	104.9	96.1	100.8	106.7	108.5	107.6	87.8	96.5	91.9	89.5	112.7	100.3	111.8	76.4	95.3	114.9	76.7	97.1
9.5	ACOUSTICAL WORK	116.1	106.8	111.1	108.9	115.0	112.2	80.0	85.7	83.1	99.6	112.9	106.8	97.3	78.2	87.0	105.2	78.8	90.9
9.6	FLOORING	96.5	108.4	99.7	115.1	107.0	112.9	89.1	111.0	95.0	111.6	105.4	110.0	89.3	93.5	90.4	101.3	67.7	92.3
9.8	PAINTING	105.3	104.3	104.5	112.5	109.3	110.0	106.9	107.6	107.5	106.9	107.6	107.5	92.3	81.7	83.9	109.4	63.9	72.3
9	FINISHES	103.8	100.1	101.8	109.1	108.9	109.0	89.4	101.7	95.9	98.3	110.6	104.9	103.2	79.7	90.7	109.5	71.8	89.4
10-14	TOTAL DIV. 10-14	100.0	101.2	100.3	100.0	107.5	102.1	100.0	104.3	101.2	100.0	106.6	101.9	100.0	97.9	99.3	100.0	84.5	95.5
15	MECHANICAL	97.6	102.7	100.2	101.0	87.8	94.5	99.6	97.2	98.4	103.5	110.3	106.9	99.0	94.8	96.9	98.1	77.6	87.9
16	ELECTRICAL	95.4	96.5	96.2	97.2	107.5	104.3	99.3	104.0	102.5	100.3	106.8	104.8	99.3	98.6	98.8	97.9	59.9	71.8
1-16	WEIGHTED AVERAGE	99.6	103.8	101.8	102.6	109.2	106.1	97.5	101.8	99.8	100.9	110.6	106.1	100.0	93.1	96.3	101.5	76.1	87.9

MICHIGAN / MINNESOTA / MISSISSIPPI

		KALAMAZOO			LANSING			SAGINAW			DULUTH			MINNEAPOLIS			JACKSON		
DIVISION		MAT.	INST.	TOTAL	MAT.	INST.	TOTAL	MAT.	INST.	TOTAL	MAT.	INST.	TOTAL	MAT.	INST.	TOTAL	MAT.	INST.	TOTAL
2	SITE WORK	96.2	91.7	94.2	115.5	92.9	105.4	114.0	92.7	104.5	123.5	90.6	108.9	119.4	104.3	112.7	101.5	80.5	92.1
3.1	FORMWORK	115.2	75.3	84.3	116.5	88.9	95.1	106.2	80.7	86.4	120.8	80.4	89.5	90.8	103.4	100.6	102.6	61.3	70.6
3.2	REINFORCING	117.7	79.2	102.1	117.7	109.2	114.3	109.2	109.2	109.2	105.3	93.2	100.4	97.2	97.7	97.4	103.9	61.9	86.9
3.3	CAST IN PLACE CONC.	98.0	94.0	95.6	95.0	84.7	88.7	91.2	94.9	93.5	95.7	97.9	97.0	98.8	97.5	98.0	96.9	87.4	91.1
3	CONCRETE	105.9	85.4	92.9	104.4	88.5	94.3	98.2	90.6	93.4	102.8	90.6	95.1	96.8	99.8	98.7	99.6	75.0	84.0
4	MASONRY	98.0	74.0	79.6	106.9	76.9	83.9	98.7	73.2	79.2	110.2	86.0	91.6	100.4	100.4	100.4	101.7	60.0	69.8
5	METALS	101.4	85.6	95.8	110.4	104.0	108.2	100.1	107.9	102.8	105.3	95.2	101.8	95.9	96.7	96.2	85.9	71.1	80.7
6	WOOD & PLASTICS	96.8	73.9	84.1	114.4	90.6	101.1	107.2	79.0	91.5	100.9	79.1	88.8	99.6	102.1	101.0	93.1	64.5	77.2
7	MOISTURE PROTECTION	89.3	74.4	84.6	111.0	80.2	101.2	93.3	84.2	90.4	94.6	86.3	91.9	90.3	106.6	95.5	83.4	59.7	75.9
8	DOORS, WINDOWS, GLASS	95.1	71.7	83.0	109.1	88.9	98.6	98.5	80.8	89.3	94.7	80.5	87.3	105.0	100.5	102.6	98.3	60.8	78.9
9.1	LATH & PLASTER	89.1	72.9	76.8	90.2	81.7	83.8	99.9	79.5	84.5	105.8	82.9	88.5	90.0	104.8	101.2	97.1	59.5	68.6
9.2	DRYWALL	107.9	73.7	92.0	116.5	83.5	101.1	94.4	76.4	86.0	107.2	78.3	93.7	99.7	103.1	101.3	114.9	61.9	90.2
9.5	ACOUSTICAL WORK	99.6	73.8	85.6	99.6	90.2	94.5	104.5	78.2	90.3	96.0	78.4	86.4	97.8	102.5	100.4	88.3	62.7	74.4
9.6	FLOORING	97.2	68.4	89.5	102.4	75.4	95.2	101.3	70.3	93.0	93.4	83.8	90.8	92.3	98.3	93.9	98.8	61.7	88.9
9.8	PAINTING	110.2	74.8	82.1	111.8	83.4	89.2	121.2	70.8	81.2	119.8	96.0	100.9	108.1	104.4	105.1	114.9	53.6	66.3
9	FINISHES	104.7	73.7	88.2	110.9	83.3	96.2	99.6	74.4	86.1	104.5	85.1	94.2	98.5	103.3	101.0	108.8	58.9	82.2
10-14	TOTAL DIV. 10-14	100.0	90.4	97.2	100.0	99.3	99.8	100.0	95.7	98.7	100.0	88.2	96.5	100.0	98.0	99.4	100.0	69.0	91.0
15	MECHANICAL	97.9	83.1	90.6	99.6	84.2	92.0	99.6	79.3	89.5	98.9	87.6	93.3	101.0	100.0	100.5	99.6	60.0	79.9
16	ELECTRICAL	94.1	77.9	82.9	100.5	84.6	89.6	99.3	92.3	94.4	97.9	86.6	90.1	93.7	103.9	100.8	95.5	62.1	72.5
1-16	WEIGHTED AVERAGE	99.1	80.5	89.2	105.4	86.3	95.2	99.9	84.8	91.9	102.2	87.5	94.4	99.6	101.0	100.4	97.9	65.3	80.5

MISSOURI / MONTANA / NEBRASKA

		KANSAS CITY			ST LOUIS			BILLINGS			GREAT FALLS			LINCOLN			OMAHA		
DIVISION		MAT.	INST.	TOTAL	MAT.	INST.	TOTAL	MAT.	INST.	TOTAL	MAT.	INST.	TOTAL	MAT.	INST.	TOTAL	MAT.	INST.	TOTAL
2	SITE WORK	91.4	103.9	96.9	83.4	97.3	89.6	99.4	94.4	97.2	94.0	93.6	93.8	103.8	87.4	96.5	113.7	94.1	105.0
3.1	FORMWORK	100.2	96.0	96.9	94.6	109.8	106.3	104.3	76.0	82.3	124.7	76.5	87.4	116.3	61.2	73.6	102.9	71.9	78.9
3.2	REINFORCING	78.0	95.1	84.9	82.3	102.7	90.5	112.5	77.5	98.3	112.7	77.5	98.4	107.4	67.6	91.3	96.7	67.6	84.9
3.3	CAST IN PLACE CONC.	95.2	94.5	94.8	81.7	101.5	93.8	103.3	93.8	97.5	115.0	93.2	101.7	104.6	91.5	96.6	93.6	93.1	93.3
3	CONCRETE	92.3	95.1	94.1	84.4	104.9	97.4	105.5	85.4	92.8	116.4	85.4	96.7	107.6	77.5	88.5	96.2	82.6	87.5
4	MASONRY	101.3	94.5	96.1	94.9	107.1	104.2	109.1	82.8	88.9	120.1	86.6	94.4	100.7	53.1	64.2	105.9	71.2	79.3
5	METALS	104.5	94.3	100.9	101.1	100.4	100.9	95.5	82.4	90.9	88.7	82.4	86.5	85.6	75.6	82.1	108.5	75.7	97.0
6	WOOD & PLASTICS	110.4	95.3	102.0	85.2	107.9	97.8	90.0	74.6	81.5	99.0	75.2	85.8	103.6	58.8	78.7	101.1	72.1	85.0
7	MOISTURE PROTECTION	102.3	99.8	101.5	93.5	106.2	97.5	100.1	73.6	91.7	102.3	72.8	92.9	88.5	61.7	80.0	105.7	70.5	94.6
8	DOORS, WINDOWS, GLASS	93.1	97.9	95.6	101.4	115.5	108.7	89.8	69.4	79.3	101.3	70.1	85.1	98.6	66.1	81.8	111.6	72.2	91.2
9.1	LATH & PLASTER	106.0	92.2	95.6	113.2	104.9	106.9	104.5	70.5	78.7	109.5	71.4	80.7	96.6	66.6	73.9	91.1	74.9	78.8
9.2	DRYWALL	97.6	94.7	96.3	95.0	108.2	101.2	103.6	72.1	88.9	106.1	72.7	90.5	90.1	64.4	78.1	91.9	72.6	82.9
9.5	ACOUSTICAL WORK	106.9	94.8	100.4	101.0	108.3	105.0	100.2	73.7	85.9	100.9	74.3	86.5	88.7	57.4	71.7	93.9	71.1	81.6
9.6	FLOORING	95.8	99.8	96.8	102.5	104.9	103.1	97.9	71.9	91.0	105.4	83.0	99.4	110.2	57.4	96.0	103.9	69.4	94.7
9.8	PAINTING	74.7	95.3	91.1	99.5	108.6	106.7	121.0	70.9	81.2	115.8	72.6	81.5	101.3	65.6	72.9	89.0	65.6	70.4
9	FINISHES	95.8	95.1	95.4	98.0	107.9	103.3	103.8	71.7	86.7	106.6	73.4	88.9	95.7	63.9	78.8	94.4	70.0	81.4
10-14	TOTAL DIV. 10-14	100.0	97.2	99.2	100.0	100.7	100.1	100.0	85.0	95.6	100.0	85.1	95.6	100.0	77.7	93.5	100.0	80.0	94.2
15	MECHANICAL	102.6	97.3	100.0	97.7	110.8	104.2	99.2	91.8	95.5	98.8	86.0	92.4	99.7	59.8	79.9	100.0	72.6	86.4
16	ELECTRICAL	104.4	98.6	100.4	103.1	115.7	111.8	98.8	84.4	88.9	98.3	83.7	88.2	96.5	72.9	80.2	92.3	75.4	80.7
1-16	WEIGHTED AVERAGE	99.2	96.6	97.8	95.5	108.1	102.2	100.1	83.6	91.3	102.4	83.1	92.1	98.5	67.8	82.1	101.4	76.0	87.8

NEVADA / NEW HAMPSHIRE / NEW JERSEY

		LAS VEGAS			RENO			MANCHESTER			NASHUA			JERSEY CITY			NEWARK		
DIVISION		MAT.	INST.	TOTAL	MAT.	INST.	TOTAL	MAT.	INST.	TOTAL	MAT.	INST.	TOTAL	MAT.	INST.	TOTAL	MAT.	INST.	TOTAL
2	SITE WORK	90.4	107.2	97.9	90.4	105.7	97.2	91.7	88.2	90.1	98.6	92.7	96.0	112.9	105.2	109.5	107.2	105.5	106.4
3.1	FORMWORK	112.7	109.4	110.2	109.2	102.4	103.9	113.1	77.0	85.1	111.9	77.6	85.3	113.7	109.7	110.6	123.7	109.7	112.9
3.2	REINFORCING	118.4	125.1	121.1	77.3	126.1	97.1	117.7	83.4	103.8	117.7	83.4	103.8	109.2	137.3	120.6	109.2	137.3	120.6
3.3	CAST IN PLACE CONC.	98.8	108.2	104.5	109.3	105.9	107.2	93.1	84.8	88.0	100.4	95.4	97.4	98.9	104.0	102.0	105.6	98.2	101.1
3	CONCRETE	106.0	110.2	108.6	102.1	106.3	104.8	102.6	81.6	89.3	106.6	87.4	94.4	104.2	109.1	107.3	110.0	106.2	107.6
4	MASONRY	108.8	94.1	97.6	123.7	88.1	96.4	103.3	76.2	82.6	107.3	76.2	83.5	106.8	96.4	98.8	110.5	98.4	101.2
5	METALS	110.8	119.2	113.7	97.8	120.2	105.7	100.6	83.6	94.6	88.0	87.2	87.7	100.9	125.0	109.3	98.3	123.6	107.2
6	WOOD & PLASTICS	88.5	107.7	99.2	88.0	100.8	95.2	105.1	75.6	88.7	110.8	77.1	92.1	111.7	107.2	109.2	109.5	107.4	108.3
7	MOISTURE PROTECTION	88.8	110.5	95.7	102.8	106.8	104.1	96.6	110.1	100.9	102.2	110.1	104.7	117.9	113.3	116.5	105.1	109.6	106.5
8	DOORS, WINDOWS, GLASS	94.5	108.7	101.8	99.9	107.7	103.9	101.9	78.7	89.9	96.5	78.7	87.3	98.8	115.6	107.5	100.7	114.2	107.7
9.1	LATH & PLASTER	86.8	105.6	101.1	101.5	104.5	103.8	110.5	76.3	84.6	110.4	76.1	84.4	98.6	108.3	105.9	99.7	99.5	99.5
9.2	DRYWALL	92.6	107.8	99.7	99.5	110.7	104.7	113.5	72.6	94.4	114.7	72.6	95.1	108.1	105.5	106.9	109.9	105.6	107.9
9.5	ACOUSTICAL WORK	110.1	108.0	108.9	101.3	100.9	101.1	106.3	74.7	89.2	106.3	74.7	89.2	100.6	107.5	104.3	100.6	107.7	104.4
9.6	FLOORING	118.5	98.0	113.0	98.4	94.5	97.4	83.7	77.7	82.1	88.5	77.7	85.6	105.2	100.0	103.8	98.7	101.8	99.5
9.8	PAINTING	105.5	110.7	109.6	114.8	97.1	100.8	126.8	61.9	75.3	105.7	61.9	70.9	92.7	110.1	106.5	89.7	110.1	105.9
9	FINISHES	100.9	108.0	104.7	101.0	103.7	102.4	107.6	69.7	87.4	107.2	69.7	87.2	105.1	107.0	106.1	104.4	106.7	105.6
10-14	TOTAL DIV. 10-14	100.0	102.8	100.8	100.0	134.3	109.9	100.0	88.9	96.7	100.0	88.9	96.7	100.0	104.0	101.1	100.0	103.7	101.0
15	MECHANICAL	100.1	104.8	102.4	102.8	104.8	103.8	97.1	81.5	89.4	98.6	81.6	90.1	99.9	102.5	101.2	97.2	99.4	98.3
16	ELECTRICAL	103.3	105.2	104.6	108.5	106.5	107.1	101.9	74.4	82.9	97.9	74.4	81.7	96.6	110.4	106.1	97.4	107.7	104.5
1-16	WEIGHTED AVERAGE	100.7	105.8	103.4	101.9	105.1	103.6	100.3	80.0	89.5	100.5	81.7	90.4	103.4	106.8	105.2	102.4	105.2	103.9

| DIVISION | | NEW JERSEY | | | | | | NEW MEXICO | | | NEW YORK | | | | | | | | |
| | | PATERSON | | | TRENTON | | | ALBUQUERQUE | | | ALBANY | | | BINGHAMTON | | | BUFFALO | | |
		MAT.	INST.	TOTAL	MAT.	INST.	TOTAL	MAT.	INST.	TOTAL	MAT.	INST.	TOTAL	MAT.	INST.	TOTAL	MAT.	INST.	TOTAL
2	SITE WORK	116.1	104.4	110.9	107.0	106.3	106.7	108.5	90.6	100.6	103.3	103.3	103.3	92.8	86.7	90.1	99.7	98.9	99.4
3.1	FORMWORK	109.5	106.4	107.1	124.3	108.0	111.7	125.8	74.0	85.6	117.3	102.0	105.5	111.8	80.9	87.9	121.7	114.7	116.3
3.2	REINFORCING	108.8	137.3	120.3	109.2	109.3	109.2	117.7	73.0	99.6	80.1	98.8	87.7	80.1	83.9	81.6	96.2	102.6	98.8
3.3	CAST IN PLACE CONC.	102.2	101.4	101.7	89.1	102.5	97.3	101.9	100.0	100.7	77.7	102.0	92.5	91.1	97.8	95.2	108.8	99.6	103.2
3	CONCRETE	105.1	106.5	106.0	100.6	105.2	103.5	110.2	87.4	95.7	86.1	101.7	96.0	92.7	90.0	91.0	108.5	105.8	106.8
4	MASONRY	109.9	125.3	121.7	105.2	98.2	99.8	102.9	73.0	80.0	87.8	101.0	97.9	100.4	79.9	84.7	99.6	114.7	111.1
5	METALS	96.1	125.0	106.3	98.1	108.1	101.6	107.4	81.9	98.4	97.1	100.9	98.5	99.2	88.2	95.3	104.4	101.7	103.4
6	WOOD & PLASTICS	117.5	107.4	111.9	120.2	108.4	113.6	100.4	76.4	87.0	96.9	100.6	98.9	103.8	78.1	89.5	112.4	116.0	114.4
7	MOISTURE PROTECTION	114.4	108.2	112.4	108.2	122.3	109.0	97.4	64.9	87.1	105.8	102.5	104.8	97.5	85.2	93.6	100.8	107.9	103.0
8	DOORS, WINDOWS, GLASS	98.7	120.1	109.8	104.8	110.5	107.8	99.6	71.8	85.2	104.2	94.4	99.2	99.5	74.5	86.6	96.2	107.3	102.0
9.1	LATH & PLASTER	99.6	106.0	104.4	115.3	102.7	105.8	119.0	74.0	84.9	106.1	101.5	102.6	109.1	81.8	88.4	111.0	105.8	107.1
9.2	DRYWALL	113.1	105.6	109.6	109.4	105.3	107.5	85.5	75.5	80.8	106.0	100.3	103.4	113.6	77.2	96.6	123.4	116.8	120.3
9.5	ACOUSTICAL WORK	100.6	107.7	104.4	96.9	107.1	102.4	92.1	75.6	83.1	111.3	100.6	105.5	110.6	77.3	92.6	115.5	116.9	116.3
9.6	FLOORING	90.7	125.6	100.0	101.1	101.5	101.2	103.9	67.7	94.2	86.1	91.8	87.6	99.5	81.3	94.6	104.1	107.8	105.1
9.8	PAINTING	100.0	110.1	108.0	96.5	110.1	107.3	110.2	68.4	77.0	116.9	94.3	99.0	103.5	77.3	82.7	112.4	108.4	109.2
9	FINISHES	105.5	108.8	107.2	105.4	106.7	106.1	93.4	72.4	82.2	103.1	97.8	100.3	109.1	77.8	92.4	117.1	112.6	114.7
10-14	TOTAL DIV. 10-14	100.0	104.5	101.3	100.0	118.0	105.2	100.0	82.1	94.8	100.0	94.7	98.4	100.0	89.9	97.0	100.0	102.5	100.7
15	MECHANICAL	99.9	101.6	100.8	99.8	109.9	104.8	100.4	88.2	94.3	95.9	97.8	96.9	100.1	74.6	87.4	97.4	95.0	96.2
16	ELECTRICAL	96.6	118.6	111.7	94.7	120.3	112.3	94.9	81.7	85.8	92.3	102.3	99.2	92.1	76.4	81.3	99.0	103.5	102.1
1-16	WEIGHTED AVERAGE	103.4	111.5	107.7	101.6	108.7	105.4	101.7	81.2	90.8	96.9	99.9	98.5	98.7	81.4	89.5	102.6	105.2	104.0

| DIVISION | | NEW YORK | | | | | | | | | | | | | | | NORTH CAROLINA | | |
| | | NEW YORK | | | ROCHESTER | | | SYRACUSE | | | UTICA | | | YONKERS | | | CHARLOTTE | | |
		MAT.	INST.	TOTAL	MAT.	INST.	TOTAL	MAT.	INST.	TOTAL	MAT.	INST.	TOTAL	MAT.	INST.	TOTAL	MAT.	INST.	TOTAL
2	SITE WORK	118.1	123.9	120.7	108.1	96.0	102.7	97.1	93.9	95.7	116.4	93.0	106.0	123.3	107.1	116.1	115.3	83.8	101.3
3.1	FORMWORK	111.0	153.6	144.0	107.9	104.3	105.1	110.5	86.6	92.0	113.4	73.6	82.6	115.5	131.8	128.2	109.8	59.6	70.9
3.2	REINFORCING	100.6	178.0	132.0	106.5	100.8	104.2	106.5	103.7	105.3	106.5	80.8	96.1	80.1	125.6	98.5	87.7	64.0	78.1
3.3	CAST IN PLACE CONC.	152.5	111.3	127.3	126.7	98.5	109.5	103.0	77.7	87.6	84.0	95.6	91.1	116.4	95.1	103.4	110.7	94.0	100.5
3	CONCRETE	132.6	133.7	133.3	118.4	101.0	107.4	105.3	83.5	91.4	94.9	85.7	89.0	108.0	112.2	110.7	105.4	77.9	87.9
4	MASONRY	104.2	148.9	138.4	102.2	105.9	105.0	102.2	98.1	99.0	100.2	76.9	82.3	122.6	143.1	138.3	89.9	47.6	57.5
5	METALS	104.6	151.2	121.0	102.0	99.0	101.0	103.1	93.9	99.9	103.8	87.5	98.1	97.5	118.7	105.0	100.3	77.7	92.3
6	WOOD & PLASTICS	111.6	152.4	134.3	98.2	104.2	101.6	107.2	84.2	94.4	112.6	72.8	90.5	103.5	151.3	130.1	104.9	62.1	81.1
7	MOISTURE PROTECTION	110.1	155.4	124.5	97.4	106.8	100.4	96.6	98.2	97.1	97.5	94.1	96.5	105.7	140.9	116.9	88.7	45.0	74.8
8	DOORS, WINDOWS, GLASS	98.2	155.5	127.8	97.1	98.6	97.8	98.0	85.3	91.4	103.6	73.5	88.0	105.2	145.3	126.0	96.1	58.7	76.7
9.1	LATH & PLASTER	87.8	134.1	122.9	107.7	95.4	98.4	106.2	95.6	98.1	109.4	78.5	86.0	93.6	113.4	108.6	99.9	51.2	63.0
9.2	DRYWALL	119.1	144.9	131.2	94.2	98.3	96.1	108.8	83.5	97.0	110.4	71.7	92.4	97.5	140.4	117.5	88.2	60.7	75.3
9.5	ACOUSTICAL WORK	102.6	154.6	130.8	115.7	104.5	109.7	98.7	83.6	90.5	115.7	71.8	92.0	115.7	153.1	136.0	92.0	60.7	75.1
9.6	FLOORING	98.8	133.4	108.1	93.6	99.3	95.1	86.3	90.2	87.4	87.9	71.6	83.5	102.1	120.8	107.1	90.7	46.7	78.9
9.8	PAINTING	118.1	142.4	137.4	98.2	102.6	101.7	103.2	92.2	94.5	108.6	88.4	92.5	107.6	145.8	137.9	96.8	60.2	67.7
9	FINISHES	112.5	143.4	128.9	96.5	100.2	98.5	102.3	87.7	94.6	105.6	77.9	90.9	100.9	140.3	121.9	90.2	58.9	73.5
10-14	TOTAL DIV. 10-14	100.0	114.1	104.0	100.0	103.1	100.9	100.0	97.3	99.2	100.0	93.3	98.0	100.0	115.9	104.6	100.0	71.7	91.8
15	MECHANICAL	99.4	148.4	123.7	97.3	104.1	100.6	100.7	88.3	94.5	99.2	83.4	91.4	96.6	126.4	111.4	97.2	63.6	80.5
16	ELECTRICAL	95.6	151.4	134.1	98.3	99.6	99.2	98.6	87.6	91.0	95.7	76.1	82.2	104.1	130.7	122.4	98.0	58.2	70.6
1-16	WEIGHTED AVERAGE	107.9	143.3	126.8	102.0	101.9	101.9	101.0	89.5	94.8	101.0	82.1	90.9	103.9	127.9	116.7	98.6	64.3	80.2

| DIVISION | | NORTH CAROLINA | | | | | | OHIO | | | | | | | | | | | |
| | | GREENSBORO | | | RALEIGH | | | AKRON | | | CANTON | | | CINCINNATI | | | CLEVELAND | | |
		MAT.	INST.	TOTAL	MAT.	INST.	TOTAL	MAT.	INST.	TOTAL	MAT.	INST.	TOTAL	MAT.	INST.	TOTAL	MAT.	INST.	TOTAL
2	SITE WORK	90.5	88.2	89.5	99.3	91.5	95.8	117.0	99.1	109.0	105.0	97.3	101.6	92.7	102.1	96.9	121.9	107.1	115.3
3.1	FORMWORK	99.9	59.7	68.7	104.3	59.6	69.7	111.4	102.8	104.7	110.2	97.7	100.5	100.0	99.8	99.9	120.4	117.3	118.0
3.2	REINFORCING	82.2	66.3	75.8	93.8	66.3	82.7	98.2	110.4	103.1	98.2	91.1	95.3	105.7	94.4	101.2	84.7	110.4	95.1
3.3	CAST IN PLACE CONC.	99.4	91.7	94.7	107.2	96.6	100.7	89.7	100.7	96.4	89.7	100.2	96.1	87.5	98.0	93.9	93.7	111.6	104.6
3	CONCRETE	95.6	76.9	83.8	103.6	79.4	88.2	95.9	102.4	100.0	95.6	98.4	97.4	94.1	98.4	96.8	97.0	113.7	107.6
4	MASONRY	102.7	47.6	60.5	89.2	47.6	57.4	93.6	100.0	98.5	107.6	96.0	98.7	74.6	93.8	89.3	96.3	114.5	110.2
5	METALS	91.3	78.3	86.7	91.2	79.9	87.2	99.0	104.9	101.1	99.0	93.3	97.0	98.6	95.0	97.3	104.9	109.1	106.4
6	WOOD & PLASTICS	91.8	62.1	75.3	95.1	62.1	76.7	104.7	103.0	103.7	103.4	98.2	100.5	109.1	98.1	103.0	142.5	115.1	127.3
7	MOISTURE PROTECTION	86.7	45.0	73.5	87.2	45.0	73.8	99.0	102.9	100.2	99.0	102.0	100.0	95.9	100.8	97.5	108.3	120.1	112.1
8	DOORS, WINDOWS, GLASS	91.0	60.2	75.0	85.4	60.2	72.4	101.0	108.4	104.8	88.8	86.8	87.8	96.9	94.0	95.4	94.2	113.5	104.2
9.1	LATH & PLASTER	97.8	62.0	70.7	106.6	55.8	68.1	116.0	101.8	105.3	112.2	87.9	93.8	103.0	97.3	98.7	104.9	116.0	113.3
9.2	DRYWALL	86.0	60.7	74.2	95.6	60.7	79.3	113.0	103.0	108.3	111.0	95.1	103.6	98.1	97.8	98.0	101.3	115.7	107.6
9.5	ACOUSTICAL WORK	97.6	60.7	77.6	108.8	60.7	82.8	82.6	103.1	93.7	101.5	98.2	99.7	96.7	98.1	97.5	98.2	115.5	107.6
9.6	FLOORING	90.0	46.7	78.4	101.9	46.7	87.1	82.3	97.9	86.5	114.1	91.2	108.0	91.1	96.0	92.4	85.0	115.0	93.0
9.8	PAINTING	91.0	60.2	66.5	90.4	60.2	66.4	107.7	101.8	103.0	101.4	94.3	95.8	103.0	91.5	93.9	104.5	114.7	112.6
9	FINISHES	88.6	59.6	73.1	97.8	59.2	77.2	103.3	102.1	102.7	110.0	94.4	101.7	97.0	95.5	96.2	97.9	115.2	107.1
10-14	TOTAL DIV. 10-14	100.0	72.1	91.9	100.0	72.9	92.1	100.0	104.3	101.2	100.0	101.3	100.3	100.0	93.5	98.1	100.0	111.2	103.2
15	MECHANICAL	95.6	63.9	79.8	97.0	63.8	80.5	99.2	96.2	97.7	99.2	83.5	91.4	99.3	92.4	95.8	101.3	106.2	103.8
16	ELECTRICAL	97.5	62.0	73.1	99.4	62.0	73.7	94.4	93.6	93.9	93.3	85.6	88.0	99.9	90.8	93.7	96.3	105.6	102.7
1-16	WEIGHTED AVERAGE	94.3	65.0	78.6	96.4	65.7	80.0	99.9	100.2	100.1	99.6	92.7	95.9	96.5	95.0	95.7	101.9	111.2	106.9

CITY COST INDEXES

OHIO / OKLAHOMA

DIVISION		COLUMBUS MAT.	INST.	TOTAL	DAYTON MAT.	INST.	TOTAL	LORAIN MAT.	INST.	TOTAL	TOLEDO MAT.	INST.	TOTAL	YOUNGSTOWN MAT.	INST.	TOTAL	OKLAHOMA CITY MAT.	INST.	TOTAL
2	SITE WORK	89.6	104.9	96.4	91.3	99.0	94.7	104.8	100.5	102.9	124.2	99.0	113.0	97.2	99.9	98.4	122.7	86.1	106.4
3.1	FORMWORK	105.2	95.6	97.8	118.8	100.4	104.5	110.5	99.1	101.6	116.5	106.7	108.9	108.5	99.8	101.8	116.0	71.2	81.3
3.2	REINFORCING	115.5	97.7	108.3	98.2	95.1	96.9	98.2	110.4	103.1	98.2	96.8	97.6	98.2	100.1	99.0	94.2	63.9	82.0
3.3	CAST IN PLACE CONC.	96.4	96.3	96.4	105.9	99.8	102.1	95.4	98.7	97.4	92.8	99.8	97.1	82.4	100.9	93.7	100.5	90.0	94.1
3	CONCRETE	102.5	96.2	98.5	106.7	99.6	102.2	99.0	99.9	99.6	98.7	102.2	100.9	91.1	100.4	97.0	102.2	80.3	88.3
4	MASONRY	90.2	94.7	93.7	86.3	97.4	94.8	101.7	91.7	94.0	109.6	92.3	96.3	96.1	96.5	96.4	99.8	80.7	85.2
5	METALS	101.7	97.4	100.1	101.0	96.1	99.3	98.8	107.0	101.7	98.8	96.6	98.0	98.3	99.0	98.6	94.2	73.3	86.9
6	WOOD & PLASTICS	107.3	95.4	100.7	108.8	100.9	104.4	99.6	98.3	98.9	119.3	109.3	113.7	106.6	98.0	101.8	115.1	74.9	92.7
7	MOISTURE PROTECTION	98.8	94.9	97.6	100.8	96.7	99.5	97.6	110.8	101.8	91.5	100.3	94.3	101.1	100.1	100.8	94.4	70.4	86.8
8	DOORS, WINDOWS, GLASS	97.2	92.8	94.9	107.9	90.0	98.6	96.3	105.8	101.2	100.3	98.4	99.3	97.7	97.2	97.5	103.0	70.4	86.2
9.1	LATH & PLASTER	91.3	92.6	92.3	110.8	98.8	101.7	110.5	92.0	96.5	115.0	105.5	107.8	115.5	96.5	101.1	99.4	76.7	82.2
9.2	DRYWALL	105.7	95.5	100.9	112.7	97.7	105.7	108.0	98.1	103.4	109.8	108.8	109.4	96.3	97.9	97.0	108.5	73.9	92.4
9.5	ACOUSTICAL WORK	101.8	95.6	98.5	86.7	100.9	94.3	91.8	98.2	95.3	115.0	109.0	111.7	95.5	98.0	96.8	86.7	74.0	79.8
9.6	FLOORING	95.8	89.1	94.0	113.6	92.5	108.0	118.5	94.5	112.1	99.8	91.6	97.6	88.2	95.2	90.1	107.2	78.1	99.4
9.8	PAINTING	105.6	97.8	99.4	84.3	93.0	91.2	103.4	114.7	112.4	115.1	90.4	95.4	74.6	99.3	94.2	105.7	75.4	81.6
9	FINISHES	102.8	95.7	99.0	107.9	96.0	101.6	108.7	103.2	105.8	108.7	101.0	104.6	92.7	98.1	95.5	106.0	74.9	89.4
10-14	TOTAL DIV. 10-14	100.0	96.5	98.9	100.0	93.2	98.0	100.0	108.0	102.3	100.0	104.0	101.1	100.0	98.3	99.5	100.0	79.3	94.0
15	MECHANICAL	102.0	96.5	99.3	98.8	92.4	95.6	99.0	87.7	93.4	99.0	96.0	97.5	99.8	92.3	96.0	94.3	76.7	85.5
16	ELECTRICAL	92.5	91.7	92.0	98.2	93.7	95.1	97.3	87.6	90.6	99.2	94.8	96.2	97.3	91.2	93.1	102.8	78.9	86.3
1-16	WEIGHTED AVERAGE	99.4	95.6	97.4	100.9	95.8	98.2	100.2	96.6	98.2	102.0	98.2	99.9	97.3	96.6	96.9	100.7	78.0	88.6

OKLAHOMA / OREGON / PENNSYLVANIA

DIVISION		TULSA MAT.	INST.	TOTAL	EUGENE MAT.	INST.	TOTAL	PORTLAND MAT.	INST.	TOTAL	ALLENTOWN MAT.	INST.	TOTAL	ERIE MAT.	INST.	TOTAL	HARRISBURG MAT.	INST.	TOTAL
2	SITE WORK	90.5	91.6	91.0	96.6	107.2	101.3	110.2	103.7	107.3	126.4	99.7	114.6	120.3	96.5	109.7	97.6	97.4	97.5
3.1	FORMWORK	125.2	70.4	82.8	120.8	101.2	105.6	122.6	102.5	107.0	110.6	109.0	109.4	90.5	93.2	92.5	113.7	88.7	94.3
3.2	REINFORCING	91.4	63.9	80.2	101.2	103.5	102.2	101.2	103.5	102.2	114.7	114.0	114.4	117.6	88.5	105.8	114.7	102.8	109.9
3.3	CAST IN PLACE CONC.	88.4	94.7	92.3	100.7	126.3	116.3	116.7	100.3	106.7	103.1	100.2	101.4	83.4	98.6	92.7	104.4	103.2	103.7
3	CONCRETE	96.3	82.5	87.5	104.8	114.5	110.9	114.4	101.4	106.2	107.2	104.9	105.7	92.5	95.6	94.5	108.5	97.5	101.5
4	MASONRY	95.8	70.1	76.1	114.6	91.6	97.0	126.6	96.8	103.7	86.2	103.3	99.3	105.6	92.8	95.8	82.8	89.9	88.2
5	METALS	105.0	74.7	94.3	103.4	111.4	106.2	106.9	102.5	105.4	104.7	107.8	105.8	100.9	91.1	97.5	96.7	103.3	99.0
6	WOOD & PLASTICS	116.8	74.0	93.0	92.2	99.5	96.3	89.1	101.1	95.8	104.2	111.0	108.0	89.4	92.8	91.3	110.1	89.9	98.9
7	MOISTURE PROTECTION	96.7	69.2	88.0	86.6	88.1	87.1	86.6	90.4	87.8	96.8	120.6	104.4	97.4	96.2	97.0	79.2	91.6	83.1
8	DOORS, WINDOWS, GLASS	100.5	68.7	84.1	101.9	97.6	99.7	107.6	98.4	102.8	102.5	98.9	100.6	92.3	88.7	90.4	104.1	87.8	95.6
9.1	LATH & PLASTER	99.7	72.7	79.2	113.1	85.0	91.8	117.8	91.3	97.7	99.4	94.7	95.9	110.3	87.0	92.7	96.5	90.0	91.5
9.2	DRYWALL	111.7	73.0	93.7	112.2	91.9	102.7	110.3	93.2	102.4	107.0	99.3	103.4	94.1	92.3	93.2	107.3	89.3	98.9
9.5	ACOUSTICAL WORK	86.7	73.1	79.4	110.5	99.5	104.6	110.5	101.1	105.4	100.6	111.2	106.4	100.3	92.4	96.0	98.8	89.4	93.7
9.6	FLOORING	111.1	80.5	102.9	123.1	94.7	115.5	102.6	97.9	101.4	97.0	101.5	98.2	102.6	86.5	98.3	96.9	92.7	95.8
9.8	PAINTING	79.1	75.1	75.9	119.8	78.7	87.2	97.5	78.7	82.6	104.6	97.0	98.5	88.7	77.9	80.1	91.5	79.4	81.9
9	FINISHES	106.0	74.3	89.1	113.5	87.7	100.0	107.5	89.1	97.7	103.8	94.9	101.4	96.3	86.6	91.1	102.4	86.2	93.8
10-14	TOTAL DIV. 10-14	100.0	78.2	93.6	100.0	101.6	100.4	100.0	102.4	100.7	100.0	101.2	100.3	100.0	93.7	98.1	100.0	92.9	97.9
15	MECHANICAL	96.8	81.7	89.3	98.7	91.9	95.3	98.2	94.5	96.3	100.4	104.4	102.4	99.8	88.6	94.2	101.1	94.8	98.0
16	ELECTRICAL	100.3	74.4	82.4	102.6	84.2	89.9	102.6	100.6	101.3	98.5	79.6	85.5	90.3	92.1	91.5	92.9	80.5	84.4
1-16	WEIGHTED AVERAGE	99.1	77.3	87.5	101.9	97.7	99.6	104.5	97.9	101.0	102.7	100.8	101.7	98.7	91.9	95.1	98.8	91.9	95.1

PENNSYLVANIA / RHODE ISLAND / SOUTH CAROLINA

DIVISION		PHILADELPHIA MAT.	INST.	TOTAL	PITTSBURGH MAT.	INST.	TOTAL	READING MAT.	INST.	TOTAL	SCRANTON MAT.	INST.	TOTAL	PROVIDENCE MAT.	INST.	TOTAL	CHARLESTON MAT.	INST.	TOTAL
2	SITE WORK	103.1	110.0	106.2	128.6	99.8	115.8	87.2	98.3	92.1	102.7	105.8	104.1	83.6	103.7	92.5	121.9	84.6	105.3
3.1	FORMWORK	93.4	125.8	118.5	103.2	97.1	98.5	110.9	93.4	97.4	107.5	91.2	94.9	116.0	115.9	115.9	104.4	56.1	67.0
3.2	REINFORCING	82.5	116.1	96.1	82.3	105.5	91.7	114.7	110.1	112.8	114.7	113.3	114.1	117.7	114.2	116.3	95.9	64.0	83.0
3.3	CAST IN PLACE CONC.	92.5	106.9	101.3	111.1	95.6	101.6	98.0	98.3	98.2	92.4	124.4	111.9	87.5	105.3	98.4	100.8	88.6	93.3
3	CONCRETE	90.4	115.1	106.1	103.0	97.0	99.2	104.3	97.4	99.9	100.4	110.4	106.7	100.0	110.2	106.5	100.4	73.7	83.4
4	MASONRY	91.0	119.6	112.9	100.2	102.2	101.7	90.4	86.3	87.2	105.4	85.7	90.3	101.3	94.0	95.7	92.2	49.8	59.7
5	METALS	103.2	122.4	110.0	100.8	102.3	101.3	96.3	105.4	99.5	98.3	117.0	104.8	111.5	106.6	109.8	98.9	73.2	89.8
6	WOOD & PLASTICS	92.0	125.7	110.7	111.0	97.8	103.7	100.3	93.3	96.4	101.8	87.0	93.6	91.4	109.1	101.3	98.8	57.4	75.8
7	MOISTURE PROTECTION	102.7	133.2	112.4	96.4	103.3	98.6	88.0	116.3	97.0	98.7	95.8	97.8	106.2	101.4	104.7	91.1	57.7	80.5
8	DOORS, WINDOWS, GLASS	94.1	126.5	110.9	101.0	104.0	102.5	103.8	90.0	96.6	93.0	86.1	89.4	110.3	101.5	105.8	101.0	55.7	77.6
9.1	LATH & PLASTER	97.5	124.1	117.7	106.4	95.1	97.8	95.1	83.4	86.2	100.9	86.1	89.7	106.2	94.4	97.3	94.0	54.0	63.8
9.2	DRYWALL	110.6	126.7	118.1	113.3	97.8	106.1	109.3	84.0	97.5	104.5	83.2	94.5	96.0	102.9	99.2	108.0	56.0	83.8
9.5	ACOUSTICAL WORK	93.0	127.0	111.4	103.0	97.9	100.2	98.8	92.9	95.6	99.6	86.3	92.4	103.9	109.5	106.9	95.0	55.8	73.8
9.6	FLOORING	96.9	113.5	101.4	112.9	100.8	109.7	103.0	83.2	97.7	100.2	81.8	95.3	88.6	95.8	90.6	89.0	50.6	78.7
9.8	PAINTING	121.8	120.3	120.6	82.6	105.0	100.4	89.9	88.1	88.5	109.4	86.7	91.4	105.4	95.1	97.2	112.0	60.2	70.9
9	FINISHES	107.0	123.4	115.8	109.1	100.3	104.4	104.8	86.0	94.8	103.6	84.7	93.5	96.2	99.7	98.1	102.9	56.9	78.4
10-14	TOTAL DIV. 10-14	100.0	115.4	104.4	100.0	98.3	99.5	100.0	97.7	99.3	100.0	94.9	98.5	100.0	106.3	101.8	100.0	71.7	91.8
15	MECHANICAL	96.2	115.5	105.8	98.4	96.6	97.5	99.3	104.4	101.9	99.1	89.2	94.2	99.8	89.7	94.8	98.4	59.9	79.3
16	ELECTRICAL	98.1	123.7	115.7	95.3	90.8	92.2	94.4	86.9	89.2	98.3	87.4	90.8	99.4	91.9	94.2	98.5	54.0	67.8
1-16	WEIGHTED AVERAGE	98.1	119.2	109.4	102.5	98.2	100.2	98.3	95.1	96.6	99.9	94.8	97.1	100.6	99.5	100.0	100.1	62.2	79.9

	DIVISION	SOUTH CAROLINA COLUMBIA			SOUTH DAKOTA SIOUX FALLS			TENNESSEE CHATTANOOGA			TENNESSEE KNOXVILLE			TENNESSEE MEMPHIS			TENNESSEE NASHVILLE		
		MAT.	INST.	TOTAL	MAT.	INST.	TOTAL	MAT.	INST.	TOTAL	MAT.	INST.	TOTAL	MAT.	INST.	TOTAL	MAT.	INST.	TOTAL
2	SITE WORK	111.8	90.7	102.4	93.4	83.0	88.8	98.4	85.8	92.8	102.4	92.9	98.2	86.5	90.4	88.2	84.1	88.7	86.2
3.1	FORMWORK	89.6	60.2	66.8	105.2	57.7	68.4	106.3	70.3	78.4	107.0	64.6	74.1	88.5	69.4	73.7	88.1	69.9	74.0
3.2	REINFORCING	91.4	64.0	80.3	104.9	62.4	87.7	98.2	68.6	86.2	98.2	63.1	84.0	96.2	66.8	84.3	96.2	65.3	83.6
3.3	CAST IN PLACE CONC.	81.1	93.3	88.6	98.6	81.3	88.0	81.7	90.2	86.9	85.8	88.4	87.4	93.3	92.4	92.7	85.2	85.2	85.2
3	CONCRETE	85.1	77.7	80.4	101.3	70.4	81.7	90.3	80.5	84.1	92.8	76.8	82.6	93.0	81.1	85.5	88.3	77.5	81.4
4	MASONRY	87.3	50.1	58.8	106.8	66.5	75.9	78.9	76.4	77.0	79.9	67.4	70.3	86.4	71.5	75.0	92.4	73.1	77.6
5	METALS	104.6	75.2	94.3	107.1	69.2	93.8	91.5	75.3	85.8	103.3	71.9	92.3	91.0	74.4	85.2	99.2	72.5	89.8
6	WOOD & PLASTICS	86.0	62.4	72.9	98.0	56.8	75.1	105.9	72.9	87.5	107.2	66.9	84.8	83.8	71.0	76.7	107.9	72.9	88.5
7	MOISTURE PROTECTION	102.0	56.8	87.6	98.4	60.2	86.2	103.2	66.1	91.4	87.8	59.9	79.0	96.7	71.8	88.8	101.6	67.1	90.6
8	DOORS, WINDOWS, GLASS	102.4	58.0	79.5	103.4	53.4	77.5	103.7	67.1	84.8	107.4	60.4	83.1	95.2	73.7	84.1	94.1	70.1	81.7
9.1	LATH & PLASTER	104.7	53.7	66.1	110.2	63.6	74.9	93.7	69.5	75.4	91.9	67.2	73.2	90.4	77.9	80.9	89.7	67.8	73.1
9.2	DRYWALL	83.5	59.2	72.2	105.9	54.7	82.0	109.6	71.9	92.0	107.6	66.2	88.3	96.3	79.3	88.4	102.8	70.8	87.9
9.5	ACOUSTICAL WORK	93.9	61.1	76.1	101.9	54.7	76.4	97.7	72.0	83.8	89.9	65.8	76.8	99.6	70.1	83.6	99.5	71.8	84.5
9.6	FLOORING	94.5	51.2	82.9	111.8	66.3	99.6	109.1	77.5	100.6	100.9	67.6	92.0	100.8	64.6	91.1	127.9	74.5	113.7
9.8	PAINTING	121.2	60.2	72.8	112.8	52.7	65.1	90.9	65.5	70.7	94.6	70.5	75.5	91.4	77.9	80.7	94.3	68.1	73.5
9	FINISHES	91.1	58.8	73.9	107.7	55.4	79.8	106.3	69.9	86.9	103.0	67.8	84.3	96.9	76.9	86.3	107.0	70.0	87.3
10-14	TOTAL DIV. 10-14	100.0	71.8	91.8	100.0	73.1	92.2	100.0	75.2	92.8	100.0	73.3	92.2	100.0	79.8	94.1	100.0	74.3	92.5
15	MECHANICAL	99.7	61.2	80.6	99.2	56.6	78.0	99.0	71.5	85.3	99.9	68.7	84.4	100.2	79.2	89.8	97.2	71.1	84.3
16	ELECTRICAL	104.3	56.9	71.7	96.7	65.0	74.9	94.5	72.0	79.0	96.2	71.9	79.4	108.7	85.6	92.8	100.0	67.3	77.5
1-16	WEIGHTED AVERAGE	97.8	64.5	80.0	101.2	64.3	81.5	97.3	74.6	85.2	98.1	71.2	83.7	96.2	78.7	86.8	96.9	73.0	84.1

	DIVISION	TEXAS AMARILLO			TEXAS AUSTIN			TEXAS BEAUMONT			TEXAS CORPUS CHRISTI			TEXAS DALLAS			TEXAS EL PASO		
		MAT.	INST.	TOTAL	MAT.	INST.	TOTAL	MAT.	INST.	TOTAL	MAT.	INST.	TOTAL	MAT.	INST.	TOTAL	MAT.	INST.	TOTAL
2	SITE WORK	117.4	84.5	102.8	88.3	83.4	86.1	127.0	93.1	111.9	112.9	85.2	100.5	114.3	86.9	102.2	107.6	99.8	104.1
3.1	FORMWORK	97.8	68.9	75.4	108.0	74.7	82.2	94.4	86.2	88.0	106.2	60.9	71.1	82.8	74.9	76.7	106.1	51.9	64.1
3.2	REINFORCING	113.6	70.7	96.2	97.1	75.2	88.2	93.8	83.3	89.6	104.4	63.6	87.9	96.2	68.2	84.8	117.7	72.8	99.5
3.3	CAST IN PLACE CONC.	114.1	90.4	99.7	96.6	91.2	93.3	111.2	94.1	100.8	106.1	88.8	95.5	97.8	92.6	94.7	92.1	80.4	84.9
3	CONCRETE	110.8	80.3	91.4	99.0	83.3	89.0	104.0	90.0	95.1	105.7	75.6	86.6	94.5	83.5	87.5	100.6	68.6	80.3
4	MASONRY	106.5	66.3	75.7	104.2	78.8	84.7	111.7	94.7	98.7	107.4	65.2	75.1	102.5	73.0	79.9	95.9	54.2	64.0
5	METALS	99.7	78.7	92.3	90.9	81.5	87.6	97.8	87.7	94.2	97.7	73.7	89.3	91.2	77.8	86.5	104.3	75.8	94.3
6	WOOD & PLASTICS	94.1	71.2	81.4	96.5	78.4	86.4	94.5	90.3	92.2	101.9	63.5	80.5	96.6	79.4	87.0	91.7	48.5	67.7
7	MOISTURE PROTECTION	94.0	56.4	82.0	97.5	64.5	87.0	95.9	74.5	89.1	97.1	59.6	85.2	102.8	69.9	92.4	98.2	54.6	84.4
8	DOORS, WINDOWS, GLASS	92.5	64.3	77.9	96.7	74.2	85.1	102.7	81.8	91.9	101.1	55.8	79.2	103.6	79.3	91.0	99.8	52.2	75.2
9.1	LATH & PLASTER	92.6	76.8	80.6	80.9	78.3	78.9	86.0	80.1	81.6	92.1	68.6	74.3	90.5	81.4	83.6	90.6	52.3	61.6
9.2	DRYWALL	84.9	73.0	79.4	84.9	77.5	81.4	90.3	85.2	87.9	87.1	64.5	76.6	78.6	78.1	78.4	87.3	47.9	68.9
9.5	ACOUSTICAL WORK	91.7	71.1	80.5	94.3	77.6	85.3	94.3	89.0	91.4	90.1	62.2	75.0	96.1	78.2	86.4	90.5	46.7	66.8
9.6	FLOORING	104.1	68.4	94.5	102.6	70.0	93.8	102.6	96.0	100.8	94.6	61.8	85.8	88.7	76.0	85.3	104.2	54.4	90.9
9.8	PAINTING	103.9	54.7	64.9	108.6	57.8	68.2	105.0	81.0	85.9	96.9	60.2	67.7	100.1	77.8	82.4	89.5	45.2	54.3
9	FINISHES	91.8	66.5	78.3	91.9	70.2	80.3	94.7	84.5	89.3	90.1	62.9	75.6	84.7	78.1	81.2	91.6	47.6	68.2
10-14	TOTAL DIV. 10-14	100.0	77.9	93.5	100.0	79.2	93.9	100.0	86.7	96.1	100.0	79.4	94.0	100.0	85.1	95.7	100.0	70.3	91.3
15	MECHANICAL	98.9	73.1	86.1	98.7	78.1	88.5	98.7	79.3	89.1	98.6	62.5	80.6	99.5	74.8	87.3	101.5	62.0	81.9
16	ELECTRICAL	104.3	68.5	79.6	102.4	71.9	81.4	105.8	85.9	92.1	100.3	60.7	73.0	97.5	80.1	85.5	94.3	62.3	72.2
1-16	WEIGHTED AVERAGE	101.2	72.4	85.8	97.2	77.6	86.7	102.0	86.7	93.8	100.4	67.5	82.8	98.1	78.7	87.7	99.6	62.6	79.8

	DIVISION	TEXAS FORT WORTH			TEXAS HOUSTON			TEXAS LUBBOCK			TEXAS SAN ANTONIO			UTAH SALT LAKE CITY			VERMONT BURLINGTON		
		MAT.	INST.	TOTAL	MAT.	INST.	TOTAL	MAT.	INST.	TOTAL	MAT.	INST.	TOTAL	MAT.	INST.	TOTAL	MAT.	INST.	TOTAL
2	SITE WORK	114.3	88.2	102.7	114.0	88.9	102.8	121.7	92.8	108.8	79.0	90.3	84.1	84.7	94.5	89.1	97.5	90.0	94.2
3.1	FORMWORK	101.7	75.1	81.1	98.5	72.9	78.7	108.8	66.5	76.0	94.4	68.4	74.3	114.0	75.9	84.5	116.0	75.4	84.6
3.2	REINFORCING	100.9	68.2	87.7	112.7	69.9	95.3	106.6	69.6	91.6	100.8	69.6	88.2	133.4	86.7	114.4	111.5	83.4	100.1
3.3	CAST IN PLACE CONC.	109.0	91.3	98.2	112.7	88.7	98.0	99.3	89.8	93.5	73.4	104.8	92.5	87.7	93.4	91.2	101.0	100.4	100.6
3	CONCRETE	105.7	83.0	91.3	109.9	80.8	91.4	102.8	78.9	87.6	83.7	87.4	86.1	103.2	86.0	92.3	106.4	89.1	95.4
4	MASONRY	108.5	70.1	79.1	114.2	79.1	87.3	113.2	62.8	74.6	97.4	69.8	76.3	97.1	83.4	86.6	110.8	59.5	71.5
5	METALS	97.1	77.2	90.1	93.3	76.3	87.3	88.8	78.0	85.0	93.3	82.4	89.5	106.5	92.7	101.6	103.5	90.9	99.1
6	WOOD & PLASTICS	99.2	79.9	88.5	104.1	74.0	87.3	91.3	70.2	79.6	88.3	71.1	78.7	83.6	75.0	78.8	119.7	75.1	94.9
7	MOISTURE PROTECTION	95.5	71.2	87.8	89.4	71.1	83.6	93.5	68.6	85.6	85.7	62.6	78.3	90.2	78.7	86.6	87.9	77.5	84.6
8	DOORS, WINDOWS, GLASS	99.2	79.3	88.9	103.2	72.8	87.4	99.0	64.5	81.1	102.8	71.0	86.4	100.2	76.0	87.7	105.6	65.0	84.6
9.1	LATH & PLASTER	101.5	82.9	87.4	92.9	79.4	82.6	92.1	68.4	74.1	95.1	70.2	76.3	117.9	71.2	82.5	91.2	64.2	70.7
9.2	DRYWALL	101.9	78.1	90.8	105.7	75.1	91.4	86.0	68.0	77.6	87.6	70.0	79.4	92.3	73.5	83.5	108.6	71.2	91.1
9.5	ACOUSTICAL WORK	91.7	78.2	84.4	96.9	73.0	84.0	91.7	69.2	79.5	98.0	70.3	83.0	99.3	74.2	85.7	98.9	74.5	85.7
9.6	FLOORING	105.3	74.5	97.0	94.1	74.9	89.0	106.7	65.0	95.5	91.1	72.5	86.1	100.1	67.4	91.3	109.1	62.8	96.7
9.8	PAINTING	98.9	85.6	88.4	103.6	78.2	83.4	102.7	53.9	63.9	96.0	56.3	64.5	97.5	75.9	80.3	110.1	63.3	73.0
9	FINISHES	101.5	80.7	90.5	101.9	76.2	88.2	92.9	63.0	77.0	90.2	65.5	77.0	95.7	73.8	84.0	107.7	67.7	86.4
10-14	TOTAL DIV. 10-14	100.0	85.2	95.7	100.0	80.3	94.3	100.0	82.0	94.8	100.0	77.0	93.3	100.0	86.5	96.1	100.0	80.1	94.2
15	MECHANICAL	99.8	70.6	85.3	100.4	75.0	87.8	99.6	70.3	85.1	100.4	83.5	92.0	103.9	79.6	91.8	100.8	71.7	86.3
16	ELECTRICAL	98.8	75.1	82.5	103.5	86.9	92.1	101.4	70.4	80.0	108.4	68.3	80.8	102.3	88.6	92.9	101.5	65.7	76.8
1-16	WEIGHTED AVERAGE	101.5	77.1	88.5	102.4	79.1	89.9	100.0	71.9	85.0	94.3	77.1	85.1	99.6	83.3	90.9	102.4	74.4	87.5

CITY COST INDEXES

VIRGINIA / WASHINGTON

DIVISION		NEWPORT NEWS MAT.	INST.	TOTAL	NORFOLK MAT.	INST.	TOTAL	RICHMOND MAT.	INST.	TOTAL	ROANOKE MAT.	INST.	TOTAL	SEATTLE MAT.	INST.	TOTAL	SPOKANE MAT.	INST.	TOTAL
2	SITE WORK	120.7	83.8	104.3	110.9	82.3	98.2	83.0	86.3	84.5	100.4	84.2	93.2	101.1	99.8	100.5	108.0	100.7	104.8
3.1	FORMWORK	102.5	64.5	73.0	106.9	64.4	74.0	99.3	63.3	71.4	105.1	57.5	68.2	83.0	103.5	98.9	106.3	96.3	98.6
3.2	REINFORCING	100.9	69.4	88.1	100.9	69.4	88.1	96.9	75.2	88.1	105.8	76.6	94.0	108.6	103.5	106.5	112.9	103.5	109.1
3.3	CAST IN PLACE CONC.	115.4	85.6	97.2	113.2	88.0	97.8	109.5	87.8	96.3	111.6	87.1	96.6	96.7	110.4	105.1	108.3	98.9	102.5
3	CONCRETE	109.6	75.9	88.2	109.2	77.1	88.8	104.6	77.1	87.2	109.0	74.6	87.2	96.6	107.1	103.3	108.9	98.3	102.2
4	MASONRY	103.9	64.8	73.9	103.4	64.8	73.8	100.8	76.6	82.3	102.2	59.3	69.4	125.4	107.5	111.7	115.6	90.5	96.4
5	METALS	86.7	75.0	82.6	86.6	75.5	82.7	96.0	74.5	88.5	96.7	80.5	91.0	105.9	104.9	105.5	102.9	101.2	102.3
6	WOOD & PLASTICS	103.8	67.7	83.7	103.5	67.7	83.6	103.2	66.2	82.6	102.6	59.0	78.3	76.7	101.6	90.5	102.0	95.1	98.2
7	MOISTURE PROTECTION	85.5	49.6	74.1	86.8	49.6	75.0	108.7	48.6	89.6	85.8	49.4	74.2	111.9	108.7	110.9	96.2	99.0	97.0
8	DOORS, WINDOWS, GLASS	90.5	65.8	77.7	89.5	65.8	77.2	90.0	62.1	75.6	94.6	59.5	76.4	94.3	98.3	96.4	100.2	93.5	96.8
9.1	LATH & PLASTER	113.7	65.4	77.1	98.3	65.8	73.7	105.5	66.3	75.8	107.4	54.3	67.2	99.5	105.8	104.3	115.5	91.2	97.1
9.2	DRYWALL	87.4	65.1	77.0	87.4	65.1	77.0	95.7	65.6	82.8	91.8	57.3	75.7	80.8	102.1	90.8	96.4	95.3	95.9
9.5	ACOUSTICAL WORK	97.4	66.3	80.6	94.7	66.3	79.3	104.5	65.2	83.2	96.7	57.3	75.4	99.1	101.5	100.4	105.2	95.4	99.9
9.6	FLOORING	98.7	62.6	89.0	94.5	62.6	86.0	88.0	76.7	85.0	103.3	55.7	90.6	90.1	104.7	94.0	106.3	94.4	103.1
9.8	PAINTING	91.0	57.4	64.3	101.7	60.6	69.1	96.7	55.3	63.8	90.8	52.2	60.2	83.7	100.5	97.0	100.9	93.3	94.9
9	FINISHES	91.6	62.4	76.1	91.3	63.5	76.5	96.2	62.8	78.5	95.0	55.2	73.8	85.0	101.9	94.0	100.2	94.3	97.1
10-14	TOTAL DIV. 10-14	100.0	72.2	91.9	100.0	72.5	92.0	100.0	74.0	92.4	100.0	71.7	91.8	100.0	104.8	101.4	100.0	100.6	100.1
15	MECHANICAL	102.7	64.2	83.6	102.5	66.1	84.4	101.6	62.7	82.2	103.2	70.7	87.0	99.6	110.0	104.8	103.1	104.3	103.7
16	ELECTRICAL	106.1	57.4	72.5	106.1	57.4	72.5	103.2	67.7	78.7	102.3	58.3	72.0	106.5	98.1	100.7	104.1	99.5	100.9
1-16	WEIGHTED AVERAGE	100.2	67.2	82.6	99.5	67.9	82.6	99.5	70.1	83.8	100.2	66.5	82.2	100.4	105.0	102.9	103.5	98.1	100.6

WASHINGTON / WEST VIRGINIA / WISCONSIN / WYOMING

DIVISION		TACOMA MAT.	INST.	TOTAL	CHARLESTON MAT.	INST.	TOTAL	HUNTINGTON MAT.	INST.	TOTAL	MADISON MAT.	INST.	TOTAL	MILWAUKEE MAT.	INST.	TOTAL	CHEYENNE MAT.	INST.	TOTAL
2	SITE WORK	104.7	97.8	101.7	119.2	98.5	110.0	124.2	101.8	114.2	83.4	98.1	89.9	75.3	100.1	86.3	109.6	89.5	100.6
3.1	FORMWORK	113.4	103.3	105.6	117.8	92.0	97.8	99.2	95.7	96.5	108.9	81.9	88.0	105.1	99.0	100.4	110.4	69.9	79.0
3.2	REINFORCING	100.7	103.5	101.8	120.8	92.0	109.1	116.3	94.0	107.3	106.2	80.9	96.0	104.5	104.6	104.5	103.2	73.1	91.0
3.3	CAST IN PLACE CONC.	100.5	101.9	101.4	117.7	96.8	104.9	114.3	97.6	104.1	100.1	102.8	101.7	80.5	95.3	89.5	98.2	91.0	93.8
3	CONCRETE	103.1	102.6	102.8	118.4	94.5	103.2	111.8	96.6	102.1	103.2	92.7	96.5	90.8	97.6	95.1	101.8	81.2	88.7
4	MASONRY	123.2	95.2	101.7	90.6	89.8	90.0	102.5	87.1	90.7	101.6	79.4	84.6	103.8	98.5	99.7	112.0	63.6	74.9
5	METALS	99.0	101.5	99.9	111.6	93.3	105.2	102.3	94.5	99.6	96.6	88.3	93.7	95.5	100.8	97.4	87.4	80.0	84.8
6	WOOD & PLASTICS	102.0	101.4	101.7	118.0	93.6	104.4	103.6	96.5	99.7	100.0	81.3	89.6	95.4	97.0	96.3	96.6	70.8	82.3
7	MOISTURE PROTECTION	110.4	107.2	109.4	89.7	90.0	89.8	89.2	91.9	90.1	87.4	81.9	85.6	103.3	97.8	101.6	89.1	68.0	82.4
8	DOORS, WINDOWS, GLASS	103.2	98.3	100.7	106.7	86.4	96.2	112.3	88.8	100.1	97.7	80.4	88.7	95.2	97.2	96.2	108.4	72.9	90.0
9.1	LATH & PLASTER	108.2	105.1	105.8	118.6	92.0	98.4	116.0	92.2	97.9	105.4	77.3	84.1	103.3	93.5	95.9	107.9	91.2	95.3
9.2	DRYWALL	99.9	102.1	100.9	100.6	92.1	96.6	98.1	94.8	96.5	109.2	80.5	95.8	95.8	97.2	96.4	95.5	79.2	87.9
9.5	ACOUSTICAL WORK	105.2	101.5	103.2	108.8	92.4	99.9	110.6	97.0	103.2	108.6	80.6	93.4	100.0	97.3	98.5	95.6	69.8	81.6
9.6	FLOORING	98.1	87.5	95.3	85.4	92.1	87.2	81.3	89.5	83.5	96.6	78.3	91.7	102.7	94.0	100.4	93.3	65.8	85.9
9.8	PAINTING	109.9	100.5	102.4	124.0	78.2	87.6	114.6	83.8	90.1	99.0	79.9	83.9	103.7	93.5	95.6	100.3	90.3	92.3
9	FINISHES	101.1	100.6	100.9	100.7	87.3	93.6	97.4	90.6	93.8	105.2	80.0	91.8	98.6	95.5	96.9	95.8	82.0	88.5
10-14	TOTAL DIV. 10-14	100.0	104.8	101.3	100.0	93.7	98.1	100.0	93.1	98.0	100.0	87.0	96.2	100.0	91.0	97.4	100.0	82.4	94.9
15	MECHANICAL	101.6	103.5	102.5	101.6	84.0	92.9	104.1	92.0	98.1	101.7	86.7	94.2	100.2	92.5	96.3	102.9	69.7	86.4
16	ELECTRICAL	104.1	106.6	105.8	95.5	86.6	89.4	98.5	88.0	91.2	89.0	89.8	89.6	93.7	102.0	99.4	98.6	74.1	81.7
1-16	WEIGHTED AVERAGE	103.5	101.7	102.5	104.5	89.7	96.6	104.0	92.1	97.7	98.3	86.7	92.1	96.3	97.0	96.7	100.1	75.0	86.7

CANADA

DIVISION		EDMONTON MAT.	INST.	TOTAL	MONTREAL MAT.	INST.	TOTAL	QUEBEC MAT.	INST.	TOTAL	TORONTO MAT.	INST.	TOTAL	VANCOUVER MAT.	INST.	TOTAL	WINNIPEG MAT.	INST.	TOTAL
2	SITE WORK	107.3	101.9	104.9	96.4	99.6	97.8	101.6	74.8	89.7	117.2	107.5	112.9	116.2	109.5	113.2	109.2	101.5	105.8
3.1	FORMWORK	120.9	99.1	104.0	120.8	106.2	109.5	114.0	106.2	107.9	119.0	126.5	124.9	109.6	125.5	121.9	108.1	93.8	97.0
3.2	REINFORCING	119.7	105.3	113.9	119.4	92.0	108.3	117.7	92.0	107.3	80.1	111.4	92.8	100.1	116.8	106.9	117.9	85.9	104.9
3.3	CAST IN PLACE CONC.	119.1	99.0	106.8	109.0	101.0	104.1	145.6	76.6	103.5	164.1	107.2	129.4	117.1	108.6	111.9	108.5	111.2	110.1
3	CONCRETE	119.6	99.6	106.9	113.7	102.2	106.4	133.0	89.5	105.4	136.3	115.2	122.9	111.8	116.0	114.4	110.5	102.1	105.2
4	MASONRY	112.8	87.7	93.5	117.8	105.2	108.1	108.8	105.2	106.0	123.3	120.8	121.4	125.7	119.0	120.5	128.6	92.3	100.8
5	METALS	102.0	103.5	102.5	88.9	100.3	92.9	86.6	91.5	88.3	103.5	110.8	106.1	104.6	111.7	107.1	108.3	100.6	105.6
6	WOOD & PLASTICS	93.7	98.0	96.1	109.9	106.1	107.8	106.8	106.1	106.4	104.7	124.2	115.5	93.4	121.0	108.7	94.7	96.5	95.7
7	MOISTURE PROTECTION	99.1	98.0	98.7	92.9	107.0	97.3	92.3	107.0	96.9	97.6	120.9	105.0	106.6	126.1	112.8	103.8	91.1	99.8
8	DOORS, WINDOWS, GLASS	100.7	98.0	99.3	101.4	89.9	95.4	102.1	101.8	101.9	95.8	119.9	108.3	107.8	119.4	113.8	111.5	89.0	99.8
9.1	LATH & PLASTER	109.2	97.0	100.0	98.9	105.2	103.7	94.9	105.2	102.7	100.2	103.6	102.7	113.3	116.5	115.2	111.9	89.9	95.2
9.2	DRYWALL	108.6	97.9	103.6	111.7	106.2	109.2	114.6	106.2	110.7	107.4	109.4	108.3	112.4	116.2	114.2	105.7	92.3	99.5
9.5	ACOUSTICAL WORK	80.0	98.0	89.8	101.5	106.4	104.1	101.5	106.4	104.1	101.5	125.4	114.5	83.1	121.8	104.0	100.6	95.8	98.0
9.6	FLOORING	99.7	93.2	98.0	92.1	107.6	96.3	88.9	107.6	93.9	95.7	115.5	101.0	100.0	120.1	105.4	109.4	91.3	104.5
9.8	PAINTING	112.4	104.0	105.8	124.3	105.3	109.2	134.3	105.3	111.3	117.7	145.9	140.1	121.2	124.8	124.0	115.8	89.8	95.2
9	FINISHES	104.8	99.6	102.0	107.5	106.0	106.7	109.4	106.0	107.6	105.2	123.4	114.9	108.2	119.9	114.4	107.3	91.5	98.9
10-14	TOTAL DIV. 10-14	100.0	95.7	98.7	100.0	102.3	100.6	100.0	102.3	100.6	100.0	102.6	100.7	100.0	114.9	104.3	100.0	98.2	99.4
15	MECHANICAL	99.4	97.5	98.4	101.1	96.3	98.7	100.2	96.2	98.2	103.4	110.5	106.9	97.9	109.8	103.8	98.8	95.5	97.1
16	ELECTRICAL	108.3	93.1	97.8	104.2	97.8	99.8	106.0	97.8	100.4	104.5	111.1	109.1	100.8	114.0	109.9	111.4	98.1	102.3
1-16	WEIGHTED AVERAGE	104.8	96.6	100.4	102.6	100.7	101.6	105.0	97.2	100.8	108.9	114.9	112.1	105.6	115.5	110.9	106.0	96.3	101.0

ABBREVIATIONS

A	Area Square Feet; Ampere	C/C	Center to Center	Demob.	Demobilization
ABS	Acrylonitrile Butadiene Styrene;	Cab.	Cabinet	d.f.u.	Drainage Fixture Units
	Asbestos Bonded Steel	Cair.	Air Tool Laborer	D.H.	Double Hung
A.C.	Alternating Current;	Calc	Calculated	DHW	Domestic Hot Water
	Air Conditioning;	Cap.	Capacity	Diag.	Diagonal
	Asbestos Cement	Carp.	Carpenter	Diam.	Diameter
A.C.I.	American Concrete Institute	C.B.	Circuit Breaker	Distrib.	Distribution
Addit.	Additional	C.C.A.	Chromate Copper Arsenate	Dk.	Deck
Adj.	Adjustable	C.C.F.	Hundred Cubic Feet	D.L.	Dead Load; Diesel
af	Audio-frequency	cd	Candela	Do.	Ditto
A.G.A.	American Gas Association	cd/sf	Candela per Square Foot	Dp.	Depth
Agg.	Aggregate	CD	Grade of Plywood Face & Back	D.P.S.T.	Double Pole, Single Throw
A.H.	Ampere Hours	CDX	Plywood, grade C&D, exterior glue	Dr.	Driver
A hr	Ampere-hour	Cefi.	Cement Finisher	Drink.	Drinking
A.H.U.	Air Handling Unit	Cem.	Cement	D.S.	Double Strength
A.I.A.	American Institute of Architects	CF	Hundred Feet	D.S.A.	Double Strength A Grade
AIC	Ampere Interrupting Capacity	C.F.	Cubic Feet	D.S.B.	Double Strength B Grade
Allow.	Allowance	CFM	Cubic Feet per Minute	Dty.	Duty
alt.	Altitude	c.g.	Center of Gravity	DWV	Drain Waste Vent
Alum.	Aluminum	CHW	Chilled Water	DX	Deluxe White, Direct Expansion
a.m.	Ante Meridiem	C.I.	Cast Iron	dyn	Dyne
Amp.	Ampere	C.I.P.	Cast in Place	e	Eccentricity
Anod.	Anodized	Circ.	Circuit	E	Equipment Only; East
Approx.	Approximate	C.L.	Carload Lot	Ea.	Each
Apt.	Apartment	Clab.	Common Laborer	E.B.	Encased Burial
Asb.	Asbestos	C.L.F.	Hundred Linear Feet	Econ.	Economy
A.S.B.C.	American Standard Building Code	CLF	Current Limiting Fuse	EDP	Electronic Data Processing
Asbe.	Asbestos Worker	CLP	Cross Linked Polyethylene	E.D.R.	Equiv. Direct Radiation
A.S.H.R.A.E.	American Society of Heating,	cm	Centimeter	Eq.	Equation
	Refrig. & AC Engineers	CMP	Corr. Metal Pipe	Elec.	Electrician; Electrical
A.S.M.E.	American Society of	C.M.U.	Concrete Masonry Unit	Elev.	Elevator; Elevating
	Mechanical Engineers	Col.	Column	EMT	Electrical Metallic Conduit;
A.S.T.M.	American Society for	CO$_2$	Carbon Dioxide		Thin Wall Conduit
	Testing and Materials	Comb.	Combination	Eng.	Engine
Attchmt.	Attachment	Compr.	Compressor	EPDM	Ethylene Propylene
Avg.	Average	Conc.	Concrete		Diene Monomer
A.W.G.	American Wire Gauge	Cont.	Continuous; Continued	Eqhv.	Equip. Oper., heavy
Bbl.	Barrel	Corr.	Corrugated	Eqlt.	Equip. Oper., light
B.&B.	Grade B and Better;	Cos	Cosine	Eqmd.	Equip. Oper., medium
	Balled & Burlapped	Cot	Cotangent	Eqmm.	Equip. Oper., Master Mechanic
B.&S.	Bell and Spigot	Cov.	Cover	Eqol.	Equip. Oper., oilers
B.&W.	Black and White	CPA	Control Point Adjustment	Equip.	Equipment
b.c.c.	Body-centered Cubic	Cplg.	Coupling	ERW	Electric Resistance Welded
BE	Bevel End	C.P.M.	Critical Path Method	Est.	Estimated
B.F.	Board Feet	CPVC	Chlorinated Polyvinyl Chloride	esu	Electrostatic Units
Bg. Cem.	Bag of Cement	C. Pr.	Hundred Pair	E.W.	Each Way
BHP	Boiler Horse Power	CRC	Cold Rolled Channel	EWT	Entering Water Temperature
	Brake Horse Power	Creos.	Creosote	Excav.	Excavation
B.I.	Black Iron	Crpt.	Carpet & Linoleum Layer	Exp.	Expansion
Bit.; Bitum.	Bituminous	CRT	Cathode-Ray Tube	Ext.	Exterior
Bk.	Backed	CS	Carbon Steel	Extru.	Extrusion
Bkrs.	Breakers	Csc	Cosecant	f.	Fiber stress
Bldg.	Building	C.S.F.	Hundred Square Feet	F	Fahrenheit; Female; Fill
Blk.	Block	CSI	Construction Specifications	Fab.	Fabricated
Bm.	Beam		Institute	FBGS	Fiberglass
Boil.	Boilermaker	C.T.	Current Transformer	F.C.	Footcandles
B.P.M.	Blows per Minute	CTS	Copper Tube Size	f.c.c.	Face-centered Cubic
BR	Bedroom	Cu	Cubic	f'c.	Compressive Stress in Concrete;
Brg.	Bearing	Cu. Ft.	Cubic Foot		Extreme Compressive Stress
Brhe.	Bricklayer Helper	cw	Continuous Wave	F.E.	Front End
Bric.	Bricklayer	C.W.	Cool White; Cold Water	FEP	Fluorinated Ethylene
Brk.	Brick	Cwt.	100 Pounds		Propylene (Teflon)
Brng.	Bearing	C.W.X.	Cool White Deluxe	F.G.	Flat Grain
Brs.	Brass	C.Y.	Cubic Yard (27 cubic feet)	F.H.A.	Federal Housing Administration
Brz.	Bronze	C.Y./Hr.	Cubic Yard per Hour	Fig.	Figure
Bsn.	Basin	Cyl.	Cylinder	Fin.	Finished
Btr.	Better	d	Penny (nail size)	Fixt.	Fixture
BTU	British Thermal Unit	D	Deep; Depth; Discharge	Fl. Oz.	Fluid Ounces
BTUH	BTU per Hour	Dis.; Disch.	Discharge	Flr.	Floor
BX	Interlocked Armored Cable	Db.	Decibel	F.M.	Frequency Modulation;
c	Conductivity	Dbl.	Double		Factory Mutual
C	Hundred; Centigrade	DC	Direct Current	Fmg.	Framing

ABBREVIATIONS

Fndtn.	Foundation	I.P.	Iron Pipe	Mat; Mat'l.	Material
Fori.	Foreman, inside	I.P.S.	Iron Pipe Size	Max.	Maximum
Fount.	Fountain	I.P.T.	Iron Pipe Threaded	MBF	Thousand Board Feet
FPM	Feet per Minute	I.W.	Indirect Waste	MBH	Thousand BTU's per hr.
FPT	Female Pipe Thread	J	Joule	MC	Metal Clad Cable
Fr.	Frame	J.I.C.	Joint Industrial Council	M.C.F.	Thousand Cubic Feet
F.R.	Fire Rating	K	Thousand; Thousand Pounds;	M.C.F.M.	Thousand Cubic Feet
FRK	Foil Reinforced Kraft		Heavy Wall Copper Tubing		per minute
FRP	Fiberglass Reinforced Plastic	K.A.H.	Thousand Amp. Hours	M.C.M.	Thousand Circular Mils
FS	Forged Steel	K.D.A.T.	Kiln Dried After Treatment	M.C.P.	Motor Circuit Protector
FSC	Cast Body; Cast Switch Box	kg	Kilogram	MD	Medium Duty
Ft.	Foot; Feet	kG	Kilogauss	M.D.O.	Medium Density Overlaid
Ftng.	Fitting	kgf	Kilogram force	Med.	Medium
Ftg.	Footing	kHz	Kilohertz	MF	Thousand Feet
Ft. Lb.	Foot Pound	Kip	1000 Pounds	M.F.B.M.	Thousand Feet Board Measure
Furn.	Furniture	KJ	Kiljoule	Mfg.	Manufacturing
FVNR	Full Voltage Non-Reversing	K.L.	Effective Length Factor	Mfrs.	Manufacturers
FXM	Female by Male	Km	Kilometer	mg	Milligram
Fy.	Minimum Yield Stress of Steel	K.L.F.	Kips per Linear Foot	MGD	Million Gallons per Day
g	Gram	K.S.F.	Kips per Square Foot	MGPH	Thousand Gallons per Hour
G	Gauss	K.S.I.	Kips per Square Inch	MH; M.H.	Manhole; Metal Halide; Man-Hour
Ga.	Gauge	K.V.	Kilovolt	MHz	Megahertz
Gal.	Gallon	K.V.A.	Kilovolt Ampere	Mi.	Mile
Gal./Min.	Gallon per Minute	K.V.A.R.	Kilovar (Reactance)	MI	Malleable Iron; Mineral Insulated
Galv.	Galvanized	KW	Kilowatt	mm	Millimeter
Gen.	General	KWh	Kilowatt-hour	Mill.	Millwright
G.F.I.	Ground Fault Interrupter	L	Labor Only; Length; Long;	Min.; min.	Minimum; minute
Glaz.	Glazier		Medium Wall Copper Tubing	Misc.	Miscellaneous
GPD	Gallons per Day	Lab.	Labor	ml	Milliliter
GPH	Gallons per Hour	lat	Latitude	M.L.F.	Thousand Linear Feet
GPM	Gallons per Minute	Lath.	Lather	Mo.	Month
GR	Grade	Lav.	Lavatory	Mobil.	Mobilization
Gran.	Granular	lb.; #	Pound	Mog.	Mogul Base
Grnd.	Ground	L.B.	Load Bearing; L Conduit Body	MPH	Miles per Hour
H	High; High Strength Bar Joist;	L. & E.	Labor & Equipment	MPT	Male Pipe Thread
	Henry	lb./hr.	Pounds per Hour	MRT	Mile Round Trip
H.C.	High Capacity	lb./L.F.	Pounds per Linear Foot	ms	Millisecond
H.D.	Heavy Duty; High Density	lbf/sq in.	Pound-force per Square Inch	M.S.F.	Thousand Square Feet
H.D.O.	High Density Overlaid	L.C.L.	Less than Carload Lot	Mstz.	Mosaic & Terrazzo Worker
Hdr.	Header	Ld.	Load	M.S.Y.	Thousand Square Yards
Hdwe.	Hardware	LE	Lead Equivalent	Mtd.	Mounted
Help.	Helper average	L.F.	Linear Foot	Mthe.	Mosaic & Terrazzo Helper
HEPA	High Efficiency Particulate Air Filter	Lg.	Long; Length; Large	Mtng.	Mounting
Hg	Mercury	L. & H.	Light and Heat	Mult.	Multi; Multiply
HIC	High Interrupting Capacity	L.H.	Long Span High Strength Bar Joist	M.V.A.	Million Volt Amperes
H.O.	High Output	L.J.	Long Span Standard Strength	M.V.A.R.	Million Volt Amperes Reactance
Horiz.	Horizontal		Bar Joist	MV	Megavolt
H.P.	Horsepower; High Pressure	L.L.	Live Load	MW	Megawatt
H.P.F.	High Power Factor	L.L.D.	Lamp Lumen Depreciation	MXM	Male by Male
Hr.	Hour	lm	Lumen	MYD	Thousand yards
Hrs./Day	Hours per Day	lm/sf	Lumen per Square Foot	N	Natural; North
HSC	High Short Circuit	lm/W	Lumen per Watt	nA	Nanoampere
Ht.	Height	L.O.A.	Length Over All	NA	Not Available; Not Applicable
Htg.	Heating	log	Logarithm	N.B.C.	National Building Code
Htrs.	Heaters	L.P.	Liquefied Petroleum;	NC	Normally Closed
HVAC	Heating, Ventilating &		Low Pressure	N.E.M.A.	National Electrical
	Air Conditioning	L.P.F.	Low Power Factor		Manufacturers Association
Hvy.	Heavy	LR	Long Radius	NEHB	Bolted Circuit Breaker to 600V.
HW	Hot Water	L.S.	Lump Sum	N.L.B.	Non-Load-Bearing
Hyd.; Hydr.	Hydraulic	Lt.	Light	NM	Non-Metallic Cable
Hz.	Hertz (cycles)	Lt. Ga.	Light Gauge	nm	Nanometer
I.	Moment of Inertia	L.T.L.	Less than Truckload Lot	No.	Number
I.C.	Interrupting Capacity	Lt. Wt.	Lightweight	NO	Normally Open
ID	Inside Diameter	L.V.	Low Voltage	N.O.C.	Not Otherwise Classified
I.D.	Inside Dimension;	M	Thousand; Material; Male;	Nose.	Nosing
	Identification		Light Wall Copper Tubing	N.P.T.	National Pipe Thread
I.F.	Inside Frosted	m/hr; M.H.	Man-hour	NQOB	Bolted Circuit Breaker to 240V.
I.M.C.	Intermediate Metal Conduit	mA	Milliampere	N.R.C.	Noise Reduction Coefficient
In.	Inch	Mach.	Machine	N.R.S.	Non Rising Stem
Incan.	Incandescent	Mag. Str.	Magnetic Starter	ns	Nanosecond
Incl.	Included; Including	Maint.	Maintenance	nW	Nanowatt
Int.	Interior	Marb.	Marble Setter		
Inst.	Installation				
Insul.	Insulation				

OB	Opposing Blade
OC	On Center
OD	Outside Diameter
O.D.	Outside Dimension
ODS	Overhead Distribution System
O & P	Overhead and Profit
Oper.	Operator
Opng.	Opening
Orna.	Ornamental
O.S.&Y.	Outside Screw and Yoke
Ovhd.	Overhead
OWG	Oil, Water or Gas
Oz.	Ounce
P.	Pole; Applied Load; Projection
p.	Page
Pape.	Paperhanger
P.A.P.R.	Powered Air Purifying Respirator
PAR	Weatherproof Reflector
Pc.	Piece
P.C.	Portland Cement; Power Connector
P.C.M.	Phase Contrast Microscopy
P.C.F.	Pounds per Cubic Foot
P.E.	Professional Engineer; Porcelain Enamel; Polyethylene; Plain End
Perf.	Perforated
Ph.	Phase
P.I.	Pressure Injected
Pile.	Pile Driver
Pkg.	Package
Pl.	Plate
Plah.	Plasterer Helper
Plas.	Plasterer
Pluh.	Plumbers Helper
Plum.	Plumber
Ply.	Plywood
p.m.	Post Meridiem
Pord.	Painter, Ordinary
pp	Pages
PP; PPL	Polypropylene
P.P.M.	Parts per Million
Pr.	Pair
Prefab.	Prefabricated
Prefin.	Prefinished
Prop.	Propelled
PSF; psf	Pounds per Square Foot
PSI; psi	Pounds per Square Inch
PSIG	Pounds per Square Inch Gauge
PSP	Plastic Sewer Pipe
Pspr.	Painter, Spray
Psst.	Painter, Structural Steel
P.T.	Potential Transformer
P. & T.	Pressure & Temperature
Ptd.	Painted
Ptns.	Partitions
Pu	Ultimate Load
PVC	Polyvinyl Chloride
Pvmt.	Pavement
Pwr.	Power
Q	Quantity Heat Flow
Quan.; Qty.	Quantity
Q.C.	Quick Coupling
r	Radius of Gyration
R	Resistance
R.C.P.	Reinforced Concrete Pipe
Rect.	Rectangle
Reg.	Regular
Reinf.	Reinforced
Req'd.	Required
Resi	Residential
Rgh.	Rough
R.H.W.	Rubber, Heat & Water Resistant; Residential Hot Water
rms	Root Mean Square
Rnd.	Round
Rodm.	Rodman
Rofc.	Roofer, Composition
Rofp.	Roofer, Precast
Rohe.	Roofer Helpers (Composition)
Rots.	Roofer, Tile & Slate
R.O.W.	Right of Way
RPM	Revolutions per Minute
R.R.	Direct Burial Feeder Conduit
R.S.	Rapid Start
RT	Round Trip
S.	Suction; Single Entrance; South
Scaf.	Scaffold
Sch.; Sched.	Schedule
S.C.R.	Modular Brick
S.D.	Sound Deadening
S.D.R.	Standard Dimension Ratio
S.E.	Surfaced Edge
S.E.R.; S.E.U.	Service Entrance Cable
S.F.	Square Foot
S.F.C.A.	Square Foot Contact Area
S.F.G.	Square Foot of Ground
S.F. Hor.	Square Foot Horizontal
S.F.R.	Square Feet of Radiation
S.F.Shlf.	Square Foot of Shelf
S4S	Surface 4 Sides
Shee.	Sheet Metal Worker
Sin.	Sine
Skwk.	Skilled Worker
SL	Saran Lined
S.L.	Slimline
Sldr.	Solder
S.N.	Solid Neutral
S.P.	Static Pressure; Single Pole; Self Propelled
Spri.	Sprinkler Installer
Sq.	Square; 100 square feet
S.P.D.T.	Single Pole, Double Throw
S.P.S.T.	Single Pole, Single Throw
SPT	Standard Pipe Thread
Sq. Hd.	Square Head
Sq. In.	Square Inch
S.S.	Single Strength; Stainless Steel
S.S.B.	Single Strength B Grade
Sswk.	Structural Steel Worker
Sswl.	Structural Steel Welder
St.; Stl.	Steel
S.T.C.	Sound Transmission Coefficient
Std.	Standard
STP	Standard Temperature & Pressure
Stpi.	Steamfitter, Pipefitter
Str.	Strength; Starter; Straight
Strd.	Stranded
Struct.	Structural
Sty.	Story
Subj.	Subject
Subs.	Subcontractors
Surf.	Surface
Sw.	Switch
Swbd.	Switchboard
S.Y.	Square Yard
Syn.	Synthetic
Sys.	System
t.	Thickness
T	Temperature; Ton
Tan	Tangent
T.C.	Terra Cotta
T & C	Threaded and Coupled
T.D.	Temperature Difference
T.E.M.	Transmission Electron Microscopy
TFE	Tetrafluoroethylene (Teflon)
T. & G.	Tongue & Groove; Tar & Gravel
Th.; Thk.	Thick
Thn.	Thin
Thrded	Threaded
Tilf.	Tile Layer Floor
Tilh.	Tile Layer Helper
THW	Insulated Strand Wire
THWN; THHN	Nylon Jacketed Wire
T.L.	Truckload
Tot.	Total
T.S.	Trigger Start
Tr.	Trade
Transf.	Transformer
Trhv.	Truck Driver, Heavy
Trlr.	Trailer
Trlt.	Truck Driver, Light
TV	Television
T.W.	Thermoplastic Water Resistant Wire
UCI	Uniform Construction Index
UF	Underground Feeder
U.H.F.	Ultra High Frequency
U.L.	Underwriters Laboratory
Unfin.	Unfinished
URD	Underground Residential Distribution
V	Volt
V.A.	Volt Amperes
V.A.C.	Vinyl Composition Tile
VAV	Variable Air Volume
Vent.	Ventilating
Vert.	Vertical
V.F.	Vinyl Faced
V.G.	Vertical Grain
V.H.F.	Very High Frequency
VHO	Very High Output
Vib.	Vibrating
V.L.F.	Vertical Linear Foot
Vol.	Volume
W	Wire; Watt; Wide; West
w/	With
W.C.	Water Column; Water Closet
W.F.	Wide Flange
W.G.	Water Gauge
Wldg.	Welding
W. Mile	Wire Mile
W.R.	Water Resistant
Wrck.	Wrecker
W.S.P.	Water, Steam, Petroleum
WT, Wt.	Weight
WWF	Welded Wire Fabric
XFMR	Transformer
XHD	Extra Heavy Duty
XHHW; XLPE	Cross-Linked Polyethylene Wire Insulation
Y	Wye
yd	Yard
yr	Year
Δ	Delta
%	Percent
~	Approximately
Ø	Phase
@	At
#	Pound; Number
<	Less Than
>	Greater Than

INDEX

INDEX

INDEX

INDEX

Avoiding and Resolving Construction Claims

by Barry B. Bramble, Esq.
Michael F. D'Onofrio, PE
John B. Stetson, IV, RA

1st Edition
NEW! Over 240 Pages • Illustrated • Hardcover

As a member of the construction management team, chances are that sooner or later you'll be involved in a construction claim.

It may relate to a design error, subsurface condition, defective installation, material, delay—any number of unexpected events which cost extra money.

How such claims are avoided and resolved is the subject of this penetrating new guide.

Each chapter addresses a different step of the process and includes a summary checklist for action, sample problems, graphic illustrations, and a few case histories to show how courts have decided disputes.

ISBN 0-87629-180-9
Book No. 67275

$59.95/copy
U.S. Funds

Means Estimating Handbook

1st Edition
NEW! Over 900 Pages • Illustrated • Hardcover

Means Estimating Handbook simplifies the task of evaluating construction plans and specs to obtain reliable quantities for pricing.

The Handbook reflects the tremendous variety of technical data required for estimating . . .

. . . questions relating to sizing, productivity, equipment requirements, codes, design standards, engineering, pressures, stresses, loads, coverages . . . and hundreds of similar factors.

The guide provides the busy estimator with every technical reference imaginable . . . tables, illustrations, definitions, examples . . . any construction factor which will help answer the key questions:

How many, how much, and how long will it take?

ISBN 0-87629-177-9
Book No. 67276

$89.95/copy
U.S. Funds

Survival in the Construction Business: Checklists for Success

by Thomas N. Frisby

1st Edition
NEW! 300 Pages • Illustrated
Index • Appendix • Hardcover

If you're wondering how to guide your company to success and profit in the decade ahead, *Survival in the Construction Business: Checklists for Success* can show you how.

In just eight concise chapters this guide details a step-by-step approach to construction company mangement. It gives you the practical tools you need to ensure that every decision you make will help you meet your goals.

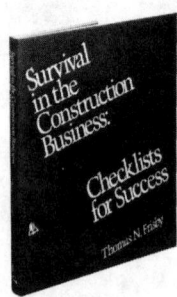

Survival's method is based on a comprehensive series of construction "Checklists" that will steer you through every phase of company evaluation, for both long-range planning and day-to-day operation. Put these checklists to use, and you'll start seeing results right away!

ISBN 0-87629-153-1
Book No. 67274

$59.95/copy
U.S. Funds

HVAC Systems Evaluation

• Comparing Systems
• Solving Problems
• Efficiency & Maintenance

by Harold R. Colen, PE

1st Edition
NEW! Over 500 Pages • Illustrated • Hardcover

A virtual encyclopedia of experienced know-how for comparing, selecting, installing, and fixing HVAC equipment and systems.

HVAC Systems Evaluation is a convenient, straight-talk way for you to understand and select HVAC systems for new or retrofit construction.

It gives you direct, right-to-the-point comparisons of how each type of system works . . . installation costs . . . operating costs . . . applications by type of building.

You get experienced advice for fixing operational problems in existing HVAC systems—*everything is covered.*

ISBN 0-87629-182-5
Book No. 67281

$57.95/copy
U.S. Funds

Means Construction Cost Indexes 1991

(Individual and back issues available at $42.25 per copy)

The index service providing updated cost adjustment factors

Whether updating construction costs from Means cost manuals or data from other sources, the construction cost index service is the efficient way to ensure 90-day cost accuracy.

Published quarterly (January, April, July, October), this handy report provides cost adjustment factors for the preparation of more precise estimates no matter how late in the year. It's also the ideal method for making continuous cost revisions on ongoing projects as the year progresses.

The report is organized in four unique sections:

- breakdowns for 209 major cities
- national averages for 30 key cities
- five large city averages
- historical construction cost indexes

Book No. 60141   **$169.00/year**   **U.S. Funds**

Means Scheduling Manual
By F. William Horsley

2nd Edition
Over 200 Pages • Illustrated • Hardcover

For experienced schedulers and project managers as well as the uninitiated

Fast, convenient expertise for keeping your scheduling skills right in step with today's cost-conscious times.

This concisely written scheduling handbook shows you the entire scheduling process far faster than any reference of its kind.

You'll benefit from all the information provided on traditional bar charts, Pert and CPM . . . and home in on the precision offered by *Precedence Scheduling*.

The book guides you through all aspects of scheduling, with fold-out spread sheets, charts, sample schedules.

ISBN 0-911950-36-2   **$49.95/copy**
Book No. 67152   **U.S. Funds**

Means Forms for Building Construction Professionals

2nd Edition
Over 325 Pages • Three-Ring Binder

Don't waste time trying to compose forms — we've done the job for you!

- Forms can be customized with your company name and reproduced on your copier or at your local instant printer.

- Forms for all primary construction activities—estimating, designing, project administration, scheduling, appraising.

- Many optional variations— condensed, detailed versions.

- Forms compatible with Means annual cost books and other typical user systems.

- Ideal for standardizing estimating and project management functions at low cost.

- Full size forms on durable reproduction paper presented in a sturdy three-ring binder.

ISBN 0-911950-87-7   **$76.95/copy**
Book No. 67231   **U.S. Funds**

Means Labor Rates for the Construction Industry 1991

18th Annual Edition • Over 325 Pages
CITY • STATE • NATIONAL

- Detailed wage rates by trade for over 300 U.S. and Canadian cities
- Forty-six construction trades listed by local union number in each city
- Base hourly wage rates plus fringe benefit package costs gathered from reliable sources
- Dependable estimates for the trade wage rates not reported at press time

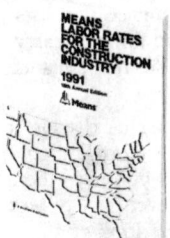

- Effective dates for newly negotiated union contracts for both 1990 and 1991
- Factors for comparing each trade rate by city, state, and national averages
- Historical 1989–1990 wage rates also included for comparison purposes
- Each city chart is alphabetically arranged with handy visual flip tabs for quick reference

ISBN 0-87629-191-4   **$145.00/copy**
Book No. 60121   **U.S. Funds**

Planning and Managing Interior Projects

by Carol E. Farren

1st Edition
Over 300 Pages • Illustrated • Hardcover

NOW, a clearly defined working guide to project management functions of interior installations

You can rely on the expertise in *Planning and Managing Interior Projects* because it's been tested and proved out *on the job!* The author draws upon the fruits of experience to give you a better working knowledge of each area of interior project management.

If your situation requires you to be knowledgeable in *any or all phases of carrying out interior projects*— working with the client, working with building managers, contractors, movers, telephone installers and suppliers, as well as preparing designs and plans—this book will be *the best investment you can make for doing a better, more professional job!*

ISBN 0-87629-097-7
Book No. 67245

$59.95/copy
U.S. Funds

The Facilities Manager's Reference

• Management • Building Audits
• Planning   • Estimating
by Harvey H. Kaiser, Ph.D.

1st Edition
Over 240 Pages with Prototype Forms and Line Art Graphics • Hardcover

Your one-step source for the latest facilities management methods used successfully by both large and small operations

One of the nation's leading management authorities describes how to manage facilities in the "real world" of modern organizations.

The author explains the diverse "hats" a facilities manager must wear as organizer, planner, estimator, supervisor, motivator, and decision-maker, and how these roles can be managed most effectively.

Now you can have access to a huge collection of new concepts, methods, and tools that you can use *immediately* to build up your level of performance and funding!

ISBN 0-87629-142-6
Book No. 67264

$59.95/copy
U.S. Funds

Means Facilities Maintenance Standards

1st Edition
Over 575 Pages • 205 Tables, Checklists, and Diagrams • Hardcover

A definitive reference addressing thousands of facilities maintenance problems

Means Facilities Maintenance Standards is a working encyclopedia that provides solutions to almost every kind of maintenance and repair dilemma.

The book guides you to the underlying causes of material deterioration, shows you how to analyze its effects, and steps you through the appropriate methods of repair. A

Man-Hours section lists estimated times to carry out various maintenance tasks.

All of the checklists in this reference are organized in the order you need them. You'll never have to worry about overlooking an important consideration or crucial step in repairs again!

ISBN 0-87629-096-9
Book No. 67246

$131.95/copy
U.S. Funds

Facilities Maintenance Management

by Gregory H. Magee, PE

1st Edition
Over 250 Pages • Illustrated • Hardcover

Provocative—instructive—you'll benefit from new ideas for planning and managing maintenance functions in your organization

This comprehensive reference will guide you through important aspects of facilities maintenance management.

It gives you ideas for staffing, estimating, budgeting, scheduling, and controlling work to produce a more efficient, cost-effective maintenance department.

No matter what your need or problem, you're sure to find direction for solving it in this authoritative book—written by a professional who's conquered every kind of challenge—*including the one on your desk right now!*

ISBN 0-87629-100-0
Book No. 67249

$59.95/copy
U.S. Funds

Estimating for the General Contractor
by Paul J. Cook

1st Edition
Over 225 Pages • Illustrated • Hardcover

For general contractors . . . breathe new life into your estimates and profits

Light on theory, heavy on practical estimating methods and ideas, here's powerful help for the contractor/estimator who wants to evaluate and polish every aspect of estimating procedures.

Estimators at all levels of experience will appreciate this comprehensive package of indispensable methods and procedures.

Through the use of clear explanations as well as detailed examples, tables, and graphs, the reader is shown how estimating may be done with *maximum efficiency*—without sacrificing *quality*.

ISBN 0-87629-110-8
Book No. 67160

$59.95/copy
U.S. Funds

Business Management for the General Contractor
by Paul J. Cook

1st Edition
Over 200 Pages • Illustrated • Hardcover

It's direct. It's current. And it's relevant to your needs.

Business Management for the General Contractor has one overriding priority—to give the contractor a basic working knowledge of all aspects of business management. With this dependable reference, you won't need bundles of cash or battalions of experts to solve your everyday problems.

You'll refer to this remarkable book again and again because the ideas it provides will be of continuous value to your career and your company's progress! You'll find efficient methods and strategies to help you handle almost any kind of management problem typical to contractors.

ISBN 0-87629-098-5
Book No. 67250

$54.95/copy
U.S. Funds

Bidding for the General Contractor
by Paul J. Cook

1st Edition
Over 225 Pages • Illustrated • Hardcover

The techniques of successful bidding and how to apply them in your own construction business

Now you can have a truly comprehensive guide for making competitive bids—methods and guidelines for all job sizes, up to multimillion-dollar projects.

It sheds new light on bidding procedures and techniques . . . at last you can see and compare your approach with other successful bidders. You'll have in-depth discussion and illustrations covering every step of the bid management process, beginning the first moment you get the sponsor's bid package.

ISBN 0-911950-77-X
Book No. 67180

$54.95/copy
U.S. Funds

Superintending for the General Contractor
Field Project Management
by Paul J. Cook

1st Edition
Over 220 Pages • Illustrated • Hardcover

A landmark guide to the effective on-site management of construction projects

At last there is a guide to on-site construction management that goes beyond textbook theory and simplistic discussion. Paul J. Cook's *Superintending for the General Contractor* is a fully developed, well organized working handbook. It delves

deeply into every area of the superintendent's job, probing, elaborating . . . pointing out the do's and don'ts of managing manpower, materials, equipment, and paperwork.

Contains hundreds of valuable insights for dealing with clients, subcontractors, foremen, suppliers, and workers . . . anyone who works on the project . . . so it gets done at targeted quality, cost, and deadlines.

ISBN 0-87629-063-2
Book No. 67233

$54.95/copy
U.S. Funds

HVAC:
Design Criteria, Options, Selection

by William H. Rowe, III, AIA, PE

1st Edition
Over 380 Pages • Illustrated • Hardcover

For a total understanding of HVAC systems . . .

If you're new to HVAC systems design, or simply want to make certain you're right in step with the newest HVAC design concepts and equipment, you'll benefit from this highly recommended resource.

HVAC: *Design Criteria, Options, Selection* is a masterful book, providing a thorough understanding of modern heating, ventilating, and air conditioning (HVAC) systems.

It gives you a comprehensive overview of the basic functions of HVAC, with emphasis on the design and costs of effective integrated climate control systems.

ISBN 0-87629-102-7
Book No. 67251

$59.95/copy
U.S. Funds

Hazardous Material and Hazardous Waste:
A Construction Reference Manual

by F.J. Hopcroft, P.E., D.L. Vitale, M. Ed., D.L. Anglehart, Esq.

1st Edition
Over 260 Pages • Illustrated • Hardcover

With OSHA, EPA and other agencies stepping up compliance inspections at construction sites, contractors can no longer afford to put off installing programs for informing and training their workers on the proper handling and disposal of hazardous materials.

You're given a full explanation of the laws as they pertain to construction work and what you—as an employer—are expected to do.

This includes suggested methods for storing and using hazardous materials, and then disposing of hazardous wastes, all in accordance with federal and state regulations.

ISBN 0-87629-136-1
Book No. 67258

$76.95/copy
U.S. Funds

Means
Electrical Estimating
Standards and Procedures

1st Edition
Over 300 Pages • Illustrated • Hardcover

Even experienced estimators applaud the countless ways this superb guidebook helps them!

This eye-opening book breaks electrical installations down into modules and explains how to estimate each of them using the *Means Electrical Cost Data* manual.

- gives you instant electrical estimating help for specific types of installations
- gives you the combined experience of the Means staff, plus top designers and electrical contractors
- covers electrical estimating in full—from takeoff to overhead and profit

80 electrical estimating modules illustrate and explain each step in the estimating process . . . units of measure for labor and materials, typical job conditions, takeoff procedures, cost modifications . . . and a great deal more!

ISBN 0-911950-83-4
Book No. 67230

$54.95/copy
U.S. Funds

Means
Mechanical Estimating
Standards and Procedures

1st Edition
Over 370 Pages • Illustrated • Hardcover

For HVAC/Plumbing/Fire Protection estimating . . . from takeoff through pricing, bidding and scheduling

Means Mechanical Estimating examines each mechanical contracting activity, pointing out the best ways to predict material and installation costs. It evaluates, analyzes, and integrates every key estimating procedure from interpreting contract documents to the final minutes of the bid.

This is a comprehensive, nononsense reference that offers frank discussion of how to evaluate mechanical plans and estimate all cost components . . . right down to the last installed pipe hanger.

It will help you estimate better, understand mechanical work better, and develop the most cost efficient approaches to achieve your goals.

ISBN 0-87629-066-7
Book No. 67235

$54.95/copy
U.S. Funds

Quantity Takeoff for the General Contractor

by Paul J. Cook

1st Edition

Over 225 Pages • Illustrated • Hardcover

How to put more speed and accuracy into your quantity takeoff work

If you're new to quantity takeoff for estimating, or simply want to be sure you're using the latest and best takeoff techniques, this book is "must" reading.

It gives you a concise overview of the process—the quantity estimator's role, the tools he or she uses, how the project is broken down, and the rules to follow which help to ensure accuracy.

You will see how to evaluate plans for the site, footings, foundation, slab, floor, wall, framing, and roof, to calculate the quantity of materials and installing labor.

If you've ever had any doubts about how to do a better, faster job making quantity takeoffs—this book will help you do just that—by pointing out ways to make error-free takeoff estimates for every project.

ISBN 0-87629-141-8
Book No. 67262

$59.95/copy
U.S. Funds

Means Unit Price Estimating

1st Edition
Over 350 Pages • Hardcover

Direct, immediate help for preparing better unit cost estimates—no matter how much experience you have!

Indispensable for strengthening your unit cost estimating

Means Unit Price Estimating directs you to the right answers to your unit cost procedural questions—and directs you fast! It describes the most productive, universally-accepted ways to

estimate, and uses checklists and charts to unearth shortcuts and time-savers. The strategy of bidding is explained, and up-to-date guidance is provided to assist in evaluation of your own approach.

A model estimate for a multi-story office building is included to demonstrate procedures. The book provides proven systems and special pointers to guide you through the building process.

ISBN 0-87629-027-6
Book No. 67232

$54.95/copy
U.S. Funds

Means Square Foot Estimating

by Billy J. Cox and F. William Horsley

1st Edition
Over 300 Pages • Hardcover

A new generation of techniques for conceptual and design-stage cost planning

Doing an effective job at the drawing board and estimating desk takes time. Too often, the time to carefully explore alternatives and evaluate different ideas is limited.

Means Square Foot Estimating is devoted to helping you accomplish more in less time. It steps you through the entire square foot cost process, pointing out faster, better ways to relate the design to the budgets.

In all, *Means Square Foot Estimating* provides clearer, better knowledge of how to greatly upgrade the *efficiency* and *effectiveness* of square foot cost estimating in the project's early stages.

ISBN 0-87629-090-X
Book No. 67145

$57.95/copy
U.S. Funds

Means Repair and Remodeling Estimating

by Edward B. Wetherill

Expanded New Edition
Over 450 Pages • Illustrated • Hardcover

- • Authoritative
- • Easy to understand
- • Follows CSI format
- • Sample estimates

Means Repair and Remodeling Estimating focuses on the unique problems of estimating renovations of existing structures. It helps you determine the true costs of remodeling through careful evaluation of architectural details and a site visit.

This, coupled with a CSI division-by-division discussion of potential pitfalls, and two sample estimates, gives you a real foundation for estimating remodeling work.

Although designed primarily for contractors and architects, the concepts in *Means Repair and Remodeling Estimating* apply to anyone who wants to enhance their renovation estimating skills.

ISBN 0-87629-144-2
Book No. 67265

$57.95/copy
U.S. Funds

Means Interior Estimating
by Alan E. Lew
1st Edition
Over 370 Pages • Illustrated • Hardcover

Four complete estimating manuals in one comprehensive reference

This book provides in-depth discussion and illustrations covering every type of interior construction . . .

- **Readable.** Tightly written, easy to read.
- **Illustrated.** Dozens of easy-to-follow prototype estimating forms and diagrams of interior assemblies and building plans.
- **Answers.** Each fact-filled chapter is a gold mine of answers to your questions.

- **Comprehensive.** Virtually anyone whose work involves interior estimating will benefit from this book.
- **Authoritative.** Authored and edited by a team of interior estimating experts, this information is based on actual experience and proven techniques.
- Hands-on application of *Means Interior Cost Data.*

ISBN 0-87629-067-5
Book No. 67237

54.95/copy
U.S. Funds

Means Landscape Estimating
by Sylvia H. Fee
1st Edition
Over 275 Pages • Illustrated • Hardcover

Answers all your questions about preparing competitive landscape construction estimates . . .

Here's the important landscape estimating reference that gives you the tools you need to solve your landscape pricing problems.

Means Landscape Estimating is a thorough, easy-reading, well organized working guide that "talks you through" every step of preparing effective bids and estimates—in a minimum of time. Written by a highly respected landscape designer and contractor.

Everything you want to know about landscape estimating is right here—including marketing your services, performing the takeoff, bidding and planning the job.

ISBN 0-87629-064-0
Book No. 67239

$57.95/copy
U.S. Funds

Home Improvement Cost Guide
2nd Edition
Over 225 Pages • Illustrated • Softcover

How to plan and price home improvements quickly, easily—projects large and small

Planning a home improvement? Maintenance project? Read this.

Prior to planning your next home improvement, wouldn't it make sense for you to talk with several expert builders who have done the same project many times before?

Suppose these professionals happily shared their insights, warned you of potential dangers, pointed out ways for you to get the

most for your money . . . and then provided you with a **reliable estimate** of what you'd expect to pay for the improvement—even if you did some of the work yourself?

You'd welcome this kind of advice with open arms, of course. Anyone would.

Well, that's exactly what the **Home Improvement Cost Guide** does for you.

ISBN 0-87629-173-6
Book No. 67280

$32.95/copy
U.S. Funds

Roofing: Design Criteria, Options, Selection
by R.D. Herbert, III
1st Edition
Over 200 Pages • Illustrated • Hardcover

This authoritative new guide is overflowing with fresh, practical know-how for professionals involved in roof construction—architects, contractors, owners, facilities managers, and roofing installers.

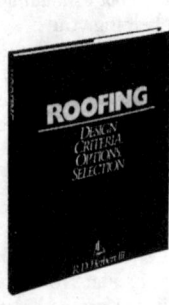

Now you can avoid roofing problems and choose the most cost-effective system—this easy-reading book helps you see the opportunities and pitfalls of modern roofing construction.

All types of roofing are covered . . . built-up, single-ply, modified bitumens, metal, sprayed-in-place, slate, tile, shingles, and shakes, as well as all types of seals and accessories.

ISBN 0-87629-104-3
Book No. 67253

$59.95/copy
U.S. Funds

Project Planning and Control for Construction

by David R. Pierce, Jr.

1st Edition
Over 275 Pages • Illustrated • Hardcover
Includes Sample Project

Here for the first time is a comprehensive action-guide to results-oriented project management

Project Planning and Control for Construction is designed specifically to help construction project managers work more effectively . . . and profitably. It is a guide to the best, most up-to-date techniques for:

- pre-construction planning
- determining activity sequence
- scheduling
- monitoring and controlling
- managing submittal data
- managing resources
- accelerating the project
- cost control

This guide allows you to compare your current methods with those recommended by one of the nation's most highly respected scheduling and project management consultants.

ISBN 0-87629-099-3
Book No. 67247

$58.95/copy
U.S. Funds

Cost Control in Building Design

by Roger A. Killingsworth

1st Edition
Over 280 Pages • Illustrated • Hardcover

Using Square Foot and Assemblies estimating techniques to predict and control costs

Direct from the practical experience of a highly respected estimating consultant and educator, *Cost Control in Building Design* examines every step of the early construction planning and estimating process . . . pointing out reliable new ways to budget and control costs as the design moves into sharp focus.

It explains—evaluates—integrates every key step in a total COST CONTROL SYSTEM that can help you bring projects to completion within 5% of budget.

Sample project includes mechanical and electrical estimates.

ISBN 0-87629-103-5
Book No. 67252

$58.95/copy
U.S. Funds

Cost Effective Design/Build Construction

by Anthony J. Branca, PE

1st Edition
Over 420 Pages • Illustrated • Hardcover

For contractors, architects, engineers . . . any professional who wants to get into the design/build business

If you've been thinking about moving into design/build construction work, this book can be invaluable for helping you make your decision.

As a contractor, architect, engineer or other construction professional, perhaps you've been intrigued by the prospect of getting into design/build construction—maybe even founding your own firm.

Cost Effective Design/Build Construction takes you through every step in the business . . . including what it takes to start your own firm. Learn how to market yourself, understand client needs, to prepare presentations, and how to manage projects using the design/build method. You'll find everything needed to understand exactly how design/build works.

ISBN 0-87629-088-8
Book No. 67242

$54.95/copy
U.S. Funds

The Building Professional's Guide to Contract Documents
(formerly Plans, Specs and Contracts for Building Professionals)

by Waller S. Poage, AIA, CSI, CCS

New! 2nd Edition
Over 400 Pages • Illustrated • Hardcover

Includes coverage of the latest AIA documents.

- detailed discussions of contract components . . . addenda, conditions, supplements, specifications, drawings
- instructions for preparing specifications and technical requirements . . . description writing, proprietary and performance specs, standards
- pointers for contract administration . . . duties and responsibilities in the OPC (owner, design professional, contractor) relationship, meetings, transmittals, shop drawings, mock-ups, the punch list, payments, and more.

ISBN 0-87629-210-4
Book No. 67261

$54.95/copy
U.S. Funds

Bidding and Managing Government Construction

by Theodore J. Trauner, Jr., PE, PP, and Michael H. Payne, Esq.

1st Edition
Over 200 Pages • Hardcover

Learn how to win and administrate government construction contracts.

Why this book will be of immense value to you in your government construction work.

- It provides you with plain language explanations of the laws that control government construction and the complex and often inconsistent requirements of government contracts.

- It explains the bidding process, pointing out the reasons for disqualification based on nonresponsive or irresponsible submissions.

- It clarifies how various government agencies administer and manage construction, pointing out your rights, responsibilities and interests.

- It describes in detail the procedures and pricing methods for changes in construction contracts.

ISBN 0-87629-111-6
Book No. 67257

$58.95/copy
U.S. Funds

Understanding the Legal Aspects of Design/Build

by Timothy R. Twomey, Esq., AIA

1st Edition
Over 370 Pages • Illustrated • Hardcover

This practical design/build guide will help you:

- **Evaluate and compare** various design/build methods
- **Analyze** the strengths and weaknesses of design/build systems
- **Understand** your role, responsibilities and liabilities as contractor, designer or client

- **Gain insight** into the types of clients most likely to use these services
- **Survey** the legal issues surrounding errors, acts, and omissions
- **See how** traditional legal concepts apply differently to design/build
- **Use** prototype design/build contracts
- **Answer** insurance, bonding and licensing questions

ISBN 0-87629-137-X
Book No. 67259

$71.95/copy
U.S. Funds

Risk Management for Building Professionals

by Thomas E. Papageorge, RA

1st Edition
Over 200 Pages • Illustrated • Hardcover

How well do you manage the complex risks of building design and construction?

If you've suffered losses of time, money, reputation … harm to workers, or damage to projects because risks went unnoticed— **we have very important information for you.**

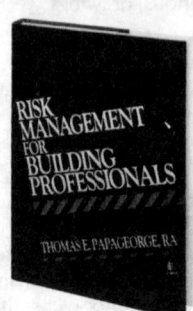

Risk Management for Building Professionals provides you with a master plan for phasing risk management functions into every department of your firm.

The goal is to provide a practical framework for recognizing potential risks and planning out how they will be handled *before* they occur.

ISBN 0-87629-106-X
Book No. 67254

$54.95/copy
U.S. Funds

Contractor's Business Handbook

- Accounting   • Tax Management
- Finance   • Cost Control
by Michael S. Milliner

1st Edition
Over 300 Pages • Illustrated • Hardcover

Attain new growth and profits through streamlined financial controls

This handbook is designed for contractors who want to put their business on a firmer financial footing and plan for future growth.

In plain language, the book describes and demonstrates how to use the contractor's most powerful financial management tool— *an efficient financial control system.*

The book delves deeply into accounting systems, financial analysis and forecasting, asset and debt management, tax reduction strategies, and computer-based control.

A convenient glossary defines financial terminology used in the text.

ISBN 0-87629-105-1
Book No. 67255

$58.95/copy
U.S. Funds

Means Man-Hour Standards for Construction

2nd Edition
Over 750 Pages • Hardcover

The "professional's choice" for uncompromised trade and labor productivity data

Here is the working encyclopedia of labor productivity information for construction professionals ... *Means Man-Hour Standards for Construction*, revised 2nd Edition.

The efficient format permits rapid comparisons of labor requirements for thousands of construction functions in the CSI MASTERFORMAT.

You'll find every bit of labor data you may need ... *all in a superb layout with "quick-find" indexes and handy visual flip tabs.*

ISBN 0-87629-089-6
Book No. 67236

$131.95/copy
U.S. Funds

Means Illustrated Construction Dictionary

1st Edition
Over 575 Pages • Illustrated • Hardcover

A working handbook of over 12,000 construction terms — explained and illustrated

Here's the on-the-job reference even the most experienced professionals can turn to for immediate answers about construction terms.

The Means Dictionary is packed with thousands of up-to-date explanations of construction terminology. It covers every area of construction from design right on through everyday lingo used by tradesmen. It's "alive" with entries covering every conceivable new technique, product.

Its no-nonsense guidance, illustrations, abbreviations, and easy-to-use format will serve you for years to come

ISBN 0-911950-82-6
Book No. 67190

$87.95/copy
U.S. Funds

Means Graphic Construction Standards

1st Edition
Over 500 Pages • Illustrated

Create construction concepts/designs Gain new insights and approaches

Means Graphic Construction Standards assists you in making preliminary "audits" of your designs. It simplifies the review of construction methods, helping you sort out potential problems. You decide visually which elements are essential and which offer you the most cost-effective alternatives.

The book illustrates and discusses the relationships between various building systems. Because each construction assembly gives you extensive design data, you're able to leap from rough concepts to workable plans quickly. You don't waste time working backwards from costs to designs. The book gives you the freedom to maximize your creativity within time and budget goals.

ISBN 0-911950-79-6
Book No. 67210

$109.95/copy
U.S. Funds

Fundamentals of the Construction Process
by K.K. Bentil, AIC

1st Edition
Over 450 Pages • Illustrated • Hardcover

The first and only book designed specifically to introduce the basics of building construction

Fundamentals of the Construction Process has been prepared for executives, facilities managers, and others whose responsibilities include overseeing, budgeting, or other involvement in

building or facilities construction. While construction processes are often highly technical, the book focuses on providing simplified, accessible, information.

It provides extensive coverage of the pre-construction phase, including an overview of project types, contract documents, cost estimating, contract procurement, bonding, scheduling and mobilization.

The greater part of the book is devoted to describing and illustrating the components of actual construction—materials, building methods, installation techniques—the differences between various construction "assemblies."

ISBN 0-87629-138-8
Book No. 67260

59.95/copy
U.S. Funds

Construction Delays:

- Documenting Causes
- Winning Claims
- Recovering Costs

by Theodore J. Trauner, Jr., PE, PP

1st Edition
NEW! 200 Pages • Illustrated • Hardcover

Learn what to look for—and what to look out for—in a construction delay case.

Whether you're a contractor or subcontractor, an owner, architect, or designer, *Construction Delays* is a detailed, fact-filled manual that can help you every step of the way—from working with your attorney to draft delay clauses in construction contracts—to learning how to document for your protection and advantage—to preparing a claim for court.

And at a time when losing a delay case could cost you thousands—even millions—of dollars, that's information you simply can't afford to be without.

Written in clear, non-legal language for construction professionals like you, *Construction Delays* takes you all the way from basic definitions and delay identification, to analysis, assessment of costs, and risk management.

ISBN 0-87629-174-4
Book No. 67278

$54.95/copy
U.S. Funds

Means Legal Reference for Design & Construction

by Charles R. Heuer, Esq., AIA

1st Edition
Over 450 Pages • Hardcover

At last—authoritative help for recognizing and understanding the legal issues in design and construction

Design and construction law is a large topic to grasp. With this in mind, the *Means Legal Reference For Design & Construction* has been prepared to enable its users to sort out and see the practical legal implications of each stage of project delivery.

Each section is illustrated and explained with examples and visuals. These make it easy for you to compare various legal documents, contracts, and situations to better understand the issues.

The large Appendix and Glossary include a comprehensive list of legal terms, names and addresses of trade associations, and samples of standard form contracts commonly used in the industry.

ISBN 0-87629-145-0
Book No. 67266

$109.95/copy
U.S. Funds

Construction Paperwork

An Efficient Management System

by J. Edward Grimes

1st Edition
Over 380 Pages • Illustrated • Hardcover

For project managers, office managers, facilities managers, design/build professionals, subcontractors . . .

Anyone responsible for using and keeping large amounts of construction documents will be delighted with this guide to controlling paperwork.

The author, J. Edward Grimes, is a project manager with 20 years of experience, and an arbitrator of construction disputes for the American Arbitration Association.

Construction Paperwork: An Efficient Management System cuts through theory to give you practical, instantly usable techniques—with immediate payback. It is the first book designed specifically to help project administrators manage paperwork more efficiently and profitably.

This book includes both manual and computer-generated formats for nearly every facet of construction documentation.

ISBN 0-87629-147-7
Book No. 67268

$65.95/copy
U.S. Funds

Insurance Repair:

Opportunities, Procedures and Methods

by Peter J. Crosa, AIC

1st Edition
Over 200 Pages • Illustrated • Hardcover

Advice for getting into the business of insurance repair—based on first-hand experience

As a contractor or subcontractor interested in expansion, you might be considering working with claims adjusters and insurance companies to repair damaged properties.

The fact is that billions of dollars are paid out annually by insurance companies . . . mostly without fierce competitive bidding. The business is also highly stable—going on in both good times and bad.

Insurance Repair: Opportunities, Procedures and Methods gives you the information you need to evaluate insurance repair, including the benefits and pitfalls—and explains what is required to be a success in this profitable business.

ISBN 0-87629-146-9
Book No. 67267

$54.95/copy
U.S. Funds

Consulting

Introducing . . .

MEANS CONSTRUCTION CONSULTING SERVICES

For corporate facilities departments, government agencies, architecture and engineering firms, developers, financial and investment companies

•Project estimating and planning
•Construction feasibility studies/research
•Development and maintenance of cost databases

R.S. Means Company now offers you the benefits of expert consulting services for individual and ongoing projects in these key areas:

Building construction estimating—planning

Control expenses, avoid cost overruns, hit budgets with Means consulting help . . . including reliable estimates . . . guidelines for expenses . . . project duration . . . cash flow projections . . . resource allocation . . . change order negotiations, and more.

Feasibility studies—analysis of construction options

Let the Means staff help you review and clarify the best construction approaches in terms of time, cost, use. These services include detailed analysis of technologies, lead times, materials use, and other important construction alternatives to provide facts and unbiased advice for management decision-making.

Creation of cost databases for unique or specialized construction applications

Utility companies, communications firms, energy companies, government agencies . . . all can benefit from Means expertise in creating construction cost databases for unique types of construction.

For complete information—call Robert Gair, Director—
1-800-448-8182

The new electronic tool that lets you access over 20,000 lines of Means construction cost data right at your PC.

Means DATA Source®

Accessing Means data has never been easier... or quicker. New **Means Data Source®** allows you to combine the power of your PC with the proven reliability of R.S. Means construction cost data. You're able to call up the information you need in mere seconds. Data Source enables you to:

Electronically search through the Means data base — over 20,000 cost lines in CSI MASTERFORMAT.

Select the appropriate lines for a job — either quantifying them in the Means program, or easily transferring them in ASCII format to any compatible program (including Lotus 1-2-3, dBase III and many others).

Feel the security and reassurance of using data that's formatted exactly like the data in our popular cost books.

Quickly and easily learn how to access and use Means data (with many useful on-line Help Screens available).

Generate professional reports — selecting from a menu of standard report formats in our program, or using your own format when you transfer the data to your spreadsheet.

Use the built-in quantifying program to extend quantities and calculate costs.

Perform your operations on all IBM and 100% IBM-compatible PC's (MS-DOS 2.0 or better).

With advantages like these, you'll want to be sure to order Means **Data Source** today. Just call **1-800-448-8182**, or mail the form below.

IBM is a trademark of International Business Machines Inc.
Lotus 1-2-3 is a registered trademark of Lotus Development Corp.
dBASE is a trademark of Ashton-Tate.
Means Data Source is a registered trademark of R.S. Means Co., Inc.

Please send me the following Data Source product(s):*

Quantity	Price		Total
_____ x	$395	Building Construction	$ _____
_____ x	$395	Mechanical	_____
_____ x	$395	Electrical	_____
_____ x	$395	Repair & Remodeling	_____
_____ x	$195	Light Commercial (10,000 line items)	_____
		Massachusetts residents add state sales tax:	_____
		Total	_____

Prices are for U.S. Delivery only. Canadian customers should write for current prices.

☐ Yes, please tell me more about the computer services exclusively designed for the construction industry.

I have a _____ computer.   Disk Size: ☐ 3½"   ☐ 5¼"
　　　　　　 make/model

Your Name: _____

Company Name: _____

Address: _____

City/State/Zip _____

Telephone _____

* ☐ Check enclosed payable to R.S. Means Company, Inc.
　　 (Save shipping & handling)
☐ Charge my order to:
☐ American Express   ☐ Visa   ☐ MasterCard

Account # _____

Expiration Date _____

☐ Bill me (P.O. #) _____

Call toll free 1-800-448-8182 or mail this request today to:
R.S. Means Co., Inc.
100 Construction Plaza
P.O. Box 800
Kingston, MA 02364-0800

*Discounts on Multiple Orders . . . starting at 25%
If you would like to order 4 or more copies of Data Source, be sure to call us at 1-800-448-8182, ext. 39. We'll provide you with complete information on multiple discounts that start at 25%.

Plug into the Means database.

PULSAR

The most significant advance in construction management systems since Means invented the cost book.

A key advantage of Pulsar is its extensive database! It is derived from the *Means Facilities Cost Data* book. You can use it to estimate all types of jobs, from simple remodeling to complex new construction. A special section is included for facilities maintenance cost estimates.

You can find the unit prices you need using either of two different methods. Use CSI numbers and the convenient windowing systems take you right to the items.

You can also key in the word or phrase. It's quick and easy. For example: type "drywall" and you get a list of items associated with drywall. Make your selection and the software takes you right to that item. It's just like using the index of a book—only simpler!

Pulsar includes over 400 pre-built Assemblies. They give you the accuracy of a line-by-line estimate without the work. For example: a "wet pipe sprinkler system" lists all the components you need, quantities, materials costs, installation costs, the cost per square foot and more! You can use Assemblies as they were developed, modify them to suit your conditions, or create new ones.

"Pages" are a real Pulsar benefit! You can create, store and reuse your own original Pages — combinations of unit prices and Assemblies that form specific portions of your project. A Page can consist of just about anything you choose.

You can develop and reuse your own Assemblies for specific tasks.

And you can insert and reuse individual line items that are special to your estimate.

Your estimate can be a combination of Means data and your own costs that are adjusted for local conditions.

Means' standardized form of estimating improves your negotiating powers with outside vendors and makes it easier to work with other people in your company.

You can print reports by CSI divisions, subdivisions, and detailed line-by-line estimates. "Sort" features allow you to segment reports by floor, room, job responsibility — basically any criteria you wish to include. You decide whether to burden or unburden the job and when to add sales tax, markups, markdowns, contingencies, and bonds.

Using the Means City Cost Index, you can quickly see what a job will cost in over 200 cities in the U.S. and Canada.

You can use Pulsar estimates with popular spreadsheet, database, scheduling, project management, job cost control, and graphics packages.

"Test drive" Pulsar and experience for yourself how this comprehensive software system saves you time and gives you the tools you need for fast, reliable construction cost estimating.

Complete the order form below or call to order your Diskette Demo Package.

Software Order/Information Form

I am interested in: ☐ Astro II   ☐ Pulsar

☐ Please send me the Multiple Diskette Demo Package for $50. (This cost can be credited toward your purchase of a system).*MA residents, please add state sales tax.

☐ Please tell me more about the computer services exclusively designed for the construction industry.

I have a _____ computer.   Disk Size: ☐ 3½″   ☐ 5¼″
make/model

* ☐ Check enclosed (save shipping & handling)
☐ Charge my Demo Package to:
☐ American Express   ☐ Visa   ☐ MasterCard

Your Name: _____

Company Name: _____

Address: _____

City/State/Zip _____

Account #
_____

Expiration Date
_____

☐ Bill me (P.O.#) _____

Call toll free 1-800-448-8182, or mail this request today to:
R.S. Means Co., Inc.
100 Construction Plaza
P.O. Box 800
Kingston, MA 02364-0800

Software

Means Software Systems Complement the Expertise of the Professional Estimator

No computer will ever replace years of experience and hard work! But the demands on your time are considerable. There is always pressure to get more accomplished in less time. Means Software Systems are the answer! They are the powerful tools you need for fast, accurate construction cost estimating.

Takeoff is efficient and easy! Programs are menu-driven and based on familiar "scroll, cut and paste" techniques. You quickly access specific portions of the multilevel database with the easy-to-use window and scroll operation. A single key stroke then marks ("cuts") only the items you need and sends them to a buffer file until you are ready to "paste" them into your estimate.

You can use the reliable Means figures to build your estimate, and you can adjust cost factors to fit actual job conditions. Costs can be modified in numerous ways to prepare estimates—from the smallest remodeling or repair projects to multi-story new construction.

Everything you need is readily available. Material costs, crews, equipment costs, man-hours, daily output, overhead, profit, and more! It's all in the databases included in the software.

Computerized estimating is dynamic! When costs change, even complex estimates can be rapidly adjusted. As soon as you enter a new figure, the software automatically changes all dependent calculations. With more flexible estimating procedures, you have guidelines to control costs and make more informed decisions.

You already know a lot about this outstanding software because it is arranged in the industry-standard CSI MASTERFORMAT. You have two ways to search the databases—using CSI numbers, or words and phrases.

With Means Estimating Software you build standardized estimates that are consistent and easy to understand. Standardized estimates improve communications with outside vendors and other departments within your company. They can even speed the budget approval process!

Reports cover as much detail as you need. You can get reports based on major CSI divisions and subdivisions, on a detailed item-by-item basis, for parts of the job, or for the specific job criteria that you choose.

For even more extensive manipulation and reporting, you can export all or part of your estimate to many popular spreadsheet, database, and graphics packages.

With our many years of experience in both software and publishing, *we understand the importance of support!* You can take advantage of Means superb documentation, step-by-step tutorials, on-screen help, toll-free information hot line, and optional on-site training.

With Means Annual Software Service Support Agreement, you receive updates on all your cost files for each year, continued hot line support and software enhancements as they occur.

Hardware requirements: IBM® XT™, AT®, or 100% compatible (AT is preferred; the program runs faster). Color monitor preferred, but not required. 640K main memory. One 5¼″ or 3½″ floppy disk drive. Hard disk with 15 mb available, MS™-DOS 2.0 or higher. IBM compatible printer.

ASTRO II
Powerful Construction Software That's Easy to Use . . . Even If You've Never Used a Computer for Estimating!

Manual estimating is cumbersome and time-consuming. Each item must be entered separately, and when costs or conditions change, all the figures must be recalculated. With Astro II, when you change a figure, all items it affects are automatically changed. Easy-to-read, on-screen "windows" guide you from the CSI MASTERFORMAT divisions, to the subdivisions, to the unit items you need. You can also search for items using words and phrases—it's just like using the index of a book—only simpler! Then you "mark" the items you need and "paste" them to your estimate.

R.S. Means is known for its timely and comprehensive construction cost data. With Astro II, you choose the Means Data File which best suits your estimating requirements.

As your company grows and changes, you can add other Means Cost Data Files to your existing Astro II package.

It's easy. You don't have to be a programmer. When there are items that are specific to your estimate, you can create your own user files. You can then save, change, update, and reuse them for other estimates.

Labor and equipment costs change constantly—so you can adjust specific costs where necessary to suit your area and your project.

When elements of the estimate change, you don't have to redo the estimate. You just enter new costs and line items to update the estimate as often as you like.

You may want a very detailed estimate for your use, and an overview for your customer. Reports can be generated by CSI divisions, subdivisions or as line-by-line estimates.

You can plan a construction project and then use Means City Cost Indexes to quickly see what it would cost to build in over 200 cities in the United States and Canada.

For additional reports, to test "what-if" scenarios, and to create job schedules, you can export your estimate to popular spreadsheet and project management software packages.

You can take advantage of on-screen help, a toll-free 800 hot line and superbly written documentation. The software is logical and works like you do.

Call 1-800-448-8182 for more information

432

Means Construction Seminars 1991

During the year, R.S. Means offers a series of 2-day seminars oriented to a wide range of construction-related topics. All seminars include comprehensive workbooks, current Means Cost Books, plus proven techniques for estimating and scheduling.

REPAIR AND REMODELING ESTIMATING

Repair and remodeling work is becoming increasingly competitive as more professionals enter the market. Recycling existing buildings can pose difficult estimating problems. Labor costs, energy use concerns, building codes, and the limitations of working with an existing structure place enormous importance on the development of accurate estimates. Using the exclusive techniques associated with Means' widely-acclaimed *Repair & Remodeling Cost Data*, this seminar sorts out and discusses solutions to the problems of building alteration estimating. Attendees will receive two intensive days of eye-opening methods for handling virtually every kind of repair and remodeling situation . . . from demolition and removal to final restoration.

MECHANICAL AND ELECTRICAL ESTIMATING

This seminar is tailored to fit the needs of those seeking to develop or improve their skills and to have a better understanding of how mechanical and electrical estimates are prepared during the conceptual, planning, budgeting and bidding stages. Learn how to avoid costly omissions and overlaps between these two interrelated specialties by preparing complete and thorough cost estimates for both trades. Featured are order of magnitude, assemblies, and unit price estimating. In combination with the use of *Means Mechanical Cost Data*, *Means Plumbing Cost Data* and *Means Electrical Cost Data*, this seminar will ensure more accurate and complete Mechanical/Electrical estimates for both unit price and preliminary estimating procedures.

CONSTRUCTION ESTIMATING– THE UNIT PRICE APPROACH

This seminar shows how today's advanced estimating techniques and cost information sources can be used to develop more reliable unit price esimates for projects of any size. It demonstrates how to organize data, use plans efficiently, and avoid embarrassing errors by using better methods of checking.

You'll get down-to-earth help and easy-to-apply guidance for:

- making maximum use of construction cost information sources
- organizing estimating procedures in order to save time and reduce mistakes
- sorting out and identifying unusual job requirements to prevent underestimates.

SQUARE FOOT COST ESTIMATING

Learn how to make better preliminary estimates with a limited amount of budget and design information. You will benefit from examples of a wide range of systems estimates with specifications limited to building use requirements, budget, building codes, and type of building. And yet, with minimal information, you will obtain a remarkable degree of accuracy.

Workshop sessions will provide you with model square foot estimating problems and other skill-building exercises. The exclusive Means building assemblies square foot cost approach shows how to make very reliable estimates using "bare bones" budget and design information.

SCHEDULING AND PROJECT MANAGEMENT

This seminar helps you successfully establish project priorities, develop realistic schedules, and apply today's advanced management techniques to your construction projects. Hands-on exercises familiarize participants with network approaches such as Critical Path or Precedence Diagram Methods. Special emphasis is placed on cost control, including use of computer based systems. Through this seminar you'll perfect your scheduling and management skills, ensuring completion of your projects *on time* and *within budget*. Includes hands-on application of *Means Scheduling Manual* and *Building Construction Cost Data*.

THE CONSTRUCTION PROCESS– METHODS & MATERIALS

This course provides an overview of the basic construction process from the time an idea/need is identified until the final move-in. Participants will learn about the components of construction and how they are assembled to build a project.

Also provided are guidelines for reading the plans and specifications identify the key elements of a building. Attendees will achieve a thorough understanding of the construction process, from concept through completion, and with a practical overview of the methods and materials used in construction. If you are new to the construction industry or just want to know more about the complex process of converting a need for a building project into a reality, this course is for you!

AVOIDING AND RESOLVING CONSTRUCTION CLAIMS

Construction claims are a common and costly problem in the construction industry. This course provides you with insight and practical recommendations for dealing with claims. The hands-on approach focuses on recognizing and responding to problems as they arise, preparing documentation and formal reports, analyzing causation, calculating damages, evaluating responsibility and defending against unmeritorious claims. Special emphasis will be given to determining schedule and cost impact of project delays. You'll perfect your skills in negotiation to resolve disputes in order to avoid the risks of trial or arbitration. In addition, the program provides realistic approaches to minimizing and avoiding the problems which lead to claims. This course is a must for those responsible for the management and administration of construction projects.

1991 Means Seminar Schedule

SPRING & SUMMER			
New York City, NY	January 28–31	Denver, CO	May 13–16
Ft. Lauderdale, FL	February 4–7	Plymouth, MA	May 13–16
Las Vegas, NV	February 18–21	Los Angeles, CA	May 20–23
San Jose, CA	February 25–28	Hartford, CT	June 3–6
Atlanta, GA	March 11–14	Tarrytown, NY	June 10–13
Atlantic City, NJ	March 18–21	Indianapolis, IN	June 17–20
Washington, DC	April 15–18	Baltimore, MD	June 24–27
Seattle, WA	April 22–25	Niagara Falls, NY	July 22–25
Chicago, IL	April 29–May 2	Minneapolis, MN	August 26–29
FALL			
Hyannis, MA	September 9–12	Philadelphia, PA	October 28–31
Washington, DC	September 16–19	Long Beach, CA	October 28–31
Rutherford, NJ	September 23–26	San Francisco, CA	November 4–7
Kansas City, MO	October 7–10	San Antonio, TX	December 2–5
Raleigh, NC	October 21–24	Orlando, FL	December 9–12

For a complete schedule of courses offered at each location, call our Seminar Registrar at 1-800-448-8182.

Registration Information

How to Register

To register, call our Seminar Registrar, Marcia Crosby, today. Means toll-free number for making reservations is:

1-800-448-8182.

Registration Fees

- One seminar registration $845 per person
- Two to four seminar registrations $745 per person
- Five or more seminar registrations $695 per person
- Ten or more seminar registrations   call for pricing

Special Offers

Early Registration:

Sign up 45 days prior to the seminar/save $50 off each seminar registration. Payment must be in advance.

One Individual:

Sign up for two seminars in a one-year period and pay only $1345. This is a saving of 40% on the additional seminar. Payment must be received at least ten (10) days prior to seminar date to confirm this offer.

Cancellations

Cancellations will be accepted up to ten days prior to the seminar start. After that time a 2-day seminar cancellation is subject to a $150 cancellation fee. The fee may be applied to any Means seminar within one calendar year of cancellation. Substitutions can be made at any time before the session starts. No-shows are subject to the full seminar fee.

AACE Approved Courses

The R.S. Means Construction Estimating and Management Seminars described and offered to you here have each been approved for 14 hours (1.4 CEU) of credit by the American Association of Cost Engineers (AACE) Inc. Certification Board toward meeting the continuing education requirements for re-certification as a Certified Cost Engineer/Certified Cost Consultant.

Daily Course Schedule

The first day of each seminar session begins at 8:30 A.M. and ends at 4:30 P.M. The second day is 8:00–4:00. Participants are urged to bring a hand-held calculator since many actual problems will be worked out in each session.

Seminar Tuition Includes

Current Means cost manuals, appropriate Means reference manuals and seminar workbooks. Continental breakfast and coffee breaks are also provided. Each participant receives a certificate of course completion.

Hotel/Transportation Arrangements

R.S. Means has arranged to hold a block of rooms at each hotel hosting a seminar. To take advantage of special group rates when making your reservation be sure to mention that you are attending the Means Seminar. You are of course free to stay at the lodging place of your choice. (Hotel reservations and transportation arrangements should be made directly by seminar attendees.)

Important

Class sizes are limited, so please register as soon as possible.

Please register the following people for the Means Construction Seminars as shown here. Full payment or deposit is enclosed, and we understand that we must make our own hotel reservations if overnight stays are necessary.

☐ Full payment of $ _____ enclosed.

☐ Deposit of $ _____ enclosed.

    Balance due is $ _____
                     U.S. FUNDS

Name of Registrant(s)
(To appear on certificate of completion)

_____

_____

Registration Form

Firm Name _____

Address _____

City/State/Zip _____

Telephone Number _____

☐ Charge our registration(s) to:

        ☐ American Express   ☐ Visa   ☐ MasterCard

Account No. _____ Expiration Date _____

_____
               CARDHOLDER'S SIGNATURE

Seminar Name	City	Dates

Please mail check to: R.S. MEANS COMPANY, INC., 100 Construction Plaza, P.O. Box 800, Kingston, MA 02364-0800 USA

434

Means In-House Training 1991

Discover the cost savings and added benefits of bringing Means proven training programs to your facility . . .

ATTENDEES

Facility, Design, and Construction Professionals who want to improve their estimating, project management, and negotiation skills

CONTENT

One or more MEANS Training Seminars (as described in Means Construction Seminar, 1991)

A two-day course running from 8 AM to 4 PM on two consecutive days

The intensive training includes conceptual work and practice exercises.

In-House Seminar (IHS) students benefit from discussion and practical work problems that are common to their particular environment.

COMPARISON

The In-House Seminar conducted at the client's facility will eliminate travel expenses and minimize time away from work.

An In-House Seminar program is more flexible and responsive to specific client needs and requirements.

IHS students improve their skills and broaden their knowledge. Students return immediate benefits to their work place.

IHS provides a standard and consistent method for preparing cost estimates.

CLASS SIZE

The cost effective break-even point is 10–12 students; most seminars have from 15 to 25 students; class size is limited to 35 students.

Occasionally, to satisfy their needs or to reduce per student training cost, clients invite outside personnel with a similar background to attend their MEANS Seminars.

SCHEDULING

Consult with Means' professional training advisor to determine the seminar that best suits your needs.

Set your training schedule immediately to assure dates that fit your calendar.

Means will make every effort to accommodate your schedule.

COMPREHENSIVE MATERIALS

Each participant receives a comprehensive workbook, Means reference books that are appropriate to the seminar, and a current Means cost book. The materials reinforce program content and serve as a valuable reference for the future.

CERTIFICATES & CEU's

A certificate of completion is awarded by R.S. Means Company, Inc. to those who complete the program. Continuing Education Units (CEU's) are also awarded through the American Association of Cost Engineers (AACE).

GUARANTEE

R.S. Means stands behind its In-House Seminars.

You will deal with real problems, not theoretical situations. If your In-House Seminar does not give you the tools you need to grow in the subject covered, just tell us why and we will give you credit toward another seminar.

Partial list of companies and organizations that have brought Mean's Training in-house:

Army Corps of Engineers
AT&T
Bell Research Lab
Bell Operating Companies
Boston Edison
Digital Equipment Corporation
Eastman Kodak
General Motors
General Services Administration
Housing and Urban Development
IBM
Internal Revenue Service
Jacksonville Electric Authority
Marine Corps
National Guard
Penn State University
Port Authority of NY/NJ
State of Missouri
State of Virginia
U.S. Air Force
U.S. Army
U.S. Navy
University of Massachusetts
Westinghouse Electric Authority

How to find out MORE—Call Joan M. Ward at (800) 448-8182

Means Facilities
Cost Data
1991

6th Annual Edition
Over 950 Pages • Illustrated

A cost planning tool for facilities construction and maintenance!

More reasons than ever to use
Means Facilities Cost Data 1991
for your facilities estimating projects!

If you are involved in facilities renovations, new construction, or maintenance as a contractor bidding for jobs or a manager planning them, there is tremendous pressure on you to justify your expenditures and to make your estimates reliable.

Means Facilities Cost Data 1991 can help you do just that. This *ultra-complete reference source* makes your estimating more precise and cuts down dramatically on the time you have to spend checking other job costs and calling subs and vendors.

You can use this book to adjust national average data to specific locations.

There are hundreds of other ways you can use this guide. With well over 40,000 unit prices, assembly costs, and square foot costs, it is actually four books wrapped into one.

Here is a brief look at what it contains:

New Facilities Construction:
Unit prices of thousands of line items not found anywhere else ... not even in our other books! These unit prices are broken down into crew, daily crew output, man-hours, material, labor, equipment, overhead and profit.

Everything imaginable is covered, from site work, drainage, and caissons to HVAC assemblies, computer rooms, electrical components, and furnishings.

Renovation, Retrofitting and Restoration:
Often you have to take an existing space and redesign it for a different use. These extensive sections can make this job easier and more efficient. They cover everything from demolition and asbestos removal to furnishings and carpets.

Facilities Maintenance:
Both interior and exterior maintenance are covered, with frequency tables and dozens of money-saving tips.

ISBN 0-87629-194-9
Book No. 60201

$173.95/copy
U.S. Funds

Means Light Commercial Cost Data 1991

10th Annual Edition
Over 500 Pages • Illustrated

For the special estimating needs of light commercial contractors and architects

Easy-to-use, reliable price information about light commercial cost planning for adjusters, architects, contractors, planners, and project managers.

No better resource for estimating commercial construction with current price information

If you're still estimating commercial construction with hard-to-follow price sheets, bills for old jobs, or "guesstimates," we have good news for you.

Means new *Light Commercial Cost Data 1991* is totally dedicated to your special estimating needs as a commercial designer or builder.

It's simple to use, provides instantaneous price quotes, and is reliable for your 1991 cost planning. With a copy on your desk, you'll have the price data you need in *one convenient place*—not stacks of file folders, bills, or wrinkled computer printouts.

Yes, here in *one manual* is all the 1991 cost data you're likely to need in three estimating formats—unit prices, assemblies prices, and square foot costs.

Means Light Commercial Cost Data 1991 is completely dedicated to estimating light commercial construction. Full range cost coverage in the most useable form ever!

- CSI MASTERFORMAT
- Quantity, unit, man-hour data
- Detailed descriptions
- Material and installation costs per square foot, assembly and unit price.
- Nine primary estimating divisions for each structure
- Man-hour data, materials and labor costs
- Over 20 appropriate cost modifications for each model
- Step-by-step listing of assembly components allows substitutions to fit unique needs

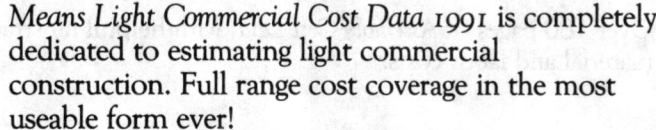

Means Light Commercial Cost Data 1991 covers all construction components

General requirements	Moisture	Special construction
Hourly and daily	thermal control	Conveying systems
crew costs	Doors, windows	Mechanical
Site work	and glass	Electrical
Concrete	Finishes	Square foot costs
Masonry	Specialties	Repair and remodeling
Metals	Equipment	City cost indexes
Wood and plastics	Furnishings	Comprehensive index

ISBN 0-87629-201-5
Book No. 60181

$65.95/copy
U.S. Funds

Means Residential Cost Data 1991

10th Annual Edition
Over 470 Pages • Illustrated

Residential costs—from a simple one-story "economy" house to a "luxury" 3-story dwelling

More valuable to you than ever with detailed S.F. models—and SUPER-ACCURATE adjustments for local price conditions

Features in this edition

- *Square Foot Cost Section* dramatically cuts down on the time you spend working on rough and ballpark estimates.

- Ultra-complete and up-to-date *Unit Price* costs for over 8,500 items, calculates all pricing data you need for reliable estimates including crews, daily output, man-hours, material, labor, equipment, overhead and profit.

- *Reference Section* with estimating data, alternate pricing methods, cost derivations and much, much more.

- *Means Residential Cost Data 1991* conforms to the CSI MASTERFORMAT, making it easier to cross-reference other MASTERFORMAT sources and estimate projects based on the MASTERFORMAT, such as many local and federal government projects.

- Over 180 pages of *Assembly Cost Data* with helpful illustrations and itemized lists of component material and labor costs.

Thousands of Unit Prices
From stonework and insulation to painting and wall coverings, these unit prices are individually figured so you can select the items you need for very exacting estimates. All in the CSI MASTERFORMAT.

Over Eighty Building Assemblies
Fully illustrated and described assemblies can save you hours of estimating time. Each assembly lists the materials and labor needed and the total costs. They give you an outstanding overview of what is needed to complete each job.

Square Foot Costs
Square Foot Costs help you verify your figures and "ballpark" estimates for all popular types of residential construction. PLUS ... Reference Notes section with added information on how unit costs were figured.

ISBN 0-87629-200-7
Book No. 60171

$65.95/copy
U.S. Funds

Means Plumbing Cost Data 1991

Piping • Fixtures • Fire Protection

14th Annual Edition
Over 400 Pages • Illustrated

Comprehensive unit and assemblies prices for every imaginable plumbing installation

Here's your all-inclusive price guide for efficiently and accurately estimating plumbing work and fire protection installations.

Tanks, filtration equipment, pipe, fittings, valves, fixtures, pumps, appliances, sprinkler and standpipe systems . . . *all related construction*

60 Illustrated Plumbing and Fire Protection Assemblies Costs

This estimating manual gives you instant access to current materials and trade labor prices for virtually all types of contemporary plumbing and fire protection installations.

One glance through this massive guide—containing thousands of "one stop" costs for piping, fittings, valves, support structures, fire protection systems, appliances—will convince you of its great value to your pricing work.

You can have plumbing prices quoted to you instantly.

Whether you use it for a complete unit price plumbing estimate, plumbing design concept, or for a sprinkler system quick price check, it will give you the estimating data you need.

Thousands of up-to-date plumbing and fire protection unit prices including:

- fixtures
- pumps
- site irrigation systems
- pipe valves and fittings
- swimming pool systems
- lawn and golf course sprinkler systems
- septic, fire hydrant and water service systems
- battery assemblies for urinals, lavatories, and water closets
- drainage and vacuum type fittings
- fire protection systems—including residential
- washing machine hook-ups
- water heaters—including point of use

Plus more graphics and illustrations of equipment and assemblies

ISBN 0-87629-204-X
Book No. 60211

$67.95/copy
U.S. Funds

Means Mechanical
Cost Data 1991
HVAC · Controls

14th Annual Edition
Over 440 Pages • Illustrated
Gives you comprehensive prices for HVAC and mechanical estimating

More valuable to you than ever—focused exclusively on HVAC, controls . . . all related piping, ductwork, accessories, and construction

Offered in 1991

- **Many prices for key materials and components used in HVAC have gone down as well as up**
- **Extensive HVAC CLASSIFICATIONS including:**
  - purging scoops
  - cocks and drains
  - liquid drainers—several types
  - flow check controls
  - suction diffusers
  - multi-purpose valves
  - mono-flow tee fittings
  - vacuum breakers
  - direct digital controls— expanded section
  - flue heat reclaim systems
  - commercial kitchen ventilation
- **Plus price coverage for:**
  - polyethylene foam closed cell sheet and tubing
  - gas appliance regulators
  - types of steam traps
  - air control vents
  - control system components
  - water and steam coils
  - silent check valves
  - FM approved fiberglass reinforced plastic fume duct

For contractors and designers who must estimate costs for mechanical installations, Means "mechanical book" is a cost tool of *increasing value* each year.

Now, the latest cost information for heating, ventilation, air conditioning and related mechanical construction is available in *Means Mechanical Cost Data* 1991.

Plumbing and fire protection cost information is published separately in *Means Plumbing Cost Data* 1991.

There are many benefits from using this storehouse of mechanical cost information.

Means Mechanical Cost Data 1991 enables you to answer every kind of HVAC and related mechanical cost question. Use it to:
- estimate better
- compare and analyze methods/costs
- check ratings, outputs, sizing information
- assist pre-bid scheduling

ISBN 0-87629-195-7
Book No. 60021

$67.95/copy
U.S. Funds

Means Square Foot Costs 1991
Commercial/Residential Industrial/Institutional

12th Annual Edition
Over 400 Pages • Illustrated

Quoting accurate square foot prices is vital to construction planning and budgeting

Take a close look at what you get in *Means Square Foot Costs 1991*

- Highly reliable adjustments for local construction costs
- Costs for Common Additives in the commercial/industrial/institutional sections
- HVAC assemblies costs and HVAC options in the Unit-in-Place (assemblies) section
- 1991 square foot prices for all types of building construction
- Four levels of residential construction quality: economy, average, custom, and luxury
- 70 types of typical commercial, industrial and institutional buildings with illustrations
- Illustrated "building assemblies"—costs broken down by materials and installation
- Regional cost adjustment factors for local labor and materials price differences by zip code
- Helpful prototype estimating forms and worksheets
- Abundant illustrations, worked out examples

You can have reliable square foot prices for any building construction you're planning in minutes

Means Square Foot Costs 1991 is the annually updated source of cost facts that covers every situation you're likely to handle—residential, commercial, industrial and institutional construction of all types.

The convenient tables enable you to simply flip to the building type under consideration, and select an appropriate square foot price in just a few minutes.

The "Unit-in-Place" assemblies style construction prices give you the ability to prepare more detailed square foot costs through the analysis of individual construction assemblies. These pages are suited perfectly to comparing and selecting various design schemes according to predetermined budgets.

ISBN 0-87629-197-3
Book No. 60051

$76.95/copy
U.S. Funds

Means Heavy Construction Cost Data 1991

5th Annual Edition
Over 380 Pages • Illustrated

Heavy construction costs for utilities, public works, earthwork, roadways, airports, pipelines, sewerage, railroads, marine work, and more!...

Now you have cost data for heavy construction in the convenient Means format

Means Heavy Construction Cost Data 1991 is designed for contractors and engineers responsible for estimating heavy construction projects with high reliability.

It's an estimating tool that provides prices based on painstaking research into the actual costs for heavy construction throughout North America in the past 12 months. The heavy construction unit and assemblies cost entries are supplemented with preparation, mobilization and finishing costs for most projects.

The book can be used to price out an entire estimate or to quickly verify estimates with cost data based on national averages, adjusted for locations in the U.S. and Canada.

The unit price entries are organized in the popular Means line item system containing crew make-up, crew productivity, man-hour data, and bare costs for materials, labor, and equipment ... with and without overhead and profit.

You'll find it more than equal to the task of answering your heavy construction cost questions in 1991.

Here's why the prices quoted in *Means Heavy Construction Cost Data 1991* can be trusted

Prices shown in *Means Heavy Construction Cost Data 1991* are a national average for heavy construction work throughout the U.S. and Canada.

The costs are those reported from the actual experience of participating heavy construction contractors and subcontractors in the latest 12-month period.

The City Cost Index supplies price adjustment factors for city-to-city labor and materials price differences.

ISBN 0-87629-202-3
Book No. 60161

$71.95/copy
U.S. Funds

Means Electrical Cost Data 1991

14th Annual Edition
Over 400 Pages • Illustrated
Complete in every way—your trusted price book for electrical estimating

Why *Means Electrical Cost Data 1991* is the "tool of choice" for your electrical estimating . . .

Means Electrical Cost Data has become a trusted and well-used guide by your fellow electrical estimating, contracting and design professionals.

This year is no exception. Particularly in light of the important price changes in recent months.

In fact, 1991 electrical estimating can hold some pricing surprises. Here's where Means value really shines. The Means staff of experts have done their homework—so you can steer clear of the pitfalls *and take advantage* of the opportunities ahead.

This is a powerful, easy-to-use resource dedicated solely to electrical estimating.

With *Means Electrical Cost Data* 1991 you'll make estimates with a price source you can trust.

Many important electrical unit price changes for 1991. You'll have them all in *Means Electrical Cost Data 1991*

Complete Unit Prices

- Crews, daily output, man-hours, bare costs for material, labor, equipment, O&P and more. Over 10,000 in all.
- Includes general requirements, site work, specialties, architectural equipment, special construction.

Electrical Systems Costs Included

- 40 Electrical assemblies with illustrations

Helpful Electrical Estimating References

- Mechanical, electrical, KW value/cost determination, lighting, H.V. line poles, workers comp, overtime (lost productivity), cost indexes, and 28 unit cost procedure models.
- City cost modifications, square foot costs, repair and remodeling factors.

ISBN 0-87629-193-0
Book No. 60031

$67.95/copy
U.S. Funds

Means Concrete Cost Data 1991

9th Annual Edition
Over 450 Pages • Illustrated

How to have the up-to-date concrete/masonry prices you need—fast!

**Comprehensive concrete/masonry cost information
in the Means fast-moving format**

This is your all-inclusive price guide for estimating concrete and related construction in 1991.

Up-to-date costs for all types of concrete/masonry work are provided in exhaustive detail. The popular Means **CSI MASTERFORMAT** layout directs you to the right data instantly.

The large, 10,000+ **UNIT PRICE** section provides detailed crew, man-hours, bare costs for material, labor, and equipment, and installing contractor's overhead and profit.

The easy-to-use **ASSEMBLIES** costs actually "picture" each assembly for you with prices for the base assembly and a variety of alternative material costs on the same or adjoining pages.

Here are just a few ways professionals use *Means Concrete Cost Data*

- **estimate concrete/masonry accurately**—rather than hunting through old cost records or making endless calls, use the guide for instant unit price quotations.

- **compare/analyze formwork better**—serving the combined functions of estimating, checking and scheduling, the guide helps you make clear, head-to-head formwork comparisons. You see alternatives, select better options.

- **check proportional quantities, mixes, labor**—use the extensive unit prices and supplemental references to locate needed concrete sizing and capacity information.

- **use it for value engineering**—perfect for designers, architects, engineers. The manual supplies every needed fact for relating concrete designs to budgets.

- **use it for scheduling**—the exhaustive crew, man-hour, and equipment productivity data is ideal for working out preliminary schedules.

ISBN 0-87629-189-2
Book No. 60111

$65.95/copy
U.S. Funds

Means Open Shop Building Construction Cost Data 1991

7th Annual Edition
Over 450 Pages • Illustrated

With so many important price changes in the last 12 months, this is "must have" information for your estimating.

You depend on BCCD for union labor . . . now you can have the same comprehensive prices for open shop work.

One flip through your copy of *Means Open Shop Building Construction Cost Data* 1991 will convince you of its value to your estimating process.

Here at last are over 20,000 reliable unit cost entries based on non-union skilled and trade labor costs. No longer will you need to use time-consuming, inefficient ways to get these prices. The Open Shop BCCD gives you every price you need in the familiar Means layout.

The labor cost data is broken down into man-hours, daily output, crew, bare labor, equipment, overhead and profit.

You can break out labor cost facts in any way you need for substitutions, comparisons, and adjustments.

Consider the many benefits of this open shop cost data:

- eliminates chasing down unit prices based on open shop labor
- helps you estimate and bid on open shop projects
- enables you to analyze and compare subcontractor price quotes
- instantly provides prices on individual items you're not familiar with
- helps you work out realistic work schedules
- enables you to access the national average for materials and open shop labor costs
- prices are totally current for your 1991 estimating

The open shop prices are the actual experience of contractors and subs.

Prices quoted in *Means Open Shop Building Construction Cost Data* 1991 are derived from an average of major urban areas in the U.S. and Canada.

The information is painstakingly prepared from the actual experience of hundreds of open shop builders and subcontractors in the latest 12-month period.

The City Cost Index provides factors for region-to-region cost differences. 162 cities are included in this table. Factors for adjusting other costs have been furnished where appropriate.

ISBN 0-87629-198-1
Book No. 60151

$76.95/copy
U.S. Funds

Annual Publications

Building Construction Cost Data 1991
Western Edition

4th Annual Edition
Over 500 Pages • Illustrated

For contractors, estimators, architects, engineers, builders, facilities professionals

Major price changes ahead for your 1991 construction estimating

You can depend on R.S. Means Company, the nation's foremost construction cost authority, to supply them to you in the *Western Edition* of *Building Construction Cost Data 1991*.

More reliable adjustment factors for local price conditions.

The unit price line items in *Building Construction Cost Data 1991 — Western Edition* are based on averages for 11 western states and western Canada.

For 1991, the City Cost Index covering 30 western metropolitan areas will reflect price adjustment factors based on the addition of thousands of new construction price reporting sources in the West.

This increased amount of localized price data will result in your ability to adjust for local price conditions with much higher precision than previously possible.

Consider the great amount of cost information you'll have for your 1991 projects

- **Up-to-date unit prices**—over 20,000—for estimating western construction
- **Price data for materials and construction approaches** common to the west
- **Labor and equipment productivity** information by man-hours and crews
- **100 pages of estimating references,** guidance for solving estimating problems
- **More reliable local cost adjustments** for your area of the state
- **Helpful examples, directions** for using the data effectively
- **CSI MASTERFORMAT** arrangement of price data for convenient use of plans and specs written in industry standards
- **500 pages of data in all!**

ISBN 0-87629-188-4
Book No. 60221

$76.95/copy
U.S. Funds

446

Means Interior Cost Data 1991

8th Annual Edition
Over 460 Pages • Illustrated

For estimating partitions, ceilings, floors, finishes, and furnishings

CSI MASTERFORMAT

Due to constantly changing styles and new design directions, thousands of new materials and hardware items for interior construction are introduced each year. Few designers and estimators can keep up with this avalanche of price options for finish work.

But now, you can keep up with—and stay ahead of—the infinite varieties of specialized materials and installation expenses involved in commercial and industrial interior construction ... thanks to *Means Interior Cost Data 1991*.

Here, in one all-inclusive resource, are all the prices and help you could ever hope to have for making accurate interior work estimates.

Comprehensive unit prices, illustrated assemblies costs, supplemental references—complete guidance for your interior estimating.

Estimating interior construction can be a lot easier with this new guide

13,000+ up-to-date unit price entries—with bare costs for labor, equipment, and materials.

Extensive labor and equipment productivity data—every key labor productivity, crew and cost fact helps you calculate prices for *any interior project*.

Materials, products, building techniques—updated cost information includes the latest materials and labor-saving installation methods for interior work.

Helpful 3-D illustrations where possible—materials, products, and assemblies are illustrated.

Quick-find index covers everything—nearly 5,000 listings in the index ensures that you'll find the item you're looking for fast.

Unit prices in CSI MASTERFORMAT—to speed estimates along, prices are arranged in the Construction Specifications Institute's MASTERFORMAT.

City Cost Index—enables you to adjust the prices based on costs in your local area of the U.S. or Canada.

ISBN 0-87629-190-6
Book No. 60091

$65.95/copy
U.S. Funds

Means Site Work
Cost Data 1991

10th Annual Edition
Over 400 Pages • Illustrated

Dedicated to the special problems
of site work estimating

Much current data to help you . . .

- Accurate local materials, labor and equipment price adjustment factors
- Site work assembly cost tables for conceptual estimating
- Reference tables for calculating quantities, pricing site work

Thumb through the new 1991 site work cost guide and you'll be convinced instantly you have the tool that's equal to today's demanding estimating problems.

It's super-inclusive, perfectly organized, and, of course, fully reliable in the Means tradition.

Its goal is to give you a cost resource that *actually does cover every aspect of modern site work estimating* . . . exploration, pollution controls, soil retention . . . underground utilities, security, site enhancement . . . *you name it!*

You'll be delighted with the clarity of descriptions, notes, and helpful guidance facts touching upon just the right information you need.

Here are just a few of the ways
Means Site Work Cost Data 1991
will help you

- estimate crews, equipment, and man-hours required for any site work operation . . . earthwork hauling, underground installations
- use the price data to compare and select designs and specifications in the project's conceptual stages
- verify your prices and estimates with the manual's national averages
- use it for the hard-to-find costs and site work installations unfamiliar to you
- double-check on subcontractor bids and estimates
- get expert guidance for complicated site work estimating problems

ISBN 0-87629-196-5
Book No. 60071

$67.95/copy
U.S. Funds

Means Landscape Cost Data 1991

4th Annual Edition
Over 350 Pages • Illustrated
Estimate landscape and related hard construction faster, easier—and with complete reliability

Means Landscape Cost Data 1991 is for busy landscape designers, contractors, facilities managers and other professionals needing accurate prices for landscape projects. Cost data is arranged in CSI MASTERFORMAT.

Here you have a landscape cost book that will give you direct, reliable answers for your landscape and outdoor improvements estimating.

Means Landscape Cost Data 1991 includes labor productivities and Standard Landscape Crew Tables with hourly and daily costs. Also includes a 162 City Cost Index for the U.S. and Canada showing price variations for materials and labor from region to region, hardiness zones, special plant uses and adaptability, scientific and common plant names, unit conversions, and complete subject index and appendix.

"One Stop" answers to your landscape price estimating questions

- Materials unit prices for shrubs, trees, seeding, ground covers, roads, walks, walls, fencing, underground utilities.
- Unit labor costs with crew, daily productivity, and man-hours.
- Equipment costs with equipment productivity including rental and operating cost data.
- Illustrated landscape assemblies prices with material and installation costs broken out.
- Helpful design and cost planning suggestions.
- Landscape site maintenance prices.
- Figures for the latest landscape methods, styles, construction approaches.
- Cost adjustment factors for 162 metropolitan areas in the U.S. and Canada.
- Easy-to-follow instructions for using the data.

ISBN 0-87629-203-1
Book No. 60191

$71.95/copy
U.S. Funds

Means®
A Southam Company
R.S. Means Company, Inc.
100 Construction Plaza, P.O. Box 800, Kingston, MA 02364-0800

1991 ORDER FORM

CALL TOLL FREE
1-800-448-8182

QTY.	BOOK NO.	COST ESTIMATING BOOKS	UNIT PRICE	Total	QTY.	BOOK NO.	REFERENCE BOOKS (cont'd)	UNIT PRICE	Total
	60061	Assemblies Cost Data 1991	$105.95			67231	Forms for Building Construction Professionals	$ 76.95	
	60011	Building Construction Cost Data 1991	65.95			67260	Fundamentals of the Construction Process	59.95	
	60221	Building Constr. Cost Data–Western Edition 1991	76.95			67210	Graphic Construction Standards	109.95	
	60111	Concrete Cost Data 1991	65.95			67258	Hazardous Material & Hazardous Waste	76.95	
	60141	Construction Cost Indexes 1991	169.00			67280	Home Improvement Cost Guide	32.95	
	60141A	Construction Cost Index–January 1991	42.25			67251	HVAC: Design Criteria, Options, Selection	59.95	
	60141B	Construction Cost Index–April 1991	42.25			67281	HVAC Systems Evaluation	57.95	
	60141C	Construction Cost Index–July 1991	42.25			67190	Illustrated Construction Dictionary	87.95	
	60141D	Construction Cost Index–October 1991	42.25			67267	Insurance Repair	54.95	
	60231	Electrical Change Order Cost Data 1991	72.95			67237	Interior Estimating	54.95	
	60031	Electrical Cost Data 1991	67.95			67239	Landscape Estimating	57.95	
	60201	Facilities Cost Data 1991	173.95			67266	Legal Reference for Design & Construction	109.95	
	60161	Heavy Construction Cost Data 1991	71.95			67236	Man-Hour Standards for Construction–2nd Ed.	131.95	
	60091	Interior Cost Data 1991	65.95			67235	Mechanical Estimating	54.95	
	60121	Labor Rates for the Const. Industry 1991	145.00			67245	Planning and Managing Interior Projects	59.95	
	60191	Landscape Cost Data 1991	71.95			67247	Project Planning and Control for Construction	58.95	
	60181	Light Commercial Cost Data 1991	65.95			67262	Quantity Takeoff for the General Contractor	59.95	
	60021	Mechanical Cost Data 1991	67.95			67265	Repair & Remodeling Estimating	57.95	
	60151	Open Shop Building Constr. Cost Data 1991	76.95			67254	Risk Management for Building Professionals	54.95	
	60211	Plumbing Cost Data 1991	67.95			67253	Roofing: Design Criteria, Options, Selection	59.95	
	60041	Repair and Remodeling Cost Data 1991	67.95			67152	Scheduling Manual	49.95	
	60171	Residential Cost Data 1991	65.95			67145	Square Foot Estimating	57.95	
	60071	Site Work Cost Data 1991	67.95			67241	Structural Steel Estimating	65.95	
	60051	Square Foot Costs 1991	76.95			67233	Superintending for the General Contractor	54.95	
		REFERENCE BOOKS				67274	Survival in the Construction Business	59.95	
	67275	Avoiding and Resolving Construction Claims	59.95			67259	Understanding the Legal Aspects of Design/Build	71.95	
	67257	Bidding & Managing Government Construction	58.95			67232	Unit Price Estimating	54.95	
	67180	Bidding for the General Contractor	54.95						
	67261	Building Profess. Guide to Contract Documents	54.95						
	67250	Business Management for the General Contractor	54.95						
	67278	Construction Delays	54.95						
	67268	Construction Paperwork	65.95						
	67255	Contractor's Business Handbook	58.95						
	67252	Cost Control in Building Design	58.95						
	67242	Cost Effective Design/Build Construction	54.95						
	67230	Electrical Estimating	54.95						
	67160	Estimating for the General Contractor	59.95						
	67276	Estimating Handbook	89.95						
	67249	Facilities Maintenance Management	59.95						
	67246	Facilities Maintenance Standards	131.95						
	67264	Facilities Manager's Reference	59.95						

MA residents add state sales tax.

TOTAL (U.S. Funds)

Prices are subject to change and are for U.S. delivery only. Canadian customers should write for current prices.

Postage and handling extra when billed. Send your check with your order to save shipping and handling charges! 1991 editions available December 1990. Means will bill you or accept MasterCard, Visa, and American Express.

BC6

SEND ORDER TO:

Name (PLEASE PRINT) _____

Company _____

☐ Company
☐ Home Address _____

_____

City/State/Zip _____

Phone # ( ) _____

P.O. # _____

(MUST ACCOMPANY ALL ORDERS BEING BILLED)

Building Construction Cost Data 1991

49th Annual Edition
Over 450 Pages • Illustrated

America's foremost construction cost information guide with over 20,000 unit prices for labor, materials, and installation

Building Construction Cost Data, now in its 49th year of publication, offers the reliability of over 20,000 thoroughly researched unit prices, plus the efficient, easy-to-use CSI MASTERFORMAT. Even the most complicated estimates can be accurately prepared in less time than with comparable references.

Astute construction professionals prefer *Building Construction Cost Data* as their primary estimating resource. It's the original, most sought-after book of its kind ... known and depended upon by its users for unparalleled accuracy and versatility ... a cost resource prepared with you—the user—in mind.

Unit prices are based on the actual experience of contractors ... coast to coast.

Price information shown in *Building Construction Cost Data* 1991 is carefully gathered from the recent, actual experience of contractors throughout the United States and Canada.

Labor prices are based on union negotiated trade rates for the year.

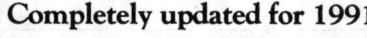

Completely updated for 1991

- materials, fixtures, hardware, and equipment items included in each section
- all items updated to reflect 1991 costs and construction techniques
- unit prices divided into material, labor, and equipment costs with overhead and profit shown separately
- hourly and daily wage rates for installation crews and crew sizes, equipment, and average daily crew output listed
- city cost adjustment factors for material and labor costs in 19 categories for each of 162 major U.S. and Canadian metro areas
- factors to compute historical costs for all cities dating back to 1947
- square foot and cubic foot cost section showing range and medium costs for 59 common building types with plumbing, HVAC, and electrical percentages tabulated separately
- easy-to-use CSI format
- costs are the national average and can be easily adjusted for local wage scales
- over 19 index pages for quick item location and cross-reference
- helpful examples, instructions, illustrations, and explanations of how costs were developed for each major division

ISBN 0-87629-187-6
Book No. 60011

$65.95/copy
U.S. Funds

Means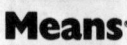
A Southam Company

R.S. Means Company, Inc.
100 Construction Plaza, P.O. Box 800, Kingston, MA 02364-0800

1992 ORDER FORM

CALL TOLL FREE
1-800-448-8182

QTY.	BOOK NO.	COST ESTIMATING BOOKS	UNIT PRICE	Total	QTY.	BOOK NO.	REFERENCE BOOKS (cont'd)	UNIT PRICE	Total
	60062	Assemblies Cost Data 1992	$115.95			67231	Forms for Building Construction Professionals	$ 84.95	
	60012	Building Construction Cost Data 1992	69.95			67260	Fundamentals of the Construction Process	64.95	
	60222	Building Constr. Cost Data–Western Edition 1992	79.95			67210	Graphic Construction Standards	114.95	
	60112	Concrete Cost Data 1992	67.95			67258	Hazardous Material & Hazardous Waste	79.95	
	60142	Construction Cost Indexes 1992	180.00			67280	Home Improvement Cost Guide	35.95	
	60142A	Construction Cost Index–January 1992	45.00			67251	HVAC: Design Criteria, Options, Selection	64.95	
	60142B	Construction Cost Index–April 1992	45.00			67281	HVAC Systems Evaluation	62.95	
	60142C	Construction Cost Index–July 1992	45.00			67293	Illustrated Construction Dictionary, Unabridged	99.95	
	60142D	Construction Cost Index–October 1992	45.00			67267	Insurance Repair	58.95	
	60232	Electrical Change Order Cost Data 1992	79.95			67237	Interior Estimating	58.95	
	60032	Electrical Cost Data 1992	72.95			67239	Landscape Estimating	62.95	
	60202	Facilities Cost Data 1992	179.95			67266	Legal Reference for Design & Construction	114.95	
	60162	Heavy Construction Cost Data 1992	76.95			67236	Man-Hour Standards for Construction–2nd Ed.	144.95	
	60092	Interior Cost Data 1992	71.95			67235	Mechanical Estimating	62.95	
	60122	Labor Rates for the Const. Industry 1992	160.00			67245	Planning and Managing Interior Projects	68.95	
	60192	Landscape Cost Data 1992	74.95			67247	Project Planning and Control for Construction	64.95	
	60182	Light Commercial Cost Data 1992	68.95			67262	Quantity Takeoff for the General Contractor	64.95	
	60022	Mechanical Cost Data 1992	72.95			67265	Repair & Remodeling Estimating	62.95	
	60152	Open Shop Building Constr. Cost Data 1992	76.95			67254	Risk Management for Building Professionals	54.95	
	60212	Plumbing Cost Data 1992	71.95			67253	Roofing: Design Criteria, Options, Selection	59.95	
	60042	Repair and Remodeling Cost Data 1992	72.95			67291	Scheduling Manual	56.95	
	60172	Residential Cost Data 1992	68.95			67145	Square Foot Estimating	62.95	
	60072	Site Work Cost Data 1992	71.95			67241	Structural Steel Estimating	71.95	
	60052	Square Foot Costs 1992	84.95			67233	Superintending for the General Contractor	54.95	
		REFERENCE BOOKS				67274	Survival in the Construction Business	59.95	
	67275	Avoiding and Resolving Construction Claims	59.95			67259	Understanding the Legal Aspects of Design/Build	72.95	
	67257	Bidding & Managing Government Construction	62.95			67232	Unit Price Estimating	58.95	
	67180	Bidding for the General Contractor	56.95						
	67261	Building Profess. Guide to Contract Documents	56.95				MA residents add state sales tax.		
	67250	Business Management for the General Contractor	58.95				**TOTAL (U.S. Funds)**		
	67278	Construction Delays	61.95						
	67268	Construction Paperwork	70.95						
	67255	Contractor's Business Handbook	59.95						
	67252	Cost Control in Building Design	59.95						
	67242	Cost Effective Design/Build Construction	59.95						
	67230	Electrical Estimating	61.95						
	67160	Estimating for the General Contractor	61.95						
	67276	Estimating Handbook	89.95						
	67249	Facilities Maintenance Management	66.95						
	67246	Facilities Maintenance Standards	144.95						
	67264	Facilities Manager's Reference	67.95						

Prices are subject to change and are for U.S. delivery only. Canadian customers should write for current prices.

Postage and handling extra when billed. Send your check with your order to save shipping and handling charges! 1992 editions available December 1991. Means will bill you or accept MasterCard, Visa, and American Express.

BC6

SEND ORDER TO:

Name (PLEASE PRINT) _____

Company _____
☐ Company
☐ Home Address _____

_____

City/State/Zip _____

Phone # ( ) _____

P.O. # _____

(MUST ACCOMPANY ALL ORDERS BEING BILLED)